Monograph Series in

UNDERWATER ACOUSTICS

Sponsored by Office of Naval Research

Monograph Series in

UNDERWATER ACOUSTICS

In 2001 the Office of Naval Research began a program to publish a selected monograph series in the field of underwater acoustics. The series publishes in-depth reviews and analysis on the state of understanding of the physics of sound in selected areas of research.

The monograph series covers the most important topics in underwater sound in an in-depth, critical, scholarly, and comprehensive fashion. It is intended to provide researchers with an understanding of the physics of underwater sound, based on both theory and experiment. The discussion of important unanswered questions is encouraged. Topics of particular importance in the field are low-frequency sound in the deep ocean, shallow-water acoustics, and ambient noise.

Sponsored by:

Office of Naval Research
One Liberty Center
875 North Randolph Street, Suite 1425
Arlington, VA 22203-1995
E-mail: onrpao@onr.navy.mil

Series Editors:

Ralph R. Goodman, Ph.D.
Research Professor
USM Department of Marine Science
1020 Balch Boulevard
Stennis Space Center, MS 39529
E-mail: ralph.goodman@usm.edu

Homer P. Bucker, Ph.D.
808 Moana Drive
San Diego, CA 92106
E-mail: bucker@nosc.mil

Ira Dyer, Ph.D.
Weber Shaughness Professor of Ocean Engineering, Emeritus
Massachusetts Institute of Technology,
Cambridge, MA 02139

Jeffrey A. Simmen, Ph.D.
Applied Physics Laboratory
University of Washington
1013 NE 40th Street
Seattle, WA 98105-6698
E-mail: simmen@apl.washington.edu

High-Frequency Seafloor Acoustics

Darrell R. Jackson
University of Washington
Seattle, Washington

and

Michael D. Richardson
Naval Research Laboratory
Stennis Space Center, Mississippi

Springer

Darrell R. Jackson
Applied Physics Laboratory
University of Washington
Henderson Hall
1013 NE 40th Street
Seattle, WA 98105-5640
e-mail: drj@apl.washington.edu

Michael D. Richardson
Marine Geosciences Division
Naval Research Laboratory
Stennis Space Center, MS 49529-5004
e-mail: mike.richardson@nrlssc.navy.mil

Library of Congress Control Number: 200692906

ISBN-10: 0-387-34154-4 e-ISBN-10: 0-387-36945-7
ISBN-13: 978-0-387-34154-5 e-ISBN-13: 978-0-387-36945-7

Printed on acid-free paper.

9 8 7 6 5 4 3 2 1

springer.com

To Linda and Jeanne for their patience and support during the lengthy preparation of this book

Series Preface

The efficacy of sound to penetrate the seas made acoustic systems in the past century the leading tools for sensing objects in and measuring properties of the seas. For over sixty years The United States Office of Naval Research (ONR) has been a major sponsor of undersea research and development at universities, national laboratories, and industrial organizations. Appropriately ONR is the sponsor of this monograph series.

The intent of the series is to summarize recent accomplishments in, and to outline perspectives for, underwater acoustics in specific fields of research. The general field has escalated in importance and spread broadly with richness and depth of understanding. It has also, quite naturally, become more specialized. The goal of this series is to present monographs that critically review both past and recent accomplishments in order to address the shortcomings in present understanding. In this way, these works will bridge the gaps in understanding among the specialists and favorably color the direction of new research and development. Each monograph is intended to be a stand-alone advanced contribution to the field. We trust that the reader will also find that each is a critical introduction to related specialized topics of interest as well.

ONR has sponsored the series through grants to the authors. Authors are selected by ONR based on the quality and relevance of each proposal and the author's experience in the field. The Editorial Board, selected by ONR, has, at times, provided independent views to ONR in this process. Its sole official role, however, is to judge the manuscripts before publication and to assist each author at his request through the process with suggestions and broad encouragement.

Ralph R. Goodman, Ph.D.
Ira Dyer, Ph.D.
Homer P. Bucker, Ph.D.
Jeffrey A. Simmen, Ph.D.

Acknowledgments

The support of the U.S. Office of Naval Research (ONR) is gratefully acknowledged by both authors. Additional support was provided to DRJ by the Applied Physics Laboratory of the University of Washington and to MDR by 6.1 Base funding from the Naval Research Laboratory. The authors are especially indebted to the ONR Editorial Panel, Ralph Goodman (Chair), Homer Bucker, Ira Dyer, and Jeffrey Simmen for their expert guidance and patience during the preparation of this monograph. Colleagues at the Applied Physics Laboratory, the Naval Research Laboratory, and other institutions contributed generously of their time, research results, figures, and informal reviews of selected material. In this regard, special thanks are due Bernie Beaudreau, Richard Bennett, Angus Best, Fred Bowles, Kevin Briggs, Joe Gettrust, Alex Hay, Marcia Isakson, Anatoliy Ivakin, Jin-wook Kim, Dennis Lindwall, Anthony Lyons, Eric Pouliquen, Richard Ray, Allan Reed, Dajun Tang, Eric Thorsos, Kevin Williams, and Mike Zimmer.

Contents

1 Introduction

The most widespread application of high-frequency underwater sound is active sonar, the acoustic equivalent of radar. In oceanography, underwater sound has found increasing use because it can provide large-scale coverage not possible with traditional sampling instruments. Likewise, underwater acoustic imagery can extend to much greater ranges than optical imagery, which is limited by high values of absorption and scattering, especially in turbid coastal waters. When sonar is used to detect and classify targets such as submarines, mines, underwater structures, animals, etc., sound scattering by the seafloor is a source of interfering reverberation that degrades the desired target echo signal. On the other hand, when sonar is used to observe or characterize the seafloor, scattering provides the desired signal. In either case, an understanding of scattering by the seafloor can provide benefits in sonar design and application.

Active sonar employs widely different frequencies, in the range 100 Hz to a few megahertz. The corresponding wavelengths range from a few meters to less than a millimeter. Acoustic wavelength is not the only spatial scale of interest in the seafloor scattering problem; equally important is the distance that acoustic energy can propagate without undue attenuation. In water, this can be hundreds of kilometers at the lowest frequencies and a few tens of meters at the highest frequencies. As for propagation in seafloor sediments, maximum propagation distances are much shorter, ranging from hundreds of meters at lower frequencies to millimeters at the higher frequencies. This monograph will emphasize small spatial scales and "high" frequencies, very roughly, frequencies from 10 kHz to 1 MHz, with corresponding wavelengths from 15 cm to 1.5 mm. With this restriction, one need only consider properties of the upper portion of the seafloor, a few meters at lower frequencies and a few centimeters at higher frequencies. Thus, by restricting attention to high frequencies, the need to consider geological stratigraphy (structure) on scales larger than a few meters is avoided. A better reason for this restriction is that lower frequencies have received more scientific attention [Jensen et al. 1994, Frisk 1998], largely because of the interest in deep-water antisubmarine warfare during the cold war. More recently, there has been a greater interest in shallow water, the so-called littoral regions (this includes regions from the continental shelf inward to shore as well as bays, sounds, and estuaries), and the research pendulum has swung accordingly. The littoral environment is more complex and varied than the deep-water environment, and the scientific problems (measurement, modeling, and prediction) are correspondingly more difficult. Recent research has resulted in significant progress, and this motivates the review given in this monograph.

1.1 Applications

In most sonar applications, operating frequency is determined by a compromise between range, which decreases as frequency increases, and angular resolution, which improves as frequency increases. While there are several military applications for high-frequency sonar, two applications in particular have motivated research on high-frequency seafloor scattering: mine countermeasures and acoustic homing torpedos. A mine countermeasure (MCM) sonar often must find and then classify proud (lying on the seafloor) or buried targets against a background of seafloor scattering. Frequencies of tens of kilohertz allow detection at ranges of order 1 km with angular resolution in the neighborhood of 1–10°. Good angular resolution restricts the sonar beam to a small patch of seafloor thereby reducing the level of interfering reverberation. For classification of targets by imaging, still better angular resolution is needed. Frequencies of hundreds of kilohertz may be used to obtain angular resolution of a fraction of 1° with a corresponding reduction in operating range compared to lower-frequency detection sonar. Similar levels of resolution can be obtained at tens of kilohertz using synthetic-aperture sonar, with techniques adapted from synthetic-aperture radar. In shallow water, bottom scattering is often the dominant source of interference for acoustic homing torpedos. Torpedos have severe restrictions on the allowable size of the sonar array, which degrades angular resolution compared to larger ship-deployed sonar. As a compromise, operating frequencies of tens of kilohertz are used to provide moderate angular resolution and range.

Military, commercial, and scientific presence in the ocean has fostered the development of acoustic communication as an important discipline. In shallow water, propagation from transmitter to receiver is complicated by multiple paths involving repeated "bounces" from the seafloor and sea surface. This causes a smearing in the time domain of transmitted signals with resultant interference between successive transmitted symbols. Schemes to reduce intersymbol interference and increase data transmission rates benefit from accurate propagation models which necessarily include acoustic interaction with the seafloor.

Nonmilitary use of high-frequency sonar has increased steadily in the last few decades. The applications that relate most immediately to this monograph are those in which sonar is used to map the seafloor or to determine its physical properties. High-frequency sonar employing multiple downward-pointing beams is used to measure seafloor relief (bathymetry). Side-scan sonar is used to image the seafloor, and some side-scan sonars can also measure bathymetry by means of an interferometric technique. In analogy with radar, which is used to remotely sense certain properties of the earth's surface, recent research has attempted to utilize the output signal of multiple-beam and side-scan sonar operating in the 10–300 kHz frequency range to measure physical properties of the seafloor. Normal incidence sonar (10–50 kHz) is often used to infer sediment physical properties, often using inversion of acoustic impedance as measured from seafloor reflection. Still higher frequencies (a few megahertz) are used in pencil-beam and mechanically scanned fan-beam sonars to observe dynamic processes at the seafloor including bedform development, sediment transport, and biological processes. Such applications fall within the realm of Acoustical Oceanography [Medwin and Clay 1998], in which acoustics is used to measure the ocean environment. Measurements of this type are dependent on acoustic interaction with the seafloor, and accurate models for this interaction can be beneficial.

1.2 Research on Seafloor Scattering

Sonar was used in World War II in submarine and antisubmarine warfare and for bottom sounding. These sonars operated at high frequencies, and important research on acoustic–environmental interactions was conducted during the war [Bergmann and Spitzer 1946, Urick 1983]. At this early stage, scattering by the seafloor was recognized as an important source of interference, and the concept of "scattering strength" was developed to quantify scattering by the seafloor and the sea surface. Measurements during the war and in the two decades following were made in an attempt to categorize the scattering strength of various seafloor types, designating them by names such as "rock, gravel, sand, silt" [Urick 1954, Urick 1956, Urick 1960, McKinney and Anderson 1964, Wong and Chesterman 1968, Flowers and Hurdle 1972]. These measurements helped to identify the main characteristics of seafloor scattering, but subsequent work has shown that simple categorizations are inadequate, as apparently similar seafloor types may have quite different scattering strengths. This issue is discussed in Ch. 12 of this monograph.

Motivated by applications in acoustic homing torpedos and minehunting, programs of seafloor scattering research were initiated by U.S. agencies, NATO, the European Union, and others. These measurements have added greatly to the existing data on seafloor scattering and, more importantly, have often been accompanied by environmental measurements intended to support physical modeling. Most of the acoustic measurements have been dedicated to backscattering (also called *monostatic* scattering), as this is the dominant seafloor reverberation mechanism in the most common sonar applications in which the source and receiver are colocated. Recent interest in sonar employing receivers spatially separated from the source has prompted so-called *bistatic* scattering measurements. Also, interest in targets buried in the seafloor (e.g., mines, pipelines, cables) has motivated measurement of acoustic penetration of the seafloor. The results of such measurement programs appear in refereed journals as well as in the proceedings of specialized conferences. An effort is made in this monograph to identify special issues and proceedings of particular interest. This relatively recent work is a primary source of material included in this monograph.

An acoustic measurement system used as part of the Coastal Benthic Boundary Layer (CBBL) Program is illustrated in Fig. 1.1. This setup was used to determine seafloor monostatic and bistatic scattering strengths. Figure 1.2 illustrates an experimental arrangement used by investigators at the NATO Undersea Research Centre to measure both monostatic scattering and penetration into the seafloor. Results of such measurements are given in Chs. 12 and 15 and in Appendix G.

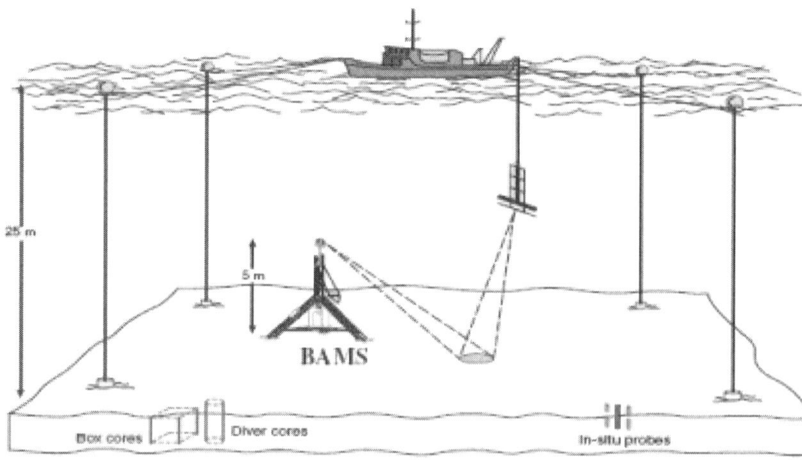

Fig. 1.1. Benthic Acoustic Measurement System (BAMS) and a ship-deployed receiving array used to measure acoustic scattering by the seafloor during the CBBL Program. Seafloor properties were measured by a variety of techniques including stereophotography, in situ probes, and sediment coring.

1.3 Organization of Monograph

It is expected that many readers will not wish to read the chapters of this monograph in serial order but will instead begin with a chapter or section of particular interest. To assist these readers, Ch. 2 provides a summary of basic concepts and connects the geophysical and acoustic aspects of this monograph. Together with the list of symbols given in the first appendix

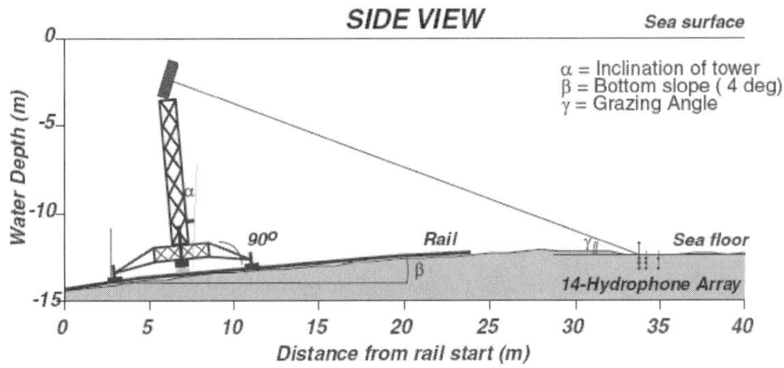

Fig. 1.2. Apparatus used to measure acoustic penetration into the seafloor. A parametric source on a movable tower was used in conjunction with a buried hydrophone array. [From Maguer et al. 2000b]

of this monograph, it is hoped that this approach will reduce the amount of preparatory reading required in the pursuit of any given subject. The remainder of the monograph is divided into chapters covering oceanographic and geophysical aspects of the subject (Chs. 3–7) and chapters on acoustic topics (Chs. 8–16).

Chapter 3, "The Nature of Marine Sediments," is a descriptive introduction to shallow-water sediments including the regional distribution of sediment types and morphology, types of sediment particles and their structure, free gas in sediments, and the effects of biological and hydrodynamic processes on sediment structure. This chapter supports following geophysical chapters but can be read independently. It is hoped that this material will provide a realistic picture of the seafloor and its dynamic nature for those attempting to develop acoustic theory and models and perhaps even stimulate interdisciplinary research connecting acoustics with oceanographic processes. The next chapter, "Physical Properties," describes sediment types and the physical properties that are of importance to high-frequency acoustics. Examples include grain size, sediment density, porosity, and permeability. The following chapter, "Geoacoustic Properties," treats parameters whose meaning derives from acoustics, e.g., wave speeds, attenuation, and impedance, and presents empirical relationships to predict those parameters from sediment physical properties. The remaining geophysical chapters, "Seafloor Roughness" and "Sediment Heterogeneity," describe the methods used to measure and characterize these random properties in the manner required for acoustic scattering models.

The acoustic chapters begin with "Fluid Theories," providing the background needed for fluid–sediment scattering models, which constitute the majority of models in current use. The next acoustic chapter, "Elastic Theories," provides the background for scattering models incorporating shear effects, and Ch. 10 does the same for Biot (poroelastic) scattering models. These theoretical chapters are followed by two chapters on acoustic data, "Reflection" and "Seafloor Scattering Experiments." The remaining acoustic chapters cover models for roughness scattering, volume scattering, acoustic penetration of the seafloor, and reverberation statistics. Model-data comparisons are presented where available. Several important and current topics are omitted or only discussed briefly. These include numerical methods for solving acoustic problems, time-domain methods, and use of acoustic data to infer seafloor properties.

The first appendix is a list of mathematical symbols, and the following seven appendices cover basic material that may be of general interest. The remaining appendices are intended for the specialist and are not essential reading for those more interested in applications than research. In part, these appendices contain model derivations that were deemed too complicated for the main text. Most of the discussion of scattering in these appendices employs the T-matrix ("transition matrix") formalism. While this approach may be less intuitive than others, once the essential definitions and relations (Appendix J) are understood, most of the models in this monograph can be derived with efficiency and within a common framework.

2 Basic Concepts and Definitions

For our purposes, three topics within underwater acoustics are of primary interest: the acoustic environment, theories for wave propagation in sediments, and models derived from these theories. This chapter will give an introductory account of each of these topics, giving concepts and definitions essential to later chapters. The intent is to provide more-or-less immediate access to later chapters with minimal introductory reading.

2.1 Geophysical Properties

An understanding of the environment is essential for solution of many of the problems in high-frequency, shallow-water acoustics. This requires that properties such as sediment mass density, sound speed, and acoustic attenuation be known.It also requires knowledge of seafloor volume heterogeneity and surface roughness. Given this information, virtually any acoustic problem can be solved, in principle, because the equations of motion and boundary conditions for the acoustic field are understood. Of course, computational difficulties may arise; more significantly, there is some uncertainty as to the applicability or accuracy of current theories for wave propagation in sediments. It is also important that sediment properties reflect seafloor conditions at the time of acoustic measurements or modeling applications. As will be shown in later chapters, seafloor properties in shallow water are not only spatially variable (heterogeneity) but can also vary with time. For example, seafloor morphology (roughness or ripple structure) changes rapidly in response to wave action during storms and can be subsequently degraded due to biological processes (Sect. 3.4).

The list of geophysical properties, or parameters, of interest is quite long, due to the continual refinement of sediment acoustic theory and models. Acoustic models can be broadly classified according to the wave theory employed (e.g., fluid, elastic, or poroelastic). Such theories sometimes require knowledge of sediment structure at the microscopic level, and often require knowledge of the centimeter-scale spatial dependence of the material parameters required by the underlying theory. Such spatial dependence includes the random spatial variation responsible for scattering and the average variation with depth below the seafloor due to stratification. The various wave theories will be discussed briefly in Sect. 2.2 and then more fully in Chs. 8 – 10.

For the sake of convenience, geophysical properties are divided into two categories in this monograph: physical properties and geoacoustic properties. Physical properties include those characteristics of sediment, sediment grains, and pore fluid that are of interest to geologists and geotechnical engineers as well as having applications to acoustics. These properties include sediment mass density, porosity, permeability, and many other parameters associated with sediment grains, pore fluid, the sediment frame, and the pore space. Geoacoustic properties include speed and attenuation for both compressional and shear waves. Acoustic impedance, the product of sediment mass density and sound speed, is another important geoacoustic property. The properties most commonly employed by current acoustic models will be introduced, and methods developed to characterize the random roughness and heterogeneity of the seafloor will be described.

The division of geophysical properties into physical and geoacoustic components is motivated in part by efforts to predict geoacoustic properties given measured values of one or more physical properties. This is one of the subjects of Ch. 5. Typically, this involves the development of empirical regressions giving wave speed or attenuation in terms of properties such as grain size or porosity.

2.1.1 Physical Properties

Some sediment physical properties, such as mass density (hereafter referred to as "bulk density"), are employed directly as input parameters in acoustic theory and models. Others, such as sediment type or mean grain size, are used indirectly as empirical predictors of acoustic behavior. For example [McKinney and Anderson 1964] relate scattering strength versus grazing angle to bottom type, expressed in descriptive terms such as "sand" and "mud." In many applications of this type of model, a value of mean grain size is all that is required. Grain size is one of the most commonly measured sediment properties, and is usually given as equivalent particle diameter, either in millimeters or in logarithmic units. Logarithmic units are convenient, because many grain-size distributions are approximately lognormal. If the grain size is d mm, the grain size is translated into base-two logarithmic units as follows:

$$\phi = -\log_2 d \ . \tag{2.1}$$

A table is provided in Ch. 4 to quickly compare different grain size scales. As defined in this table, clay-sized particles have grain size less than 0.0039 mm (8 ϕ), silt-sized particles have grain size between 0.0039 mm and 0.0625 mm (8 ϕ to 4 ϕ), and sand-sized particles have grain size between 0.0625 mm and 2.0 mm (4 ϕ to -1 ϕ). Particles larger than sand-sized are often referred to as "gravel," while clay- and silt-sized particles are collectively referred to as "mud." Several measures of logarithmic grain-size statistics are used. The most common is the "mean grain size," denoted by the symbol M_z. Various definitions are used for mean grain size, but the "graphical mean" definition given in Sect. 4.1 will be used in this monograph. As the logarithmic scale is used, the units of M_z are denoted ϕ.

A very broad classification into "siliciclastic" and "carbonate" sediments is often used to define both origin and mineralogy, the former composed of particles (e.g., quartz sand grains and clay minerals) derived from the breakdown of primarily silicon-bearing rocks that have been transported some distance from their origin, and the latter composed of calcite or aragonite particles formed from the in situ breakdown of shells and reef material. The term "clay" may be confusing to the casual reader as it refers both to the size class defined above and to a large class of silicate minerals. Chapter 3 provides a brief description of the origin, classification, and structure of particles that make up most shallow-water sediments.

Bulk density, ρ, is defined as the mass, M, of a suitably chosen small volume of the sediment divided by the volume, V:

$$\rho = \frac{M}{V} \ . \tag{2.2}$$

Bulk density is also considered a geoacoustic property by many (e.g., [Hamilton 1971b, Hamilton 1980]).

Sediments are porous media, and one commonly used measure is *fractional porosity*,

$$\beta = \frac{V_w}{V} \ , \tag{2.3}$$

where V_w is the volume of water in the total sediment volume, V, assuming the sediment has no gas (is "fully saturated"). Porosity is often expressed in percent, denoted by the symbol $\eta = 100\beta$ in this monograph. There are numerous other sediment bulk properties, such as void ratio, water content, degree of saturation, permeability, as well as grain shape, density, and bulk modulus and pore fluid density, viscosity and bulk modulus that are described in Ch. 4 and Appendix B. Many of these physical properties were neither defined nor measured with high-frequency acoustic applications in mind. However, Ch. 4 should provide the acoustic modeler insight into how sediment physical properties are measured and reported, and how values of these sediment properties should be interpreted relative to acoustic theory and models.

2.1.2 Geoacoustic Properties

Geoacoustic parameters, such as speed and attenuation of waves propagating in sediment, are the essential inputs to sediment wave propagation theories. It follows that they are also needed for theory-based acoustic models. Acoustic impedance is another frequently encountered geoacoustic parameter whose utility stems from its ease of measurement and high degree of correlation with other parameters. While the model builder has the luxury of choosing a certain set of geoacoustic parameters as inputs, the model user faces the problem of obtaining values for these parameters. In a few cases (e.g., Biot theory), some of the geoacoustic parameters can be calculated in terms of physical parameters such as the densities and elastic moduli of the constituents (particles and pore water) as well as sediment bulk properties such as porosity and permeability. Generally, however, one has neither a sufficient set of measured physical parameters nor a sufficiently complete and accurate theory that this approach can be used. Failing this, direct measurement of bulk density, and in situ measurement of wave properties is the preferred method of obtaining geoacoustic parameters. As wave speeds generally depend on frequency, it is desirable that the measurement frequency be similar to the frequency of application. Unfortunately, such detailed geoacoustic measurements are rare, as they require instrumentation and expertise usually available only in carefully planned field experiments. It is more likely that a few physical properties, such as mean grain size and porosity, might be measured. In these cases, one can use empirical regressions to estimate values of geoacoustic parameters (Sect. 5.1.2). Acoustic inversion is another approach to this problem, but this subject falls outside the scope of this monograph.

Fluid theory requires that the sediment be described in terms of three parameters, usually taken to be the bulk density, compressional wave speed, and compressional wave attenuation. Additional geoacoustic parameters are needed if the seafloor is modeled as an isotropic elastic solid supporting shear waves. In this case sediment bulk density and a set of four elastic properties must be known. One such set consists of the compressional and shear wave speeds and corresponding attenuations. Alternately, elastic moduli can be used. Hamilton [Hamilton 1971a] presents a version of elasticity in which the elastic constants (shear modulus and bulk modulus) are replaced with complex constants with the real parts governing wave speeds and the imaginary parts governing wave attenuations. This is the "linear viscoelastic" formalism, but it will simply be called "elastic theory" in this monograph. Hamilton assumes that the complex moduli are independent of frequency. As a result, attenuation increases linearly with frequency and wave speeds are independent of frequency (negligible speed dispersion). The proportionality constant, k, connecting attenuation and frequency is widely encountered in the literature and is usually given in dB $\mathrm{m}^{-1}\mathrm{kHz}^{-1}$. The assumption that both k and sound speed are independent of frequency is an approximation at best, as it violates causality (Appendix I).

Sediments, particularly sands, can be treated as poroelastic media that allow independent movement between the pore fluid and the "frame," comprised of sediment particles in contact. Modeling scattering from a poroelastic seafloor requires several physical parameters, many of which can be inferred from geoacoustic parameters. This is a role reversal compared to fluid modeling, where geoacoustic parameters are the model inputs and are often inferred from physical parameters. The physical parameters required by poroelastic theory can be broken into four sets characterizing the (1) fluid, (2) pore, (3) grain, and (4) frame properties. The fluid parameters are density, viscosity, and bulk modulus. The porometric parameters are porosity, tortuosity, pore size, and permeability, and the grain parameters are density and bulk modulus. The frame parameters are the bulk and shear moduli (real and imaginary parts). Obtaining practical values for the several poroelastic parameters is an unsolved problem that is approached with a combination of empiricism and theoretical approximation.

The acoustic theories described above assume marine sediment is a single-phase (fluid or solid) or two-phase (fluid–water) medium. In fact, sediments are quite complex assemblages of a variety of particle types, pore fluid, organic matter, and, sometimes, free gas. Even a small amount of free gas can greatly alter sediment compressional wave speed (compressibility), compressional wave attenuation, and acoustic scattering (Sects. 3.3 and 14.1.7). However, sediment density and shear wave speed and attenuation are not affected as much by small amounts of free gas.

2.1.3 Characterization of Gradients and Randomness

Random roughness of the sediment–water interface causes scattering of sound as does small-scale variability (heterogeneity) of sediment properties. Consequently, scattering models require statistical characterization of this randomness (Chs. 6, 7). This characterization is also needed to understand and model the effect of randomness on acoustic reflection.

Statistical characterization of the randomness of the seafloor can vary from a simple measure such as a variance to a more detailed measure such as a power spectral density. The shorter term "spectrum" will be frequently used in this monograph. Many roughness scattering models require a 2D spectrum whose argument is the "wave vector" $\mathbf{K} = (K_x,\ K_y)$. Following a convention used throughout this monograph, the two-dimensional vector, $\mathbf{K}$, is denoted by a boldface uppercase symbol. Three-dimensional vectors will be denoted by boldface lowercase. The magnitude, K, of the wave vector is called the wavenumber and is measured in radians/length. Wavenumber is analogous to the temporal radian frequency, $\omega = 2\pi f$, where f is frequency in cycles/s. Often, "spatial frequency" is used in preference to wave vector or wavenumber with an analogous factor of 2π involved.

An example of a 2D roughness spectrum (in spatial frequency) is shown in Fig. 2.1. The seafloor in this case had been raked to produce artificial ripples, and is an example of anisotropic roughness. Natural roughness may likewise be anisotropic, especially if ripples are present, but is often approximately isotropic. In fact, many natural roughness spectra are rather featureless, and can therefore be adequately characterized by two parameters, a "spectral strength" and a "spectral exponent" (Sect. 6.1). It is usually necessary to

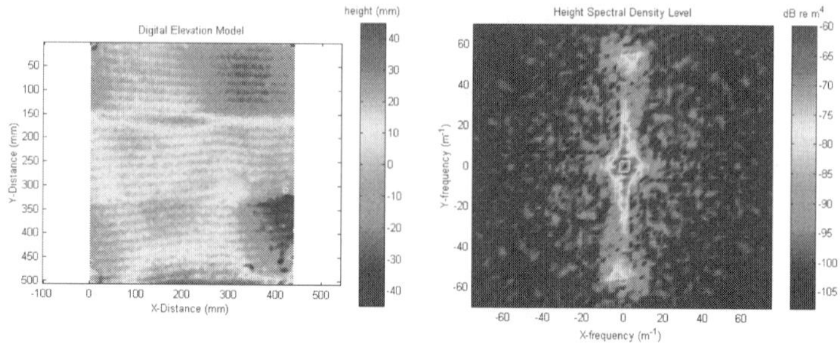

Fig. 2.1. False-color map of relief of a rippled sand seafloor (left panel) and resulting two-dimensional power spectrum (right panel). Courtesy of A. P. Lyons. (see first color insert)

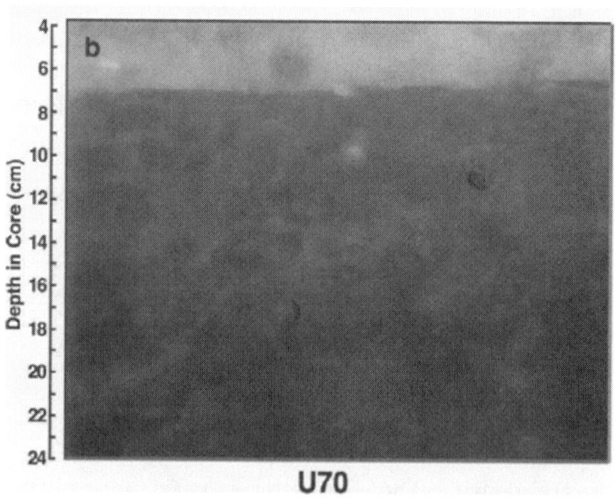

Fig. 2.2. Positive X-radiograph of core sample taken near the mouth of the Eel River, CA, USA, showing continuous heterogeneity as well as discrete (shell) inclusions [Richardson et al. 2002b].

know the roughness spectrum over a wavenumber range that includes and extends beyond the acoustic wavenumber in water, $k_w = 2\pi/\lambda_w$, where λ_w is the acoustic wavelength in water. In experimental tests of roughness scattering models, it is important that roughness and acoustic measurements be temporally and spatially concurrent. As will be seen in later chapters, biological and hydrodynamic processes can cause significant temporal variability in seafloor roughness.

Most models for scattering by sediment heterogeneity (Ch. 14) require statistical characterization of the volume heterogeneity of compressional wave speed and sediment bulk density (Ch. 7). This typically includes determina-

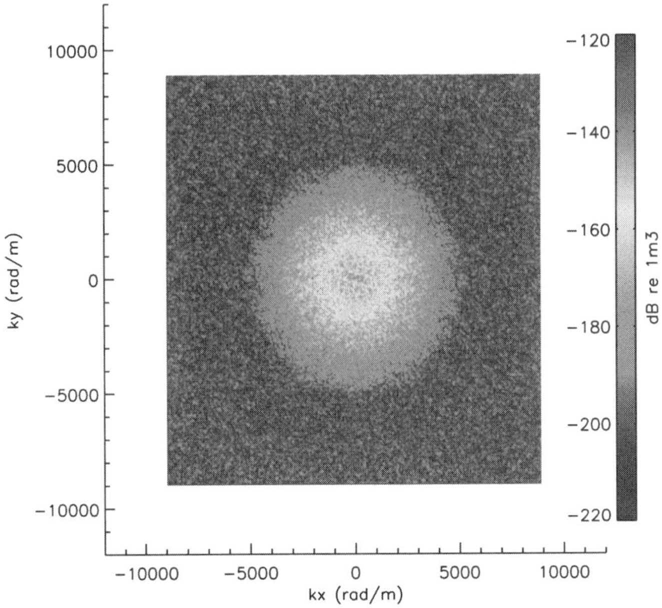

Fig. 2.3. A 2D slice through a 3D spectrum of density fluctuations obtained by X-ray tomography [Pouliquen and Lyons 2002]. (see first color insert)

tion of correlation functions or spectra characterizing the spatial fluctuations of sound speed and density.

An example of sediment heterogeneity is shown in Fig. 2.2. Scattering models usually require 3D spectra for fluctuations, and Fig. 2.3 shows a 2D slice through such a 3D spectrum. In the three-dimensional case, the argument of the spectrum is the wave vector $\mathbf{k} = (k_x,\ k_y,\ k_z)$.

Other forms of sediment heterogeneity such as layering or inclusions such as rocks or shells may be characterized either stochastically or deterministically depending on acoustic frequency and on the requirements of the acoustic model. Work is needed in obtaining statistical descriptions for the heterogeneity of poroelastic media, which would require spectra for several material parameters. Hard seafloors such as those composed of solid rock or coral pose another challenge. There has been no attempt in high-frequency acoustic experiments to characterize the random heterogeneity of rock or coral seafloor, yet models predict that this could be an important cause of scattering (Sect. 14.2.3).

Gradients and/or layering are common in sediments; therefore density, as well as compressional and shear wave speed and attenuation, must be measured as functions of depth in the sediment (Chs. 4, 5, and 7). Although this falls within the capabilities of present methods, the corresponding issue of layering in poroelastic seafloors has not been addressed, at least not with regard to a complete characterization of the several parameters of interest.

2.2 Sediment Wave Theories

Theories for the acoustic behavior of sediments will be placed into three classes. As noted in the previous section, these treat the sediment as a fluid, an elastic solid, or a poroelastic medium. Some theories do not fit this simple classification, for example, the theory of Buckingham [Buckingham 2000] has elements of both fluid and elastic behavior. Each class of theory has particular equations of motion and boundary conditions, to be discussed in later chapters. The older, simpler fluid models are by no means outmoded, as there are problems for which the fluid approximation for the seafloor is sufficiently accurate. A short, nonmathematical description of these three theories will be given here with technical details left to Chs. 8–10. Before beginning this description, it should be noted that this monograph does not discuss numerical methods for solving the field equations, such as ray approximations, modal expansions, fast-field methods, parabolic approximations, and finite difference methods [Jensen et al. 1994]. Such numerical methods take into account the spatial dependence of sound speed and other parameters in the water column and seafloor, treating a number of interesting phenomena, such as lateral waves and modal cutoff effects. Also omitted is any discussion of nonlinearity (so-called finite-amplitude effects). In most cases of interest, the acoustic field is weak enough that acoustic stresses are sufficiently small to be within the linear (Hooke's law) range.

For convenience in presentation, a distinction will be made between the terms "theory" and "model." The terms "wave theory," "propagation theory," or, simply, "theory" will be applied to the field equations and boundary conditions that determine the acoustically excited stresses and strains within the sediment. Thus, the wave equation for pressure and the boundary conditions for pressure and displacement at an interface between two fluids constitute a "wave theory," which will be denoted "fluid theory." There are analogous elastic and poroelastic theories. While use of the terms "wave theory," "fluid theory," etc., in this sense is by no means universal in sediment acoustics, it will be employed to avoid confusion with other products that the acoustic modeler produces. Thus, one may combine a fluid theory with an approximate rough-interface boundary condition to produce a scattering model. One may conduct measurements to test either the fundamental theory or a derived model. The theories to be discussed fall into the class of "effective medium" theories, because they treat the granular sediment as a continuum with properties such as density and elastic moduli obtained as averages over their microscopically varying values. It follows that such theories must fail at frequencies sufficiently high that the acoustic wavelength is comparable to the grain or pore size.

2.2.1 Fluid Theory

Seawater is a fluid, so it is not surprising that the first seafloor acoustic theories treated sediments as fluids [Kuo 1964], [Urick 1983, Ch. 5]. Fluid theory is discussed in detail in Ch. 8. The term "fluid" will be used in a very limited and well-defined sense: it is assumed that the acoustic stresses in the sediment can be described adequately by a pressure field and accompanying wave equation. This definition might seem to admit only a single fluid theory, but this is not the case. The differences lie in the dissipation mechanism(s) assumed in the acoustic wave equation.

In the fluid models to be encountered later in this monograph, the essential parameters, bulk density, compressional wave speed, and attenuation, are often represented by the dimensionless parameters,

$$a_\rho = \frac{\rho}{\rho_w} , \tag{2.4}$$

$$\nu_p = \frac{V_p}{c_w} , \tag{2.5}$$

$$\delta_p = \frac{\alpha_p \nu_p c_w \ln(10)}{40\pi f} . \tag{2.6}$$

In (2.4), a_ρ is called the "density ratio," ρ is the sediment bulk density, and ρ_w is the density of the overlying water. Similarly, in (2.5), ν_p is called the "sound speed ratio" and will also be represented by the notation V_pR. This ratio involves V_p, the phase speed of the compressional ("P") wave in the sediment, and c_w, the sound speed of the overlying water. The attenuation of the compressional wave in sediment is denoted α_p and usually has units dB m^{-1}. Equation (2.6) provides a means of converting this parameter into the dimensionless "loss parameter," δ_p. The acoustic frequency in Hz is denoted f. Several other, equivalent, parameters are used in the literature to characterize attenuation (Sect. 8.3).

Sound speed and attenuation generally depend on frequency, and these two dependencies are related by the constraint of causality (Appendix I) which simply states that the acoustic wave generated by a source cannot occur before the excitation. Frequency dependence of wave speed results in "dispersion," distortion of the shape of a transmitted pulse as it propagates through the medium.

Very fine sediments composed mostly of clay- or silt-sized particles ($< 62\mu$m) are often "slow," with sound speeds 1–5% less than the overlying water. Sediments composed mostly of sand-sized particles ($> 62\mu$m) are usually "fast," with sound speeds 10–15% greater than water. Even though these differences seem modest, there can be a marked difference in acoustic behavior between slow and fast sediments. In a slow sediment, sound is refracted downward toward the vertical, while in a fast sediment, it is refracted toward the horizontal and tends to penetrate to a lesser depth. This gives rise to large differences in the amount of acoustic energy penetrating the seafloor or redirected into the water column.

2.2.2 Elastic Theory

Sediments generally support static shear stresses, unlike true fluids, so it is logical to apply the theory of elasticity in underwater acoustics (Ch. 9). The relevant theory has largely been developed by geophysicists, who consider both "P" (denoting "primary," but to be understood as meaning "compressional") and "S" (originally denoting "secondary," but now meaning "shear") waves in the earth's crust [Telford et al. 1990, Aki and Richards 2002]. Compressional waves are analogous to acoustic waves in fluids and in typical unlithified surficial sediments travel at speeds (1450 to 1850 m s^{-1}) comparable to the water sound speed, while shear waves travel at much lower speeds (5 to 200 m s^{-1}). In lithified sediments, both compressional and shear speeds are faster than the water sound speed. In an isotropic elastic medium, compressional waves have particle motion parallel to the direction of propagation, while shear waves have particle motion perpendicular to the direction of propagation. Sediments are somewhat anisotropic owing to anisotropy in the stresses caused by overburden. Nonetheless, the theory and models treated in this monograph assume isotropy as an approximation. For an isotropic medium, particle motion for a shear wave can be decomposed into "vertical" and "horizontal" polarizations. These terms must be used with caution, as particle motion may not be in the indicated directions unless the wave propagates horizontally. The terms refer to the fact that particle motion is in either a vertical plane (with the plane parallel to the direction of propagation) or a horizontal plane.

An isotropic elastic medium is described by five parameters. Three parameters are analogous to those of the fluid model: bulk density, compressional wave speed, and compressional wave attenuation coefficient. The two additional parameters are shear wave speed and shear wave attenuation. These can be expressed in the dimensionless form of (2.5) and (2.6) with the replacement of compressional wave speed and attenuation by the corresponding shear values. The resulting parameters are denoted ν_t and δ_t in this monograph. The parameter ν_t is defined as the ratio of sediment shear speed to water sound speed, *not* the ratio of sediment shear speed to water shear speed, as water does not support shear waves. The subscript "s" would be more mnemonic, but "t" (for transverse) is employed because the subscript "s" will be used to indicate "scattering." As in the fluid case, elastic models differ in their description of dissipation, with constraints imposed by causality. As noted in Sect. 2.1.2, in Hamilton's approach [Hamilton 1971a], causality is violated to a degree, though this violation may not be serious in applications where the frequency band of interest is narrow. The assumed frequency independence of the elastic moduli is equivalent to frequency independence of the dimensionless parameters ν_p, ν_t, δ_p and δ_t.

2.2.3 Poroelastic Theory

Sediments are (at least approximately) two-phase mixtures of particles and water and are candidates for the application of Biot's theory for poroelastic media [Biot 1956a, Biot 1956b, Biot 1962a, Biot 1962b, Deresiewicz and Skalak 1963, Stoll 1974, Stoll 1989]. In this theory (Ch. 10), the medium consists of a shear supporting, porous "frame" with the pore space saturated with fluid. For sediments, the grains constitute the frame, with grain-to-grain contacts providing a degree of rigidity and with seawater constituting the fluid. Motion of the fluid relative to the frame provides an extra degree of freedom compared to elastic theory. This extra degree of freedom is manifested as a splitting of the compressional wave into two waves, the "fast" wave (essentially an acoustic wave) and the "slow" wave. As in elastic theory, shear waves of two polarizations are possible. While fluid theory requires three parameters to characterize the medium, and elastic theory requires five to characterize an isotropic medium, Biot theory requires a minimum of 13 parameters to characterize an isotropic medium. As noted in Sect. 2.1.2, measurement of all the required parameters is a difficult task, although it is usually agreed that fluid properties can be assigned handbook values. Typical parameter values for sandy sediments can be used to show that sands behave approximately as fluids, that is, the "fast" wave is dominant, and both the shear and "slow" waves have speeds of a few hundred meters per second and are not excited to an appreciable degree in typical circumstances. One of the most striking predictions of Biot theory is that dispersion will be significant. In particular, the fast wave speed in sands should increase by roughly 10% as frequency increases from a few hundred hertz to a few tens of kilohertz. Over this same frequency range, attenuation is expected to vary nonlinearly with frequency, in contrast to Hamilton's assumption. In addition, poroelastic effects are predicted to be significant in reflection [Stoll and Kan 1981, Chotiros et al. 2002b] and scattering [Williams et al. 2002b] in sands.

2.3 Reflection and Scattering

Following common practice, "reflection" and "scattering" will be treated as separate topics, although subsequent discussion will show that a clean separation is not usually possible. The term "reflection" will be used to imply the interaction of sound with a nonrandom, flat seafloor, in which case the seafloor behaves as a partially reflecting acoustic mirror, reflecting sound (with some energy loss) at an angle equal to the angle of the incident sound, which defines the "specular direction." Reflection is discussed in Chs. 8–11 and Appendix G. The term "scattering" refers to the equivalent process when the roughness (Chs. 6 and 13) and/or heterogeneity (Chs. 7 and 14) of the seafloor are important and result in a redistribution of acoustic energy over angles other than that of the incoming sound. Seafloor roughness on scales relevant to high-frequency acoustics is usually measured by means of stereophotography and summarized in a power spectrum estimate, as in Fig. 2.1. High-porosity sediments, such as silts and clays, have little acoustic impedance contrast in comparison to the overlying water, consequently, the sediment–water interface may not be a strong scatterer of acoustic energy. This lack of contrast allows sound to penetrate deeper into soft sediments than into sands. As a result, sound scattering by heterogeneity in physical properties such as bulk density and sound speed is expected to be stronger in mud than scattering by roughness.

In discussing reflection and scattering by the seafloor it is useful to define a coordinate system that provides a means of specifying the directions of incident and scattered acoustic energy. Figure 2.4 defines the angular coordinate system to be used, following the usual practice in underwater acoustics of measuring angles from the horizontal rather than from the vertical, thus departing from the convention used in electromagnetics and optics. When measured from the horizontal, angles will be referred to as *grazing angles*, with the direction of incoming acoustic energy specified by the *incident grazing angle*, θ_i, and the direction of the outgoing acoustic energy specified by the *scattered grazing angle*, θ_s, and the *bistatic angle*, ϕ_s. For a flat, non-random seafloor, the reflection is in the specular direction, for which the scattered grazing angle is equal to the incident grazing angle, and the bistatic angle is zero. In scattering, the monostatic case in which the scattered sound field is observed at the same location as the source is quite common. This "backscattering" direction has $\theta_s = \theta_i$, $\phi_s = \pi$.

2.3.1 Reflection

Although the seafloor is never perfectly flat and homogeneous, it is useful to consider this ideal case in which scattering is absent. This idealization is sometimes a reasonable approximation to the true situation (Sect. 13.1) and, in any case, allows definition of reflection and transmission coefficients appearing in *both* reflection and scattering models. As a further idealization, it will be assumed that the incident pressure field is a unit-amplitude plane wave at the angular frequency ω. A source at great distance produces a plane pressure wave with spatial dependence that can be approximated at the interface as

$$P_i = P_{i0} e^{i\mathbf{k}_i \cdot \mathbf{r}} , \tag{2.7}$$

where P_{i0} is the complex pressure amplitude at the origin, and $\mathbf{k}_i$ is the wave vector giving the direction of propagation of the plane wave. As indicated in Fig. 2.4, the z-coordinate will be taken perpendicular to the seafloor. For convenience, the coordinate system will be chosen so that the direction of propagation of the plane wave lies in the x–z plane. Then the wave vector has $(x,\ y,\ z)$ components

$$\mathbf{k}_i = \frac{\omega}{c_w}(\cos\theta,\ 0, -\sin\theta) , \tag{2.8}$$

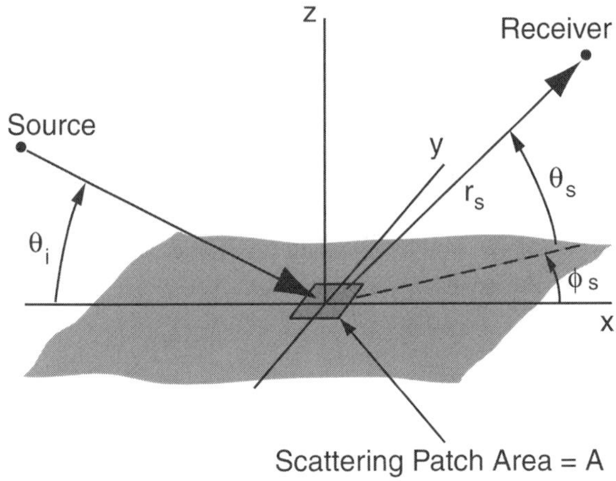

Fig. 2.4. Definition of angular coordinates used in treating reflection and scattering. If the seafloor has a preferred direction, for example due to a ripple field, it may be desirable to assign an arbitrary azimuthal angle, ϕ_i, to the incident sound as well.

where c_w is the speed of sound in water. For simplicity, the subscript i has been removed from the grazing angle, as there is only one angle of interest. The reflected wave will be a plane wave having the same grazing angle, θ, but generally will have reduced amplitude and shifted phase. That is, the reflected pressure field is of the form

$$P_r = V_{ww}(\theta)P_{i0}e^{i\mathbf{k}_r\cdot\mathbf{r}} , \tag{2.9}$$

where

$$\mathbf{k}_r = \frac{\omega}{c_w}(\cos\theta,\ 0, \sin\theta) . \tag{2.10}$$

The complex parameter, $V_{ww}(\theta)$, is the reflection coefficient, with the subscripts ww added to indicate that the incident and reflected fields are both measured in the water. In later chapters, transmission into the sediment will be considered. The reflection coefficient is dimensionless and is a function of grazing angle. Energy conservation imposes an upper limit of unity on the magnitude of the reflection coefficient. The "bottom loss," L, is usually defined as the positive decibel quantity

$$L = -20\log_{10}(|V_{ww}(\theta)|) . \tag{2.11}$$

Bottom loss is sometimes denoted "BL." Seafloor reflection is discussed in Ch. 11 and Appendix G, and the use of decibels is discussed in Appendix H.

Figure 2.5 shows two theoretical examples of bottom loss typical of sand and clay seafloors. For the sand sediment, bottom loss is small for grazing angles less than the "critical angle," θ_{crit}, where

$$\theta_{crit} = \cos^{-1}(1/\nu_p) , \tag{2.12}$$

and where ν_p is the compressional/water speed ratio (2.5) in the sediment. In this example, $\nu_p = 1.1$, giving a critical angle of 24.6°. As the grazing angle increases to values larger than the critical angle, bottom loss increases owing to transmission of an increasing fraction of the incident energy into the sediment. For the clay seafloor example, a sound speed ratio $\nu_p = 0.98$ was chosen, and loss becomes large at the "angle of intromission," about 10.5° in this case. These two examples show that sediment sound speed plays an important role in determining bottom loss. Using language employed in Sect. 2.2.1, the sand example is a "fast" seafloor, because the sediment–water sound speed ratio is greater than unity. In this case, according to Snell's law, sound is refracted away from the interface normal. At the critical grazing angle, sound energy is refracted into a horizontal direction. At still smaller grazing angles, the sound field does not propagate in the sediment, so little energy is lost due to transmission into the sediment. For a "slow" seafloor, such as the clay example given above, refraction is toward the normal. There is no critical angle in this case, but, at the angle of intromission, transmission of sound across the interface is nearly perfect, resulting in a small reflection coefficient and large bottom loss.

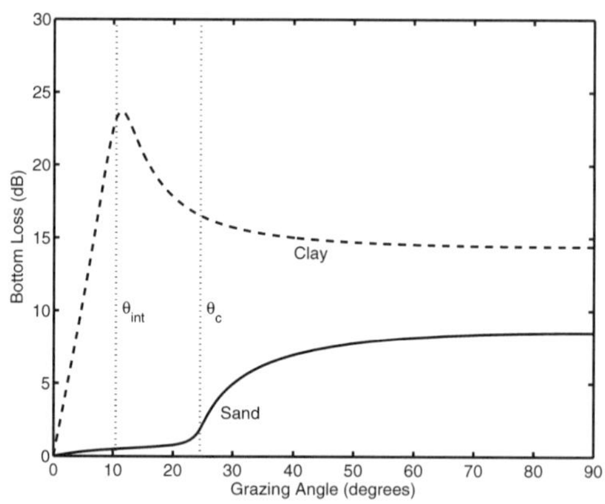

Fig. 2.5. Theoretical bottom loss for representative sand and clay seafloors. The acoustic parameters are the same as those used in computing the curves of Fig. 8.3.

These examples employ the simplest possible fluid model for sediment acoustics and results have been given without derivation. Chapter 8 will give a more complete account of reflection by the fluid–fluid boundary.

2.3.2 Scattering

Acoustic waves are scattered randomly by irregularities in the seafloor, including roughness of the water–sediment interface, spatial variations in sediment physical properties, and by discrete inclusions such as shell pieces or bubbles. These scattering processes are suggested by the highly schematic Fig. 2.6. A seafloor that is perfectly stratified (flat with properties varying only in the vertical direction) would exhibit no random scattering, only reflection (albeit with the possibility of multiple, interfering reflections from buried interfaces between different sediment types). At high frequencies, all seafloors have substantial irregularities on the scale of the acoustic wavelength, thus scattering is ubiquitous. Published treatments of seafloor scattering at high frequencies typically employ statistical methods, predicting various moments of the acoustic field. The most commonly encountered statistical quantity is the scattering cross section, which is proportional to the variance of the scattered field. Scattering by random boundaries and random heterogeneities is a large subject, and several books go into considerable depth on this subject [Beckmann and Spizzichino 1963, Bass and Fuks 1979, Ogilvy 1991, Voronovich 1994, Ishimaru 1997]. Figures 1.1 and 1.2 show two experimental arrangements used for making seafloor scattering measurements.

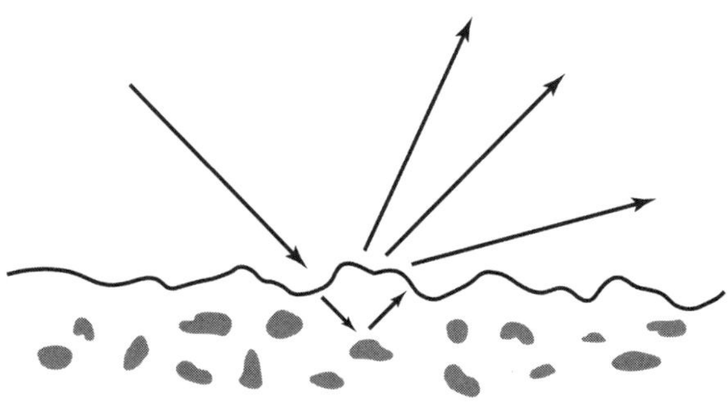

Fig. 2.6. Sketch showing acoustic scattering due to the roughness of the water–sediment interface and heterogeneity of the sediment.

It is typical in the formal treatment of scattering by random media to discuss the mean, $< P >$, of the complex pressure field, and fluctuations about this mean. The total field is decomposed as follows:

$$P =< P > +P_s \; . \tag{2.13}$$

The mean-square fluctuation is of primary interest in discussions of scattering. This is equal to the variance of the field and can be expressed as follows:

$$< |P_s|^2 >=< |P|^2 > -|< P >|^2 \; . \tag{2.14}$$

The "incident" field, that is, the field produced by the source in a hypothetical unbounded medium, is non-random and can either be included in P or treated separately. The subscript "s" is attached to the fluctuating part of the field, as it is the *scattered* part of the field. The mean field can be described in terms of the coherent reflection coefficient, V_{wwc}, defined in a manner equivalent to the reflection coefficient of a flat surface. The magnitude of the coherent reflection coefficient will be smaller than the magnitude of the reflection coefficient of the corresponding nonrandom case (in which the seafloor has physical properties equal to the mean of the properties of the heterogeneous medium). This is because a portion of the field that is reflected from the nonrandom medium is converted to an incoherent, scattered field in the random case.

The averaging symbol, $<>$, requires some explanation. It refers to an idealized average over an infinite ensemble of different seafloors, using identical experimental equipment and identical geometry. Each member of the hypothetical ensemble of seafloors has different roughness and/or sediment heterogeneity, but all are "drawn" from a statistically homogeneous population. In this monograph, this approach will be referred to as "formal averaging." This type of averaging exists only in the theorist's imagination, of course, so one may ask whether there is some practical approximation to this hypothetical averaging process. Experimentalists face the task of "estimating" (a statistician's word) quantities such as the mean-square scattered pressure, $< |P_s|^2 >$, from a finite statistical population. Usually, each member of this population is an echo from a different portion of seafloor, but at the same altitude and vertical pointing direction for the source and receiver ("depression angle"). For this averaging to be meaningful, the statistical properties of the seafloor must be "stationary," that is, independent of position over some reasonably well-defined region. If this is the case, an average can be achieved using a ship-mounted sonar, a towed sonar, or a rotating sonar mounted on the seafloor. With a fixed, rotating sonar, a complication enters if the scattering properties of the seafloor depend on the azimuthal angle of the sonar beam. This may happen if the seafloor has a well-defined field of directional sand ripples. Such seafloors are "anisotropic," but are within the capabilities of current scattering models.

Scattering by the seafloor is usually quantified in terms of the "scattering strength," whose definition follows from the situation depicted in Fig. 2.4, in which a small patch of seafloor having area A is situated in the far field of a source (see Appendices F, J). If this patch is rough and/or heterogeneous on scales comparable to the acoustic wavelength, it will scatter sound energy over a distribution of directions. In Fig. 2.4, the scattered pressure field is measured at one particular point, hence a particular direction. If this measurement is performed many times using the same geometry, but for an ensemble of statistically equivalent seafloor patches, then the mean-square pressure fluctuation (averaged over the ensemble of measurements) will be proportional to both the area of the patch and the squared incident pressure,

$|P_i|^2$, and will be inversely proportional to the square of the distance, r, from the patch. Thus,

$$< |P_s|^2 >= |P_i|^2 A\sigma \frac{1}{r_s^2} , \tag{2.15}$$

where attenuation and refraction in the seawater are neglected. Examination of (2.15) shows that the proportionality factor, σ, is dimensionless. It is sometimes referred to as the "scattering cross section per unit area per unit solid angle" because the integral of σ over the upper solid angle hemisphere yields the total mean scattered power, U_s, through the relation

$$U_s = \frac{A|P_i|^2}{2\rho_w c_w} \int_{2\pi} \sigma(\theta_i,\ \theta_s,\ \phi_s) d\Omega_s \ . \tag{2.16}$$

In (2.16), ρ_w is the mass density of seawater (about 1025 kg/m^2), and the product $\rho_w c_w$ is the acoustic impedance (Ch. 8) of seawater. The factor 2 in the denominator of (2.16) appears because time averages of squared sinusoidally oscillating functions are one-half the square of the peak magnitude. For brevity, $\sigma(\theta_i,\ \theta_s,\ \phi_s)$ will be simply referred to as the "scattering cross section," although this conflicts with its dimensionless nature. The more widely used nomenclature, "bottom scattering strength" [Urick 1983] is applied to the decibel equivalent,

$$S_b = 10 \log_{10} \sigma \ . \tag{2.17}$$

It is important to remember that the scattering cross section is defined here as a statistical average. Although ping-to-ping fluctuations in the scattered pressure are large (Ch. 16), the scattering strength itself is not a random quantity. The units of scattering strength are dB (Appendix H), although the units are sometimes incorrectly given as dB per unit area or dB re 1 m. These incorrect assignments of units imply that a change in the choice of length unit, say a change from meters to feet, would result in a different numerical value for S_b. Such is not the case, provided the same length units are used in measuring area and range. The above discussion is intended to define scattering strength and is inadequate to suggest measurement techniques, which are outlined in Appendix G.

For the sake of brevity, the term "scattering strength" will be used rather than "bottom scattering strength" or "seafloor scattering strength." There is some danger of misinterpretation here, because a "volume scattering strength" is often used to quantify scattering from within the ocean water column or from within the volume of the sediment. The definition just given applies only to scattering from an approximately planar interface, and the volume scattering concept will be needed in later treatment of scattering from within the sediment volume. Even in this case, however, it is possible to treat the scattering as if it were from a planar interface, if attenuation or refraction limit acoustic penetration of the sediment to small depths as often happens.

As defined here, scattering strength depends on properties of the seafloor and the water immediately above it, acoustic frequency, and the angles that define the directions of the incident and scattered acoustic energy. It is important to note those parameters that scattering strength does *not* depend on. These include measurement geometry and measurement system parameters such as source level and pulse length. Water column properties are eliminated if proper account is taken of propagation from the source to the seafloor and back to the receiver. With reference to Fig. 2.4, scattering cross section depends on three angular variables: a grazing angle for the incident field (θ_i) and grazing and azimuthal angles for the scattered field (θ_s and ϕ_s). These dependencies can be shown explicitly by writing the scattering cross section as $\sigma(\theta_s,\ \phi_s,\ \theta_i)$. The angle ϕ_s is often referred to as the "bistatic angle." If the seafloor has a preferred direction, as would be the case if a ripple field were present, an azimuthal angle, ϕ_i, is also required for the incident field. In either case, σ is referred to as the "bistatic" scattering cross section. A simpler and more common case is backscattering, the "monostatic" case, in which the transmitter and receiver are situated at the same point in space. For that case $\theta_s = \theta_i = \theta$ and $\phi_i = \phi_s + \pi = \phi$, and S_b is referred to as the "backscattering strength". Only two angular variables, θ and ϕ, are needed. If the random seafloor has no preferred direction, then the variable ϕ can be eliminated.

Two important constraints are placed on the scattering cross section by the fundamental principle of reciprocity [Morse and Ingard 1968, Pierce 1989, Kinsler et al. 1999] and energy conservation. Reciprocity demands that the cross section be unchanged if the source and receiver are interchanged. This imposes the condition

$$\sigma(\theta_i,\ \phi_i,\ \theta_s,\ \phi_s) = \sigma(\theta_s,\ \phi_s,\ \theta_i,\ \phi_i)\ . \tag{2.18}$$

It is necessary in (2.18) to use the most general form of the angular arguments for the bistatic cross section in order to clearly state the reciprocity condition. Energy conservation requires that the sum of the coherently reflected power, U_c, and the incoherently scattered power, U_s (2.16), be less than or equal to the incident power. The power incident on the scattering patch is

$$U_i = \frac{A|P_i|^2}{2\rho_w c_w} \sin\theta_i\ , \tag{2.19}$$

where the factor $\sin\theta_i$ is required to project the area A in the direction of the incident energy. Requiring $U_c + U_s < U_i$ gives the inequality

$$\int_{2\pi} \sigma(\theta_i,\ \theta_s,\ \phi_s) d\Omega_s + |V_{wwc}|^2 \sin\theta_i < \sin\theta_i\ , \tag{2.20}$$

where V_{wwc} is the coherent reflection coefficient mentioned earlier. The inequality results from energy dissipated in the seafloor. Expression (2.20) can be used to set bounds on the backscattering cross section, useful in assessing the validity of experimental data. In high-frequency acoustics, the coherent reflection coefficient is often expected to be much smaller than unity, essentially vanishing in cases where the RMS relief of the seafloor is comparable to or larger than the acoustic wavelength. In such cases, the inequality

$$\int_{2\pi} \sigma(\theta_i,\ \theta_s,\ \phi_s) d\Omega_s < \sin\theta_i \tag{2.21}$$

provides a more useful bound than (2.20).

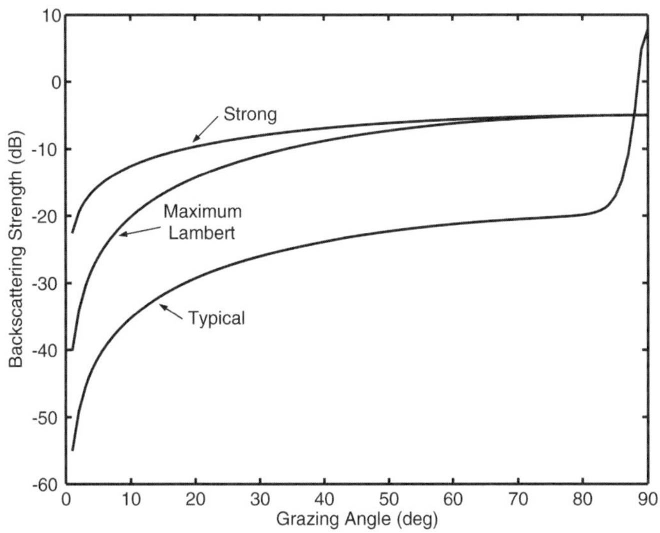

Fig. 2.7. Hypothetical backscattering strength curves. Although the curve labeled "Typical" exceeds 0 dB over a small range of angles, it does not violate energy conservation. The curve labeled "Maximum Lambert" is the highest level that maximally diffuse scattering can attain. The "Strong" case exhibits suspiciously high scattering strengths that can be sustained only if scattered energy is preferentially directed back toward the source ("backscatter enhancement").

The inequality (2.21) does not place a strict upper limit on scattering cross section, only a bound on its integral over all bistatic scattering directions. Thus, the curve labeled "Typical" in Fig. 2.7 is perfectly legal, even though it exceeds 0 dB near vertical incidence ("nadir"). Referring to Fig. 2.8(a), the large values of cross section only occupy a small lobe in angular space and do not contribute greatly to the integral in 2.21. The tendency for the scattering cross section to have a near-specular peak does not imply a focusing phenomenon, rather, it simply represents a sort of reflection degraded by the randomness of the seafloor (Sect. 13.1). One may inquire whether there is a

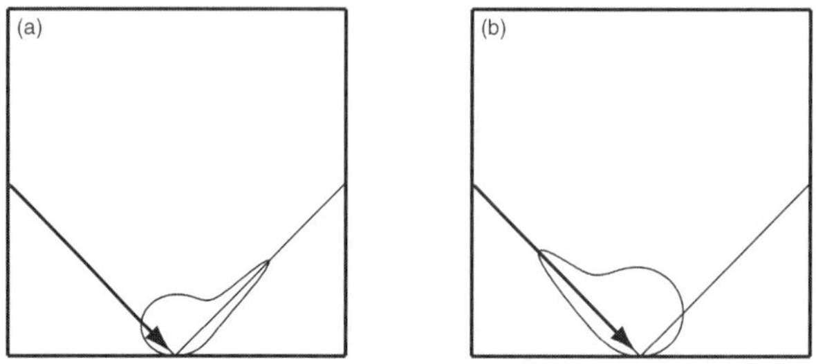

Fig. 2.8. Polar plots of the bistatic scattering cross section for two hypothetical cases with a plane wave incident from the left. (a) "Typical" seafloor scattering with a peak in the specular direction, representing degradation of the flat-interface reflection. (b) Strong scattering, with no peak in the specular direction and a peak in the backward direction, so-called "backscatter enhancement."

bound on scattering cross section for cases where angular dependence is weak, so-called "diffuse" scattering. The most well-known such case is described by Lambert's law

$$\sigma(\theta_i, \phi_i, \theta_s, \phi_s) = \mu \sin\theta_s \sin\theta_i \ . \tag{2.22}$$

Note that reciprocity is obeyed and that there is no dependence on azimuth. For backscatter, $\theta_s = \theta_i$, and the expression reduces to the form most common in the ocean acoustics literature, $\sigma = \mu \sin^2\theta_i$. Integrating the bistatic form (2.22) over the upper 2π solid angle as in (2.21), one obtains [Lurton 2002, p. 91],

$$\mu < \frac{1}{\pi} \ . \tag{2.23}$$

The case of strongest-possible Lambert scattering is shown in Fig. 2.7. Although the case labeled "Strong" exhibits a backscattering cross section that is everywhere greater than the maximal Lambert case, this need not imply violation of energy conservation if the scattering is not diffuse, but is concentrated in the backscatter direction, as illustrated in Fig. 2.8(b). Such "backscattering enhancement" (at levels relative to scattering in other directions of 3 dB and less) has been reported in optics [Knotts et al. 1993, Maradudin et al. 1994] and ultrasonics [Bayer and

Niederdraenk 1993,Ye and Alvarez 1998], and has been treated theoretically in the context of seafloor volume scattering [Ivakin 1982]. Although this interesting behavior is a possibility in extremely hard, rough seafloors, it has not been observed to date.

On occasion, the measured backscattering strength seems to approach a constant as grazing angle approaches zero [Wong and Chesterman 1968]. Such behavior indicates measurement error, most likely due to interfering multipath arrivals or noise. This can be asserted with some confidence because the scattering strength bounds given above force the bistatic cross section to approach zero as the incident grazing angle approaches zero. Reciprocity forces the same limiting behavior on the scattered grazing angle. Thus, the backscattering cross section must approach zero at least as rapidly as $\sin^2 \theta_i$. This argument is not rigorous, as a very narrow and strong backscattering enhancement peak could elevate the backscattering strength without contributing greatly to the integral in (2.21). But all experimental and theoretical evidence points away from this possibility, so any measured backscattering strength that does not fall off rapidly as the grazing angle becomes small should be viewed with suspicion.

3 The Nature of Marine Sediments

The discussion in this chapter will be restricted to surficial sediments (upper few meters) found on the continental shelf (beach to the continental shelf-slope break, approximately to 100–130 m water depth) and in other nearshore environments (lagoons, estuaries, and bays). This coastal area is approximately 29 $\times 10^6$ km^2, or 8% of the world's oceans and is characterized by high spatial and temporal variability in both morphology and sediment distribution [Reineck and Singh 1973]. Short-term temporal changes in the seafloor, which occur at time scales of minutes to decades, are the result of the interaction of hydrodynamic (waves, currents and tides, and deposition from rivers), biological (bioturbation), and biogeochemical (mineral dissolution, precipitation, and in situ gas bubble formation) processes acting on the seafloor. Longer-term changes in seafloor characteristics result from global changes in sea level which alter regional sediment erosion, transport, and depositional patterns and occur over geological time scales. These coastal areas are the likely locations for high-frequency sonar operations and high-frequency acoustic experiments. The general trends in sediment morphology and characteristics described herein should reflect corresponding trends in high-frequency acoustic behavior.

This chapter will include a general discussion of coastal morphology and will describe global patterns of sediment distribution. Next, the structure of sediments is discussed with the hope of providing developers of acoustic propagation theory with a realistic picture of the morphology, fabric, and particle interactions of sediments. This structure provides the building blocks for coastal marine sediment and has a great influence on the sediment physical and geoacoustic properties discussed in Chs. 4 and 5. Bubbles in sediments are also considered, including their morphology and the conditions that promote their existence. The chapter concludes with a discussion of the hydrodynamic and biological processes that alter sediment morphology, structure, and relief. Other diagenetic (post-depositional) changes that occur in near-surface sediments related to consolidation and biogeochemical processes are discussed within appropriate sections on sediment structure. Together these environmental processes are responsible for vertical gradients in sediment physical properties, create much of the random spatial variability treated in Chs. 6 and 7, and cause temporal changes in sediment properties and seafloor morphology.

3.1 The Origin and Classification of Coastal Sediments

For roughly the last two million years, the larger-scale morphology and depositional patterns on continental shelves have been shaped by repeated advances (transgressions) and retreats (regressions) of the oceans across the shelves, as sea level rose and fell in response to changes in global forcing during glacial and interglacial stages of the Pleistocene Period. The most recent sea level transgression (sea level rise) occurred during the Holocene Period between about 18,000 and 6000 years BP (Before Present). During the last period of low sea level stand, extensive areas of the inner and middle continental shelves were exposed to subaerial processes, while rivers crossed the exposed shelf depositing their sediment loads on the outer shelf and slope environments. During that period, much of the inner continental shelf received little or no sediment. The subsequent transgression during the Holocene represented a 100- to 130-m rise in sea level [Kennett 1982] that resulted in a landward migration of the seashore and the development of estuaries, lagoons, and drowned river channels that trapped sediment. As a result, the middle and, in particular, the outer areas of many continental shelves (e.g., east coast of the United States) were deprived to some degree of recent (last 6000 years) sedimentation. The seafloor in these areas consists of relict sediments (70% of shelf sediments may be relict), areas of exposed bedrock, and reef material that are not in equilibrium with present-day hydrodynamic processes. Areas located near major river systems (e.g., Mississippi and Amazon rivers) have developed significant deltas often composed of fine-grained silts and clays. It is of interest that most high-frequency acoustic measurements and experiments (Chs. 11–14) have been conducted on sediments that are in equilibrium with present hydrodynamic processes, although most of the sand-sized sediment particles are relict in origin.

Sediments are often characterized descriptively and texturally by type (e.g., sand, silty clay) or by mean grain size (see Sect. 4.1). One of the advantages to the geologist is that such textural classification often reflects the composition and events that occurred during erosion, transport, and deposition. The other advantage is that textural types can be determined from disturbed sediment samples, such as grab samples, whereas bulk properties such as density, permeability, and sound speed are altered by these collection methods. Methods to determine sediment grain size and other textural and bulk properties of sediment are discussed in Chs. 4 and 5. As a general rule, depositional patterns tend to reflect distance from shore (i.e., water depth). Coarse sediments (gravel and sand), for example, tend to accumulate near shore, whereas fine sediments (silt and clay) tend to accumulate farther from shore. However, this simple distribution pattern is modified by the existence of relict sediments (usually sands) on some shelves, and ... port and reworking of recently deposited sediments by hydrodynamic factors (waves, tides, currents) and large storms (e.g., hurricanes). Global climate also plays a significant role [Hayes 1967]. For example, sands, silts, and clays derived from terrestrial erosion are ubiquitous on continental shelves and in shallow-water seas, however, coral and coralline debris are primarily restricted to shelf areas in tropical waters where mean temperatures exceed 21°C [Bathurst 1975]. Rocky, gravelly sediments are more abundant on high-latitude (polar and subpolar) shelves where glacial and meltwater transport are common, or where they now exist as relict sediments and larger morphological features (e.g., terminal moraines) left behind by the retreat of the last Pleistocene ice sheets. The genesis of clay minerals (kaolinite, illite, and smectite or montmorillonite) which make up the greater part of most fine-grained marine sediments is highly dependent on climate. Kaolinite is common in shelf sediments where high temperatures and abundant rainfall on land contribute to intense chemical weathering of parent rock material. In cooler, more temperate climates, chemical weathering is less intense and kaolinite tends not to develop. Illite and smectite form in either environment with illite often the more ubiquitous. Clay minerals can also be derived from the erosion of older sedimentary rocks (e.g., shales and mudstones). The structure and importance of clay-mineral types to seafloor properties will be discussed later in this chapter. The term "clay" is often used (loosely) when referring to clay mineral composition. The term refers both to a size category ($> 3.9\ \mu$m; see Sect. 4.1) and to a group or class of minerals. Throughout this monograph the term "clay" will be used as a grain size category unless followed by the term "mineral."

Sediments also can be classified by origin (hydrogenous, biogenous, or lithogenous). Hydrogenous (derived from water) sediment deposits consist of minerals such as salt, palygorskite, sepiolite, aragonite, etc. that form by chemical reactions within the water column or as precipitates from supersaturated oceanic or coastal waters. Quantitatively, hydrogenous sediments are less important than biogenous (derived from organisms) and lithogenous (derived from terrestrial, volcanic, and cosmic sources) sediments.

Lithogenic sediments are derived primarily from the mechanical and chemical weathering of continental igneous, metamorphic, and sedimentary rock consisting mostly, but not exclusively, of silicate minerals. The nature of the weathering products is largely controlled by climate (temperature and rainfall), topography, and parent rock material. Hot, humid climates facilitate intense chemical weathering. In cold climates, chemical reactions are slow and the weathering is mechanical, e.g., expansion and contraction due to differential heating causing rock surfaces to exfoliate, rock fragmentation due to ice formation in cracks and crevasses, and pulverizing of rock material under the weight of glaciers. Eighty-five to ninety percent of weathering products are transported to coastal areas and to the open shelves by rivers and streams but only about 7% by ice. Transport by wind accounts for less than 1% of surficial sediment particles and is quantitatively important only in deep-sea sediments, where other sedimentary processes are slow [Bryant and Bennett 1998]. It is important to note that, during transport, particle size and shape can be mechanically altered by impact and abrasion (during bed load transport in rivers and streams) and by grinding (due to the movement of glaciers).

Biogenic materials in coastal and shelf sediments consist primarily of calcite (crystalline calcium carbonate) which includes the skeletal remains of animals and plants that live on or within the sediments and the debris from breakdown of coral reef communities. The dominant taxonomic groups that contribute to carbonate sediments include Mollusca (gastropods, bivalves and scaphopods), Echinodermata (sea urchins, sand dollars, and crinoids), corals, and calcareous algae. Approximately 50% of near-shore sediments contain a significant percentage of calcareous remains, especially in tropical regions (between 30°N and 30°S). Differences in the physical properties of siliciclastic (composed of silicon dioxide) and carbonate sediments are to be expected. Carbonate sediments (also called calcareous sediments) are often composed of weak, irregularly shaped, porous or hollow particles (as is illustrated in Fig. 3.12). As a result, porosity can be higher for a given mean grain size because of the presence of both intra- and interparticulate porosity. This characteristic in particular affects other sediment parameters (e.g., bulk density, sound speed, attenuation, impedance). In addition, cementation at grain contacts can increase sediment rigidity, resulting in high shear and compressional wave speeds.

In summary, nearshore and continental shelf areas are complex environments for conducting sonar operations and high-frequency acoustic experiments. This is due primarily to the wide spatial and temporal variability of the sediments which are overwhelmingly siliciclastic and calcareous in nature. Recently deposited sediments and shelf morphologies are in equilibrium with present-day hydrodynamic processes and global climates; however, many shelf areas are characterized by relict sediments that are a result of depositional and physical processes during the last lower sea stand (18,000 years BP).

3.2 Sediment Structure

Sediments are primarily composed of solid particles (minerals), pore fluid, free gas, and organic matter. Much of the discussion that follows will treat sediments as two-phase media (solid particles and pore fluid). Sediment structure includes fabric, the structural arrangement and orientation of particles; interactions between particles, mechanical and physiochemical; and interactions of particles with organic matter or the pore fluid. These, in turn, control sediment physical properties (bulk density, porosity, and permeability) and sediment behavior (acoustic and geotechnical). Sediment is generally fully saturated (i.e., no free gas) except for nearshore sandy sediment where air is entrained from breaking waves or in organic-rich sediment where methane bubbles form in supersaturated pore water from the bacterial degradation of organic matter [Martens et al. 1998, Richardson and Davis 1998]. Gas bubbles are discussed in Sect. 3.3, and their acoustic effects are treated in Sect. 14.1.7. Organic matter, common in fine-grained sediments, can also play an important role, adhering to clay minerals and affecting sediment structure and behavior (mechanical and acoustic). Sediment structure discussed in this section should be distinguished from larger-scale heterogeneity, which includes the larger-scale laminations, feeding voids, burrows, and shells which are discussed in Ch. 7, "Sediment Heterogeneity." The discussion of sediment structure will be divided into clay microstructure, fabric of granular sand and gravel, and structure of carbonates, because the structure, behavior, and properties of clay, sand and gravel, and calcium carbonate sediments can be quite different. Mixtures of granular material and clay are common, and their behavior is often determined by whether the fabric is matrix or particle/grain dominated. "Matrix-supported" refers to predominately clay-size sediment with embedded sand-size particles which are not in contact. "Grain-supported" refers to larger sand- or gravel-sized particles in contact with clay-sized particles within the pore structure.

Traditional methods of characterizing sediment structure include imaging thin sections of resin- or wax-impregnated sediments using optical microscopy. Recent observations of sediments using Transmission (TEM) and Scanning (SEM) Electron Microscopy and high-resolution Computed Tomography (CT) have greatly increased knowledge and thus improved models of the fabric of both fine- and coarse-grained marine sediments [Bennett et al. 1991a, Mees et al. 2003]. High-resolution CT provides the best opportunity to characterize sediment fabric without disturbing the original fabric. This approach allows simultaneous measurement of acoustic, electrical, or thermal properties during consolidation experiments with fine-grained mud or packing experiments with sand. The aforementioned techniques combined with more traditional techniques of mineral identification including optical, chemical, or X-ray diffraction analyses provide the basis for characterizing sediment structure. In addition, advances in theories of particulate mechanics and computational methods, such as discrete element modeling, suggest that it is possible to predict both the low-strain geoacoustic and high-strain geotechnical behavior of sediments directly from knowledge of the size and shape of particles, their arrangements, and the forces between them and with organic matter. In this monograph, the more traditional effective medium methods (Chs. 8–10) are used to describe wave propagation in these complex and not yet fully understood media. However, knowledge of sediment structure is important for current efforts to determine which wave propagation theories are appropriate for the variety of clay, granular sand and gravel, carbonate, and mixed sediments common to coastal marine environments.

3.2.1 Clay Microstructure

Much of the description presented here is derived from a review by [Bennett and Hulbert 1986], a collection of papers edited by [Bennett et al. 1991b], and from the text of [Mitchell 1993]. As indicated previously, clay microstructure includes both clay microfabric and physiochemical interactions. The term "clay microfabric" refers to the spatial distribution, orientation, and particle-to-particle relationships of clay particles and aggregates in sediments. Physiochemical interactions are the expression of electrostatic and electrodynamic forces among clay particles, organic matter, and the surrounding water-hydrated ions. Together, both control the behavior (acoustic propagation and geotechnical strength properties) and physical properties of clayey marine sediments. Clay refers both to a type of sediment particle (phyllosilicates) and to a size class ($< 3.9\ \mu$m). Most clay-mineral and clay-sized particles are derived from weathering of terrestrial rocks and soils that are transported to the oceans by major river systems, and then deposited on the seafloor where the clay structure is reworked by biological, chemical, hydrodynamic (waves and currents), and consolidation processes (see Sect. 3.1). Clay mineral types not only depend on the source deposit or parent rock but also can be altered by a variety of physiochemical and biological processes during weathering, transport, deposition, and subsequent diagenesis. Diagenesis refers to all of the chemical, physical, and biological changes or transformations undergone by sediment after initial deposition. All of these processes contribute to the temporal evolution of clay mineralogy and microstructure. A discussion of the structure and properties of clay minerals is required to understand the evolution of clay microstructure in marine sediments.

Clay minerals belong to the phyllosilicate family and are hydrous aluminosilicates that have a unique layered or plate-like structure (Fig. 3.1). The most common clay minerals found in marine sediment are illite, kaolinite, smectite, chlorite, and mixed-layer clays (clay minerals that consist of alternating layers of two distinct clay types). Montmorillonite is the most common species of smectite-group clay minerals, but the more general term "smectite" is used in this monograph. Other common groups of phyllosilicates found in marine sediments include talc, mica, and vermiculite. Common clay minerals are composed of two simpler structural units, the silicon tetrahedron and the aluminum or magnesium octahedron. The manner in which tetrahedron and octahedron sheets are stacked and the ions, cations, and water layers that glue the sheets together, provide the taxonomy and control the characteristics of the different clay mineral groups. Mineralogy determines the size, shape, and surface characteristics of clay plates, and the interactions of clay particles with the fluid phase. The basic characteristics of the idealized forms of these clay mineral groups are described below.

Kaolinite [$Si_4Al_4O_{10}(OH)_8$] is a two-layer, 1:1 (silica:alumina), clay mineral that consists of aluminum octahedral and silicon tetrahedral sheets with strong O-OH hydrogen bonding and little isomorphic cation substitution

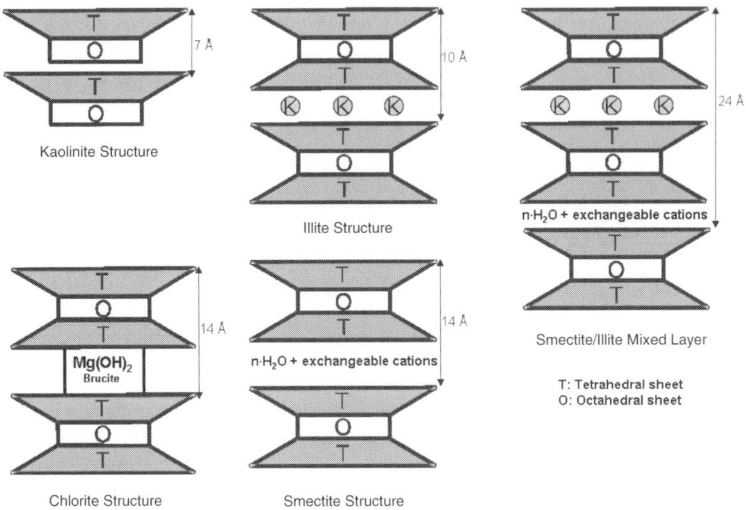

Fig. 3.1. Arrangement of silicon tetrahedron (trapezoids) and aluminum or magnesium octahedron (rectangles) for common clay minerals found in marine sediments: kaolinite, smectite, illite, chlorite, and smectite/illite mixed layer clays, adapted from [Mitchell 1993]. Octahedral sheets include gibbsite and brucite where the cations are mainly aluminum or magnesium, respectively. Kaolinite is a two-layer (1:1) or (T:O) clay mineral, whereas illite and smectite are three-layer clay minerals (2:1) or (2T:O), and chlorite is a four-layer clay mineral (2:1:1) or (2T:O:brucite).

within layers. Kaolinite plates occur as six-sided flakes with a 7.2-Å basal spacing, a lateral dimension of $\sim$ 0.1 to 4 μm, a thickness of 0.05 to 2 μm, and kaolinite can occur in 3000 to 4000 sized stacks. The specific gravity is 2.60–2.68 with a specific surface area of 10–20 $m^2\ g^{-1}$ and a cation exchange capacity of 3–15 milliequivalent (mEq) per 100 g dry weight of clay.

Smectite $[Si_8(Al_{3.34}Mg_{0.66}/Na_{0.66})O_{20}(OH)_4{\cdot}nH_2O]$ is a three-layer (2:1) clay mineral that consists of an aluminum octahedral and magnesium and sodium tetrahedral sheets with weak interlayer bonding and isomorphic substitution by magnesium or other cations for aluminum in the octahedra and aluminum for silicon in the tetrahedra. Smectite plates occur as equal-dimensional flakes with basal spacing from 9.6 Å to almost complete separation (swelling to 18 Å) and have thicknesses of > 10 Å and lengths to 1–2 μm. The specific gravity is 2.35–2.70 with a primary specific surface area of 50–120 $m^2\ g^{-1}$, a secondary specific surface area including the between layers that are exposed by the swelling of 700–840 $m^2\ g^{-1}$, and a cation exchange capacity of 80–150 mEq per 100 g.

Illite $[(K,H_2O)_2Si_8(Al,Mg,Fe)_{4-6}O_{20}(OH)_4]$ is a three-layer (2:1) clay mineral that consists of (Al,Mg,Fe) octahedral and (Al,Si) tetrahedral sheets with strong potassium bonding and some isomorphic substitution of silicon with aluminum balanced by potassium within the tetrahedra. Illite plates occur as flakes with a basal spacing of 10 Å, a size of 0.1 to 10 μm, and a thickness of 0.003–1 μm. The specific gravity is 2.60–3.00 with a specific surface area of 65–100 $m^2\ g^{-1}$, and a cation exchange capacity of 10–40 mEq per 100 g.

Chlorite $[(SiAl)_8(MgFe)_6O_{20}(OH)_4]$ with $(MgAl)_6(OH)_{12}$ interlayer is a four-layer (2:1:1) clay mineral that consists of alternating (Mg,Fe) octahedral and (Al,Si) tetrahedral sheets separated by a (Mg,Al) octahedral sheet and isomorphic substitution of aluminum for silicon in the 2:1 layer and aluminum for magnesium in the interlayer. Chlorite occurs as flakes with a basal spacing of 14 Å with a size of approximately 1 μm. The specific gravity is 2.60–2.96, and the cation exchange capacity is 10–40 mEq per 100 g.

Of the characteristics listed above, the ability of smectite to adsorb water (designated by $\cdot nH_2O$ in the chemical formula), greatly expanding the basal spacing between sheets, may be of greatest importance to predicting differences in wave propagation in different clay mineral assemblages. This expansion, or swelling, not only leads to low values of bulk density but to an increase in compressibility and a decrease in hydraulic conductivity (Sect. 4.4.1) compared to illite or kaolinite. However, it must be remembered that these are idealized characteristics of the pure clay minerals and that mixed-layer clays with varied properties are also common. Smectite–illite with alternating layers of expanded water-bearing and contracted non-water-bearing layers is the most common mixed-layer clay mineral. In addition, coastal marine clays usually contain mixtures of smectite, illite, kaolinite, and chlorite. Interactions of clay minerals with their environment during weathering, transport, exposure to cation- and anion-rich seawater, deposition, and subsequent diagenesis can alter the structure of the octahedral, tetrahedral, or interlayer cations changing not only clay properties but also changing one clay species or group to another. This is especially true for smectite which has the highest cation exchange capacity. Therefore it is not surprising that clay mineralogy is, at most, only indirectly taken into consideration in acoustic modeling of fine-grained marine sediments [Anandarajah and Lavoie 2002].

Clay particles, because of their small size and platy nature, have a very large surface area over which surface force interactions in the clay–seawater electrolyte system can act. These forces influence the flocculation behavior of clay particles in suspension and the consolidation behavior, compressibility, and strength properties of natural sedimentary deposits. The physiochemical interactions between clay plates are primarily electrical and include, in decreasing order of importance: Born repulsive forces, covalent bonds, electrostatic interactions, hydrogen bonds, and van der Waals forces. These forces control the structure and properties of the early developmental stages of clay microfabric (particles in suspension). The clay plates form multiplate particles or domains which are the fundamental particle units or building blocks of clay microfabric in marine sediments (Fig. 3.2). Together the mineralogy and domain structure of clays determine the plasticity or cohesion, swelling,

Fig. 3.2. Fabric models for aggregates or domains of clay plates from [Bennett et al. 1991b]. These aggregates and domains form the fundamental particle units or building blocks of clay microfabric in marine sediments. A–C are multiplate domains in plan view and in cross section as would be observed using TEM; D includes plan and cross-sectional views of shingle-type arrangements of domains with offset plates; F is a plan view of edge-to-edge contacts of domains forming long chains; G is a cross-sectional view of edge-to-face domains; E and H include cross-sectional and plan views of clay plates formed by face-to-face arrangements.

compressibility, strength, and fluid conductivity in fine-grained marine sediments [Mitchell 1993, pp. 37–38]. In organic-rich sediments, organic material may also play an important role by controlling the number and strength of interparticle bonds [Bennett et al. 1999].

When clay minerals, transported by rivers, reach salt water, they aggregate into larger sizes of particles (floccules) held together by van der Waals forces after the electrostatic repelling forces are neutralized by cations and anions in saline water. This flocculation provides the initial fabric or random arrangement of domains common to clay floccules recently deposited on the seafloor (Fig. 3.3). Floccules can combine with organic or inorganic ma-

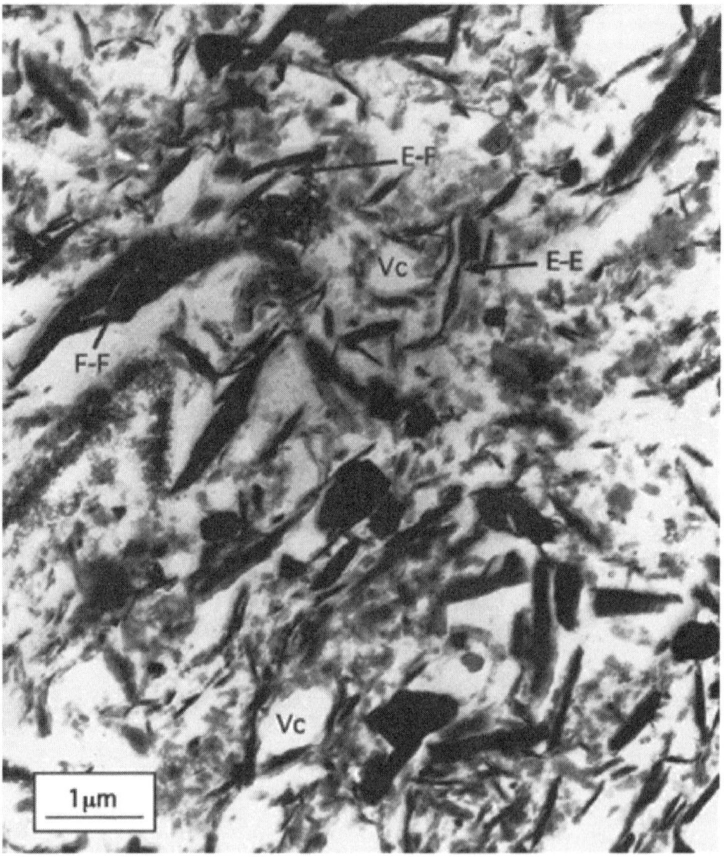

Fig. 3.3. Fabric of recently settled organic-rich clay aggregates from Eckernförde Bay, Baltic Sea (TEM image from [Anandarajah and Lavoie 2002]). Note the very high porosity (90%) or void ratio for these recently deposited sediments. Clay particles are arranged in domains in edge-to-edge (E-E) and edge to face (E-F) configuration. A tangential arrangement of clay particles is found around some void spaces (Vc). Smectite is characterized by its amorphous appearance.

terial to form chains, aggregates, agglomerates, or fecal pellets (collectively called marine snow when in suspension) which are more rapidly transported to the seafloor. These larger particles, rich in microbes, make up 90% of suspended material in the ocean and, by inference, most of the deposited clay minerals on the seafloor [Pierce 1991]. These aggregates mix with sediments already at the seafloor creating organic-rich aggregates containing a mixture of microbes, filamentous extracellular mucopolysaccharides, and randomly arranged clay domains in the form of the chains and aggregates depicted in Fig. 3.2 and provide the initial conditions important in the development of sediment microstructure [Ransom et al. 1998a, Ransom et al. 1998b,

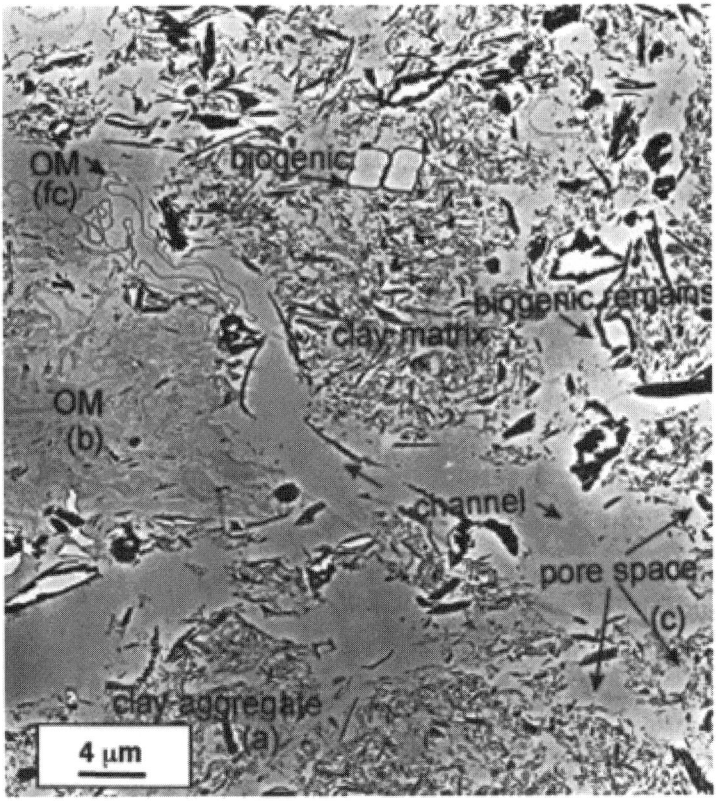

Fig. 3.4. Fabric of recently settled organic-rich clay aggregates from Eckernförde Bay, Baltic Sea (TEM image from [Anandarajah and Lavoie 2002]). Note the large void spaces, channels within and especially between clay aggregates, and abundant organic matter within channels (OM fc), as discrete blobs (OM b), and as biogenic remains.

Bennett et al. 1999, Hulbert et al. 2002]. The result is often a surficial sediment with large void spaces, very high porosity (> 95%) and low density (< 1.11 g cm^{-3}) (Fig. 3.4). The importance of organic matter in binding clay particles into larger aggregates may be greatly underestimated, and organic matter, in the form of long polymer strings, may be the dominant bonding mechanism in recently deposited sediments. Subsequent reworking by abundant benthic fauna, hydrodynamic effects of waves and currents, biogeochemical changes, and consolidation by dewatering in the upper meter of sediment can restructure chains, aggregates, or pellets and more tightly pack the clay domains, producing a surficial mud with porosity of 65–75%. The rate of these processes depends on the initial microstructure of the deposited material, clay mineralogy, percent and types of organic matter, rates of de-

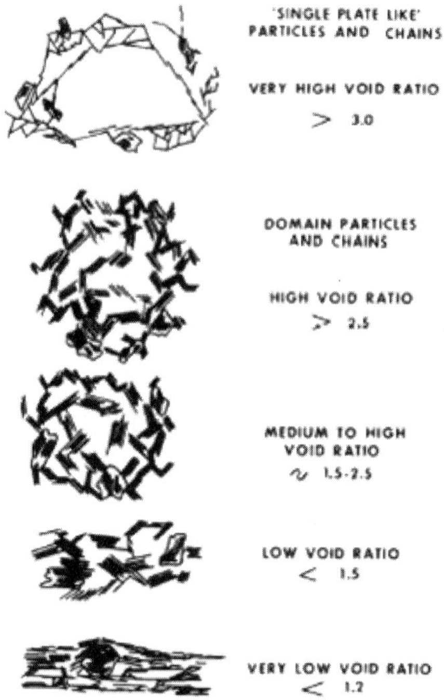

Fig. 3.5. Changes in void ratio of clay aggregates during consolidation by dewatering (from [Bennett et al. 1991a]).

position, and the abundance and depth distribution of benthic fauna. Dewatering or consolidation and the resultant clay microstructure depend strongly on electrostatic forces between particles, clay surface area, hydration state, and organic matter–clay particle interactions. These factors control both the permeability and compressibility of sediments and provide input to models of sediment consolidation [Toorman 1999, Boudreau and Bennett 1999]. Under additional stress from consolidation, the random arrangement of domains becomes reordered, with particle orientation normal to the vertical effective stress, and the pore space, as measured by void ratio or porosity, is further reduced (Fig. 3.5). The overall effect is an increase in sediment bulk density, a decrease in porosity and permeability, an increase in speed of compressional and shear waves, and a decrease in attenuation of both. Consolidation produces gradients in sediment physical and geoacoustic properties. When clay domains have a preferential orientation, consolidation may cause vertical/horizontal anisotropy in permeability and wave speed as well.

In spite of considerable advances in SEM and TEM characterization of the microfabric of naturally occurring and laboratory-consolidated clayey sediments, physical models of sediment geotechnical and geoacoustic behavior and properties which utilize that microfabric have lagged behind. Measurement of sediment porosity or void ratio from images of thin sections of embedded clay sediments is well developed. Changes in clay microstructure during consolidation are well documented and have been correlated with changes in the shape, orientation, and spatial arrangement or distribution of clay domains and aggregates. It is even possible to predict sediment permeability based on sizes of pore throats, pore bodies, and number and degree of interconnectivity of pore bodies [Vaughan et al. 2002]. Physiochemical interactions between clay particles and between clay minerals and the pore fluid are reasonably well known for the clay minerals commonly found in marine sediments. The cation exchange capacity, effective surface area, and surface potential for various clay minerals have been measured. However, physical models that predict values of compressional or shear wave speed or attenuation from clay microfabric and physiochemical interactions between particles and pore fluid have not been highly successful. Yet, propagation theories that treat clay sediment as a collection of noninteracting particles (similar to granular material) and neglect (a) the electrostatic forces between particles, (b) the interactions between particles and the pore fluid, and (c) the cohesiveness associated with organic matter often abundant in these surficial sediments cannot capture all of the physics of sediment compressibility and shearing associated with wave propagation through these complex sediments. Clay microstructure studies coupled with acoustic research may provide a "springboard" for successful understanding and modeling of fine-grained sediment behavior.

3.2.2 Fabric of Siliciclastic Sands

Sediments comprised of predominantly sand-size particles are common in shallow-water, high-energy environments (waves and currents) and in offshore areas where relict sands predominate. The types, size distribution, sorting, density, bulk modulus, shape, and roughness features of individual particles, together with packing, affect the fabric or the spatial arrangement of sand grains in natural settings. Discussion of the morphology and physical properties of individual grains is found in Ch. 4 "Physical Properties." In this section, methods to characterize the spatial arrangement of grains in naturally occurring sandy sediments and the effects of these particle arrangements (fabric) on sediment physical properties relevant to wave propagation in cohesionless sediments will be presented. The extensive literature on laboratory studies of artificially created or packed sediments will not be reviewed. The term "cohesionless" (noncohesive) applies to assemblages of particles that have little or no surface attraction, one to another, and is in contrast to cohesive clayey sediments where physiochemical attractions among clay particles, organic matter, and pore water dominate stress–strain behavior. In cohesionless sediments, friction between individual particles tends to dominate stress–strain behavior. This fundamental difference in the low-strain behavior between cohesive and noncohesive sediments should be taken into account when developing or using physical models of wave propagation in sediments.

The fabric of sand-sized assemblages of particles (0.0625 to 2.0 mm in diameter) traditionally has been characterized using either optical methods (petrographic microscopy) [Curry et al. 2004] or SEM [Reed et al. 2002] applied to thin sections of resin-impregnated sediments. More recently, CT methods [Mees et al. 2003] have been used with whole samples. Fundamental to this characterization of fabric is particle orientation, contacts, and packing, as well as the size, shape, and orientation and connectivity of pore space. Two-dimensional (2D) SEM or optical images can resolve pore-scale (10s of micrometers) and grain-scale (10s to 100s of micrometers) structures in marine sands and have been used to determine bulk sediment properties, such as porosity and density. Some inherent assumptions of 2D image analysis are that the sediment system is homogeneous and isotropic, and that the images are large enough that small-scale, grain-size variations do not dominate measurements (e.g., a minimum elemental volume is achieved). Provided the system is two-phase (i.e., water and sand), 2D imaging and simple thresholding techniques enable quick determinations of porosity and bulk density. Bulk density is readily computed in terms of porosity, fluid density, and average grain density (see Table 4.5). Examples of 2D high-resolution SEM images of resin-impregnated sand from the SAX99 experiment conducted off Ft. Walton Beach Florida are shown in Fig. 3.6. As can be seen from these images, the assumptions of homogeneity and isotropy in the distribution of sand grains and pore space are sometimes violated, complicating the calculation of porosity, void ratio, and density. Two-dimensional images, such as in Fig. 3.6, also complicate the determination of particle size, morphology, contacts, and arrangement, as well as pore size and connectivity. Note that most of the grains appear to be free-floating in a matrix of pore space. That is not the case, either in nature or in the resin-impregnated samples depicted in Fig. 3.6. The number of contacts is typically 5–10 per grain. As will be seen next, properties of the fabric are often used to determine parameters important to wave propagation in sediments such as permeability, tortuosity, pore size, and grain contacts.

The CT methods can provide volumetric (3D) characterization of grain arrangements and pore space from either resin-impregnated sediment or samples in the natural state [Reed et al. 2005]. High-energy linear accelerators or synchrotron CTs also have been used to image unconsolidated sediments but, because of their limited use and small sample size at high resolution, will not be discussed here [Mees et al. 2003]. Typical medical CT imaging systems have voxel (3D volume) resolution of approximately 0.5 mm^2 with 1.0-mm-thick slices, which is inadequate for fabric characterization of typical marine sands [Mees et al. 2003]. Recent advances in micro-CT scanning (Fig. 3.7) provide the resolution (5–20 μm) needed to determine values of sediment

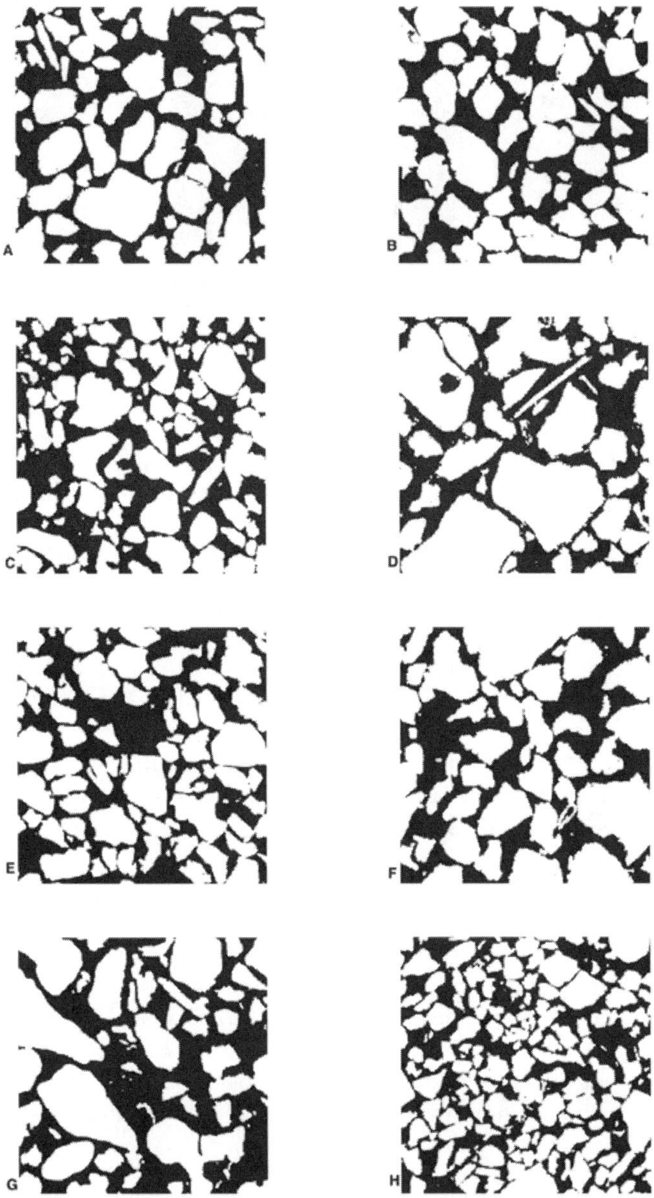

Fig. 3.6. Representative images of the microstructure of surficial sediments (upper 10 cm) collected from near Ft. Walton Beach, Florida [Reed et al. 2002]. The SEM images (3.5 mm^2) are from polished sections of resin-impregnated sand and depict cases of homogeneity (A, B) and heterogeneity (C, D) from horizontally oriented images. Examples of spatial isotropy (E, F) and anisotropy (G, H) are depicted from pairs of vertical and horizontal images sliced from the same section. The mean grain size of these predominately quartz sand grains as determined by conventional bulk methods is 420 μm, the porosity is 36.6%, and the bulk density is 2074 kg m^{-3}.

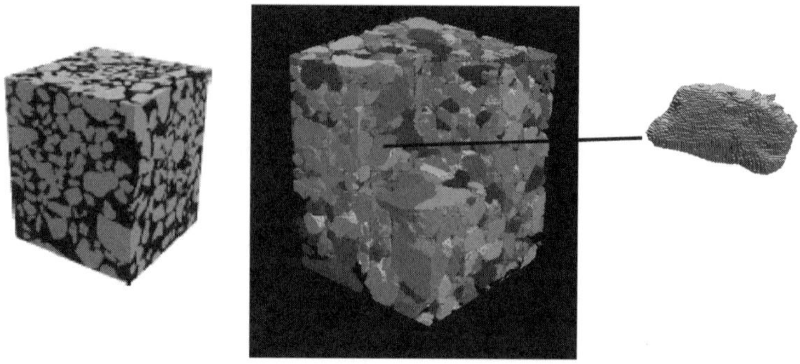

Fig. 3.7. CT scan of resin-impregnated sediment collected with a diver core during SAX99 (from A.H. Reed in [Richardson et al. 2005]). The volume of scanned sediment is 82 mm^3 with a calculated porosity of 39%. The mean inscribed radius of the pores is 58 μm, the mean inscribed radius of the pore throats is 42 μm, and the mean grain size is 375 μm. The mean pore coordination number (number of interconnected pores) is 6.0, and the average grain coordination number (number of grain contacts) is 7.1. The individual grain to the right has a volume of 2.5×10^8 μm^3, surface area of 2.0×10^6 μm^2, and aspect ratio of 2.2. (see first color insert)

properties in sand relevant to acoustic propagation (Chs. 8–10), including porosity, bulk density, permeability, tortuosity, and pore size parameter.

Another difference between low-resolution medical and high-resolution microfocal CT images is the relationship between the source, detector, and the sample to be imaged. In typical medical CT imaging systems, the X-ray source and detectors rotate around the sample and data are collected from multiple angles. For high-resolution CT scanners, the sample rotates between a fixed source and receiver. CT scanners record the differences in the degree of attenuation of X-rays, which is both material- and energy-dependent. X-ray attenuation is a complex function of two processes: (1) the photoelectric effect or photoelectric absorption, which is more prevalent at lower energies (i.e., below 100 keV) and correlates with atomic number, and (2) Compton scattering, which is more prevalent at high energies and correlates with material density. A direct correlation between X-ray attenuation and particle density does not exist; therefore correlations between attenuation data (i.e., voxel values) with sample density for different CT scanners have been developed for a variety of materials including quartz and carbonate sediments [Wellington and Vinegar 1987, Orsi and Anderson 1999]. The differences in the relationships clearly display the importance of atomic number, as well as density, to X-ray attenuation [Orsi and Anderson 1999].

Once the differences in X-ray attenuation are determined, the images can be segmented into two or more components, and the distributions of grain and pore space can be determined. This can simply include the quartz grain and pore space (resin) as depicted in Fig. 3.8, or the distribution of free gas or other minerals can be determined if they are present. The simple thresh-

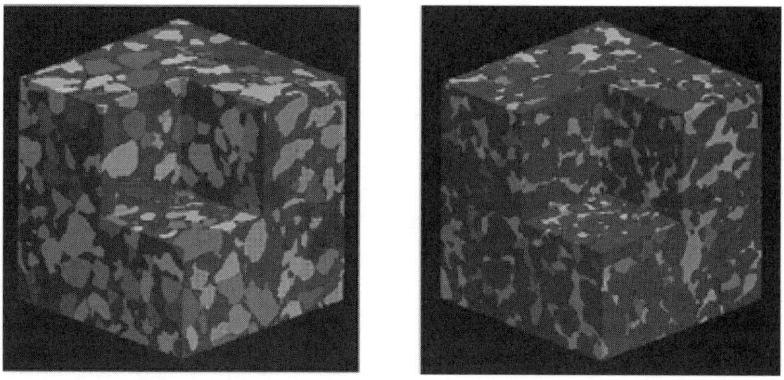

Fig. 3.8. Volumetric CT image of quartz sand (SAX04) from which porosity and bulk density were determined. Grains are false-color coded in the left panel, and pores are false-color coded in the right panel. Courtesy of A. H. Reed. (see first color insert)

olding techniques used to segment the 2D images described earlier sometimes are inadequate because grain and pore phases are not clearly distinct. This occurs as a result of volume averaging [Ketcham and Carlson 2001]. A better technique is indicator kriging [Oh and Lindquist 1999] which relies on a nearest-neighbor approach to assign ambiguous voxels to one or more of the phases (e.g. grain or pore space in a two-phase medium). From these images, bulk properties such as porosity, void ratio, and density can be determined.

3.2.3 Carbonate Structure

Carbonate structure differs from the siliciclastic structure (clay minerals and sands) discussed in the previous section not only because of the differences of chemical structure (carbonates are mostly composed of aragonite, low-magnesian calcite, and high-magnesian calcite), but also because of differences in origin (mostly biological versus terrigenous), transport (produced mostly in the vicinity, versus transported long distances by rivers), and shape and structure (irregular shapes, often pitted or with hollow cavities). Other differences may result because carbonates are often subject to higher levels of subaerial exposure, and biochemical reactions (especially related to dissolution, precipitation, and cementation), which affect compaction, permeability, and other physical properties [Bathurst 1975, Bennett et al. 1990b, Rezak and Lavoie 1993, Richardson et al. 1997]. In the following discussion, attention is restricted to modern shallow-water carbonates, as shallow water is the primary realm of interest in high-frequency acoustics. Different types of carbonate fabric will be illustrated with the use of photographic, TEM, SEM, and CT images. This monograph will not consider the substantial literature on microstructure and physical and geoacoustic properties of deep-sea deposits including foraminifera and coccolith and pteropod oozes (for example, see [Hamilton et al. 1982, Bachman 1984, Briggs et al. 1985]).

Most shallow-water carbonates occur in tropical and subtropical waters where water temperatures remain above 18°C, and coral reefs are common. The main exceptions are hard grounds composed of shell material and relict reefs that are essentially devoid of sediments (Fig. 3.9). These regions occur mostly on sediment-starved coasts where the finer-grained sediments are winnowed out leaving shell and shell hash as lag deposits or in bays and sounds where biological productivity is very high, and shell material produced in situ is mixed with mud. Carbonate sediment is primarily derived from the debris of skeleton-building plants and animals. Carbonate reef material includes the skeletal remains from a wide variety of invertebrates (gastropods, bivalves, echinoids, ophiuroids, barnacles, corals, alcyonarians, ascidians, bryozoans, and benthic foraminifers), polychaete worm tubes, calcareous (in particular Halimeda) and encrusting algae. This material is broken and transported by hydrodynamic (waves and currents) and biological processes (especially feeding and creation of fecal pellets), then deposited in the vicinity of a reef in a bewildering array of sizes and shapes. The result is often sediments that are very poorly sorted (having a variety of particle sizes) and composed of mixtures of odd-shaped particles combined with living organisms (larger macrofauna to bacteria-size organisms) and abundant organic matter (Fig. 3.10).

The techniques used to characterize sand-sized fabric in carbonate sediment are the same as used for clay mineral and siliciclastic sediments: optical, SEM, and TEM imaging of thin sections or polished surfaces of resin-impregnated sediments and CT scanning of whole samples. Much of the early work on carbonate sediments is summarized in [Bathurst 1975] and brought up to date in [Bathurst 1993]. The importance of cementation in preserving carbonate fabric and resisting consolidation is emphasized in those references. In surficial carbonate sediments, the pore water is in contact with the overlying ocean water, and carbonate sediments are exposed to a mixture of postdepositional biological and biogeochemical processes that control fragmentation, dissolution, precipitation, and subsequent cementation of grains and matrix material. Cementation of grains increases the rigidity of the sediment frame resulting in higher values of compressional and shear wave speeds.

Rezak and Lavoie [Rezak and Lavoie 1990, Rezak and Lavoie 1993] characterize fabric in carbonate sediments using: (1) grain-to-grain relationships (shapes, orientations, and the nature of grain-to-grain contacts), (2) grain-to-matrix relationships (grain-supported versus matrix-supported), and (3) matrix component relationships (shapes, sizes, orientations, and the nature of particle-to-particle contacts). Carbonate sediments are considered matrix-supported (Fig. 3.11, left panel) when larger grains are suspended within a fine-grained matrix and grain-supported (Fig. 3.11, right panel) when grains are mostly in contact with finer-grained material found between

Fig. 3.9. Upper: Photograph of carbonate hard grounds on the west coast of Florida north of Tampa Bay. Courtesy of R. I. Ray. These hard grounds are sediment-starved and only a very thin layer of sand-sized carbonate sediment covers the exposed reef material. Lower: Image of shell material within a muddy matrix collected from Chesapeake Bay using the REMOTS interface camera system. Courtesy of J. D. Germano.

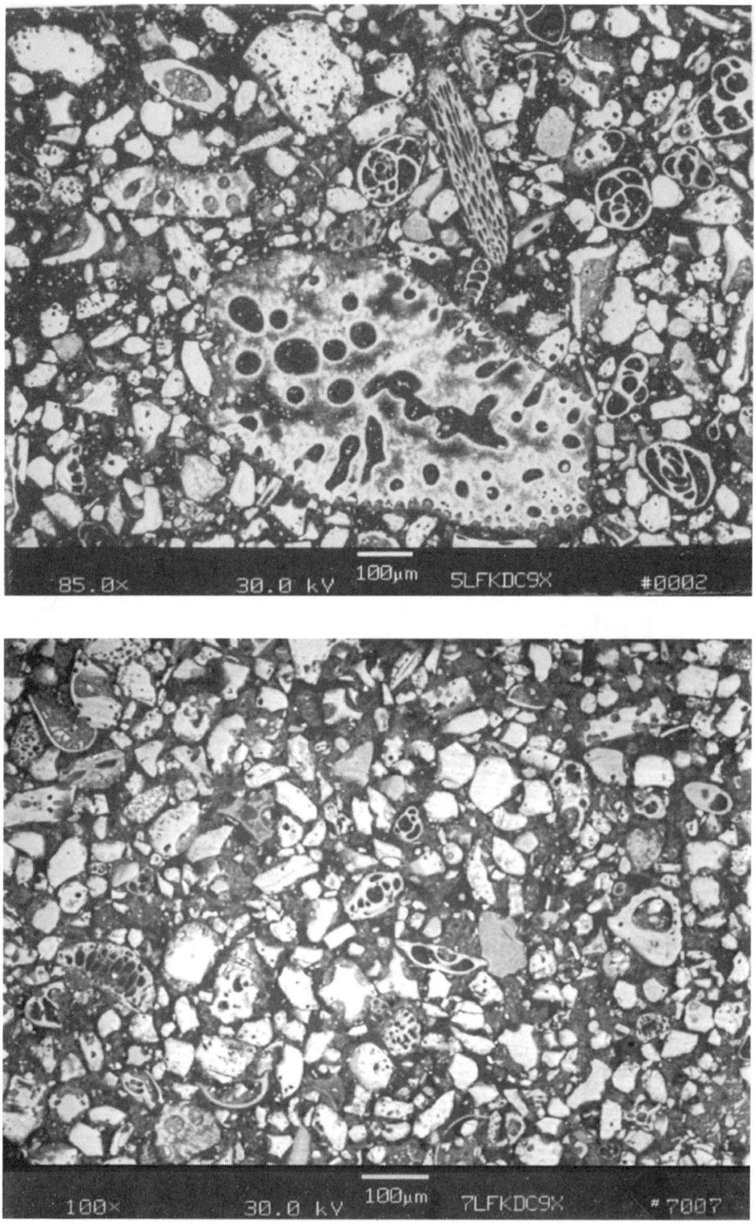

Fig. 3.10. SEM images of polished thin sections of embedded carbonate sediments from the Dry Tortugas. Note the great variety of sizes and shapes of these mostly sand-sized carbonate particles. Some particles are whole skeletal sections of reef-dwelling plants or animals; others are highly degraded carbonate fragments and other debris. These images are approximately 0.95 by 1.34 mm. Courtesy of D. Lavoie.

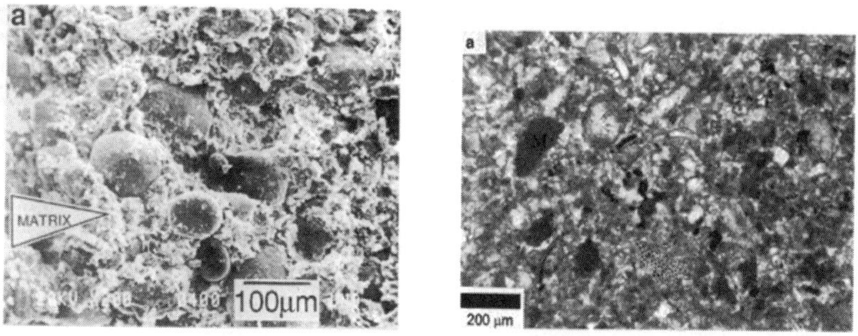

Fig. 3.11. Images of matrix-supported (left) and grain-supported (right) carbonate fabric. The matrix-supported image is a SEM photomicrograph of sand-sized nearly spherical ooids embedded within a matrix of silt- and clay-sized aragonite needles and needle clusters from the Bahama Islands [Bennett et al. 1990b]. The grain-supported image is from carbonate sediments from the Dry Tortugas. Courtesy of D. Lavoie.

grains. Behavior, including geoacoustic behavior, of grain-supported and matrix-supported carbonates can be quite different [Rezak and Lavoie 1990, Bennett et al. 1990a, Bennett et al. 1990b]. In general, sediments with grain-supported fabric are more resistant to compaction and have higher porosity and permeability than matrix-supported sediments [Bennett et al. 1989].

The partitioning of porosity within (intraparticulate porosity) and between (interparticulate porosity) carbonate particles presents problems for both empirical and physical models predicting sediment geoacoustic properties (wave speeds and attenuations) based on sediment porosity or bulk density (see Ch. 5). In the example of Fig. 4.12 (Ch. 4), fragments or plates of the calcareous algae, Halimeda, contained approximately 30% intraparticulate porosity or 10–15% of the total porosity [Richardson et al. 1997]. The shear wave speed in these carbonate sediments was higher than for siliciclastic sediments of comparable porosity, whereas the compressional wave speed was lower than for siliciclastic sediments of comparable mean grain size. The sound speed in the carbonate sands shown in Fig. 3.12 followed similar trends [Fu et al. 2004].

An exception to the very poorly sorted carbonate sediments derived from reef debris are the well-sorted oolitic sands common in the Bahama Islands. Ooids are spherical accretionary grains formed in situ by precipitation of supersaturated carbonate (mostly aragonite) when cold water upwells along energetic coastal tropic shallow-water environments (Fig. 3.13). Because of their mechanism of formation, ooids are well-sorted, medium-to-coarse sand-sized particles that occur in mobile beds. Models for physical, acoustic, and geotechnical properties that treat particles as spheres may be especially suited to oolitic sands except when these beds lose mobility and ooids begin to be cemented together.

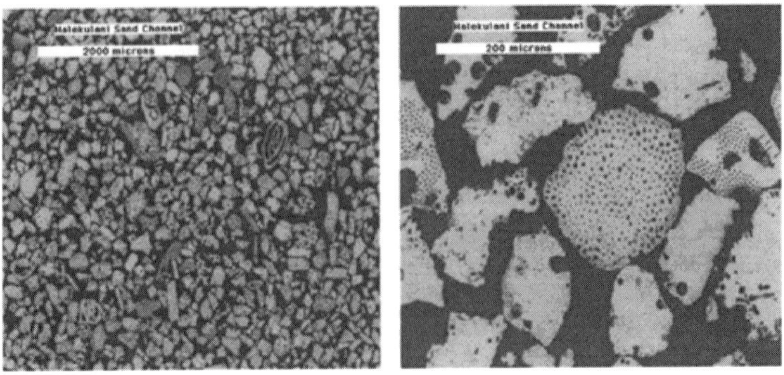

Fig. 3.12. SEM images of polished sections of resin-impregnated coral sand from Waikiki, Hawaii [Fu et al. 2004]. These sands show no evidence of postdepositional cementation within the upper 10 m of sediment. The range of mean grain size (0.19 to 0.25 mm), porosity (49–57%), bulk density (1780–1930 kg m^{-3}), and sound speed (1620 m s^{-1}) are typical for coralline sands but values of porosity are higher and values of sound speed are lower than for quartz sand of comparable mean grain size.

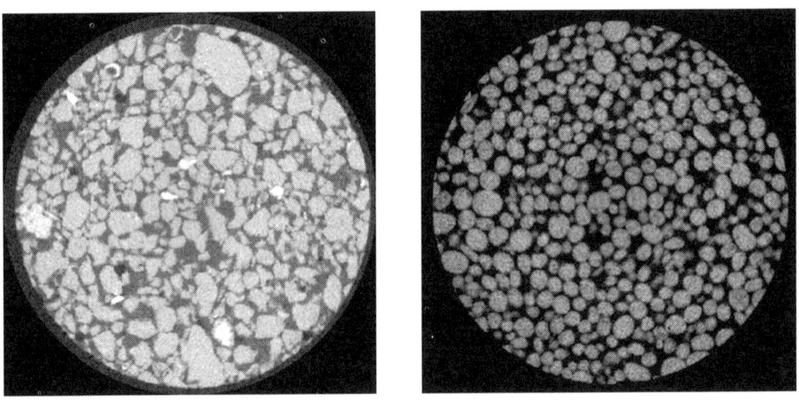

Fig. 3.13. A comparison of high-resolution CT images of resin-impregnated quartz sand (left) collected from near Ft. Walton Beach, Florida, and aragonite ooid sand (right) collected south of Bimini in the Bahama Islands. Both images are 7.7 mm in diameter and have roughly the same mean grain diameter and porosity: 0.375 mm and 37% for the more angular quartz sand; 0.325 mm and 37% for the more spherical ooid sand. However, the sound speed was higher (1836 m s^{-1}) and the attenuation was lower (120 dB m^{-1} at 400 kHz) in the well-sorted ooid sand compared to the moderately well-sorted quartz sand (1775 m s^{-1}, 175 dB m^{-1}). Courtesy of A. H. Reed.

3.3 Bubbles in Sediment

Gas bubbles and associated methane seeps are ubiquitous in organic-rich, muddy sediments of coastal waters and shallow adjacent seas [Judd and Hovland 1992, Richardson and Davis 1998, Fleischer et al. 2001, Judd 2004]. The depth and horizontal distributions of these gassy regions are often determined from seismic profiling. One of the first examples of such profiles (Fig. 3.14) comes from early sound profiling using an echo sounder in the Baltic Sea [Schüler 1952]. The presence of gas bubbles often impedes acoustic characterization of sediments below the gas horizon (Fig. 3.15) and terms such as acoustic masking or blanking, acoustic turbidity, bright spots, wipe-outs, and pulldowns are used to characterize these gas-charged sediments [Fleischer et al. 2001]. Gas bubbles also produce anomalously high acoustic backscattering from the seafloor [Tang et al. 1994, Boyle and Chotiros 1995b, Tang 1996b, Lyons et al. 1996, Chu et al. 1997, Fonseca et al. 2002] degrading the effectiveness of high frequency sonar (see Sect. 14.1.7). Sound speed and attenuation in gassy sediments have unique frequency dependencies with very high attenuation at frequencies near the acoustic resonance of bubbles and low sound speed at frequencies significantly below the bubble resonance [Anderson and Hampton 1980a, Anderson and Hampton 1980b, Bedford and Stern 1983, Wilkens and Richardson 1998, Gardner and Sills 2001, Best et al. 2004].

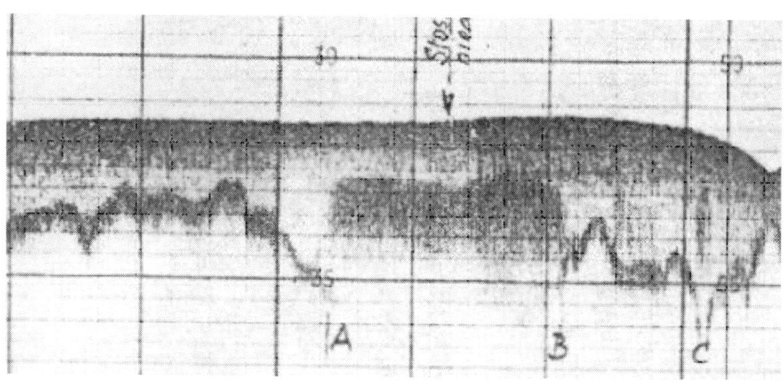

Fig. 3.14. One of the first echo soundings showing acoustic turbidity caused by free gas bubbles in coastal marine sediments. The area between A and B was first referred to as acoustic turbidity by [Schüler 1952] and masks the deeper higher impedance subbottom glacial horizon found in Flensburger Förde in the Baltic Sea.

Methane is the most abundant hydrocarbon gas in marine sediments and can originate from biogenic or thermogenic degradation of organic matter or by degassing and/or cooling of magma or igneous rocks in the mantle [Judd 2003]. Most methane in shallow-water sediments is of biogenic origin,

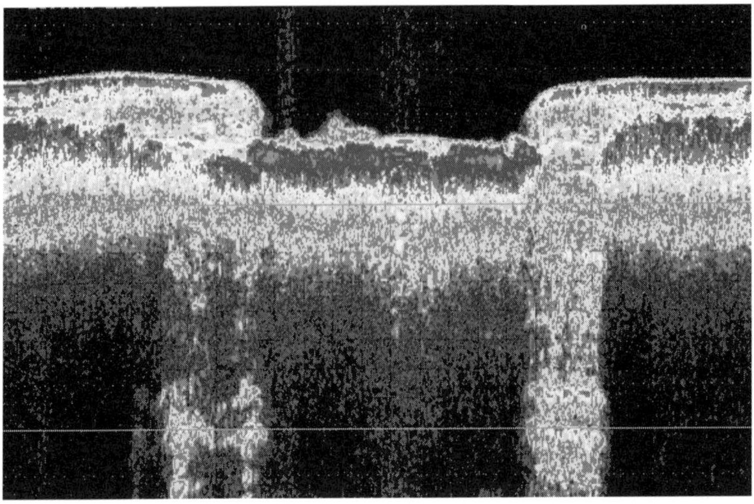

Fig. 3.15. Acoustic image obtained from 12 kHz echo sounder data during experiments on gassy sediments in Eckernförde Bay in the Baltic Sea [Richardson and Davis 1998]. A 50-m-wide "pockmark" is evident. At 5 m depth (indicated by the horizontal white line), strong acoustic returns from glacial deposits are seen just outside the edges of the pockmark. These are masked elsewhere by a layer of methane gas whose uppermost surface is visible as dark red outside the pockmark and violet inside. Bubbles rising through the water column are visible over the pockmark. (see first color insert)

whereas methane in gas seeps can be thermogenic (formed by thermocatalytic degradation of organic matter at high temperature and pressure) or biogenic in origin. Methane derived from abiotic sources in the mantle is thought to be mostly associated with hydrothermal vents and is not considered here. In addition to methane, benthic microalgae (e.g., diatoms, dinoflagellates, cyanobacteria) can produce sufficient oxygen during respiration that pore waters are supersaturated, forming oxygen bubbles that are trapped in surficial sands, especially in the presence of algae mats [Holliday et al. 2004]. The presence, concentrations, and thus importance of these types of bubbles to scattering from the seafloor are currently a subject of conjecture [Boyle and Chotiros 1998, Greenlaw et al. 2004]. Concurrent measurements of acoustic properties and gas characteristics could resolve this issue. Air can be found trapped in intertidal sediments even during high tide when the sediments are fully covered with seawater. Breaking waves in nearshore regions can also inject air in the form of bubbles into otherwise fully saturated sediments. The importance of this process to high-frequency acoustics and the concentration or even presence of air in subtidal sediments are largely unknown.

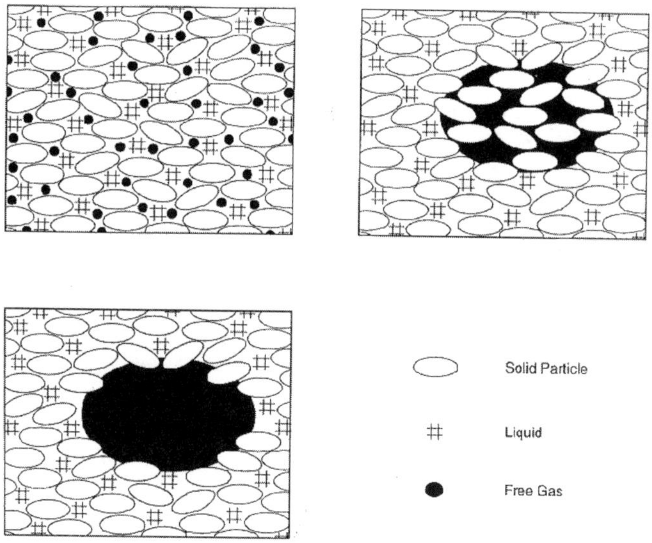

Fig. 3.16. Types of seafloor bubbles. Top left: Type I, interstitial bubbles. Top right: Type II, reservoir bubbles. Bottom left: Type III, sediment displacement bubbles. From [Anderson et al. 1998].

Descriptions of three types of bubbles that commonly occur in marine sediments (Fig. 3.16) have been given by [Anderson et al. 1998]. Type I, interstitial bubbles, are small bubbles that are entirely contained within the pore space between sediment particles. These bubbles are typically spherical and usually occur in sand-size sediments. Type II, reservoir bubbles, are comprised of free gas in pockets that are larger than the pore space and include sediment grains. This type of bubble is common in petroleum deposits and usually contains methane, ethane, and higher-molecular-weight organic gases. Type II bubbles are also typical of air trapped in intertidal or shallow subtidal sands. Type III, displacement bubbles, are larger than the normal pore space but do not contain sediment particles. This type of bubble is surrounded by a wall of sediment that has been displaced as the gas bubble grows. Smaller Type III bubbles may be spherical but larger bubbles tend to be coin-shaped or even irregularly shaped (resembling a corn flake) [Anderson et al. 1998, Boudreau et al. 2005]. Type III bubbles are most often composed of methane and are the most abundant type of bubble encountered in high-frequency acoustic experiments [Richardson and Davis 1998]. The size, shape, and type of bubbles as well as bubble concentration have a major effect on bubble resonance, sound speed, attenuation, and scattering characteristics of the sediment (see Sect. 14.1.7). The remainder of this section will concentrate on biogenic methane in the form of Type III bubbles.

The distribution of shallow-water gassy sediments generally correlates with the distribution of fine-grained, organic-rich sediments, and these types of sediments are most common in estuaries, bays, deltas, drowned coastlines and on the continental shelf near areas of major river discharges (Fig. 3.17). The preponderance of reported cases of gassy sediment are in and around North America and Europe and reflect more the distribution of field studies than any fundamental differences in sediment or gas distribution. Most gas in surficial sediments originates from the generation of methane as a by-product of metabolism by methanogenic bacteria [Floodgate and Judd 1992, Martens et al. 1998, Whiticar 2002]. In mud with high organic content, aerobic respiration is usually restricted to the upper few centimeters, or even the upper few millimeters, of the seafloor, an oxygen-rich zone that is often brown in color. Below this zone, sulfur bacteria produce black, odoriferous sediment rich in hydrogen sulfide. Still deeper, after all of the sulfate is used as a terminal electron acceptor, methanogenic bacteria further reduce simple organic compounds, producing methane. Pore-water methane concentrations increase until they exceed saturation levels and free methane bubbles form in the sediment [Abegg and Anderson 1997]. Bubble volume in shallow-water sediments is typically less than 1% but volumes as high as 6–9% have been reported from pockmarks (Fig. 3.18) [Lyons et al. 1996] and as high as 12% from sites of very high anaerobic mineralization (Fig. 3.19) [Martens and Klump 1980].

Fig. 3.17. Global view of reported sites of free gas bubbles in shallow-water sediments based on [Fleischer et al. 2001]. (see first color insert)

A kinetic model has been developed by [Martens et al. 1998] for bacterial methane production which includes: (a) methane consumption at the base of the hydrogen-sulfide-reduced horizon, (b) advective and diffusive trans-

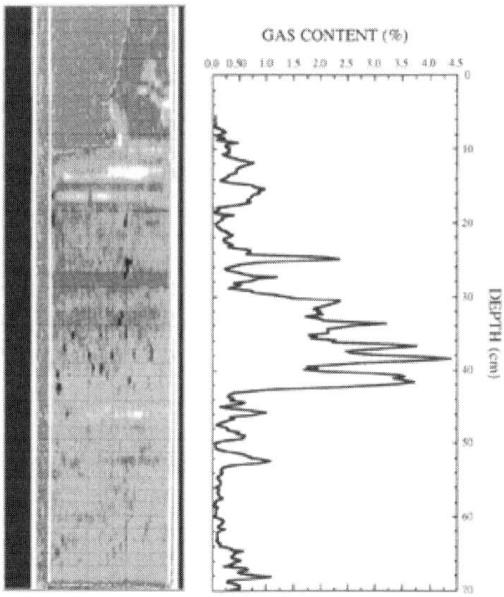

Fig. 3.18. The left panel is a CT-scan showing methane gas bubbles from a sediment core collected from center of a pockmark in Eckernförde Bay, Baltic Sea (see Fig. 3.15). False colors, based on X-ray densities, represent water (blue), sediment (brown), core liner (white) and methane bubbles (black). The bubble volume was as high as 4% as shown in the plot on the right. The core had been maintained at in situ temperature and pressure (see [Abegg and Anderson 1997] for details). (see first color insert)

Fig. 3.19. X-radiograph of a 2.5-cm-thick section of sediment (approximately 10 by 20 cm) collected by divers during summer months from the gassy sediments of Cape Lookout, North Carolina [Martens and Klump 1980]. Large methane bubbles appear as white (less dense) objects in this positive X-ray. Note the open bubble tubes (light gray) leading to the sediment–water interface.

port processes, (c) organic supply, and (d) sedimentation rates. They have successfully used this model to predict methane and sulfate concentration profiles, rates of biogeochemical reactions, and the depth distribution of dissolved and free gas in the sediments of Eckernförde Bay, Baltic Sea, and Cape Lookout, North Carolina. The flux rate and quality (reactivity) of organic matter deposited at the seafloor are the most important factors controlling the distribution of methane in these sediments.

Methane saturation in sediments is controlled by temperature, salinity, and pressure and has been modeled in Eckernförde Bay by [Wever and Fiedler 1995, Abegg and Anderson 1997, Wever et al. 1998] and in Chesapeake Bay by [Hagen and Vogt 1999]. Gas bubble horizons migrate with seasonal changes in sediment temperature and with rapid local changes in hydrostatic conditions due to storm or tidal conditions [D. Jackson et al. 1998]. Gas ebullition has been observed during rapid pressure changes [Martens and Klump 1980, D. Jackson et al. 1998] and during changes in pore water salinity resulting from increased ground water transport [Bussmann and Suess 1998].

Knowing total bubble volume alone is insufficient to model the acoustic behavior of gassy sediments. The bubble size distribution, shape and spatial and temporal distributions are important to both propagation (sound speed and attenuation) and scattering. Individual bubble volume, size, shape, and spatial and temporal distributions have been measured using a variety of techniques including: seismic profiling, geochemical methods, standard X-radiography, and medical and high-resolution X-ray computed tomography [Martens and Klump 1980, Richardson and Bryant 1996, Abegg and Anderson 1997, Albert et al. 1998, Anderson et al. 1998, Boudreau et al. 2005]. Perhaps the most comprehensive study of the spatial distribution of bubbles including size distribution and shape was conducted using medical CT scanning of sediments collected from Eckernförde Bay, Baltic Sea, by [Anderson et al. 1998]. These sediments were placed in pressure-tight containers by divers at the seafloor and were retained at in situ pressure and temperature until imaged. Most bubbles were Type III bubbles and ranged in size between 0.42 mm equivalent radius (the best resolution possible with the medical CT system) and 10 mm (Figs. 3.18 and 3.20). The equivalent radius is the radius of a sphere having the same volume as the measured bubble. The number of bubbles smaller than 0.42 mm is unknown, but acoustic measurements suggest significant numbers as small as 0.25 mm [Wilkens and Richardson 1998]. The size and number of bubbles varied considerably both with depth and between cores collected within a few meters of each other. Gas concentrations ranged from less than 0.1% to as high as 9% with high volumes associated with larger bubbles. Bubble shapes ranged from almost spherical for the smaller bubbles to highly eccentric for the larger bubbles. The most common shape was an oblate spheroid with the long axis oriented vertically (Fig. 3.20). More recent work using a high-resolution CT scanner [Boudreau et al. 2005] shows bubbles with a shape similar to a corn flake (Fig. 3.21). Growth of Type III displacement bubbles in soft sediment can be modeled by fracture mechanics [Johnson et al. 2002] resulting in the highly eccentric shapes shown in Figs. 3.18–3.21.

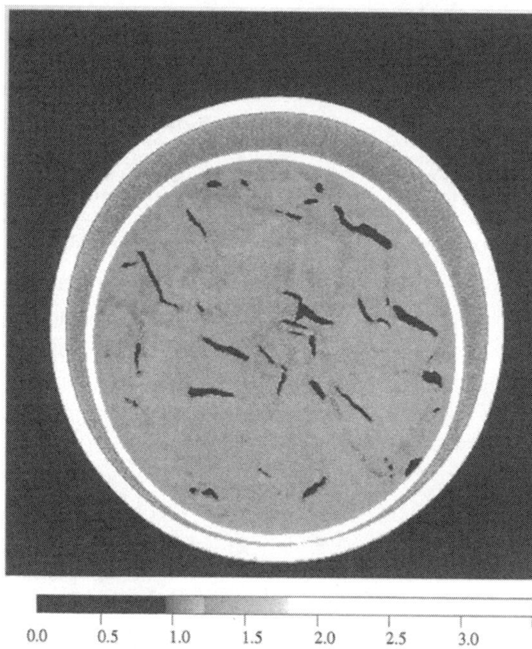

Fig. 3.20. A reconstructed slice from a medical CT scan showing the distribution of methane gas bubbles from a sediment core collected from center of a pockmark in Eckernförde Bay, Baltic Sea (see Figs. 3.15 and 3.18). False colors, based on X-ray densities, represent water (blue), sediment (brown), core liner (white), and methane bubbles (black). The core has been maintained at in situ temperature and pressure (see [Abegg and Anderson 1997] for details). (see first color insert)

3.4 Effects of Hydrodynamic and Biological Processes

Predicting how high-frequency acoustic energy interacts with the seafloor often requires a thorough knowledge of sediment physical properties and morphology of the seafloor. As will be shown in this section, the spatial distribution and temporal changes in sediment bulk properties and seafloor morphology are controlled by a combination of biological, hydrodynamic, and depositional processes [Richardson and Young 1980, Bentley and Nittrouer 2003]. The links between hydrodynamic and biological processes, sediment physical

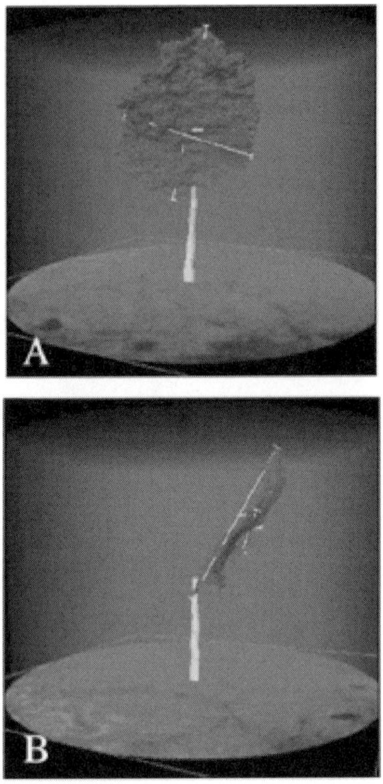

Fig. 3.21. A 3D reconstruction of a bubble injected into fine-grained sediment using high-resolution X-ray computed tomography [Boudreau et al. 2005]. The bubble is blue and the injection capillary is yellow. The bubble is 20 mm across (A) and 7 mm thick (B) with a resulting volume of 0.3 cm^3. (see first color insert)

and geoacoustic properties, and acoustic scattering, propagation, and penetration are often overlooked during high-frequency acoustic experiments. Understanding these relationships may provide insights into understanding and thus predicting temporal changes and spatial variability of high-frequency acoustic–seafloor interactions. For example, winter storms can create periodic roughness features (ripples) in a matter of hours that can increase seafloor acoustic scattering strengths. These highly anisotropic features can in turn be degraded by biological processes in a matter of days yielding a much reduced isotropic pattern of scattering. It will be shown how modeling and predicting acoustic–seafloor interactions will benefit from an understanding of how biological and hydrodynamic processes create, destroy, or alter seafloor characteristics. This is an important area of research that is still in its infancy.

Theories for acoustic propagation (fluid, elastic, viscoelastic, or poroelastic) and models for scattering (roughness and volume) require extensive characterization of sediment bulk properties (density, porosity, permeability, and pore-scale properties such as tortuosity and pore size distribution). In addition, poroelastic theory requires knowledge of the properties of the pore fluid (fluid density, bulk modulus, and viscosity) and solid grains (density and bulk modulus). Buckingham's theory (Sect. 9.8) also requires knowledge of grain shape, number and types of grain contacts, and surface roughness. Many of the aforementioned physical properties are estimated from grain size statistics. For some propagation theories the mean or effective values of these sediment properties are sufficient; but for predicting volume scattering the statistical distributions of these properties (especially bulk density and sound speed) are required. Larger features such as mud inclusions, shells, and gas bubbles are discrete scatterers of acoustic energy. Knowledge of the morphology of the seafloor (roughness and slope), including deterministic or statistically defined rough-interface conditions, are required by many scattering and penetration models. Biological, hydrodynamic, erosional, and depositional processes all play major roles in controlling mean values and statistical distributions of seafloor properties. Acoustical oceanographers use acoustic–seafloor interactions to determine the magnitude and rates of biological and hydrodynamic processes. This is especially true for studies of sediment transport (for a recent review see [Thorne and Hanes 2002]) and is beginning to be important for rate studies in benthic ecology [Jumars et al. 1996, Self et al. 2001]. High-frequency sonar, including side-scan, normal-incidence echo sounders, and multibeam systems, are used to determine sediment provinces, biotopes, and estimate sediment physical properties [Diaz et al. 2004, Schock 2004].

There is a long history of studies of the interaction between hydrodynamic and biological processes, sediment physical properties, sediment structure, and seafloor morphology, but these different disciplines have approached the problem from different perspectives. Most benthic biological studies have concentrated on the effects of sediment properties on the distribution and abundance of benthic fauna. It was not until recently that the importance of the behavior of benthic fauna on sediment structure and seafloor morphology was emphasized. By contrast, most studies concerned with the interaction of hydrodynamic processes and sediment structure have concentrated on the effects of waves and currents on seafloor erosion and transport, deposition, sediment distribution, and morphology. Recently the feedback of the effects of seafloor physical properties and morphology on surface gravity waves was recognized [Tolman 1994, Ardhuin et. al. 2003, Sheremet et al. 2005]. Both of these fields will be reviewed in this section with emphasis on their influence on the sediment properties and structure important to acoustic–seafloor interactions. For a more comprehensive coverage of interactions between biological processes and the seafloor, the reader is directed to reviews by [Rhoads 1974, Carney 1981, Rhoads and Boyer 1982, Murray et al. 2002]. Hydrodynamic–seafloor interactions are covered in several textbooks in the *Advanced Series on Ocean Engineering* from World Scientific; of particular interest are [Fredsøe and Deigaard 1992] and [Nielsen 1992]. In addition, [Middleton 1984, Julien 1995, Winterweap and van Kestern 2004] are excellent sources of information on sediment erosion, transport, and deposition.

3.4.1 Hydrodynamic Processes

Bottom currents, surface gravity waves, and storms have major and often predictable effects on the distribution of sediment types, formation of beach profiles [Inman et al. 1993], types and shapes of sand ripples [Clifton 1976], distribution of sand banks and ridges [Dyer and Huntley 1999], and seafloor microtopography [Briggs 1989]. As a general rule, muddy sediments are more common in low-energy environments because of the slow settling rates of clay- and silt-sized particles. Biological processes, which will be discussed in the next section, tend to control sediment physical properties and microroughness in these environments, often destroying layering generated by floods or depositional events [Richardson and Young 1980, Bentley and Nittrouer 2003]. Sands are more common in high-energy environments where offshore bars and berms, sand ridges, and sand ripples are the most obvious features. Where storms or tidal currents are common, these features, which are in equilibrium with hydrodynamic processes, are predictable [Traykovski et al. 1999]; but, where storms are rare, larger features are relict or degraded by biological processes, and the time history of hydrodynamic processes must be considered to predict sediment properties and roughness [Briggs et al. 2001]. An excellent review of the types of depositional features typical of nearshore sedimentary regimes is provided by [Reineck and Singh 1973]. These features include wave and current ripples, megaripples, sand ridges or dunes, scour and markings caused by objects on the seafloor, beach profiles, and a variety of bedding types. The following paragraphs will concentrate on the development of tide-, storm-, and wave-induced sand ripples which are of a size to scatter high-frequency acoustic energy (e.g., [Williams et al. 2002b]) and influence acoustic penetration into the seafloor (e.g., [D. Jackson et al. 2002]). It is these features that are of most interest to current high-frequency acoustic research.

Ripples are small bed features (wavelengths 10–200 cm) that form on sandy beds with grain sizes smaller than 0.8 mm (coarse sand) when the flow speeds exceed the threshold of sediment motion but are below that which can wash out the ripples (sheet flow, Fig. 3.22). Current-generated ripples are asymmetrical with steeper slopes on the downstream side of crest. They form irregular patterns when viewed from above. Sediment transport or ripple migration is in the direction of dominant flow, either downstream or in the direction of the usually stronger ebb tidal flow. Wave-generated ripples are often symmetrical about the crest in cross section, with the long crests being relatively sharp. In water sufficiently shallow that waves interact with the seafloor, the waves become asymmetrical and the ripples tend to migrate inshore in response to the asymmetry of the bottom wave velocity distribution.

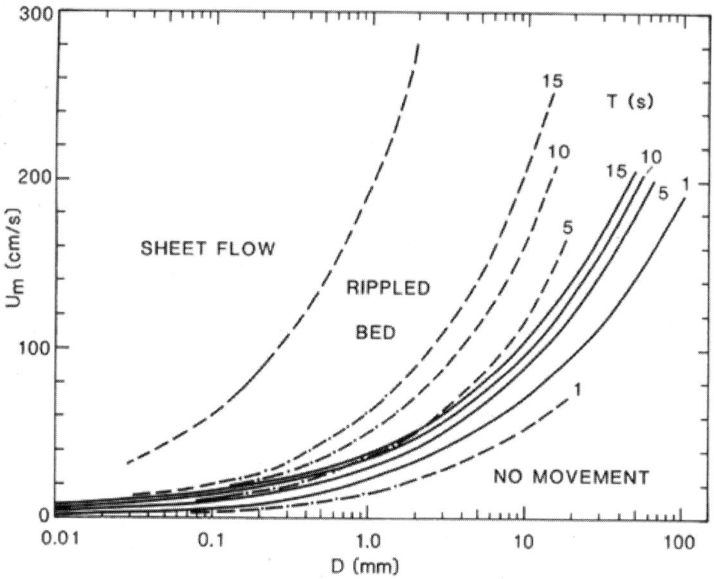

Fig. 3.22. Velocity thresholds versus grain size for different wave periods, T, based for grain movement, formation of ripple beds and sheet flow [Clifton and Dingler 1984].

In coastal regions, wave-generated ripples tend to dominate except where tidal currents are strong. All ripple types degrade with time due to physical and biological processes. A conceptual model of wave-generated bedform states in nearshore sandy environments has been developed by [Clifton 1976]. This model is based on extensive observations by SCUBA divers after storms (Fig. 3.23), and depends on sediment grain size, wave period, wave amplitude, and the statistics (energy and skewness or asymmetry) of the incident wave field (Fig. 3.24). Nearshore is defined in Clifton's model as shoreward of the depth at which asymmetric waves appear or shoaling first begins to develop. Bedforms evolve in response to increasing wave energy (wave orbital velocity) in a more-or-less predictable manner: from irregular ripples, to cross ripples, to linear transition ripples, and finally to a flat bed at maximum wave energy. Various bed states are illustrated in Fig. 3.25, which shows images from a rotary-scanning sonar operating at 2.25 MHz [Hay and Mudge 2005]. The large features, "lunate megaripples," appearing at high wave energy, occur infrequently. It is interesting that asymmetry (skewness) of surface waves and the strength or direction of longshore currents appears uncorrelated with bedform type. The surface wave nonlinearity does, however, generate sand ripple asymmetry in which shoreward-facing ripple slopes are steeper than seaward-facing slopes for cross- and linear-transition ripple states. This is consistent with the observed shoreward ripple migration and onshore net

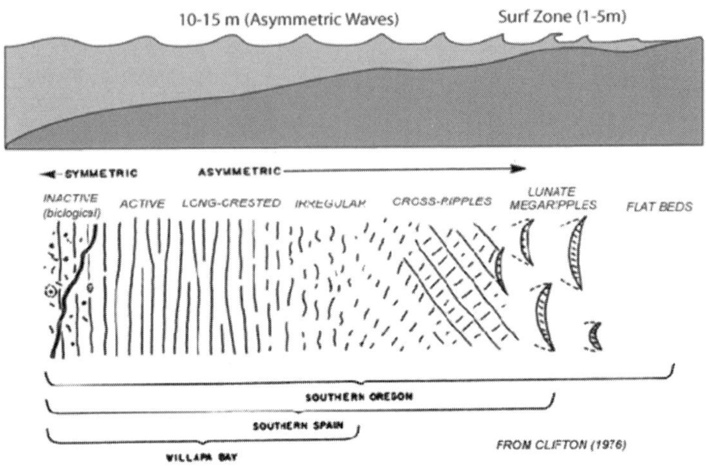

Fig. 3.23. A sequence of wave-formed ripple structures commonly observed within high-energy sandy environments. This depth-related sequence of bedforms was based on observations made by scuba divers after storm events along the southern Oregon coast, southern Spain, and Willapa Bay [Clifton 1976]. The similarity with the bedforms observed by [Hay and Mudge 2005] using high-frequency sonar (Fig. 3.25) demonstrates that these bed states are strong functions of incident wave energy and have temporal as well as spatial dependence. Figure adapted by P. Traykovski from [Clifton 1976].

sediment transport [Clifton 1976, Hay and Mudge 2005]. Each bed type has its own high-frequency (MHz) acoustic scattering characteristics. The flat and irregular rippled bedforms are nearly isotropic (uniform spectral densities), with much lower scattering levels from the flat bedforms. The linear transition rippled bedforms exhibit the greatest anisotropy with strips of higher scattering perpendicular to ripple strike (offshore and onshore). Cross-rippled bedforms are intermediate with respect to symmetry and energy. Both cross-rippled and linear transition ripples exhibit higher backscatter strengths in the seaward direction, a result of asymmetry in ripple slopes. It was suggested [Hay and Mudge 2005] that a spectrum-based anisotropy index could be developed for automatic bed-state recognition from rotary sonar imagery.

A set of equations was developed by [Wiberg and Harris 1994] to predict ripple type and geometry based on mean grain size, wave orbital diameter, and estimated anorbital ripple height, η. Wave orbital diameter can be estimated from surface gravity wave height and period and the water depth, using linear wave theory. For orbital ripples (Fig. 3.24) where the ratio of orbital diameter, d_0, to mean grain size, D, is less than 2000, the ripple

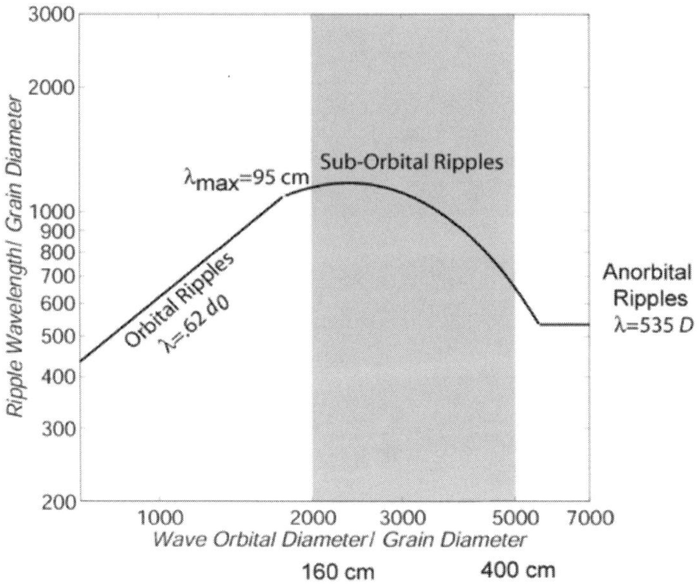

Fig. 3.24. The canonical two-dimensional parameter space for wave-formed ripples based on [Clifton and Dingler 1984].

wave length, λ, is a function of the wave orbital diameter ($\lambda = 0.62 d_0$), and the ratio of ripple height to wavelength is constant ($\eta/\lambda = 0.17$). For anorbital ripples where $d_0/D > 5000$, the ripple wavelength is a function of the mean grain size ($\lambda = 500D$), and the ratio of wavelength to ripple height is less for orbital ripples and decreases with wave height. Suborbital ripples are intermediate between the other two ripple types. More recent long-term, continuous observations using sector scanning sonar (Fig. 3.25) tend to support the models of Clifton and Wiberg and Harris [Traykovski et al. 1999, Crawford and Hay 2001, Hay and Mudge 2005]. These instruments are also called rotary or fan-beam sonars. It is interesting that most studies of evolving bedforms have been conducted in high-energy environments where ever-changing bedform types are in equilibrium with recent wave or current events. In less energetic environments, bedform roughness degrades with time to a state of low-amplitude isotropic roughness.

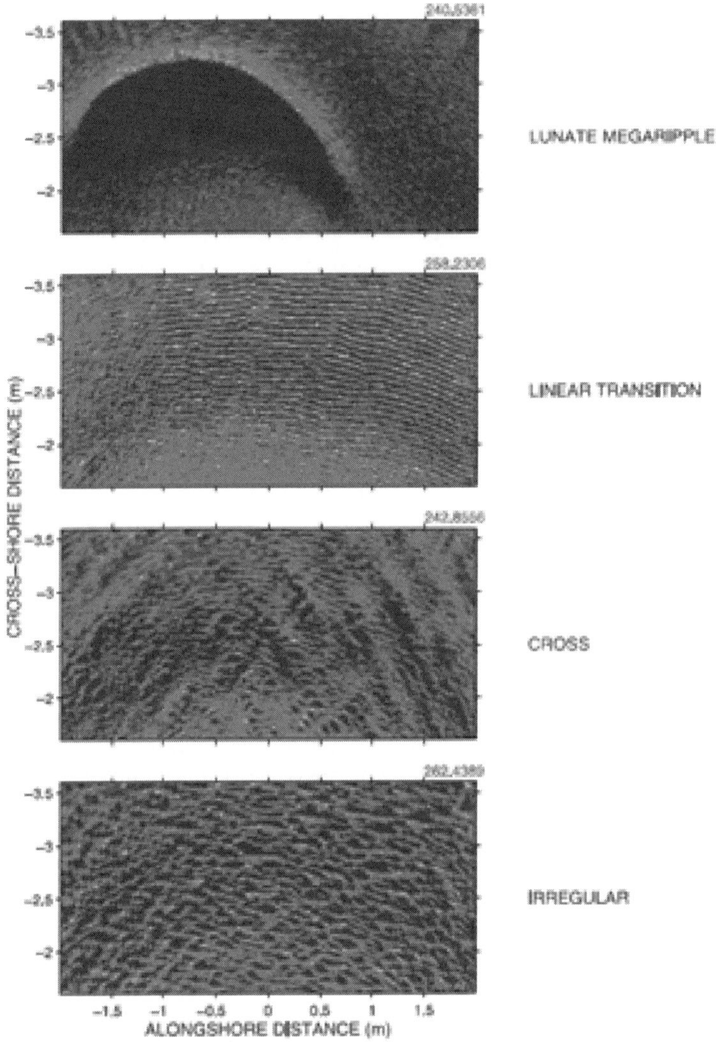

Fig. 3.25. Bed state images obtained by [Hay and Mudge 2005] using a rotary sonar with a "fan-beam," narrow in the horizontal and wide in the vertical. Four principal seafloor states are illustrated by images (2 x 4 m) collected during Sandy-Duck97 near Duck, North Carolina. Wave energy increases moving from the lower to the upper panels of the figure. The fifth bed state is featureless and therefore not shown. Color scale indicates backscatter intensity, with blue corresponding to low intensity and red to high intensity. (see first color insert)

3.4.2 Biological Processes

Bioturbation of surficial coastal sediments can have profound effects on sedimentary characteristics [Rhoads 1974, Richardson and Young 1980, Carney 1981, Rhoads and Boyer 1982, Wheatcroft 1990, Murray et al. 2002, Crusius et al. 2004]. Bioturbation in this context refers to the churning, stirring, mixing, or reworking of sediments by organisms during such activities as feeding, locomotion, or home building [Richter 1952]. The types of bioturbation are as diverse as the fauna that inhabit the seafloor. The result of this biological activity is an ever-changing, often heterogeneous, landscape that is called the "benthic boundary layer."

Deposit feeding is one of the dominant modes of feeding in fine-grained sediments and includes the bulk ingestion and defecation of particles as animals move through the sediment. Two examples include conveyer-belt species that feed at depth depositing sediments at the surface (Fig. 3.26) and tentacular feeders that ingest particles recently deposited on the sediment creating fecal pellets deposited on the surface or at depth. Also abundant are filter feeders which collect organic-rich material from the water column, depositing it at or below the sediment surface. In fine-grained sediments, much of the sediment surface can be transformed into sand-sized fecal pellets by these deposit- or suspension-feeding processes (Fig. 3.27). Other animals selectively feed on either large or small particles creating local heterogeneity in grain size or graded bedding. Home building by benthic fauna can create a variety of temporary or permanent structures such as tubes, burrows, or galleries. Some animals collect larger particles incorporating them into their tubes or burrows. Mobile surface deposit feeders create a variety of surface features including trails, tracks, mounds, and pits that cover the seafloor. Others, such as migrating sand dollars or sea cucumbers, destroy surface features at an alarming rate (Fig. 3.28). Demersal fish, decapods, and other larger mobile epifauna that feed near the sediment floor destroy sediment surface features such as ripples and create new structures such as small pits, mounds, and other feeding traces (Fig. 3.29). Other larger mobile benthic fauna, such as octopi, collect and concentrate shells near their home. The length scales of the features range from meters (ray pits) to decimeters (mounds created by larger conveyer-belt feeders) in diameter, down to millimeter (worm trails; amphipod tubes), to submillimeter scales (fecal pellets).

The activities of benthic fauna and the structures they construct have a major effect not only on local sedimentary conditions but also impact larger-scale geomorphological features and processes at scales of 50 meters to hundreds of kilometers [Murray et al. 2002]. Interactions between benthic fauna and sediments at the millimeter-to-meter scales (microscale interactions) influence sediment properties and processes associated with sediment stability, erosion, transport, sedimentation, and consolidation, and can change larger-scale sedimentary characteristics. Tubes and burrows increase large-scale permeability, which could increase dewatering and promote consolidation. How-

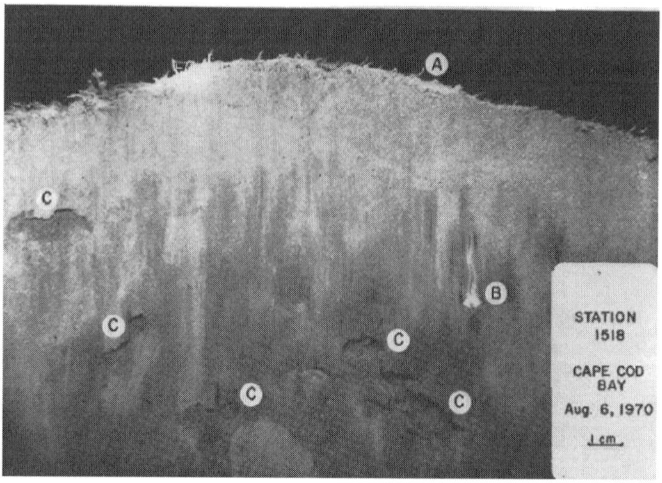

Fig. 3.26. Sediment-profile photographs of sediments in Cape Cod Bay, showing the cone-shaped mound created by *Molpadia oolitica*, a common conveyer-belt species found in these silty-clay sediments [Rhoads and Young 1971]. The mound is densely populated by tubedwelling sabellid polychaetes (A) which tends to stabilize the sediment surface. The areas of intensive feeding by the holothurian *M. oolitica* are easily recognized as sand-filled voids (C). A free-living (errant) deposit-feeding polychaete (B) also occurs in the photograph. Feeding by this conveyer-belt species thus creates considerable topographic roughness and a high degree of volume heterogeneity in sediment grain size, porosity, and bulk density.

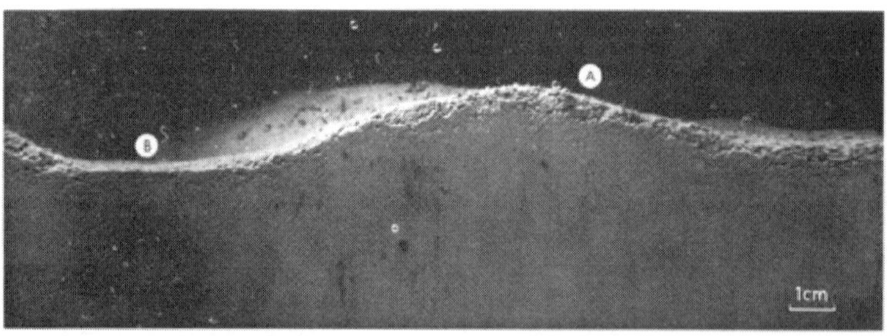

Fig. 3.27. Sediment-profile photograph of sediments in Cape Cod Bay showing a stable fecal cone populated by tube-dwelling suspension feeders (A) [Rhoads and Young 1971]. The effective grain size of the near-surface fecal pellets is much larger than the sediments below. Also note that the adjacent depression (B) is filled with unconsolidated fecal-rich mud which is more easily resuspended by tidal flow than the larger-size fecal pellets on top of the mound.

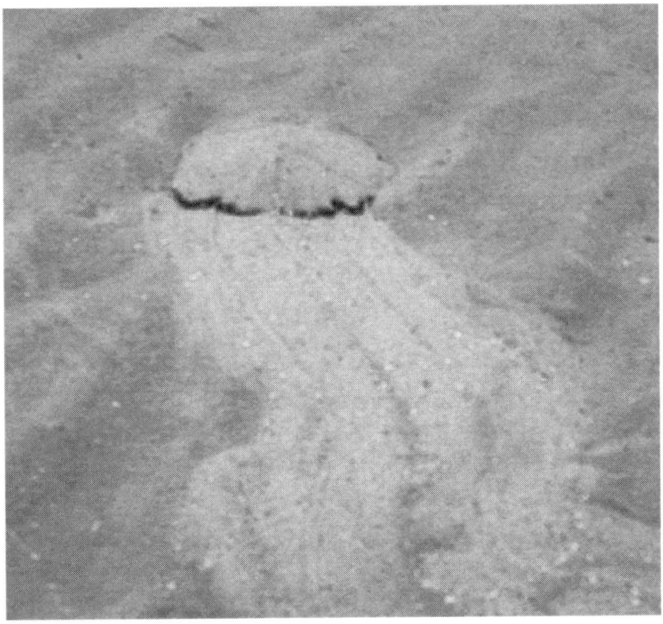

Fig. 3.28. A sand dollar creating track in fine sand in shallow water off the western Florida coast. Note the destruction of the wave-induced ripples. Photograph by R. I. Ray.

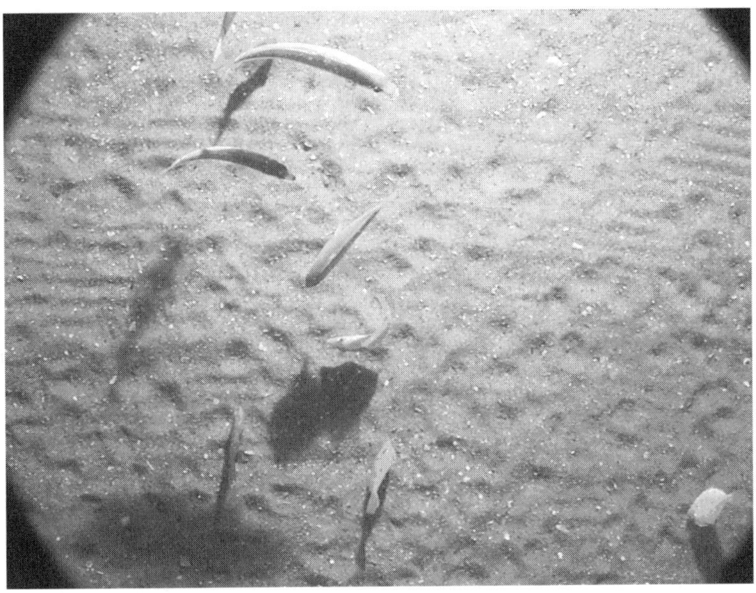

Fig. 3.29. Pockmarks created by the feeding of pinfish during the SAX99 experiments off in the northeastern Gulf of Mexico. Photograph by A.P. Lyons.

ever, this is a point currently in debate, as tubes with more-or-less impermeable linings may actually resist dewatering. The tubes of benthic fauna also help stabilize the sediment surface, increasing its resistance to erosion and promoting sedimentation. Tubes, mucus, cementation, algae filaments, and consolidation can increase sediment shear strength. Mixing of sediments by benthic fauna can obliterate layering or structures created by hydrodynamic processes or, alternatively, create layering, such as graded bedding, by selective feeding. Suspension-feeding benthic fauna can completely filter the water column in a matter of a few days, reducing suspended matter, increasing visibility, and temporarily increasing rates of sediment deposition, often in the form of sand-size fecal pellets or more loosely compacted pseudofaeces. Burrowing benthic fauna have been shown to compact sediment by increasing density and decreasing porosity compared to self-weight consolidation. Benthic fauna have also been shown to fluff, dilate, or decrease density in more compacted sediments. Changes in bed roughness (microtopography) can modify boundary layer flow near the sediment surface, either increasing or decreasing turbulent flow near the bed depending on the biogenic roughness scales (size and density). Changes in biogenic roughness can alter sediment stability, erosion, transport, and sedimentation patterns. Given the omnipresence and wide variety of benthic fauna–seafloor interactions, it is not surprising that attempts have been made to correlate high-frequency acoustic bottom interactions to bioturbation or other biogenic processes.

A search of the term "bioturbation" from peer-reviewed papers on the ISI Web of Science resulted in 1620 hits, most of which were from the marine environment. At least 50 dealt directly or indirectly with acoustics-related studies. In the following paragraphs, these relationships are explored. From the perspective of the acoustician, the prediction of acoustic phenomena from sediment physical properties or roughness (the forward problem) is of greatest interest but, from a benthic ecologist's perspective, acoustic remote characterization of benthic biotopes (benthic physical habitats and their associated biota) and high-frequency acoustic characterization of the spatial and temporal effects of benthic fauna on sediment characteristics (inverse problems) are of greater interest.

One of the first attempts to relate bioturbation by benthic animals to sediment geoacoustic properties was provided by [Richardson and Young 1980]. In a review of the literature, they demonstrate how bioturbation profoundly affects the physical properties of sediments. Activities such as burrowing, ingestion/digestion/defecation, tube building, biodeposition, cementation, and metabolic activities modify sediment porosity, grain size, bulk density, rigidity, compressibility, and microroughness properties important to high-frequency acoustics. Using an example of dewatering or compaction of a fine-grained sediment they, perhaps naively, calculated the resulting changes in sound speed and attenuation, impedance, reflection coefficient, and bottom loss. Later experiments reported by [Rhoads and Boyer 1982, Richardson et al. 1983, Richardson 1983, Jones and Jago 1991, Jones and Jago 1993, Briggs and Richardson 1997, Rowden et al. 1998, Richardson et al. 2002b] all confirmed a relationship between bioturbation, sediment physical properties, and sediment geoacoustic properties. These studies and others have shown that bioturbation alters sediment physical properties including grain size and sorting, porosity or water content, bulk density, permeability, packing, tortuosity, electrical resistivity, and consolidation behavior. In addition, bioturbation has been shown to alter the sediment fabric, affecting sediment shear strength, shear modulus, and shear wave speed [Jones and Jago 1993]. Bioturbation can alter sediment density and porosity thus directly affecting sound speed [Richardson 1983], often creating strong gradients in sound speed and attenuation. As demonstrated by [Richardson et al. 1983], intense bioturbation by surface-dwelling and deeper-living deposit feeders also can thoroughly mix the sediment, destroying layering from previous storm deposits thus creating a spatially and temporally homogeneous deposit. Sound speed and attenuation in these homogeneous deposits exhibit little vertical or horizontal variability. Other species can create considerable variability in sediment physical and geoacoustic properties as well as in sediment impedance [Richardson et al. 2002b]. Most of the aforementioned studies correlated values of sediment physical, geoacoustic, and geotechnical properties to the presence, absence, abundance, or vertical distribution of fauna that are known to alter seafloor properties. Therefore, in spite of the considerable work relating bioturbation to sediment physical properties and to geoacoustic and geotechnical behavior, predictive models relating bioturbation to sediment properties and to behavior are rare (see [Rhoads and Boyer 1982] for one of the few examples). In contrast, many bioturbation models have been developed to describe the sediment and pore water mixing process and those will be described later in this section. These models have rarely been used to predict the mean values or spatial distribution of values of sediment properties.

High-frequency acoustic systems (side scan or normal-incidence echo sounders) have been used to map the distribution of benthic communities or assemblages based on the community association with particular sediment characteristics (grain size, compactness, density, or bottom roughness) or from the distinctive scattering from hard parts such as shells or from habitat structures such as tubes or burrows (recent examples include [Brown et al. 2002, Freitas et al. 2003, Diaz et al. 2004]). However, the general conclusion from these acoustic habitat mapping studies is that acoustic classification techniques currently lack sufficient discrimination of benthic habitats unless supplemented by considerable ground-truth, either in the form of benthic surveys or sediment characterization [Diaz et al. 2004].

High-frequency (265 kHz and higher) acoustic detection of the diurnal vertical migration patterns of benthopelagic fauna has been reported by Holliday, Greenlaw, and McGehee in [Richardson et al. 2001a] and by [Kringel et al. 2003] for coastal and estuarine environments. Small crustaceans, typically mysids or decapod shrimp, emerge from the seafloor at night, presumably feeding in the water column, then reenter the seafloor for protection from predation just before dawn. This behavior may be more common than predicted by emergence traps [Kringel et al. 2003] and can provide another mechanism of sediment mixing or bioturbation.

High-frequency (40 kHz) bottom backscattering measurements made using a bottom-mounted scanning system [Dworski and Jackson 1994] have shown evidence of the presence and movement of benthic fauna and the temporal effects of bioturbation on seafloor scattering. Experiments have included natural environments [Jumars et al. 1996, D. Jackson et al. 1996b, Jones 1999, Jones and Jackson 2000] as well as artificial manipulations of live fauna, fauna replicates, and seafloor roughness [Self et al. 2001, Williams 2001b, Richardson et al. 2002b] and Jumars and Self in [Richardson et al. 2001b] The objectives of these experiments were to determine if high-frequency acoustics could be used to quantify benthic faunal behavior and/or qualitatively measure the effects of benthic processes on sediment physical structure including micro-roughness. The results to date have been partially successful. Acoustic scattering from larger fauna with hard parts was easily determined from acoustic scans [Jumars et al. 1996] and vertical (diurnal) and horizontal migrations have been quantified [Holliday et al. 2004]. This should not be surprising as numerous high-frequency acoustic surveys have noted strong backscatter strengths from hard-bodied benthic fauna living at or near the sediment surface (e.g., [Fenstermacher et al. 2001]). Increased values of backscatter strength have also been noted with the deliberate introduction of larger numbers of smaller bivalves during acoustic experiments [Self et al. 2001].

Temporal changes in backscattering, measured using correlation techniques, have demonstrated changes in seafloor roughness and volume heterogeneity attributed to bioturbation by benthic fauna. Most of the temporal decorrelation is attributed to phase changes in the acoustic signal rather than changes in mean scattering level. This may indicate spatial reordering of bottom microroughness or bulk density by benthic fauna rather than changes in average sediment properties or in changes in roughness or volume heterogeneity spectra (see Chs. 6, 7). In sandy sediment, changes in seafloor roughness or movement of discrete scatterers such as shells led to a rapid temporal decorrelation of scattering strength (Jackson and Richardson, unpublished). Some natural or artificial manipulations of fine-grained sediments having low bulk density yield no measurable changes in high-frequency backscatter strengths. The lack of sound speed contrast between the overlying water and high-porosity sediments suggests even significant changes in bottom roughness may not be detectable using high-frequency acoustics [Self et al. 2001].

Most seafloor microroughness in fine-grained sediments is biogenic in origin [Wheatcroft 1994]. In sand, roughness in the form of quasiperiodic ripples is created by hydrodynamic processes (waves and currents) only to be destroyed by sediment mixing by a variety of organisms that inhabit the benthic boundary layer. The rapid decay in backscatter strength of artificially created roughness was demonstrated by [Richardson et al. 2001b], the result of active destruction of quasiperiodic roughness by bioturbation. Both [Briggs et al. 2001] and [Pouliquen et al. 2004] have demonstrated fluctuations in backscattering strength due to changes in roughness spectra. Both attributed these changing roughness spectra to modification of seafloor roughness by benthic fauna. These experiments will be discussed in more detail in Ch. 6.

Mathematical models have been developed to describe the effects of bioturbation on both sediment particles and pore water. The most common representation of bioturbation is the 1D biodiffusion model where $D_b(z)$ is the bioturbation or mixing coefficient, a diffusion coefficient analogous to the heat diffusion coefficient appearing in (4.24) [Guinasso and Schink 1975]. The mixing coefficient is often determined from the depth distribution of impulsive or continuous tracers such as naturally occurring radionuclides (Pb-210, Th-235, Cs-137, Be-7) or radionuclides associated with fallout from nuclear testing (Pu-239, Pu-240). Vertical mixing coefficients and the maximum depth of mixing, L_b, have been determined for many coastal sites and generally range between 10 and 100 cm^2 yr^{-1} and 10–30 cm, respectively [Boudreau 1998, Wheatcroft and Drake 2003, Crusius et al. 2004]. Values of D_b increase with increased sedimentation rate and decreased water depth, perhaps in response to increasing availability of reactive/labial organic matter [Boudreau 1994]. Sediment mixing by benthic fauna tends to be advective rather than diffusive, but, averaged over sufficient time, can be approximated by a diffusive process. In general D_b, as measured with tracers, appears to decrease with depth and is often modeled by a variable mixing coefficient [Boudreau 1986] or as discrete layers, each with a different value for the mixing rate [Crusius et al. 2004]. At the site of the STRATAFORM experiments off the northern California coast, high rates of sediment mixing (D_b = 10–100 cm^2 yr^{-1}) tend to fully mix or obliterate the signatures of thick (6–8 cm) depositional layers within 3–15 years [Wheatcroft and Drake 2003] and destroy thinner layers or smaller structures even more rapidly. This suggests that much of the heterogeneity described in Ch. 7 is of recent origin. Only after the sediment structure is buried below the depth of mixing by benthic fauna ($L_b > 20$ cm) will the sedimentary layer or biogenic structure be preserved [Bentley and Nittrouer 2003]. Data also exist suggesting horizontal mixing rates may be as much as 10 times greater than vertical mixing rates [Wheatcroft 1991] providing higher rates of structural degradation.

More complex sediment mixing models that allow quantification of non-diffusive mixing by individual processes have been developed and successfully tested by [Boudreau and Imboden 1987, François et al. 1997, Shull 2001], and reviewed by [Meysman et al. 2003] but are well beyond the scope of this monograph. In an attempt to treat sediment mixing in a more mechanistic manner, Boudreau and colleagues developed a lattice-automaton bioturbation simulator (LABS)[Choi et al. 2002]. The model treats sediment as a 2D random lattice of particles, pore space, and organisms and allows the organisms, as operators or automatons, to move particles according to well-defined rules. Simulations with small deposit-feeding automatons mimic the diffusive behavior of 1D biodiffusion models when allowed to operate over sufficient time intervals. LABS should provide considerable insight into 3D bioturbation processes, even allowing modeling of ripple decay due to bioturbation at the sediment–water interface, which is important for seafloor roughness studies (Ch. 6).

An attempt to combine a stochastic model of bioturbation with an acoustic volume scattering model (Ch. 14) was made by [Jones 1999, Jones and Jackson 2000] to predict time-dependent decorrelation of backscattering strength. The mixing coefficient, D_b, and mixing depth, L_b, of macrofauna and meiofauna were used to predict the temporal evolution of bulk density spectra (Ch. 7). Nonlocal mixing, for example feeding by a conveyer-belt species, gave rise to spatial heterogeneity, while fine-scale mixing by smaller animals resulted in diffusive decay of heterogeneity. This approach was used to model temporal decorrelation of backscattering strength from silty-clay sediment in West Sound, Orcas Island in Puget Sound, Washington. At this site volume scattering rather than roughness dominated backscattering. The general shape of the measured and predicted decorrelation of backscattering were similar, suggesting it is possible to model changes in high-frequency acoustic scattering due to the effects of bioturbation on sediment structure. Equally important to benthic ecologists, it may be possible to measure rates of biological processes using high-frequency acoustics.

3.5 Research Issues

Although this chapter has emphasized nonacoustic topics, the discussion reveals some challenging areas for future acoustic research. Foremost is the use of knowledge of hydrodynamic, biological, and biogeochemical processes to predict acoustic behavior. As an example, one might hope to build predictive models for interface roughness incorporating the hydrodynamic growth and biological decay of sand ripples (Fig. 3.30). Similarly, it might be possible to develop a predictive model for sediment heterogeneity based on the lines of existing models for sediment mixing due to bioturbation. This is the forward problem, but its solution should suggest new inverse methods for acoustically measuring hydrodynamic and biological benthic activity. Equally challenging (and equally rewarding) is the prospect of using knowledge of sediment microstructure in wave-theoretic models with the goal of solving problems apparent in current theory (e.g., poroelastic theory, see Ch. 10).

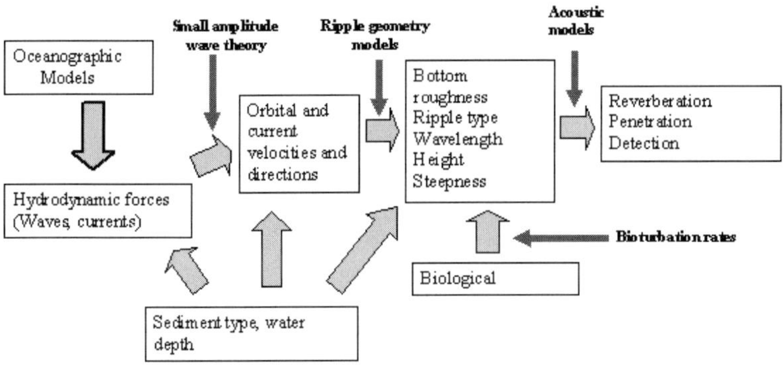

Fig. 3.30. A scheme for coupling hydrodynamic, biological, and acoustic models to predict ripple structure and resultant acoustic behavior.

4 Physical Properties

In this chapter, surficial sediment physical properties relevant to high-frequency acoustics will be discussed. The term "physical properties" denotes such things as porosity, permeability, mean grain size, particle, pore fluid and pore space characteristics, and sediment type. These properties will be distinguished in this monograph from "geoacoustic properties," to be discussed in Ch. 5. Geoacoustic properties include such things as compressional and shear wave speeds, the corresponding attenuations, and impedance. The division into physical and geoacoustic properties is made for convenience and is not followed universally in the literature. For example, compressional and shear wave speed and attenuation are often considered sediment physical properties by geologists and geophysicists but are of sufficient importance to high-frequency acoustics to be treated separately. Empirical relationships between sediment physical and geoacoustic properties are given in Ch. 5. These allow prediction of sediment geoacoustic properties based on measured values of sediment physical properties presented in this chapter. In many high-frequency acoustic applications, values of sediment geoacoustic properties are not available, and these relationships provide a reliable method to obtain estimates of the values of geoacoustic parameters needed in acoustic theories and models (Chs. 8–15).

Methods used to measure sediment grain size distribution and define common descriptors used in sediment databases (such as "fine sand" or "silty clay") will be described first. Bulk properties associated with the relative percentages of solid, fluid, and gas phases in sediments will be described next, followed by sediment properties associated with fluid, electrical, and heat conductance. This chapter concludes with discussions of the characteristics of sediment pore space followed by the characteristics of sediment particle and pore water properties. The purpose of this chapter is not to provide an exhaustive documentation of all methods used to characterize sediment physical properties but instead give the acoustic modeler insight into how these properties are measured and reported, how values of these sediment properties should be interpreted relative to acoustic theory and models, and how to sort through the bewildering array of jargon used by sedimentologists and geotechnical engineers.

4.1 Sediment Grain Size Distribution

The size distribution of particles in marine sediments covers almost eight orders of magnitude from boulders hundreds of centimeters in size to particles having sizes smaller than one micron. The type, size, shape, and orientation of these particles provide useful information to the sedimentary geologist about sediment particle origin, history during transport and deposition, and changes during postdeposition processes. Other bulk properties of sediments, such as porosity, permeability, and compressional and shear wave speeds, are either dependent on, or at least empirically correlated with grain size. Particle size distribution can be measured on sediment samples collected by grabs, cores, or other nonquantitative collection tools, where the intactness of the sediment is not requisite for analysis. Sediment mean grain size and sediment type (e.g., carbonate fine sand) are the most common descriptors found in sediment databases and often are the only sediment physical descriptions available to the acoustic modeler. More measurements of grain size distribution have been reported in the literature than for all other physical properties combined. It has been estimated [Syvitski 1991] that earth scientists perform 800,000 grain size analyses each year.

Classic methods for particle size analysis in the earth sciences are described in texts on sediment petrology [Krumbein 1938, Griffiths 1967, Folk 1980] and sediment marine geotechnology [Lamb and Whitman 1979]. The field of powder technology, principally involving the chemical industry, has developed many of the automated methods of particle size analysis. More recently, acoustical, optical, and electrical methods have been developed to measure in situ sediment particle size at or near the seabed [Syvitski 1991, Crawford and Hay 1993, Schaafsma and Hay 1997, Agrawal and Pottsmith 2000, Thorne and Hanes 2002]. In one way or another almost all methods of particle size analysis relate back to classical sieving and particle settling techniques. Most of the emphasis in this section is therefore devoted to these classic techniques. An accurate interpretation of particle-size data requires an understanding of the nature and morphology of sediment particles, sediment collection methods, dispersion of sediment samples prior to size analysis, accuracy and precision of the size analysis techniques, and statistical methods used to report the results. The terms "grain size" and "particle size" tend to be used interchangeably, but it should be remembered that all sediment particles are not grains (i.e., not discrete particulate or crystalline masses) and can be composed of organic matter or other material that is neither a mineral nor a rock fragment.

Grain size (ϕ, or "phi") has been traditionally reported in the base-2 logarithmic scale devised by [Krumbein 1934], where $\phi = -\log_2 d$, and d is the particle size in millimeters. If one assumes the sediment size distribution is lognormal, the phi-transformed size distribution becomes normal (Gaussian), condensing a very broad range of grain diameters into a more restricted range of values of phi, expanding the spectrum of very small grain sizes, thereby facilitating the calculation of grain size statistical parameters and the comparison of statistical size distributions of different sediments. Natural processes, such as weathering and sediment transport, break down and abrade small sediment grains more rapidly than larger grains tending to create lognormal distributions of particles. Natural processes that sort grains according to size also tend to create lognormal distributions. Deviations from a lognormal distribution (parameterized as sorting, skewness, kurtosis, or the presence of more than one mode) are included in the statistical summary of grain size distribution. A visual method to equate phi, diameter in microns and millimeters, U.S. standard sieve mesh size, and the descriptive Udden–Wentworth size classes is given in Table 4.1. Acoustic modelers can thus easily convert the rather cryptic (at least to them) logarithmic phi size-distributions reported by sedimentologists to more familiar metric units or descriptive sediment size classes. Acoustic modelers should realize that the phi transformation is not unlike that of the decibel, proportional to the base-10 logarithm of the ratio of measured sound pressure to a reference sound pressure [McManus 1982]. Analogously for grain size analysis, phi is the ratio of the negative logarithm (base 2) of measured grain size to a reference grain size of 1.0 mm. The Udden-Wentworth grain-size scale may be the most widely used description of sediment particles but is not the only such scale (for other examples see [Krumbein 1938]). For example, geotechnical researchers often use 2 μm (9.0 ϕ), rather than the traditional 3.9 μm (8.0 ϕ), as the break between silt and clay thus adding a category of very fine silt (8–9 ϕ) [Blott and Pye 2001]. If one employs a scale in which the size-class descriptors differ from the standard Udden–Wentworth scale, reference to that other scale should be given.

In the classical methods of particle size analysis devised by sedimentary petrologists, the weight of each size class of sand- and gravel-sized particles is determined after particle disaggregation by dry or wet sieving using the appropriate U.S. Standard screen sizes. Silt- and clay-sized particle distribution is determined by pipette methods by which particles segregate according to diameter during settling as described by Stokes' law. The size-range classes reported for marine sediment are usually restricted to -2 ϕ (large granules, 4 mm equivalent diameter) to 12 ϕ (0.24 μm). Particles larger than granules are rare and not statistically represented in most small geological samples. The weight frequency distribution of each size class is plotted, and graphical methods are used to determine the statistics of the sediment grain size distribution (mean, mode, sorting or standard deviation, skewness, and kurtosis). There are many modifications to this general approach and the devil is in the details. Replacement of the classic techniques (sieving and pipette methods) with modern automated and digital methods has been advocated [Syvitski et al. 1991b]. The reader is directed to the excellent source of information on modern laboratory particle size analysis techniques, [Syvitski 1991]. Specifics associated with measurement problems, calibrations, and comparisons of the wide

Table 4.1. Sediment grain size classification based on [Folk 1980]. The grade scale proposed by [Udden 1914] and modified by [Wentworth 1922] is logarithmic where each size class is twice as large as the next smaller size class. Phi (ϕ) is $-\log_2(d)$ where d is the equivalent spherical diameter in millimeters [Krumbein 1938]. Equivalent diameter is also expressed in millimeters and microns. The U.S. standard wire mesh (No. of wires/in^2) for sieving granules, sands and coarse silt-size particles and inches/opening for particles $\geq$ 6.3 mm is also given.

U.S. Standard Wire Mesh No.	Millimeters	Micron	Phi (ϕ)	Udden-Wentworth Size Class
	1024		−12	Boulder (−8 to −12 ϕ)
—	256	—	−8	—
	128		−7	Cobble (−6 to −8 ϕ)
$2\ ^1/_2$	64	—	−6	—
$1\ ^1/_4$	32		−5	
$^5/_8$	16		−4	
0.530	13.2		−3.75	
$^7/_{16}$	11.2		−3.50	
$^3/_8$	9.5		−3.25	Pebble (−2 to −6 ϕ)
$^5/_{16}$	8		−3.00	
$^1/_4$	6.3		−2.75	
3.5	5.6		−2.50	
4	4.75		−2.25	
5	4.00	—	−2.00	—
6	3.36		−1.75	
7	2.83		−1.50	Granule (−1 to−2 ϕ)
8	2.38		−1.25	
10	2.00	—	−1.00	—
12	1.68		−0.75	
14	1.41		−0.50	Very coarse sand
16	1.19		−0.25	
18	1.00	—	0.00	—
20	0.84		0.25	
25	0.71		0.50	Coarse sand
30	0.59		0.75	
35	0.50	500	1.00	—
40	0.42	420	1.25	
45	0.35	350	1.50	Medium sand
50	0.297	300	1.75	
60	0.250	250	2.00	—

Table 4.1. Continued

US Standard Wire Mesh No.	Millimeters	Micron	Phi (ϕ)	Udden-Wentworth Size Class
70	0.210	210	2.25	
80	0.177	177	2.50	Fine sand
100	0.149	149	2.75	
120	0.125	125	3.00	
140	0.105	105	3.25	
170	0.088	88	3.50	Very fine sand
200	0.074	74	3.75	
230	0.0625	62.5	4.00	
270	0.053	53	4.25	
325	0.044	44	4.50	Coarse silt
400	0.037	37	4.75	
450	0.031	31	5.00	
500	0.023	23.4	5.50	Medium silt
635	0.0156	15.6	6.00	
	0.0117	11.7	6.50	Fine silt
	0.0078	7.8	7.00	
	0.0045	4.5	7.50	Very fine silt
	0.0039	3.9	8.00	
	0.0020	2.0	9.00	
	0.00098	0.98	10.00	
	0.00049	0.49	11.00	Clay
	0.00024	0.24	12.00	
	0.00012	0.12	13.00	
	0.00006	0.06	14.00	

variety of automated techniques are beyond the scope of this monograph, but a discussion of more classic techniques adequately describes many of the issues associated with particle size analysis techniques and will allow the acoustician to interpret grain size data. In addition, these classic techniques have provided much of the worldwide database on sediment grain size and are still used in many laboratories today. Most of the automated laboratory methods of determination of grain size distributions are based, in part, on these classic methodologies.

Sediment samples are generally collected remotely using grabs, gravity cores, piston cores, or box cores, or collected in situ by divers using a variety of hand tools including diver cores. Care must be taken to include the entire sample and not allow fine-grained, silt- and clay-size particles to winnow out of the sample. If the sample is too large for analysis, subsamples must be prepared. Sediments are often vertically and horizontally quite variable, therefore a sufficient number of samples must be collected to adequately characterize vertical gradients and horizontal spatial heterogeneity [Richardson 1986, Richardson and Briggs 1996]. In most cases the acoustic footprint or patch size for acoustic scattering measurements is much larger than the sediment sample and larger particle size classes, such as shell material and rocks, may be missed by typical grain-size analysis techniques. For longer acoustic wavelengths, especially in fine-grained sediments, propagation and scattering can occur from depths up to a few meters, therefore the depth distribution of grain size statistics should match the depth of acoustic penetration. Sediments used for grain size analysis can be somewhat disturbed and stored for long periods of time without adverse effects, although drying and disaggregation of clay particles could cause some problems. Consequently, laboratory determination of grain size is usually the last step in analysis of sediment physical properties.

The sand and mud (in this monograph, "mud" refers collectively to silt- and clay-sized particles) fractions are usually segregated by wet sieving and analyzed separately using the techniques described below. Preparation of sediment samples often includes some form of mechanical disaggregation of sand-, silt-, and clay-sized particles and subsequent dispersion of the suspension of silt- and clay-sized particles [Matthews 1991a]. Unless sand-sized particles are cemented, as in some carbonates, mechanical agitation will usually separate unconsolidated sediments into individual particles. Silt- and clay-sized particles are dispersed using mechanical agitation, such as stirring or ultrasonic energy; chemical reagents to remove organic binding material; or ionic dispersants such as sodium hexametaphosphate (Calgon®). This provides analysis of the size distribution of individual clay particles rather than clay mineral floccules or aggregates. For example, fecal pellets that are commonly in the size range of 500 to 3000 μm in muddy sediments are broken down to clay-size particles in the 1 to 5 μm range. In fact, most clay-sized particles found in surficial sediments are bound in some form of aggregate, chain, crumb, or pellet (see Sect. 3.2.1) by electrostatic forces and organic films. Whether characterization of the size distribution of individual dispersed particles or of larger naturally occurring aggregates is most appropriate for high-frequency acoustics may depend on the specific application but has received little attention to date. It should be noted, however, that in situ acoustic or optical methods are used to measure particle size distribution of naturally occurring aggregates in the water just above the seafloor [White 1998, Thorne and Hanes 2002].

The sand fraction is routinely separated into 0.25-ϕ intervals using stacks of U.S. standard screens (ASTM standard E-11), and, after shaking, the particles retained on each screen are weighed. Other techniques to determine sand-sized distribution include settling tubes [Syvitski et al. 1991a] and image analysis. Recent advances in sieving techniques include automated methods for wet sieving and sonic, electromagnetic, and air-jet particle agitation

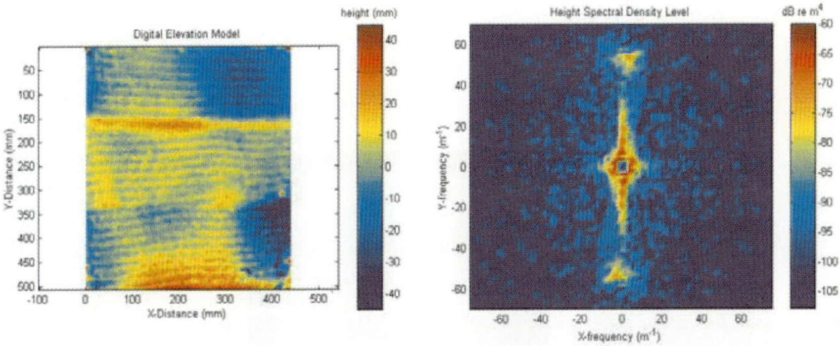

Fig. 2.1. False-color map of relief of a rippled sand seafloor (left panel) and resulting two-dimensional power spectrum (right panel). Courtesy of A. P. Lyons.

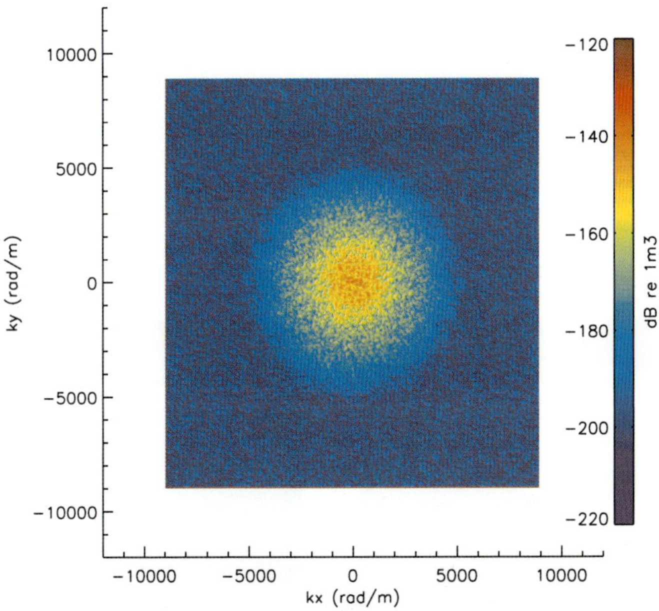

Fig. 2.3. A 2D slice through a 3D spectrum of density fluctuations obtained by X-ray tomography [Pouliquen and Lyons 2002].

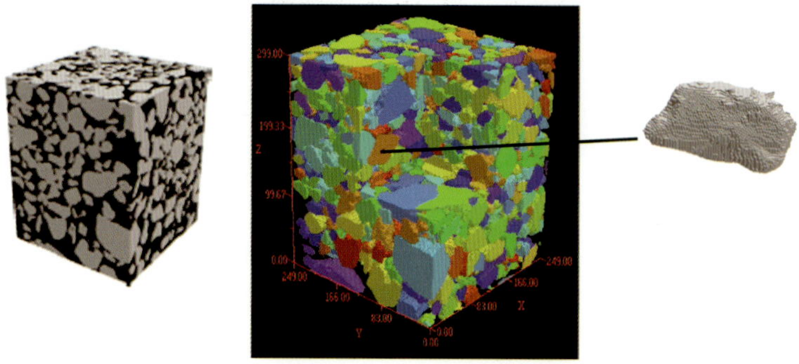

Fig. 3.7. CT scan of resin-impregnated sediment collected with a diver core during SAX99 (from A.H. Reed in [Richardson et al. 2005]). The volume of scanned sediment is 82 mm^3 with a calculated porosity of 39%. The mean inscribed radius of the pores is 58 μm, the mean inscribed radius of the pore throats is 42 μm, and the mean grain size is 375 μm. The mean pore coordination number (number of interconnected pores) is 6.0, and the average grain coordination number (number of grain contacts) is 7.1. The individual grain to the right has a volume of 2.5×10^8 μm^3, surface area of 2.0×10^6 μm^2, and aspect ratio of 2.2.

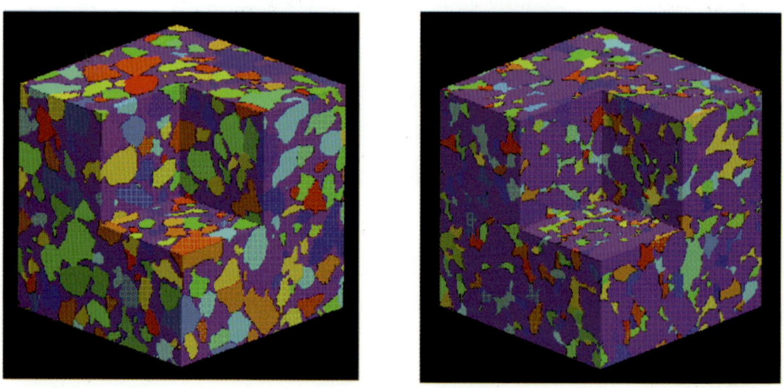

Fig. 3.8. Volumetric CT image of quartz sand (SAX04) from which porosity and bulk density were determined. Grains are false-color coded in the left panel, and pores are false-color coded in the right panel. Courtesy of A. H. Reed.

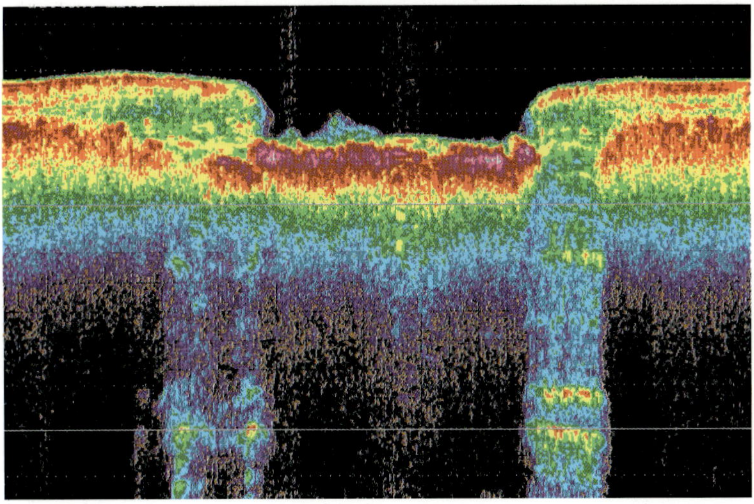

Fig. 3.15. Acoustic image obtained from 12 kHz echo sounder data during experiments on gassy sediments in Eckernförde Bay in the Baltic Sea [Richardson and Davis 1998]. A 50-m-wide "pockmark" is evident. At 5 m depth (indicated by the horizontal white line), strong acoustic returns from glacial deposits are seen just outside the edges of the pockmark. These are masked elsewhere by a layer of methane gas whose uppermost surface is visible as dark red outside the pockmark and violet inside. Bubbles rising through the water column are visible over the pockmark.

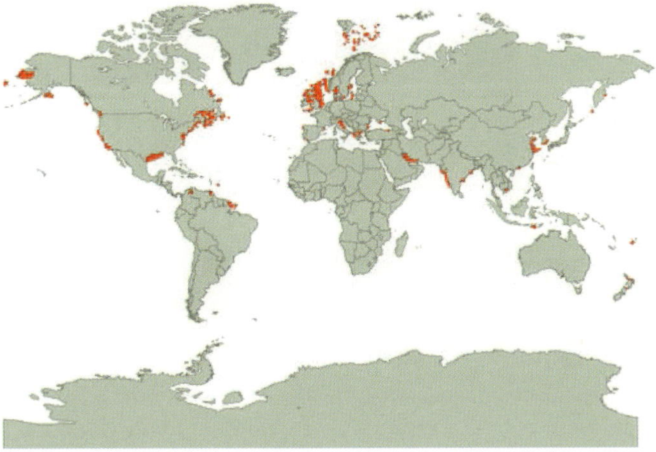

Fig. 3.17. Global view of reported sites of free gas bubbles in shallow-water sediments based on [Fleischer et al. 2001].

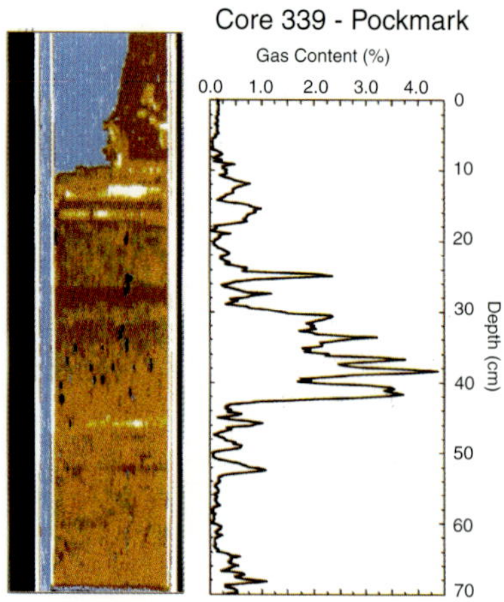

Fig. 3.18. The left panel is a CT-scan showing methane gas bubbles from a sediment core collected from center of a pockmark in Eckernförde Bay, Baltic Sea (see Fig. 3.15). False colors, based on X-ray densities, represent water (blue), sediment (brown), core liner (white) and methane bubbles (black). The bubble volume was as high as 4% as shown in the plot on the right. The core had been maintained at in situ temperature and pressure (see [Abegg and Anderson 1997] for details).

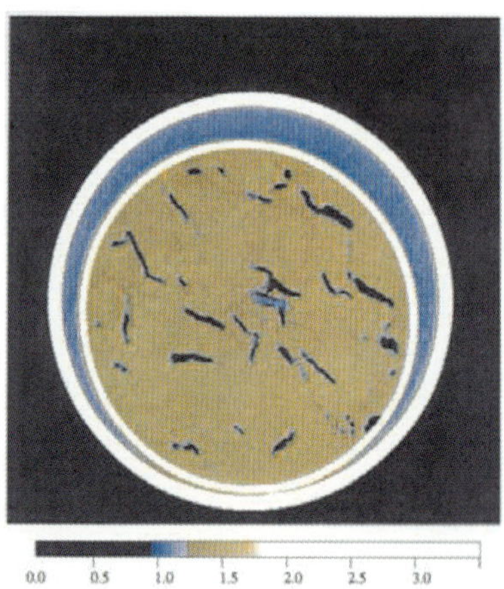

Fig. 3.20. A reconstructed slice from a medical CT scan showing the distribution of methane gas bubbles from a sediment core collected from center of a pockmark in Eckernförde Bay, Baltic Sea (see Figs. 3.15 and 3.18). False colors, based on X-ray densities, represent water (blue), sediment (brown), core liner (white), and methane bubbles (black). The core has been maintained at in situ temperature and pressure (see [Abegg and Anderson 1997] for details).

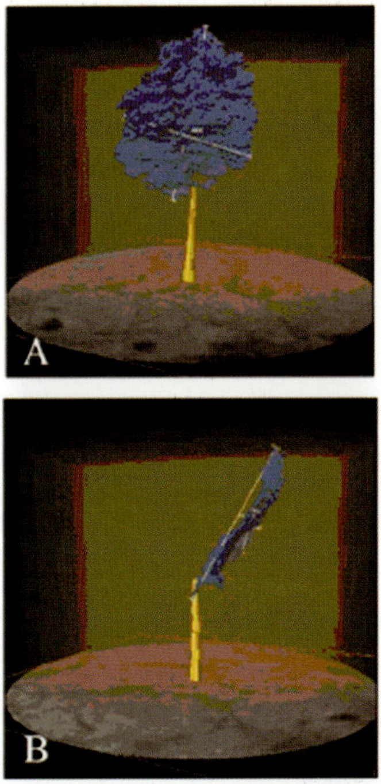

Fig. 3.21. A 3D reconstruction of a bubble injected into fine-grained sediment using high-resolution X-ray computed tomography [Boudreau et al. 2005]. The bubble is blue and the injection capillary is yellow. The bubble is 20 mm across (A) and 7 mm thick (B) with a resulting volume of 0.3 cm^3.

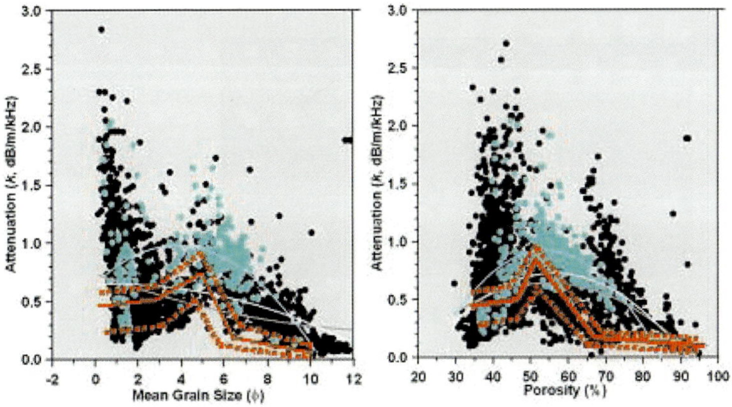

Fig. 5.3. Attenuation factor, measured at 400 kHz, as a function of porosity and mean grain size. The lighter symbols represent carbonate sediments and overlay the darker symbols which represent siliciclastic sediments. Equations for regressions are given in Table 5.3. The relationships between attenuation factor and porosity and mean grain size summarized by [Hamilton 1980] are in red.

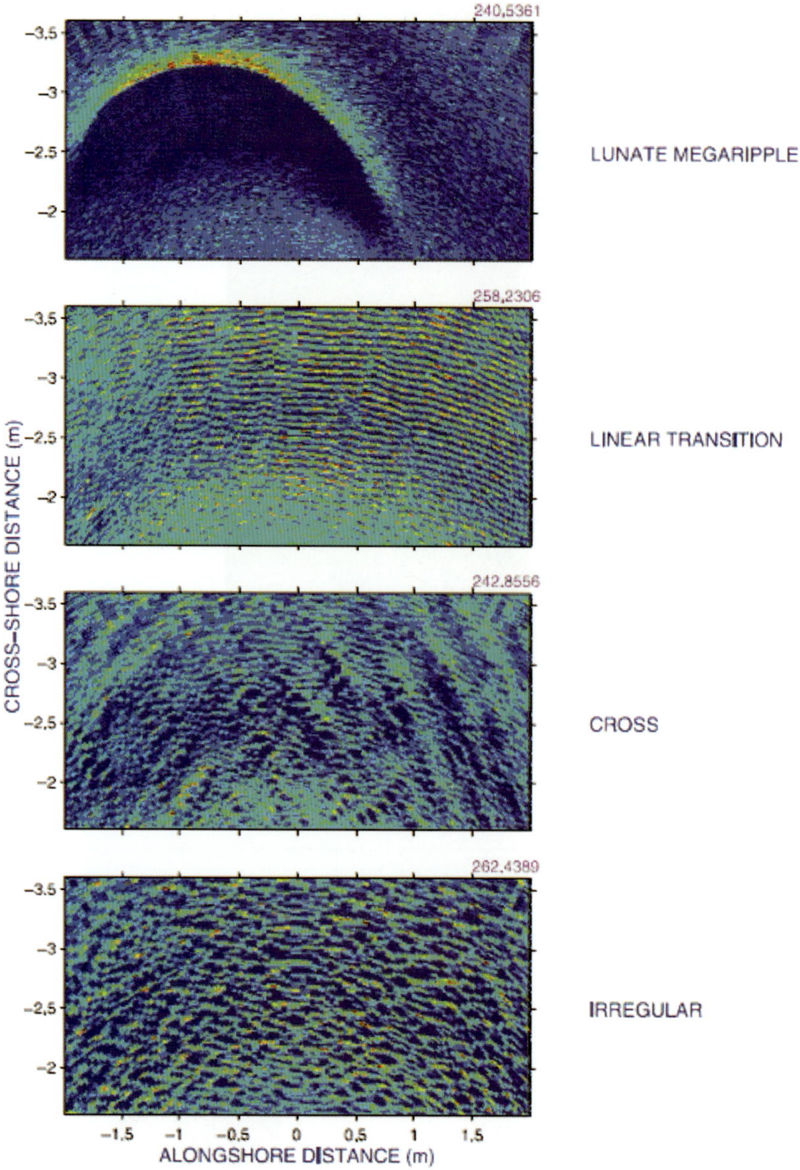

Fig. 3.25. Bed state images obtained by [Hay and Mudge 2005] using a rotary sonar with a "fan-beam," narrow in the horizontal and wide in the vertical. Four principal seafloor states are illustrated by images (2 x 4 m) collected during SandyDuck97 near Duck, North Carolina. Wave energy increases moving from the lower to the upper panels of the figure. The fifth bed state is featureless and therefore not shown. Color scale indicates backscatter intensity, with blue corresponding to low intensity and red to high intensity.

Fig. 6.1. Examples of seafloor roughness: (A) Fresh storm-generated ripples in a medium sand sediment in 19-m water depth. (B) A fine sand with biogenic roughness features such as mounds, trails, and track marks. (C) A sand dollar destroying small 10-cm wavelength ripples. (D) A highly bioturbated muddy sand seafloor with many examples of biogenic features such as fresh mounds, trails, and tracks. (E) A shallow water ooid sand with a mixture of ripple types. (F) A field of 30- to 40-cm wavelength ripples in ooid (carbonate) sands from the Ocean Cay, Bahama Islands (note the smaller set of secondary ripples formed 45° to the primary ripple field). (G) Carbonate hard ground which is nearly devoid of sediments and covered with sponges and corals. (H) A shelly seafloor with sea urchins. Note that hydrodynamically induced roughness tends to be anisotropic whereas biologically induced roughness tends to be isotropic. Courtesy of R. I. Ray.

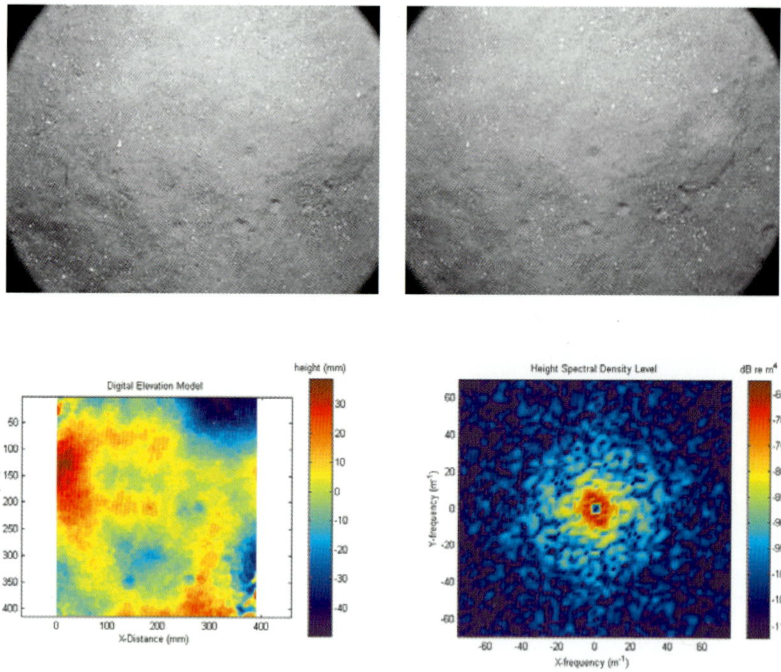

Fig. 6.4. Upper: Overlapping stereophotographs (photo: SAX27) collected by A. P. Lyons from the SAX99 experiment in the northeastern Gulf of Mexico. Lower left: Digital elevation map. Lower right: Calculated two-dimensional power spectral density calculated from the digital elevation map. Note the isotropic nature of the microtopography. See text or [Lyons et al. 2002b] for description of methodology. The spectrum color scale is for $10\log_{10}[S(F_x,\ F_y)]$.

instead of the traditional shaking and tapping. More recent techniques associated with settling tubes include measuring the accumulating weight of sand-sized particles at the bottom of the settling tube, or the use of X-ray or laser diffraction to determine particle settling rates. Most methods used to determine silt- and clay-size distribution are based on particle settling velocity in a dilute dispersant solution, with the particles assumed to have spherical shape. Settling methods are generally restricted to particles in the 0.5 to 70 μm size range. Particles larger than 70 μm in the confined space of a settling tube create turbulence thus accelerating settling velocities; particles smaller than 0.5 μm are affected by Brownian motion and could theoretically remain in suspension indefinitely [Griffiths 1967]. In the classic pipette method, the dispersed silt- and clay-sized particles are distributed uniformly by vigorous stirring in a cylinder of the dispersant and small aliquots (samples) are collected at specific times and depths using a calibrated pipette. The samples are dried, weighed, and corrected for dispersant weight and related to the volume of the suspension from which each aliquot is drawn. The result is a percent weight of each size class which can be combined with the weight of gravel- and sand-size classes (Figs. 4.1–4.5). Silt-sized particles are usually divided into ten 0.5-ϕ size bins and clay-sized particles into four 1.0-ϕ bins down to 12 ϕ. The remaining particles finer than 12 ϕ are often divided evenly between 13 ϕ and 14 ϕ.

Recently, a variety of optical attenuation, optical imaging (counting and sizing), laser diffraction spectroscopy, X-ray attenuation (SediGraph), electrical resistivity, photon correlation spectroscopy, and acoustic automated techniques have been developed to measure grain size distribution of silt- and clay-sized particles [McCave and Syvitski 1991]. These automated methods provide continuous measurements at greater resolution, are less prone to operator eccentricities, and are much faster than classic pipette or sieving methods. Proper calibration, however, is required to maintain precision and accuracy. Both the U.S. Naval Oceanographic Office and The Naval Research Laboratory use a Micromeritics Sedigraph to measure size distribution of silt- and clay-sized particles down to 1 μm [Coakley and Syvitski 1991]. Sieves are still used for the dried sand and gravel fractions.

The classic size distribution data are plotted as a histogram with both size class (bin) frequency and cumulative frequency (see examples in Figs. 4.1–4.5). Each size class represents the weight of particles between two size bin limits with the size class designated by the smaller-size bin limit. Thus, in sieving, particles are assigned to the class corresponding to the screen that retains them. Summary statistics (mean, median, standard deviation or sorting, skewness, and kurtosis) are based on either analytical (arithmetic based on first to fourth moments) or graphical analysis of the frequency histograms of the mid-point of each size class using ϕ-percentile or metric-percentile units. The most commonly used graphical measures were devised by [Folk and Ward 1957] and include median,

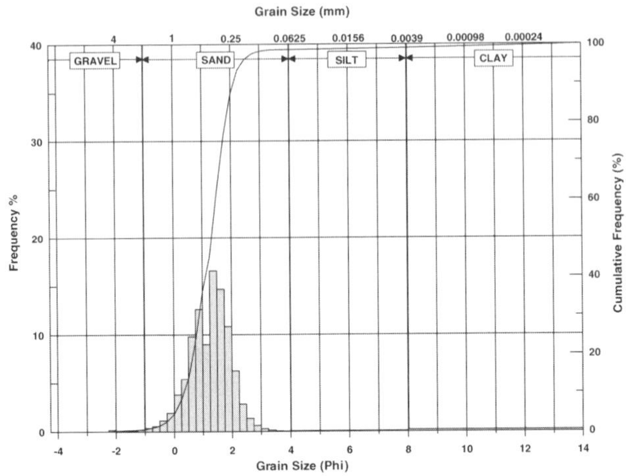

Fig. 4.1. Histogram of grain size distribution for sandy sediment collected during the SAX99 high-frequency acoustic experiment (station 20-1) in 19-m water depth in the northeastern Gulf of Mexico, October 1999 [Richardson et al. 2001a]. Grain size statistics are given in Table 4.2. The sediment is moderately well sorted, coarse-skewed, mesokurtic, medium sand. The silt- and clay-sized fractions accounted for less than 2% of the total particulate weight and do not affect graphic grain size statistics.

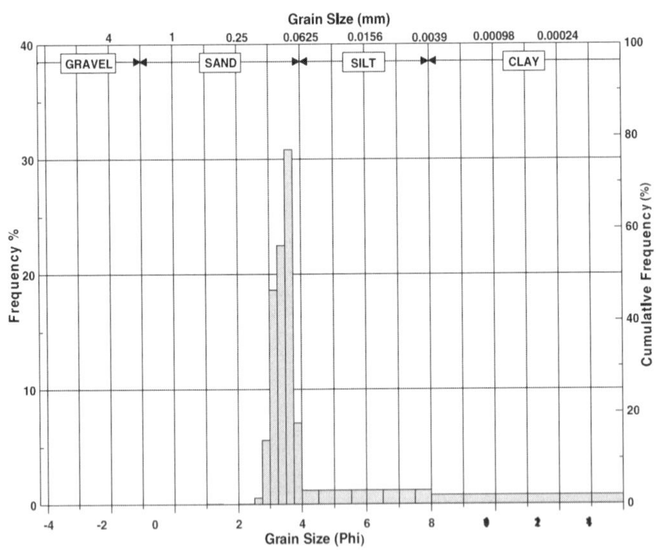

Fig. 4.2. Histogram of grain size distribution for sandy sediments collected during high-frequency acoustic scattering experiments (station T1-2) in 8-m water depth off northern Italian coast (Tirrenia) in September 1997. Grain size statistics are given in Table 4.2. The sediment is moderately sorted, strongly fine-skewed, leptokurtic very fine sand. Courtesy of K. B. Briggs.

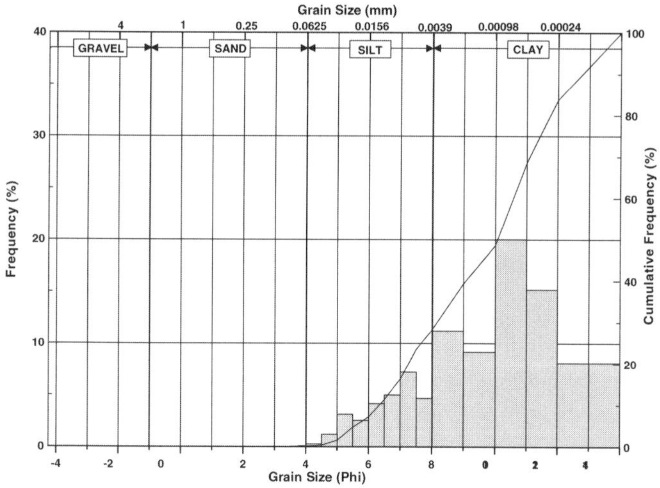

Fig. 4.3. Histogram of grain size distribution for muddy sediment collected off the California coast (Eel River) during the STRATAFORM experiments (station 75-1) at 70-m water depth, June 1996 [Richardson et al. 2002b]. Grain size statistics are given in Table 4.2. The sediment is very poorly sorted, coarse-skewed, platykurtic silty clay.

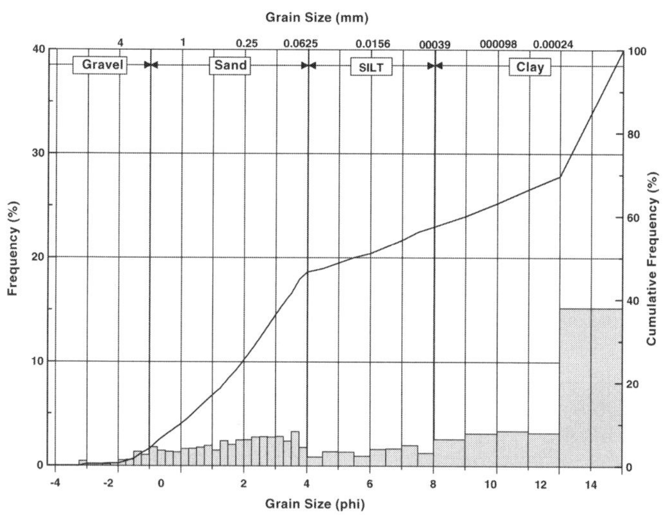

Fig. 4.4. Histogram of grain size distribution for shelly mud sediment collected in the Arafura Sea (station 18-3), north of the Gulf of Carpentaria off the Australian coast in 70-m water depth, May 1984 [Briggs et al. 1989]. Grain size statistics are given in Table 4.2. The sediment is extremely poorly sorted, fine-skewed, very platykurtic sandy clay.

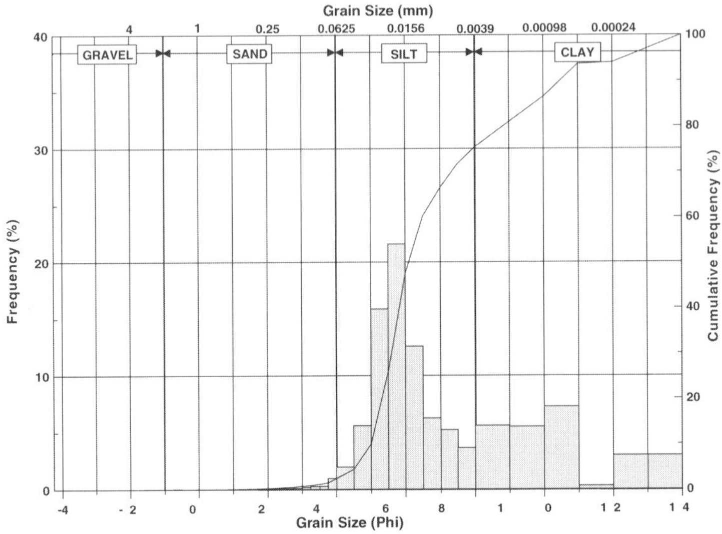

Fig. 4.5. Histogram of grain size distribution for muddy sediment collected off the California coast (Eel River) during the STRATAFORM experiments (station 41-1) at 70-m water depth, June 1996 [Richardson et al. 2002b]. Grain size statistics are given in Table 4.2. The sediment is very poorly sorted, strongly fine-skewed, leptokurtic clayey silt.

Table 4.2. Summary statistics for moderately well-sorted medium sand (SAX99), moderately sorted very fine sand (Tirrenia), very poorly sorted silty clay (Eel River), extremely poorly sorted mixtures of sandy mud (Arafura Sea), and clayey silt from the Eel River. Frequency histograms for the grain size distributions for these sediments are presented in Figs. 4.1–4.5, with the order of the figure citation corresponding to the order in the table. Grain size statistics are based on the graphical methods of [Folk and Ward 1957].

Statistic	Gravel %	Sand %	Silt %	Clay %	Mean	Sorting	Skewness	Kurtosis
Location								
SAX99	0.55	98.08	0.39	0.99	1.27	0.70	−0.11	1.20
Tirrenia	0.00	85.31	9.93	4.76	3.45	0.96	0.41	4.30
Eel River 75	0.00	0.10	28.14	71.74	9.69	2.45	−0.19	0.85
Arafura Sea	4.12	42.42	10.92	42.54	6.43	5.23	0.18	0.57
Eel River 41	0.00	2.44	72.86	24.71	6.95	2.27	0.59	1.28

$$M_d = \phi_{50} \ , \tag{4.1}$$

graphic mean,

$$M_z = (\phi_{16} + \phi_{50} + \phi_{84})/3 \ , \tag{4.2}$$

inclusive graphic standard deviation,

$$\sigma_I = (\phi_{84} - \phi_{16})/4 + (\phi_{95} - \phi_5)/6.6 \; , \tag{4.3}$$

inclusive graphic skewness,

$$Sk_I = (\phi_{16}+\phi_{84}-2\phi_{50})/[2(\phi_{84}-\phi_{16})]+(\phi_5+\phi_{95}-2\phi_{50})/[2(\phi_{95}-\phi_5)] \; , \tag{4.4}$$

and graphic kurtosis,

$$K_G = (\phi_{95} - \phi_5)/[2.44(\phi_{75} - \phi_{25})] \; . \tag{4.5}$$

Graphic mean grain size, M_z, is one of the measures most often used and will be simply referred to as "mean grain size" in this monograph. Additional measures of the statistics for graphic grain size frequency distributions are given in [Griffiths 1967]. Descriptive terminology for ranges of the grain size statistics (sorting, skewness, and kurtosis) proposed by [Folk and Ward 1957] are given in Table 4.3.

Other common statistical (arithmetic and geometric using either metric or phi units) and descriptive measures of sediment-size distribution are reviewed by [Krumbein 1938, Griffiths 1967, Folk 1980, Blott and Pye 2001]. The consensus of marine geologists and soil scientists is that the graphical methods of [Folk and Ward 1957] provide the most robust basis for comparisons of variable sediment size distributions. This logarithmic approach defines statistical distributions of the central tendency of the data rather than outliers in the tails of the distribution which are often ill-defined in standard grain size analysis. The values and ranges for the descriptive terminology vary considerably depending on which mathematical or graphical method is used to determine the grain size statistics. Therefore a comparison between values of statistical distributions derived from different statistical techniques is not recommended [Blott and Pye 2001].

The statistical distribution of grain size has many uses in high-frequency acoustics. Mean grain size is used as an empirical predictor of other sediment physical and geoacoustic properties, such as porosity and sound speed (Ch. 5) or used as a descriptor predicting scattering strength. This approach is subject to large errors (Ch. 12), but is sometimes the only physical description of the sediment available. Mean grain size, M_z, as defined by Folk and Ward, is a suitable descriptor for empirical predictions of other sediment properties (Ch. 5), as these relationships tend to be logarithmic with respect to grain diameter. For several other acoustic applications, such as scattering by discrete particles and characterization of volume heterogeneity, a size frequency histogram, such as those presented in Figs. 4.1–4.5, is the preferred form of data. For applications in which a physical relationship connects grain size to the desired parameter, such as scattering from particles suspended in the water column, grain size statistics derived from the arithmetic method of moments may be preferred. A program is available [Blott and Pye 2001] to calculate

Table 4.3. Descriptive terminology for the grain size statistics based on the graphic measures of [Folk and Ward 1957]. Inclusive graphic standard deviation (σ_I) is a measure of sorting, inclusive graphic skewness (Sk_G) is a measure of symmetry, and graphic kurtosis (K_G) is a measure of peakedness of the grain size distribution.

Sorting (σ_I)	Descriptive terms
<0.35	very well sorted
0.35 to 0.50	well sorted
0.50 to 0.71	moderately well sorted
0.71 to 1.0	moderately sorted
1.0 to 2.0	poorly sorted
2.0 to 4.0	very poorly sorted
>4.0	extremely poorly sorted
Skewness (Sk_G)	
1.00 to 0.30	strongly fine-skewed
0.3 to 0.1	fine-skewed
0.1 to −0.1	near symmetrical
−0.1 to −0.3	coarse-skewed
−0.3 to −1.0	strongly coarse-skewed
Kurtosis (K_G)	
> 0.67	very platykurtic
0.67 to 0.90	platykurtic
0.90 to 1.11	mesokurtic
1.11 to 1.50	leptokurtic
1.50 to 3.00	very leptokurtic
< 3.00	extremely leptokurtic

mean, standard deviation, skewness, and kurtosis from size-frequency data using arithmetic, geometric, and logarithmic methods of moments with either metric or phi units, thus allowing acousticians to calculate the appropriate grain size statistics.

Another common approach to sediment classification is characterization of sediments with respect to the relative percentages (by weight) of sand and mud (silt and clay combined) or in a ternary phase diagram, where sediments are characterized according to percentages of three classes of particles (e.g., sand:silt:clay or gravel:sand:mud). In the scheme of [Flemming 2000], sediments are divided into six textural classes based on the relative percentages of sand and mud (Table 4.4). Numerous similar binary approaches exist.

Three of the most commonly used ternary diagrams were suggested by [Shepard 1954] and by [Folk 1980] for sand, silt, and clay mixtures (Figs. 4.6 and 4.7), and by [Folk 1980] for the ratios of gravel, sand, and mud (silt and clay combined, Fig. 4.8). There appears to be no standardized terminology

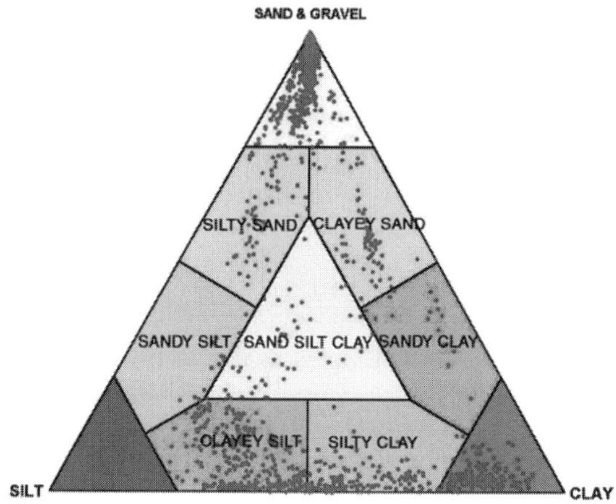

Fig. 4.6. Ternary diagram of [Shepard 1954] based on sand–silt–clay ratios.

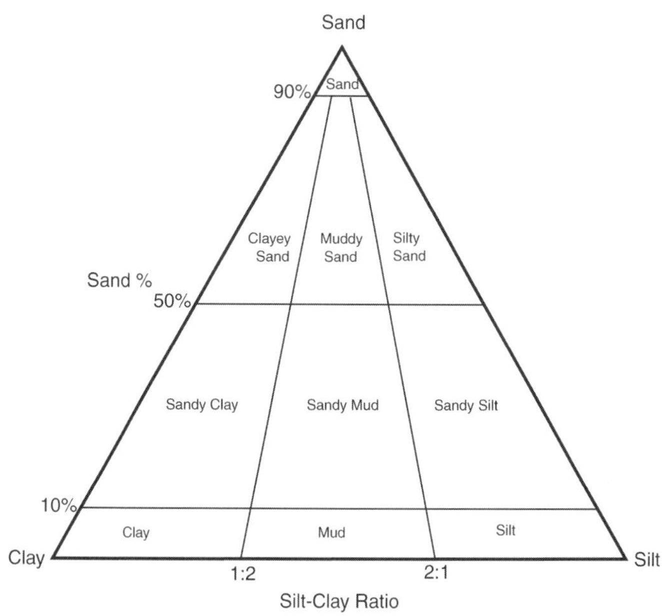

Fig. 4.7. Ternary diagram of [Folk 1980] based on sand–silt–clay ratios.

Table 4.4. Descriptive terminology for the six textural classes of sediments based on sand–mud ratios [Flemming 2000].

Mud Content %	Textural Class	Mud Content %	Textural Class
< 5	Sand	50-75	Sandy mud
5-25	Slightly muddy sand	75-95	Slightly sandy mud
25-50	Muddy sand	> 95	Mud

Fig. 4.8. Ternary diagram of [Folk 1980] based on gravel–sand–mud ratios.

for these commonly used ternary diagrams. The papers of [Shepard 1954, Folk 1980, Flemming 2000] describe over 10 different ternary phase diagrams for various combinations of gravel-sand-silt-clay with identical descriptors covering much different ratios of the three components. Some authors have reported sediment textural characteristics in a form that implies quantitative analysis using ternary diagrams but instead employ qualitatively estimated relative percentages of gravel, sand, silt, and clay. Therefore, when reporting sediment textural characteristics, reference should be given to the form of the ternary diagram employed.

Based on quantitative analysis of sediment grain size, several descriptive approaches to sediment classification have been proposed. The approach of [Folk 1980] combines the description based on a ternary diagram with the graphical statistics. For example, the four sediments having the distributions shown in Figs. 4.1–4.5 are described as: (a) a medium sand, moderately well sorted, coarse-skewed mesokurtic; (b) a very fine sand, moderately sorted, strongly fine-skewed, leptokurtic; (c) a silty clay, very poorly sorted, coarse-skewed, platykurtic; (d) a sandy mud, extremely poorly sorted, fine-skewed, very platykurtic; and (e) a clayey silt, very poorly sorted, fine-skewed, leptokurtic. The U.S. Naval Oceanographic Office combines textural information with modifiers denoting sediment origin (for example, carbonate clayey sand or terrigenous sandy clay) giving over 400 sediment types. In this monograph, sediment size distribution is classified using the Shepard ternary diagram (Fig. 4.6), Wentworth size classification (Table 4.1), and Folk and Ward (Table 4.3) graphical grain size statistics and descriptors.

Reported statistical values of sediment grain size distribution should be used with careful consideration of their limitations. First of all, sediments exhibit considerable variability on spatial scales of centimeters to meters to kilometers and over temporal scales of hours to years. Because of this variability, a few small sediment samples may not represent the larger footprint of the acoustic system of interest. Second, most sediment analyses assume sediment particles are spherical; they are not. Sieving techniques measure the smallest cross-sectional area of particles, and plate-shaped clay minerals rarely settle according to Stokes' law for equivalent spheres. Particle density and shape affect calculated size distributions for both sieve and settling techniques, and particle interactions during settling may also affect the calculated size distributions [Matthews 1991b]. Furthermore, size characterization of particles in the clay size range is controlled by the effectiveness of the disaggregation procedure and the concentration of the dispersant. Often, when the clay particles are completely dispersed, the characterization does not reflect the in situ particle size distribution. Different techniques cover different size classes, usually requiring more than one technique to characterize the entire particle size distribution of many types of marine sediment. Accuracy or precision of grain size analyses are rarely reported, and different techniques often give very different results [Syvitski 1991]. Statistical analyses of particles with no lower size cutoff can present analytical problems. Two different sediments with the same mean grain size can be physically diverse and acoustically behave quite differently when one is a mixture of sand and mud (Fig. 4.4) and the other is mostly silt (Fig. 4.5). For example, the clayey sand from the Arafura Sea (mean grain size 6.46 ϕ) has a porosity of 82% and sound speed of 1512 m s^{-1}, whereas the clayey silt sediment from the Eel River (mean grain size 6.95 ϕ) has a porosity of 56% and sound speed of 1565 m s^{-1}.

4.2 Sand Grain Morphology

Grain morphology, as well as the grain size distribution described above, is an important factor controlling packing, porometric properties associated with fluid flow [Berryman and Blair 1986], electrical resistivity [P. Jackson et al. 1978], consolidation under loading, low-strain geoacoustic behavior described by contact mechanics [Buckingham 2000], and perhaps acoustic scattering when wavelengths are comparable to grain size [Boyle and Chotiros 1995b, Greenlaw et al. 2004]. However, grain morphology is rarely taken into account in high-frequency acoustic applications. Numerous methods have been developed to characterize natural grain morphologies. These methods range in complexity from visual evaluations summarized by [Folk 1980] to computer-enabled and mathematically complex methods, such as Fourier transforms [Ehrlich and Weinberg 1970], fractal analyses [Kaye 1978], two-point correlation functions [Berryman and Blair 1986], and other less commonly used approaches, such as mathematical morphology [Pirard 1994] and wavelet transforms [Drolon et al. 2000]. These methods have been used to characterize the morphology or surface of grains to provide information about potential source locations [Ehrlich and Weinberg 1970], transport mechanisms [Bagnold 1956, Drake 1990], settling rates [Le Roux 2002], and depositional history [Hudson and Ehrlich 1980].

Naturally occurring sediment particles that are closest to idealized smooth-surfaced, spherically shaped particles are carbonate ooids (formed by precipitation, see Sect. 3.2.3) and well-rounded beach sands. Most other sediment particles deviate considerably from idealized spheres both in shape and in surface roughness (Fig. 4.9). If all sediment particles were spherical, a single parameter would fully characterize shape (e.g., radius, diameter, circumference, or volume). A summary of much of the early work relating to shape and surface morphology of sediment particles has been given by [Folk 1980]. Measures of particle morphology were divided into form, sphericity, roundness, and surface features. Form is a measure of the relationship of the three dimensions of an object and may qualitatively be classed as compact (three axes having nearly equal lengths), elongated (rod-like), or platy (disk-like). Sphericity is a measure of the equality of the dimensions of a particle. Roundness is a comparison of the curvature of a particle surface to the curvature of the largest inscribed sphere or circle within that particle. Particle surface characterization includes subjective visual observations and employs terms such as frosted, polished, or dull. Subjective determination of particle form, sphericity, and roundness is often made by visual comparison to standard charts [Griffiths 1967]. Development in computer technology has paved the way for more efficient and exacting methods to analyze particle morphology. Fourier analysis techniques [Ehrlich and Weinberg 1970] have been used to characterize grain morphology (i.e., surface characteristics and markings) and to determine source locations, method of transport, and depositional history for specific sediment deposits [Hudson and Ehrlich 1980, Ehrlich and Chin 1980]. Fractal dimensionality has been used to differentiate grains from different sources or provinces [Orford and Whalley 1983]. Grain morphology characterization in terms of two-point correlation functions has been used to determine specific surface area of grains or exterior boundaries of the pore

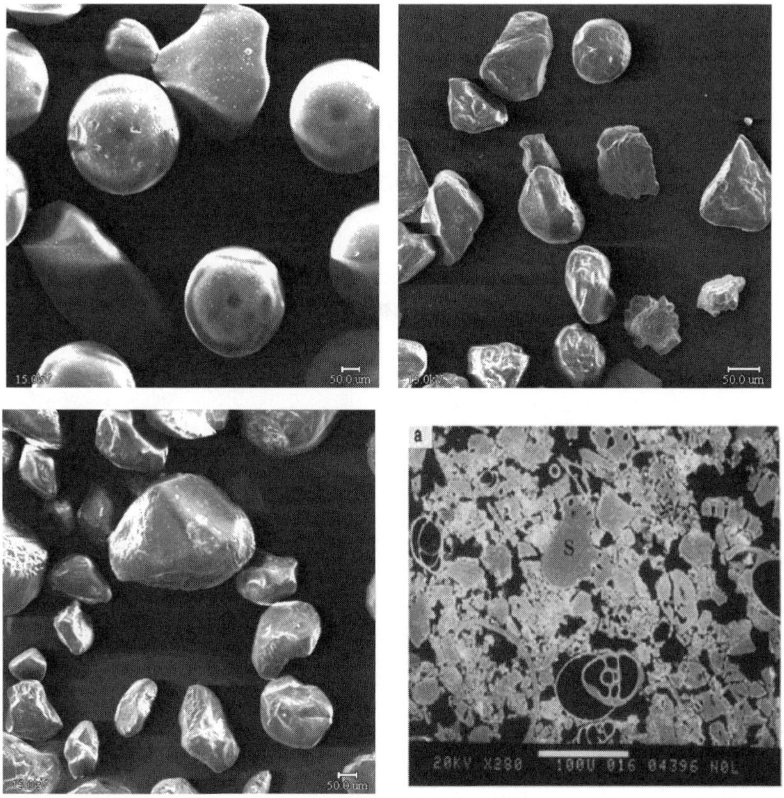

Fig. 4.9. SEM micrographs of quartz and carbonate grains. Top left: glass beads. Top right: Ottawa sand. Bottom left: sand grains collected during SAX99 off Fort Walton Beach Florida. Bottom right: carbonate skeletal remains from the Dry Tortugas. A micron scale bar can be found in each image. Courtesy of R. I. Ray.

walls using high-magnification scanning electron microscope (SEM) images [Krinsley and Takahashi 1962, Berryman and Blair 1986, Blair et al. 1996].

Sediment morphology reflects the mineralogy of sediment particles as well as weathering, transport history, and depositional processes associated with terrigenous and biogenic sediments. Mineral hardness affects the rate at which particles become rounded during transport (carbonate shell material abrades and rounds more easily than quartz); particles in high-energy environments, such as beach and bar regimes, tend to be more rounded and spherical. Grain morphology and roughness (roundness) have long been known to affect grain packing, relative density (4.11) and to control porosity, bulk density, and permeability in sediment [Fraser 1935]. It has been shown that particle angularity influences the relationship between porosity and formation factor [P. Jackson et al. 1978] (see Sect. 4.4.1 of this monograph). Grain morphology also influences the number and area of grain contacts in sands. In spite of these obvious relationships between grain morphology and sediment physical properties, no high-frequency acoustic study has directly utilized grain shape or morphology to determine wave propagation properties. Conceptually, measures of grain roughness are part of the theory of wave propagation proposed by [Buckingham 2000], but shearing which he predicts will occur at grain contacts during the passage of a wave may be quite difficult to measure or predict based on the number of grain contacts and grain microroughness. However, it is safe to assert that grain shape and morphology affect wave propagation through sediment by their influence on porosity, density, permeability, and the number and types of grain contacts and should be an area of active research.

4.3 Phase Relationships: Porosity, Void Ratio, Water Content, and Bulk Density

Sediments are composed primarily of solid minerals, pore fluid, free gas, and organic matter, although most marine sediments are totally saturated (containing no gaseous phase) and are often treated as two-phase media (solid particles and pore fluid). Porosity, water content, void ratio, and bulk density are measures of the relationship between the mass and/or volume of sediment particles and pore water in fully saturated sediments.

Several approaches have been devised to directly or indirectly measure these properties in fully saturated sediments. The two most direct laboratory methods are to measure the total weight, W, of a measured volume, V, of sediment or to measure the weight loss, or water weight, $W_w = W - W_s$, of a known weight, W, of sediment dried for 24 hours at 105–110°C where W_s is the weight of the dried solid fraction. For acoustic applications, it is appropriate to correct the water weight, W_w, for pore water salt content [Hamilton 1971b]. The volume-mass method provides the sediment saturated bulk density, ρ, and the weight-loss technique (thermogravimetric) provides the water content, w, from which the fractional porosity, β, or void ratio, e, can be calculated, assuming the density of the solid particles, ρ_s, and pore fluid, ρ_w, are known. The density of pore water is often calculated from the temperature, salinity, and pressure (see Appendix B). The average particle density can be measured on dried samples with a pycnometer (NRL uses a Quantachrome Ultrapycnometer) or derived from handbook values for pure crystalline forms of minerals (Table 4.7). It has been shown by [Lee and Chough 1987] that the volume-weight method can underestimate bulk density by as much as 2–6% compared to the weight-loss method, primarily due to the inclusion of air in the measurement tube.

The various measures of porous, water-saturated sediment are defined as follows: bulk density is the total mass, $M = W/g$, of sediment divided by the total sediment volume,

$$\rho = \frac{M}{V} , \tag{4.6}$$

water content is the weight of water divided by weight of solids,

$$w = \frac{W_w}{W_s} , \tag{4.7}$$

fractional porosity is the ratio of pore water volume, V_w, to total volume,

$$\beta = \frac{V_w}{V} , \tag{4.8}$$

and void ratio is the ratio of pore water volume to solid particle volume, V_s,

$$e = \frac{V_w}{V_s} = \frac{\beta}{1-\beta} . \tag{4.9}$$

Porosity is also commonly reported as a percentage ($\eta = 100\beta$). Table 4.5 provides useful relationships between porosity, water content, void ratio, and bulk density. For example, if the porosity is measured and the densities of solid particles and pore water are known, bulk density can be calculated.

Table 4.5. Relationships among sediment fractional porosity, β, water content, w, void ratio, e, and bulk density, ρ, for saturated marine sediments.

	Porosity, β	Water Content, w	Void Ratio, e	Bulk Density, ρ
β	—	$w/(\rho_w/\rho_g + w)$	$e/(e+1)$	$(\rho_g - \rho)/(\rho_g - \rho_w)$
w	$(\rho_w/\rho_g)\beta/(1-\beta)$	—	$e(\rho_w/\rho_g)$	$(1 - \rho/\rho_g)/((\rho/\rho_w) - 1)$
e	$\beta/(1-\beta)$	$\rho_g w/\rho_w$	—	$(\rho_g - \rho)/(\rho - \rho_w)$
ρ	$\rho_w\beta + \rho_g(1-\beta)$	$\rho_w(1+w)/(w + \rho_w/\rho_g)$	$e(\rho_w - \rho_g)/(e+1)$ $+\rho_g$	—

Most wet-weight, dry-weight measurements of water content are not corrected for the residue dry salts (see, e.g., [Bennett and Lambert 1971, Richards et al. 1974]). Methods to correct the calculated seawater and measured sediment weights to account for the residue of dried salts that is included with the dried sediment grains have been provided by [Hamilton 1971b]. These corrections are made prior to calculations of void ratio, porosity, water content, or bulk density. Hamilton also suggests that uncorrected values of porosity can be multiplied by 1.012 to derive approximate values of salt-free porosity. This approximation is only valid for high-porosity sediments when the salinity is near 35 ppt. More accurate calculations of true porosity, β, can be obtained from

$$\beta = \beta_0 + 0.036 \left(\frac{S_p}{35}\right) - 0.0224 \left(\frac{S_p}{35}\right) \beta_0 , \tag{4.10}$$

where S_p is the pore water salinity in parts per thousand, and β_0 is the uncorrected fractional porosity. All values of porosity reported in this monograph have been corrected for salt content.

The bulk density of granular sediment (sands) can also be characterized in terms of relative density, D_r, defined as

$$D_r = 100 \frac{e_{max} - e}{e_{max} - e_{min}} , \tag{4.11}$$

where e_{max} is the maximum value of void ratio, measured in laboratory testing with minimum packing, and e_{min} is the void ratio measured at maximum packing. The measured void ratio before laboratory testing is denoted e (4.9). Bulk density of sandy sediments studied during SAX99 averaged 2074 kg m^{-3} based on laboratory testing [Richardson et al. 2001a], which amounts to a relative density of 75%. Relative density in most sandy environments is high (D_r = 65–85%) due to dense packing in response to hydrodynamic forces (waves and currents) on surficial sediments. However, during prolonged periods of calm (lack of storm events), bioturbation may lead to a moderately packed state with a corresponding decrease in relative density. The state of relative density in sandy sediments has been shown to have a profound effect on sediment bearing capacity and liquefaction by the geotechnical community [Youd et al. 2001]. Sandy sediments with low values of relative density have much lower values of bearing strength relative to similar sediments with higher values of relative density and are more prone to liquefy under stresses associated with earthquakes or pressure fluctuations from surface gravity waves (i.e., high sea states). Shear wave speeds (Ch. 5) are often used to predict liquefaction potential with lower sound speeds associated with lower values of relative density [Robertson et al. 1995]. The effects of relative density on wave propagation have not received the attention they deserve. However, it should be expected that sound speed will increase and attenuation decrease with increasing values of relative density.

If free gas in the form of air, methane, or oxygen bubbles is present in sediment, the degree of saturation, S, can be calculated as the percentage of the pore volume, V_v, filled with water, with the water volume denoted V_w:

$$S = 100 \frac{V_w}{V_v} . \tag{4.12}$$

For fully saturated sediments the degree of saturation is 100%.

Methane gas bubbles are common in organic-rich muddy, coastal sediments [Fleischer et al. 2001] and rarely exceed 1% of the pore volume but, on rare occasions, have been reported as constituting as high as 4–5% of the pore volume [Martens et al. 1998]. Air can be entrained in shallow-water sandy sediments by breaking waves, and the degree of saturation can descend to less than 75% in the swash zone where waves roll up on the beach but rarely falls below 95% in subtidal sediments. Algae mats generate oxygen as a metabolic by-product, and oxygen bubbles have been observed trapped in the upper few millimeters of sediment [Nilsson et al. 1991]. A more detailed discussion of free gas in sediments is presented in Sect. 3.3. Calculation of the degree of saturation makes little sense in the case where oxygen bubbles are present only at the sediment surface. The effects of gas in sediments on acoustic scattering are discussed in Sect. 14.1.7.

Determination of sediment density, porosity, void ratio, or water content from sediments collected with cores is subject to errors associated with sediment disturbance during collection, transport, and storage of samples and from loss of pore water during the volume-weight or weight-loss measurement techniques. Fine-grained, muddy sediments may be compacted during collection, transport, and subsequent storage, reducing porosity and increasing density. The relative density of sandy sediments can be either increased (tighter packing) or decreased (looser packing) during core collection and transport. Some water loss is inevitable during laboratory manipulations. Granular sediments (sands and gravels) collected with gravity or piston corers are disturbed to the extent that determination of sediment porosity or density is not recommended. However, careful collection of these granular sediments by divers or with large (0.25 m^2) box corers may ameliorate most, but not all, of these effects. The bow-wave effect of gravity or piston corers may also blow away high-porosity near-surface sediments creating additional problems in fine-grained sediments. Again, careful collection of sediments by divers or using large box corers can preserve the sediment–water interface and is the preferred method of sediment collection for high-frequency acoustic applications where properties of the upper few centimeters are important. Recently, sediments collected with diver cores were evaluated to determine the degree of alteration of sediments due to transport, storage, and laboratory manipulations [Briggs and Richardson, unpublished]. The volume of sediment was measured immediately after collection (the sediment–water interface was marked on the outside of the sediment cores) and compared to the sediment volume just before laboratory measurement (days to months after collection). In most cases, the sediment volume changed less than 1% during transport and storage for both muddy and sandy sediments, suggesting changes in porosity or bulk density of less than 1% for carefully handled sand and mud sediments. A comparison of values of porosity measured on sectioned cores and entire (unsectioned) cores showed differences of less than 0.003 (0.3%), suggesting manipulations during weight-loss measurements were insignificant.

Weight-loss methods used to determine porosity of sediments comprised primarily of clay minerals (most marine mud) present additional issues associated with the physical association of water with clay minerals. Water present between clay multiplate particles or domains is driven off by heating at temperatures below 70°C, water intercalated between crystalline folia or plates is driven off at 100°C, and formation water bound to the crystals is driven off at temperatures above 200°C [Bourbié et al. 1987]. The amount of water driven off at different temperatures depends on mineralogy and domain structure, which differs among the dominant clay minerals (kaolinite, smectite, illite) found in marine sediments (see Sect. 3.2.1). Standard weight-loss methods call for drying sediments at 105–110°C for 24 to 48 hours which may drive off interparticulate water as well as water between crystalline folia or plates. It is not known what role, if any, the water between folia or plates plays during acoustic propagation in sediments nor is there general agreement on what types of water should be considered part of an individual clay particle. Future weight-loss determinations of porosity should use an oven thermostat setting no higher than 70°C if the water between crystalline folia is considered to be part of the clay particle. The importance of vicinal water (water in the immediate vicinity of the surfaces of sediment particles) on acoustic propagation in clay-rich fully saturated sediments is generally unknown. However, vicinal water, with its higher viscosity and density, may affect fluid flow (permeability), and the swelling of clays may affect the sediment bulk modulus [Murad and Cushman 1997].

In order to avoid disturbance issues associated with destructive types of laboratory sediment density and porosity measurements, less invasive techniques have been developed including electrical resistivity, X-ray attenuation or scattering, and X-ray tomography. Only electrical resistivity and gamma-ray attenuation techniques have developed to the point where in situ measurements can be routinely made, allowing quantitative characterization of the undisturbed spatial distribution of sediment bulk density and porosity. The use of these nondestructive techniques to measure the spatial heterogeneity of sediment density associated with acoustic volume scattering is extensively reviewed in Ch. 7. All other techniques require sediments be removed from the seafloor, and issues associated with sediment disturbance during core collection and transport still apply. Laboratory and in situ techniques used to measure sediment bulk density from larger bulk samples including vertical gradients will be emphasized in the following paragraphs. Electrical resistivity measurement techniques will be discussed in Sect. 4.4.2, but attempts to develop predictive relationships between electrical resistivity and sediment porosity will be presented in this section. Several attempts have been made to provide empirical relationships that predict porosity from values of electrical resistivity in a variety of sediment types [Boyce 1968, Kermabon et al. 1969, Taylor-Smith 1971, P. Jackson et al. 1978, Lovell 1985a, Lovell 1985b, Bennett et al. 1990a, Gerland et al. 1993, Erickson and Jarrard 1998a]. The empirical relations are based on the power-law relationship between "formation factor" and porosity first suggested by [Archie 1942]. Formation factor, F, is the ratio of sediment and pore water resistivities (4.20), or, inversely, the ratio of pore water to sediment conductivity. The power-law relationship between formation factor and fractional porosity developed in [Winsauer et al. 1952] is displayed in (4.21), where a and m are constants based on best-fit polynomial regressions specific to different sediment types. Sediment bulk density can be estimated from porosity if pore water and grain density are known or estimated (1000 kg m^{-3} and 2650 kg m^{-3} provide adequate estimates of water and grain densities).

Winsauer's version of Archie's law in the specific form $F = 1.30\beta^{-1.45}$ was found by [Boyce 1968] to fit data collected by gravity cores from the Bering Sea. A third-order polynomial was used by [Kermabon et al. 1969] to describe the relationship between percent porosity and formation factor of unconsolidated muddy sediments from the Tyrrhenian Sea ($\eta = 171.25 - 105.4F + 40.04F^2 - 5.9F^3$). The relationships $F = \beta^{-1.5}$ for sands and $F = \beta^{-2.0}$ for clays have been suggested by [Taylor-Smith 1971]. Using laboratory experiments, [P. Jackson et al. 1978] reported that values of the Archie exponent, m, were primarily controlled by particle shape and varied from 1.2 to 1.9 for spheres to platy shell, from 1.4 to 1.9 as the ratio of shell hash in sand increased, and from 1.4 to 1.6 for natural sands (Fig. 4.10). In all cases Jackson set $a = 1.0$ to satisfy conditions where the fractional porosity and formation factor are both 1.0. Further discussion of measurements of this type is given in Sect. 4.4.2.

Using the same techniques and equipment as [P. Jackson et al. 1978], [Lovell 1985a] found either a third-order polynomial ($\beta = 1.386 - 0.463F + 0.0833F^2 - 0.0073F^3$) or power law ($F = 1.29\beta^{-1.42}$) fit the results for 9 samples of deep-sea mud. Lovell also found a third-order polynomial ($\beta = 1.4154 - 0.4799F + 0.067F^2 - 0.0033F^3$) fit the results for nine samples of deep-sea mud. The fits $F = 1.27\beta^{-1.11}$ for oolitic sand and $F = 1.48\beta^{-1.11}$ for muddy oolitic sand on the Grand Bahama Bank were obtained by [Bennett et al. 1990a]. A formation factor–porosity relationship of $F = 2.56\beta^{-2.51}$ was reported by [Erickson and Jarrard 1998a] for muddy sands collected from the Amazon Fan. When these data, restricted to the ranges of porosity and formation factor from the original measurements, are plotted together, it is obvious that values of formation factor decrease with increasing porosity and increase with sediment bulk density and mean grain size, in more or less the power-law form first suggested by [Archie 1942]. However, special care must be exercised not to use these regressions outside of their original range of measurement, because unrealistic predictions do occur.

The use of a single power law or polynomial regression to describe the formation factor–porosity relationship over the entire range of shallow-water sediments is not recommended. Factors such as mineralogy, tortuosity, pore connectivity, grain morphology, orientation, sorting, packing, and cementation all complicate the relationship. Differences in laboratory consolidation and natural packing and differences in the statistical methods of representing this relationship also complicate the problem. Clean sands and muddy sediments seem to have different power-law relationships, and the effects of particle shape, percent fine-grained material, and sediment matrix cementation yield different values of the coefficients a and m. Electrical resistivity measurements might best be used to measure and quantify small-scale spatial

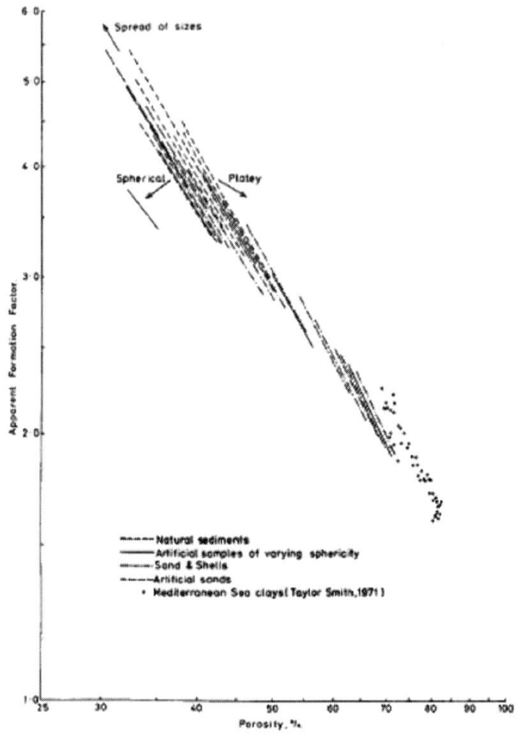

Fig. 4.10. Formation factor versus porosity for a range of artificially packed natural and artificial clean sands. Changes in porosity–formation factors cover packing state from minimum to maximum density for each sediment sample. Adapted from [P. Jackson et al. 1978].

heterogeneity of porosity and density in sediments (see Ch. 7) rather than provide a universal measure of porosity and density.

Nondestructive measurement of sediment density using gamma-ray attenuation or scattering has a long history dating back to the 1950s and 1960s. Transmission techniques have generally been preferred because they can be used on smaller samples such as sediments collected with cores and can characterize the fine-scale vertical gradients in sediment density that are of interest to geologists [Preiss 1968a]. In situ methods to measure density by gamma-ray scattering [Keller 1965] and attenuation [Preiss 1968b] were developed in the 1960s. One of the more widely used systems is the Gamma Ray Attenuation Porosity Evaluator (GRAPE) which was routinely used to measure sediment density as part of the Ocean Drilling Program [Boyce 1973]. Density resolution of the GRAPE is reported to vary from 0.015 g cm^{-3} to as much as 0.04 g cm^{-3} for typical deep-sea sediments [Herbert and Mayer 1991]. Gamma-ray attenuation is one of the three sediment measurements made by core logger systems [Weaver and Schultheiss 1990]. Other logger measurements include sediment compressional wave speed and magnetic susceptibility. Core logger systems provide a nearly continuous, nondestructive, automated, rapid measurement of sediment properties without opening cores. The gamma-ray source is Cs-137 and gamma-ray intensity is measured with a scintillation counter. Gamma rays are attenuated by Compton scattering with the attenuation factor proportional to density [Best and Gunn 1999]. One of the major difficulties of using this methodology is obtaining accurate calibration of the system.

The use of X-radiography to image sediment structure within a laboratory setting has a long history in geology, geotechnical engineering, and soil science [Bouma 1964, Howard 1968, Holyer et al. 1996]. However, most traditional X-radiographic studies of sediments tend to be qualitative in nature and do not provide accurate estimates of sediment bulk density. Problems of interpretation of photographic film occur because of the loss of image detail when three-dimensional structures (sediments in cores or slabs) are projected onto two-dimensional photographic film, because of the non-uniformity of typical X-ray films, and the effects of out-of-plane scattering [Lyons and Pouliquen 2004]. Some, but not all, of these problems can be overcome with newer digital X-ray imaging systems which offer expanded contrast or resolution of gray scales, uniform quality, and are amenable to modern digital image processing algorithms [Migeon et al. 1999, Lofi and Weber 2001]. The problems associated with both analog and digital X-radiographic density characterization hve led several investigators to use 3D reconstructions associated with medical or high-resolution CT-scanning [Orsi et al. 1994, Orsi et al. 1996, Orsi and Anderson 1999, Reed et al. 2002, Muzi et al. 2004, Lyons and Pouliquen 2004]. Recent reviews of the use of medical and high-resolution CT scanning to estimate sediment bulk density or porosity in sediments [Lyons and Pouliquen 2004] and in sedimentary rocks [Van Geet et al. 2001] exist and will not be repeated here. The main problems in measuring sediment density with CT scanning are associated with eliminating artifacts, such as beam hardening, and with calibration of the so-called CT numbers to actual values of density. CT number or the attenuation (absorption) coefficient is a function of both sediment density and atomic number. Thus separate calibration regressions are needed for different sediments (Fig. 4.11) and for different types of CT scanners (monochromatic versus polychromatic). CT scanning is especially useful for characterizing density heterogeneity which is covered in Ch. 7.

4.4 Sediment Conductance

Water, heat, electrical current, and soluble material flow through sediments are governed by Darcy's law, Fourier's law, Ohm's law, and Fick's law, respectively, all of which have the same general form. Hydraulic, thermal, and

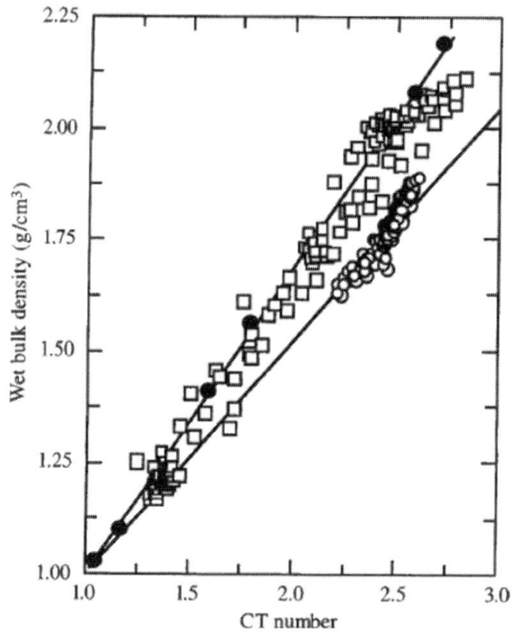

Fig. 4.11. Regressions between CT number and bulk density for siliciclastic sediments (squares) and carbonate sediments (circles) based on a comparison of colocated laboratory measurement of bulk density and average CT numbers [Orsi and Anderson 1999]. The average percent difference between measured and bulk density predicted from these regressions is 2.7%.

electrical flow rates are a product of, respectively, the hydraulic, thermal, and electrical conductivities and the hydraulic, thermal, and electrical gradients. All these flow phenomena are in some way correlated to sediment fabric, the grain-to-grain arrangements and pore characteristics of sediments that are described in Sect. 3.2.

4.4.1 Permeability

The hydraulic flow rate through a porous medium is equal to the product of the hydraulic conductivity and the negative of the gradient of the pressure as expressed in Darcy's law:

$$\mathbf{V} = -\frac{K}{g\rho_f}\nabla P \,, \tag{4.13}$$

where g is the acceleration of gravity, ρ_f is the fluid density, and P is the pressure. The hydraulic flow rate, $\mathbf{V}$, in volume/unit area/unit time is the flow velocity averaged over the region of flow. The hydraulic conductivity,

K, is also known as Darcy's coefficient of permeability, and has dimensions of length/time. The flow rate will depend on the viscosity of the fluid, so K will depend on the fluid type. This dependence is removed by defining the intrinsic coefficient of permeability, κ, such that

$$K = \kappa \frac{g\rho_f}{\mu} . \tag{4.14}$$

Here, μ is the dynamic (or absolute) viscosity, denoted η in Ch. 10. The dynamic viscosity has units kg m^{-1} s^{-1}, and the coefficient of permeability has dimensions (length)^2. The flow rate can be written in terms of the dynamic viscosity as

$$\mathbf{V} = -\frac{\kappa}{\mu} \nabla P . \tag{4.15}$$

The hydraulic conductivity has a greater range (10 orders of magnitude) and exhibits larger spatial variability than any sediment physical property discussed in this chapter. It also exhibits a scale dependence that is not often appreciated [Richardson et al. 2002b]. The largest values of permeability occur in highly porous coarse gravel and sand, and the smallest values are found in practically impermeable clays (Table 4.6). Sediment hydraulic conductivity or permeability can be measured on sediment samples in the laboratory using falling- or constant-head methods, measured in situ with probes using flow rate or pressure drop methods, calculated from mean grain size or from descriptions of sediment fabric, or predicted empirically from other sediment properties. In clays, permeability has been determined indirectly from consolidation or triaxial laboratory tests. Generally, values of permeability derived from consolidation testing are one to two orders of magnitude less than values measured directly (see Sect. 4.4.4). The dissipation of excess pore pressure developed from rapid insertion of penetrometers into the sediment has also been used to measure in situ permeability in low-permeability clays [Robertson et al. 1992, Bennett et al. 1996]. However, the dissipation of excess pore pressure in sand is too rapid to use this method except when mud content reduces permeability. Permeability has also been predicted from sound speed [Taylor-Smith 1971, Fang et al. 1993] and electrical resistivity [Lovell 1985a].

During SAX99, four different methods were used to estimate sediment hydraulic conductivity of a medium sand [Richardson et al. 2001a], including a laboratory constant-head flow method (1.8 to 5.1 $\times$ 10^{-2} cm s^{-1}), an in situ constant-head flow pump (0.3 to 5.4 $\times$ 10^{-2} cm s^{-1}), an in situ falling-head permeameter (0.99 to 1.23 $\times$ 10^{-3} cm s^{-1}), and effective medium theory calculations from sediment microstructure images (3.5 $\times 10^{-3} \pm 2.3 \times 10^{-3}$ cm s^{-1}). This example illustrates the variability as well as the difficulties and uncertainties associated with permeability measurement. Permeability strongly depends on sediment fabric, therefore any disturbance associated with (1) emplacement of in situ probes, (2) high flow rates, or (3) collection, transport, or manipulation of cored sediments can alter the permeability.

Table 4.6. Range of values for sediment intrinsic permeability and porosity.

Sediment Type	Grain Size, d (mm)	Permeability, κ (m^2)	Porosity, β
Clayey	< 0.002	10^{-18} – 10^{-15}	0.5 - 0.9
Silty	0.002 - 0.05	10^{-16} – 10^{-12}	0.35 - 0.5
Sandy	0.05 - 2	10^{-14} – 10^{-10}	0.35 - 0.5
Gravelly	> 2	10^{-10} – 10^{-7}	0.25 - 0.4

The intrinsic coefficient of permeability, κ, is dependent on particle size and shape, porosity or void ratio, mineralogy, and pore topology (size and length of pore throats, size of pore bodies, coordination number, and tortuosity). This is often expressed in the form of the Kozeny–Carman equation [Carman 1956],

$$\kappa = \frac{e^3}{k_0 T^2 S_0^2 (1+e)} , \tag{4.16}$$

where T is the hydraulic tortuosity factor, k_0 is a pore shape factor, S_0 is the specific surface area (having dimensions area/volume = length^{-1}) of a sediment particle, and e is the void ratio [Mitchell 1993, Eq. 12.32]. The hydraulic conductivity has also been expressed as

$$K = \frac{g C D_s^2 \rho_f e^3}{\mu (1+e)} , \tag{4.17}$$

where D_s is some characteristic grain diameter and C is a shape factor related to sediment fabric [Mitchell 1993, Eq.12.33]. These relationships apply to fully saturated, well-sorted sand- and silt-sized assemblages of grains (cohesionless sediments) but may not be appropriate with an appreciable fraction of clay-sized particles. Numerous experimental observations (as reviewed in [Mitchell 1993]) indicate that hydraulic conductivity and intrinsic permeability vary directly with $e^3/(1+e)$ and some measures of characteristic grain or pore size supporting the functional relationships in (4.16) and (4.17). Relationships of this type have proven useful in predicting permeability based on grain size distribution and packing.

Carbonates exhibit special problems with respect to relationships between sediment fabric, porosity or void ratio, grain size, and permeability. Carbonate sediments often contain a significant percentage of intraparticulate porosity due to naturally occurring vugs (closed pores), thus not all void space that contributes to porosity contributes to permeability [Bennett et al. 1990a] or to acoustic propagation (Fig. 4.12). Carbonates that are matrix-supported (larger grains embedded in smaller-grained matrix) can have a much lower permeability (3–4 orders of magnitude) than grain-supported carbonates (smaller particles within sand-size pore space) yet both having nearly the same porosity or void ratio [Bennett et al. 1989].

Clay microfabric rarely exhibits a uniform distribution of pore sizes but instead consists of chains, aggregates, agglomerates, or fecal pellets with large

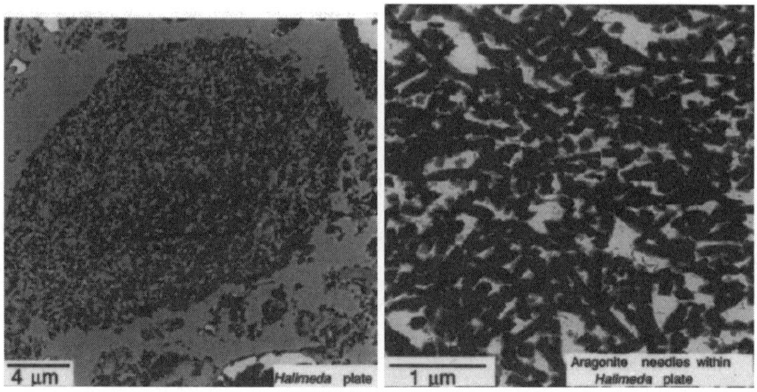

Fig. 4.12. TEM micrographs of fragments of Halimeda plates (calcareous algae) collected from the Dry Tortugas [Richardson et al. 1997]. Left: entire fragment showing both inter- and intraparticulate porosity; right: high magnification reveals the porous nature of aragonite needles that made up the fragments of Halimeda plates. The silt-sized plates contained approximately 30% intraparticulate porosity and contributed 10–15% of the total porosity measured by weight-loss methods. From [Richardson and Briggs 1996].

variations in pore body sizes and pore throat lengths and diameters. Sediments usually contain a mixture of sand-, silt-, and clay-sized particles. A small volume of fines can effectively clog the pores of an otherwise coarse material with an open pore network (grain-supported sediment) greatly reducing the permeability. This is especially true for carbonate sediments which are often very poorly sorted. The presence of coarser grains in an otherwise clayey sediment (matrix supported) has little effect on permeability.

It has been suggested by [Mitchell 1993] that there are three scales of clay fabric that have much different effects on permeability: *microfabric*, which consists of particle aggregations through which little flow will occur; *minifabric*, which includes larger aggregates and fecal pellets with much greater rates of flow; and *macrofabric*, which includes cracks, fissures, tubes and burrows for which the flow rate is so great as to obscure the other rates of flow. This behavior has been observed in a fine-grained sediment for which permeability measured in a bioturbated sample was much higher that than predicted from grain size, mostly because of the large number of open tubes and burrows (Fig. 4.13) which conducted water. However, these high values of permeability were not consistent with values of sound speed and attenuation measured on the same sediments [Richardson et al. 2002b]. A much lower permeability that is characteristic of the sediment matrix is more consistent with the measured sound speeds.

It can be concluded that permeability in clays and some muds is very scale-dependent, therefore difficult to measure. The scale-dependent require-

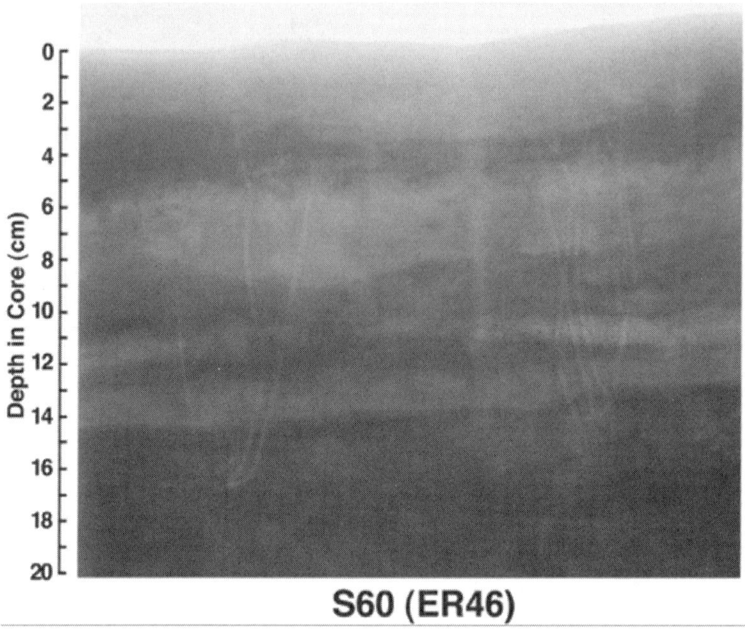

Fig. 4.13. X-radiograph positive image of slab subcore from a box core collected at site S60 off the northern California coast [Richardson et al. 2002b]. Permeability measured on sediment cores also collected from this site were much higher than predicted from mean grain size based on the Kozeny–Carman equation. Thirteen vertical burrow conduits (some are connected, continuous burrows) are evident in this image. Burrows are lighter vertically oriented features. Remnants of buried ripples are evident at 8 to 14-cm sediment depth.

ments for permeability may be much different for predicting sediment acoustic propagation than for predicting wave-driven ventilation flows in sands or consolidation behavior in mud. Thus, care should be exercised when using large-scale permeability measurements in propagation theory. Permeability is less scale-dependent in well sorted, cohesionless sediments (gravel-, sand-, and silt-sized sediment) and best measured using in situ techniques; permeability in both cohesionless and cohesive sediments appears to be predictable from sediment micro- and macrofabric characteristics.

4.4.2 Electrical Resistivity

The electrical flow rate through a conducting medium is governed by the point form of Ohm's law,

$$\mathbf{J} = -\frac{1}{\rho_e}\nabla V \, , \tag{4.18}$$

where **J** is current density in amperes m^{-2}, ρ_e is the electrical resistivity in ohm m, and V is the electric potential in volts. Most of the electrical resistivity techniques now in use to characterize surficial sediments were developed as part of the well-logging techniques used for geophysical prospecting, where electrical resistivity is used to characterize the porosity, hydrocarbon and water saturation, and permeability of hydrocarbon reservoirs. Electrical resistivity is routinely measured on cores collected as part of the Ocean Drilling Program [P. Jackson et al. 1993]. Two common in situ approaches for measuring sediment electrical resistivity in near-surface sediments involve penetrating the sediment with an electrode array [Wheatcroft 2002] or using a planar array to inject a focused current into the sediment from the sediment surface [P. Jackson 1975, P. Jackson et al. 1978, P. Jackson et al. 1981, Davis et al. 1989, Davis et al. 1996, P. Jackson et al. 1996, Evans 2001, Richardson et al. 2001a]. Both types of arrays have also been used for the laboratory measurement of electrical resistivity on both natural sediments (collected with coring devices) and artificially created sediments, where consolidation or packing provided a range of values of electrical resistivity and porosity for samples with the same grain morphology. A single-electrode arrangement has been used by [Tang et al. 2002] to measure fine-scale variabilty of porosity (Ch. 7).

Many of the probe systems use an array of four equally spaced electrodes (Wenner array), where the outer pair of electrodes is used for current injection and the inner pair is used to measure potential. An alternating current (100 Hz) is used to avoid electrode polarization or electrolysis from prolonged use. The resistivity as determined by the Wenner array is

$$\rho_e = 2\pi a \frac{\Delta V}{I} , \tag{4.19}$$

where a is the distance between equally spaced electrodes, I is the current passed between the outer two electrodes, and the potential difference (ΔV) is measured across the inner electrodes [P. Jackson 1975]. Electrical resistivity is often expressed in terms of the formation factor, F, which is the ratio of values of sediment resistivity, ρ_{es}, to pore fluid resistivity, ρ_{ew},

$$F = \frac{\rho_{es}}{\rho_{ew}} . \tag{4.20}$$

In this type of analysis, it is assumed that values of electrical resistivity of the overlying water and pore water are identical [P. Jackson 1975]. Electrical resistivity of sediments is primarily controlled by the resistivity of the pore water, the porosity, and the pore morphology. Pore water resistivity is a function of temperature, salinity, and pressure and can be calculated from algorithms of [Fofonoff and Millard 1983] or from tabular values given in Appendix B when pore water or the overlying water resistivity is not measured. However, the use of formation factors provides sediment characterization that is independent of pore water resistivity and array configurations.

The relationship between sediment fractional porosity, β, and formation factor was described by [Archie 1942] as a power law, where the exponent, m, is a function of grain shape and resultant pore morphology. The following more general form has been used by [Winsauer et al. 1952, P. Jackson et al. 1978]:

$$F = a\beta^{-m} , \tag{4.21}$$

in which the constant factor a is added to Archie's power-law relationship. This form cannot be used in the limit as porosity approaches unity, where the formation factor must also approach unity.

The values of the exponent m are obtained from packing or consolidation experiments for individual sediment samples, where electrical resistivity increases and porosity decreases with increased packing or consolidation. Using laboratory experiments, [P. Jackson et al. 1978] found that values of the exponent, m, were primarily controlled by particle shape (see Fig. 4.10). In all cases he set $a = 1.0$ to satisfy conditions where the fractional porosity and formation factor are both 1.0. Values as high as $m = 2$ have been calculated for high-porosity mud in Eckernförde Bay by P.D. Jackson (in [Briggs et al. 1998]) and $m = 1.9$ to 2.6 for the carbonate sediments of the Dry Tortugas [P. Jackson et al. 2002], in each case assuming $a = 1.0$. The values $F = \beta^{-1.5}$ for sands and $F = \beta^{-2.0}$ for clays have been suggested by [Taylor-Smith 1971], very similar to Jackson's laboratory results for a variety of sediment types and shapes (Fig. 4.14).

4.4.3 Thermal Conductivity

The thermal conductivity of sediments has an indirect effect on high-frequency acoustics, as sediment temperature gradients cause corresponding gradients in geoacoustic properties. The point form of Fourier's law for heat flow is

$$\mathbf{Q} = -k_b \nabla T , \tag{4.22}$$

where $\mathbf{Q}$ is heat flux in W m^{-2}, k_b is the thermal conductivity in W m^{-1} K^{-1}, and T is the temperature in °C.

Most measurements of thermal conductivity in marine sediments have been made in conjunction with heat flow investigations where both the thermal conductivity and thermal gradient are needed to model heat flow [Von Herzen and Maxwell 1959, Ratcliffe 1960, Sclater et al. 1969]. Laboratory measurements have been made using transient needle probe techniques in which the temperature increase of a needle inserted into sediment is measured while being heated at a constant rate [Von Herzen and Maxwell 1959]. The transient needle method compares well (within 3–4%) with the more direct and time-consuming, steady-state hot-plate method, where the temperature difference across a sediment specimen (usually a sediment core) is measured during the application of a steady-state heat flow [Ratcliffe 1960]. The transient needle probe techniques have also been used in situ to measure

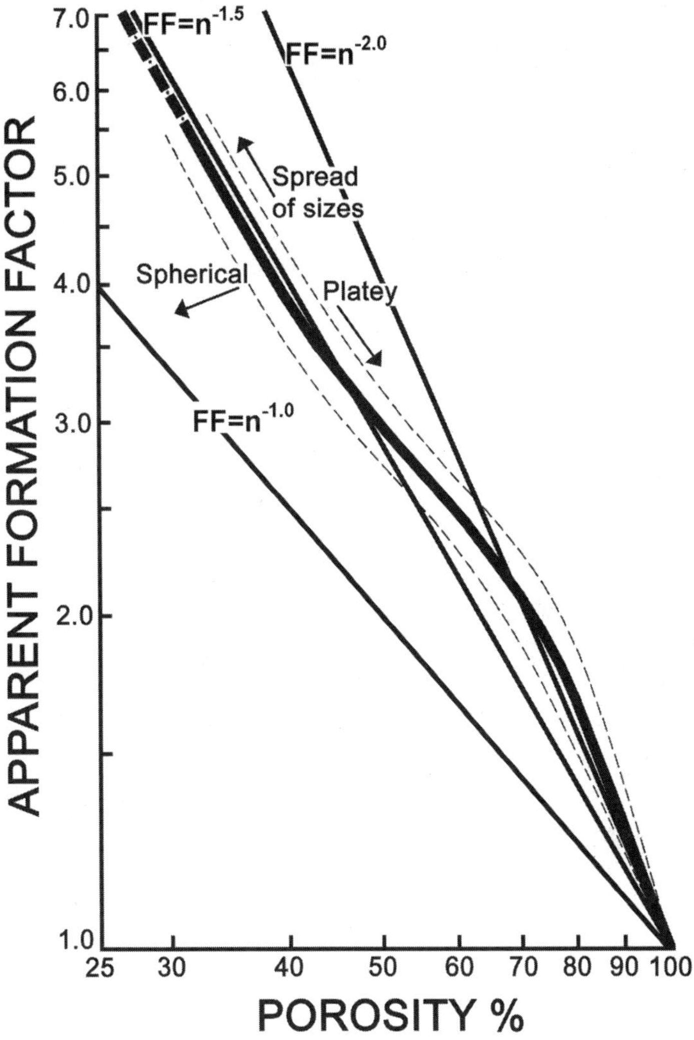

Fig. 4.14. A summary of laboratory packing experiments by [P. Jackson et al. 1978] giving formation factor versus porosity. The solid line is the mean for all measurements, and the dashed lines represent an envelope containing most of the experimental data. Also included are the formation factor versus porosity relationships for mud ($F = \eta^{-2.0}$) and sand ($F = \eta^{-1.5}$) proposed by [Taylor-Smith 1971]. Figure adapted from [P. Jackson et al. 1978].

thermal conductivity in clay-rich deep-sea sediments [Sclater et al. 1969]. The decay of temperature elevations due to the friction associated with probe insertion has also been used to measure thermal conductivity in sediments with low conductivity [Lee and Von Herzen 1994]. Typical values of thermal conductivity of deep-sea sediments range between 1.6 to 2.8 x 10^{-3} cal cm^{-1} s^{-1} $°C^{-1}$ (0.67 to 1.17 W m^{-1} $°C^{-1}$). A clear relationship between thermal conductivity and porosity has been demonstrated by numerous investigators for deep-sea sediments [Ratcliffe 1960]. However, this relationship is complicated by differences in conductivity of the solid phase in sediments of different mineralogical composition (e.g., carbonate versus siliciclastic clays). Studies of thermal conductivity in shallow-water or in sandy sediments have been mostly neglected with the exception of [Lovell 1985a, Lovell 1985b], although several studies of temperature gradients in intertidal sandy beaches and mud flats have been published (e.g., [Piccolo et al. 1993]). Studies of intertidal temperature gradients are complicated by rapid temperature fluctuations associated with diurnal solar heating and radiant cooling, tidal inundation, and the three-phase nature (air, water, sediment grains) of intertidal sediments and are not discussed in this monograph.

It has been demonstrated using laboratory experiments [Lovell 1985a, Lovell 1985b] that sediment thermal conductivity increases with decreasing porosity for both sand and mud. Lovell proposed a simple geometric model to predict sediment thermal conductivity, k_b, from sediment fractional porosity, β, and thermal conductivities of the pore fluid, k_f, and solids, k_s,

$$k_b = k_s^{(1-\beta)} k_f^{\beta} \; . \tag{4.23}$$

A least-squares fit of his measurements yielded k_s = 8.58 W $m^{-1}°C^{-1}$ and k_f = 0.64 W $m^{-1}°C^{-1}$.

Sediment thermal diffusivity, α^2, is a function of sediment thermal conductivity, k_b, specific heat at constant pressure, S_p, and bulk density, ρ,

$$\alpha^2 = \frac{k_b}{S_p \rho} \; . \tag{4.24}$$

This parameter can be used in the heat diffusion equation to determine temperature as a function of position and time:

$$\frac{\partial T}{\partial t} = \alpha^2 \nabla^2 T \; . \tag{4.25}$$

Temperature depth-time series measured in the upper 2 meters of the sandy sediments during SAX99 [Richardson et al. 2001a] were compared with solutions of the heat diffusion equation by [D. Jackson and Richardson 2002]. These authors obtained a sediment thermal diffusivity of 0.006 cm^2 s^{-1}, smaller than the value 0.012 cm^2 s^{-1} obtained using (4.23) and (4.24) and a handbook value $S_p \rho_g$ = 1.97×10^6 J m^{-3} $°C^{-1}$ for the specific heat of quartz. These results raise the question as to whether advective processes such as ventilation from wave-induced pressure fluctuations, mixing by bioturbation, or an upward migration as a result of a hydraulic head produced by groundwater aquifers could alter temperature gradients. Mixing by ventilation or bioturbation would decrease the temperature gradient—the opposite of what is observed. Furthermore, there is no evidence for freshwater advection. The difference between measured and modeled temperature gradients in their work thus remains unsolved pending additional study.

Seasonal variations in gradients of sediment temperature and sound speed and their effects on acoustic scattering have been investigated [Rajan and Frisk 1992, D. Jackson and Richardson 2002]. In temperate regions, bottom water temperature fluctuations are nearly sinusoidal, often exceeding 20°C, corresponding to sound speed fluctuations of 50–70 m s^{-1}. It is often assumed that sediment temperature is nearly the same as the overlying bottom water temperature, but a lag in heat transport due to the finite thermal diffusivity of sediments can create fluctuating gradients in sediment temperature and sound speed (Fig. 4.15). Such gradients can have a substantial effect on sound propagation at low acoustic frequencies (50 and 200 Hz) [Rajan and Frisk 1992], but the effects on scattering strength or reflection diminish at higher frequencies [D. Jackson and Richardson 2002].

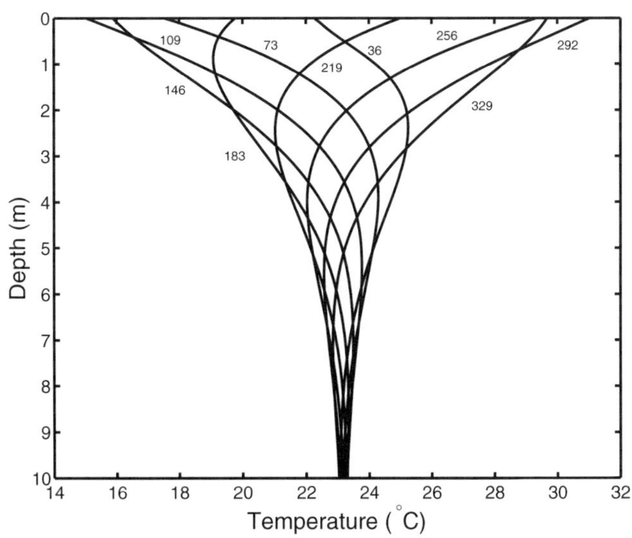

Fig. 4.15. Synthetic temperature profiles obtained by solving the one-dimensional heat diffusion equation with diffusivity 0.006 cm^2 s^{-1} for sand sediment (SAX99) from the northeastern Gulf of Mexico [D. Jackson and Richardson 2002]. The curves are labeled by the time (in days) after 0000 GMT, 5 October 1999.

4.4.4 Conductance Measured in Consolidation and Packing Experiments

Several of the laboratory studies of fluid flow (permeability), electrical resistivity, and heat flow cited in the previous sections were conducted with clay sediments that were artificially consolidated or with sandy sediments for which packing states were altered. Laboratory consolidation testing occurs over short time scales (seconds to days) and at high applied stress compared to the much longer times scales (decades to millennia) and low applied stress of natural consolidation. It is not known if the micro- or macrostructure of artificially consolidated sediments mimics that of naturally consolidated sediment. As a result, conclusions about sediment conductance phenomena (electrical resistivity, heat flow, permeability, and chemical flow) derived from laboratory experiments may be questioned when applied to natural conditions. For example, laboratory-consolidation-derived permeability values are between one and two orders of magnitude less than those measured directly [Lovell 1985a]. In addition, experiments over the full range of packing states for sand may include conditions that do not occur in nature, especially at the tenuous minimum-density/maximum-porosity state. Laboratory consolidation and packing experiments also exclude the effects of biological processes (Sect. 3.4.2) on sediment structure and conductance phenomena. This does not mean that the resultant relationships between sediment physical properties and conductance are not applicable to nature, but that the results should be viewed with a bit of skepticism.

4.5 Grain Properties

The density and bulk modulus of sediment grains are important parameters controlling compressional wave speed in sediments. This is true for the suspension theory proposed by [Wood 1964], the elastic theory of [Gassmann 1951], the viscoelastic theory used by [Hamilton 1971a], the poroelastic theory preferred by [Stoll 1989], or the theory proposed by [Buckingham 1997, Buckingham 1998, Buckingham 2000].

4.5.1 Density of Sediment Particles

In general, the density of the sediment particles (or grains) contributes to the overall bulk density of sediments. Particle density (also called grain density) can be determined by weight-volume methods if the specimen is large enough, but most constituents of sediments are much too small, consequently mean density of sediment grains is often determined using a helium pycnometer. Using this technique, average grain density is calculated from the dry weight of the grains and the pressure difference of a known volume of helium introduced into a sample chamber with and without the mineral sample. This method has an accuracy of about 1 kg m^{-3}. Other techniques to measure grain bulk density include X-ray crystallography, buoyancy in heavy liquids, liquid displacement, and gas volumenometry [Johnson and Olhoeft 1984].

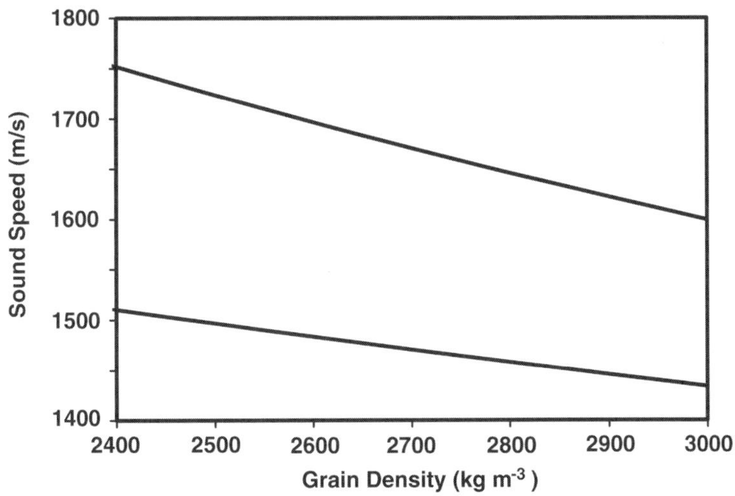

Fig. 4.16. Sound speed (m s^{-1}) as a function of grain density (kg m^{-3}) for typical sand of 37% porosity (top curve) and mud with 75% porosity (bottom curve) calculated using Wood's equation (8.72). Grain and water bulk modulus are assumed to be 36 GPa and 2.4 GPa, respectively, and water density is set at 1023 kg m^{-3}. The calculations represent the relative minimum sound speeds as the effects of frame modulus, shear modulus, or any fluid–skeletal interactions were excluded. The relationship nevertheless shows the relative decrease in sound speed with increasing grain density (increasing bulk density) for both quartz sands and fine-grained sediment dominated by clay minerals.

A compilation of densities of constituents commonly found in sediments taken from handbooks and the peer-reviewed literature is presented in Table 4.7. Although there is a range of bulk densities, even within individual mineral assemblages, most values of mineral or rock density are between 2500 and 2800 kg m^{-3}. Based on sound speed calculations using the Wood equation (8.72), this range of density translates into sound-speed differences in typical sediments of up to 50 m s^{-1} (Fig. 4.16). The range in values of mineral density reported in physical-properties handbooks often reflects differences in quality or purity of the specimens analyzed, or the varying chemical compositions of the mineral populations rather than differences in measurement techniques. The density of quartz, the dominant mineral in most nearshore terrigenous sands (35–50% of the terrigenous component), varies little from 2650 kg m^{-3}. Other constituents of sediments include metamorphic rock fragments (e.g.,

Table 4.7. Typical values for density and bulk modulus of common constituents found in coastal marine sediments. Numbers in parentheses are estimated percentages of the terrigenous component (excluding biogenic or carbonate components) found worldwide in coastal terrigenous sediments.

Particle Type	Density ($g\ cm^{-3}$)	Bulk Modulus (GPa)
Quartz (SiO_2, 35-50%)	2.65	36-40
Feldspars (5-15%)	2.50-2.80	37.5
K-Feldspars (Orthoclase)	2.54-2.57	
Na-Ca-Feldspars (Plagioclase)	2.62-2.76	
Metamorphic Rock Fragments (5-15%)		
Slate, Phyllite, Schist, Gneiss, Quartzite	2.50-3.00	
Clay Minerals (25-35%)		
Kaolinite	2.60-2.68	6-50
Smectite	2.35-2.70	6-50
Illite	2.60-3.00	6-50
Coarser hydrous phyllosilicates (0.1-0.4%)		
Chlorite	2.60-3.30	86.9
Muscovite $K_2Al_4(Si_3AlO_{10})_2(OH,F)_4$	2.70-3.10	40-60
Biotite ($K_2Mg_4(Si_3AlO_{10})_2(OH,F)_4$	2.70-3.30	40-60
Heavy Minerals (0.1-1.0%)		50-230
Magnetite (Fe_3O_4)	5.18-5.20	180-217
Limonite (FeO - FeOH)	3.20	60
Pyrite (FeS_2)	4.95-5.03	130-155
Hematite (Fe_2O_3)	5.25-5.28	200-230
Zircon ($ZrSiO_4$)	4.68-4.70	200-210
Tourmaline	3.00-3.25	
Rutile (TiO_3)	4.18-4.25	210
Sphene ($Ca\ Ti\ Si\ O_5$)	3.45-3.55	
Ilmenite ($FeTi\ O_4$)	4.70-4.78	210
Carbonates		
Calcite ($CaCO_3$)	2.71-2.72	64-75
Aragonite ($CaCO_3$)	2.93-2.95	45
Dolomite ($CaMg(CO_3)_2$)	2.72-2.87	70-90

slate, phyllite, schist, quartzite, and gneiss) which comprise 5–15% of the terrigenous components of sediments, and feldspars derived from weathering of igneous rocks which comprise roughly 5–15% of terrigenous components of sediments. These low-porosity sand- or gravel-sized rock fragments have values of density that reflect the density of the parent rock, 2730 kg m^{-3}, with igneous rocks having a slightly higher mean density than metamorphic rocks [Johnson and Olhoeft 1984]. Carbonates derived from the breakdown of skeletal material of a variety of plants and animals are only slightly denser than quartz (2700–2950 kg m^{-3}) and comprise nearly all the biogenic components of nearshore sediments. Differences in density in the calcite group are related to the crystalline structure, with minerals in the calcite and dolomite groups having a hexagonal structure and the denser aragonite group an orthorhombic structure. Coarser (> 20 microns) mica-rich (muscovite and biotite) or chlorite-rich particles, although abundant in source rock, are easily broken down during transport and generally make up less than 0.5% of the terrigenous fraction of sediments. Heavy minerals, as the name implies, all have densities greater than quartz, carbonates, and clay minerals, but are chemically or physically unstable under normal transport and sedimentary conditions and thus typically comprise less than 1% of the terrigenous component of coastal marine sediments [Louis and McConchie 1994]. Several heavy minerals such as Ti and Fe oxides and hydroxides and other metallic minerals that are common in sedimentary rocks are listed in Table 4.7.

Clay minerals are finely crystalline hydrous aluminum layer silicates (phyllosilicates) that have layered or sheet-like structure that is described in Sect. 3.2.1. Clay minerals are usually clay-sized (< 2 microns) but some silt-size phyllosilicates exist in rocks and sediments. The density of clay minerals, which make up 25 to 35% of the terrigenous component of sediments, ranges from 2600 to 3000 kg m^{-3} (Table 4.7). However, issues associated with both measurement techniques and the differences in the structure of these hydrated clay minerals make application of these values to high-frequency acoustic studies questionable. Measurement of the density of clay minerals, such as those presented in Table 4.7, typically represent densities measured on larger dried rock specimens or on sediment grains after the pore water between clay clusters and between crystalline folia has been eliminated by drying and only the formation water of the crystals remains (see Sect. 4.3). The "effective density" of hydrated clay particles in saturated marine sediments may, in many cases, be less than reported in rock physics handbooks or dry-measured using a helium pycnometer, as some of the structural water may have been driven off before density determination. This especially applies to the smectite and vermiculite groups, because these groups have hydrous exchangeable cations and water within the interlayers which are very sensitive to changes in temperature or the ionic composition of the pore fluid. The illite and kaolinite groups, on the other hand, have an organized octahedral sheet or fixed potassium layer, respectively, within interlayers and are therefore much less susceptible to changes in the ionic composition of the pore fluid, or structural changes during drying, even at high temperature. Many clay-sized particles in sediments are mixed-layer clays including both expandable water-bearing and contracted non-water-bearing layers; smectite–illite is the most common example [Mitchell 1993]. Swelling in the presence of seawater can change the "effective density" of clay particles in marine sediments. The amount of swelling varies between different clays depending on crystalline structure which affects the types and amount of bound water, the cation exchange capacity, and the specific surface area [Mitchell 1993].

In the NRL data set, comprised of over 1500 dried samples of siliciclastic and carbonate sediments, including clay-sized to sand-sized particles, the bulk density of the solid fraction of coastal sediments ranges between 2350 and 2800 kg m^{-3}. The lowest values of grain density are found in organic-rich mud and the highest values in carbonates, reflecting the trends presented in Table 4.7. Predominately quartz sand sediments have values of particle density very near 2650 kg m^{-3}, whereas values of average particle density in carbonate sediments are slightly higher, 2700–2800 kg m^{-3}. Clay minerals have the greatest range in values of density because of differences in chemical composition and the effects of temperature and pore water ionic composition on their structure. The measured density values cited above and the representative values given in Table 4.7 provide bounds for acoustic modeling but, given the natural variability, may not be a substitute for laboratory determination of sediment grain density using a helium pycnometer.

4.5.2 Bulk Modulus of Sediment Particles

Sediment-grain dynamic bulk modulus for a representative sediment sample is often calculated from bulk moduli corresponding to the crystalline minerals that make up sediment grains of that sample [Hamilton 1971a, Christensen 1996, Wang et al. 2001], by using either the Voigt–Reuss–Hill, weighted Hashin–Shtrikman, or Kroner approximation averaging methods [Hashin and Shtrikman 1962, Hill 1963, Kroner 1967]. Laboratory measurements of the bulk modulus of small rock specimens or polycrystalline aggregates are often based on static compression or dynamic (resonance bar or ultrasonic) acoustic techniques or, more recently, by X-ray diffraction, Brillouin scattering, and optical techniques. The values tabulated for the common minerals that make up lithospheric rock [Simmons and Wang 1971, Sumino and Anderson 1984, Mavko et al. 1998] are mostly based on measurements of crack-free, gem quality crystals [Christensen 1996]. For the crystalline materials that make up most of the sediment particles in coastal regions, such as quartz, calcite, aragonite, feldspars, silicates, etc., static and dynamic methods yield about the same results at either 1 atmosphere or for conditions of elevated pressure and temperature [Simmons and Brace 1965].

Using the Voigt–Reuss–Hill averaging method for determination of bulk modulus in a mixture of mineral species [Hamilton 1971a] calculated the following values of aggregate bulk modulus for typical marine sediments: nearshore fine sand, 51.2 GPa; clayey silt, 54.4 GPa; deep-sea silty clays, 50 GPa. Stoll, in a series of papers (1969 through 2002), consistently used 36 GPa, which is the average bulk modulus for crystalline quartz, for the average bulk modulus of all sand-, silt-, or clay-sized mixtures of particles in sediments. Many other investigators, following Stoll's approach, have adopted the handbook values of quartz for both grain density (2560 kg m^{-3}) and grain bulk modulus (36 GPa) as standard or default input values in theories for wave propagation in sediments. Values of bulk modulus for minerals and rocks that make up most sedimentary particles are given in Table 4.7. Figure 4.17 shows sound speed versus porosity calculated using Wood's equation (8.72) for the range of values of grain bulk modulus presented in Table 4.7. For these calculations, values of grain and water density are assumed to be 2650 and 1023 kg m^{-3}, respectively, and the pore water bulk modulus is set at 2.4 GPa. These calculations represent the relative minimum sound speeds as the effects of frame modulus, shear modulus, or any fluid–skeletal or grain-to-grain interactions are excluded. Nevertheless, Fig. 4.17 demonstrates the relative effects of grain bulk modulus on sound speed–porosity relationships, where small changes in grain bulk modulus have the greatest effect at low values of sediment porosity and for low values in grain bulk modulus. The effects on sound speed of the differences in average grain bulk modulus recommended by Hamilton (50 GPa) and used by Stoll (36 GPa) presented in Fig. 4.17 are not great. However, values of grain bulk modulus (about 7 GPa) that are much lower than those recommended by Hamilton or Stoll have been proposed. When inserted in Wood's equation (see Fig. 4.17), these lower values give much lower sediment sound speeds than obtained using handbook values. The validity of the lower values for grain bulk modulus will be discussed in the next paragraph. Given the obvious importance of grain modulus to sound speed determination in sediments it is not surprising that these recent controversies should arise.

Static measurements of compressibility of a mixture of sand-sized siliciclastic sediment grains reported in an abstract by [Molis and Chotiros 1992] suggest that values of bulk modulus of grains naturally occurring in surficial marine unconsolidated sediments may be much lower than values typically reported for crystalline minerals in rock physics handbooks. They proposed that imperfections or microfractures, typical of rock specimens, might contribute to the higher compressibility. These lower than tabulated handbook values of grain bulk modulus (0.7×10^{10} Pa versus 3.6 to 4.0×10^{10} Pa for crystalline quartz in handbooks) have subsequently been quoted and/or used in acoustic modeling of sediments by several investigators [Boyle and Chotiros 1995b, Chotiros 1995a, Chotiros 1995b, Buckingham 1997, Chotiros et al. 1997, Hickey and Sabatier 1997, Boyle and Chotiros 1998, Buckingham 1998, Chotiros 1998]. The effect on sound speed of one or the other of these two values of grain bulk modulus is evident in Fig. 4.17 where differences as great

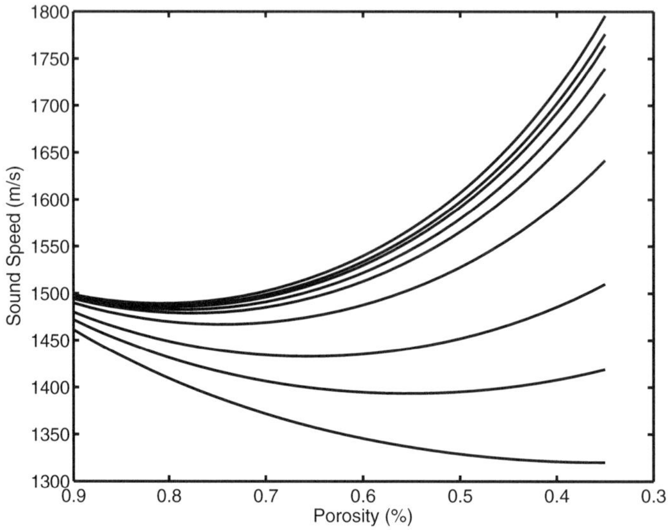

Fig. 4.17. Sound speed versus porosity calculated using Wood's equation (8.72) over a range of values of sediment grain bulk modulus (5–200 GPa). Sound speed increases with increasing values of grain bulk modulus found in Table 4.7: 5, 7, 10, 20, 36, 50, 75, 100, 200 GPa.

as 150 m s^{-1} are predicted. The merit of these two values of quartz grain bulk modulus was debated in comments in the *Journal of the Acoustical Society of America* [Stoll 1998, Chotiros 1998].

More recent low-strain acoustic measurements by [Richardson et al. 2002a], using a suspension of quartz grains and beads suspended in a heavy liquid, have shown that the bulk modulus of Ottawa sand and quartz sand grains collected from the coastal sediments of the northeast Gulf of Mexico ranged between 3.8 and 4.7 $\times 10^{10}$ Pa, with 95% confidence limits between 3.0 and 5.7 $\times 10^{10}$ Pa. These measured values of bulk modulus are consistent with the range of handbook values for polycrystalline quartz (3.6 to 4.0 $\times 10^{10}$ Pa). The authors suggest that until more accurate low-strain measurements of sediment grain bulk modulus are made, it is best to use the traditional handbook values derived from static compression, or dynamic (resonance bar or ultrasonic) acoustic measurements on small rock specimens or polycrystalline aggregates as inputs to sediment acoustic models.

The variations in bulk modulus reported for heavy minerals (Table 4.7) may reflect the gem quality of the mineral being measured rather than differences or inaccuracies in techniques. The relatively low concentrations of heavy minerals in coastal marine sediments suggest that small variations in values of bulk modulus, especially at these high values, would have little effect on predictions of acoustic propagation theories. The same might be stated for carbonate particles which have a higher crystalline bulk modulus than quartz or other metamorphic minerals such as feldspars. However, most carbonate particles are biogenic, being skeletal fragments of plants and animals often degraded by chemical and physical processes, and may not behave as their crystalline constituents. Measurements of bulk modulus using the suspension techniques suggested by [Richardson et al. 2002a] on carbonate sediment types archived from past sediment acoustic propagation studies could provide insights into possible effects of biological and chemical degradation on grain structure and compressibility and their resultant effect on acoustic propagation in sediments. This methodology is especially appropriate for examining the anomalous acoustic behavior associated with biogenic carbonate sediments possessing intraparticulate porosity.

Measurements of the bulk modulus of clay minerals are rare because of the small size of individual sheets ($< 2\ \mu$m) and because of their intrinsic properties such as swelling during hydration, reactions with organic polar compounds, low permeability, and their varied and changing microstructure given differences in temperature, pressure, and ionic composition of the pore fluid. The manipulations required during measurements can change the clay structure and significantly alter the bulk modulus [Vanorio et al. 2003]. These possible changes can vary depending on the types of clay and the ionic composition of the bound fluid or pore water. As stated before, Hamilton, Stoll and several others have generally assumed a value of bulk modulus for clays between 36 and 50 GPa. However, more recent measurements are not all in agreement with these values. Extrapolations from bulk modulus measured on shale suggest a clay bulk modulus near 20 GPa [Castagna et al. 1995]. Measurements on clay mixed with an epoxy resin by [Wang et al. 2001] suggest somewhat higher values of bulk modulus for kaolinite (45 GPa) and illite (60 GPa) and a lower value for smectite (10 GPa). Using theoretical considerations, [Berge and Berryman 1995] arrive at a bulk modulus of 10–12 GPa for clay. The values of bulk modulus for clay minerals were determined by [Vanorio et al. 2003] using a combination of acoustic pulse transmission techniques on cold-pressed clay powders at different hydrostatic confining pressures and during uniaxial stress compaction, and on suspensions of clay particles. They concluded that the dynamic bulk moduli of clay particles, including kaolinite, montmorillonite, and smectite, lay between 6 and 12 GPa with little difference among clays. In summary, the bulk modulus of clay particles, including theoretical calculations, direct measurement, measurements in clay–epoxy mixtures, and empirical extrapolations from measurements on shale, show little agreement, with values ranging between 6 and 50 GPa.

As stated at the beginning of the previous paragraph, hydration state and ionic composition of pore water may have an important effect on clay structure, possibly leading to a wide variation in measured values of bulk modulus. In addition, different clay types may react differently to changes in temperature, pressure, and the ionic composition of the surrounding water, resulting in changes in crystalline structure which probably result in changes in values of bulk modulus. This applies to both the laboratory measurement conditions and the in situ conditions within marine sediments. Recent laboratory experiments [Kim et al. 2003, Kim et al. 2004] have shown that biological activity (bacterial reduction) can alter the clay structure at ambient pressure and temperature. This includes significant changes in lattice spacing and ionic composition of interlayers. Therefore one of the difficulties arises in matching the laboratory-measured bulk modulus to the state of clay minerals in marine sediments. The values of bulk modulus suggested by [Vanorio et al. 2003] may seem more appropriate for marine sediments, because the sample state is closer to the natural hydrated state of clays than in other measurements. However, if these low values of bulk modulus (6–12 GPa) are assumed, values of sound speed calculated by Wood's equation (Fig. 4.17) are lower than typically measured in marine sediments, especially in lower-porosity mud. Shear speeds in mud, even at lower porosity, are generally less than 50 m s^{-1}, suggesting the contribution of shear modulus to sound speed is almost negligible. The contribution of frame modulus to sound speed, using the Gassmann equations (Sect. 10.1.1), must be much larger than usually assumed or the "effective bulk modulus" is higher than 6–12 GPa. It is also possible that clay minerals in marine sediments cannot be treated as discrete particles as required by elastic and poroelastic theories, thus these models do not adequately describe acoustic propagation in high-porosity marine clay sediments. Other approaches, such as discrete element modeling, might better capture the physics of low-strain acoustic wave propagation and dissipation in marine sediments where electrostatic attraction and repulsion forces, organic binding, and small-scale deformation of clay particles are significant [Yao and Anandarajah 2003]. Other approaches include treating the grain bulk modulus as a frequency-dependent complex number accounting for movement of interlayer water into and out of the pore space in response to low-strain acoustic wave propagation through the sediment [Leurer 1997].

4.6 Factors Affecting Fluid Motion

In the Biot theory of propagation (Ch. 10), acoustic displacements of the fluid are taken as averages over a suitably chosen local volume such that details of flow within individual pores are lost. Four variables control average fluid motion: pore water dynamic viscosity, μ, permeability, κ, structure factor, α (also called tortuosity), and pore size parameter, a. Direct methods of measuring permeability were discussed in Sect. 4.4.1, and pore water viscosity will be adequately covered in Sect. 4.7.2. The remaining two variables are related to the geometry associated with sediment fabric discussed in Sect. 3.2. It is obvious that permeability, tortuosity, and structure factor are interrelated (see (4.16) and (10.16)), but these variables can be difficult to physically relate to the complex structure of marine sediments. Although it is appealing to theoretically determine the permeability, structure factor, and pore size parameter from microgeometry of the sediment fabric, [Stoll 1989] and many others have instead estimated these parameters from experimental data (the phenomenological approach).

Tortuosity is a term used to describe the sinuosity and interconnectivity of pore space as it affects hydraulic transport through porous media. Definitions of tortuosity are not universally accepted and are complicated by their use in different transport fields [Clennell 1997]. Geometric tortuosity is often defined as the ratio of the shortest path of interconnected points in pore fluid space to the straight-line distance between those points. In practice, tortuosity is an average property of a given volume of sediment and may vary with direction and scale. The hydraulic tortuosity, T, appearing in the Kozeny-Carman equation (4.16) is the ratio of the effective hydraulic path length to the straight-line distance in the direction of flow. It is important to note that the tortuosity factor, α, used in Biot theory is equal to the *square* of the hydraulic tortuosity, $\alpha = T^2$. This is because the overall factor by which the flow is retarded is proportional to the effective path length squared. Tortuosity can be calculated from high-resolution images of pore structure (Figs. 3.6 - 3.8), measured using conductance experiments, or calculated from pore size distribution. Stoll [Stoll 1989] set the structure factor to 1.25 for sand and 3.0 for fine-grained sediments. However, theoretically, the structure factor can vary between 1.0 for parallel capillaries and 3.0 for capillaries with random orientation.

The pore size parameter, a, with units of length, is the size of parallel channels or slits through which fluid flows during the passage of an acoustic wave. This is equivalent to the pore geometry (geometry and shape of pores and pore tubes) described earlier in Sect. 4.4.1. Based on experimental data [Stoll 1989] set the pore size parameter to 1/6 to 1/7 of the mean grain size for sediments composed of predominately sand or silt. Other investigators have related the pore size parameter to the mean and standard deviation of the grain size distribution [Chotiros 1995a], mean and standard deviation of the pore size distribution [Yamamoto and Turgut 1988], porosity and wetted surface area of the grains [Hovem and Ingram 1979], and porosity, permeability, and pore size parameter [Johnson et al. 1987].

4.7 Pore Water Properties

Pore water physical properties are usually assumed to be the same as those of the overlying water and are not usually measured as part of the environmental characterization during high-frequency acoustic experiments. In a rare exception, [Richardson et al. 2001a] found no difference in salinity and viscosity of pore and overlying waters during the SAX99 experiments conducted in sandy sediments in the northeastern Gulf of Mexico. However, as will be shown below, the assumption that pore water has the same properties as the overlying seawater may not be accurate. Seawater properties are discussed in some detail in Appendix B.

4.7.1 Pore Water Density

Pore water is usually assumed to have the same density as the overlying water, with the density calculated from temperature, salinity, and water depth using the Knudsen's Hydrographic Tables, a polynomial regression adopted by the UNESCO Joint Panel on Oceanographic Tables and Standards in 1980 [Fofonoff and Millard 1983], or the Gibbs thermodynamic potential function proposed by [Feistel 2003] (see Appendix B for details). These precise polynomial regressions are required for dynamical oceanographic calculations in open-ocean waters where the dominant seawater constituents maintain relatively constant ratios. In coastal waters, values of density calculated from temperature, salinity, and pressure may deviate as much as 0.05 kg m^{-3} (5×10^{-5} g cm^{-3}) for actual conditions, a result of variations in the composition of dissolved salts in the seawater [Fofonoff 1985, Millero 2000]. Pore water with abundant inorganic matter may be significantly denser than the overlying seawater. However, these differences may not be important for many high-frequency acoustic applications, and a pore water density of 1000 kg m^{-3} is often assumed (e.g., [Stoll 1989]).

4.7.2 Pore Water Viscosity

Viscosity is the measure of a fluid's internal, or molecular, resistance to shear stress (velocity shear). The thicker or more consistent the fluid, the more viscous is the fluid. Dynamic (or absolute) viscosity, μ, is the tangential force per unit area required to produce unit flow velocity at unit distance from a flat boundary where the velocity is zero. Thus, it has dimensions force$\times$(length)$^{-2}$, or mass$\times$(length)$^{-1}\times$(time)$^{-1}$. Kinematic viscosity is defined as the ratio μ/ρ_w.

Viscosity of open-ocean seawater is well documented, and tables can be found in a variety of physical handbooks. Moreover, empirical formulas are available to calculate seawater viscosity at a given water temperature, salinity and pressure (see Appendix B). Seawater viscosity decreases with temperature, increases with salinity, but only slightly increases with pressure with a range of 0.0007 to 0.0019 kg m^{-1} s^{-1} for most oceanic conditions. For application of Biot theory, a dynamic viscosity of 0.001 kg m^{-1} s^{-1} is typically used for pore water [Stoll 1989]. Using sensitivity analysis, [Williams et al. 2002a] determined that sound speed and attenuation at high frequencies are not sensitive to pore water viscosity in sandy sediments such as found in SAX99, at least over the range predicted for filtered seawater (9.5 to 11.5×10^{-4} kg m^{-1} s^{-1}).

Algae and bacteria secrete copious amounts of mucus exopolymers (slime) into the water column, changing the rheological properties (i.e., flow characteristics, including viscosity) of seawater [Jenkinson 1993]. Increased seawater viscosity may inhibit particle sinking, may slow the locomotion and feeding rates of plankton (especially those microorganisms that use cilia or flagella), and may change flow dynamics [Jenkinson and Biddanda 1995, and references cited therein]. Excess viscosity (that which is greater than predicted for seawater at a given temperature and salinity) has been correlated to the concentration of mucus in the water [Jenkinson 1993]. Jenkinson also found that the seawater behaved as a non-Newtonian fluid in that the viscosity increases with the rate of shear. At the lowest measured shear rate (0.0021 s^{-1}), viscosity averaged 3.5 times that calculated for seawater at the same temperature and salinity for samples collected from the Mediterranean and averaged 52 times the calculated viscosity for samples collected in the German Bight. These high-viscosity samples were collected intentionally during plankton blooms where abundant mucus was expected, but the increased viscosity in seawater due to biological productivity has been noted unexpectedly by others. Sediment pore waters, which are also rich with bacteria and other microorganisms that secrete all sorts of organic material, may also have elevated viscosity. The possible effects of this elevated viscosity on permeability, sound speed, and attenuation should be investigated.

4.7.3 Sound Speed and Attenuation

Precise algorithms to calculated sound speed from seawater temperature, salinity, and pressure have been developed (see Appendix B). The Del Grosso [Del Grosso 1974] algorithm has 19 coefficients compared to the 41 coefficients of the algorithm of [Chen and Millero 1977], which is accepted as part of the international seawater equation of state. The differences in values of sound speeds calculated from these algorithms is insignificant (less than 0.2 m s^{-1}) in shallow water conditions (< 100 m) where high-frequency acoustics often finds application. Pore water sound speed may be more than 10 m s^{-1} different than the overlying water, because of temperature differences or gradients, effects of the abundant inorganic and organic matter typically present in pore water, and diagenetic alteration of ionic concentrations in pore fluids.

A simplified algorithm to calculate sound speed from temperature, T, salinity, S, and depth in meters, D, has been suggested by [Medwin 1975]

$$V_p = 1449.2 + 4.6T - 0.055T^2 + 0.00029T^3 + (1.34 - 0.01T)(S - 35) + 0.016D \,. \tag{4.26}$$

Sound speed from this algorithm is less than 1 m s^{-1} different than sound speed calculated using the Chen and Millero algorithm over the range of environmental conditions in which high-frequency experiments are usually conducted (0–40°C, 0–40 PSU, 0–500 m). Sound absorption in seawater over the frequencies considered here (10–500 kHz) is dominated by magnesium sulfate absorption and can be predicted by the algorithms given by [Francois and Garrison 1982a, Francois and Garrison 1982b] which are given in Appendix B. Graphical representations of attenuation over the high-frequency band can also be used (Appendix B). Attenuation in turbid waters has been shown to degrade the performance of high-frequency sonar [Richards et al. 1996, Richards 1998, Richards et al. 2003]; therefore scattering from particles found in pore waters must also increase attenuation, but the effect on predictions of propagation theories should be slight.

4.8 Research Issues

Measurement of sediment physical properties is a fairly mature area of research because of the many applications to geology, geophysics, and geotechnical engineering. Sediment type and mean grain size are routinely measured and included in large databases. However, it can be questioned whether the destructive techniques used in traditional grain size analysis are appropriate for high-frequency acoustic applications or if the in situ, non-destructive techniques developed by the sediment transport community are more appropriate. The effects of grain shape and surface morphology on packing and conductance have been well documented, but their direct effects on wave propagation (compressional and shear wave speed and attenuation) are at best subjects of speculation and could provide ideal topics for laboratory studies. The relative density of typical marine sediments (sands) is generally unknown and can have an important impact on sediment porosity and all of the conductance phenomena (permeability, electrical resistivity, thermal conductivity). The effects of sediment collection, transport, and destructive manipulations during measurements of porosity, bulk density, or water content and the determination of packing (sand) or consolidation state (silts and clays) are all hotly debated, which suggests improvements to nondestructive and in situ measurements of these properties are needed. For example, small differences in porosity (1%) can produce significant differences in sound speed. Additional studies on the effects of sediment structure on sediment conductance properties are also needed. The appropriate values for high-frequency acoustics of particle density and bulk modulus, especially for clay minerals and perhaps for carbonate skeletal material, are unknown. The assumption that the physical properties of pore water are the same as the overlying seawater should be investigated.

5 Geoacoustic Properties

The geoacoustic properties to be discussed in this chapter include speed and attenuation of both compressional and shear waves, acoustic impedance (expressed as the "index of impedance"), and bulk density (which is included in both the physical and geoacoustic categories in this monograph). The speed and attenuation of waves propagating through seafloor sediments (body waves) or at the sediment–water interface (interface waves) have been measured and modeled for over half a century. The types of measurements and models are as varied as the motivations of the investigators (Exploratory Geophysics, Sedimentology, Soil Engineering, Acoustic Modeling, Sediment Dynamics). Measurements using widely different techniques make comparisons difficult, especially for wave attenuation. The frequency dependence of wave speed (dispersion) and attenuation makes prediction of wave speed or attenuation difficult when the measured and required acoustic frequencies are substantially different. In spite of these problems, some generalizations can be made, and values of wave speed and attenuation as well as their gradients can be measured or predicted using physical and empirical models. This chapter will concentrate on measurement and prediction of near-surface (upper few meters), high-frequency wave speed and attenuation. Three types of body waves will be considered: compressional, shear, and the Biot "slow wave." Interface waves (Love waves and Stonely waves) will only be discussed where these techniques provide data on speed and gradients of shear waves. Following common practice, the compressional wave speed in sediments will often be referred to as the "sound speed."

In fluids, which lack rigidity, only compressional (or longitudinal) waves are present, whereas both compressional and shear waves propagate through all known sediments, demonstrating that all sediments have some rigidity, and that elastic theory (Ch. 9) or poroelastic theory (Ch. 10) may be required in many cases. Shear waves traveling at speeds as low as 1.2 m s^{-1} have been detected in slurries of natural clay-size marine sediment which have been allowed to settle in laboratory cylinders [McDermott 1991], and shear wave speeds as low as 5 m s^{-1} have been measured in very high porosity (90%) mud in Eckernförde Bay, Baltic Sea [Richardson and Briggs 1996]. Poroelastic theory [Biot 1962b, Stoll 1989], presented in Ch. 10 of this monograph, predicts the existence of a second bulk longitudinal wave ("slow wave"). Although the slow wave has not been observed in field experiments, this is in keeping with its expected low speed and very high attenuation. This does not mean that poroelastic effects are insignificant in unconsolidated sediments. As explained in Ch. 10 and later chapters, poroelasticity results in frequency-dependent sound speed, nonlinear dependence of attenuation on frequency, and reductions in the strength of reflection and scattering. In summary, although elastic and poroelastic theory provide important refinements of fluid acoustic theory, the ordinary compressional wave remains the most important excitation in unconsolidated sediments, and its speed and attenuation are the subject of most geoacoustic measurements. Sound speed and attenuation embody poroelastic effects and, together with physical properties such as density, provide the inputs needed for the majority of high-frequency reflection and scattering models described later.

In this chapter, a review will be given of the various laboratory and in situ methods that have been used to measure wave speed and attenuation for shear and compressional waves in near-surface marine sediments. The term "speed" rather than "velocity" will be used, as most of the measurements are reported as scalar rather than vector quantities. It is assumed that all measurements involve excitation levels small enough to ensure that the resulting strain levels are below the maximum for which linearity can be assumed. This maximum strain level is sometimes taken to be 10^{-6} [Hamilton 1971a, Hamilton 1972, Hamilton 1980, Stoll 1989], but the less conservative number 10^{-5} is reasonable. An even higher maximum level of 10^{-4} is given for shear strains by [Davis and Bennell 1986]. The shear modulus decreases as strain increases to values above this maximum. As a point of reference, a pressure of 1 atmosphere (101.3 kPa) causes a strain (divergence of displacement) in water of about 4×10^{-5} (Sect. 8.1) and strain of similar magnitude in sediment. Modeling the sediment as a poroelastic medium does not alter this statement, as solid and fluid displacements are comparable in magnitude (Sect. 10.1.5), hence the strains are similar. Thus, in assuming linearity, one is assuming that acoustic pressures are substantially smaller that 1 atmosphere. The measurements reported here were conducted at acoustic pressure levels in the linear region.

Measured attenuation includes all forms of energy loss, including the intrinsic attenuation modeled by effective medium wave propagation theories, wave conversion at boundaries and other heterogeneities, and scattering. The range and variability of compressional and shear wave speed and attenuation for a variety of sediment types will be given as well as examples of near-surface gradients. Empirical regressions for prediction of sediment acoustic properties from easily measured physical properties are presented. In addition, regressions that can be used to predict sediment physical properties from wave speeds and impedance will be given. The limited evidence for the existence of the Biot slow wave is then discussed. The literature on the frequency dependence of attenuation and sound speed (dispersion) is also reviewed.

5.1 Compressional Wave Speed and Attenuation

Speed and attenuation of waves propagating through unconsolidated sediments have been measured since the 1950s in order to (1) constrain inversions of low-frequency geophysical data, especially related to oil and gas exploration, (2) provide geoacoustic models in support of antisubmarine warfare, (3) determine relationships between geotechnical and acoustic properties in support of acoustic surveying techniques, (4) investigate the validity of theoretical formulations of wave propagation, (5) predict geotechnical behavior of sediments, and more recently, (6) provide the necessary inputs to high-frequency scattering and penetration models.

The compressional wave speed (or sound speed) will be denoted V_p, following common usage in the geoacoustic literature. The symbol c_p is used for compressional wave speed in the purely acoustic chapters and appendices of this monograph, following standard practice in the underwater acoustics literature. There is a difference in meaning, however, in these two symbols. The parameter c_p is complex, as its definition incorporates attenuation. The parameter V_p is real, being the phase speed of the compressional wave. In Ch. 8, the phase speed is denoted c_{pphase} and is given by (8.22). Similar dual notation is used for shear wave speeds.

5.1.1 Measurement Techniques

Early laboratory and in situ measurements of sound speed and attenuation by [Hamilton 1956, Hamilton et al. 1956, Shumway 1956, Laughton 1957, Nafe and Drake 1957, Sutton et al. 1957, Shumway 1958, Shumway 1960, Hamilton 1963, Nolle et al. 1963, Wood and Weston 1964, Buchan et al. 1967, Hampton 1967, McCann and McCann 1969, Hamilton et al. 1970, McCann 1972] provided the database necessary to determine the range of sound speed and attenuation in surficial marine sediments and to develop empirical predictive relationships among sediment acoustic and physical properties. The speed of sound in sediment was found to be correlated with sediment porosity and mean grain size with a sound speed minimum (3% lower than seawater) in muddy sediments (70–90% porosity) and a maximum in medium to coarse sands (porosity 35–40%). Attenuation was low and highest in sediments with intermediate grain size (30–125 microns) and porosity (40–60%). Hamilton and Bachman, in a series of review papers, provided the most widely used regressions to predict sound speed and attenuation from easily measured sediment physical properties (porosity, bulk density, mean grain size) [Hamilton 1972, Hamilton 1980, Hamilton and Bachman 1982, Bachman 1985, Hamilton 1987, Bachman 1989].

Sediment sound speed and attenuation in unconsolidated sediments have been measured in situ, in the laboratory, and remotely using a wide variety of techniques and over a frequency band from a few hertz to the megahertz range. Sound speed and attenuation measurements using remote lower-frequency seismic techniques, including seismic reflection, refraction, and various inversion approaches, are beyond the scope of this monograph. In general, seismic techniques provide estimates of average sound speed and attenuation over large patch sizes and to greater depths than are of interest for high-frequency acoustic bottom interactions. Therefore the discussion in this chapter will be mostly restricted to in situ and laboratory techniques used to measure compressional wave speed and attenuation at frequencies greater than 10 kHz (wavelengths less than 20 cm). This allows measurement and prediction of near-surface sound speed and attenuation at the frequencies of interest and provides information on the spatial variability and gradients of wave propagation properties needed for high-frequency acoustic models.

In situ measurements have been made using a variety of probes pushed into the sediment by divers, from submersibles, or by remotely operated systems [Hamilton 1956, Richardson and Briggs 1996, Best et al. 1998, Kraft et al. 2002]. Sound speeds are usually determined directly from measurement of time-of-flight and distance between probes or by comparison to waves propagating through water. Attenuation is often measured from the loss of energy (waveform amplitude) over different probe separations by comparison to a standard such as water. These and most other in situ techniques assume spherical spreading. Other direct methods include impedance tubes [Dunlop 1992], bottom-mounted hydrophones and geophones using a variety of sources within the water column and sediment [Best et al. 2001, Stoll 2002], and downhole [Fu et al. 1996, Wilkens and Richardson 1998, Gorgas et al. 2002] or crosshole [Yamamoto 1995, Rapids et al. 1998, Chu et al. 2001] tomographic methods.

Most laboratory measurements rely on pulse techniques, although higher-strain resonance techniques are popular in geotechnical engineering. The U.S. Naval Oceanographic Office developed one of the first pulse techniques to measure sediment sound speed within cores [Winokur and Chanesman 1966]. This approach with some modification is still used at the Naval Research Laboratory [Richardson 1986]. Sound speed has also been measured using thin slices of sediments cut from core samples using a Hamilton frame [Abernethy 1965, Boyce 1973]. These latter data have provided values of sound speed for most of the shallow-water database used by Hamilton. One of the most popular recent laboratory methods to measure sediment sound speed in cores is the multi-sensor core logger [Weaver and Schultheiss 1990, Schultheiss and Weaver 1992, Weber et al. 1997, Gunn and Best 1998, Best and Gunn 1999]. This technique uses two rolling 500-kHz transducers that maintain contact with the outside of the core liner. A water-filled core liner serves as a standard to allow time-of-flight through the core liner and other electronic time delays to be determined. Graduated aluminum slugs are used for calibration of sediment bulk density measurements.

Any laboratory technique used to measure sound speed or attenuation requires very carefully collected and maintained sediment samples. Removing sediments from cores to make acoustic measurements often results in changes in packing in sand and consolidation state in mud and loss of water by drying or draining in both types of sediments. Altering the packing of sandy sediment or the consolidation of muddy sediment can greatly change the sediment structure and values of sound speed and attenuation measured on these sediments may not reflect in situ conditions. Sound speed in sediment varies with temperature, pressure, and pore water salinity and should be corrected to a common set of conditions before regressing against sediment physical properties. Hamilton and co-workers used laboratory conditions of 23°C, atmospheric pressure, and a salinity of 35 ppt to report sound speed in their earlier work. Hamilton later instituted the term "velocity ratio," V_pR, which is the ratio of measured sound speed of sediment to the sound speed of water at the same temperature, salinity, and pressure. This ratio will be referred to here as the "sound speed ratio" and will be denoted ν_p in the acoustic chapters of this monograph. The sound speed of water, c_w, can easily be calculated from temperature, salinity, and depth or pressure (see Appendix B) and can be used to calculate in situ sound speed from the sound speed ratio ($V_p = V_pR \times c_w$). The validity of this approach has been demonstrated in laboratory measurements by [Shumway 1956, Laughton 1957, Shumway 1958, Hamilton 1971b, Bennell 1979, Bell and Shirley 1980, Carbo and Molero 2002], where the differences in sound speed in sediments for different conditions of temperature, pressure, and salinity were shown to be approximately proportional to the changes in pore water sound speed over the same range of conditions. This relationship was later extended by [Richardson and Briggs 1993, Richardson and Briggs 2004b] to regressions between sediment impedance and seafloor physical properties. The index of impedance, IOI, which is defined as the product of sound speed ratio and density, is used to account for the dependence of impedance on pore water temperature, pressure, and salinity (see Sect. 5.1.8). The effects of temperature or pressure on attenuation in sediments have been rarely documented except under laboratory conditions at very high frequencies (500 to 1000 kHz). Under these conditions, attenuation in sands decreases with increasing temperature, whereas attenuation increases with increasing temperature in muddy sediments [Bennell 1979, Carbo and Molero 2002]. Pressure appears to have little effect on attenuation, at least over the range of pressure found in shallow littoral regions. Whether these measurements can be extended to lower frequencies is an open question, however, as any variations in attenuation caused by seasonal fluctuations in temperature or with depth are likely to be masked by the high variability of attenuation in naturally occurring sediments.

The technique developed by Richardson and Briggs [Richardson 1986] is used for one of the primary data sets to be discussed in this chapter, and will be described here in detail. The method uses travel time differences (Δt) between 400-kHz signals (5- to 10-cycle pulsed sine wave) transmitted through sediment cores and identical reference cores filled with distilled water to measure compressional wave phase speed (V_p, m s^{-1}) and attenuation (α_p, dB m^{-1}),

$$V_p = \frac{c_w}{1 - c_w \Delta t/d} , \tag{5.1}$$

$$\alpha_p = \frac{20}{d} \log_{10} \left(\frac{e_w}{e_s} \right) , \tag{5.2}$$

where c_w (m s^{-1}) is the sound speed of distilled water, d (m) is the inside diameter of the core liner, and e_w/e_s is the ratio of the received voltages from waveforms transmitted through sediments and distilled water. Phase speed, rather than group speed, is measured in this method, as time shifts are measured as shifts of signal peaks, that is, of the cycles making up the signal. Group speed would result if the shift in the signal envelope were measured. If dispersion is large, however, differential time shift between the envelope and the signal cycles could make it difficult to make a correspondence between cycles in the water and sediment signals. This is not a problem, because, experimentally and theoretically, there is a negligible difference in time delays for phase and group speed. Comparison of cross-correlation, envelope, and peak-picking techniques showed no significant difference in calculated sound speeds [Briggs and Richardson, unpublished]. Using parameters from the SAX99 site and a frequency of 400 kHz, the theoretical ratio of group speed to phase speed is 1.007 using Buckingham's theory (Sect. 9.8) and 1.0014 using Biot theory (Sect. 10.1.3). These ratios correspond to differential envelope-cycle shifts less than one-tenth of a cycle, too small to cause confusion in measurement.

The core tubes used for sediments and water standards were all taken from a single extrusion with carefully controlled tolerances to reduce the effects of differences in core diameter and core wall thickness. Transducers are maintained at a constant distance in oil-filled rubber housings (Fig. 5.1). Sound speed and attenuation are measured at 1-cm intervals along the core tubes, usually within 24 hours of collection, using time-of-flight and amplitude techniques (5.1) and (5.2). The resulting acoustic wavelengths are 0.35 to 0.45 cm which are about five times larger than the diameter of coarse sand. This satisfies the requirement given in the introduction of Ch. 8 for treating the sediment as a continuous medium. Following [Hamilton 1971b], sound speed is reported as the dimensionless sound speed ratio, V_pR, of measured sediment sound speed to the sound speed of pore water at the same temperature, salinity and pressure. Sound speed values are also reported for in situ conditions and for the standard laboratory conditions favored by Hamilton (23°C, 35 ppt, and atmospheric pressure). These conditions correspond to a water sound speed of 1529.4 m s^{-1} and a water density of 1024 kg m^{-3}. Attenuation is expressed as the "attenuation factor," $k = \alpha_p/f_{kHz}$, where f_{kHz} is the acoustic measurement frequency in kHz [Hamilton 1971b]. The units of k are dB m^{-1} kHz^{-1}.

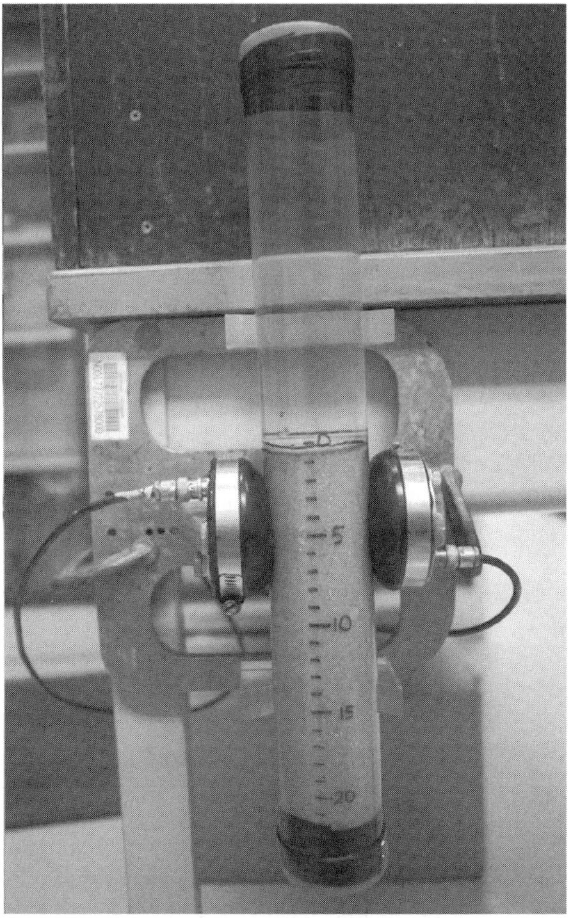

Fig. 5.1. Transducer setup used to measure sediment sound speed and attenuation. Two identical transducers (transmit and receive) are maintained at a constant distance across a 5.9-cm (inside diameter) sediment core by a rigid frame. The transducers are disk-shaped ceramics mounted on brass slugs within oil-filled, rubber housings. Travel-time and amplitude measurements (400 kHz) are made at 1-cm intervals.

After acoustic measurements on cores are concluded, sediments are extruded and sectioned at 2-cm intervals to determine sediment porosity and grain size distribution (Ch. 4). Porosity is determined from weight loss of sediments dried at 105°C for 24 hours and corrected for residual salt. Grain density is determined using a pyncnometer. Sediment bulk density is calculated from the measured porosity and from the densities of pore water and sediment grains. Sediment grain size is determined from disaggregated samples by dry sieving for sand-sized particles and by either pipette methods or Micromeritics Sedigraph for silt- and clay-sized particles.

The other primary data set to be discussed in this chapter was obtained using various versions of an in situ sediment acoustic measurement system (ISSAMS). In situ sediment geoacoustic measurements complement laboratory measurements allowing independent confirmation of results as well as covering frequencies and spatial scales not accessible in the laboratory. The in situ data presented here were measured remotely using a hydraulically operated platform that drives geoacoustic probes into sediments [Griffin et al. 1996] or a diver-deployed version of the same system [Barbagelata et al. 1991]. Compressional wave measurements use identical radially poled ceramic cylinders for both transmission and reception. Compressional wave speed and attenuation are measured over path lengths ranging from 30 to 100 cm at depths of 5–30 cm below the sediment–water interface. Transmissions utilize 5- to 10-cycle pulsed sine waves, and the time delays and voltages of received signals are used to determine values of sound speed and attenuation. Measurements with the diver-operated system were made at 58 kHz while the remotely operated system employed a frequency of 38 kHz. The resulting acoustic wavelengths are between 2.5 and 4.5 cm. Compressional wave speed and attenuation are calculated by comparison of received signals transmitted through the sediment with those transmitted through seawater overlying the sediments, using (5.1) and (5.2). Sound speed (c_w) and signal amplitude (e_w) in the overlying water are substituted for their respective values in distilled water. The attenuation measurement technique used here does not account for differences in transducer sensitivity in water and sediment. Comparison of attenuation measurements made by transposition, which eliminates the effects of differences in sensitivity and insertion loss, with the methods used in this analysis suggests these effects are small (0–10% of measured k) and within the natural variability of surficial sediments [Richardson 1997b]. Recalculation of attenuation based on transposition techniques was only possible for a small subset of these data and did not significantly change the results.

5.1.2 Data Sets and Regression Relations

Both empirical and physical models have been used to predict relationships between seafloor geoacoustic and physical properties, but the empirical approach has significant advantages. Some of the parameters required by the wave theories described in Chs. 8–10 are either very difficult (pore size, tortuosity, grain roughness) or nearly impossible to measure (i.e., frame modulus). The predictive capability of these theories when applied to naturally occurring sediments is accordingly limited. Often the empirical relationships between acoustic and physical properties have been used to validate wave theories or derive unknown constants (e.g., [Buckingham 2005]). For these reasons, empirical regressions are the preferred means of relating geoacoustic and physical properties. Regressions among sediment physical properties (porosity, bulk density, mean grain size) and geoacoustic properties (sound speed and attenuation) have been developed by numerous authors [Nafe and Drake 1963, Buchan et al. 1972, Hamilton 1972, Akal 1974, Hamilton 1980, Hamilton and Bachman 1982, Bachman 1985, Hamilton 1987, Bachman 1989, Richardson and Briggs 1993, Richardson 1997a, Richardson 1997b, Richardson et al. 1997, Kraft et al. 2002, Richardson and Briggs 2004a, Richardson and Briggs 2004b, Goff et al. 2004]. Many of these regressions are based on a limited range of sediment types or on data collected using a variety of techniques. To eliminate issues associated with intercomparison with different techniques, two of the largest available compilations of measured values of near-surface compressional wave speed, attenuation, and sediment physical properties will be presented here. Each compilation is internally consistent in that the data were obtained by identical or similar techniques. One compilation includes nearly 800 cores collected from 67 shallow-water sites around the world (12 carbonate and 55 siliciclastic sites) [Richardson and Briggs 2004a].The other compilation consists of in situ measurements made at 87 sites using various versions of an in situ geoacoustic measurement system [Richardson 1997a].

5.1.3 Core Data Set

All the laboratory measurement values presented in Tables 5.1 and 5.2 were made using the Richardson–Briggs technique, over a wide variety of sediment types, on carefully collected and maintained sediments, and all within the upper 30 cm of the seafloor [Richardson and Briggs 2004a, Richardson and Briggs 2004b]. The data include nearly 4500 data points resulting from measurements or calculations (3922 siliciclastic and 621 carbonate) of the following parameters: sediment sound speed (V_p, m s^{-1}), sediment sound speed ratio (V_pR, no units), attenuation (α_p, dB m^{-1}; k, dB m^{-1} kHz^{-1}), mean grain size (M_z, ϕ), sediment porosity (η, %), sediment bulk density (ρ, g cm^{-3}), index of impedance (IOI, g cm^{-3}, see Sect. 5.1.8), and sediment type. Each of these parameters has been determined for each core, so that empirical relationships between parameters are based on precise colocated sampling (the same 1- cm to 2-cm slice of sediment). The sediments in these data sets include most types commonly found in coastal regions. Most measurements were made at the sites of high-frequency acoustic scattering experiments that often include in situ measurements of compressional and shear wave speed and attenuation. The disadvantages of this data set include (1) measurement at a single high frequency (400 kHz) that may not be applicable to lower frequencies because of dispersion, (2) possible disturbance of the sediments during collection, transport and measurement, and (3) a limited number of samples from coarser-grained sites (coarse sand and gravel). Similar regressions based on in situ sound speed and attenuation measurements made at lower frequencies, presented later in this chapter, overcome some of these disadvantages and provide a method to evaluate some of these problems.

Table 5.1. Summary of sediment physical and geoacoustic properties from 57 siliciclastic sites. These properties consist of sound speed (V_p, m s^{-1}), sound speed ratio (V_pR, no units), attenuation (α_p, dB m^{-1}), mean grain size (M_z, ϕ), porosity (η, %), bulk density (ρ, g cm^{-3}), attenuation factor (k, dB m^{-1} kHz^{-1}), index of impedance (IOI, g cm^{-3}, see Sect. 5.1.8), and sediment type. Values of all acoustic parameters were determined at 400 kHz.

Site	V_p	V_pR	α_p	M_z	η	ρ	k	IOI	Sediment
SABay	1518.9	0.993	38.7	10.94	89.14	1.170	0.097	1.162	clay
Diga	1480.4	0.968	58.0	10.05	69.12	1.506	0.145	1.458	silty clay
Eck93	1515.5	0.991	72.3	9.88	87.40	1.188	0.181	1.177	silty clay
Portovenere	1501.7	0.982	66.2	9.45	68.30	1.546	0.166	1.518	silty clay
Viareggio	1511.3	0.988	99.5	8.98	61.74	1.634	0.249	1.615	silty clay
STeresa	1502.4	0.982	122.3	8.78	66.98	1.569	0.306	1.541	silty clay
JDF7	1507.2	0.985	114.2	8.50	83.43	1.313	0.285	1.294	silty clay
CLBight	1521.9	0.995	114.0	8.10	86.50	1.223	0.285	1.216	silty clay
Orcas	1511.9	0.988	179.1	8.08	75.22	1.403	0.448	1.387	clayey sand
LISound	1503.1	0.982	—	7.64	76.64	1.411	—	1.386	clayey silt
EelRiver	1554.6	1.016	190.7	7.17	57.32	1.745	0.477	1.773	clayey silt
JDF4	1521.7	0.995	206.8	6.93	74.35	1.470	0.517	1.462	glacial till
RussRiver	1545.5	1.010	231.8	6.35	64.35	1.597	0.579	1.613	clayey sand
Tellaro	1614.4	1.055	184.7	6.08	50.70	1.820	0.462	1.921	sand-silt-clay
Arafura	1511.4	0.988	347.8	5.24	71.63	1.494	0.869	1.476	clayey sand
Monasteroli	1652.4	1.080	220.2	5.12	46.62	1.891	0.550	2.042	sand-silt-clay
Eck94	1609.7	1.052	210.7	4.59	59.38	1.659	0.527	1.745	sand-silt-clay
JDF1	1617.6	1.057	238.5	4.37	55.37	1.800	0.596	1.903	silty fine sand
VAzzura	1686.4	1.102	156.5	4.14	45.17	1.911	0.391	2.106	muddy sand
Misby/fine	1682.4	1.100	195.8	3.77	—	—	0.489	—	v. fine sand
Tirrenia	1683.1	1.100	127.6	3.72	45.76	1.906	0.319	2.097	v. fine sand
JDF6	1668.2	1.090	314.3	2.94	47.56	1.922	0.786	2.096	fine sand/s-s-c
Quinault	1709.3	1.117	177.2	2.94	41.76	1.971	0.443	2.202	fine sand
TBay/fine	1746.0	1.141	206.1	2.92	40.16	2.013	0.515	2.297	fine sand
PC84	1742.9	1.139	241.7	2.61	40.08	1.998	0.604	2.276	fine sand
ATB/G40	1651.9	1.080	219.8	2.56	56.61	1.716	0.549	1.853	fine sand
LTB	1716.8	1.122	317.1	2.54	43.57	1.929	0.793	2.165	fine sand
Duck	1758.8	1.150	116.2	2.53	39.54	2.051	0.291	2.357	fine sand
MVCO	1755.1	1.147	154.5	2.52	38.49	2.028	0.386	2.327	fine sand
PCB I&II	1755.1	1.147	176.1	2.34	39.72	2.018	0.440	2.315	fine sand
JDF5	1701.5	1.112	213.8	2.31	45.44	1.946	0.534	2.164	fine sand/s-s-c

Table 5.1. Continued

Site	V_p	V_pR	α_p	M_z	η	ρ	k	IOI	Sediment
PCB99	1764.2	1.153	133.5	2.24	39.33	2.020	0.334	2.329	fine sand
SWEAT	1747.6	1.142	213.3	2.23	40.38	2.007	0.533	2.292	fine sand
ATB/B14	1752.6	1.146	107.2	2.15	39.52	2.006	0.268	2.298	fine sand
SG98-8	1747.1	1.142	265.7	2.14	39.65	2.026	0.664	2.314	shelly fine sand
MonPt	1744.4	1.140	92.1	2.04	37.21	2.045	0.230	2.332	fine sand
JDF2	1771.6	1.158	179.5	2.03	39.10	2.039	0.449	2.361	medium sand
Charl/fine	1728.4	1.130	281.0	1.97	39.94	2.001	0.703	2.260	fine sand
NoSea	1779.0	1.163	155.7	1.93	37.56	2.054	0.390	2.388	med/fine sand
TOSSEX	1762.7	1.152	161.8	1.93	35.64	2.075	0.404	2.391	med/fine sand
NS	1735.0	1.134	226.1	1.87	41.07	2.046	0.565	2.320	medium sand
IRB	1745.2	1.141	281.2	1.77	40.63	2.023	0.703	2.307	medium sand
SG98-10	1752.1	1.145	164.1	1.62	40.69	1.979	0.410	2.266	medium sand
SG98-9	1747.1	1.142	206.7	1.56	39.45	2.010	0.517	2.295	medium sand
Charl/crse	1729.1	1.130	308.1	1.44	39.63	2.006	0.770	2.267	medium sand
TBay/crse	1754.2	1.147	610.2	1.36	44.85	1.966	1.526	2.254	coarse/fine sand
Hood Canal	1767.1	1.155	184.6	1.34	36.46	2.108	0.462	2.435	medium sand
KB/bar	1758.2	1.149	254.4	1.33	37.28	2.047	0.636	2.352	medium sand
PE99	1770.7	1.157	153.0	1.28	37.08	2.052	0.383	2.375	medium sand
SAX99	1766.3	1.154	177.5	1.27	37.27	2.066	0.444	2.385	medium sand
PE00	1774.1	1.160	149.5	1.21	37.32	2.050	0.374	2.377	medium sand
PC93	1708.5	1.117	404.0	0.98	40.93	2.008	1.010	2.242	coarse sand
Misby/crse	1762.4	1.152	145.4	0.95	—	—	0.357	—	coarse sand
KB/lyn	1709.2	1.117	586.9	0.90	40.14	2.020	1.467	2.256	shell hash
PCII	1716.4	1.122	391.2	0.85	41.09	2.000	0.978	2.244	c. sand/sh. hash
SG98-1	1713.0	1.120	430.2	0.84	40.66	2.053	1.076	2.299	shell hash
SG98-6	1649.6	1.078	632.5	0.08	43.47	2.001	1.581	2.158	shell/coral hash

Sediment geoacoustic and physical property measurements were made on sediments collected with 45-cm-long, 5.9-cm-inside-diameter, clear, polycarbonate coring tubes. Most sediments were collected by divers, but sediments collected from eight sites (Montauk Point, Quinault Range, Arafura Sea, Russian River, Eel River, North Sea, TOSSEX, and Straits of Juan de Fuca), which were too deep for diving operations, were subsampled from 0.25 m^2 spade box cores. Cores were capped at both ends immediately after collection to retain the overlying water and kept in an upright position during transport to the laboratory for analysis. Collection, measurement, and handling procedures were designed to minimize sampling disturbance and to maintain an intact sediment–water interface within the coring tube.

In Tables 5.1 and 5.2, the siliciclastic and carbonate sites are arranged in order of increasing mean grain size (decreasing values of phi), from clay to coarse sand or shell hash. Sound speed ranges from values less than the

Table 5.2. Summary of sediment physical and geoacoustic properties from 12 carbonate sites. Geoacoustic and physical properties consist of sound speed (V_p, m s^{-1}), sound speed ratio (V_pR, no units), attenuation (α_p, dB m^{-1}), mean grain size (M_z, ϕ), porosity (η, %), bulk density (ρ, g cm^{-3}), attenuation factor (k, dB m^{-1} kHz^{-1}), index of impedance (IOI, g cm^{-3}, see Sect. 5.1.8), and sediment type. Values of all acoustic parameters were determined at 400 kHz.

Site	V_p	V_pR	α_p	M_z	η	ρ	k	IOI	Sediment
Hawaii/mud	1495.3	0.977	68.6	8.67	84.02	1.296	0.171	1.267	calc. silty clay
DTortugas	1561.8	1.021	343.0	6.62	59.00	1.755	0.858	1.792	calc. s-s-clay
MarqKeys	1555.6	1.017	391.3	6.15	59.66	1.726	0.978	1.755	calc. s-s-clay
SG98-5	1560.8	1.020	322.3	5.85	59.59	1.748	0.806	1.783	calc. s-s-clay
LFK/fine	1581.3	1.034	365.8	5.40	57.19	1.759	0.914	1.818	calc. s-s-clay
Hawaii-4	1609.7	1.052	246.2	3.88	56.42	1.771	0.615	1.864	calc. silty sand
Hawaii-2	1671.6	1.093	438.3	2.33	47.68	1.933	1.096	2.112	calc. med. sand
SG98-3	1777.3	1.162	236.7	1.66	40.92	2.067	0.592	2.401	ooid/skel. sand
SG98-2	1669.4	1.091	383.1	1.57	49.47	1.921	0.958	2.096	crse. skel. sand
RebShoal	1733.1	1.133	279.1	1.26	43.85	2.022	0.698	2.290	carbonate sand
Hawaii/crse	1639.4	1.072	695.2	0.74	45.18	1.960	1.738	2.100	crse. coral sand
LFK/crse	1704.7	1.114	488.9	0.54	41.97	2.054	1.222	2.289	crse. coral sand

water sound speed in high-porosity muddy sediment to values almost as high as 1800 m s^{-1} in low-porosity fine-to-medium sand.

5.1.4 Core Data Regressions

Figure 5.2 shows data and second-order regressions for sediment sound speed and sound speed ratio. The regressions and coefficients of determination (r^2) are given in Table 5.3. Sound speed ratio is highly correlated with both bulk density and porosity, and to a lesser degree with mean grain size. A similar second-order polynomial regression has been used by [Hamilton and Bachman 1982]. Note that in all regressions, the sound speed minimum can be lower than the water sound speed by 3–4%. This lower speed is the result of the mass loading due to the grains in sediments that have low frame bulk modulus and loosely follows the trend of Wood's equation (8.72) [Hamilton and Bachman 1982]. Regressions for siliciclastic and carbonate sediments are not significantly different, and [Richardson and Briggs 2004b] suggest using the regression for the sediment types combined (Table 5.3). The fact that regressions between sound speed and sediment physical properties are similar for carbonate and siliciclastic sediments and differ very little from regressions based on a much smaller data set presented by [Richardson and Briggs 1993] suggests a universal applicability of the sound-speed regressions presented here.

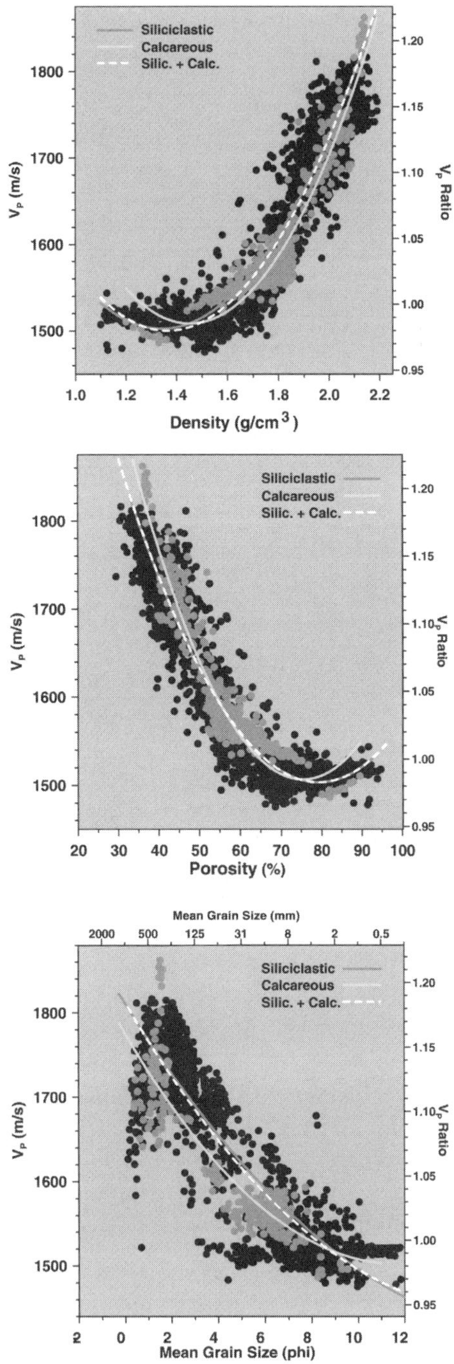

Fig. 5.2. Sound speed and sound speed ratio as functions of bulk density, porosity, and mean grain size. The lighter symbols represent carbonate sediments and overlay the darker symbols which represent siliciclastic sediments. Regression equations are given in Table 5.3.

Table 5.3. Regressions for sediment physical and geoacoustic properties for siliciclastic and carbonate sites. Geoacoustic and physical properties: sound speed ratio (V_pR, no units), mean grain size (M_z, ϕ), porosity (η, %), bulk density (ρ, g cm^{-3}), and attenuation factor (k, dB m^{-1} kHz^{-1}). These regressions should not be used outside of the range of data found in Figs. 5.2 and 5.3: $1100 < \rho$ (kg m^{-3}) < 2200, $30 < \eta$ (%) < 95, $0 < M_z$ (ϕ) < 12.

Sediment Type	Regression	No. of Points	r^2
Siliciclastic	$V_pR = 1.603 - 0.0156\eta + 0.0001\eta^2$	3905	0.95
Carbonate	$V_pR = 11.760 - 0.0206\eta + 0.0001\eta^2$	609	0.91
All Sediments	$V_pR = 1.606 - 0.0158\eta + 0.0001\eta^2$	4514	0.95
Siliciclastic	$V_pR = 1.585 - 0.8991\rho + 0.3352\rho^2$	3905	0.94
Carbonate	$V_pR = 1.878 - 1.2289\rho + 0.4232\rho^2$	609	0.90
All Sediments	$V_pR = 1.649 - 0.9807\rho + 0.3595\rho^2$	4514	0.93
Siliciclastic	$V_pR = 1.184 - 0.0288M_z + 0.0008M_z^2$	2392	0.82
Carbonate	$V_pR = 1.161 - 0.0308M_z + 0.0013M_z^2$	371	0.82
All Sediments	$V_pR = 1.184 - 0.0307M_z + 0.0010M_z^2$	2763	0.82
All Sediments	$k = 0.74 - 0.07M_z - 0.02M_z^2$	2653	0.10
All Sediments	$k = -1.121 + 0.066\eta + 0.0006\eta^2$	4391	0.19

Attenuation is generally higher in the carbonate than siliciclastic sediments but is very poorly correlated with mean grain size or porosity (Fig. 5.3, Table 5.3). Attenuation as measured by the technique employed for the data set represented here, however, includes both intrinsic attenuation and the effects of scattering from both grains (such as shells) and larger-scale heterogeneities. The attenuation relationships proposed by [Hamilton 1972, Hamilton 1980, Hamilton 1987] and included in Fig. 5.3 are from a compilation of in situ and laboratory values of attenuation. Hamilton's attenuation data set was obtained under a greater variety of conditions and generally lower frequencies than his sound speed data set, which was collected under laboratory conditions at a single frequency (200 kHz). The lower bounds of attenuation measured at 400 kHz may come close to representing intrinsic attenuation and, as such, closely mimic the attenuation values in the scatter plots of [Hamilton 1972] which have the lowest values of attenuation in coarse-to-medium sand and clay and higher values of attenuation in the fine-sand to silt-sized range. Based on the data presented here, empirical relationships among attenuation and sediment physical properties have little predictive value at this high measurement frequency (400 kHz).

5.1.5 Gradients

A large number of studies have used naturally occurring or reconstituted sediment samples to study the effects of consolidation (up to the equivalent of 100s of meters below the sediment surface) on wave propagation. As the pri-

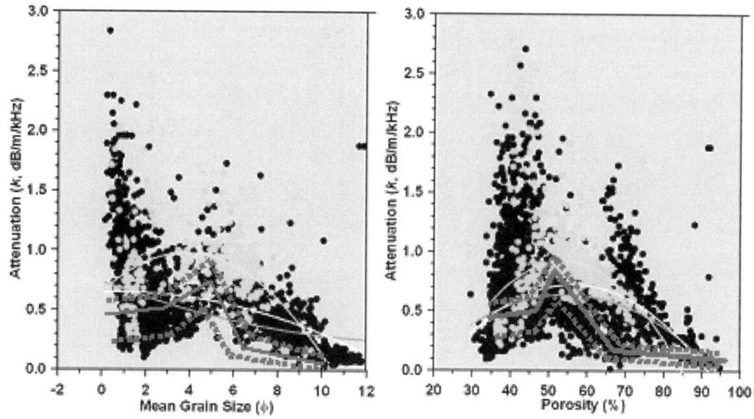

Fig. 5.3. Attenuation factor, measured at 400 kHz, as a function of porosity and mean grain size. The lighter symbols represent carbonate sediments and overlay the darker symbols which represent siliciclastic sediments. Equations for regressions are given in Table 5.3. The relationships between attenuation factor and porosity and mean grain size summarized by [Hamilton 1980] are in red. (see first color insert)

mary interest in this monograph is in predictions of near-surface in situ sound speed and attenuation, these studies are not emphasized, and the reader is directed to [Hamilton 1976b, Hamilton 1979a, Hamilton 1987, Kibblewhite 1989, Bowles 1997] for excellent reviews of gradients of sound speed and attenuation expected in marine sediments, based on laboratory consolidation and lower-frequency seismic studies. Based on very limited data, [Hamilton 1979a, Hamilton 1987] concluded that near-surface gradients of increasing sound speed are higher in sand (20 s^{-1}) than in mud (1 s^{-1}). The scarcity of appropriate high-resolution, near-surface data and variability make gradients in attenuation difficult to predict, but attenuation generally decreases with depth in sand and increases with depth in mud at least in the upper 100 meters of sediment [Hamilton 1976b, Hamilton 1987]. The sound speed and attenuation profiles (Figs. 5.4–5.6) reported for the upper 20–30 cm of both siliciclastic and carbonate sandy and muddy sediment often have higher gradients than suggested by Hamilton [Richardson 1986, Richardson and Briggs 1996, Briggs and Richardson 1997]. In sand, both sound speed and attenuation exhibit strong positive gradients in the upper 20–30 cm. Sound speed gradients of 50 s^{-1} or higher are common, and attenuation may double over this small depth interval (tens of centimeters). In high-porosity mud, sound speed and attenuation gradients can be either positive or negative. In general, these gradients are difficult to predict without knowledge of the biological, physical, and sometimes biogeochemical processes that control sediment structure and properties. The reader is cautioned not to extrapolate the gradients reported here beyond the measurement interval (0–30 cm).

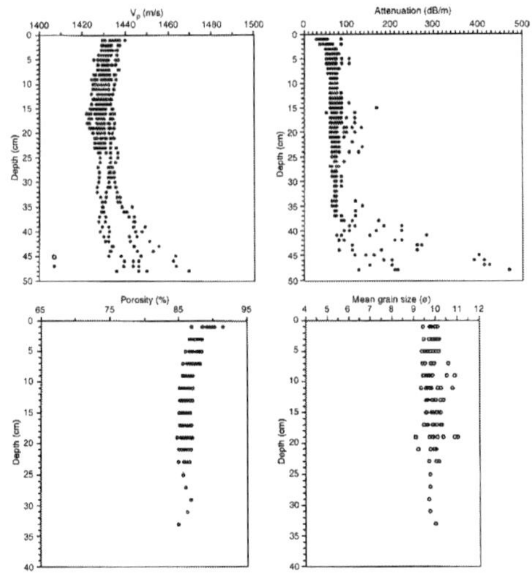

Fig. 5.4. Profiles of sediment sound speed, attenuation factor, porosity, and mean grain size for high-porosity clayey sediment from Eckernförde Bay, Baltic Sea [Richardson and Briggs 1996]. Sediments below 35 cm contained methane bubbles and exhibited higher values of attenuation and slightly higher values of sound speed. All acoustic measurements were made at 400 kHz.

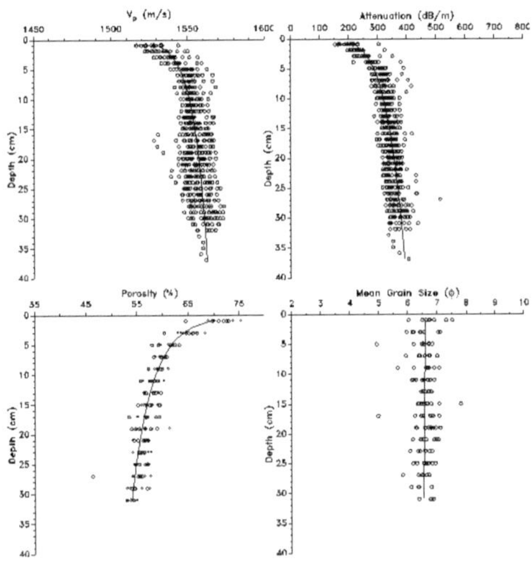

Fig. 5.5. Profiles of sediment sound speed, attenuation factor, porosity, and mean grain size for carbonate sediments collected from the Dry Tortugas [Richardson et al. 1997]. The lines are best-fit power-law regressions and demonstrate a rapid increase in sound speed in response to dewatering (reduction in porosity) of the sediments by bioturbation. Note that the mean grain size does not vary with depth, suggesting a single source of deposition. All acoustic measurements were made at 400 kHz.

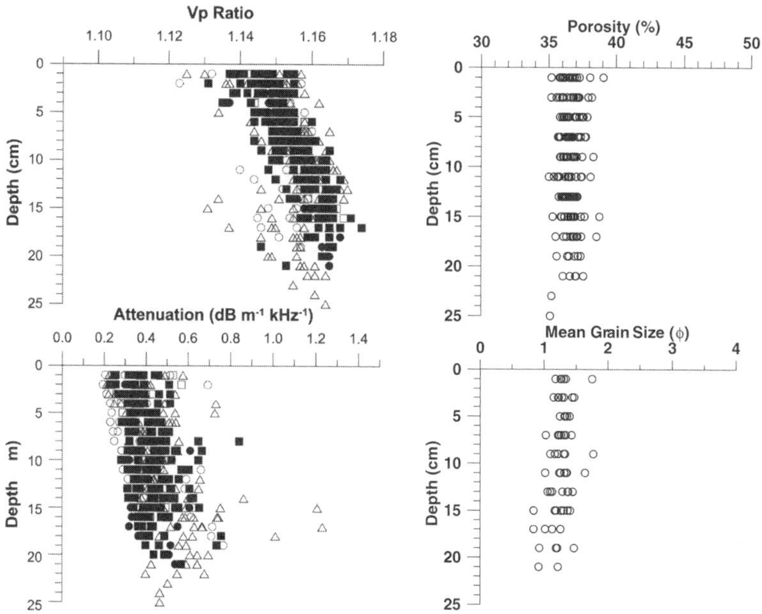

Fig. 5.6. Profiles of sound speed ratio, attenuation factor, porosity, and mean grain size for sediment collected at the SAX99 experimental site [Richardson et al. 2001a]. Values of porosity and mean grain size vary little with depth, whereas values of sound speed and attenuation increase. All acoustic measurements were made at 400 kHz.

Spatial variability (in both the vertical and horizontal), evident in the scatter of data points in Figs. 5.4–5.6, is discussed in Ch. 7 and quantified for these and many other examples in Appendix C.

Further discussion of gradients is given in Ch. 7. That chapter addresses an issue that is peculiar to the rather small depths of interest in high-frequency acoustics: spatial variability may mask vertical trends in geoacoustic parameters. Accordingly, the vertical profile of a given geoacoustic parameter is estimated after an average has been taken over the measurement ensemble.

5.1.6 In Situ Data Regressions

For the in situ measurements (made using the technique described previously) geoacoustic data are reported in the same form as the core data: sound speed ratio, V_pR, and attenuation factor, k. Core data of the type described above were taken at the same sites, providing physical property data and allowing comparison of the two methods. Sound speed, attenuation, and physical property measurements were not colocated; therefore, averages for each site were use to construct the regressions given in Table 5.4. Values of sound speed followed the same trends as in the higher-frequency laboratory measurements (Fig. 5.7 compared to Fig. 5.2). Sound speed ratios range from 0.978 (1428 m s^{-1}) in the silty-clay sediments found in Eckernförde Bay, Baltic Sea, to 1.170 (1767 m s^{-1}) in coarse sand sediment in the North Sea. The sound speed ratio was only slightly lower than determined from the higher-frequency laboratory measurements at the same locations (Fig. 5.8). At least part of this trend may be due to dispersion predicted by poroelastic theory (Ch. 10). Recent evaluation of this in situ measurement technique by [Buckingham and Richardson 2002] suggests that tone-burst time-of-flight measurements between a single source and receiver in close proximity may underestimate sound speed when compared to single-source two-receiver configurations. These differences are generally less than 15 m s^{-1} for sandy sediments, such as the SAX99 sediments evaluated by [Buckingham and Richardson 2002]. Attempts to reevaluate the entire data set used here have proven difficult, as many of the measurements exist only as analog plots, and spatial variability in sound speed over the measurement paths increases the variability in values of measured sound speed and attenuation. Of the thirty or so locations successfully re-analyzed, the same trends were evident for similar sandy environments (10–15 m s^{-1} lower sound speed for single-source receiver methods), but insignificant differences were found in more muddy sediments.

Table 5.4. Regressions for sediment physical and geoacoustic properties for in situ measurements using ISSAMS. Regressions are plotted in Figs. 5.7 and 5.9. Geoacoustic and physical properties, sound speed ratio (V_pR, no units), mean grain size (M_z, ϕ), porosity (η, %), bulk density (ρ, g cm^{-3}), and attenuation factor (k, dB m^{-1} kHz^{-1}). These regressions should not be used outside of the range of data found in Figs. 5.7 and 5.8: $1200 < \rho$ (kg m^{-3}) < 2100, $30 < \eta$ (%) < 90, $0 < M_z$ (ϕ) < 10.

Property	Regression	No. Points	r^2
Density	$V_pR = 1.705 - 1.035 \times 10^{-3}\rho + 3.664 \times 10^{-7}\rho^2$	86	0.92
Porosity	$V_pR = 1.576 - 0.015677\eta + 1.0269 \times 10^{-4}\eta^2$	86	0.91
Mean Grain Size	$V_pR = 1.190 - 0.03956M_z + 1.9476 \times 10^{-3}M_z^2$	86	0.92
Density	$k = 0.00332e^{0.00241\rho}$	87	0.45
Porosity	$k = 2.153e^{-0.0401\eta}$	87	0.43
Mean Grain Size	$k = 0.697e^{-0.183M_z}$	87	0.52

Plots of attenuation factor and sediment physical properties show an increase in the coarser-grained, lower-porosity sediments (Fig. 5.9). Attenuation, expressed as k (dB m^{-1} kHz^{-1}), is often lower than measured at 400 kHz (Fig. 5.8), and the trend of highest attenuation in sediments in the fine-sand or silt-size ranges summarized by [Hamilton 1972] is not evident, as values of attenuation appear to increase with increasing grain size and de-

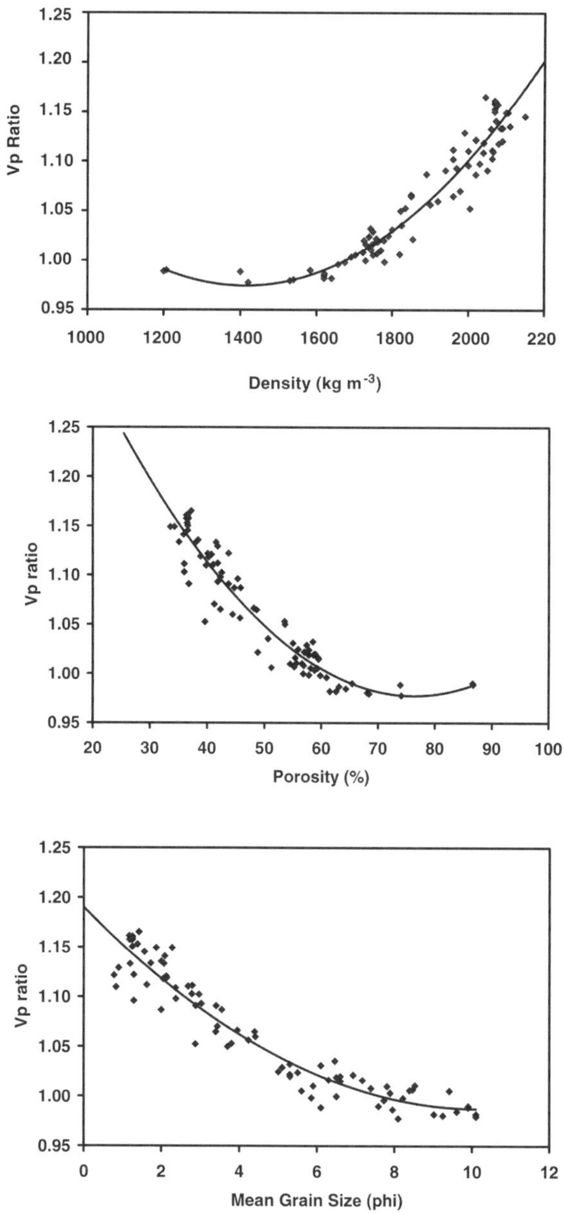

Fig. 5.7. Sound speed ratio as a function of bulk density, porosity, and mean grain size for average values of in situ sound speed ratio and average values of sediment physical properties for 88 sites where the ISSAMS systems were deployed. The data include both siliciclastic and carbonate sites. The regressions are an update from similar regressions presented in [Richardson 1997a]. Sound speed measurements were made at either 38 or 58 kHz depending on which version of ISSAMS was used.

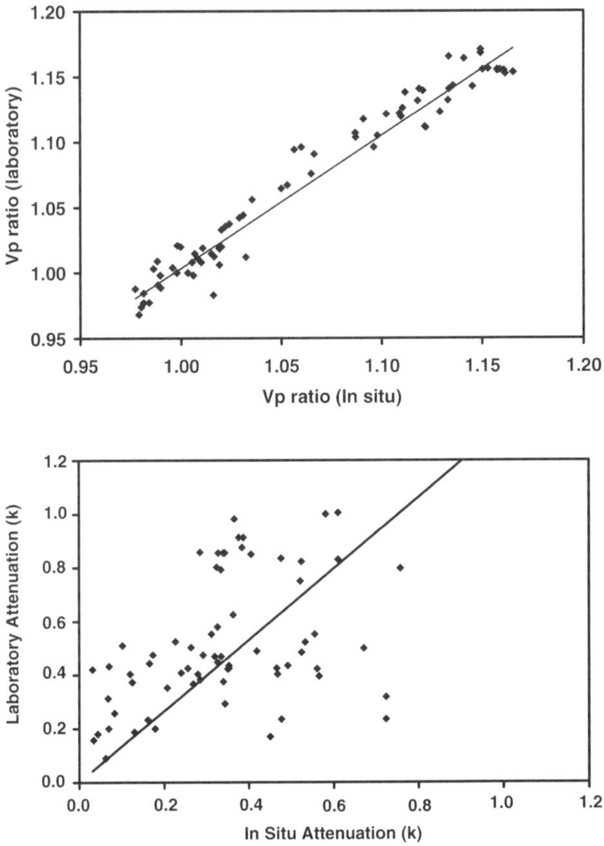

Fig. 5.8. Comparison of sound speed ratio and attenuation factor for average laboratory (400 kHz) and in situ (38 kHz) measurements at 78 sites worldwide, where both cores and in situ measurements were available. The data include both siliciclastic and carbonate sites. The regression lines are: Attenuation, k (lab) = 1.3276 × k (in situ) with r^2 = 0.16; V_p (lab) = -0.0154 + 1.0187 × V_p (in situ) with r^2 = 0.089.

creasing porosity. Although regressions are given for attenuation factor versus sediment physical properties, the coefficients of determination are low (r^2 = 0.45–0.52). These regressions yield only approximate values of attenuation and should be used with caution. Measured attenuations are much preferred.

5.1.7 Comparison with Other Regressions

The most widely used regressions of sound speed against density, porosity, and mean grain size were compiled by [Hamilton and Bachman 1982] with later versions by [Bachman 1985, Bachman 1989]. These regressions yield consis-

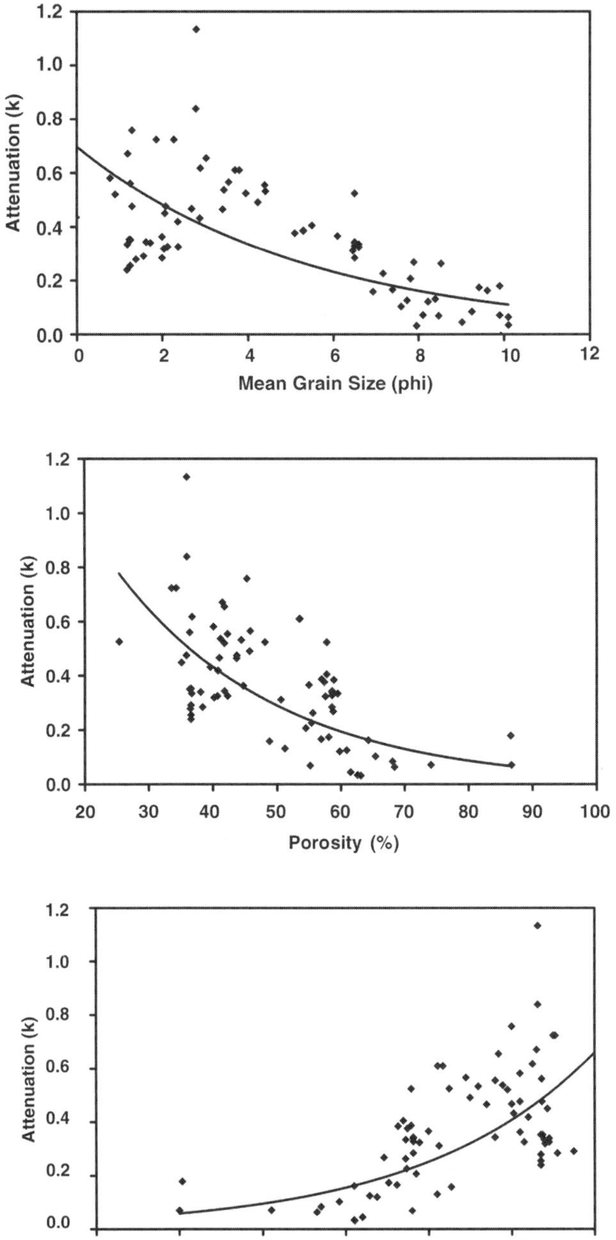

Fig. 5.9. Attenuation factor (k, dB m^{-1} kHz^{-1}) as a function of mean grain size, porosity, and bulk density for the 78 siliciclastic and carbonate sites where in situ measurements were made. Acoustic measurements were made at either 38 or 58 kHz.

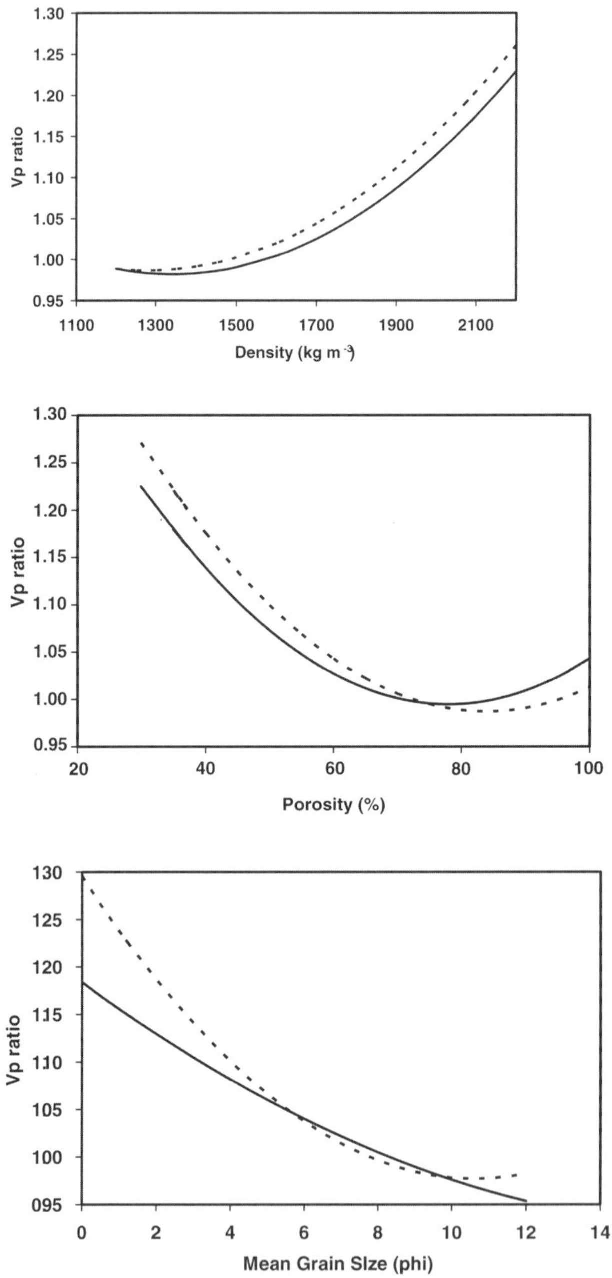

Fig. 5.10. Comparison of the predictions of sound speed ratio from sediment mean grain size, porosity, and density based on the empirical regressions given by [Hamilton and Bachman 1982] (dashed lines) and [Richardson and Briggs 2004a] (solid lines). Hamilton's measurements were made at 200 kHz, Richardson and Briggs' at 400 kHz.

tently higher sound speeds for similar values of sediment physical properties for sandy sediments than the regressions presented herein for both laboratory and in situ measurements. Figure 5.10 compares Hamilton's regressions with the regressions given here for core data at 400 kHz. The causes of these differences remain a mystery, but additional in situ measurements, using a variety of techniques at the same sites as the laboratory and in situ measurements reported here, suggest that the regressions given in this monograph are more suited to high-frequency applications. For example, sound speed measurements made during the SAX99 experiments using a variety of in situ probe techniques [Williams et al. 2002a] yielded lower sound speed values than predicted by Bachman (see Table 5.5) and were consistent with the regressions given here. Recent in situ measurements using pulse techniques (68 kHz) on the New Jersey coast [Goff et al. 2004] and near the Martha's Vineyard Coastal Observatory (Kraft, personal communication) both align better with the regressions of [Richardson and Briggs 2004a] than with those of Hamilton.

Most of the sound speed measurements that Hamilton and Bachman based their regressions on were laboratory measurements at 200 kHz using a pulse technique and a Hamilton frame. Values of sediment physical properties were obtained using techniques very similar to those of [Richardson and Briggs 2004a]. Density was measured with a weight-volume method, drying the samples at 110°C for 24–48 hours. Porosity was calculated from the density and corrected for salt content. Grain size was determined by first wet sieving the dispersed sediment samples to separate sand and mud fractions. The sand fraction was analyzed using the Emery settling tube and the fine fraction by pipette. Mean grain size was calculated using the same graphic mean as Richardson and Briggs ($M_z = (\phi_{16} + \phi_{50} + \phi_{84})/3$), Sect. 4.1). It seems unlikely that differences in physical property measurement techniques could account for the differences in the regressions. Other possible causes include differences in acoustic measurement techniques, disturbance of sediments measured with the Hamilton frame, or actual differences in the sediments. Comparison of the sand–silt–clay ratios presented as Shepard ternary diagrams (Fig. 5.11) suggests higher concentrations of silt-sized particles in the Hamilton–Bachman data set and fewer examples of sediments dominated by sand- or clay-sized particles when compared to the sediments analyzed by [Richardson and Briggs 2004a]. Regardless of the cause, the regressions of [Richardson and Briggs 2004a] and those given in this chapter should be preferred for prediction of sediment sound speed for high-frequency acoustic experiments. It follows that the regressions of [Richardson and Briggs 1993, Richardson and Briggs 2004b] that predict sediment physical properties from impedance (Sect. 5.1.8) also are preferred when using acoustic remote classification systems.

Section 11.2.1 gives "regressions" for density ratio and sound speed ratio taken from [APL-UW TR 9407]. As noted in Sect. 11.2.1, these are *not*

Table 5.5. Comparison of sound speed and sound speed ratio predictions based on the empirical regressions of [Bachman 1989] and [Richardson and Briggs 2004a] using mean values of sediment physical properties appropriate to the SAX99 site. The symbol / separates the two predictions, with Bachman's on the left and Richardson–Briggs' on the right. The predicted values should be compared with the measured values from Table 5.1, V_p = 1766.3 m s^{-1} and sound speed ratio = 1.154. The average sound speed of the pore water during SAX99 is assumed to be 1536 m s^{-1}.

Physical Property	Sound Speed (m s^{-1})	Sound Speed Ratio
Porosity = 37%	1847 / 1786	1.202 / 1.163
Density = 2074 kg m^{-3}	1830 / 1785	1.191 / 1.162
M_z = 1.25 ϕ	1880 / 1764	1.224 / 1.149

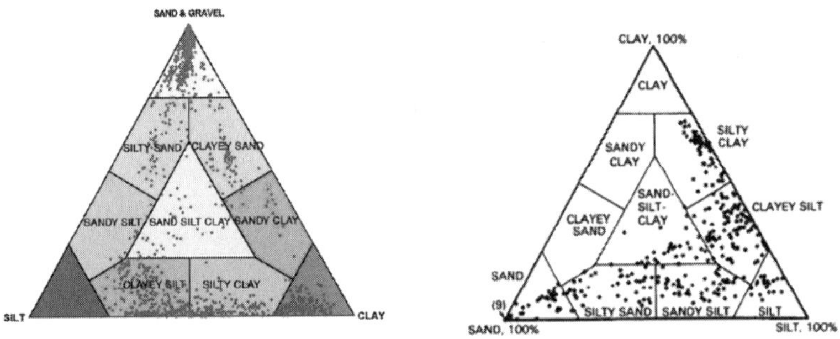

Fig. 5.11. Comparison of the ratios of sand, silt, and clay for sediment included in the acoustic-physical property regression of [Richardson and Briggs 2004a] (left) and [Hamilton and Bachman 1982] (right). Note the ternary diagrams are rotated 120° relative to each other.

geoacoustic regression relations; rather, they are ad hoc relations designed to improve acoustic model-data fits.

5.1.8 Index of Impedance Regressions

Many acoustic sediment classification systems use the amplitude of echo returns to estimate seafloor acoustic impedance [Richardson and Briggs 1993]. Empirical relationships between seafloor impedance and sediment physical properties are then used to map seafloor physical properties such as porosity, bulk density, percent sand and gravel, or mean grain size as well as geoacoustic properties such as sound speed and attenuation [Richardson and Briggs 2004b]. Sediment impedance is the product of sediment sound speed and bulk density with units kg m^{-1} s^{-1}. Like sound speed, impedance is dependent on pore water temperature and salinity and pressure (water depth). Sound speed and thus the impedance at a single site can vary up to 10% over the range of seasonal conditions expected in coastal waters. Therefore, a pore-water-independent index of impedance (IOI), which is the product of the sediment bulk density and sound speed ratio, is used to calculate empirical relationships between sediment impedance and other sediment physical properties.

The index of impedance provides excellent predictions of sound speed ratio, bulk density, and porosity for both siliciclastic and carbonate sediments (Figs. 5.12 and 5.13, Table 5.6). This is not surprising, as IOI is the product of sound speed ratio and bulk density, and both sediment bulk density and porosity are determined from the same wet loss measurements. Data on sound speed, sound speed ratio, bulk density, and porosity for carbonate and siliciclastic sediments fall within bands of width 14 m s^{-1}, 0.001, 0.04 g cm^{-3}, and 4%, respectively, over the full range of values of IOI, suggesting regressions for each parameter derived from the entire data set are appropriate (Table 5.6). The coefficients of determination (r^2) between IOI and sediment mean grain size and percent sand and gravel are lower for carbonate than siliciclastic sediments. The lower values of r^2 between IOI and grain size properties, or scatter in the data, justify combined regressions using all carbonate and siliciclastic data in spite of up to 0.6 ϕ and 17% differences in predicted mean grain size and percent sand and gravel between the two data sets. The predictions of attenuation based on the index of impedance are not better, based on values of coefficient of determination, than similar regressions between attenuation and other sediment physical properties (Table 5.3).

Most of the empirical relationships presented here were developed using sound speeds measured at 400 kHz. Many echo sounders operate at 3.5 to 30 kHz, where sound speeds and thus impedance values may be lower. Biot theory predicts that sound speed is dispersive (Ch. 10), with dispersion more pronounced in sand compared to muddy sediments. Values of sound speed ratio measured at 38 and 400 kHz for most sediments are nearly equal (Fig. 5.8). However, at the lower frequencies used with many acoustic sediment classification systems, especially in sand, these regressions may not accurately predict sediment physical properties from measured values of IOI. In the example given by [Williams et al. 2002a] for the sand sediment of the SAX99 experiments, measured sound speeds were up to 75 m s^{-1} higher at 400 kHz than at 3.5 kHz. Based on this degree of sound speed dispersion, sediment properties predicted using the higher-frequency IOI regressions would be different than if lower sound speeds predicted at 3.5 kHz were used to calculate IOI: actual porosity would be 2.6% higher, bulk density would be 50 kg m^{-3} lower, mean grain size 0.69 ϕ unit higher (finer), and sound speed 44 m s^{-1} higher. Given the current state of knowledge of dispersion in sandy sediment, corrections are premature; however, future research should attempt to both define and model the dispersive behavior of sound speed in sediments and the associated frequency dependence of attenuation.

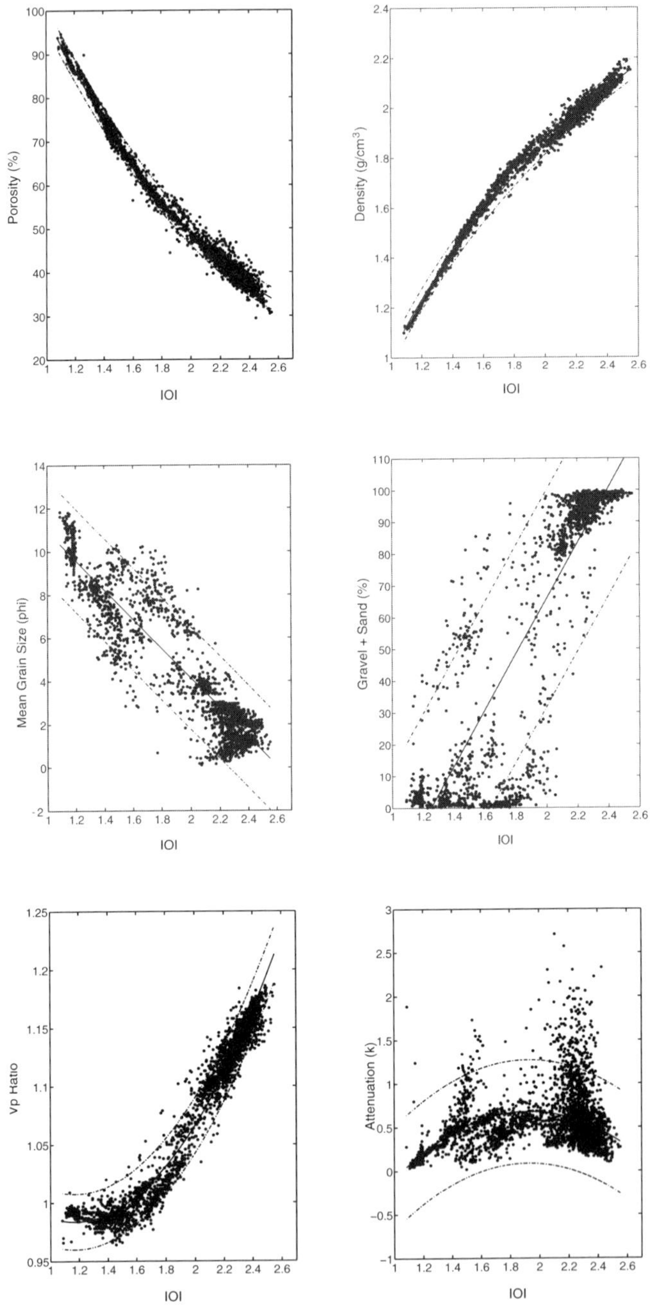

Fig. 5.12. Empirical relationships used to predict sediment physical and acoustic properties from the index of impedance (IOI) for siliciclastic sediments. Data (Table 5.1) and regressions (Table 5.6) are based on over 3900 measurements made on cores collected from 55 shallow-water sites worldwide. The dashed lines denote the 95% confidence intervals.

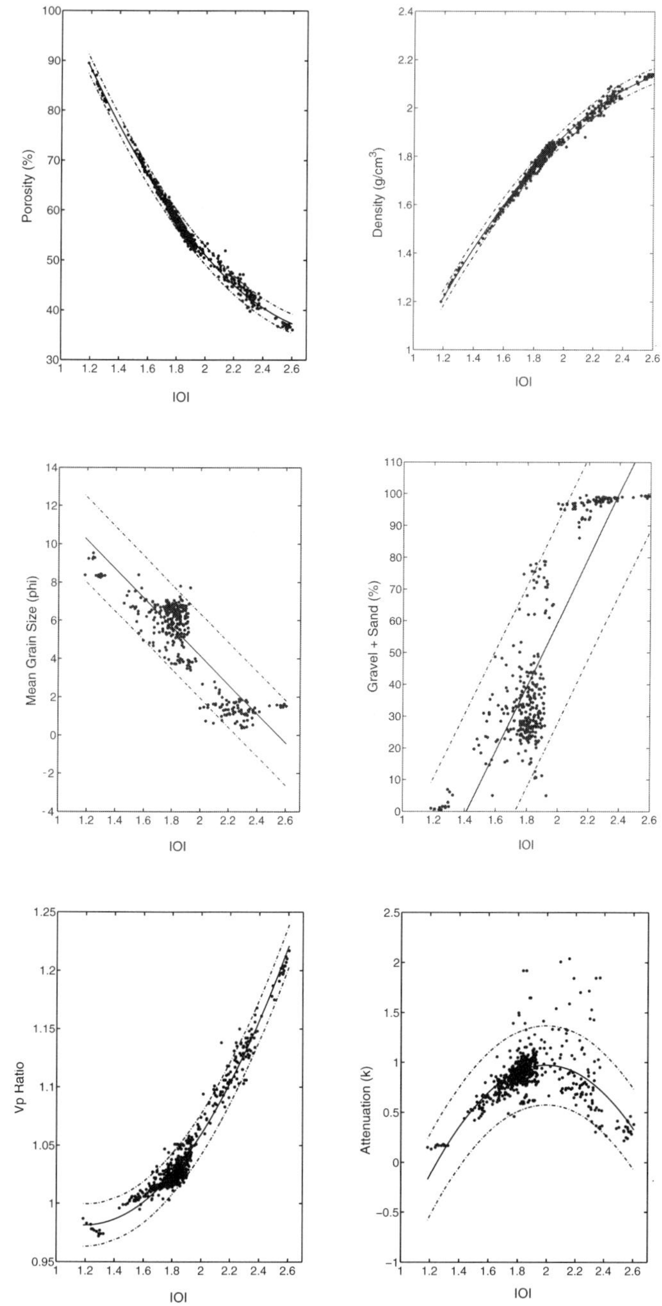

Fig. 5.13. Empirical relationships used to predict sediment physical and acoustic properties from the index of impedance (IOI) for carbonate sediments. Data (Table 5.2) and regressions (Table 5.6) are based on 69 cores collected from 12 sites around southern Florida and in the Hawaiian Islands. The dashed lines denote the 95% confidence intervals.

Table 5.6. Regressions for sediment physical and geoacoustic properties based on the index of impedance (IOI) for siliciclastic and carbonate sediments. The percentage of sand and gravel is denoted SG. The coefficient of determination (r^2) is given for each regression. These regressions should not be used outside of the range of data found in Figs. 5.12 and 5.13: $1.05 < IOI$ (kg m^{-1} s^{-1}) < 2.6.

Sediment Type	Regression	r^2
Siliciclastic	$V_pR = 1.149 - 0.2821(IOI) + 0.1203(IOI)^2$	0.97
Carbonate	$V_pR = 1.164 - 0.3001(IOI) + 0.1253(IOI)^2$	0.96
All Sediments	$V_pR = 1.164 - 0.3001(IOI) + 0.1253(IOI)^2$	0.97
Siliciclastic	$k = -2.61 + 3.41(IOI) - 0.885(IOI)^2$	0.16
Carbonate	$k = -5.96 + 6.94(IOI) - 1.174(IOI)^2$	0.43
All Sediments	$k = -3.31 + 4.33(IOI) - 1.138(IOI)^2$	0.22
Siliciclastic	$\eta = 178.60 - 94.60(IOI) + 14.86(IOI)^2$	0.99
Carbonate	$\eta = 186.18 - 102.20(IOI) + 17.29(IOI)^2$	0.99
All Sediments	$\eta = 174.16 - 89.12(IOI) + 13.37(IOI)^2$	0.99
Siliciclastic	$\rho = 1.01 + 1.22\ln(IOI)$	0.99
Carbonate	$\rho = -0.52 + 1.81(IOI) - 0.305(IOI)^2$	0.99
All Sediments	$\rho = 1.02 + 1.21\ln(IOI)$	0.99
Siliciclastic	$M_z = 17.7 - 6.8(IOI)$	0.85
Carbonate	$M_z = 19.3 - 7.6(IOI)$	0.75
All Sediments	$M_z = 17.9 - 6.0(IOI)$	0.84
Siliciclastic	$SG = -109.6 + 87.7(IOI)$	0.82
Carbonate	$SG = -143.2 + 101.4(IOI)$	0.73

The coefficient of determination, r^2, between IOI and mean grain size is much lower than the coefficient of determination between IOI and sediment bulk density or porosity. This lower coefficient of determination is reflected in the lack of a fundamental physical relationship between mean grain size and either sediment bulk density or porosity (Fig. 5.14). In muddy sediments, consolidation (dewatering) can lower porosity and increase density without a change in mean grain size, as indicated by the horizontal arrow in Fig. 5.14. This increase in density occurs with little or no change in sound speed. In sands, porosity can vary up to 10%, for a given grain size depending on packing (horizontal arrow in Fig. 5.14). Given the same packing, a uniform assemblage of spheres would theoretically achieve the same porosity regardless of grain diameter. Using values of mean grain size as an index, especially in the silt-size range, may be very misleading because of major differences in sorting (standard deviation of the particle size distribution) or due to the effects of compaction and packing. Well-sorted sediment composed of wholly silt-size particles may have the same mean grain size as poorly sorted sediment with a mixture of sand- and clay-size particles. The resultant density and sound speed of these two sediments, however, can be very different. Given the aforementioned issues, it is perhaps amazing that empirical regressions between grain size-related parameters and sediment density, porosity, sound speed, or impedance have any predictive value.

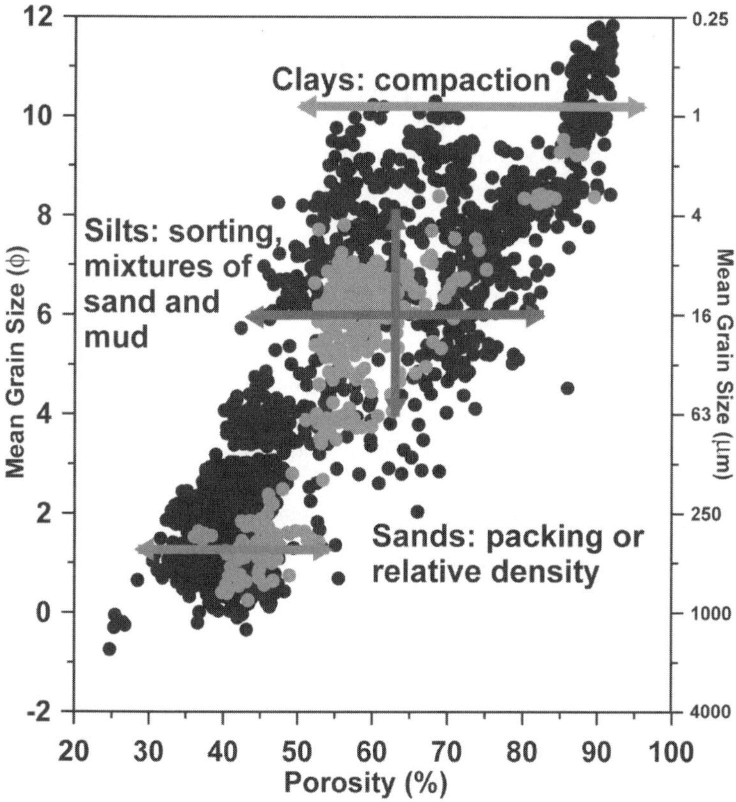

Fig. 5.14. Scatter diagram of mean grain size versus porosity. The lighter-colored symbols, which represent carbonate sediments, overlay the darker-colored symbols, which represent siliciclastic sediments. The great range of porosity for a given mean grain size indicated by the arrows reflects processes controlling sediment microstructure: compaction in clays, packing in sand, and both processes in silts. In addition, the relative percentages of clay- and sand-sized particles in silt also control porosity.

5.2 Shear Wave Speed and Attenuation

Shear wave speed and attenuation have been measured using a variety of laboratory and in situ techniques which will be reviewed in the next two sections. Laboratory measurements are generally made at higher frequencies and include both naturally occurring sediments collected with cores and artificially created sediments. The emphasis of many of these laboratory measurements is on consolidation behavior and geotechnical testing to determine values of shear modulus. In situ measurements are generally at much lower acoustic frequencies and include inversions of interface waves and seismic reflection and refraction techniques. Recent in situ probe measurements by Richardson and colleagues are used to develop the predictive relationships among shear wave and sediment physical properties presented later in this chapter. Although laboratory measurements can be converted to in situ values, issues associated with sample disturbance and maintenance of in situ effective stresses, shear strain amplitudes, and strain rates dictate a preference for in situ measurements for high-frequency acoustic applications [Richardson et al. 1991a].

5.2.1 Laboratory Measurement Techniques

Shear wave speed and attenuation have been measured using a variety of laboratory techniques including resonance columns, pulse techniques using quartz crystal transducers, piezoceramic shear plates or radial expander transducers, torsional resonance columns [Davis and Bennell 1986], torsional cyclic loading, and a variety of ceramic bimorph or bender elements (see review of [Bennell and Taylor-Smith 1991]). The bender elements can be cantilever-mounted where one end of the bender is clamped and the other end allowed to vibrate [Shirley 1978, Shirley and Hampton 1978] or mounted free in a compliant material [Bennett et al. 1991a, Barbagelata et al. 1991]. The cantilever mounting is best for laboratory measurements, and the free mounting is better for in situ applications. Bender transducers are preferred because they provide lower Q (wider bandwidth) and lower operating frequency.

Laboratory measurements of shear modulus and shear wave speed are commonly used in geotechnical investigations of foundation loading, liquefaction during earthquakes, or studies of consolidation behavior. For these studies sediment shear modulus is often correlated with sediment void ratio and "mean effective stress" or overburden pressure. Some definitions are required here. Static strain results from overburden, but these static stresses and strains must be distinguished from the much smaller "dynamic" stresses and strains due to the passage of compressional and shear waves. Dynamic stress and strain are related through dynamic moduli (Ch. 9). Returning to static stress, the mean effective stress is the stress due to the effects of gravity through self-weight of the solid particles of the sediment. There are three principal stresses, σ_{xx}, σ_{yy}, and σ_{zz} in the notation of (9.1). The pore water pressure, P_f, is subtracted from these to obtain the components of effective stress, e.g.,

$$\sigma'_{zz} = \sigma_{zz} - P_f \; . \tag{5.3}$$

The vertical (zz) component of effective stress may not be equal to the horizontal components. Defining the "coefficient of earth pressure," K_0, as the ratio of effective stress in the horizontal to that in the vertical, the "mean effective stress" is

$$\sigma_0' = \frac{1 + 2K_0}{3} P_0 , \tag{5.4}$$

where $P_0 = \sigma_{zz}'$. The coefficient of earth pressure may range from 0.4 for dense sands to 1.0 for very soft mud [Mitchell 1993].

Laboratory measurements of dynamic shear modulus with varying effective stress have been collated by [Bryan and Stoll 1988] who present the following relationship to predict sediment shear modulus, μ, from void ratio, e:

$$\mu = 2526(\sigma_0')^{0.448} e^{-1.504e} . \tag{5.5}$$

Here, the symbol e is used in two ways, as the base of the natural logarithms and, in the exponent, as the void ratio.

Shear wave speed can be determined from the shear modulus and sediment bulk density using (9.10). As for the case of compressional wave speed, the symbol c_t is used for the complex shear speed in Ch. 9, while the symbol V_s is used for the (real) shear phase speed. Here, the shear modulus will be assumed to be real, so that (9.10) becomes

$$V_s = \sqrt{\frac{\mu}{\rho}} . \tag{5.6}$$

Laboratory measurement of shear wave attenuation has been mostly neglected except in the fields of soil mechanics and foundation engineering [Stoll 1989]. Shear wave attenuation, α_s (denoted α_t in the acoustic chapters of this monograph), will be presented in terms of the attenuation factor, $k_s = \alpha_s / f_{kHz}$, and as the logarithmic decrement, Δ_s, which is preferred in the geotechnical community. Adapting (8.26) and (8.29) to the present case, the logarithmic decrement can be written in terms of the attenuation as

$$\Delta_s = \frac{\alpha_s V_s \ln(10)}{20 f} . \tag{5.7}$$

With the exception of [Brunson and Johnson 1980], most values of shear wave attenuation are based on various laboratory torsional decay techniques which by their nature are fairly low in frequency (< 1 kHz). A good review of these types of measurements is provided by [Stoll 1989]. Cantilever-mounted bender elements were used by [Brunson and Johnson 1980] to measure shear wave attenuation in laboratory sand over the frequency range of 0.45–7.0 kHz. Shear wave attenuation in sands, glass beads, and kaolinite clay sediments was measured by [Bell 1979, Bell and Shirley 1980] using bender elements. Summaries of shear wave attenuation often include laboratory measurements on artificially created or dried sediments which may not behave as fully saturated naturally occurring sediments. This may, in part, explain the poor predictability of shear wave attenuation which is discussed later.

5.2.2 In Situ Measurement Techniques

A variety of interface waves (Love, Scholte, also known as Stonely waves, Sect. 9.6) propagating at the sediment–water interface have been used to estimate shear wave speed and attenuation in marine sediments. In general, the inversion of higher-frequency interface waves (>10 Hz) allows higher-resolution characterization of near-surface shear wave properties. Most inversion schemes take advantage of the dispersive nature of shear wave propagation, i.e., strong positive gradients in shear wave speed resulting from increased effective stress from overburden pressure. However, heterogeneity of surface sediment properties may complicate interpretation of these inversions [Stoll et al. 1994, Winsborrow et al. 2003]. Interface-wave measurements have employed ocean bottom seismometers (OBS) with a variety of configurations of three-axis geophones as receivers [Bucker et al. 1964, Rauch 1980, Rauch 1986, Muir et al. 1991, Bibee 1993]. Interface waves are generated with man-made explosives, impact sources, torsional sources, or naturally occurring bottom motion. Explosive sources can be placed at or in the seafloor to directly generate interface waves, or explosions in the water column can be used to generate interface waves through conversion of compressional to shear wave energy at the seafloor. Towed sleds have been used to measure shear wave speed with an array of geophones, coupled to the seafloor using a rubber mat, towed at fixed distances behind a seismic source [Davis et al. 1989, Haynes et al. 1997, Huws et al. 2000]. Near-surface gradients of shear wave speeds have been inverted from Scholte [Brekhovskikh and Godin 1990] and Love waves [Aki and Richards 2002, Ch. 7] using impulsive and torsional sources coupled to the seafloor followed by a towed array of geophones [Bautista and Stoll 1995, Stoll 2002, and references therein]. Yamamoto and colleagues have inverted profiles of shear wave speed using seabed motion induced by surface gravity waves and measured with bottom mounted or buried seismometers [Yamamoto et al. 1989, Trevorrow and Yamamoto 1991].

Seismic reflection and refraction techniques, where both sources and receivers are in the water column, have also been used to invert for shear wave properties. However, the systems typically deployed by the petroleum industry utilize low frequency sources (< 10 Hz) and do not have the resolution required for high-frequency applications (upper few meters of sediment). Crosshole and downhole and seismic tomographic measurements have also been used to measure shear wave speed and attenuation [Yamamoto 2001]. Geotechnical engineers often use seismic cone penetrometers (SCPT, for seismic cone penetration test) to estimate gradients of shear wave speed for evaluation of liquefaction potential during earthquakes (see [Robertson and Wride 1998] for a review).

Reviews of shear wave speed and attenuation in marine sediments based on the aforementioned measurements have been given by [Hamilton 1976a, Hamilton 1976c, Hamilton 1987, Bryan and Stoll 1988, Kibblewhite 1989,

Bowles 1997]. These reviews cover such a wide range of frequencies and depth gradients that generalizations of near-surface (upper meter) shear wave properties are obscured. However, the following conclusions can be drawn from these reviews: (1) all marine sediments support shear waves, (2) shear wave speed is higher in sand than mud, (3) depth gradients in shear wave speed follow a power law (with an exponent 0.25 to 0.33), (4) in situ shear wave speeds are 15 to 60% higher than shear wave speeds derived from laboratory measurements of shear modulus, (5) shear wave attenuation is between 1 and 2 orders of magnitude higher than for compressional waves, and (6) shear wave attenuation decreases rapidly within the upper 5 m becoming nearly constant at depths below 10–20 m. However, the data reviewed by these authors are inadequate to provide the type of empirical regressions between shear wave speed or attenuation and sediment physical properties developed for compressional wave speed and attenuation (Figs. 5.2–5.7).

Over the past 20 years, measurements of shear wave speed have been made using an in situ sediment geoacoustic measurement system (ISSAMS) [Richardson 1997a]. The system utilizes "bimorph" ceramic benders to transmit and receive shear waves [Richardson et al. 1987, Richardson et al. 1991a]. These transducers have been mounted on a variety of diver-deployed and remotely operated systems [Griffin et al. 1996] allowing measurement of near surface shear wave speed at 87 locations worldwide. Shear wave speed is measured as time-of-flight over known distances. More recently a transposition method using two transmit and two receive transducers was developed that allows measurement of shear wave attenuation without standards [Richardson 1997b]. This new technique eliminates the need to know transducer sensitivity or to measure variable insertion losses. All measurements were made within the upper 30 cm of sediment, usually at 10, 20, and 30 cm below the sediment–water interface.

5.2.3 Shear Wave Data and Regressions

Empirical regressions among shear wave speed and sediment physical properties (porosity, density, and mean grain size) were previously reported by [Richardson 1997a, Richardson 1997b]. These regressions were based on a data set much smaller than used here. The number of deployments of ISSAMS where shear wave speed and sediment physical properties have been measured has increased to 87 sites, and shear wave attenuation has now been measured at 18 sites [Richardson et al. 1997, Wilkens and Richardson 1998, Richardson et al. 2001a, Richardson et al. 2002b, Richardson 2002]. These data, combined with sediment physical properties measured on carefully collected core samples, were used to develop the regressions shown in Figs. 5.15–5.21 for carbonate and siliciclastic sediments. All in situ acoustic measurements were made at 20–30 cm below the sediment–water interface, over distances of 30–100 cm, and at frequencies ranging from 1 kHz in sands to

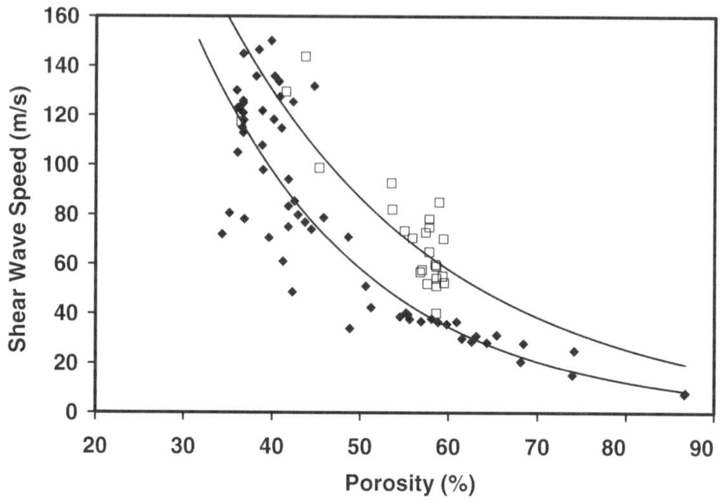

Fig. 5.15. Shear wave speed versus porosity for 24 carbonate (open squares) and 63 siliciclastic (filled diamonds) sediment sites. Regressions for carbonate ($V_s = 656e^{-0.0405\eta}$) and siliciclastic sites ($V_s = 770e^{-0.0517\eta}$) accounted for 67% and 89% of the variability. Porosity, η, is expressed as a percent.

as low as 100 Hz in high-porosity mud. Based on the combination of frequency and shear wave speed, wavelengths ranged between 10 and 20 cm. These are several times greater than the wavelengths associated with in situ compressional wave measurements discussed here. Average shear wave speeds are used for the regressions and include up to 8 separate measurements for each of 1–4 deployments of ISSAMS at each site. Sediment physical properties were measured as described in previous sections and are average values for one or more cores collected at each site with up to 15 measurements per core.

Mean shear wave speed ranged from a low of 7.7 m s^{-1} in the very high porosity sediments of Eckernförde Bay, Baltic Sea, to 150 m s^{-1} in shelly coarse sand sediments found on the southeastern Florida shelf. Shear wave speed increases with increasing density and mean grain size (expressed in millimeters) and decreases with porosity and mean grain size (expressed in phi units) for both carbonate and siliciclastic sediments. These are the same general trends as found in the sound speed–physical property relationships. Shear speed tends to be higher in carbonate sediments than in siliciclastic sediments for similar values of porosity (Fig. 5.15). However, no difference in shear wave speed was found for similar values of mean grain size. The shear speed measurements that define the carbonate relationships are mostly from the Florida Keys. It has been postulated [Richardson et al. 1997] that intraparticulate porosity (the porosity of the particles themselves, approximately 20% within Halimeda plates) at these sites in the Florida Keys could account for the differences between carbonate and siliciclastic relationships. If only interparticulate porosity (i.e., the porosity that results if the 20% intraparticulate porosity is considered part of the particulate volume) is used to determine porosity, the relationship between shear wave speed and porosity is nearly identical for the two sediment types (Fig. 5.16). Until additional measurements are made in carbonate sediments, it is suggested that the regression given in Fig. 5.16 ($V_s = 692e^{-0.0494\eta}$) be used to predict shear wave speed from porosity in %. Interparticulate porosity should be used for carbonate sediments where significant intraparticulate porosity is found. Regressions between shear wave speed and bulk density and mean grain size are given in Figs. 5.17 and 5.18.

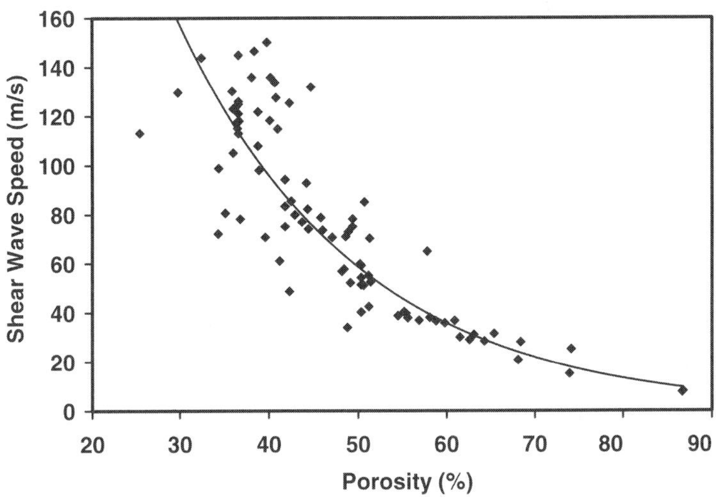

Fig. 5.16. Shear wave speed versus porosity for all 87 sediment sites. Porosity is calculated as interparticulate porosity (see text) for the carbonate sediments from the Florida Keys. Regression ($V_s = 692e^{-0.0494\eta}$, $r^2 = 0.86$) is nearly identical to the regression for the siliciclastic sediment alone. Porosity, η, is expressed as a percent.

The empirical regressions given here provide the most accurate method to predict shear wave speed in near-surface sediment, however, they are only appropriate to predict shear wave speed at 20–30 cm depth in the sediment. Near-surface measurements of shear wave speed (Fig. 5.19) show that a power-law regression provides a good fit of the observed dependence of shear wave speed upon depth for a particular site (also see [Richardson et al. 1991b, Wilkens and Richardson 1998, Richardson 2002]). A more general empirical prediction of the depth dependence was obtained, using a regression for shear

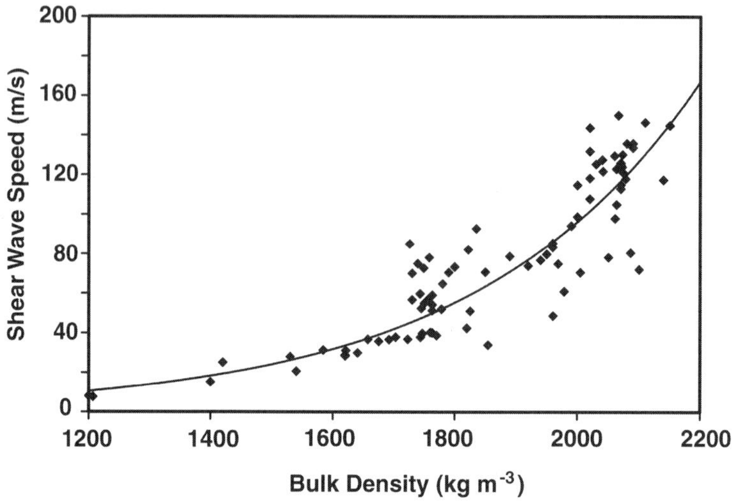

Fig. 5.17. Shear wave speed versus bulk density (kg m^{-3}) for 87 sediment sites. Regression ($V_s = 0.3823e^{0.00284\rho}$, $r^2 = 0.85$). Shear wave speed for the carbonate sediments from the Florida Keys was slightly higher than for siliciclastic sediments with the same bulk density.

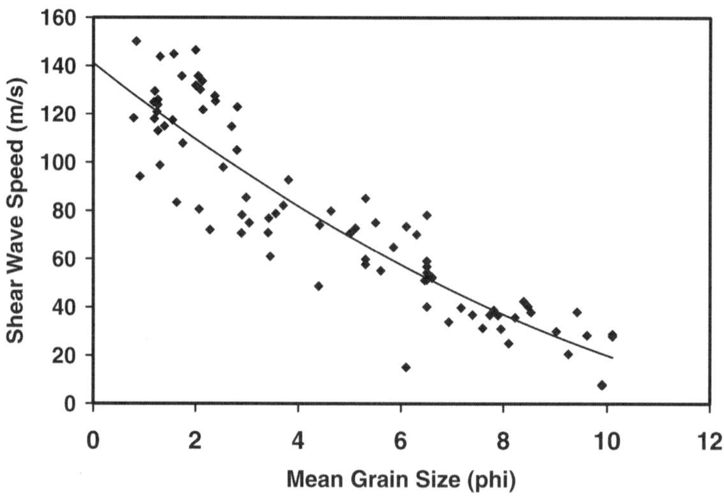

Fig. 5.18. Shear wave speed versus mean grain size for 87 sediment sites. Regression $V_s = 141 - 16.65M_z + 0.455M_z^2$. Regressions for carbonate and siliciclastic sediments were nearly identical.

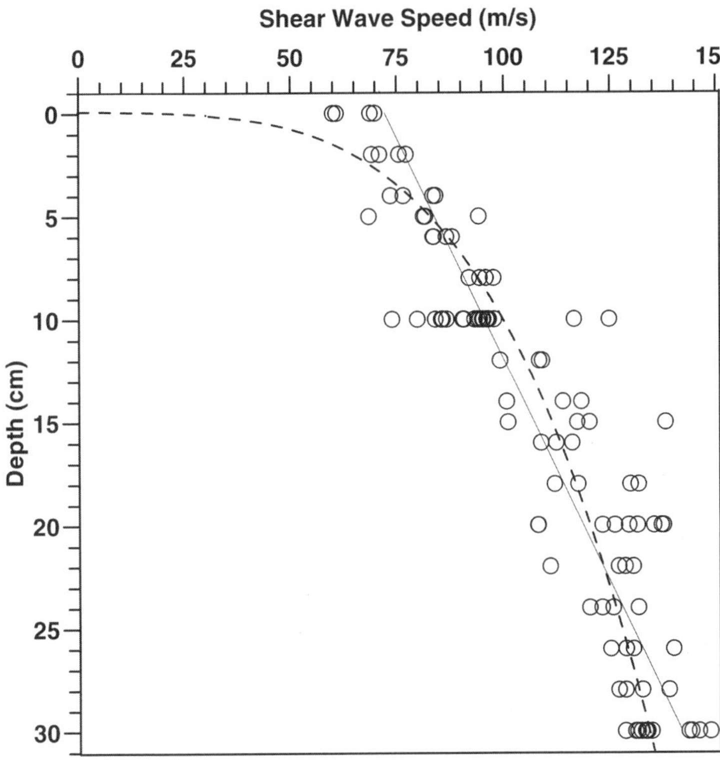

Fig. 5.19. Profile of shear wave speed measured using ISSAMS at 1 kHz for a sandy site in the North Sea (Richardson, unpublished). Best-fit linear ($V_s = 72.3 + 232d$) and power-law ($V_s = 186.5.8d^{0.27}$) regressions account for 83% and 81% of the variability, respectively, where d is the depth below the sediment–water interface in meters. The power-law relationship is preferred to the linear relationship for extrapolation to depths greater than measured here.

wave speed as a function of void ratio from [Richardson et al. 1991a], where values of shear speed and void ratio were measured between 20 and 30 cm below the sediment–water interface,

$$V_s = 55.3e^{-1.1394} \ . \tag{5.8}$$

Note that e is the void ratio, not the base of the natural logarithms. This expression was combined with appropriate depth gradients to obtain the following empirical prediction for shear wave speed in the upper meter of sediment [Richardson et al. 1991b]:

$$V_s = 79.5e^{-1.1394}d^{0.3} \ , \tag{5.9}$$

where d is depth below the sediment–water interface in meters, and e is the void ratio. Some errors in the equations, as originally published, have

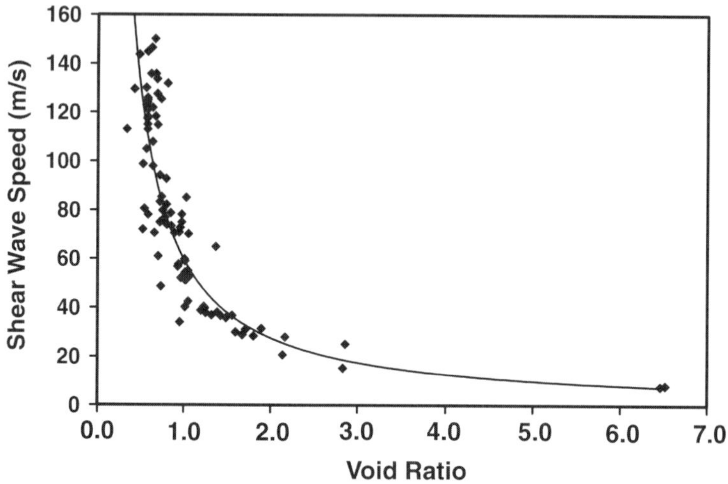

Fig. 5.20. Shear wave speed versus void ratio for 87 sediment sites. Void ratio is calculated from values of fractional porosity ($e = \beta/(1-\beta)$) using the correction for interparticulate porosity (see text) for the carbonate sediments from the Florida Keys. Regression ($V_s = 59.7e^{-1.12e}$, $r^2 = 0.86$) is only slightly different than that presented by [Richardson et al. 1991a] for a smaller data set. Note that e in the exponent is the void ratio and not the base of the natural logarithms.

been corrected here. This predictive regression matched shear wave gradients calculated from inversions of Scholte waves, measured using a geophysical sled, from a variety of sediments on the northern coast of California [Huws et al. 2000].

The following empirical predictive equations for shear wave speed in the upper meter of sediment were generated by combining the most recent empirical relationships between sediment physical properties (Figs. 5.15–5.18, 5.20) with a depth dependence $d^{0.3}$:

$$V_s = 992e^{-0.0494\eta}d^{0.3} , \tag{5.10}$$

$$V_s = 0.549e^{0.00284\rho}d^{0.3} , \tag{5.11}$$

$$V_s = [202.5 - 23.9M_z + 0.6528M_z^2]d^{0.3} , \tag{5.12}$$

$$V_s = 85.7e^{-1.12e}d^{0.3} , \tag{5.13}$$

where d is depth in meters, η is porosity in percent, ρ is bulk density in kg m^{-3}, and M_z is mean grain size expressed in phi units. In (5.13) the symbol e stands for both the base of the natural logarithms and (in the exponent) the void ratio. These four predictive equations are preferred to the more general equations given by [Hamilton 1976a, Hamilton 1987] or to the empirical formula (5.5) derived by [Bryan and Stoll 1988] from laboratory data. It should be noted that in situ measurements by [Stoll 1989] gave values of shear modulus between 1.3 and 2.5 times values obtained by laboratory techniques using (5.5).

Shear wave attenuation was measured using ISSAMS and the transposition technique at 18 sites and is regressed against average sediment physical properties from cores collected at the same sites (Fig. 5.21). The shear wave attenuation factor (k_s, dB m^{-1} kHz^{-1}) is much higher than the compressional wave attenuation factor at the same sites (see Fig. 5.3) and appears to be slightly higher in high-porosity fine-grained sediments compared to lower-porosity sandy sites.

Based on the sparse data set available at the time, [Hamilton 1976c, Hamilton 1987] concluded that the logarithmic decrement (5.7) for low-strain shear waves was 0.3 ± 0.15 for sands and 0.2 ± 0.10 for mud. Log decrements calculated for shear wave attenuation summarized here (Fig. 5.22) are only slightly higher than those summarized by [Hamilton 1987] for both sand and mud sediments. Gradients of shear wave attenuation have also been noted in near-surface sediments (Fig. 5.23). The lack of strong relationships between attenuation of both compressional and shear waves and sediment physical properties (Figs. 5.3, 5.9, and 5.21) may reflect the nature of attenuation measured at these acoustic frequencies. Attenuation of both compressional and shear waves includes a variety of intrinsic attenuation mechanisms as well as scattering and conversion to other forms of energy due to heterogeneity. Given the high variability in shear and compressional wave attenuation and the generally poor predictability based on regressions with sediment physical properties, it is interesting that a good relationship exists between the ratio of compressional and shear wave attenuations and porosity (Fig. 5.24). Similarly, compressional and shear wave speed have a strong relationship (Fig. 5.25).

5.3 Biot's "Slow Wave"

Poroelastic theory ([Biot 1962b, Stoll 1989], Ch. 10) predicts the existence of a second bulk longitudinal wave (a wave having particle and fluid displacement parallel to the direction of propagation). The speed of this wave is much lower than that of the usual compressional wave, hence the name "slow wave." For slow waves, the motion of the fluid is out of phase with that of the skeletal fame, and the attenuation is very high. The laboratory studies of [Plona 1980, Johnson and Plona 1982] are often cited to support the existence of the slow wave in water-saturated sediments. However, they were only able to detect a slow longitudinal wave in fused glass beads (shear speeds of 1400 m s^{-1}) and were unable to detect the slow wave or a shear wave in similar unconsolidated water-saturated beads. The presence of the slow wave in laboratory experiments on fused or cemented samples has been adequately documented by [Klimentos and McCann 1988, Smeulders 2005] and others, yet

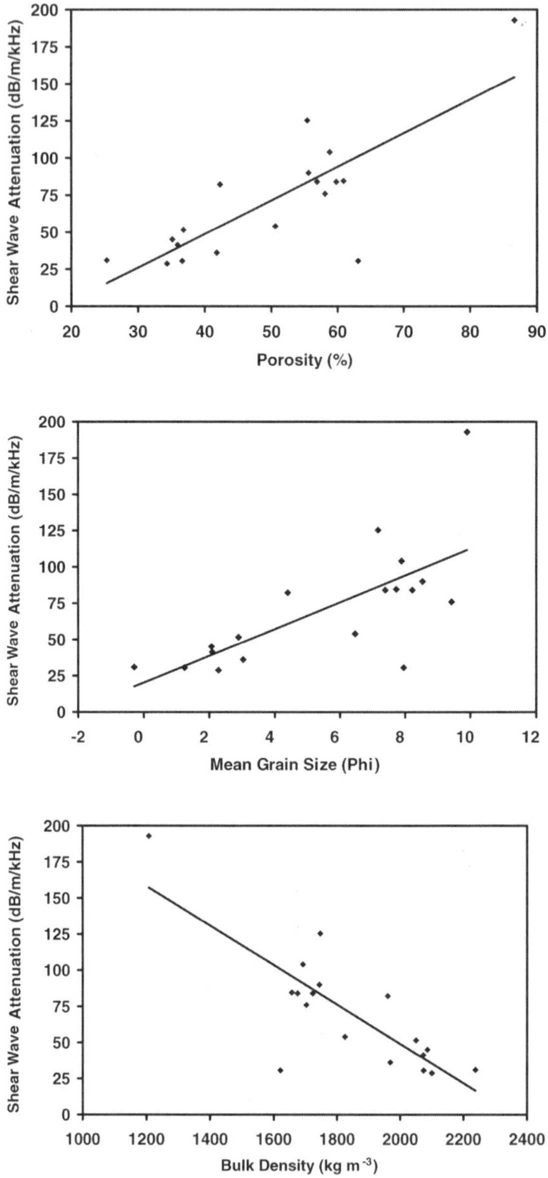

Fig. 5.21. Empirical relationships between shear wave attenuation factor (k_s, dB m^{-1} kHz^{-1}) and sediment physical properties for 19 sites including both carbonate and siliciclastic sediments. The regressions are $k_s = 42.2+2.2\eta$, $k_s = 20.2+9.23M_z$, and $k_s = 321-0.136\rho$, where η is porosity in %, M_z is mean grain size in phi units, and ρ is bulk density in kg m^{-3}.

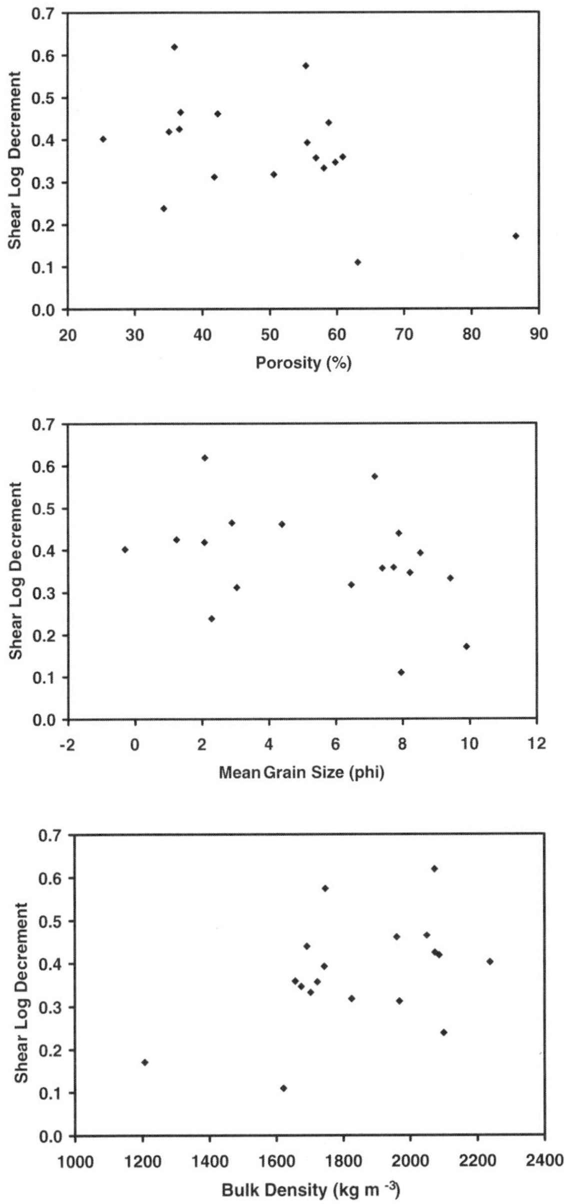

Fig. 5.22. Comparison of shear wave attenuation (shear logarithmic decrement) and sediment physical properties for 19 sites including both carbonate and siliciclastic sediments.

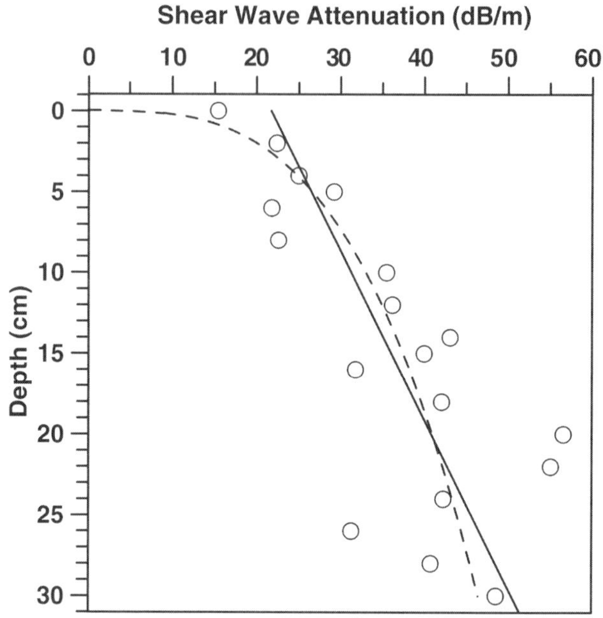

Fig. 5.23. Profile of shear wave attenuation measured using ISSAMS (1 kHz) at a sand site in the North Sea (unpublished data, M.D. Richardson). Best-fit linear line ($\alpha_s = 21.7 + 96d$) and power-law ($\alpha_s = 67.1d^{0.31}$) regressions account for 58% and 62% of the variability, respectively where d is depth in meters. Shear wave attenuation probably decreases below the depths measured here.

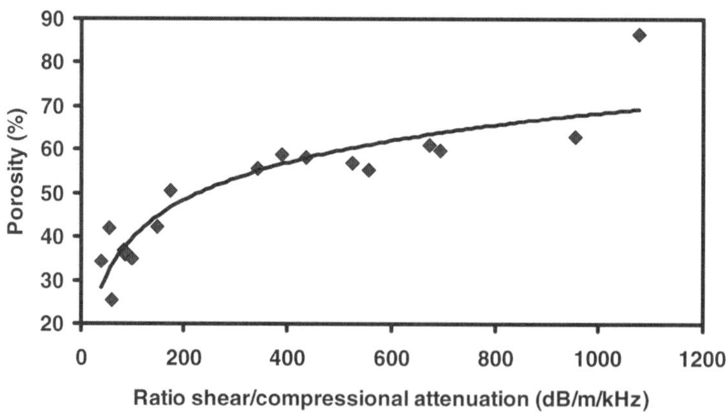

Fig. 5.24. Relationship between sediment porosity and the ratio of shear and compressional wave attenuation. All acoustic measurements were made in situ using probes deployed by ISSAMS and a transposition technique for calculation of attenuation [Richardson 1997b].

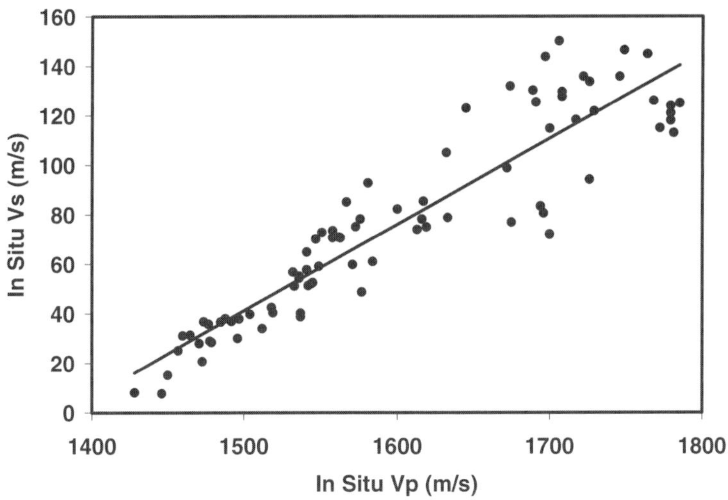

Fig. 5.25. Empirical relationship between sediment shear and compressional wave speed measured at 87 sites worldwide. All measurements were made in situ using probes deployed by ISSAMS. The regression line is $V_s = -481 + 0.348V_p$ with $r^2 = 0.85$.

the slow wave has not been directly observed in field experiments nor has its existence been unequivocally inferred from high-frequency reflection, scattering, and penetration experiments [Thorsos et al. 2000a, Simpson et al. 2003].

In addition to the fundamental question of the existence of the slow wave in unconsolidated sediments, there is the question of the importance of poroelastic effects in sands. That is, are there significant poroelastic effects that are not predicted by fluid or elastic theory even when the slow wave is unobservable? In fact, theory and experiment indicate that poroelasticty can reduce reflection and scattering levels in sands by 1–3 dB [Stoll 1980, Stoll and Kan 1981, Stoll 1985, Chotiros et al. 2002b, Williams et al. 2002b] (Ch. 10 of this monograph). In high-frequency acoustic seafloor interactions in sands, these effects are well approximated by the effective density fluid approximation ([Williams et al. 2001a], Sect. 10.2 of this monograph). This is not a conventional fluid approximation, as the density is reduced from its true value according to a prescription obtained from Biot theory.

5.4 Sound Speed and Attenuation versus Frequency

Wave speed and attenuation are not always measured at the acoustic frequency of interest, thus frequency-dependent relationships become important. Comparison of the sound speed and attenuation measurements of different investigators over a large frequency range (1 Hz to 1 MHz) is difficult given

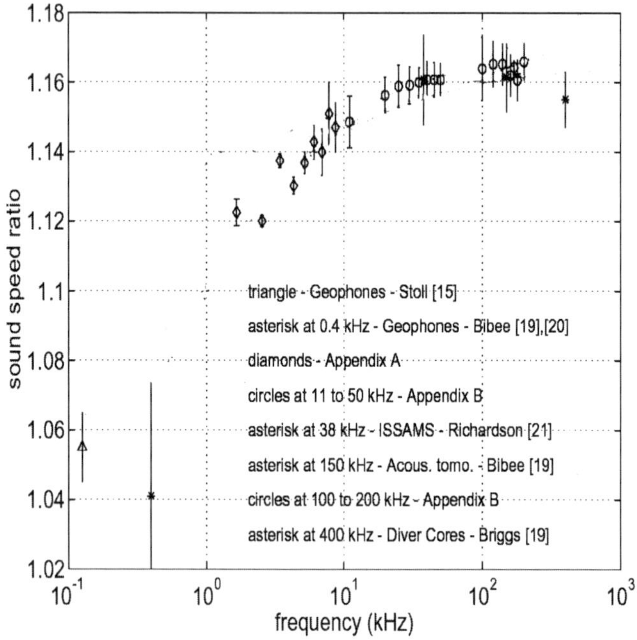

Fig. 5.26. Sediment compressional phase speed measurements from the SAX99 experiment. The figure is adapted from [Williams et al. 2002a], and the references are from that article. These data are identical to those displayed in Fig. 10.9.

the diverse number of sediment types involved as well as the diversity of laboratory and field measurement techniques. Furthermore, any conclusions are further compromised by the range of wavelengths (millimeters to 100s of meters) and varying scales of sediment heterogeneity. In spite of the numerous attempts to measure or compile data on wave speed and attenuation [Hamilton 1971a, Stoll 1977, Stoll 1980, Stoll 1985, Stoll 1989, Kibblewhite 1989, Turgut and Yamamoto 1990, Bowles 1997, Maguer et al. 2000b, Stoll 2002], these complications have prevented unambiguous validation of competing wave propagation theories. The compilation of [Williams et al. 2002a] shown in Figs. 5.26 and 5.27 is based on data from a single site spanning three decades of frequency. Although several different methods were used to obtain this frequency coverage, this data set provides a beginning for efforts to test competing theories for propagation in sediments (See Ch. 10 of this monograph). These theories (Hamilton, Biot-Stoll, Buckingham) require a large number of input parameters that are often difficult and sometimes impossible to measure, may have ambiguous meaning, and must often be applied to heterogeneous natural sediments, making prediction of seafloor geoacoustic

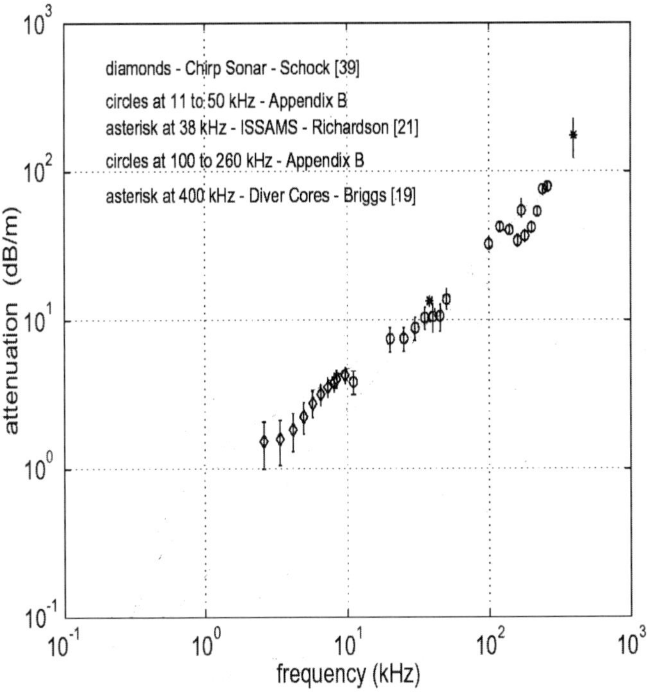

Fig. 5.27. Sediment attenuation measurements from the SAX99 experiment. The figure is adapted from [Williams et al. 2002a], and the references are from that article. These data are identical to those displayed in Fig. 10.10.

properties from physics-based models difficult. Although unsatisfying from a physics perspective, the best estimates of wave properties (speed and attenuation) come from either direct measurements or empirical relationships among seafloor properties. This topic is revisited in Sect. 10.3.

5.5 Bulk Density Regressions

Although bulk density is treated as a sediment physical property in Ch. 4, it is an essential parameter in all sediment wave theories. The regressions given here are based on the Richardson–Briggs data set and methods described in Sect. 5.1.1. Both siliciclastic and carbonate sediments are included. The regression for bulk density as a function of mean grain size is compared with data in Fig. 5.28 and is

$$\rho = 2.17 - 0.082 M_z \ . \tag{5.14}$$

The inverse regression for mean grain size in terms of density is

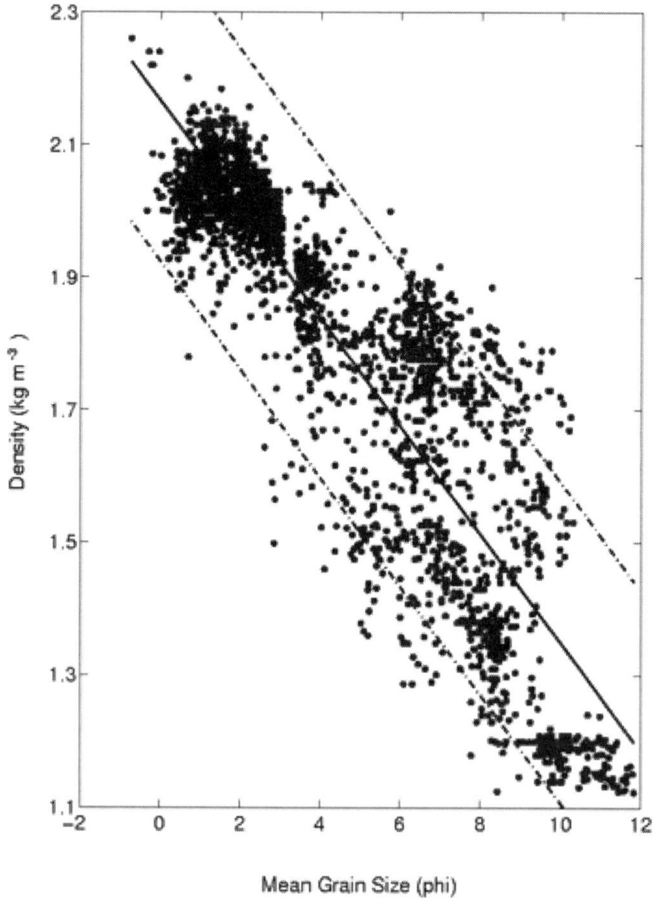

Fig. 5.28. Comparison of regression for density as a function of mean grain size with data from 12 carbonate and 55 siliciclastic sites. The dashed lines are 95% confidence bounds.

$$M_z = 22.11 - 9.82\rho \ . \tag{5.15}$$

Both regressions have a coefficient of determination, r^2, of 0.80. This reflects the high degree of scatter in the data, showing that grain size is not an accurate predictor of bulk density, as is typical of earlier regressions, e.g., [Hamilton and Bachman 1982]. Nonetheless, in many circumstances this is the only means available for estimating density at sites where density or porosity data are lacking. If porosity has been measured, the expression given in Table 4.5 can be used to obtain bulk density.

5.6 Properties of Rock

Rock or rock material in the form of large boulders or calcareous reefs is often a source of acoustic scattering at the seafloor. The bulk density and compressional and shear speeds (equivalently, bulk and shear modulus) of rocks typically found at the seafloor are functions of the chemistry and mineralogy of component minerals, the relative porosity and percent saturation (see Sect. 4.3 for definitions), and the rock microstructure. The bulk density and bulk modulus of the common constituents of coastal marine sediments as well as rock material are given in Table 4.7. Specific values for properties of different rock types have been compiled in numerous handbooks of the physical properties of rocks (e.g., [Mavko et al. 1998] and compressional- and shear-wave speeds of marine rock are given in [Hamilton 1978, Hamilton 1979b]. For typical low-porosity ($< 1\%$), saturated rock specimens, wet bulk density is higher for basalts (2870 kg m^{-3}) than granites (2660 kg m^{-3}); whereas higher-porosity (10–30%) sandstones and siltstones tend to have lower bulk density (2100–2500 kg m^{-3}) which is dependent on porosity. The bulk density of calcareous reef material and shell material is also dependent on porosity and can range from the bulk density of calcite or aragonite for shell material (2700–2950 kg m^{-3}) to as low as 2000 kg m^{-3} for higher-porosity cemented reef material (see Sect. 3.2.3). Sound speed in low-porosity saturated igneous and metamorphic rock samples is about 5–6 km s^{-1}. Sound speed in saturated sandstone and mudstones (sedimentary rocks) is a function of both porosity and the relative percentages of sand and mud and can range between 2 and 6 km s^{-1} with lower sound speeds at higher porosities [Erickson and Jarrard 1998b]. Sound speed in low-porosity carbonate material ($< 10\%$) can be as high as 5.5 to 6.0 km s^{-1} but is much lower in higher-porosity cemented reef material. Average measured shear speeds tend to be roughly 40–60% of the sound speed values. This translates into a Poisson's ratio of 0.2 to 0.4 for most rock material. In contrast, unconsolidated marine sediments have a Poisson's ratio near 0.5. Attenuation in rocks is frequency dependent and, at higher frequencies, more variable and thus difficult to measure. Attenuation mechanisms are the subject of much debate and are a topic which is well beyond the scope of this monograph.

5.7 Research Issues

A consistent theme throughout this chapter is the importance of characterizing the frequency-dependent behavior of sound speed (dispersion) and attenuation for near-surface sediments (e.g., [Williams et al. 2002a]). An internally consistent set of accurate measurements of sound speed and attenuation at the same location and time and for a homogeneous section of sediment over the frequency range of a few hundred hertz to hundreds of kilohertz does not exist. Much of the problem is associated with the difficulties of making accurate measurement of sound speed and attenuation, especially at lower frequencies. The accuracies of most sound speed and attenuation measurements are rarely reported. Understanding and predicting dispersion is not only important for making accurate predictions of sound speed from sediment physical properties (and the inverse) and for predicting sound speed at frequencies other than those measured but also for determination of which theories best account for wave propagation in near-surface sediments. Measurement of attenuation presents additional problems. Attenuation measurements include all types of energy losses including intrinsic attenuation, wave conversion at boundaries and other heterogeneities, and scattering. These effects are rarely separated, making frequency-dependent intrinsic attenuation difficult to evaluate. The higher-than-predicted values in attenuation measured at high frequencies ($>$ 25 kHz) are sometimes ascribed to various scattering mechanisms, but this has not been shown convincingly. The high variability in attenuation presented herein and in most other studies may be real or a result of measurement techniques. Comparison of frequency-dependent sound speed and attenuation can also be complicated by gradients in sound speed or attenuation or by layering.

A reasonable database of shear wave speed has recently been collected using in situ measurement techniques. It does appear that shear wave speed can be predicted from sediment physical properties. However, there is a very sparse data set from which to predict shear wave attenuation, and more measurements are needed. More data, from a variety of sediment types, on the gradients of shear wave speed and attenuation are also needed to validate Eqs. (5.9)–(5.13). The potential anisotropic behavior of shear wave propagation also needs investigation. All of the measurements used to develop the shear-wave versus physical property regressions given in this chapter were based on horizontally generated stresses propagating in a horizontal direction. The possible effects of anisotropy due to the distribution of effective stress (characterized by K_0) and by layering or particle orientation have been neglected.

The importance of poroelastic behavior to reflection and scattering levels is an area of disagreement among high-frequency investigators. It is unlikely that the "slow wave" will be detected in natural sediments by conventional measurement methods. However, the losses in energy as compressional waves are converted to "slow waves" might provide an indication of their presence.

6 Seafloor Roughness

Seafloor roughness can be a dominant contributor to sound scattering at higher acoustic frequencies, causing reverberation through backscattering and altering propagation through forward scattering. Seafloor roughness, especially that associated with sand ripples, can also diffract sound downward into the sediment promoting the seafloor penetration necessary for sonar detection of buried objects. Together with sediment sound speed, attenuation, and bulk density, a statistical characterization of seafloor roughness, often in the form of RMS roughness, RMS slope, or power spectral density, is a critical input to most scattering and penetration models described in Chs. 13 and 15. Much of the present body of knowledge, especially related to measurement techniques and statistical characterization, comes from studies in geology, geophysics, and oceanography, but an increasing effort is being directed toward specific high-frequency acoustic applications. This recent work is the primary subject of this chapter.

The morphology of the seafloor encompasses a wide range of spatial scales. Larger-scale features traditionally studied in the fields of geology and geophysics include basinwide tectonic features, like the mid-Atlantic Ridge, regional bathymetry, sand ridges, coral reefs, shell deposits and other hard grounds, and beach profiles. Smaller-scale features studied in the fields of benthic ecology, hydrodynamics, and sediment transport include sand ripples, pits, craters, and mounds created by fish and decapods; smaller biogenic features such as trails, tracks, tubes, and burrows created by worms and other centimeter-size organisms (Fig. 6.1); and, ultimately, grain-scale features. It is these smaller-scale features that are of most interest in high-frequency acoustic applications. While larger features are stable over tens to thousands of years, smaller features, such as sand ripples and biogenic features, are constantly being altered, which may change seafloor characteristics over hours or days.

Generally, the needs of the acoustic modeler differ from those of the geologist or geophysicist with regard to the characterization of roughness. While geologists and geophysicists seek information on recurring patterns in the dimension, distribution, and shape of seafloor features over a large range of spatial scales, high-frequency acoustic modelers are primarily interested in a statistical analysis of small features on the scale of the acoustic wave-

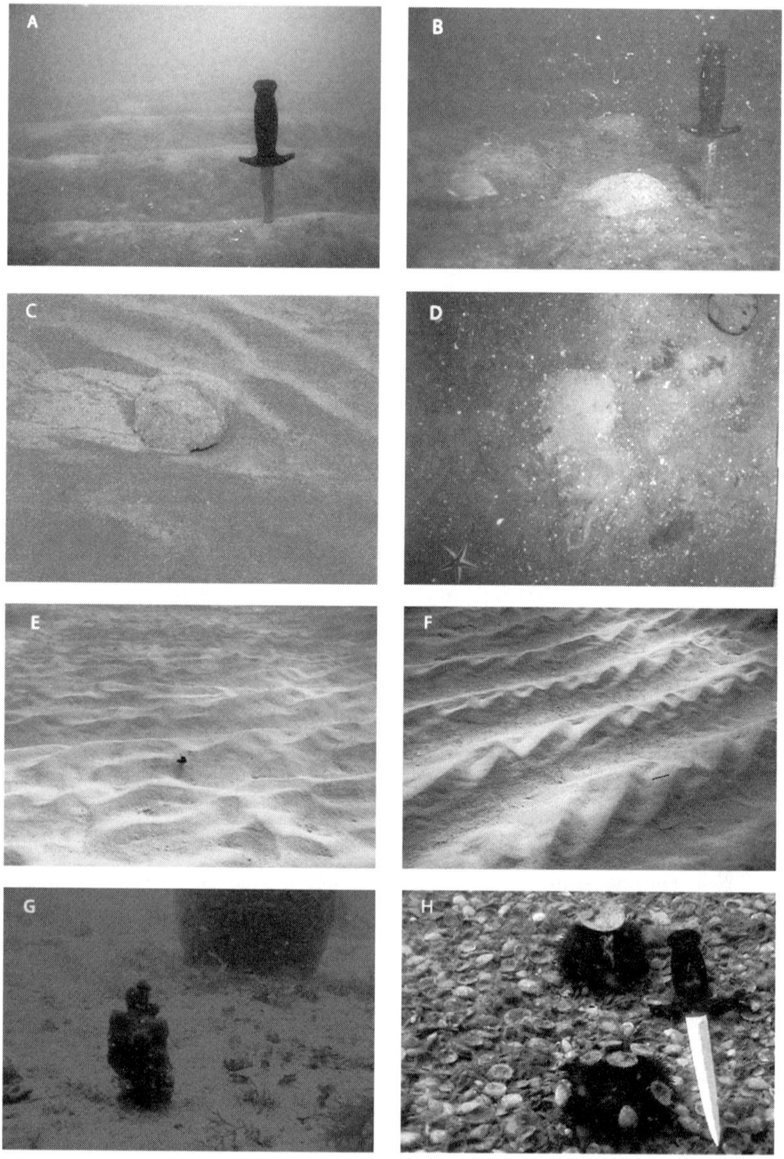

Fig. 6.1. Examples of seafloor roughness: (A) Fresh storm-generated ripples in a medium sand sediment in 19-m water depth. (B) A fine sand with biogenic roughness features such as mounds, trails, and track marks. (C) A sand dollar destroying small 10-cm wavelength ripples. (D) A highly bioturbated muddy sand seafloor with many examples of biogenic features such as fresh mounds, trails, and tracks. (E) A shallow water ooid sand with a mixture of ripple types. (F) A field of 30- to 40-cm wavelength ripples in ooid (carbonate) sands from the Ocean Cay, Bahama Islands (note the smaller set of secondary ripples formed 45° to the primary ripple field). (G) Carbonate hard ground which is nearly devoid of sediments and covered with sponges and corals. (H) A shelly seafloor with sea urchins. Note that hydrodynamically induced roughness tends to be anisotropic whereas biologically induced roughness tends to be isotropic. Courtesy of R. I. Ray. (see first color insert)

length, so-called "microtopography." For the purpose of understanding and predicting interactions of high-frequency acoustic energy with the seafloor, roughness features having scales comparable to acoustic wavelengths are responsible for sound scattering, while larger features cause changes in the grazing angle of the acoustic wave with respect to the bottom (i.e., they determine the seafloor slope). In the acoustic penetration problem, the ripple amplitude and phase both affect the detection of buried objects. Statistical characterization of seafloor roughness at spatial scales larger than the "ensonified region" adds little to the predictive nature of acoustic backscattering, however spatial variations in roughness affect the statistics of sonar echoes (Ch. 16). We therefore restrict most of our discussion to techniques used to measure and quantify roughness at resolution scales of less than a few meters to tenths of a millimeter. It must be realized that, depending on acoustic wavelength, individual features such as shells and other roughness elements at the sediment–water interface may be more appropriately characterized as discrete scatters than as microtopography.

The most commonly used statistical measure of seafloor roughness is the "power spectral density" (hereafter referred to as the "roughness spectrum"). Definitions will be given in Sect. 6.1. Roughness measurement techniques are described in Sect. 6.2, where it is noted that some techniques only provide relief measurement along one-dimensional tracks or profiles across the seafloor. In these cases, the estimated 1D spectrum must be converted to 2D, using assumptions and methods outlined in this chapter. Other issues discussed in this chapter include spatial and temporal stationarity and their effects on seafloor scattering, the use of a single power-law to characterize roughness over a range of spatial scales, relationships between sediment roughness and sediment physical properties, and the importance biological and hydrodynamic processes that create and alter seafloor roughness.

6.1 Statistical Characterization of Seafloor Roughness

Acoustic models for scattering due to interface roughness require various statistical measures of roughness. This section will give essential definitions, leaving technical details to Sect. D.1. The 2D interface relief can be expressed as

$$z = f(\mathbf{R}) \; , \qquad \mathbf{R} = (x,\; y) \; , \tag{6.1}$$

where $f(\mathbf{R})$ is the "interface relief function," with zero mean

$$< f(\mathbf{R}) >= 0 \; , \tag{6.2}$$

and mean square

$$h^2 =< f^2(\mathbf{R}) > \; . \tag{6.3}$$

The parameter h will be referred to as the "RMS roughness." The covariance of the interface relief is required for the Kirchhoff (Sect. 13.3) and small-slope (Sect. 13.4) approximations

$$B(\mathbf{R}) = < f(\mathbf{R}_0 + \mathbf{R}) f(\mathbf{R}_0) > . \tag{6.4}$$

Note that 2D vectors are denoted by uppercase boldface characters. The brackets <> in the equations above denote an average over an arbitrarily large ensemble of different seafloor relief functions. This formal averaging process can be approximated in practice by averaging over multiple realizations of seafloor relief derived from the acoustic, optical, electrical, or manual tracing techniques described in Sect. 6.2. If the covariance is divided by the mean-square roughness, the result is the "autocorrelation," which has a peak value of unity when the "lag," $\mathbf{R}$, is zero.

Equations (6.2–6.4) embody the assumption of spatial stationarity: all statistical measures such as variance, covariance, and higher-order moments are independent of the location at which they are evaluated. As a result, the covariance depends only on the difference of the two coordinate vectors defining the two positions at which relief is measured. This is an approximation whose utility depends on the scale over which the seafloor can be considered statistically similar. If the roughness statistics can be regarded as similar over an area large compared to the size of the area ensonified by the sonar of interest, then the assumption of stationarity is reasonable.

Scattering models based on small-roughness perturbation theory (Sect. 13.2) require the roughness spectrum, which is the Fourier transform of the relief covariance:

$$W(\mathbf{K}) = \frac{1}{(2\pi)^2} \int B(\mathbf{R}) e^{-i\mathbf{K}\cdot\mathbf{R}} d^2R . \tag{6.5}$$

The variable $\mathbf{K} = (K_x, K_y)$ is a two-dimensional "wave vector," whose magnitude, $K = \sqrt{K_x^2 + K_y^2}$, is called the "wavenumber." Wavenumber and wavelength, λ, are related by

$$K = \frac{2\pi}{\lambda} , \tag{6.6}$$

so that small wavenumbers correspond to large wavelengths and vice versa. The spectrum is normalized so that its integral over all K_x and K_y (positive and negative) is equal to the mean-square roughness, h^2:

$$h^2 = \int W(\mathbf{K}) d^2K . \tag{6.7}$$

The dimensions of the 2D spectrum are (length)^4. As will be seen, roughness measurements are usually presented as spectra, and, given the spectrum, the covariance can be found using an inverse Fourier transform. This step may be taken numerically or formally, after a convenient function has been fitted to the spectrum (Sect. D.1.1). Other measures of roughness can also be obtained from the spectrum, such as the RMS roughness, h (6.3), and the RMS slope (13.68).

In addition to the assumption of stationarity, it is often assumed that seafloor roughness obeys Gaussian statistics. In this case, "second-order" statistics (either the covariance or the spectrum) provide a complete characterization from which any desired statistical property (e.g., higher statistical moments) can be determined. This largely untested assumption is usually made in implementing the Kirchhoff (Sect. 13.3) and small-slope (Sect. 13.4) approximations in scattering models. Figure 6.1H is an example of seafloor roughness that cannot obey Gaussian statistics.

If the seafloor roughness exhibits no directional features, such as ripples, the spectrum will be "isotropic," dependent only on the wavenumber, K. Examples of isotropic and anisotropic roughness and corresponding spectra are shown in Sect. 6.3.

Scattering models usually employ convenient functional forms for roughness spectra. One of the simplest and most often used isotropic spectral forms is the "power law,"

$$W(\mathbf{K}) = \frac{w_2}{K^{\gamma_2}}. \tag{6.8}$$

The parameter w_2 is called the "spectral strength," and the parameter γ_2 is called the "spectral exponent." The power-law spectral form can only hold over a finite range of scales, as the extension to zero and infinite scales causes pathological behavior (Sect. D.1). If the power law holds over a sufficiently wide range of scales, the relief can be regarded as "fractal," possessing interesting scaling properties. One such property is "self-similarity," in which the roughness appears statistically the same if it is scaled by any factor, with the same scaling in both the horizontal and vertical. This behavior only occurs for $\gamma_2 = 4$. For the more general case, a different scaling is required in the horizontal than in the vertical in order to restore similarity. Such behavior is "self-affine," and the validity of this idealization of the rough seafloor is discussed in Sect. 6.5.

An important consequence following from power-law behavior is that RMS roughness is a scale-dependent parameter, in spite of its widespread use in quantifying roughness. As noted above, the mean-square roughness is the integral of the spectrum over all K_x and K_y. With a power-law spectrum, this integral is infinite, with the infinity arising from the limit $K_x\ K_y \to 0$, for which the spectrum approaches infinity. This unphysical result shows that the power law cannot persist for very long wavelengths; the spectrum must "roll over" as wavenumber becomes smaller than some finite value. The length scale corresponding to this value is sometimes referred to as the "outer scale." This outer scale is apparently rather large (Sect. 6.5), certainly larger than the field of view of typical seafloor relief measurements (Sect. 6.2). If RMS roughness is determined from stereophotography, the field of view sets a size limit to the longest wavelengths that can be included. A finite RMS roughness results from such measurements, just as one would obtain a finite value by integrating the spectrum down to the smallest resolvable wavenumber. From this, it follows that RMS roughness increases as the field of view increases in size, until the field of view becomes larger than the outer scale. This point is discussed in [APL-UW TR 9407] and Sect. D.1.1. In spite of this, comparison of RMS roughness for different sites determined from photographs is still a sensible procedure, provided the photographs all have similar-sized fields of view and provided the relative nature of RMS roughness is acknowledged.

Some authors prefer spatial frequency (cycles/unit length) to wavenumber as the spectral argument. The relation between wavenumber, K, and spatial frequency, F, is analogous to the relation between temporal frequency, f, and angular frequency, ω, thus $K = 2\pi F$. In this case, the wave vector, $\mathbf{K}$, is replaced by spatial frequencies F_x and F_y, with $K_x = 2\pi F_x$ and $K_y = 2\pi F_y$. The 2D spectra are related as follows:

$$\Phi(F_x,\ F_y) = (2\pi)^2 W(2\pi F_x,\ 2\pi F_y), \tag{6.9}$$

where $\Phi(F_x,\ F_y)$ is the 2D spectrum in terms of spatial frequency.

In many instances, roughness measurements are confined to a 1D track. The dimensions of 1D roughness spectra are $(\text{length})^3$. The resulting 1D spectra can be converted to equivalent 2D form if the roughness is isotropic. This conversion is not trivial and is described in Sect. D.1.2. Briggs' published spectra (see Table 6.1) are one-dimensional, and have spatial frequency as the argument. These spectra will be denoted $\Phi_1(F)$ in this monograph. Analogous to (6.9), the connection between the 1D spectrum in wavenumber, $W_1(K)$, and the 1D spectrum in spatial frequency is

$$\Phi_1(F) = 2\pi W_1(2\pi F)\ . \tag{6.10}$$

Measured 1D spectra are often fit by the power law

$$\Phi_1(F) = \frac{\phi_1}{F^{\gamma_1}}. \tag{6.11}$$

The parameter ϕ_1 is called the "intercept," because in a log–log plot, the power-law spectrum appears as a straight line intercepting the vertical line ($F = 1$) at the value ϕ_1 [Briggs 1989]. The slope of the spectrum in such a plot is $-\gamma_1$. That is, the spectral slope reported by Briggs is the negative of the 1D spectral exponent. Examples will be given in Sect. 6.3, and Table 6.1 lists slope and intercept values for a large number of sites. The equivalent 2D parameters, w_2 and γ_2, can be found using relations given in Sect. D.1.2. Section D.1.3 discusses the units of the parameters w_2 and ϕ_1.

6.2 Measurement of Seafloor Roughness

A variety of techniques have been used to quantitatively characterize seafloor microtopography including manual tracing [Briggs 1989], stereophotography

[Briggs 1989, Wheatcroft 1994, Lyons and Pouliquen 2004], electrical resistivity probing [Briggs et al. 2002b, Tang 2004], laser line scanning [Moore and Jaffe 2002], laser imaging (N. P. Chotiros in [Richardson et al. 2001a]), ultrasound profiling (Y. Igarashi and R. L. Allman cited in [D. Jackson et al. 1986b]), towed acoustic multibeam and sidescan sonar [Stewart et al. 1994, Briggs et al. 2005a], and bottom-mounted sector-scanning and pencil beam sonar [Irish et al. 1999]. These physical, optical, electrical, and acoustic techniques are used to develop one-dimensional (1D) linear elevation profiles or two-dimensional (2D) maps of seafloor elevation, the elemental data required for statistical characterization of seafloor topography. The roughness statistics that are often extracted from these data have been chosen to meet the input requirements of high-frequency scattering models rather than to study the evolution of seafloor microtopography or the processes that create and destroy that morphology. However, future studies may emphasize the use of statistical characterization of microtopography and spatial and temporal variability in high-frequency acoustic scattering to quantify types and rates of biological and hydrodynamic processes that alter seafloor microtopography [Wheatcroft 1994, Traykovski et al. 1999, Pouliquen and Lyons 2002, Self et al. 2001, Hay and Mudge 2005]. In the next few paragraphs a historical development of seafloor roughness measurement is presented.

Hull-mounted and towed multibeam sonar systems developed in the 1970s, and operating in the approximate frequency range of 12–30 kHz, provided a unique 2D view of the seafloor and one of the first opportunities to statistically characterize seafloor topography at resolution scales of tens of meters to hundreds of kilometers. Fractal parameters, autocorrelation, slope histograms, Gaussian statistics or other statistical methods were developed to characterize topography of a variety of oceanic ridge crest, seamount, abyssal plain, continental shelf, and continental rise terrains (e.g., [Fox and Hayes 1985, Goff and Jordan 1989a, Goff and Jordan 1989b, Goff and Tucholke 1997]). Fox and Hayes suggested that a single power-law spectrum could be used over scales ranging from centimeters to kilometers. This was based on data collected with a hull-mounted multibeam sonar, a deep-towed profiling sonar, and stereophotographic images, all collected from the continental rise along the Atlantic coast of the United States. The validity of this hypothesis will be discussed in Sect. 6.5.

One of the earliest uses of stereophotographs to statistically quantify seafloor microtopography is due to [Akal and Hovem 1978]. They developed digital elevation maps from pairs of stereo bottom photographs and processed these to obtain a two-dimensional autocorrelation function that provided information on the wavelength content and orientation of microtopography. As will be seen, modern stereophotographic techniques are not significantly different from those employed by Akal and Hovem. Several others [Swift et al. 1985, Huntley and Hazen 1988, Hollister and Nowell 1991] also resolved seafloor morphology at spatial scales appropriate for high-frequency modeling but focused on contouring sediment features rather than providing the statistical characterization needed for acoustic scattering models. All of these early studies were restricted to a few images because of the large computational time required to create 2D elevation maps at high-resolution scales.

The first serious attempts to measure seafloor microtopography for high-frequency acoustic modeling were made in the late 1970s and early 1980s. One method, described in [Briggs 1989], employed a tracing system by which divers manually drew 1.8-m-length profiles of seafloor roughness on a Mylar sheet, thus creating digitized profiles of bottom microtopography. The system was labor-intensive, but useful in conditions where the optical clarity of the water is poor. Also in the early 1980s, Igarashi and Allman (as reported in [D. Jackson et al. 1986b]) developed an acoustic microprofiler operating at 5.75 MHz to obtain microroughness spectra. Early microroughness measurements using stereophotography by investigators at a British Admiralty laboratory are also reported in [D. Jackson et al. 1986b].

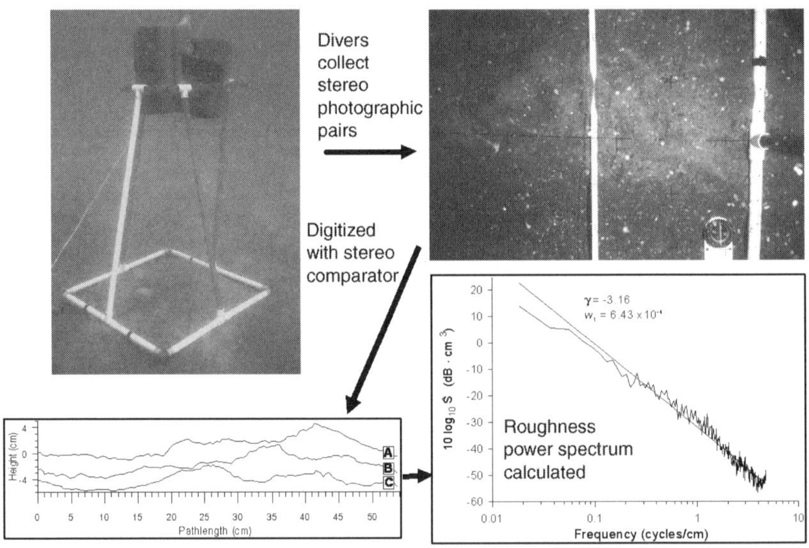

Fig. 6.2. A summary of photogrammetric techniques used by K. Briggs to measure seafloor microtopography. Stereophotographs are made with a diver-operated 35-mm camera, paired images are processed with a stereocomparator to obtain relative height profiles, and power spectra are estimated from the profiles and then averaged (see [Briggs 1989] for details).

The bulk of 1D microroughness data from stereophotography is due to [Briggs 1989], who developed diver-operated stereophotogrammetric systems for shallow-water and remotely operated systems for deeper waters. In shallow water, divers navigate a 35-mm underwater stereo camera, which is attached to a rigid frame to control focal distance (91 cm) and orientation, along the seafloor to be characterized (Fig. 6.2). At greater depths a remote system using a pair of 70-mm cameras is used to photograph the seafloor from distances of 90–200 cm, depending on water clarity. Both systems are self-contained and thus independent of surface electrical cables. Roughness height values are digitized from stereo pairs with a stereocomparator along linear profiles at intervals spaced according to the particular application and picture quality. Typical resolutions of 1 mm in the vertical and 0.5 mm in the horizontal are obtainable from high-quality images. Several profiles are normally analyzed from each pair of photographs to obtain the roughness power spectrum over path lengths of 30–100 cm. The digitized data from each profile (with samples equally spaced and totaling a power of 2) are prewhitened by taking the differences of adjacent points, and then possible leakage is eliminated by subtracting the sample mean from the prewhitened data. The resultant data are tapered with a 20% cosine bell and a fast Fourier transform is used to compute a periodogram. The periodograms are smoothed by ensemble averaging from all profiles from the same orientation or for all orientations if the roughness is isotropic. In most cases, spectra from several stereophotographs collected at each site are averaged. Normally the roughness height profiles are oriented along the azimuthal direction of the acoustic transmitters used in the experiments. Where sand ripple fields are present, profiles are oriented parallel and orthogonal to the crest-to-crest ripple train. Roughness is reported as the 1D spectral slope and spectral intercept and as RMS along the same transects (Table 6.1).

A digital stereophotogrammetric system has been used at the NATO Undersea Research Centre since 1997 [Lyons and Pouliquen 2004] to automatically characterize 2D seafloor relief over scales from about 1 m to 1 mm (Fig. 6.3). Digital cameras (1280 x 960 pixels), mounted on a rigid frame at 91-cm focal distance from the seafloor and hard-wired to either the shore or ship, are used to produce stereo images of the seafloor. A Desktop Mapping System (R-Wel, Inc.) using area-based mapping is used to construct 2D digital elevation maps from the stereo images. After data tapering, Fourier-transformed height data are used to construct 2D roughness spectral density maps. These steps are illustrated in Fig. 6.4. Either azimuthally averaged or azimuthal-specific 2D roughness power spectra can then be calculated and used as inputs to scattering models (see Sect. 13.2.4). This stereophotogrammetric system has been used to characterize seafloor roughness near Elba Island, Italy, in 1997 [Lyons et al. 2002b] and again in 2003 [Pouliquen et al. 2004], near the coast of Halifax, Canada, in 2001 [Pouliquen et al. 2004], in La Spezia Bay [Pouliquen and Lyons 2002], and in the northeastern Gulf of Mexico during SAX99 (A.P. Lyons in [Richardson et al. 2001a]). This system has since been

Fig. 6.3. The digital stereo camera system described by [Lyons et al. 2002b, Lyons and Pouliquen 2004] being deployed off the harbor of Marciana Marina, Island of Elba, Italy (left) and collecting data from a sandy seabed in 10-m water depth (right). The data acquisition, camera control, and power supply are connected to the ship using an electromechanical cable.

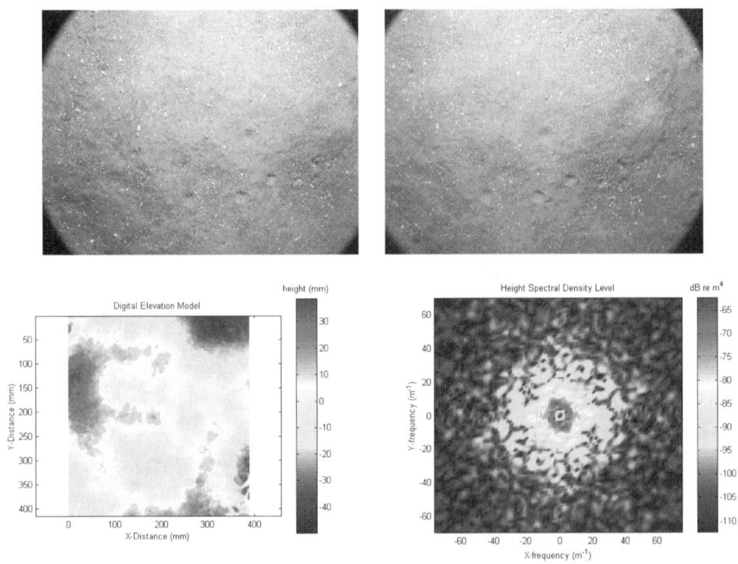

Fig. 6.4. Upper: Overlapping stereophotographs (photo: SAX27) collected by A. P. Lyons from the SAX99 experiment in the northeastern Gulf of Mexico. Lower left: Digital elevation map. Lower right: Calculated two-dimensional power spectral density calculated from the digital elevation map. Note the isotropic nature of the microtopography. See text or [Lyons et al. 2002b] for description of methodology. The spectrum color scale is for $10\log_{10}[S(F_x,\ F_y)]$. (see first color insert)

improved by updating to higher-resolution cameras and improving the digital elevation software.

Electrical resistivity probes have been used to measure seafloor microtopography [Briggs et al. 2002b, Tang 2004]. The abrupt increase in electrical resistivity (expressed as formation factor, see Sects. 4.4.2 and 7.1) as electrodes were inserted in the sediment was used to map the seafloor elevation at 1-cm horizontal resolution during SAX99 (Fig. 6.5). Roughness power spectra estimated with the same techniques used for elevation profiles from stereophotographs yielded statistically similar values of spectral strength and spectral exponent, but did not include the highest spatial frequencies obtained with stereophotography. A second electrical resistivity system (IMP2) was developed by [Tang 2004] to characterize seafloor relief along a longer path length (4 m) but at a slightly lower resolution. This instrument provides the capability to measure the relief of buried interfaces as shown in Fig. 6.6.

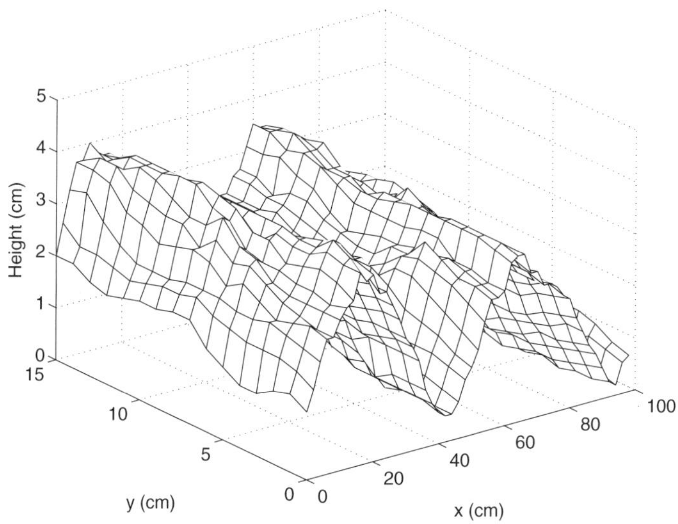

Fig. 6.5. Two-dimensional microtopography measured using IMP during SAX99. Note that all three scales are different. Adapted from [Briggs et al. 2002b].

Using a laser line scanner, [Moore and Jaffe 2002] measured elevation over a 1.35-m transect at $<$ 1-mm resolution over a 9-day period during SAX99. They reported larger-scale features such as ripples decreasing in amplitude from 2.5 cm to 2.0 cm and smaller-scale features degrading after 20–30 hours. Rapid changes in bottom features were attributed to feeding and locomotion behavior of benthic and pelagic animals. Roughness power spectral slopes and intercepts estimated from laser images were about the same as those

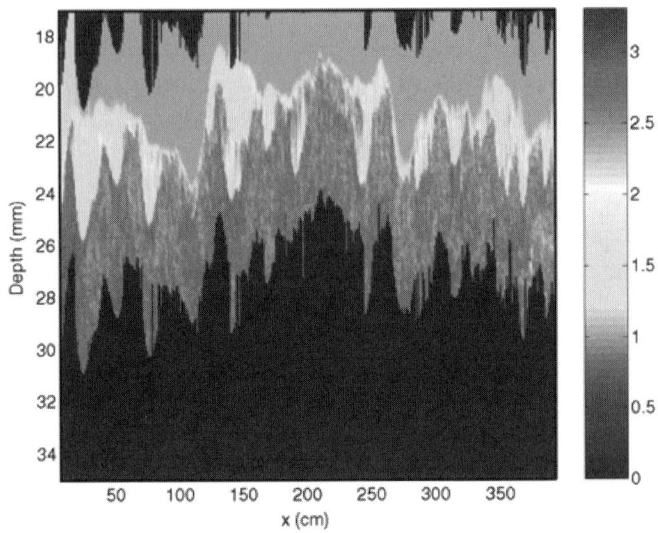

Fig. 6.6. Formation factor measured along a 1D track using IMP2 during SAX04. These data show a mud layer overlying rippled sand. The sand appears as red-orange, while the mud appears as light blue-green with yellow inclusions of sand. The light blue indicates seawater, and regions for which there are no data are dark blue. Courtesy of D. Tang. (see second color insert)

measured from stereophotographic images over the spectral frequency range 0.02 to 2 cycles cm^{-1}.

6.3 Examples of Seafloor Roughness Spectra

Figure 6.7 displays 1D spectra showing time variation of roughness at the SAX99 site [Briggs et al. 2002b] over a period of one month. The 4 October spectra are richer in long-wavelength features, which were storm-generated ripples. As time progresses, these ripples decay, and smaller roughness features grow, as evidenced by the increase in the high-frequency part of the spectrum. Also evident in Fig. 6.7 is an apparent break in the slope of the roughness spectra between 1 and 2 cycles cm^{-1} (see [Williams et al. 2002b]. Whether this change in slope is real or a function of the method of manual digitization is not certain. In computing this type of spectrum, several data manipulations are made in order to reduce bias. Several tracks are selected from each pair of photographs to create digitized elevation maps using a stereocomparator. The digital elevations maps are prewhitened and tapered prior to calculation of individual power spectra. The power spectra from multiple photographs collected on the same date are averaged. Averaging the rough-

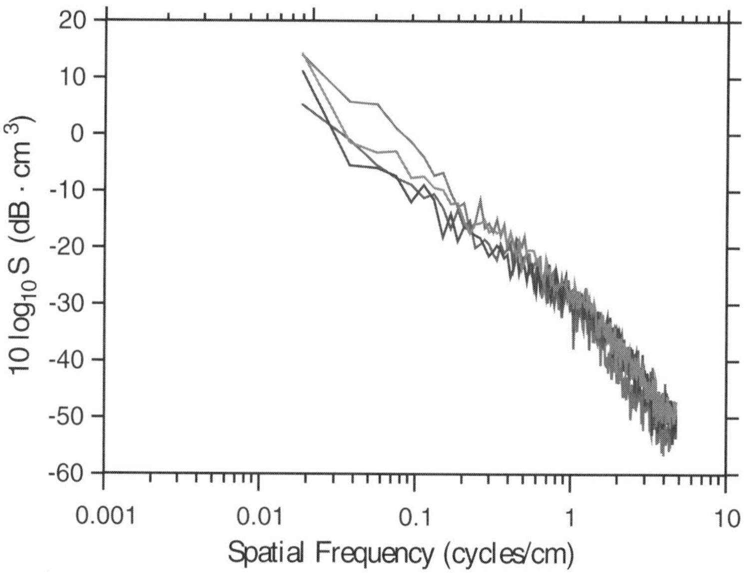

Fig. 6.7. Changes in 1D spectra derived from stereophotographs collected during the SAX99 experiments in the northeastern Gulf of Mexico in 1999: 4 October (red), 19 October (green), 23 October (blue), and 5 November (purple) [Briggs et al. 2002b]. The ordinate is $10\log_{10}[S(F)]$. (see second color insert)

ness power spectra assumes that the spatial and temporal seafloor relief is a stationary, random process, at least over the spatial and temporal scales photographed. This averaging may blur features that seem evident on casual inspection of photographs. The human brain is good at recognizing regular patterns, such as relatively faint ripple superimposed on random microtopography. If the ripples do not have a consistent wavelength and direction, they may not stand out as a peak in the roughness spectrum. An example showing a clear spectral peak due to ripple is shown in Fig. 6.8. These data were obtained over a single track during SAX04 using IMP2.

For anisotropic conditions, such as rippled seafloors, 2D spectra give more information than 1D spectra (Fig. 6.9). Alternately, 1D spectra determined along specific azimuths can be used to reveal anisotropy, but it is difficult to infer the 2D spectrum from such 1D data. It is important to realize that slices through a 2D spectrum, such as those displayed in Fig. 6.10, are *not* 1D spectra in the corresponding directions. Connections between 1D and 2D spectra are given in Sect. D.1.2.

A comparison of 2D anisotropic roughness spectra from sand ripples from Marciana Marina, Elba Island, Italy, with a well-bioturbated sandy sediment found in the Bay of La Spezia, Italy, demonstrates the importance of 2D digital elevation mapping in determining relief spectra for anisotropic and isotropic seafloors (Fig. 6.9). Two-dimensional roughness power spectra can

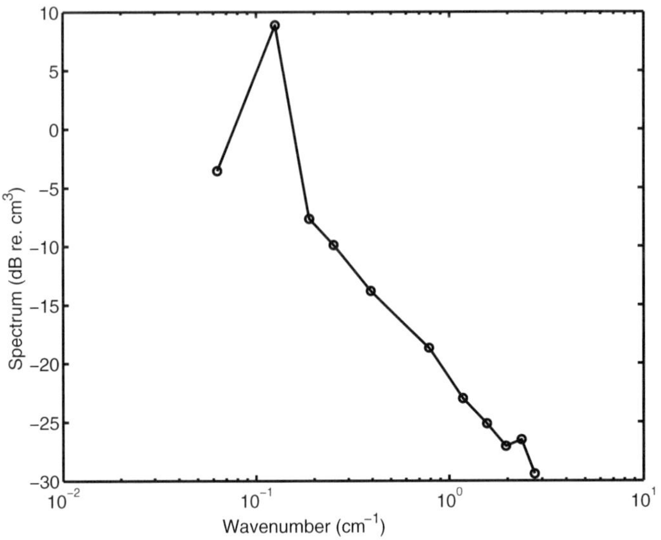

Fig. 6.8. Roughness spectrum measured using IMP2 during SAX04. The measurement track was orthogonal to the ripple crests, and a spectral peak due to the ripples is evident. The ordinate is $10 \log_{10}[W_1(K)]$. Courtesy of D. Tang.

also be used to parameterize seafloor roughness at different azimuthal orientations (Fig. 6.10).

Table 6.1 provides roughness data from sites where grain size as well as seafloor roughness were measured. The table gives RMS roughness, spectral slope, and spectral intercept. The uncertainties in these data (+2 dB to −3 dB for the 95% confidence limits) combined with typical errors in measured acoustic data (± 2 dB in [Williams et al. 2002b]) are such that model-data differences as large as 4 dB do not necessarily imply disagreement. Although values of roughness parameters may not be applicable outside the measured wavenumber range, it is not uncommon to extrapolate outside of this range, assuming that the power law holds over a wide band of spatial frequency (Sect. 6.5). The data in Table 6.1 will be used in Sect. 6.6 to demonstrate relationships among values of roughness statistics and sediment physical properties.

6.4 Temporal Variability in Seafloor Roughness

As discussed in Ch. 4, the seafloor is dynamic with bottom features being constantly altered by competing hydrodynamic and biological processes. Unfortunately, characterization of roughness during high-frequency acoustic experiments is often restricted to a single point in time. As will be

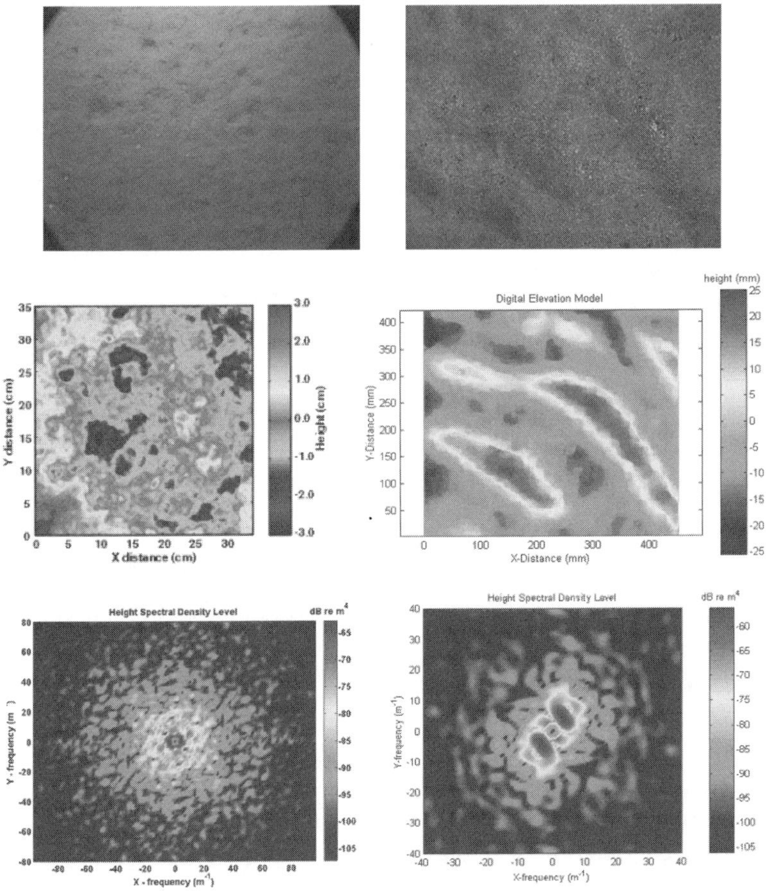

Fig. 6.9. Single images of the paired stereophotographs from a fine sand sediment near the Bay of La Spezia (left) and from a ripple seafloor near Elba Island (right) collected by A.P. Lyons. The digital stereo images were used to develop digital elevation maps (middle panels) and two-dimensional relief spectra (bottom panels) and visually demonstrate the difference between bioturbated isotropic sediments and anisotropic rippled seafloors (see [Lyons et al. 2002b, Pouliquen and Lyons 2002] for details). The color scale is for $10\log_{10}[S(F_x,\ F_y)]$. (see second color insert)

shown, the lack of temporally coincident acoustic and roughness measurements may degrade comparisons of measured and predicted scattering. In energetic sandy environments, seafloor roughness is often in equilibrium with hydrodynamic forcing, resulting in anisotropic ripple fields (Fig. 6.11a). Under these conditions, sand ripple type, height, and wavelength are generally predictable given sediment type (mean grain size), water depth, surface wave conditions, and storm or tidal bottom current speed and direction [Wiberg and Harris 1994, Traykovski et al. 1999, Hay and Mudge 2005]. In

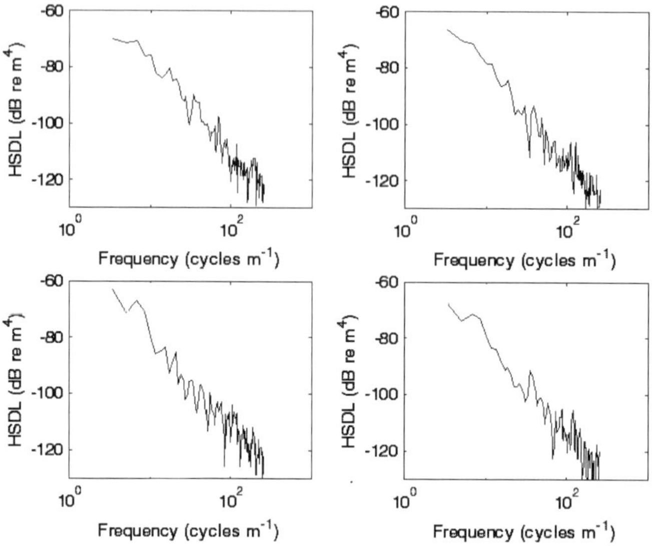

Fig. 6.10. Roughness height spectral density level slices of the 2D power spectral density presented in Fig. 6.4. The slice directions are (a) 0°, (b) 45°, (c) 90°, (d) 135°. The ordinate (HSDL) is $10\log_{10}[S(F_x,\ F_y)]$. Courtesy of A. P. Lyons.

less energetic sandy environments, ripple fields created during major storms are slowly destroyed by biological activity until an isotropic roughness regime having a less steep spectral slope is achieved [Briggs et al. 2002b] (Fig. 6.11). The rate of degradation of these wave- or current-induced features is largely unknown. In less energetic, often fine-grained environments, bottom currents and waves are rarely of such strength as to erode or resuspend sediments, and roughness is dominated by a combination of depositional events and by biological alteration of the surface. The seafloor has a smooth, often isotropic relief after depositional events. However, biological processes soon begin to create small-scale isotropic relief features, such as trails, pits, burrows, and mounds, and the equilibrium state of roughness is often achieved within only a few days [Wheatcroft 1994].

The importance of temporally concurrent roughness and acoustic measurements is demonstrated by the results of three experiments. In these experiments temporal changes in the bottom relief spectra were measured using a combination of analog and digital stereophotography coupled with the measurements of high-frequency acoustic backscattering strength. These are the only examples known to the authors of concurrent temporal characterization of seafloor roughness and measurement or prediction of high-frequency acoustic backscattering strengths. During SAX99 in the northeastern Gulf of Mexico, [Briggs et al. 2001] observed seafloor roughness features for a 30-

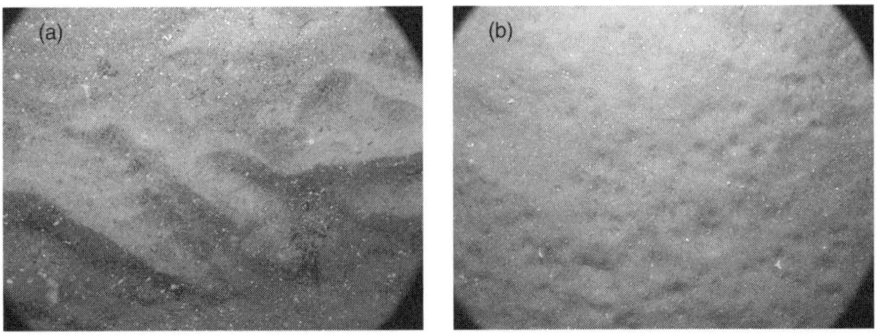

Fig. 6.11. Bottom photographs taken by A. P. Lyons during the SAX99 high-frequency experiments: (a) sharp-crested, well-defined ripples were present at the beginning of the experiment and (b) small pockmarks created by feeding fish. From [Richardson et al. 2001a].

day period (4 October to 5 November 1999), during which a storm created relatively sharp-crested, well-developed storm orbital ripples (50- to 70-cm wavelength) on the sandy seafloor. Within 15 days after the storm, biological processes reduced ripple heights to subtle undulations, resulting in a reduction in both the slope and intercept of the relief spectra (see Fig. 6.7). These differences in relief spectra led to an approximate 2-dB reduction in modeled backscatter strength at 40 kHz.

Also during SAX99, the seafloor was raked with 1.95-cm tine spacing within the field of view of an acoustic tower [Richardson et al. 2001b]. These artificial ripples were completely destroyed within 24 hours by biological processes. The 2D power spectra derived from digital stereophotographs at a raking site near the site of these acoustic experiments show the change from an isotropic roughness just before the raking to a highly anisotropic roughness just after raking and then a decay of the ripples back to the original isotropic relief within 24 hours (Fig. 6.12). This rapid decay of ripple structure was commensurate with rapid changes in acoustic backscattering (Fig. 6.13). Note that the tine spacing of the rake was set to the approximate Bragg frequency (1/2 wavelength at 40 kHz) for maximum effect (see Sect. 13.1).

In another observation of the temporal change of seafloor roughness, [Pouliquen et al. 2004] collected stereophotographs using a bottom-mounted system every 10 minutes for 10 days at a 10-m-deep site near Elba Island, Italy. They attributed the rapid changes in roughness to competing biological and hydrodynamic processes (wave-induced ripples). Figure 6.14 compares the time evolution of roughness spectral parameters with time series for current speed and wave height. Sand ripples were present between 1800 on day date 146 to 1100 on day date 147 when winds exceeded 30 kts, with significant wave heights greater than 1 m. It was during this period that the 2D relief spectrum displayed anisotropy. After day date 147, sea states were low, and the 2D relief spectrum quickly became isotropic with an almost continuous decline in RMS roughness until the end of the experiment. Diurnal changes in

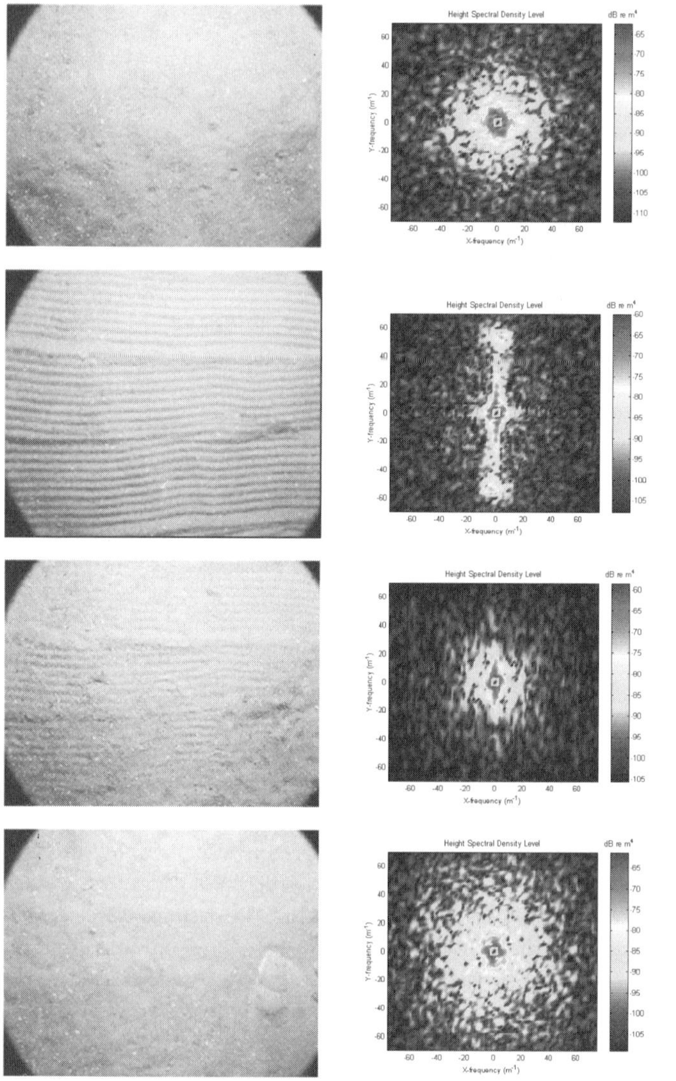

Fig. 6.12. Photographic images and corresponding 2-D power spectra from SAX99. These show temporal decay of ripples created by raking the seafloor. Courtesy of A. P. Lyons. (see second color insert)

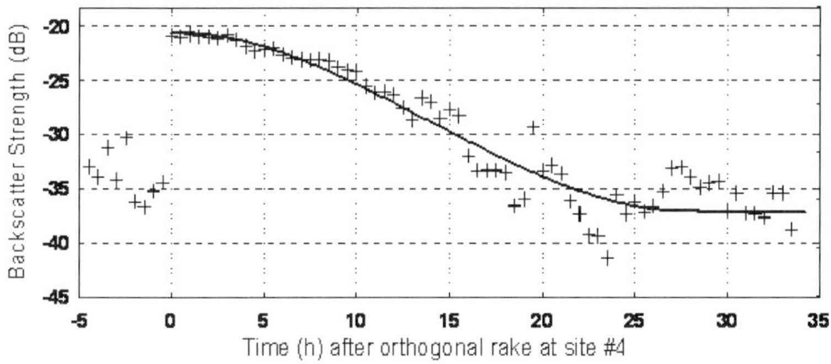

Fig. 6.13. Decay of acoustic scattering after raking orthogonal to the direction of acoustic incidence. Background scattering strengths averaged −32 dB over all measurement sites in SAX99, but were closer to −35 dB at this site [Richardson et al. 2001b].

bottom relief, a result of burrow construction (60 to 80 mounds and funnels per m^2) by the mud shrimp *Callianassa truncata*, dominated the relief spectrum. Spectral strength and intercept seemed more sensitive to these diurnal bottom relief changes than RMS roughness. Changes in calculated backscatter strength based on the diurnal changes in relief spectra were as high as 10 dB (Fig. 6.14).

The information presented in this section suggests that considerable temporal changes in backscattering strength are possible based on the temporal evolution of bottom roughness. However, care must be exercised when interpreting and extrapolating the magnitude and rates of these processes. All of the experiments described in this section, and many other observations of temporal changes in bottom roughness features were made with sensors attached to bottom-mounted platforms. The presence of these platforms may alter the rates and magnitude of the temporal changes in roughness either by altering the hydrodynamic flow around the platform, creating turbulence and scour, or by attracting benthic and pelagic fauna to an otherwise barren location, the refugia or reef effect. However, it does appear that the rate of destruction of larger-scale roughness is both a function of the size of the features (large-scale features decay more slowly because of their larger mass) and a function of the activity of the dominant fauna inhabiting the seafloor. It must also be noted that good fits between modeled (predicted) and measured backscatter strengths have not generally been achieved in these dynamic environments. This should be an area of active research, because the temporal variability of backscatter strengths may on many substrates be greater than the spatial variability.

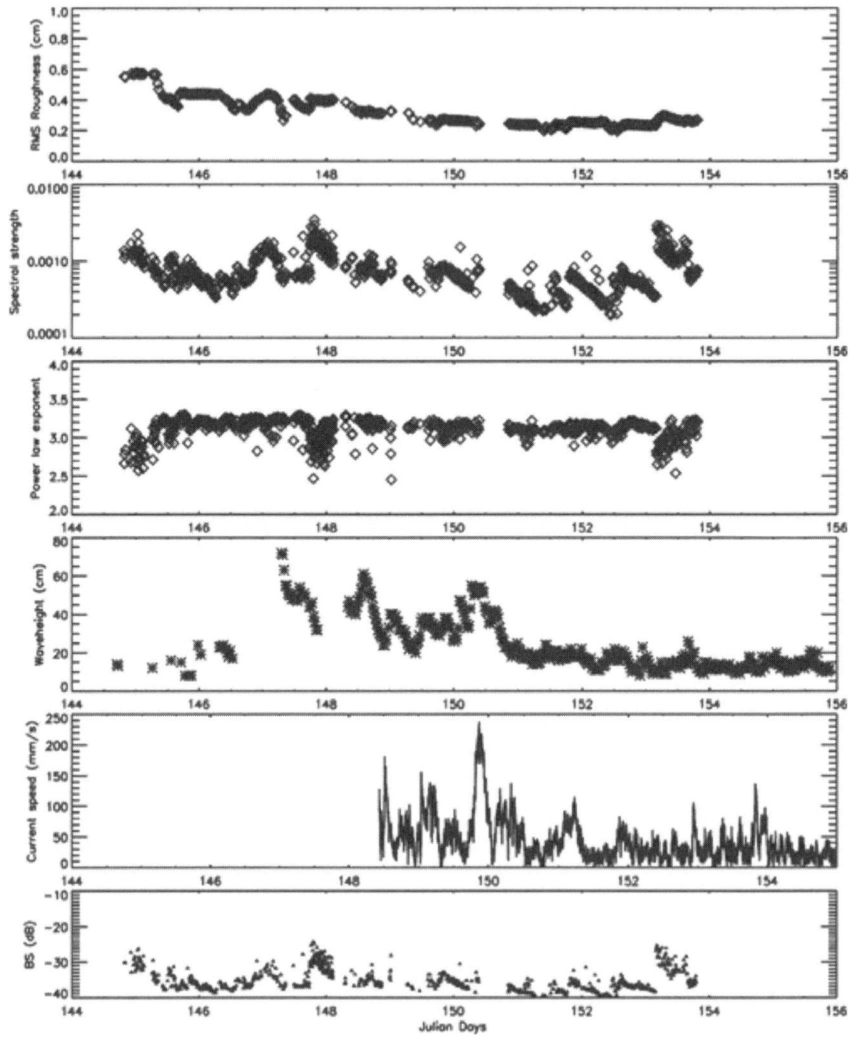

Fig. 6.14. Temporal evolution of roughness properties (RMS roughness, spectral slope, and intercept) calculated from digital stereophotographs taken every 10 minutes compared to the temporal changes in significant wave height and bottom currents. (From [Pouliquen et al. 2004].) Bottom backscatter strength (lowest panel) is calculated for 20° grazing angle at 100 kHz using first-order small-perturbation theory for a fluid–fluid interface (see Ch. 13) using a bottom sound speed ratio $\nu_p = 1.31$, loss parameter $\delta_p = 0$, and a density ratio $a_\rho = 1.92$.

6.5 Validity of Power-Law Assumption for Roughness Spectra

It is sometimes assumed that seafloor roughness power spectra have the simple power-law dependence, expressed in Eqs. (6.8) and (6.11), over a wide range of spatial scales [Berkson and Matthews 1983, Fox and Hayes 1985, D. Jackson et al. 1986a]. It has been noted that a single power law fits roughness spectra over a very wide range of scales on the Atlantic coast continental rise, combining data from a multibeam bathymetric sonar operating at 12 kHz, a deep-towed profiling sonar operating at 30 kHz, and stereophotography [Fox and Hayes 1985]. A more recent comparison of power spectra [Briggs et al. 2005a] calculated from stereophotographs and data generated from high-resolution multibeam bathymetry collected in the northeastern Gulf of Mexico (Fig. 6.15) suggests two power laws. One power law applies to wavelengths shorter than 10 cm, corresponding to features that may be biogenic in origin and having roughness that may be rather stable in time (Sect. 6.4). The other power law applies to wavelengths larger than 100 cm, corresponding to long-lived morphological features. The 10- to 100-cm wavelength portion of the spectrum corresponds to wave-generated ripple which can vary considerably with time (Sect. 6.4).

More data relevant to the power-law question is provided by the extended-track photogrammetric spectra presented in Fig. 6.16. These show some tendency to roll over for wavelengths longer than 10 cm. The power law does not seem to hold at the highest frequencies, owing to the change in slope mentioned in connection with Fig. 6.7, but this break is not seen in most spectra from other sites.

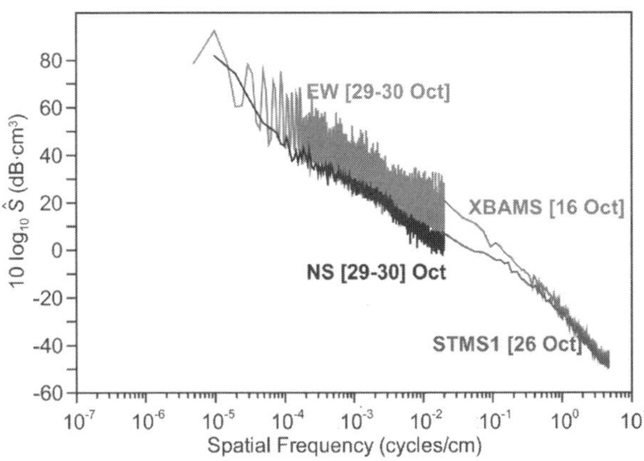

Fig. 6.15. 1D roughness power spectra, measured during SAX99, from multibeam bathymetry (EW, NS profile directions) and close-range photogrammetry from fresh and decayed ripples similar to those depicted in the spectra of Fig. 6.7. The ordinate is $10\log_{10}[S(F)]$. Modified from [Briggs et al. 2005a].

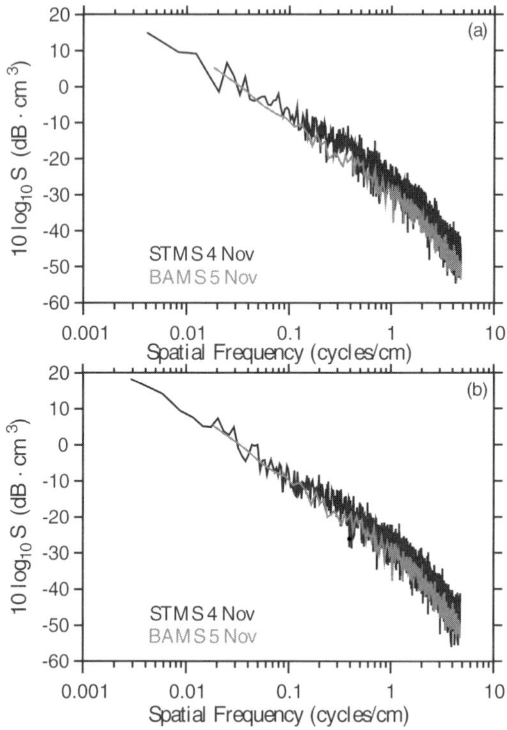

Fig. 6.16. Spectra determined from extended photogrammetric roughness profiles oriented (a) NE and (b) NW. Spectra are from the SAX99 BAMS and STMS sites. The ordinate is $10\log_{10}[S(F)]$ [Briggs et al. 2002b].

There is no a priori reason to assume self-affine behavior (Sect. 6.1) over such a wide range of spatial scales as in Fig. 6.15. In fact, the existence of this type of relationship may be counterintuitive, given the variety of scales over which biological and hydrodynamic processes alter seafloor morphology (tens of meters to millimeters). Each hydrodynamic process or animal activity operates over a limited spatial frequency band, and if these bands do not overlap greatly, there is no reason to expect scale-free, power-law behavior. Rippled seafloors provide one clear example in which a narrow-band process causes a departure from power-law behavior. On the other hand, it can be argued that multiple processes acting on multiple overlapping scales may give rise to power-law behavior. In the absence of ripple fields, it is reasonable to inquire as to the spatial frequency limits of a given power-law fit to the spectrum. Figure 6.16 suggests that extrapolation of the power law to meter scales corresponding to frequencies of a few kHz is reasonable, but more data are needed to verify this. Lacking such data, it is still possible to bound the lower spatial frequency limit of any proposed power law by computing the mean-square roughness as the integral of the spectrum down to the lower frequency limit. If the resulting RMS roughness (the square root of the mean-square roughness) is greater than the height of known or expected bathymetric features, it is clear that the power law has been overextended. Extrapolation to decameter and larger scales may be of geologic interest, but is not necessary for most high-frequency acoustic applications.

It has not been convincingly demonstrated that a single power-law regression can adequately describe seafloor roughness over all measured scales. Nonetheless, determination of roughness spectra over a wide spatial frequency band is well within the capability of present measurement techniques and deserves further attention.

6.6 Relationships between Roughness and Sediment Physical Properties

Although sediment type is often assumed to control, at least in part, the morphology of bedforms, evidence of relationships between the statistics of seafloor roughness relevant to high-frequency acoustics and sediment grain size has only recently been sought by [Pouliquen et al. 2004] and [Briggs et al. 2005a]. When seafloor roughness, as represented by RMS roughness, is plotted as a function of mean grain size, the result is a scattering of points across the range of sediment types (Fig. 6.17). In general, fine-grained sediments ($M_z < 5\phi$) have lower values of RMS roughness (< 0.75 cm), whereas coarser-grained sediments have varied values of RMS ranging from 0.2 to 2.3 cm. In fine-grained, more cohesive sediments, hydrodynamic forcing is weak, and roughness is generally thought to be the result of biological processes associated with benthic fauna feeding, locomotion, or habitat construction [Richardson and Young 1980] (also see Figs. 6.1 and 6.9 in this chapter). The result is an isotropic roughness of relatively low amplitude. In coarser-grained sediments (sand), where hydrodynamic processes such as waves and currents are stronger, sediment transport creates a variety of ripple types ranging from small-wavelength (10–50 cm) orbital ripples to larger meter-scale lunate ripples [Clifton 1976, Wiberg and Harris 1994, Hay and Mudge 2005]. The morphology of these bedforms depends on sediment grain size, water depth, and the strength of the hydrodynamic forcing at the seafloor. In many high-energy shallow-water areas, bedforms are in almost constant dynamic equilibrium with active sediment transport processes, and ripple morphology is predictable (e.g., LEO-15 off the New Jersey coast [Traykovski et al. 1999]). In other areas, where storm conditions are intermittent, bedforms may not be in equilibrium with hydrodynamic conditions (i.e., are relict) and are being degraded by biological activity [Briggs et al. 2002b]. Examples include the sites of the SAX99 and SAX04 high-frequency acoustic experiments. Bedform morphology is difficult to predict under these lower energy conditions. The overall result is a temporally varying RMS roughness and no predictive relationship between RMS roughness and sediment mean grain size. The effects of biological and hydrodynamic processes on seafloor roughness are treated in more detail in Ch. 3.

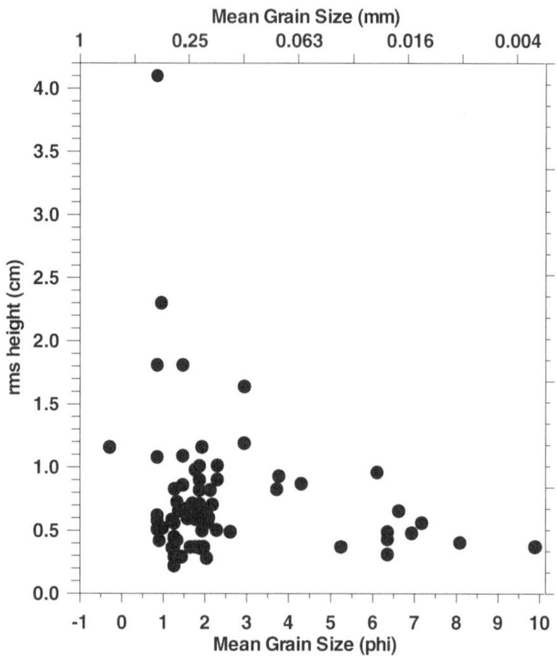

Fig. 6.17. Plot of RMS roughness as a function of mean grain size based on profiles of seafloor roughness derived from analog stereophotographs of the seafloor. The data are from Table 6.1, and the figure is modified from [Briggs et al. 2005a].

The 2D spectral characteristics (spectral strength, w_2, and spectral exponent, γ_2) are plotted against mean grain size in Fig. 6.18. The 2D spectral parameters were obtained from the 1D values in Table 6.1 under the assumption of isotropy. The relevant power law representations for 1D and 2D spectra were given earlier in this section (6.8) and (6.11) and relationships used to convert the 1D spectral slope and intercept to 2D spectral strength and spectral exponent are given in Sect. D.1.2. As with the comparison with RMS roughness (Fig. 6.17), no empirical relationships are evident between roughness spectral exponent and grain size or between roughness spectral strength and grain size. However, the measurements of spectral exponent and strength appear to cluster according to sediment type, with separate trends for sand and mud. The dashed lines in Fig. 6.18 are not regression fits but an attempted delineation of trends for the different sediment types. Measurements from sites that include roughness features such as storm ripples and biogenic mounds and pits are distributed throughout the data and do not align with the delineated trends. It is obvious that more data are needed if us-

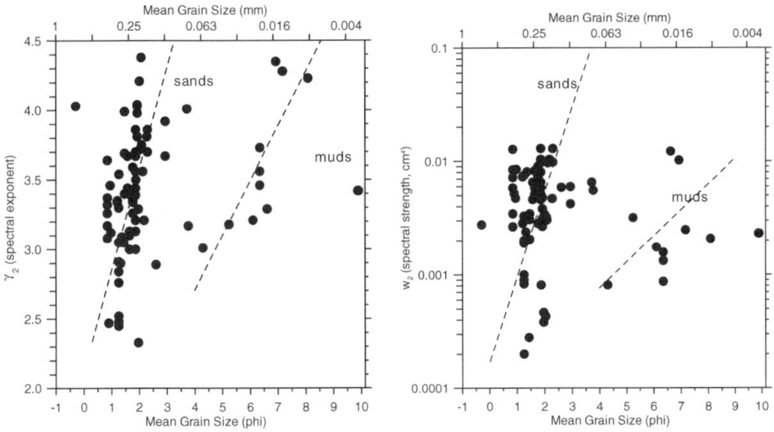

Fig. 6.18. Plots of 2D seafloor roughness spectral exponent, γ_2, and spectral strength, w_2, as functions of mean grain size. The spectral parameters are calculated from the 1D data presented in Table 6.1, and the figure is modified from [Briggs et al. 2005a].

able predictive relations are to be established. Sediment mean grain size alone may never yield the types of predictive relationships for roughness required by high-frequency acoustic modelers unless the effects and rates of hydrodynamic and biological processes that create, modify and destroy roughness features are incorporated into predictive relationships. A similar lack of predictability from grain size for 1D spectral slopes and intercepts has been reported by [Pouliquen et al. 2003].

6.7 Relationship between Spectral Parameters

One-dimensional spectral slope, $-\gamma_1$, as a function of the spectral intercept, ϕ_1, is plotted in Fig. 6.19 from a compilation of roughness data using stereophotogrammetric measurements made on continental shelf seafloors since 1983 [Briggs et al. 2001] (also see Table 6.1). The relationships inferred from this display may indicate important patterns that might help model roughness scenarios and, ultimately, predict backscattering from sediment type. Spectral slope is apparently related to intercept, with a positively sloped trend. The dashed line in Fig. 6.19 gives the relation between spectral intercept and slope for the family of 1D power-law spectra that all have the same value, $S(F_0) = 7.1 \times 10^{-5}$ cm^3, at spatial frequency $F_0 = 0.05$ cycle/cm. This frequency corresponds to a ripple wavelength of 20 cm, a relatively long wavelength in comparison to those of interest in high-frequency scattering. Thus the dashed line is the intercept–slope relationship expected if the low-frequency portion of the spectrum is held fixed and the high-frequency

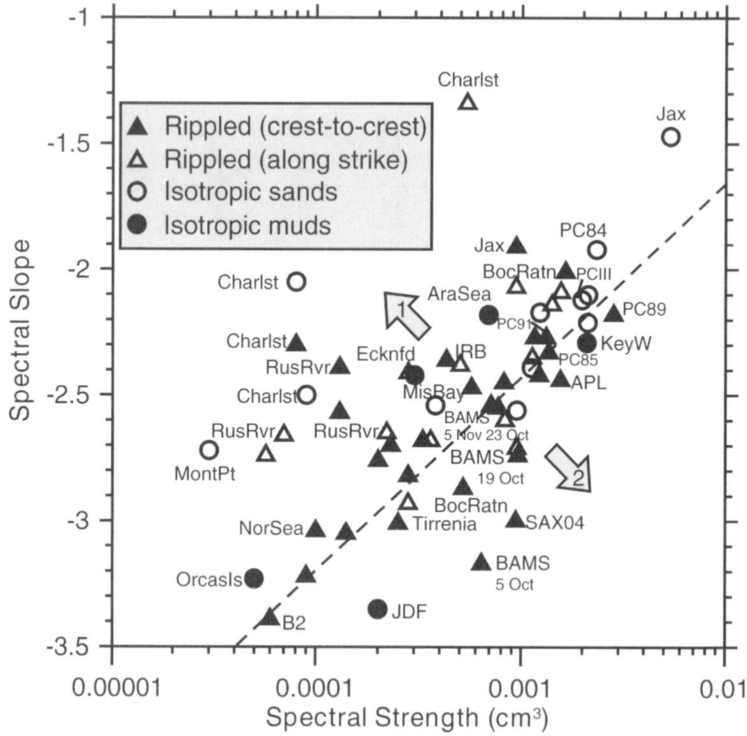

Fig. 6.19. Relationships between 1D spectral intercept and slope determined from analog stereophotographs. The data are from Table 6.1, and the figure is modified from [Briggs et al. 2005a].

portion is varied, while maintaining the power-law behavior. Values occurring in the direction of arrow "1" (to the left of the dashed line) represent roughness spectra deviating from the trend by significant decreases in low-frequency roughness relative to changes in high-frequency roughness. Values occurring in the direction of arrow "2" (to the right of the dashed line) represent roughness spectra deviating from the trend by significant increases in low-frequency roughness relative to changes in high-frequency roughness. Examples of an environment represented in the former scenario are muddy sites with little biogenic roughness (Eckernförde Bay, Arafura Sea, Orcas Island) or sands with fully decayed or low relief ripples (Charleston, Montauk Point). It is not possible at present to deduce predictive relationships for seafloor microtopography from the trends seen in Fig. 6.19, because widely different seafloor types are not segregated in the plot (e.g., isotropic and anisotropic roughness data somtimes are found in the same region). The natural decay of ripple morphology at the SAX99 (BAMS) site, which was discussed in Sect. 6.4 and is seen in the spectra of Fig. 6.7, may be tracked from a point far below and to the right of the dashed line (fresh ripples) to a point closer to the line but still to the right (decayed ripples). Longer-term decay (months) and an eventual equilibrium between biogenic and hydrodynamic processes may be responsible for the spectral parameters deviating to the left of the trend line. The data plotted in Fig. 6.19, with the notable exceptions of the SAX99 BAMS, Russian River, and PC89-91-III measurements, are only an instant temporal sample of a dynamic process. It is possible that long-term measurements showing the evolution of roughness spectra at various sites would reveal regularities not evident in the "snapshots" presently available.

6.8 Summary

A statistical description of seafloor microroughness is required for prediction of high-frequency scattering from or into the seafloor. For this purpose, digital stereophotographs provide the best description of the seafloor and allow for automated calculation of the high-resolution 2D relief spectra that are required by many seafloor scattering models. If the seafloor roughness is isotropic, 2D relief spectra can be calculated from averaged 1D relief spectra. These 1D relief spectra can be calculated from analog stereophotographs using a stereo comparator or when water clarity is poor from relief spectra determined from manual tracing, electrical resistivity probes, or very high-frequency (MHz) acoustic profiles. The statistics of 1D and 2D relief spectra derived from stereophotographs, together with sediment sound speed, attenuation, and bulk density, have proven adequate for predicting backscattering strength from sandy seafloors (e.g., [D. Jackson et al. 1996a, Williams et al. 2002b]).

Sediment microtopography is created and altered by a combination of geological, hydrodynamic, biological, and depositional processes. Hydrodynamic processes, waves and currents, tend to create anisotropic relief spectra with peaks in the direction and at the spacing of sand ripples. Biological processes tend to create isotropic relief spectra which have lower average relief amplitudes. Larger meter-scale bottom features tend to persist longer, sometimes up to weeks or months, than smaller centimeter-scale features which can be destroyed within hours. In environments dominated by hydrodynamic processes, microroughness is spatially and temporally variable and relief spectra are highly variable; whereas in environments where biological processes dominate, roughness features rapidly change (exhibit temporal decorrelation), but the relief spectra are relatively constant. Prediction of relief spectra from sediment properties, such as sediment type or mean grain size, is unreliable and direct measurement of relief spectra during high-frequency acoustic experiments is highly recommended. Preliminary data suggest that it may not be possible to approximate roughness spectra by a single power law over a wide range of scales.

6.9 Research Issues

Several issues remain relative to both the statistical approaches and measurement techniques used to characterize seafloor microroughness. Historically, the statistical requirements of roughness scattering models (Ch. 13) and the scientific objectives, methods, and statistical techniques used to study seafloor morphology in the environmental sciences often have little overlap. While most of the data in this chapter have been obtained with the acoustic problem in mind, there are still issues as to the most appropriate methods for statistical characterization of seafloor roughness. The widely used assumption that roughness is a stationary Gaussian random process has not been sufficiently tested. If this assumption is not valid, the averaging used to obtain spectra will obscure potentially important roughness features, and other statistical methods may be needed. It is not known whether a single power law is adequate to characterize seafloor roughness over spatial scales from kilometers to millimeters. If not, over what spatial frequency band will a simple power law adequately predict roughness (single values of spectral strength and spectral exponent)? The spatial and temporal stationarity of seafloor roughness is assumed but not often validated during high-frequency acoustic experiments. The temporal component of roughness varies with size of features, bottom type, water depth, and dominant roughness-creating processes and requires additional study if seafloor roughness is to be predicted. For now, concurrent (in space and time) measurements of seafloor roughness and acoustic properties are required to test high-frequency scattering models. For practical applications, there is a need for rapid remote methods for determining seafloor roughness statistics. The typical size of the patches over which the seafloor can be regarded as statistically stationary is not known. Means of statistically characterizing the patchiness of the seafloor should be developed. All the unknowns are directly related to the acoustic issue of how roughness statistics affect the sonar echo signal (see Ch. 16).

Table 6.1. Seafloor roughness measured from a variety of sediment types in shallow coastal sites by Kevin Briggs (NRL). Sediments are arranged in order of decreasing mean grain size (increasing phi units) from coarse sand to clay. Roughness is expressed as the RMS height (cm), slope, and intercept of the roughness power spectrum. Anisotropic roughness measurements have been measured crest-to-crest (c-c) and along-strike (a-s). Measurements are accomplished with diver-operated 35-mm stereo cameras, remote 70-mm stereo cameras, or a manual trace on Mylar by divers. The units of the spectral slope, $-\gamma_1$, and spectral intercept, ϕ_1, are discussed in Sect. D.1.3.

Site	M_z (ϕ)	Type	RMS (cm)	Meas. Tech.	Slope	Intercept	Reference
Panama City 1985 (c-c)	0.85	c. sand/sh. hash	1.81	35 mm	-2.32	0.00136	Unpublished data
Panama City 1985 (a-s)	"	"	0.51	"	-2.37	0.000502	"
Panama City 1989 (c-c)	"	"	4.10	"	-2.17	0.00282	"
Panama City 1989 (a-s)	"	"	0.58	"	-2.64	0.00022	"
Panama City 1991 (c-c)	"	"	1.08	"	-2.26	0.00132	[Briggs and Ray 1997]
Panama City 1991 (a-s)	"	"	0.62	"	-2.08	0.00156	"
Kings Bay/Lynch site	0.90	hash	0.42	"	-1.47	0.00534	[Stanic et al. 1989]
Mission Bay/coarse	0.95	coarse sand	2.30	Trace	-2.46	0.00570	[Richardson et al. 1983]
Panama City 1993	0.98	coarse sand	0.52	35 mm	-2.12	0.00198	[D. Jackson et al. 1996a]
SAX99 (c-c)	1.27	rip.med sand	0.83	"	-2.54	0.00077	[Briggs et al. 2001]
Kings Bay/Bart. site (c-c)	1.33	rip.med sand	0.42	"	-1.90	0.00094	[Briggs 1989]
Kings Bay/Bart. site (a-s)	"	"	0.60	"	-2.06	0.00094	"
Charleston/coarse	1.44	med sand	0.26	"	-2.05	0.00008	[Briggs et al. 1986]
SAX04 (c-c)	1.47	rip.med sand	1.81	"	-2.99	0.00014	Unpublished data
SAX04 (c-c)	"	"	1.09	Trace	-2.69	0.00023	"
SAX04 (a-s)	"	"	0.86	"	-2.40	0.00028	"
Panama City 1998	1.65	med sand	0.65	35 mm	-2.10	0.00212	"
Indian Rocks 6 (c-c)	1.77	"	0.59	"	-2.35	0.00043	[Stephens et al. 1997]
Indian Rocks 6 (a-s)	"	"	0.37	"	-2.34	0.00113	"
Indian Rocks 7 (c-c)	"	"	0.98	"	-2.26	0.00116	"
Indian Rocks 7 (a-s)	"	"	0.62	"	-2.59	0.00083	"
Boca Raton NS01	1.87	med sand	0.71	"	-2.21	0.00211	"
Boca Raton NS02	"	"	0.71	"	-2.39	0.00112	"
Boca Raton NS03 (c-c)	"	"	1.01	"	-2.86	0.00052	"
Boca Raton NS03 (a-s)	"	"	0.90	"	-2.70	0.00095	"
Boca Raton NS05 (c-c)	"	"	0.60	"	-2.44	0.00082	"
Boca Raton NS05 (a-s)	"	"	0.68	"	-2.67	0.00036	"
Boca Raton NS06 (c-c)	"	"	0.37	"	-2.00	0.00164	"
Boca Raton NS06 (a-s)	"	"	0.63	"	-2.13	0.00141	"
Boca Raton NS07	"	"	0.82	"	-2.56	0.00095	"
Charleston/fine	1.88	"	0.36	"	-2.50	0.00009	[Briggs et al. 1986]
No. Sea C1	1.93	"	0.67	70 mm	-3.04	0.00014	Unpublished data
No. Sea B2	"	"	0.58	"	-2.98	0.00013	"
No. Sea TOSSEX6	"	"	1.16	"	-3.03	0.00010	"
No. Sea TOSSEX7	"	"	0.50	"	-2.81	0.00028	"
Charleston/fine (c-c)	1.97	rip. med sand	0.37	35 mm	-2.29	0.00008	[Briggs et al. 1986]

Table 6.1. Continued

Site	M_z (ϕ)	Type	RMS (cm)	Meas. Tech.	Slope	Intercept	Reference
Charleston/fine (a-s)	"	"	0.36	"	-1.33	0.00054	"
Montauk Point	2.04	fine sand	0.28	"	-2.72	0.00003	[Richardson et al. 1983]
Panama City 1984	2.61	"	0.49	"	-1.89	0.00233	[Stanic et al. 1988]
Quinault Range (c-c)	2.94	rip. fine sand	1.65	70 mm	-2.67	0.00033	[D. Jackson and Briggs 1992]
Quinault Range (a-s)	"	"	1.19	"	-2.92	0.00028	"
Tirrenia Italy	3.72	v. fine sand	0.83	35 mm	-3.01	0.000255	[Briggs et al. 2002c]
MissionBay/fine	3.77	fine sand	0.93	Trace	-2.17	0.00123	[Richardson et al. 1983]
Arafura Sea	5.24	clayey sand	0.37	70 mm	-2.18	0.00069	[D. Jackson and Briggs 1992]
Russian River 1988	6.35	"	0.31	"	-2.46	0.00019	"
Russian River 1989 (c-c)	"	rip. clayey sand	0.43	"	-2.56	0.00013	"
Russian River 1989 (a-s)	"	"	0.49	"	-2.73	0.00006	"
Lower FL Keys	6.62	carb. s-s-clay	0.65	35 mm	-2.29	0.00209	[D. Jackson et al. 1996a]
Juan de Fuca site 4	6.93	glacial till	0.48	70 mm	-3.35	0.00020	[Briggs 1989]
Eel River	7.17	clayey-silt	0.21	"	-3.28	0.000056	[Richardson et al. 2002b]
Orcas Island	8.08	clayey sand	0.40	35 mm	-3.23	0.000052	[Self et al. 2001]
Eckernförde Bay	9.88	silty clay	0.37	"	-2.42	0.00030	[D. Jackson et al. 1996a]

7 Sediment Heterogeneity

A variety of physical, biological, and geochemical processes create heterogeneity (or inhomogeneity) in sediment physical properties, that is, values of sediment physical properties tend to be nonuniform with respect to depth and horizontal position (for example, see Fig. 7.1). This heterogeneity can manifest itself as a patchy distribution of sediment physical properties, such as worm burrow or feeding voids, layered bedding created by alternating erosion and deposition processes, or as individual scatterers such as large shells. High-frequency acoustic volume scattering (Ch. 14) is caused by this heterogeneity and is dependent on its magnitude and spatial scale. The effects of physical, biological, and biogeochemical processes on the internal 3D structure of sediments and the spatial distribution of values of sediment physical properties were reviewed in Ch. 3.

From the viewpoint of modeling acoustic volume scattering, it is convenient to divide heterogeneity into nonrandom and random parts. The nonrandom part of a given sediment physical parameter is the average of that parameter over the patch of seafloor of interest, and the random part is the fluctuation about this average. For these definitions to be useful, there must be a suitably large region of seafloor over which the random heterogeneity is statistically the same, that is, spatially stationary. The average can be taken (at least in principal) over horizontal coordinates with the result that the nonrandom part depends only on the vertical (z) coordinate. Thus, the nonrandom part of a given parameter is described by a profile or gradient while the random part is described statistically, usually in terms of a variance, covariance, or power spectrum.

The sediment properties of interest in this chapter are those that appear in models for scattering. Some volume scattering models require spectra describing the spatial fluctuations of sediment bulk density and compressibility (alternately, fluctuations of porosity and compressional wave speed). Other volume scattering models require statistical descriptions of discrete acoustic scatterers such as shells and rocks. The primary need is for statistics at scales comparable to the acoustic wavelength, but scales larger than this are important to the question of patchiness and resulting non-Rayleigh backscatter statistics (Ch. 16). For many applications, less ambitious measures of spatial variation such as mean profiles and coefficients of variation (100% $\times$ stan-

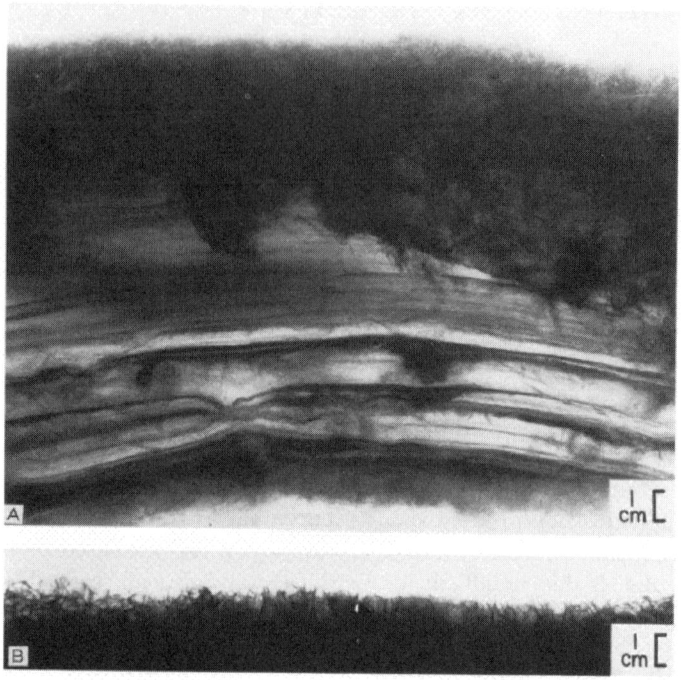

Fig. 7.1. (A) An X-radiograph of a 2-cm-thick slab of sediment collected from a muddy site in Long Island Sound [Richardson et al. 1983]. The surface top layer (3–5 cm) is well-mixed by reworking of the sediment by abundant benthic macrofauna. The deeper layers are alternate layers of coarser more dense (darker) sediments and finer less dense (lighter) sediments deposited during storms. With time, bioturbation can mix sediments to depths of at least 20 cm. (B) An X-radiograph of the upper 3 cm of the same sediment core showing a dense population of the bivalve *Mulinea lateralis* (exposure time reduced). The gray-scale density in the X-radiograph can be considered a proxy for bulk density.

dard deviation/mean) are sufficient. These properties include mean grain size, shear wave speed, compressional and shear wave attenuation, porometric parameters such as permeability, pore size and tortuosity, and pore water and grain properties such as density and bulk modulus (these properties are discussed in detail in Chs. 4, 5, and 10).

Biological (bioturbation) and physical (waves and currents; deposition and consolidation) processes create, destroy, and alter sediment structure at all scales including: large kilometer-scale facies, meter-scale sand ripples, centimeter-scale mounds and pits, millimeter-scale burrows or feeding traces, and micron-scale particle flocs and aggregates [Richardson and Young 1980]. The processes can be continuous in time (biological mixing) or episodic (storms, river flooding) as well as competing (operating at the same spatial and temporal scales).

One of the best-documented examples of the evolution of spatial heterogeneity in near-surface sediments can be found in the published results of ONR's STRATAFORM program [Nittrouer and Kravitz 1996, Nittrouer 1999, Cutter and Diaz 2000, Richardson et al. 2002b, Bentley and Nittrouer 2003]. The effects of interacting processes such as (a) biological mixing, (b) episodic sediment deposition as a result of river flooding, and (c) sediment resuspension, mixing, and transport by waves and currents on the temporal evolution and ultimate preservation of near-surface sediment structure were well documented. Episodic floods off the northern California coast produced 5- to 10-cm-thick deposits of fine-grained sediments at middepths (50–70 m) on the continental shelf. These deposits were characterized by basal layers of silt-sized sediment, grading toward the interface into clay-size particles. Within 6 months the upper 3–5 cm of these deposits were fully mixed by benthic fauna, resulting in a mottled distribution of sediment physical properties. Resuspension and deposition of sediments by waves and currents (storms) further mixed these near-surface sediments. Below ~5 cm, deeper-dwelling deposit feeders mixed sediments to depths of ~20 cm over time scales of 6–350 years. These interacting processes created a tiered distribution of sediment properties which is quite heterogeneous on both horizontal and vertical scales. This spatial heterogeneity is evident in large-scale acoustic surveys of the study site (Fig. 7.2), and in X-radiographs (Fig. 7.3) and CT scans (Fig. 7.4) from sediment collected with box corers. Many other examples of heterogeneity of modern near-surface sediments, as well as the structure preserved in the historical record, are well documented, but the time scales over which heterogeneity and gradients evolve have only recently received the attention it deserves. With regard to acoustic scattering, very slow decorrelation in backscattering over time scales on the order of one month have been reported for sites in which scattering by volume heterogeneity was dominant [Dworski and Jackson 1994, Jones and Jackson 1997]. In contrast, decorrelation times shorter than one day have been reported for sites in which roughness scattering was dominant [D. Jackson et al. 1996b, Richardson et al. 2001b].

The impacts of bioturbation and hydrodynamic processes tend to decrease with depth in the sediments. Below a meter or so physical features such as storm layers, graded bedding, and biogenic features such as tubes, burrows, and feeding voids tend to remain relatively unchanged, even in the event of major storms such as hurricanes. The area of interest for high-frequency acoustics is roughly the upper 30 cm of sandy sediment and the upper 100 cm in muddy sediments. These near-surface sediments are strongly affected by biological and hydrodynamic processes and therefore exhibit high temporal variability. Consolidation or dewatering is common in finer-grained sediment resulting in positive gradients in density and sound speed (increasing with

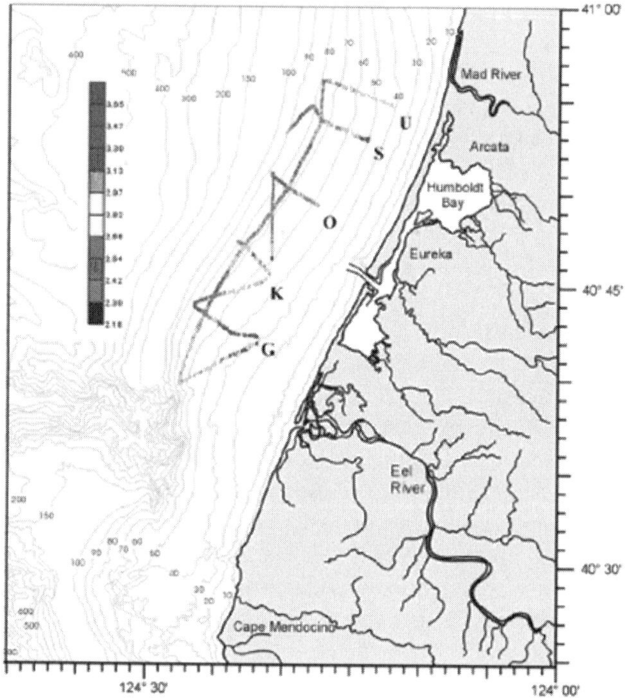

Fig. 7.2. Sediment acoustic impedance measured along tracks surveyed with the Acoustic Sediment Classification System at 15 kHz on the Eel River shelf, northern California. The red to yellow colors correspond to sand and the green to purple correspond to muddy sediments [Richardson et al. 2002b]. (see second color insert)

depth) and negative gradients in porosity (Fig. 7.5). In most sands, sound speed also increases with depth in response to increased packing (Fig. 7.6). As shown in Ch. 5, values of shear wave speed in both sand and mud sediments increase with depth as a result of increased overburden pressure (effective stress, Sect. 5.2). Given the variety of processes and scales operating within the benthic boundary layer there is no reason to assume a single statistic (e.g., mean, variance, or correlation length) would characterize the resultant structure over all scales. Even though the range of scales is broad, most scattering models only require statistical characterization of sediment physical properties over a limited range of spatial scales (acoustic wavelength to patch size of the acoustic footprint).

In this chapter, the techniques used to quantify heterogeneity of sediment properties will be reviewed: traditional laboratory analysis of sediment cores, X-radiography including CT (computed tomography), and in situ electrical, optical, and acoustic methods. The methods used to characterize spatial fluctuations in geoacoustic parameters in terms of correlation functions and

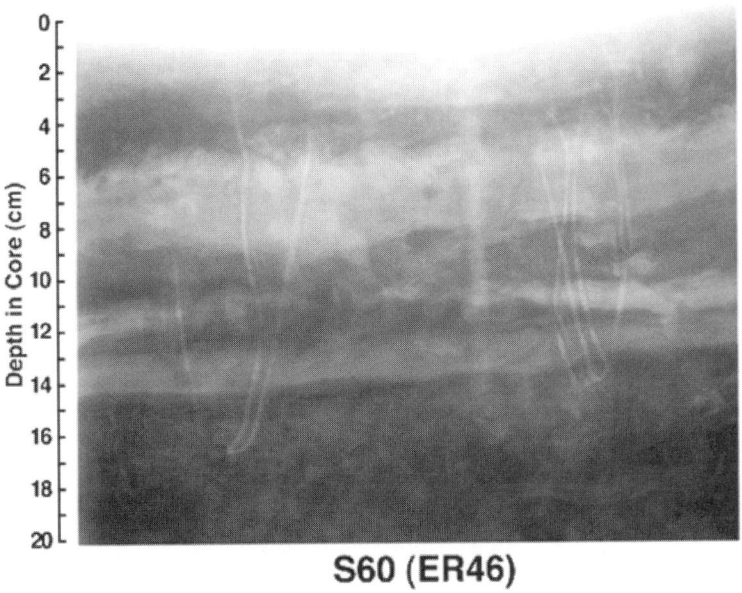

Fig. 7.3. X-radiograph positive image of slab subcore collected from box core 46 from site S60 along the Eel River shelf, northern California coast, showing 13 vertical burrow conduits (some are connected, continuous burrows). Burrows are lighter vertically oriented features. Remnants of buried ripples are evident at 8- to 14-cm sediment depth [Richardson et al. 2002b].

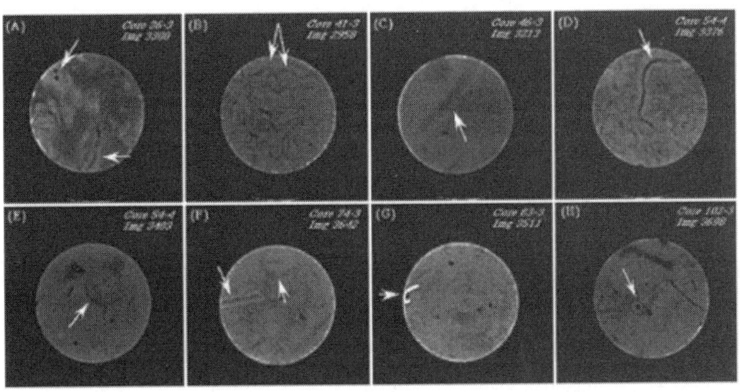

Fig. 7.4. Horizontal CT images with arrows pointing out bioturbation features found in sediments on the Eel River shelf, northern California. The gray scale indicates density, with higher-density regions being in lighter shades. Features include vertical worm burrows, A, H; horizontal worm burrows, B, C, D, F; a shell fragment G; and a burrow created by a larger animal, E [Richardson et al. 2002b].

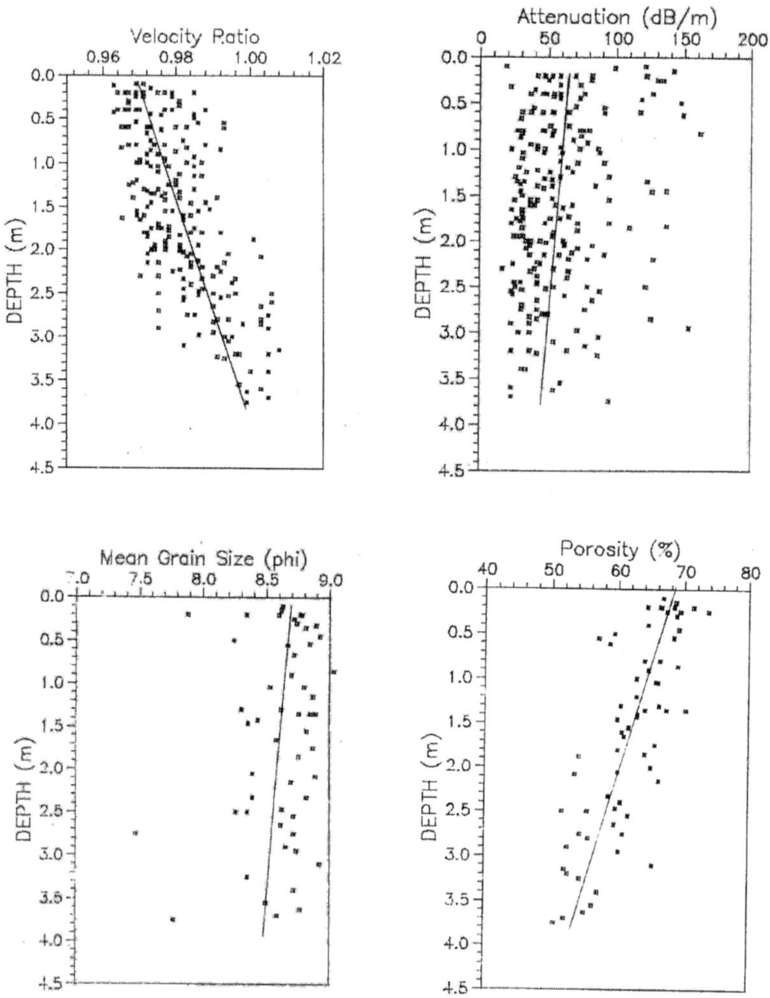

Fig. 7.5. Mean gradients in porosity, mean grain size, sound speed (velocity ratio), and attenuation for fine-grain recent (Holocene) sediment in the Adriatic [Richardson et al. 1992]. Sound speed and attenuation were measured at 400 kHz using the system described in Sect. 5.1.1.

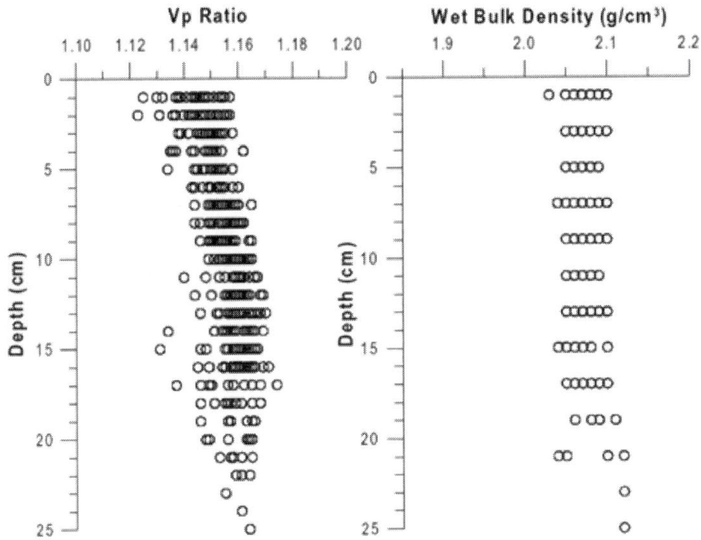

Fig. 7.6. Measurements of compressional wave speed ratio (400 kHz) and bulk density from several cores in the SAX99 experiment [Tang et al. 2002]. The precision of the bulk density measurements is such that numerous data points have the same value and therefore overlap in the plot. Note that the compressional wave speed data have a 1-cm vertical sampling interval, whereas the density data have a 2-cm vertical interval.

spectra will then be presented. This discussion includes the problem of obtaining spectra for unmeasured parameters from those that are measured (e.g., compressibility from sound speed and density from porosity). The "bias" in spectral estimates associated with the sample size (resolution) of physical property measurements is also considered. This chapter includes discussion of several recent attempts at statistical characterization of seafloor physical properties, and these examples are followed by a "Research Issues" section that summarizes areas requiring further study.

7.1 Measurement of Heterogeneity

Of the several sediment physical properties relevant to acoustics, only a few have been subjected to measurements of heterogeneity in a manner meeting the needs of high-frequency acoustic volume scattering models: bulk density, porosity, and compressional wave speed. The traditional analysis of sediment cores described in Ch. 5 provides a low-resolution means of measuring heterogeneity. These techniques include sound speed measurements and weight loss measurements on slices of extruded sediments for porosity or density [Briggs 1994]. More recently, automated core logging techniques have been employed [Schultheiss and Weaver 1992]. First- and second-moment averages provide depth-dependent profiles and covariances [Briggs 1994, D. Jackson et al. 1996a, Richardson and Briggs 1996, Briggs and Percival 1997]. Such core measurements have poor spatial resolution relative to other methods due to the finite size of core sections (typically 4–12 cm in diameter and 1–2 cm in thickness or depth). One advantage of this method is that both compressional wave speed and density can be measured on the same sample (Fig. 7.6), allowing determination of the correlation between these two parameters, at least in principle.

X-ray techniques have been used for over 40 years to provide nondestructive means of characterizing structure of near-surface marine sediments (for early examples see [Calvert and Veevers 1962, Bouma 1964, Howard and Frey 1973]. Recently these techniques have been adapted to high-frequency acoustic applications to provide a quantitative characterization of the fine-scale distribution of sediment bulk density [D. Jackson et al. 1996a, Holyer et al. 1996, Briggs et al. 1998, Lyons and Pouliquen 2004]. Two X-ray techniques are commonly used to study heterogeneity: X-radiography and computed tomography (CT). Standard X-radiography provides images of heterogeneity which can be used to characterize sediment stratigraphy or heterogeneity (e.g., [Richardson and Young 1980, D. Jackson et al. 1996a, Wheatcroft 2002, Richardson et al. 2002b]). X-radiography of sediments collected using gravity or piston cores is a standard method in the analysis of sediment stratigraphy in marine sediments (Fig. 7.3). These X-radiographs often show sediment layering and biological structure not visible in split cores and combined with core logging (density and sound speed described in Chs. 4 and 5) yield useful information to characterize past and recent sedimentary events and provide a guide for more detailed sampling of sediment layering. More recently, slabs or thin slices of sediment have been collected by divers or from box cores which provide a larger sample to create X-ray images without the distortion created by the variable thickness of cylindrical cores [Richardson et al. 1983, Briggs et al. 1985, Holyer et al. 1996, Richardson et al. 2002a]. A typical size of these sample slabs is 35 $\times$ 40 cm by 2–3 cm thick [Holyer et al. 1996]. Until recently X-radiographs from both sediment cores and slabs only allowed qualitative descriptions of sediment structure or heterogeneity. New digital X-ray imaging systems combined with modern image processing techniques provide quantitative analysis of the gray-scale images created by X-radiography [Migeon et al. 1999, Lofi and Weber 2001]. Analysis techniques such as autocorrelation and binary run length have provided means to characterize periodic structures and provide quantitative analysis of anisotropy from sediment cores or slabs [Holyer et al. 1996, Briggs et al. 1998]. These new digital techniques have eliminated difficulties associated with the subjective visual analysis of X-radiographs and other drawbacks such as the quality of film, limited range of contrast, variations within and between images, and the image blurring effects due to out-of-plane scattering within the imaged sediment [Migeon et al. 1999, Lyons and Pouliquen 2004]. However, these 2D images still suffer from the visual ambiguities created by superimposing of density features on the radiograph and a lack of constant thickness for cylindrical cores. To solve some of these problems, investigators have begun using CT techniques to characterize 3D fine-scale heterogeneity in sediment cores [Orsi et al. 1994, Lyons and Pouliquen 2004].

Imaging of sediment cores with CT provides very-high-resolution characterization of the density distribution within sediments (Figs. 7.7 and 7.4). The 3D imaging process eliminates much of the ambiguity associated with superposition of density contrast typical of standard X-ray images and provides high quality digital images that are needed for modern image processing software. The resolution of medical CT images is approximately 0.5–1.0 mm (voxel size), and resolutions of about 10 μm are possible with high-resolution systems [Reed et al. 2005]. This approach could potentially provide the high-resolution data required for the statistical characterization needed by high-frequency acoustic models. As with standard X-ray imaging, a high correlation between volumetric electron density and the bulk density of sediment is assumed. A linear relationship has been demonstrated ($r^2 > 0.98$) between the electron density determined from X-ray attenuation and bulk density measured using gravimetric methods [Orsi et al. 1994, Orsi and Anderson 1999]. This makes it possible to use CT images to determine the spatial statistics of bulk density.

There has been a significant effort to characterize the heterogeneity of sediment density using medical CT-scanning techniques [Orsi et al. 1994, Orsi et al. 1996, Orsi et al. 1997, Orsi and Anderson 1999, Tang and Orsi 2000a, Briggs et al. 2001, Pouliquen and Lyons 2002, Briggs et al. 2002a, Lyons and Pouliquen 2004]. Of these studies, only those published after 1999 attempt to provide the random and nonrandom statistics required by the volume scattering models described in Ch. 14. In addition to the statistical characterization of density fluctuations, CT provides a useful nondestructive method to characterize larger discrete scatterers, such as individual shells or rocks, layers of shell hash or other forms of graded bedding, storm layers, gradients in density due to dewatering, gas bubbles, and the variety of biological structures typical of marine sediments. Two examples are presented which demonstrate the usefulness of CT scanning. The first, from a sandy sediment collected off Panama City, Florida, demonstrates the ability to statistically characterize discrete scatterers such as shell hash (Fig. 7.8); the second demonstrates the ability to characterize the size distribution of methane bubbles in the gassy sediments of Eckernförde Bay, Baltic Sea (Fig.

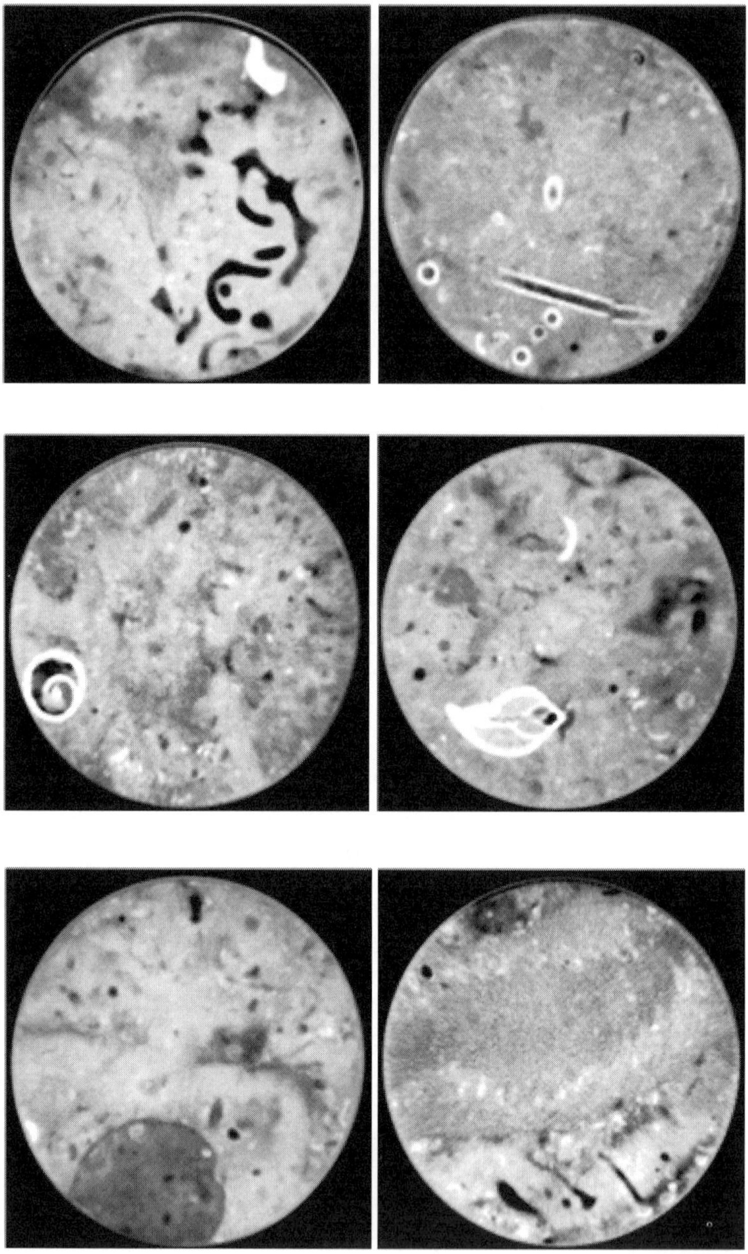

Fig. 7.7. Biological structures as seen in CT images, CBBL Tortugas site. Top row (burrows/tubes): left (low-density structures, 87-mm depth); right (high-density features, 235-mm depth). Middle row (shells/fragments): left (shell, 147-mm depth); right (shell and fragment, 191-mm depth). Bottom row (miscellaneous): left (urchin?, 165-mm depth); right (feeding structure, 149-mm depth). The outer diameter of the core image is 8.3 cm [Briggs et al. 2002a].

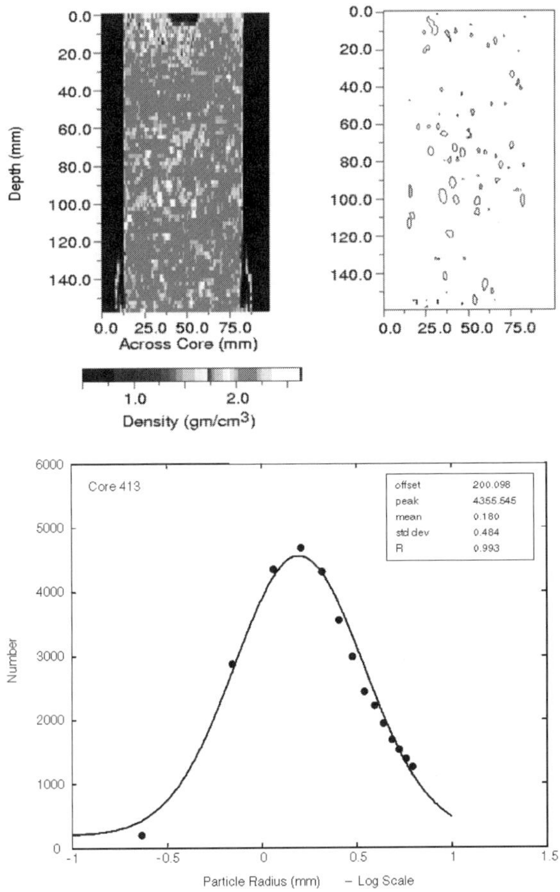

Fig. 7.8. Characterization of discrete scatterers from a shelly sand sediment collected off the coast of Panama City, Florida. Top left is a 2D density image derived from CT data; top right is a contour map of the shell hash which is more dense, white material in the image to the left; bottom is a plot of shell size distribution derived from 12 vertical images (slices) of the CT data from two cores. The particle radius is given in terms of the base-10 logarithm. From [Lyons 2005]. (see second color insert)

7.9). Some of the possible drawbacks associated with CT imaging of cores include sediment disturbance during collection, transport, and manipulation; the limited data set derived from CT imagery, primarily density contrasts; and the relatively small sample size. In spite of these shortcomings, CT images provide the best current method to statistically characterize density heterogeneity for modeling of high-frequency acoustic volume scattering.

Another technique for imaging sediments is provided by the sediment-interface profile camera. The interface profile camera was originally designed

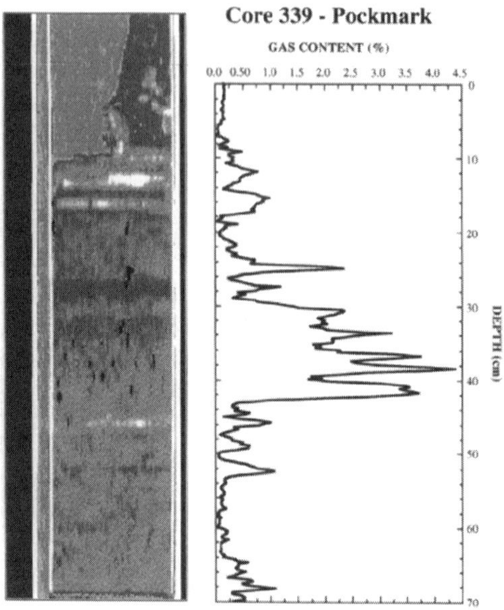

Fig. 7.9. Methane bubble distribution from a pockmark in Eckernförde Bay, Baltic Sea. Left is a 2D slice of sediment retained at in situ pressure and temperature, white is the core liner, blue is water, brown is a silty-clay sediment, and black is methane bubble; right is the methane gas bubble volume based on analysis of the entire core [Lyons et al. 1996]. (see second color insert)

as a diver-operated underwater camera attached to a freshwater-filled Plexiglas prism inserted into the sediment [Rhoads and Cande 1971]. Later, a remotely operated version, "REMOTS," was used to optically image an undisturbed vertical cross section of the upper 15–20 cm of the sediment (Fig. 7.10). This allows characterization of sediment structure, biochemical processes, sediment reworking, and community succession in a variety of estuarine, coastal, and deep-sea environments [Rhoads and Germano 1982]. One of the most obvious biochemical features is the "redox" discontinuity, the interface between oxidized and reduced sediments, marked by a change in sediment color. The profile camera has been used in both a survey mode [Cutter and Diaz 2000] and for long-term studies [Diaz and Cutter 2001]. Recent addition of digital cameras allows quantification of grain size distribution, subsurface methane bubbles, biogenic structures, etc. The potential for REMOTS to provide at least qualitative characterization of sediment hetero-

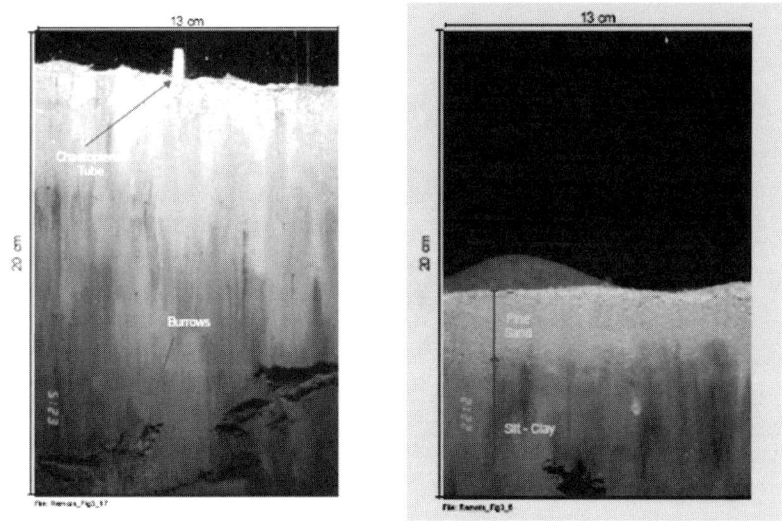

Fig. 7.10. Two images made with the REMOTS sediment interface camera in Buzzards Bay, MA. The image on the left shows a worm tube at the surface and numerous feeding voids and burrows at depth. The image at the right shows fine sand covering silty-clay sediment with a prominent feeding void at about 8 -cm depth. In both cases the structure and roughness of the sediment–water interface are intact. Courtesy of J. Germano. (see second color insert)

geneity is demonstrated in images of changes in sediment structure along a transect on the northern California coast [Cutter and Diaz 2000] during the STRATAFORM experiments. The quantification of larger discrete objects such as shells and feeding voids in these images is obvious.

Electrical resistivity probes and imaging systems have also been used to quantify the 3D distribution of sediment bulk density [P. Jackson et al. 1996, P. Jackson et al. 2002, Briggs et al. 2002a, Tang et al. 2002, Wheatcroft 2002, Tang 2004]. All of these systems exploit the relationship between sediment electrical resistivity and porosity given by Archie's law (see Sects. 4.3 and 4.4.2) to determine the spatial distribution of porosity or sediment bulk density. Fluctuations in density are calculated from fractional porosity, β, using the density–porosity relation given in Table 4.5, repeated here for convenience:

$$\rho = \beta \rho_w + (1 - \beta)\rho_g \ . \tag{7.1}$$

Density fluctuations can be inferred from porosity fluctuations assuming that the grain density, ρ_g, and water density, ρ_w, are known and constant over the measurement volume. The mean value of porosity is usually determined from a calibration, with porosity determined gravimetrically. A standard 4-probe diver-deployed Wenner array was used by [Wheatcroft 2002] to measure gradients of electrical resistivity at a sand site during SAX04 and at a carbonate

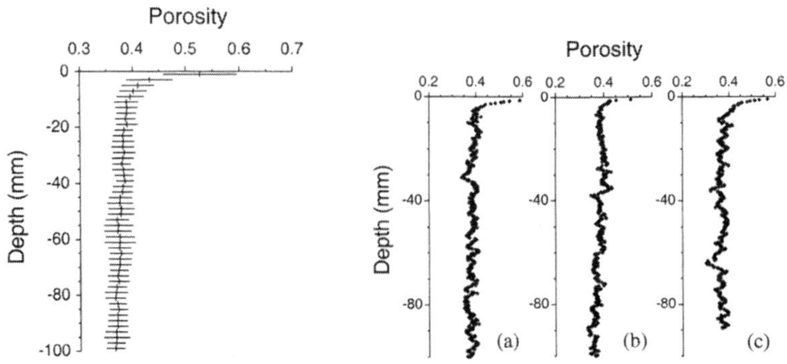

Fig. 7.11. Fluctuations in fractional porosity measured with a diver-deployed microresistivity system in a sand sediment in the northeastern Gulf of Mexico. On the left the depth gradient of average fractional porosity (with bars representing ± one standard deviation) and on the right three representative porosity profiles (a) under a ripple crest, (b) under a mound, and (c) under a sand ripple trough [Wheatcroft 2002].

site in the Bahamas (Fig. 7.11). He found evidence for strong 5- to 15-mm surficial gradients of porosity with 0.05 to 0.15 fluctuations in fractional porosity around the mean with vertical length scales of 5–15 mm. He suggested that these fluctuations were a result of grain packing caused by a combination of hydrodynamic sorting and biological mixing.

A 16-probe resistivity array was developed by [Tang et al. 2002] to measure surface roughness and 3D fluctuations in porosity within the upper 10–15 cm of the sediment, the "In-Situ Measurement of Porosity" system (IMP). The ratio of sediment to seawater resistivity is the "formation factor," F, discussed in Sects. 4.3 and 4.4.2. Fractional porosity, β, is determined from resistivity using Archie's law (4.21) with the parameter a set to unity ($F = \beta^{-m}$). Porosity can then be converted to bulk density using (7.1). Combining (4.21) with $a = 1$ and (7.1) yields

$$\rho = \rho_g - \frac{\rho_g - \rho_w}{F^{1/m}} \ . \tag{7.2}$$

The exponent of Archie's formation factor relationship, m, is usually determined by calibration with sediment of known porosity. Calibration errors of several percent are acceptable, as IMP is intended to measure fluctuations of porosity or density. As will be seen, these fluctuations are small compared to the mean, and errors as large as 10% in fluctuation amplitude are acceptable from an acoustic modeling point of view, as they correspond to 1-dB errors in scattering strength (Ch. 14). The 16-probe array can be mechanically inserted into the sediment at depth intervals as small as 1 mm. Repeated insertions after horizontal displacement allow both 2D and 3D images to be generated (Fig. 7.12). The 1D vertical spectra determined from gradients of

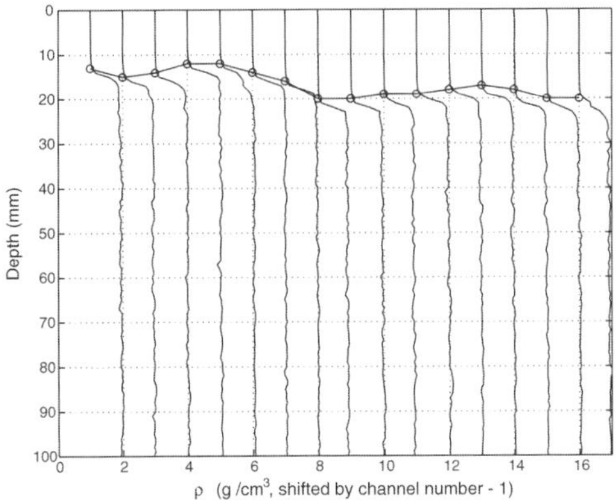

Fig. 7.12. Depth profiles of bulk density calculated from electrical resistivity measurements made using IMP-1 from a sand sediment in the northeastern Gulf of Mexico. The profiles are shifted by 1 g cm^{-3} for display purposes. The connected circles near 10-mm depth represent the estimated sediment–water interface. From [Tang et al. 2002].

electrical resistivity have a much higher resolution than those obtained with cores (1–3 mm versus 1–2 cm) and are therefore of importance to modeling high-frequency volume scattering. Resolution will be discussed further in Sect. 7.4. The horizontal resolution of resistivity measurements (about 1 cm) is not as good as the vertical resolution because of the necessary spacing between array elements and possible sediment disturbance if successive insertions have insufficient separation. The use of density profiles such as those in Fig. 7.12 to calculate 1D vertical density fluctuation spectra will be discussed in Sect. 7.5.

A second-generation electrical resistivity measurement system (IMP2) with a single resistivity probe was developed for midfrequency (1–10 kHz) applications [Tang 2004]. The system is mounted on a longer frame allowing a 4-m horizontal coverage. IMP2 was deployed in the East China Sea during the Asian Seas International Acoustic Experiment (ASIAEX). Note the continued increase with depth of bulk density shown in Fig. 7.13. The increase in density (decrease in porosity) in these silty, fine-sand sediments is probably a result of dewatering due to overburden pressure and bioturbation.

A noninvasive electrical resistivity technique has been developed by Peter Jackson [P. Jackson et al. 1996, Briggs et al. 1998, P. Jackson et al. 2002]. An array of electrodes makes contact with the sediment interface, and a focusing method is used to image sediment resistivity. Several versions of these microresistivity arrays have been tested, including a 576-electrode 2D array used to image soft muddy sediments from rectangular cores collected by divers from Eckernförde Bay, Baltic Sea [P. Jackson et al. 1996] and from

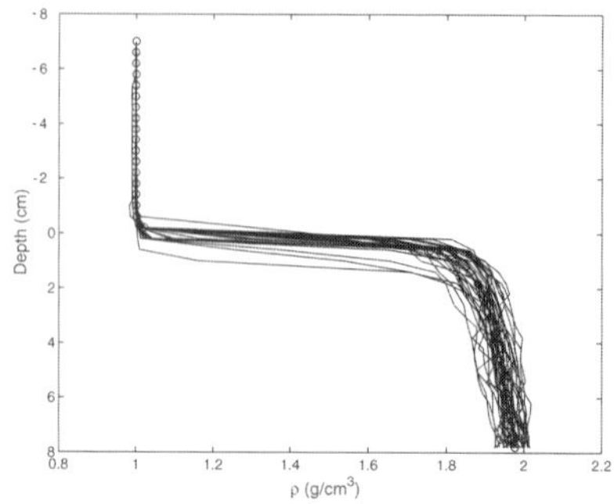

Fig. 7.13. Depth profiles of sediment bulk density measured by electrical resistivity profiling from the East China Sea. The light lines are individual profiles and the solid line with circles is the mean profile [Tang 2004].

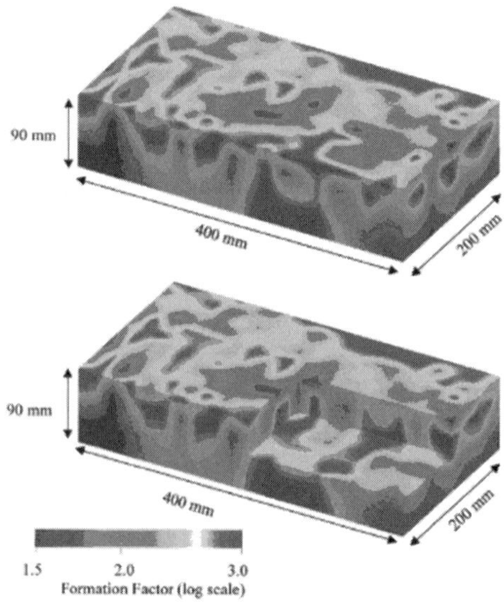

Fig. 7.14. Spatial variability of formation measured in carbonate sediment using a focused microresistivity array. The sediments were collected from the Florida Keys using a box corer [P. Jackson et al. 2002]. (see second color insert)

carbonate sediments in the Florida Keys [P. Jackson et al. 2002]. The diver cores were designed to collect rectangular slabs of relatively undisturbed sediment (36 cm wide, 44 cm long and 3 cm thick). The electrode array was placed flat against the larger (originally vertical) sediment surface. The result is a 2D image of sediment electrical resistivity in a vertical plane. Another version of the system collected 3D images of electrical resistivity from larger 0.25-m^2 box cores collected in the same carbonate sediment in the Florida Keys (Fig. 7.14). A later version was a diver-deployed focused array used to make in situ measurement of electrical resistivity of sand sediments during SAX99 [Richardson et al. 2001a]. These studies demonstrated the feasibility of using a 3D focused array to determine depth-dependent gradients in electrical resistivity. Preliminary analysis of these data yielded average values of electrical resistivity similar to those obtained using conventional 4-electrode Wenner arrays, but the observed higher variability is probably the result of the higher resolution possible with focused microresistivity imaging.

A variety of well-developed seismic exploration techniques are commonly used to determine the spatial distribution of sound speed, attenuation and other seismic attributes in marine sediments [Telford et al. 1990, Yamamoto 1995, Kearey et al. 2002, Chopra and Marfurt 2005]. These techniques employ data acquired by towed hydrophone arrays, bottom geophone arrays, and down-hole or cross-hole geophones. Ray-based inversions of amplitudes and travel times between transducers are used to produce 2D or 3D images of sound speed and attenuation [Sheriff and Geldart 1995, Yilmaz 2001]. Full waveform inversion techniques are also applied to these data providing images of sediment density and bulk modulus as well as sound speed and attenuation [Watanabe et al. 2004]. Most applications of seismic exploration techniques provide 2D or 3D images of much coarser-resolution (meter- to kilometer-scale resolution) of sediment acoustic properties than are needed for high-frequency acoustic modeling, where resolution of centimeters or less is required (Ch. 14). The spatial resolution for down-hole and cross-hole tomography is limited by the acoustic wavelengths used for acoustic mapping and the geometric configuration and size of the acoustic transducers [Chu et al. 2001]. Inversions from towed arrays are limited by size, orientation, and stability of the array [Parkinson 2001] and have not yet been used for in situ sediment characterization for high-frequency scattering studies. However, recent high-frequency scattering studies from laboratory physical models utilizing movable and fixed transducers with nonlinear full waveform inversions show some promise for field applications [Watanabe et al. 2004, Lindwall 2006]. An evaluation of seismic tomographic measurement techniques and inversions is well beyond the scope of this monograph but is adequately covered in several textbooks on seismic exploration [Sheriff and Geldart 1995, Yilmaz 2001]. Several authors have attempted to measure the spatial fluctuations of sound speed and attenuation in support of high-frequency acoustic volume scattering experiments [Yamamoto 1995, Yamamoto 1996, Rapids et al. 1998,

Rogers and Yamamoto 1999, Chu et al. 2001]. Most use two or more vertical arrays pushed into the sediments and linear inversions of travel times (using ray theory) to produce 2D images of sound speed. These efforts are reviewed below.

Yamamoto summarized velocity variability measured by cross-hole acoustic tomography [Yamamoto 1995, Yamamoto 1996]. Measurements were made at 1 kHz for sediments in Tokyo Bay and the Florida–Bahamas carbonate platform but lacked the resolution needed for most high-frequency volume scattering applications. Yamamoto also used sound speed variations determined from acoustic tomography techniques to create images of porosity, density, permeability, shear wave speed, and shear strength having similar resolution using empirical or physical models.

A tomographic system using two linear receive arrays and one source array was developed by [Rapids et al. 1998]. The arrays are inserted into the seafloor by scuba divers using water jets for liquefaction and PVC pipes as well casings. Each array contains 16 transducer elements at 10-cm increments for a total length of 1.5 m (Fig. 7.15). Experiments were conducted in a medium sand with shell fragments off Ft. Pierce, Florida. The transmitter array remained stationary while four locations ranging between 2 and 10 m distant were chosen for the receiver arrays. Transmitter elements were driven with an FM sweep centered at 7.5 kHz. A linear inversion strategy using cross-correlation of the first arrivals between all transmitter–receiver pairs was used to determine travel times and sound speed fluctuations. Two-dimensional power spectra were derived from these tomographic sections [Rogers and Yamamoto 1999] and used as inputs to volume scattering models. Anisotropy was observed in the sound speed spectra where vertical fluctuations had much shorter scale than horizontal fluctuations. This was in agreement with the lower-resolution measurements of [Yamamoto 1995, Yamamoto 1996] from Tokyo Bay and the Florida–Bahamas carbonate platform. However, the resolution analysis of [Chu et al. 2001] indicates that much of this anisotropy may have been an artifact arising from the large horizontal–vertical aspect ratio of the measurements.

A tomographic system using two identical 1-m vertical arrays and a 1-m horizontal array with transducer spacing of 5 cm was developed by [Tang 1997, Chu et al. 2001] (Fig. 7.16). Inversion of travel times and relative amplitudes between all possible pairs of transducers (90 kHz with 40-kHz composite bandwidth) were used to construct 2D images of sound speed and attenuation of a muddy sediment in Hadley Harbor, MA (Figs. 7.17 and 7.18). The system is capable of spatial resolution of approximately 10 cm and can differentiate sound speed variations greater than 2%. Tang (in [Richardson et al. 2001a]) used a high-resolution sediment acoustic imager with 60 transducers arranged in a circle of 25-cm diameter to image sediments during the SAX99 experiments on sandy sediment in the northeastern

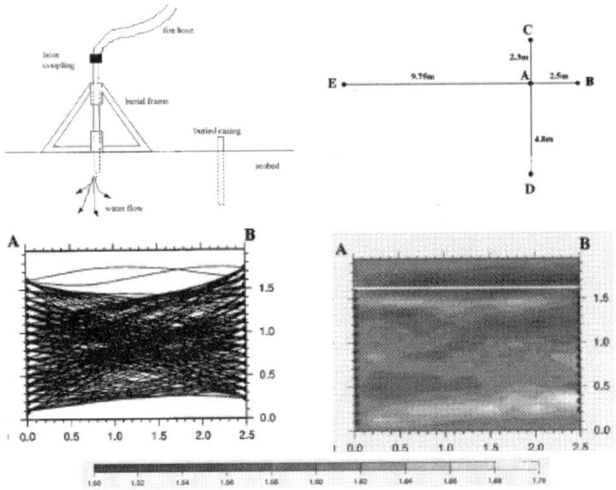

Fig. 7.15. Tomographic techniques and results of [Rapids et al. 1998]. Upper left: Cartoon depicting method used to bury arrays. Upper right: Overhead geometry of the experiment. Lower left: Side view of the ray paths between A and B. Lower right: Tomographic image obtained using the ray paths shown on the left; the gray scale is given below.

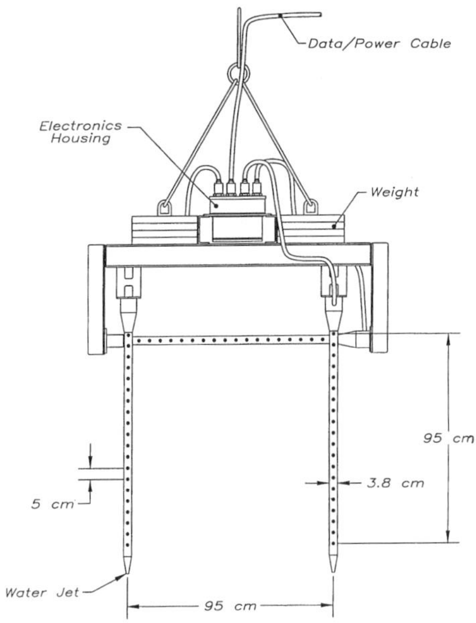

Fig. 7.16. Drawing of the fine-scale acoustic imaging system used by [Chu et al. 2001].

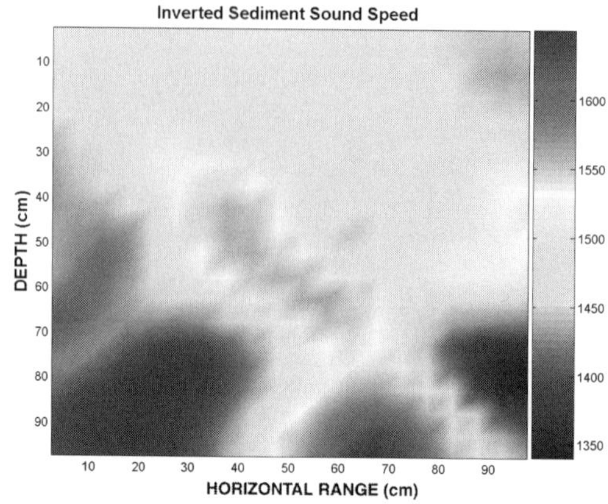

Fig. 7.17. Tomographic sound speed image [Chu et al. 2001]. (see second color insert)

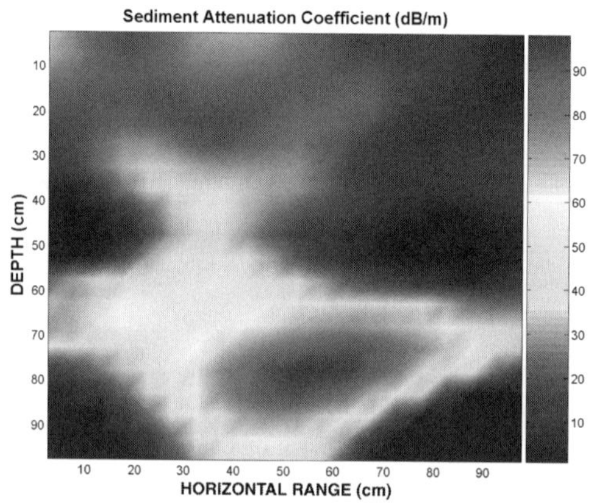

Fig. 7.18. Tomographic attenuation image [Chu et al. 2001]. (see second color insert)

Gulf of Mexico. The working principle is similar to a CT scanner and the system is capable of obtaining 3D images of sediment sound speed and attenuation with a 2-cm resolution. This is achieved by successively imaging 2D slices as the system is pushed into the sediment. The data have not yet been fully exploited.

All of the acoustic tomography techniques reviewed above should be considered preliminary but show considerable promise in characterizing fluctuations in sound speed and attenuation. Many have been validated by ground truth and are to first order correct (capture gradients and large-scale variations). All experimental tomographic techniques are affected by uncertainty in source–receiver geometry, unknown effects of layering on ray paths, multipaths, distorted waveforms from scattering by shells, artifacts of the inversion process, and poor resolution at the edges of the image. None, with the possible exception of the acoustic imaging system of Tang [Richardson et al. 2001a], provide the spatial resolution required by volume scattering models used at higher frequencies (Ch. 14).

Comparisons of microresistivity imaging with other means of determining density heterogeneity have been made by [P. Jackson et al. 2002]. The values of sediment bulk density measured from cores in the laboratory, determined from density contrasts using X-radiographs, and from inverted 2D and 3D electrical microresistivity measurements yielded much different values of correlation length. Vertical correlation lengths (Sect. 7.2) were consistently less than horizontal correlation lengths for all methods and this may reflect the natural layered nature of marine sediments and the gradients of sediment density from consolidation and rates of biological mixing with depth. Similar relationships were reported by [Briggs et al. 1998] for muddy sediments in Eckernförde Bay, Baltic Sea. Vertical correlation lengths determined from fluctuations in electrical resistivity (2D system) and from the gravimetric method applied to diver cores are on the average similar (2.1 versus 2.8 cm) but both are approximately 10 times greater than correlation lengths determined from X-ray density contrasts. These issues are discussed in Sect. 7.4. A related issue is failure of the X-radiographic methods to discern fine-scale fluctuations in density. It was suggested [P. Jackson et al. 2002] that this may be the result of the effects of the superposition of density contrasts (mainly skeletal carbonate fragments) in these 3-cm-thick slabs and out-of-plane scattering of X-rays. They conclude that the gravimetric and electrical resistivity methods provide a more accurate description of density fluctuations in these carbonate sediments and that correlation lengths are much longer in the horizontal than vertical directions.

Each of the systems described herein to measure spatial fluctuations in sediment physical properties has advantages and disadvantages. No single system simultaneously provides the range of spatial scales and accuracy required by models for high-frequency acoustic volume scattering. The accuracy of many of the newer systems has not been established, especially at very high resolutions. The X-radiographic, microresistivity ,and CT methods only provide data on bulk density fluctuation and not compressibility (alternately, sound speed) fluctuation. Vertical and horizontal fluctuation scales can be much different, suggesting that 3D measurements are required. Sound speed fluctuations measured from cores or using in situ probes or tomographic methods provide spatial fluctuations in sediment bulk compressibility but at resolution insufficient for most high-frequency applications. It is rather common to extrapolate lower-resolution measurements to smaller spatial scales by assuming a single power-law relationship over the wavenumber scales required, but this procedure is not supported by available data. It is also possible to assume that the spatial fluctuations of compressibility covary with bulk density, at least in a statistical sense. This assumption is widely used and is discussed in Sect. D.2.2, but it has not been validated. In situ measurement of both density and compressibility is much preferred because removal of sediment from the natural environment, even using very carefully collected diver core or slab samples, can disturb sediment macrostructure and microstructure, the building blocks of density and compressibility fluctuations. This is especially true when information on larger-scale fluctuations is required. Finally, acousticians typically worry about spatial stationarity, but temporal stationarity should be a concern as well. The shallow littoral regions where high-frequency experiments are conducted is dynamic, constantly changing in response to biological mixing, hydrodynamic forcing (waves and currents), sediment deposition, and biochemical processes.

7.2 Statistical Characterization of Continuous Heterogeneity

Heterogeneity in sediments may be in the form of isolated discontinuities in physical properties due to tubes, burrows, feeding voids, or shell pieces, or it may involve more extensive and continuous fluctuations due to bioturbation, sediment deposition, or hydrodynamic processes (Fig. 7.1). The continuous form of heterogeneity has received the most attention with regard to statistical characterization and is the focus of this section. Additional discussion of the statistical approach to characterization of sediment heterogeneity is given in Appendix D.

As noted in the beginning of this chapter, it is useful to break the spatial dependence of a given sediment physical parameter, denoted α for generality, into its average and fluctuating parts:

$$\alpha(x,\, y,\, z) =< \alpha(x,\, y,\, z) > [1 + \gamma_\alpha(x,\, y,\, z)] \; . \tag{7.3}$$

Note that the fluctuating quantity, γ_α, is normalized so as to be dimensionless. That is, it has been defined as the fluctuation in α divided by the mean. This definition is not universal, but is motivated by the requirements of the acoustic models introduced in Ch. 14. Neglecting large-scale variability in the horizontal, the average will be assumed to depend only on the vertical coordinate,

$$< \alpha(x,\, y,\, z) >= \bar{\alpha}(z) \; . \tag{7.4}$$

From the point of view of the acoustic models discussed in Ch. 14, the essential properties of the fluctuations are contained in the covariance

$$< \gamma_\alpha(x_1,\, y_1,\, z_1)\gamma_\alpha(x_2,\, y_2,\, z_2) >= B_{\alpha\alpha}(x_2 - x_1,\, y_2 - y_1,\, z_1,\, z_2) \; . \tag{7.5}$$

The subscripts $\alpha\alpha$ are used to specify that this is a second moment of parameter α. Acoustic models also require moments between two physical parameters (e.g., density and compressional wave speed), so one must introduce the more general covariance

$$< \gamma_\alpha(x_1,\ y_1,\ z_1)\gamma_\beta(x_2,\ y_2,\ z_2) >= B_{\alpha\beta}(x_2 - x_1,\ y_2 - y_1,\ z_1,\ z_2) \quad (7.6)$$

between two parameters, α and β. The labels α and β are arbitrary and should not be interpreted to represent particular parameters (e.g., not attenuation and fractional porosity). Equations (7.5) and (7.6) express a key assumption made in all the acoustic models of Ch. 14: the normalized fluctuations are stationary random processes with respect to the horizontal coordinates, x and y. Stationarity implies that statistical measures, such as probability densities and moments, depend only on the difference in coordinates. Stationarity is unlikely to apply in the vertical coordinate, as sediment properties, including heterogeneity, often depend on depth. Nonetheless, many acoustic models assume stationarity in z as an approximation, and this may be reasonable in many cases if attention is confined to depths below the seafloor on the order of 5–30 cm. As will be seen, statistics may be nonstationary with respect to depth in the upper 1 cm or so. Also, significant non-stationarity at depths of 1 m and greater can be found owing to layering and the increasing overburden pressure. This situation can be accommodated at least approximately by assuming that stationarity in the vertical holds over each layer in the form

$$< \gamma_\alpha(\mathbf{r}_1)\gamma_\beta(\mathbf{r}_1 + \mathbf{r}) >= B_{\alpha\beta}(\mathbf{r}) \ , \quad (7.7)$$

where $\mathbf{r} = (x,\ y,\ z)$, but where both $\mathbf{r}_1$ and $\mathbf{r}_1+\mathbf{r}$ are constrained to lie within the layer. If the means are depth-independent within each layer, the normalized fluctuations are simply proportional to the unnormalized fluctuations, e.g., the density and compressional wave speed fluctuations are

$$\gamma_\rho(\mathbf{r}) = \frac{\delta\rho(\mathbf{r})}{\bar{\rho}} \quad (7.8)$$

and

$$\gamma_p(\mathbf{r}) = \frac{\delta V_p(\mathbf{r})}{\bar{V}_p} \ . \quad (7.9)$$

In the case of 3D stationarity, the 3D power spectral density ("spectrum" for brevity) is obtained as a Fourier transform of the covariance

$$W_{\alpha\beta}(\mathbf{k}) = \frac{1}{(2\pi)^3}\int B_{\alpha\beta}(\mathbf{r})e^{-i\mathbf{k}\cdot\mathbf{r}}d^3r \ . \quad (7.10)$$

If stationarity only holds over a layer of finite thickness, a different approach must be followed. The consequences for scattering are discussed in Appendix M. For the present, it will be assumed that the covariance falls to zero on scales comparable to the layer thickness so that the limits can be extended to infinity. The 3D wave vector, $\mathbf{k}$, has components k_x, k_y, and k_z and magnitude, k. The magnitude will be referred to as the "wavenumber." The wavenumber is related to the wavelength, λ, of fluctuation features as follows:

$$k = \frac{2\pi}{\lambda} . \tag{7.11}$$

Expression (7.10) is the "cross-spectrum" between the two fluctuating parameters α and β. Appendix D discusses assumptions that modelers have used to accommodate this situation. For the purposes of this chapter, the ordinary spectrum (the case $\alpha = \beta$) is of primary interest. The definition of the spectrum is such that its integral over any given region of 3D wave vector space gives the contribution of the corresponding wavelengths to the normalized variance of the parameter in question. If the integral is carried out over all $\mathbf{k}$,

$$\int W_{\alpha\alpha}(\mathbf{k}) d^3k = \hat{\sigma}_\alpha^2 , \tag{7.12}$$

where

$$\hat{\sigma}_\alpha^2 = \frac{< [\delta\alpha(\mathbf{r})]^2 >}{\bar{\alpha}^2} \tag{7.13}$$

is the variance of the fluctuating parameter, α, divided by the square of its mean. The term "normalized standard deviation" will be applied to $\hat{\sigma}_\alpha$, and, when expressed in percent (multiplied by 100), it will be called the "coefficient of variation," a statistical parameter often used in describing fluctuations of geoacoustic parameters (Appendix C). This makes it possible to compare the fluctuation strength of parameters having different measurement units. If the fluctuations in a given random parameter are a stationary random process, the coefficient of variation of that parameter will be independent of position.

The following form for fluctuation spectra is adapted from a more general form used by [Yamamoto 1996]:

$$W_{\alpha\beta}(\mathbf{k}) = \frac{w_{3\alpha\beta}}{[\Lambda^2 k_x^2 + \Lambda^2 k_y^2 + k_z^2 + L_c^{-2}]^{\gamma_3/2}} . \tag{7.14}$$

Special cases of this spectrum have been used in scattering models and in fitting measured heterogeneity data. The parameter $w_{3\alpha\beta}$ will be called the "spectral strength," with the subscript "3" indicating that the spectrum is three-dimensional. The parameter γ_3 is the "spectral exponent," governing the rate of falloff of the spectrum at large wavenumbers. Because the normalized fluctuation parameter, γ_α, is dimensionless, the spectrum has units (length)^3. As a consequence, $w_{3\alpha\beta}$ will have dimensions (length)$^{(3-\gamma_3)}$. The parameter L_c will be referred to as the "correlation length." It is sometimes not used (set to infinity), but is necessary in principle to limit the spectrum to finite values as spatial frequency approaches zero. This parameter is sometimes referred to as the "cutoff scale," or "outer scale" of the random process.

The degree of anisotropy is determined by Λ. If $\Lambda > 1$, the fluctuations have greater high-frequency content (shorter correlation length) in the vertical direction than in the horizontal directions and vice versa. Strictly speaking, L_c is the vertical correlation length and L_c/Λ is the horizontal correlation length. If $\Lambda = 1$, the fluctuations are statistically isotropic, with equal correlation lengths in all directions.

The normalized variance is obtained by performing the integral (7.12):

$$\hat{\sigma}_\alpha^2 = \frac{\pi^{3/2} w_{3\alpha\alpha} L_c^{\gamma_3 - 3} \Gamma(\frac{\gamma_3 - 3}{2})}{\Lambda^2 \Gamma(\frac{\gamma_3}{2})} , \tag{7.15}$$

where Γ is the gamma function. Certain special cases of the spectrum (7.14) are often encountered. If the correlation length, L_c, is set to infinity, and the aspect ratio is unity, the spectrum becomes a pure power law:

$$W_{\alpha\alpha}(\mathbf{k}) = \frac{w_{3\alpha\alpha}}{k^{\gamma_3}} . \tag{7.16}$$

This form is often used in scattering models and has the advantage of requiring only two parameters. This is a reasonable approximation if the correlation length is larger than the acoustic wavelength. The pure power behavior can only persist over a finite wavenumber range, as the integral giving the normalized variance (7.12) would otherwise be infinite. It follows that the coefficient of variation is undefined if the spectrum is modeled as in (7.16). In the case $\gamma_3 = 3$, the heterogeneity described by (7.16) is self-similar, meaning that, viewed at any magnification, the random structure would look the same. This self-similar property results in the parameter $w_{3\alpha\alpha}$ being dimensionless, and, as noted in Ch. 14, implies that the sediment volume scattering contribution to seafloor backscattering strength would be nearly frequency independent. Another commonly encountered model spectrum has $\gamma_3 = 4$ and $\Lambda = 1$:

$$W_{\alpha\alpha}(\mathbf{k}) = \frac{w_{3\alpha\alpha}}{(k^2 + L_c^{-2})^2} . \tag{7.17}$$

In this case, an inverse Fourier transform can be performed analytically to show that the covariance is an exponential in the spatial lag,

$$B_{\alpha\beta}(\mathbf{r}) = \pi^2 w_{3\alpha\beta} L_c e^{-r/L_c} . \tag{7.18}$$

Note that the normalized variance in this case is $\hat{\sigma}_\alpha^2 = \pi^2 w_{3\alpha\alpha} L_c$. The exponential covariance is often assumed in fitting of core data for porosity and density. It is an attractive choice because two parameters suffice to define the covariance: the zero-lag value and the correlation length. Such a highly constrained form for the covariance suits the analysis of core data where only a few lag values are usually available and the limited sample size does not allow fitting of multiparameter functions. The exponential covariance results from the first-order autoregressive model for random processes, and this provides additional motivation for its use in modeling [Briggs 1994, Tang et al. 2002]. The covariance in the first-order autoregressive model takes the form

$$\tilde{B}_{\alpha\alpha}(n\Delta z) = \tilde{B}_{\alpha\alpha}(0)\phi^n \ , \quad n = 0,\ 1,\ 2,\ \cdots \tag{7.19}$$

where ϕ is a dimensionless positive parameter smaller than unity, the "autoregressive coefficient." The spatial interval between samples is denoted Δz and is equal to the length of the individual core sections. The symbol $\tilde{B}_{\alpha\alpha}$ is used for the covariance, because the random process of interest is not the process described by $B_{\alpha\alpha}$, rather it is the process obtained when density (or porosity) is averaged over core sections. While it is useful to estimate $\tilde{B}_{\alpha\alpha}$ from core data, it is essential to make this distinction and accept the necessity of translating properties of the section-averaged process into the properties of the random process relevant to acoustic scattering models. This topic will be discussed at some length in Sect. 7.4. If one ignores this distinction and compares (7.18) and (7.19),

$$L_c = \frac{-\Delta z}{\ln \phi} \ . \tag{7.20}$$

This differs from the definition due to [Yaglom 1987] and used in [Briggs 1994, Tang et al. 2002], $L_c = [1/2 + \phi/(1-\phi)]\Delta z$, but the numerical difference between these two definitions is negligible in the cases of interest. A means of estimating the autoregressive parameters from core data is given by [Tang et al. 2002].

Analysis of sectioned cores only yields the vertical dependence of heterogeneity. Since scattering models require the 3D dependence, relationships between 1D and 3D spectra are needed. Appendix D treats this subject in some detail and derives the 1D spectrum corresponding to (7.14)

$$W_{z\alpha\alpha}(k_z) = \frac{2\pi w_{3\alpha\alpha}}{(\gamma_3 - 2)\Lambda^2 (k_z^2 + L_c^{-2})^{(\gamma_3-2)/2}} \ . \tag{7.21}$$

Given a measured 1D spectrum, one cannot in general obtain the corresponding 3D spectrum, as there are an infinite number of possibilities. If isotropy holds, however, a unique correspondence exists. In the pure power-law case, inspection of (7.21) shows that the 1D spectrum

$$W_{z\alpha\alpha}(k_z) = \frac{w_{1\alpha\alpha}}{k_z^{\gamma_1}} \tag{7.22}$$

corresponding to the 3D spectrum (7.16) has

$$w_{1\alpha\alpha} = \frac{2\pi w_{3\alpha\alpha}}{\gamma_3 - 2} \tag{7.23}$$

and

$$\gamma_1 = \gamma_3 - 2 \ . \tag{7.24}$$

7.3 Relationships Connecting Fluctuating Parameters

Sediment volume scattering models (Ch. 14) require spectra representing the fluctuation of parameters such as compressional wave speed, bulk density, and compressibility. While porosity, sound speed, and bulk density may be measured with spatial resolution sufficient to allow useful spectral estimates, it is sometimes necessary to infer spectra for unmeasured parameters in terms of measured spectra. Volume scattering models also require cross-spectra, for example, between density and compressibility. Cross-spectra express correlation between two parameters, and very little is known about such correlations. Some of the relationships used to bridge the gap between measured and unmeasured spectra are based on physical principles, but these alone are not sufficient to meet all modeling needs, so various investigators have proposed relationships based on arguments that are plausible but not rigorous. The physically based relations will be presented here and the conjectured relations are given in Appendix D.

For porosity and density, fluctuations are usually small compared to the mean of the given parameter, allowing a useful connection to be made between these two fluctuation types. Differentiating the density–porosity relation (Table 4.5) while assuming that the water and grain densities are fixed, one finds after some algebra

$$\gamma_\rho = -\frac{1 - \rho_w/\bar{\rho}}{1 - \bar{\beta}}\delta\beta \; , \tag{7.25}$$

where, in the notation of (7.3), γ_ρ is the fluctuation in density divided by the mean density, while $\delta\beta$ is the fluctuation in fractional porosity, β, *not* divided by the mean. The assumption of spatially uniform grain density may be questioned, but the bulk density–porosity regressions shown in Fig. 7.19 support the linear relationship between density and porosity. These data (M. Richardson, unpublished) show some scatter, but most of this scatter is due to the inclusion of data from many different sites. It is still an open question, however, as to how much of density fluctuation is due to fluctuation in grain density.

Assuming that (7.25) holds, the normalized standard deviation for density, $\hat{\sigma}_\rho$, is related to the variance of porosity as follows:

$$\hat{\sigma}_\rho = \frac{1 - \rho_w/\bar{\rho}}{1 - \bar{\beta}}\sqrt{< \delta\beta^2 >} \; . \tag{7.26}$$

Some scattering models require statistical measures of compressibility fluctuations as input (Ch. 14). Compressibility is not commonly measured in sediments, but can be inferred from density and compressional wave speed. Noting that compressibility, κ, is the inverse of the bulk modulus, K_b, (8.6) can be written in the form

$$c_p = \frac{1}{\sqrt{\kappa\rho}} \; . \tag{7.27}$$

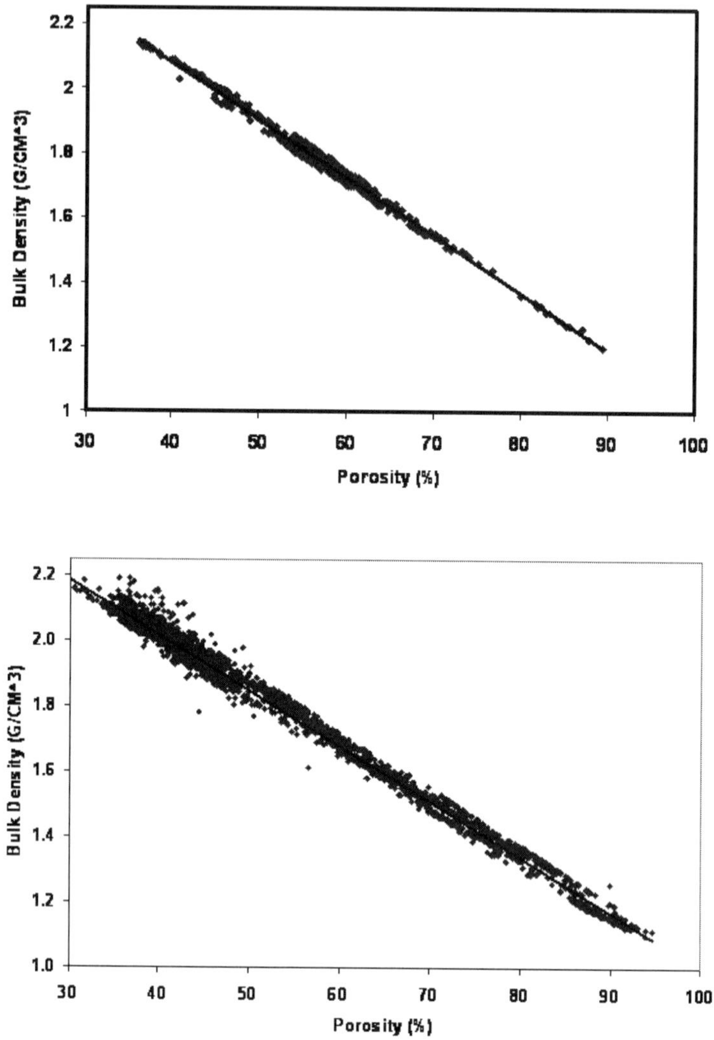

Fig. 7.19. Bulk density versus fractional porosity measurements for siliciclastic (upper) and carbonate (lower) sediments. The regression lines are $\rho = 2.695 - 0.01697\beta$ (siliciclastic) and $\rho = 2.794 - 0.01783\beta$ (carbonate) with r^2 values of 0.9936 and 0.9906, respectively.

This expression neglects shear effects, which should be an acceptable approximation unless fluctuations of the (small) shear modulus are extraordinarily large. Differentiating (7.27), the normalized fluctuations in compressibility (fluctuations divided by the mean) can be expressed in terms of the normalized fluctuations of compressional wave speed and density:

$$\gamma_\kappa = -2\gamma_p - \gamma_\rho . \tag{7.28}$$

In obtaining this expression, the mean of $1/\sqrt{\kappa\rho}$ has been approximated by replacing compressibility and density in this expression by their means. The error in this approximation is second-order in the fluctuations and negligible in the present context. Using (7.28), one can obtain the compressibility covariances in terms of experimentally accessible covariances:

$$B_{\kappa\kappa}(\mathbf{r}) = 4B_{pp}(\mathbf{r}) + 4B_{\rho p}(\mathbf{r}) + B_{\rho\rho}(\mathbf{r}) \tag{7.29}$$

and

$$B_{\rho\kappa}(\mathbf{r}) = -2B_{\rho p}(\mathbf{r}) - B_{\rho\rho}(\mathbf{r}) . \tag{7.30}$$

Identical relations exist between the corresponding spectra.

7.4 Measurement Bias Due to Finite Resolution

All techniques for measurement of heterogeneity have finite resolution, leading to bias in estimates of first and second moments (profiles, covariances, and spectra). This issue has been studied in connection with density statistics obtained from core sections [Briggs and Tang 2002]. These authors used Monte Carlo methods, creating simulated random density fields and then averaging over volumes typical of core sections to produce synthetic data series. The simulations employed an exponential correlation function with correlation length $L_c = 3.51$ cm. The averaging resulted in estimated correlation lengths greater than 15 cm, an extreme bias. This section will employ an analytical approach to the bias problem. The method used is partly adapted from unpublished work by D. Percival.

Averaging is equivalent to the application of a 3D low-pass filter to the random function, $\gamma_\alpha(\mathbf{r})$. Thus, the measured parameter, $\tilde{\gamma}_\alpha(\mathbf{r})$, is related to the actual parameter through the 3D correlation integral

$$\tilde{\gamma}_\alpha(\mathbf{r}) = \int \gamma_\alpha(\mathbf{r}')y(\mathbf{r}' - \mathbf{r})d^3r' , \tag{7.31}$$

where $y(\mathbf{r})$ is a function that specifies the volume averaging inherent in a finite-resolution measurement. While (7.31) is a correlation, it can be written as a convolution between $\gamma_\alpha(\mathbf{r})$ and $y(-\mathbf{r})$, thus as a filtering operation. As the averaging produced by this filtering is unbiased when the parameter $\gamma_\alpha(\mathbf{r})$ has no spatial dependence,

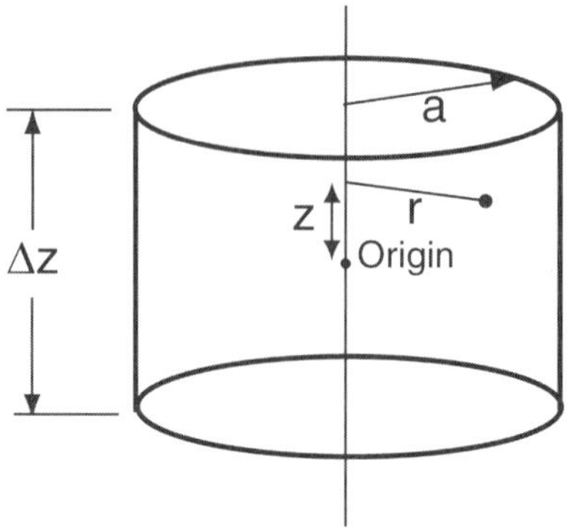

Fig. 7.20. Averaging region appropriate to core sample measurements of porosity and bulk density. The two dots show the origin of the coordinate system and an arbitrary point within the averaging volume described by cylindrical coordinates, (r, z). The azimuthal coordinate is not needed owing to symmetry.

$$\int y(\mathbf{r})d^3r = 1 . \tag{7.32}$$

The nature of $y(\mathbf{r})$ depends on the type of measurement. In measurements of bulk density through core sections, this function is, in cylindrical coordinates (Fig. 7.20),

$$\begin{aligned} y(\mathbf{r}) &= 1/V , \qquad |z| < \Delta z/2, \quad r < a, \\ &= 0 , \qquad \text{otherwise} , \end{aligned} \tag{7.33}$$

where Δz is the vertical length of the core section, a is the core inner radius, and $V = \pi a^2 \Delta z$ is the volume. This might be called a "hockey puck" function owing to the shape and proportions relevant to core sectioning. It can be considered as a sort of 3D "boxcar" function. Compressional wave speed measurements on core samples are typically made by means of transmission of sound horizontally through the core using source and receiving transducers on opposite sides of the core. In this case, the averaging function might be approximated as constant within a circular cylinder (with its axis horizontal) and zero outside. In actuality, the averaging function should be continuous.

The averaging function for resistivity measurements using a spherical electrode can be determined by assuming that the fluctuations in resistivity are small compared to the mean. In this case, the averaging function is discontinuous at the probe radius and continuous at larger radii:

$$\begin{aligned} y(\mathbf{r}) &= \frac{a}{4\pi r^4} , \qquad r > a, \\ &= 0 , \qquad \text{otherwise} . \end{aligned} \tag{7.34}$$

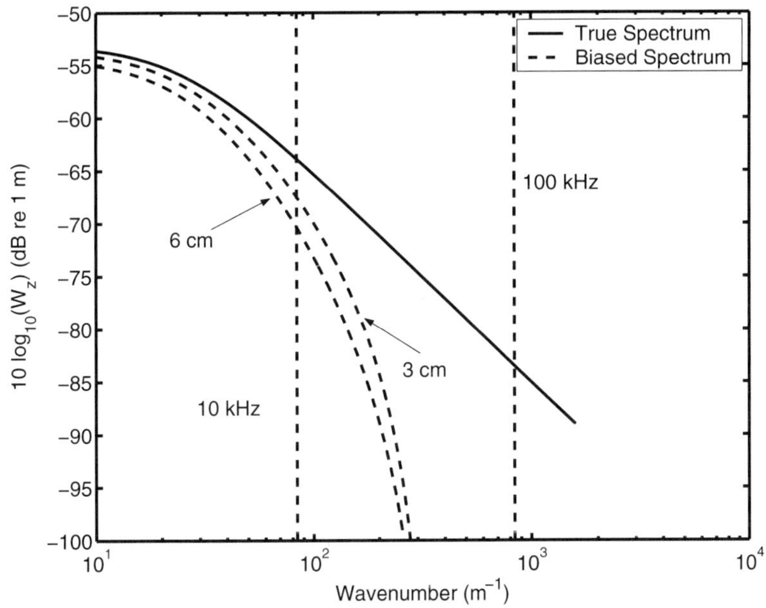

Fig. 7.21. Averaging model results comparing true 1D spectrum and 1D spectrum resulting from averaging over core sections having length 2 cm and diameters 3 cm and 6 cm. The true 3D spectrum is given by (7.17) with $w_{\alpha\alpha} = 0.001 \text{ m}^{-1}$ and $L_c = 0.04$ m. The vertical dashed lines show the wavenumber range appropriate to volume scattering models applied to the frequency range 10–100 kHz.

This expression takes no account of the disturbance of the sediment caused by probe insertion. Spherical coordinates are used in (7.34), so the meaning of r is different than in (7.33) where r is a cylindrical coordinate.

The mathematical details of the effect of finite resolution for the two cases above are worked out in Sect. D.2.3. An example of the bias in spectral estimation from core sections is shown in Fig. 7.21. This figure compares the true spectrum and the biased 1D spectrum resulting from this expression using parameters appropriate to typical core sampling. The true spectrum is taken to be the isotropic spectrum (7.17) corresponding to the exponential covariance. The true correlation length is taken to be 4 cm, comparable to the core section dimensions, radius $a = 3$ cm, length $\Delta z = 2$ cm. The bias is significant (5 dB or greater) over the wavenumber domain of interest for high-frequency acoustics. The biased spectrum is lower at all frequencies than the true spectrum.

The immediate output of core analysis is often the covariance rather than the spectrum. The covariance of the averaged process can be found by performing a multiple integral (D.43), but it is easier to perform an inverse FFT

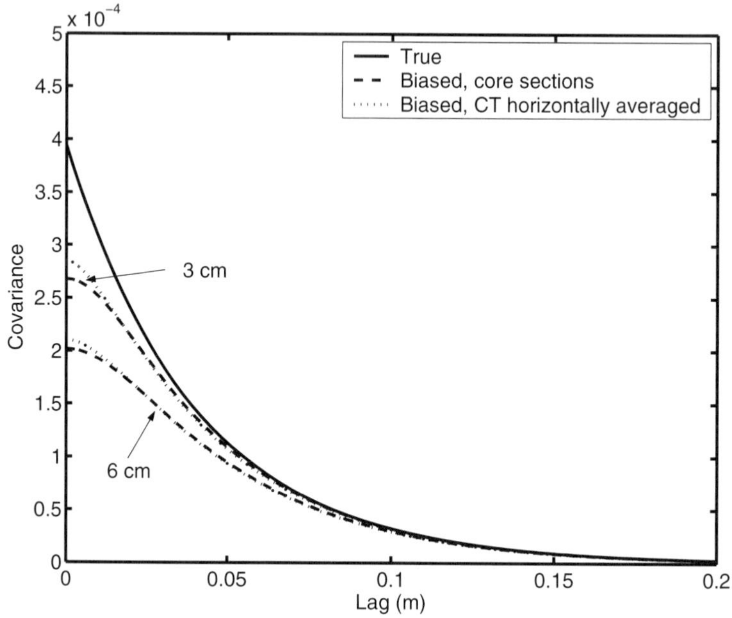

Fig. 7.22. Averaging model results comparing true covariance and covariance resulting from averaging over core sections. The dashed curves indicate the biased covariance resulting from averaging over entire sections, as in conventional analysis. The dotted curves show the biased covariance resulting from averaging over the horizontal extent of the core with no averaging over depth. The core sections are assumed to have length 2 cm and radii 3 cm and 6 cm. The spectral parameters of Fig. 7.21 are used.

on the result for the 1D spectrum. The result of this procedure is shown in Fig. 7.22, which compares the covariances of the true process and the biased process. Two different bias cases are shown. The first case is identical to that of Fig. 7.21 in which averaging is over the entire core section. In the second case, averaging is over a thin disk having radius equal to the core radius. This type of averaging was used for the CT data of Figs. 7.24 and 7.25. All parameters are the same as those used to obtain Fig. 7.22, and the covariances have been marked at 2-cm intervals, as these are the lag values typically measured in core analysis. Although the true process has an exponential covariance, the averaged processes do not. Both "hockey puck" and disk averaging cause approximately the same bias, and bias is greater at small lags. The zero-lag covariance (the variance) is reduced by about a factor of two by averaging over core sections. If one wrongly assumes the section-averaged process has exponential covariance, the value for the autoregressive coefficient would be $\phi = \ln[\tilde{B}(0)/\tilde{B}(2)] = 0.81$, where lag is given in cm, and subscripts on the covariance have been dropped for convenience. From this and the core section length of 2 cm, one would (wrongly) infer using (7.20) that the correlation length was $L_c = 9.5$ cm, when the underlying true process had $L_c = 4$ cm.

This example is realistic in the magnitudes of the parameters used, but may not be realistic in the choice of the "true" process as being first-order autoregressive. Examples will be seen later in which the first-order autoregressive process provides reasonable fits to core-averaged data, consequently, a more realistic example would replace the dashed curve in Fig. 7.22 with one having exponential shape. The question then arises as to which true process would yield an approximately exponential covariance after averaging. The true process would have a sharper spike in the covariance at zero lag, and a 3D spectrum that falls off more slowly than the k^{-4} of the autoregressive process. The expressions derived above could be used to at least partially determine the underlying random process, but this line of inquiry has not yet been followed.

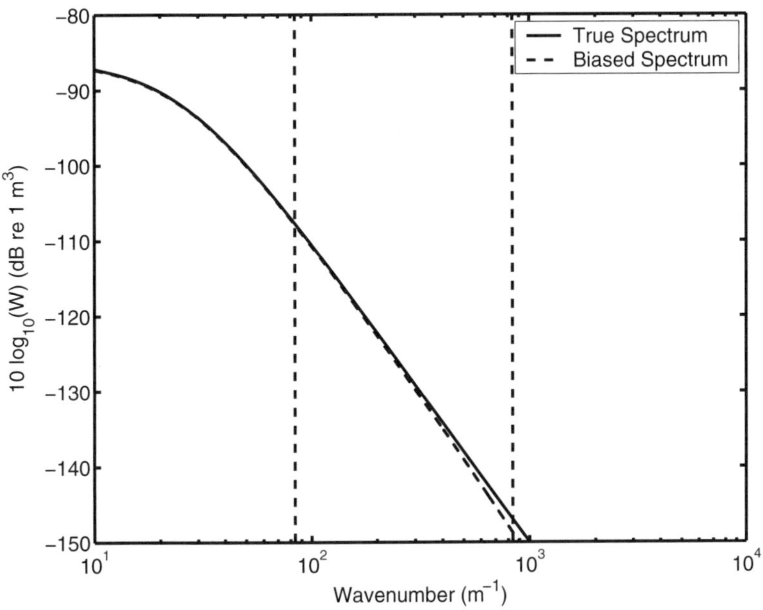

Fig. 7.23. Averaging model results comparing true 3D spectrum and slightly biased spectrum resulting from measurements with conductivity probe having radius 0.3 mm. The spectral parameters of Fig. 7.21 are used, and the vertical dashed lines show the wavenumber range appropriate to volume scattering models applied to the frequency range 10–100 kHz.

In summary, measurements on sectioned core samples produce a series of random numbers that is a strongly filtered version of the underlying random process. As the resulting bias is predictable, it may be possible to make useful corrections. Resistivity probes average over much smaller volumes, and the resulting bias is correspondingly smaller, as seen in Fig. 7.23. The bias expected in CT determinations of heterogeneity spectra has not been considered here and deserves attention.

7.5 Examples of Heterogeneity Statistics

This section gives measurement results for heterogeneity from several sources. These examples include average vertical profiles, coefficients of variation, correlation functions, and spectra. Such examples provide useful inputs for acoustic models, but apparent inconsistencies appearing in some cases also highlight the need for further research.

7.5.1 Coefficient of Variation

The coefficient of variation, defined in Sect. 7.2, provides a useful dimensionless measure of the levels of fluctuation of geoacoustic parameters. This allows comparisons of the fluctuation levels of different parameters (e.g., density and sound speed) in terms that correspond to their relative contributions to acoustic scattering. Table C.1 in Appendix C gives the coefficient of variation for sound speed, attenuation, porosity, mean grain size, bulk density, and grain density for sediment collected with cores from siliciclastic and carbonate sites [Richardson and Briggs 2004a, Richardson and Briggs 2004b]. These sites have roughly the same spatial scale as typical acoustic scattering experiments (1 km^2 or less). Most of the sediment samples were 1- to 2-cm-thick sections of cores having approximate diameter 6 cm. Because sound speed ratio and sound speed have identical values of the coefficient of variation, this coefficient is only presented for sound speed ratio. For the same reason, the coefficient of variation for the attenuation factor k (dB m^{-1} kHz^{-1}) is given with the realization that α_p (dB m^{-1}) has the same coefficient of variation. These data are mostly for sites of high-frequency acoustic experiments and demonstrate the type of variability typical for sediment physical and geoacoustic properties on scales of a kilometer or less. Note that the fine centimeter-scale, within-core variability is combined with the larger meter-scale, between-core variability in these calculations. Several general trends are noted from the values of the coefficient of variation given in Appendix C. The highest coefficients of variation are calculated for attenuation (mean, 32.53); the lowest for grain density (mean, 0.99) and sound speed (mean, 1.20). The next highest coefficients of variation are calculated for mean grain size (mean, 18.74). Bulk density (mean, 2.95) and porosity (mean, 5.84) have intermediate coefficients of variation. The coefficients of variation are higher in carbonate compared to siliciclastic sediments, especially for mean grain size and less so for porosity and density. Fine-grained sediments generally have higher values of the coefficient of variation for grain density, porosity, and bulk density and lower values of the coefficient of variation for sound speed and mean grain size. The coefficients of variation for attenuation exhibit no apparent trends with respect to grain size.

7.5.2 Early Work on Fluctuation Spectra

The earliest examples of heterogeneity spectra were obtained using large-scale data not suited to high-frequency application. Nevertheless, these techniques could be applied at smaller scales. Porosity and density fluctuations in deep-sea drilling cores have been analyzed by [Yefimov et al. 1988]. They fit the resulting spectrum over the wavenumber range 0.2 to 100 m^{-1} with a power law having $\gamma_3 = 3.5$. This method could be applied to the very large data sets available from deep-sea drilling.

The tomographic measurements of Yamamoto and collaborators are outlined in Sect. 7.1. Spectra for compressional wave speed fluctuations obtained using cross-well tomography have been presented by [Yamamoto 1995, Rogers and Yamamoto 1999]. These measurements had resolution appropriate to acoustic application below 10 kHz. Higher-resolution measurements were made by [Rapids et al. 1998], and two-dimensional power spectra were derived from these tomographic sections by [Rogers and Yamamoto 1999]. As noted in Sect. 7.1, the resolution analysis of [Chu et al. 2001] suggests that the reported horizontal–vertical anisotropy may be in part a measurement artifact.

Several examples of higher-resolution heterogeneity statistics obtained from short core samples are provided in Table 14.1 of Ch. 14. The data for the Arafura, San Francisco, and Orcas sites were obtained by Fourier analysis of conventional core data. Section averaging has presumably biased the spectral strength values downward, and comments on this are made in Ch. 14. The data for the other two sites were obtained by Fourier analysis of higher-resolution CT data and should not suffer from this bias.

7.5.3 CBBL Fine-Grained Sediment Measurements

Both CT and conventional core analysis were employed by [Briggs et al. 2002a], allowing comparison of the two techniques. Figure 7.24 compares vertical profiles obtained at the CBBL site off the Dry Tortugas in the Florida Keys where the sediment was a carbonate sand-silt-clay. Both sound speed and density data show a transition in the first 5 cm from values appropriate to low-density sediments to the higher values characteristic of the remainder of the core samples. The CT density data show a less pronounced transition, possibly due to settling of the cores in transit.

Covariance estimates from the same data set [Briggs et al. 2002a] are shown in Fig. 7.25 in terms of the correlation coefficient obtained by dividing the covariance by its zero-lag value (the variance). The figure also

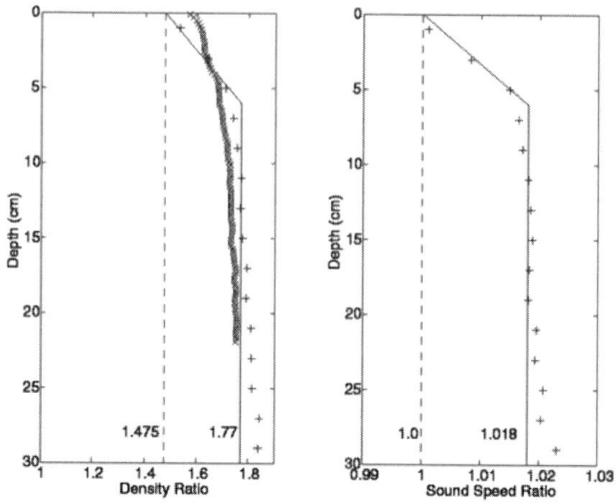

Fig. 7.24. Averaged profiles of density and compressional wave speed ratio for a site off the Dry Tortugas [Briggs et al. 2002a]. The CT data (x) show a less abrupt transition in the uppermost portion of the cores compared to the data obtained by conventional core analysis (+). The solid lines are a piecewise-linear fit suitable for acoustic models that allow for stratification, while the dashed lines represent extrapolation of surficial values as often done in models that do not allow stratification.

shows curves obtained using the first-order autoregressive model with a correlation length of 4 cm. The confidence bounds (95%) were obtained by a Monte Carlo method. The autoregressive model fits the data reasonably well for lags less than 5 cm, and the confidence bounds indicate that departures from this model at larger lags are not statistically significant. The CT and core-section data are similar which is interesting given that the CT data have higher resolution than the core-section data. In fact, the resolutions are similar, as the CT data were averaged over a square area equal to about 20% of the core cross-sectional area, and the core sections employed 23–33% of the cross-sectional area with the remainder saved for grain size analysis [K. Briggs, private communication]. There is essentially no vertical averaging of the CT data (the voxel vertical dimension is 2 mm), and the core-section data are averaged vertically over 2 cm. This situation is modeled approximately by the curves of Fig. 7.22 corresponding to a core diameter of 3 cm. On the basis of that analysis, the measured CT and core-section covariances should be similar with coefficients of variation approximately equal to 0.8 times the true value. The measured coefficients of variation for density are 1.4% and 0.85%, respectively, for the core-section and CT data.

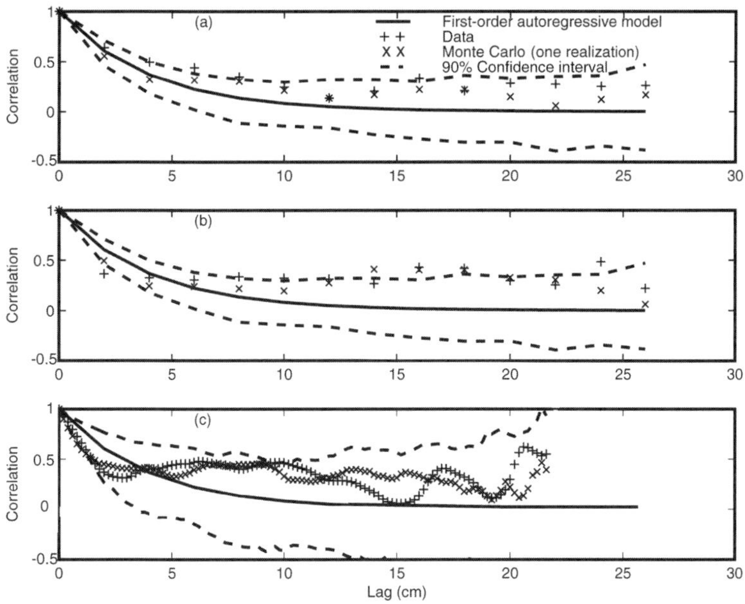

Fig. 7.25. Correlation versus lag for sound velocity (a), density ratio determined from core-section measurements (b), and density ratio determined from CT measurements (c) for a site off the Dry Tortugas [Briggs et al. 2002a]. The CT data and the data obtained by conventional core analysis show rather similar behavior. A first-order autoregressive model having correlation length 4 cm is plotted for all three cases (solid curve). The dashed curves are 95% confidence bounds obtained using a Monte Carlo method, and a single Monte Carlo realization is plotted (x). The legend of panel (a) applies to the other panels as well.

7.5.4 CBBL Coarse-Grained Sediment Measurements

An analysis of CT data has been carried out by [Tang and Orsi 2000a], using three cores from the CBBL Panama City site, where the sediments were coarse-grained quartz sand with shell fragments. The CT measurements provide high resolution with voxel size 0.25×0.25 mm in the horizontal dimensions and 2 mm in the vertical dimension. The authors find that the density fluctuations exhibit Gaussian statistics if outliers resulting from shell and mud inclusions are eliminated. One-dimensional spectra in transverse and vertical directions are shown in Fig. 7.26. There is a slight anisotropy in the vertical as compared to the horizontal, but the two horizontal spectra in orthogonal directions are essentially identical. The authors fit the data with an isotropic 1D spectrum:

$$W_{z\rho\rho}(k) = \frac{a_1}{[1 + (L_1 k)^2]^{\beta_1}} + \frac{a_2}{[1 + (L_2 k)^2]^{\beta_2}} , \tag{7.35}$$

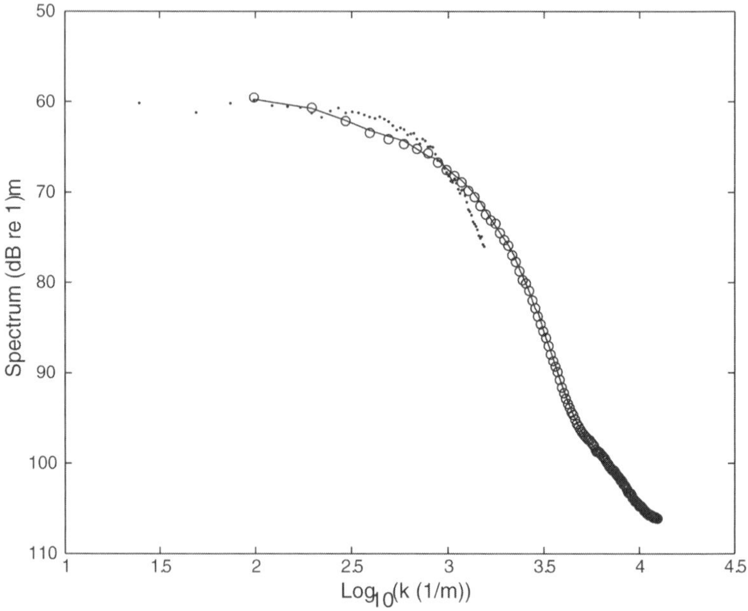

Fig. 7.26. One-dimensional spectra obtained using CT data from the CBBL Panama City site [Tang and Orsi 2000a]. The circles and solid line are the spectra in orthogonal horizontal directions, and the dots are the spectrum in the vertical.

with $a_1 = 8.59 \times 10^{-7}$ m, $a_2 = 1.96 \times 10^{-10}$ m, $L_1 = 0.09$ cm, $L_2 = 0.01$ cm, $\beta_1 = 2.8$, and $\beta_2 = 2.1$. By analyzing sample portions that do not contain shell pieces, the authors conclude that the high-frequency part of the spectrum ($k > 1000$ m^{-1}) is primarily governed by intrinsic fluctuations in sand bulk density due to sorting and packing of grains, while the lower-frequency part is largely governed by heterogeneity due to shells and shell fragments. Note that (7.35) is the sum of two terms of the form (7.21). Isotropy allows one to obtain 3D spectral exponents $\gamma_3 = 7.6$ and 6.2. These values are large, but the pure power behavior is only apparent at very large wavenumbers ($k > 1000$ m^{-1}) corresponding to acoustic frequencies greater than about 100 kHz. As expected, the correlation lengths are much smaller than those usually obtained from analysis of core sections owing to the relative lack of averaging inherent in the measurement.

7.5.5 Pouliquen–Lyons Fine-Grained Sediment Measurements

High-resolution density spectra obtained from CT data by [Pouliquen and Lyons 2002] are shown in Fig. 7.27. Each panel of the figure displays three different 1D spectra. Two spectra are in orthogonal horizontal directions and one is in the vertical direction. Except for the Tellaro site, the three spectra

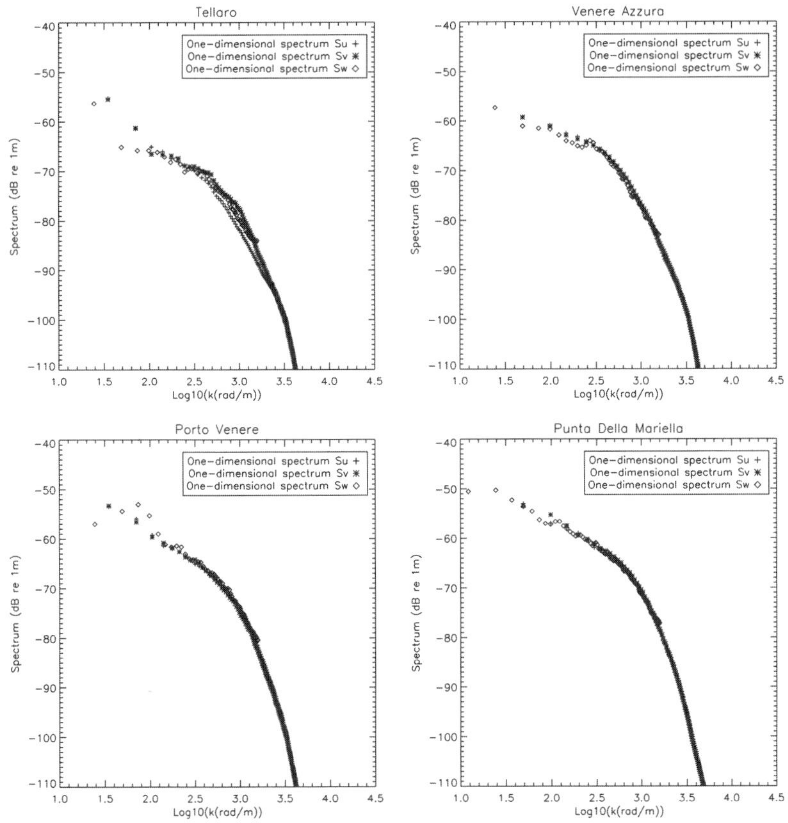

Fig. 7.27. One-dimensional spectra obtained from CT data [Pouliquen and Lyons 2002]. The + and $*$ symbols are the spectra in orthogonal horizontal directions, and the diamonds are the spectrum in the vertical.

overlap quite well indicating a high degree of isotropy. The authors fit the 3D spectra with the form (7.14) with $\Lambda = 1$. The fits are best at higher frequencies ($k > 300\ \mathrm{m}^{-1}$). At lower frequencies the measured spectra depart from the model of (7.14) by failing to level off as wavenumber becomes much smaller than the inverse of the correlation length. Table 7.1 summarizes the results from the fits and also gives the mean grain size and voxel dimensions applicable to each site. These sites had much finer-grained sediments than either the CBBL or SAX99 sites. Interestingly, the power-law exponents and correlation lengths are comparable to those of the dominant (first) term in the spectrum (7.35) characterizing the CBBL site, even though the sediment at that site was a coarse sand. The mean bulk density profile of each site was fitted by the function

$$\bar{\rho}(z) = \rho_s - \frac{\rho_s - \rho_0}{1 + az} , \tag{7.36}$$

where $z = 0$ at the sediment–water interface and z is positive downward into the seafloor. The values of the parameters in this expression are given in Table 7.1.

Table 7.1. Heterogeneity spectrum and vertical density profile parameters from [Pouliquen and Lyons 2002]. The parameters in that reference have been translated into the notation of this monograph.

Parameter	Tellaro	Venere Azzura	Porto Venere	P. D. Mariella
$M_z(\phi)$	4.3	6.1	9.1	7.2
$\hat{\sigma}_\rho$	0.021	0.021	0.023	0.031
γ_3	6.4	6.4	7.6	8.0
L_c (cm)	0.2	0.2	0.14	0.12
Voxel Dim. (mm)	0.35×0.35×2	0.25×0.25×2	0.35 x 0.35×2	0.25×0.25×2
ρ_s (g cm^{-3})	1.97	1.99	1.61	1.81
ρ_0 (g cm^{-3})	1.52	1.37	1.18	1.1
a (m^{-1})	300	800	800	100

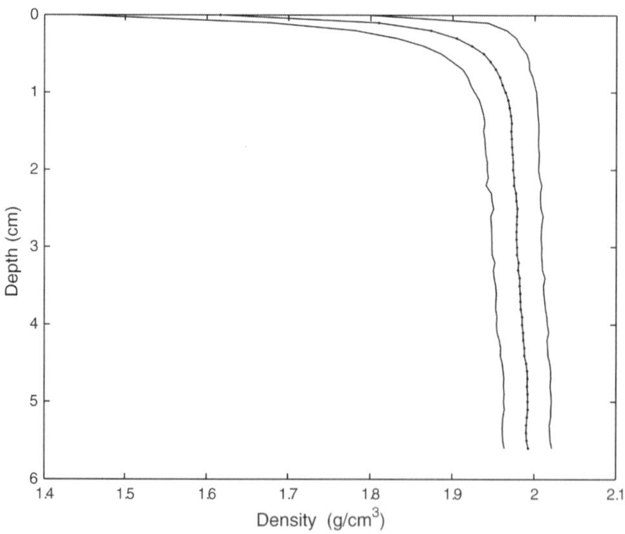

Fig. 7.28. Mean density profile (line with dots) measured using IMP at the SAX99 site [Tang et al. 2002]. Also shown are curves denoting the mean plus and minus one standard deviation.

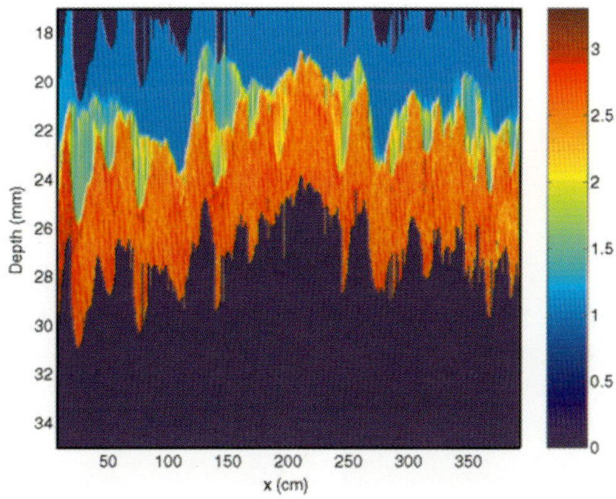

Fig. 6.6. Formation factor measured along a 1D track using IMP2 during SAX04. These data show a mud layer overlying rippled sand. The sand appears as red-orange, while the mud appears as light blue-green with yellow inclusions of sand. The light blue indicates seawater, and regions for which there are no data are dark blue. Courtesy of D. Tang.

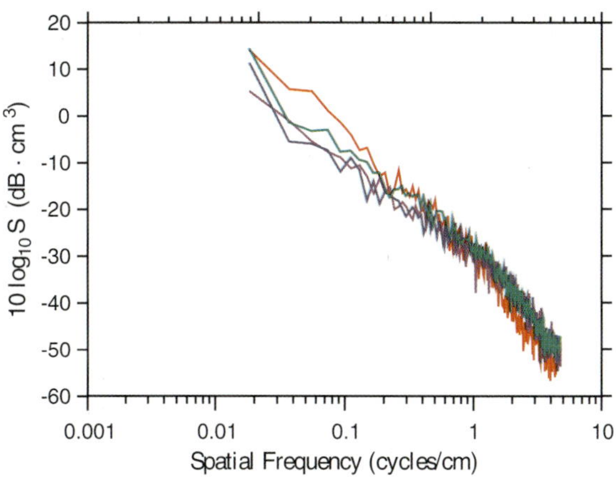

Fig. 6.7. Changes in 1D spectra derived from stereophotographs collected during the SAX99 experiments in the northeastern Gulf of Mexico in 1999: 4 October (red), 19 October (green), 23 October (blue), and 5 November (purple) [Briggs et al. 2002b]. The ordinate is $10\log_{10}[S(F)]$.

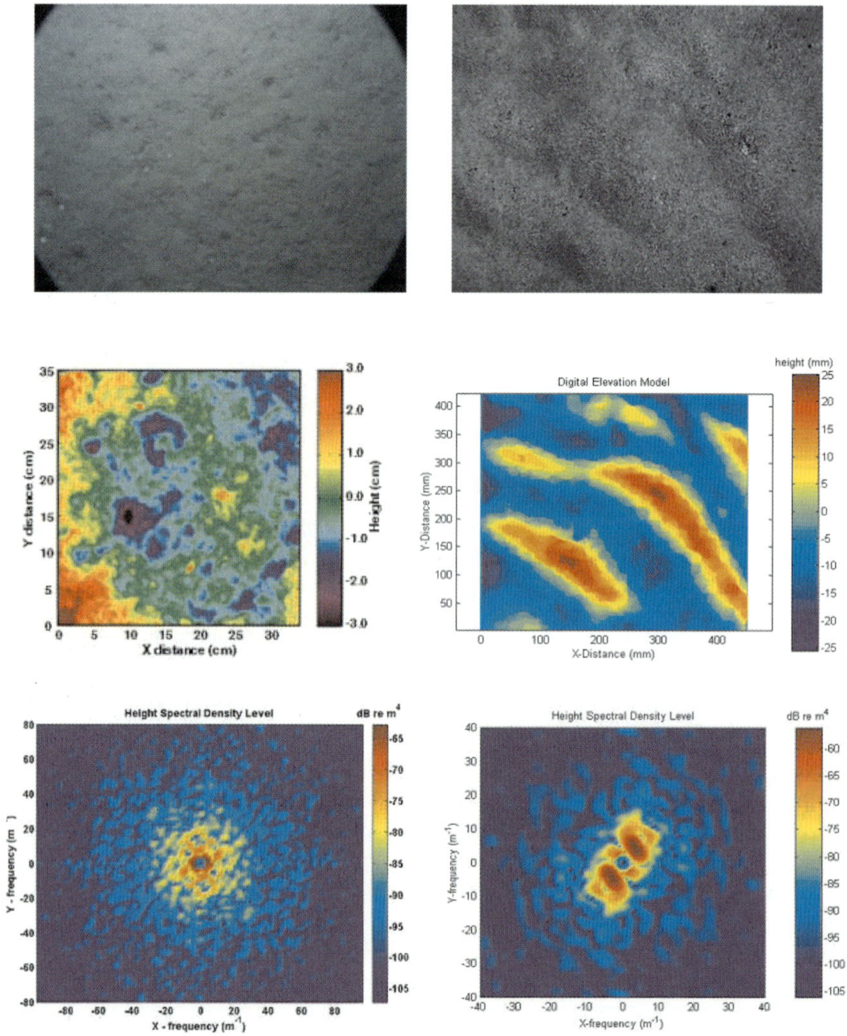

Fig. 6.9. Single images of the paired stereophotographs from a fine sand sediment near the Bay of La Spezia (left) and from a ripple seafloor near Elba Island (right) collected by A.P. Lyons. The digital stereo images were used to develop digital elevation maps (middle panels) and two-dimensional relief spectra (bottom panels) and visually demonstrate the difference between bioturbated isotropic sediments and anisotropic rippled seafloors (see [Lyons et al. 2002b, Pouliquen and Lyons 2002] for details). The color scale is for $10\log_{10}[S(F_x,\ F_y)]$.

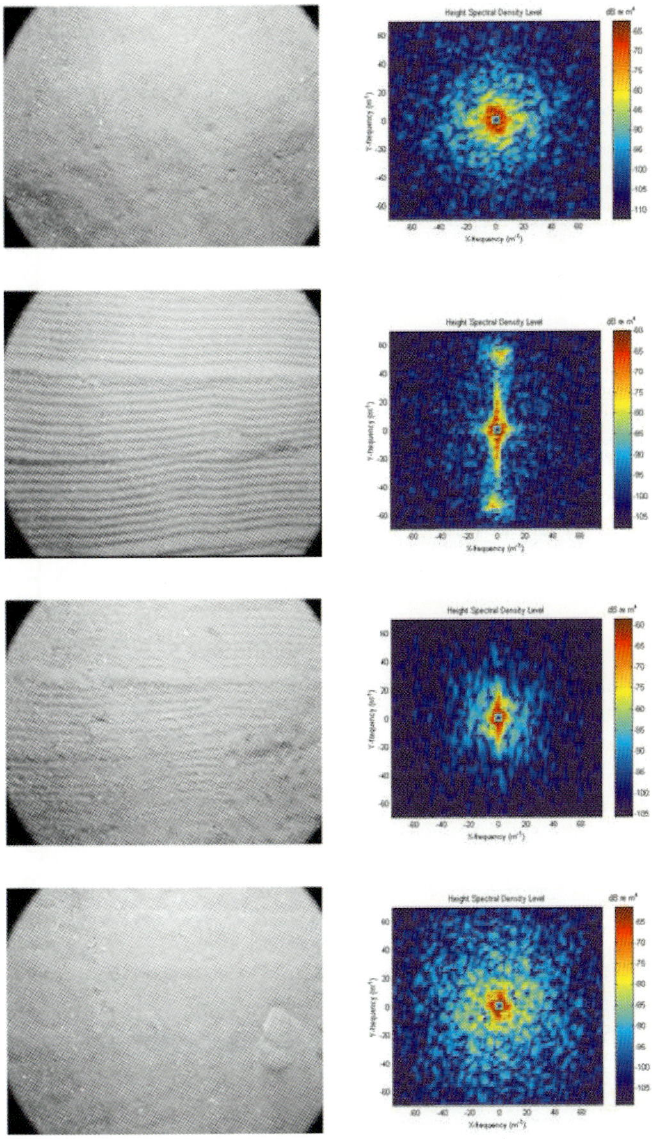

Fig. 6.12. Photographic images and corresponding 2-D power spectra from SAX99. These show temporal decay of ripples created by raking the seafloor. Courtesy of A. P. Lyons.

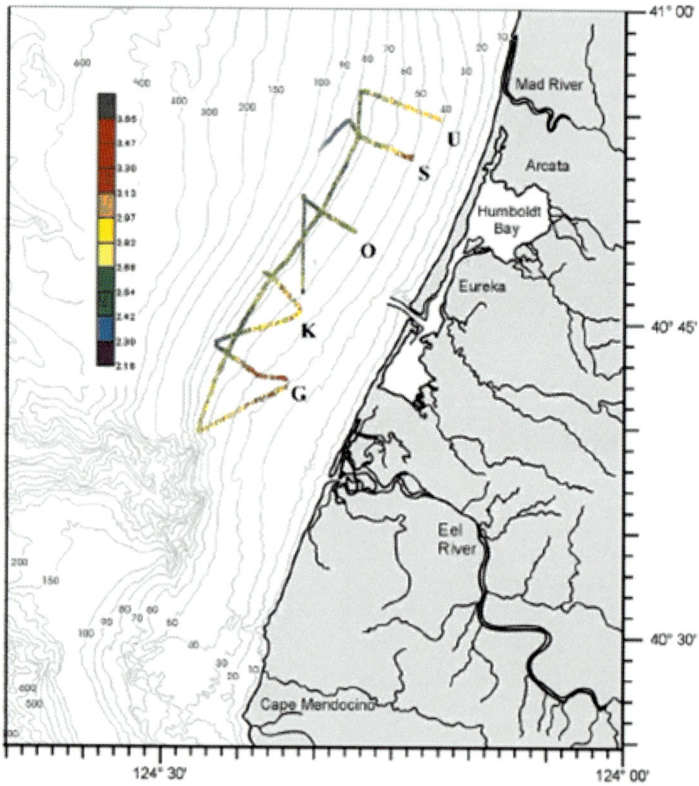

Fig. 7.2. Sediment acoustic impedance measured along tracks surveyed with the Acoustic Sediment Classification System at 15 kHz on the Eel River shelf, northern California. The red to yellow colors correspond to sand and the green to purple correspond to muddy sediments [Richardson et al. 2002b].

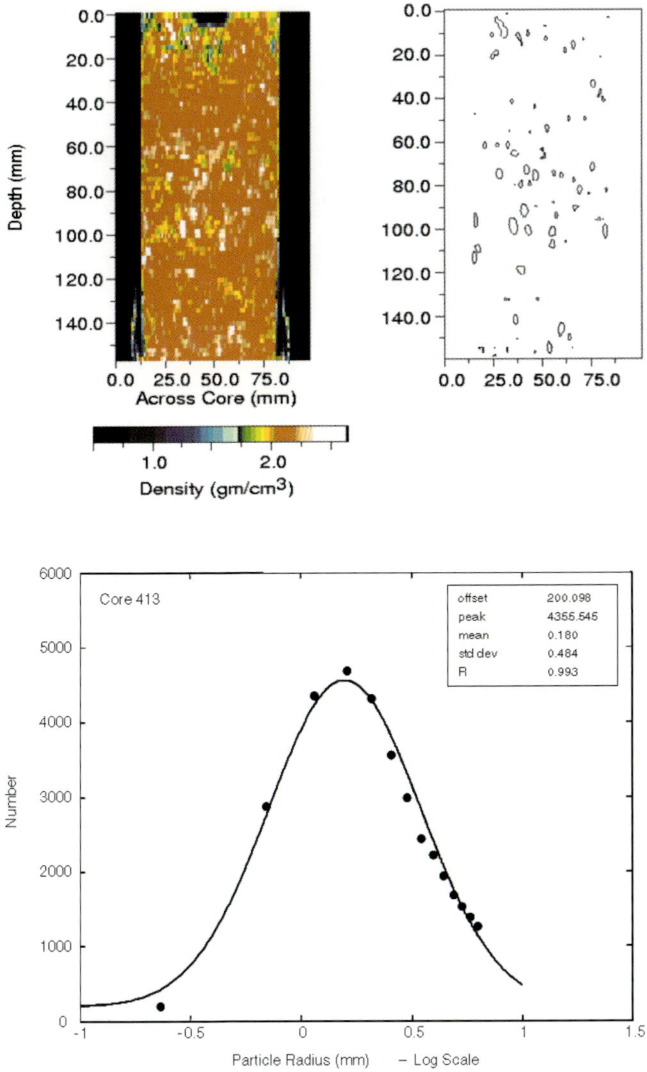

Fig. 7.8. Characterization of discrete scatterers from a shelly sand sediment collected off the coast of Panama City, Florida. Top left is a 2D density image derived from CT data; top right is a contour map of the shell hash which is more dense, white material in the image to the left; bottom is a plot of shell size distribution derived from 12 vertical images (slices) of the CT data from two cores. The particle radius is given in terms of the base-10 logarithm. From [Lyons 2005].

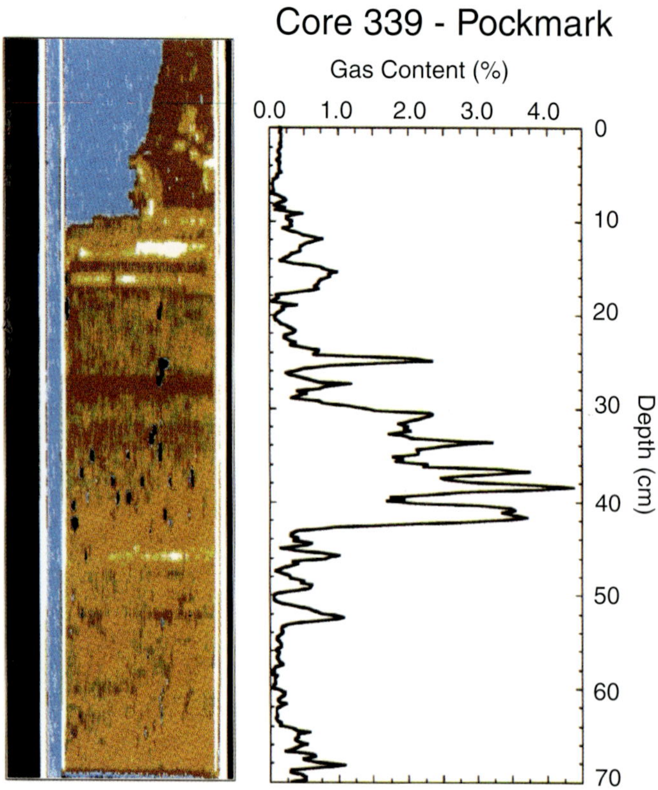

Fig. 7.9. Methane bubble distribution from a pockmark in Eckernförde Bay, Baltic Sea. Left is a 2D slice of sediment retained at in situ pressure and temperature, white is the core liner, blue is water, brown is a silty-clay sediment, and black is methane bubble; right is the methane gas bubble volume based on analysis of the entire core [Lyons et al. 1996].

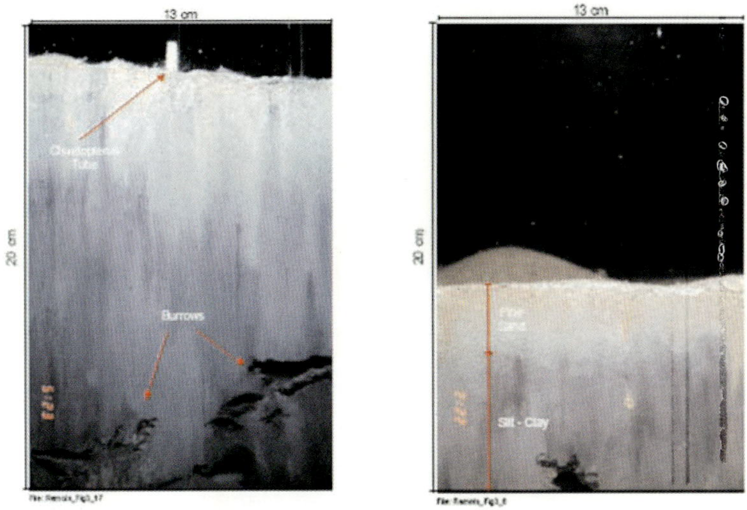

Fig. 7.10. Two images made with the REMOTS sediment interface camera in Buzzards Bay, MA. The image on the left shows a worm tube at the surface and numerous feeding voids and burrows at depth. The image at the right shows fine sand covering silty-clay sediment with a prominent feeding void at about 8 -cm depth. In both cases the structure and roughness of the sediment–water interface are intact. Courtesy of J. Germano.

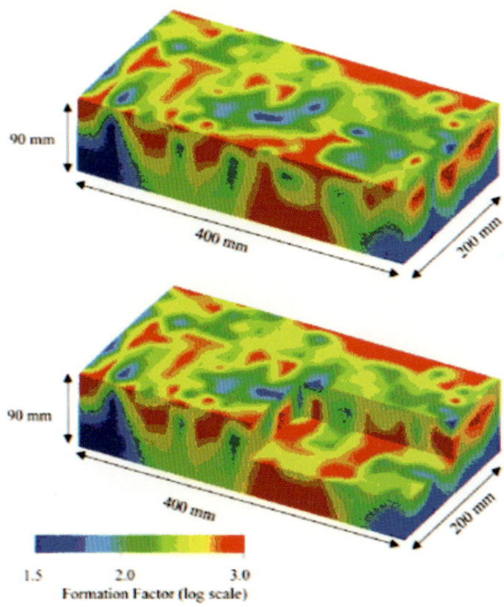

Fig. 7.14. Spatial variability of formation measured in carbonate sediment using a focused microresistivity array. The sediments were collected from the Florida Keys using a box corer [P. Jackson et al. 2002].

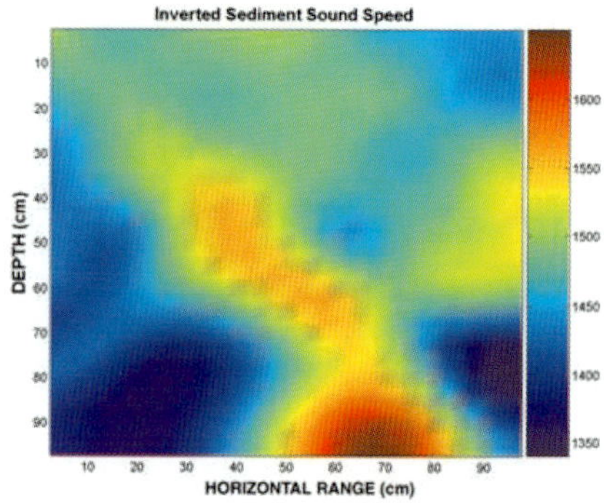

Fig. 7.17. Tomographic sound speed image [Chu et al. 2001].

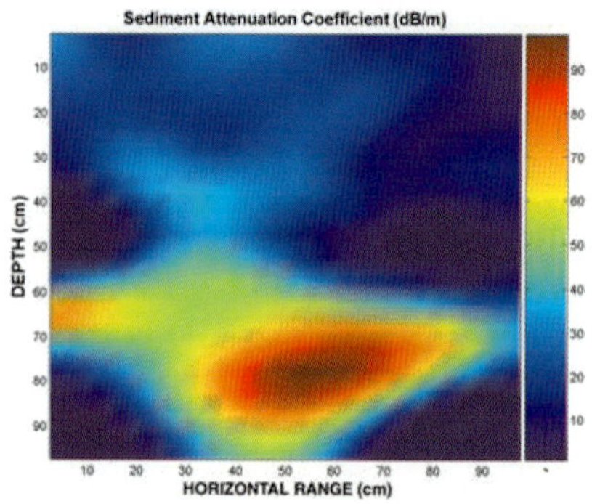

Fig. 7.18. Tomographic attenuation image [Chu et al. 2001].

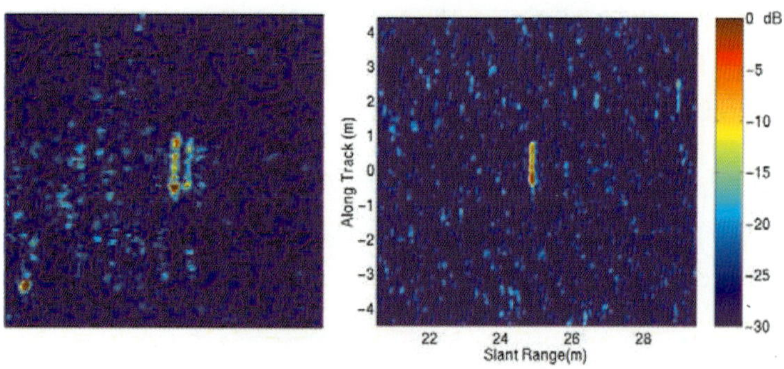

Fig. 15.1. Images of cylindrical targets buried in a sandy sediment. These images were obtained using a synthetic-aperture sonar operating at 20 kHz [Piper et al. 2002]. In the left image the top of the target is 50 cm below the interface, and the grazing angle is 3.3°–4.3°. In the right image the top of the target is 15 cm below the interface, and the grazing angle is 8°–10°. A ripple field having mean wavelength of about 70 cm was present at the site.

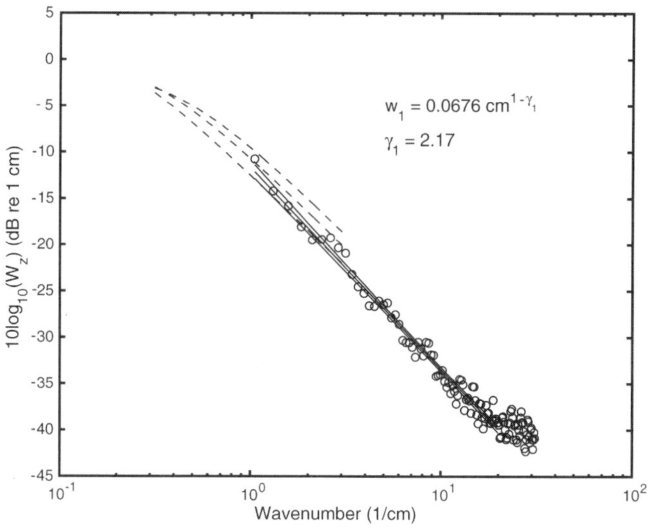

Fig. 7.29. One-dimensional spectrum in the vertical obtained from conductivity probe data at the SAX99 site [Tang et al. 2002]. The data points are open circles and the solid lines are a power-law fit (45) and corresponding 95% confidence intervals. The dashed curves correspond to a first-order autoregressive model (exponential covariance) obtained from core-section analysis with 95% confidence intervals.

7.5.6 SAX99 Sand Measurements

High-resolution density measurements obtained using IMP (conductivity probe) at the sandy SAX99 site have been presented by [Tang et al. 2002]. The IMP data provided much higher resolution than core sectioning, owing to the small probe radius of 0.3 mm (see Fig. 7.23 for a model determination of the bias expected due to finite resolution in this case). The density data are fit reasonably well by a Gaussian probability density function, and the authors suggest that departures from Gaussian statistics are due to shell fragments and organic matter. Figure 7.28 shows the mean density profile obtained using IMP along with bands on either side of the mean profile showing the depth-dependent standard deviation. The mean was fit by the closed-form expression

$$\bar{\rho}(z) = 1.98 - 0.4e^{-3.5z^{0.6}} , \tag{7.37}$$

where $z = 0$ at the sediment–water interface and z is positive downward into the seafloor. The measured standard deviation increases markedly near the interface; this is evidence that the density fluctuations, in this case, are not a stationary random process. The authors fit the depth-dependent normalized standard deviation with the function

$$\hat{\sigma}_\rho = 0.0152 + 0.096e^{-3.7z^{0.82}} . \tag{7.38}$$

The coefficient of variation is about 1.5% over most of the 6-cm measurement depth, but increases to about 10% very near the interface. In order to perform spectral analysis, $\gamma_\rho(\mathbf{r})$ was divided by this function to obtain a random variable that has depth-independent (unit) standard deviation. This random variable was then used to estimate the 1D vertical spectrum shown in Fig. 7.29 with the spectrum renormalized to yield the normalized standard deviation of the main part of the data series, $\hat{\sigma}_\rho = 0.0152$. Also shown in the figure is a power-law fit of the form (7.22).

The IMP data suggest a "transition layer" having a thickness of a few millimeters, in which density begins at a very small value (about 1.5 g cm^{-3}) at the interface and increases rapidly to the at-depth value (about 2.0 g cm^{-3}). This implies that the sand has a very tenuous structure near the interface, then exhibits a rapid increase in packing with depth. This is in opposition to the conventional view of sands as sediments that are deposited in a low-porosity state, while muds are deposited in a high-porosity state with consolidation occurring as a function of depth and time. Further work is needed to verify that this transition behavior is not a measurement artifact. In any case, the transition layer is so thin that it would have little effect on acoustic behavior at frequencies below 100 kHz (A. Ivakin, private communication)

These authors also employed conventional analysis of cores, allowing a comparison of the two techniques. The sound-speed data interval was 1 cm and that of the density data was 2 cm. The first-order autoregressive model was fit to the compressional wave speed and density covariances obtained by conventional core analysis. Correlation lengths of $2.55^{+0.95}_{-1.68}$ cm and $3.51^{+2.27}_{-1.03}$ cm for the compressional wave speed and density, respectively, were found, with the 95% confidence intervals as indicated. The coefficients of variation obtained from this analysis were 0.7% and 0.73%, respectively. Figure 7.29 compares the first-order autoregressive spectrum obtained from analysis of sectioned cores with that obtained using high-resolution IMP data. The core and IMP data overlap within the 95% confidence intervals, and the IMP data have a 1D spectral exponent similar to that ($\gamma_1 = 2$) of the first-order autoregressive process. In fact, this agreement is troubling, as the averaging model predicts that the IMP spectrum should lie above the core-section spectrum (represented by the autoregressive model). As the IMP data were obtained in situ, it is difficult to imagine a mechanism that would reduce the strength of measured fluctuations. One should not forget, however, that the conductivity probe is invasive, and may alter porosity determinations in an unknown way. One might also inquire whether there is any mechanism that could increase the measured level of fluctuation obtained in the core sample analysis. The authors argue against the possible effects of sectioning, noting that errors in porosity due to sectioning are expected to be less than 0.003. Sectioning error was evaluated by comparison of the mean of porosity values measured from sectioned cores with the porosity values measured from intact, unsectioned cores collected from the same location. Equation (7.25) shows that errors in fractional porosity will produce comparable fractional errors in the bulk density. The porosity error, however, is not comfortably smaller than the observed coefficient of variation, 0.73%. It appears that further study of the error due to core sectioning is warranted.

The IMP data of [Tang et al. 2002] were also used to determine 1D spectra in the horizontal. In keeping with the nonstationarity seen in the vertical, the horizontal spectra are dependent on depth. Power-law fits of the form (7.22) yield exponents in the range $\gamma_1 = 1.52$–2.35, with the smaller exponents appearing at the greatest depths (5–6 cm).

7.6 Research Issues

Some aspects of volume heterogeneity are understood, but much remains to be learned. The various measurement techniques tend to suggest a variety of forms for fluctuation spectra. This may indicate that there is no canonical form, or it may be symptomatic of errors that are not yet understood. CT and conductivity measurements both offer high resolution, but the CT spectra show much steeper falloff at the highest wavenumbers, with 3D spectral exponents as large as 6–8. The CT data offer support for the common assumption of isotropy on the small scales relevant to high-frequency scattering. The errors due to disturbance of the sediment by conductivity probes deserve attention and further measurements of the thin transition layer seen in measurements on sand are warranted. The bias resulting from averaging over core sections is readily understood, but the random error due to core collection and subsequent manipulations, including sectioning, deserves further scrutiny, as does the possible degradation of cores in transit from the field to a CT facility. More comparisons between techniques in laboratory-prepared samples are needed to resolve some of these issues. Regarding correlations between acoustic parameters, further study of the conventional core measurement methodology is needed, and development of new high-resolution methods should be given a high priority.

The issues outlined above relate primarily to fine-scale fluctuations in sediment structure which have received the most attention to date. Isolated fluctuations such as those due to shell fragments and bubbles have received less attention, at least as regards the type of statistical characterization needed for high-frequency acoustic modeling. Resolution of issues connected with sediment volume scattering will require laboratory measurements and field experiments in which comparable effort is dedicated to both acoustic and geoacoustic measurements.

8 Fluid Theories

The term "fluid theory" is to be interpreted in the nomenclature of Ch. 2. That is, "wave theory," "propagation theory," or, simply, "theory," refer to systems of equations governing wave motion and boundary conditions for sound pressure fields (and related fields) in sediments. The term "theory" is distinct from the term "model" which is reserved for applications of theory (often with simplifying approximations) to produce specific predictions, e.g., of sound scattering. The next three chapters will consider, in order of increasing complexity, fluid, elastic, and poroelastic theories. These theories are required for the scattering and penetration models treated in Chs. 13–15. The acoustic modeler will usually choose the simplest theory that can account for the phenomena evident in the data at hand. Because of the close relationships between the various theories, the discussion of research issues will be unified and placed at the end of Ch. 10.

The parameters used in fluid theories are discussed in detail in Chs. 4 and 5, and in Appendix B. The parameters of interest are the sound speed, attenuation, and bulk density of sediment and seawater. Methods to measure sediment bulk density in the laboratory and in situ are presented in Ch. 4. Sound speed and attenuation measurement techniques are discussed in Ch. 5, and empirical regressions between these parameters and commonly measured sediment physical properties (such as mean grain size and porosity) are also given. Empirical relations to predict the relevant parameters of the water column in terms of temperature, pressure, and salinity are given in Appendix B.

Although sediments are composed of discrete particles, the theories discussed in this chapter and in Chs. 9 and 10 treat sediments as continuous media. This approximation is reasonable provided the wavelength is much larger than the size of the sediment grains [Bourbié et al. 1987]. The theories of interest can be classified as *effective medium* theories in which physical properties such as density and elastic moduli are spatial averages of some sort. The term "bulk density" is a reminder of this averaging and serves to distinguish sediment density from the constituent grain and water densities. There is a possibility of confusion, however, because the term "bulk modulus" is used as well, but here the modifier "bulk" refers to the volume change in response to hydrostatic pressure, and is not meant to denote any sort of averaging. Thus, while the term "bulk" does not necessarily imply volume averaging when applied to moduli, such averaging is assumed for the sediment modulus and density in this chapter.

There should be no mystery surrounding effective medium theories, as fluids and solids are routinely treated as continuous media even though they are composed of discrete atoms, molecules, and/or crystal grains. For example, a quartz crystal is often described in terms of elastic moduli, optical refractive index, etc., without concern for its underlying atomic structure. This structure becomes apparent, however, when X-rays are scattered by the crystal, because the X-ray wavelength is comparable to the atomic spacing. Returning to the main topic of this monograph, the discrete structure of sediments will become apparent acoustically when the acoustic wavelength becomes comparable to sediment particle size, particle separation, or pore size. For the acoustic frequencies of interest, wavelengths (in water) range from about 15 cm to 1.5 mm. As the coarser sand sediments have grain sizes of order 1 mm, the continuum treatment should be adequate except at the highest frequencies for the coarser sediments.

This picture implies averaging over a volume containing many sediment particles, but doesn't clearly specify the size of this averaging volume. If the averaging volume is too small, containing only a few particles, the resulting average density and modulus may have an extreme variation with position and may not be physically meaningful. On the other hand, if the averaging volume is chosen to be several acoustic wavelengths in size, the heterogeneity responsible for acoustic volume scattering (Ch. 14) will be smoothed out and lost. Thus, the idea of an effective medium is not as clear as it should be, and the success of this approach is conditional (as previously noted) on the acoustic wavelength. It will be assumed in this and in the following chapters that geoacoustic measurements provide averages over suitable volumes, neither too small nor too large for the acoustic problem at hand. In Sect. 7.4 it is demonstrated that this optimistic view is not always justified, that is, the averaging volume in geoacoustic measurements may be too large for the acoustic application. One can believe, however, that as long as the acoustic wavelength is much larger than the sediment particle size, suitable averaging should be possible in principle, if not always in practice.

All the theories to be discussed in this and the two next chapters have their beginnings in Newton's second law which states that force equals mass times acceleration. For a continuous medium, one considers the forces acting on small volume elements and the resulting displacements and deformations. These deformations (quantified in terms of *strains*) are assumed to be proportional to the applied forces (quantified in terms of *stresses*). This approximation is equivalent to Hooke's law for springs and leads immediately to linear behavior in which response is proportional to the strength of excitation and superposition applies. That is to say, only linear acoustics will be considered, a good approximation provided the deformation per unit length (strain) is small (typically, strains less than 10^{-5} fall in the linear regime). Strains larger than this fall in the range of interest of sediment geotechniques [Mitchell 1993]. The fluid, elastic, and poroelastic theories to be discussed differ in the so-called constitutive relations that connect stress and strain, where stress is force divided by the area over which it is applied.

In the following discussion, all quantities will be assumed to have $\exp(-i\omega t)$ time dependence unless the time dependence is explicitly noted to be arbitrary. Fourier synthesis can be used to develop the time dependence resulting from excitations having arbitrary time dependence. Much of the physics contained in these theories is evident in the frequency dependence of parameters such as sound speed and attenuation. In this chapter, such frequency dependence is left unspecified, so the results can be applied to "dispersive" media in which sound speed depends on frequency.

Fluid theories are commonly used to describe sediment acoustic properties and have been found to be adequate for application to many of the scattering models discussed later in this monograph. The various fluid theories differ primarily in their energy loss mechanisms, an area of current research. With $\exp(-i\omega t)$ time dependence, loss mechanisms can be modeled by replacing the bulk modulus (or sometimes, the bulk density) by a complex, frequency-dependent value. The following discussion will first treat the lossless case, after which losses will be considered.

8.1 Equations of Motion

In classical fluid theory, wave motion is governed by the linearized Navier–Stokes equation [Morse and Ingard 1968, Ch.6], [Pierce 1989, Ch. 10], [Kinsler et al. 1999, Ch. 8], in which viscosity contributes a force term proportional to velocity shear. Viscosity is discussed in Sects. 4.4.1, 9.8, 10.1.2, and B.4. For the present, viscous effects will be ignored, and attention will be restricted to simpler, lossless fluid theories leading to the Helmholtz equation, for which acoustic pressure, P, is the only stress to be considered. In this case the force per unit volume, $\mathbf{f}$, is equal to the negative gradient of pressure

$$\mathbf{f} = -\nabla P \, . \tag{8.1}$$

Newton's second law can be expressed on a per unit volume basis as

$$\mathbf{f} = -\omega^2 \rho \, \mathbf{u} \, , \tag{8.2}$$

where the $-\omega^2$ factor arises from the second time derivative of displacement, $\mathbf{u}$, giving acceleration, and ρ is the density. The displacement is a position-dependent field whose three components at any point in the medium give the (small) displacement of the medium due to passage of the acoustic wave. Hooke's law enters through the relation between pressure and the resulting fractional decrease in volume, $-\nabla \cdot \mathbf{u}$:

$$P = -K_b \nabla \cdot \mathbf{u} \; , \tag{8.3}$$

where K_b is the dynamic bulk modulus of the fluid, the inverse of the dynamic compressibility. For convenience, these terms will be shortened to "bulk modulus" and "compressibility" hereafter. This equation is a typical *constitutive* relation connecting stress and strain. More complicated examples will be seen in later chapters.

Equating (8.1) and (8.2) gives

$$\nabla P = \omega^2 \rho \, \mathbf{u} \; , \tag{8.4}$$

and dividing (8.4) by ρ on both sides and taking the divergence, (8.3) can be used to eliminate the unknown $\mathbf{u}$, yielding the following equation of motion for acoustic pressure:

$$\rho \nabla \cdot (\frac{1}{\rho} \nabla P) + \frac{\omega^2}{c_p^2} P = 0 \; , \tag{8.5}$$

where

$$c_p = \sqrt{\frac{K_b}{\rho}} \tag{8.6}$$

is the sound speed. The subscript p denotes "P-wave" using seismic terminology for consistency with later chapters. Both the density and sound speed can be functions of position. Thus, (8.5) is capable of describing the reflection and scattering properties of layered or otherwise heterogeneous sediments, provided the fluid approximation is adequate.

Alternatively, an equation of motion for the displacement can be found by taking the gradient of (8.3) and combining the result with (8.1) and (8.2) to obtain

$$\omega^2 \rho \, \mathbf{u} + \nabla (K_b \nabla \cdot \mathbf{u}) = 0 \; . \tag{8.7}$$

As the unknown in this equation is a three-component vector field, the equation of motion for pressure (8.5) is usually preferred. If density (but not necessarily sound speed) is position-independent, (8.5) reduces to the Helmholtz equation,

$$\nabla^2 P + \frac{\omega^2}{c_p^2} P = 0 \; . \tag{8.8}$$

In this case, applying the curl operator ($\nabla \times$) to (8.4), one finds that the displacement field is *irrotational*, that is, $\nabla \times \mathbf{u} = 0$.

8.2 Plane Waves

If both density and sound speed are position-independent, (8.8) has plane-wave solutions of the form

$$P = P_0 e^{i\mathbf{k}_p \cdot \mathbf{r}} \ , \tag{8.9}$$

where P_0 is an arbitrary complex amplitude, and the magnitude, k_p, of the *wave vector*, $\mathbf{k}_p$, is called the *wavenumber*, equal to the ratio of angular frequency and sound speed:

$$k_p = \frac{\omega}{c_p} \ . \tag{8.10}$$

The corresponding displacement can be found by using (8.4):

$$\mathbf{u} = i \frac{\mathbf{k}_p}{k_p^2 K_b} P \ . \tag{8.11}$$

Since displacement is parallel to the direction of propagation, this is a "longitudinal wave." In dealing with horizontally stratified media such as the seafloor, the vertical (z) coordinate takes on special significance, and it will prove useful in later developments to break coordinate vectors and wave vectors into horizontal and vertical parts. The two horizontal (x and y) components will be collected in two-dimensional vectors indicated by bold uppercase letters:

$$\mathbf{r} = (\mathbf{R}, z) \ , \tag{8.12}$$

$$\mathbf{k}_p = (\mathbf{K}, \pm k\beta_p(K)) \ . \tag{8.13}$$

In (8.12), $\mathbf{R} = (x, y)$, while in (8.13), $\mathbf{K} = (k_x, k_y)$. The magnitude of the horizontal part of the wave vector is $K = \sqrt{k_x^2 + k_y^2}$. The z-component of the wave vector is written in a form that forces $k_p^2 = k_x^2 + k_y^2 + k_z^2$. That is, once k_x and k_y are known, k_z is known up to a possible sign change, depending on whether the wave vector is tilted upward or downward. This constraint is imposed by the definition

$$\beta_p(K) = \sqrt{1 - \frac{K^2}{k_p^2}} \ , \tag{8.14}$$

and the choice of (+) is made in (8.13) if the wave vector is tilted toward the positive z-direction (upward), while the choice of (−) is made if the wave vector is tilted toward the negative z-direction (downward). Note that $\beta_p(K)$ can be interpreted as the sine of the angle of propagation measured from the horizontal. As will be seen, this is a generalized sine that is sometimes a complex number.

The square root in (8.14) requires some discussion. While the square root usually has two values differing only in sign, $\beta(K)$ is defined as the positive square root, with the sign change (determining the upward or downward tilt of the wave vector) specified elsewhere (as in (8.13)). An interesting situation arises if the magnitude, K, of the horizontal component of the wave vector exceeds k_p. With a coordinate vector, the analogous situation would be impossible: the horizontal component, R, of $\mathbf{r}$ must always be less than the magnitude, r. In other words, $R \leq r$. However, if $K \geq k_p$, the relation $k_p^2 = k_x^2 + k_y^2 + k_z^2$ can still be satisfied if k_z is imaginary so that its square is negative. This means that the definition of $\beta_p(K)$ can be used without change, as the square root will produce the necessary imaginary k_z. To make the definition unambiguous, the positive sign is chosen for the imaginary part of the square root.

The situation in which k_z is imaginary is not just a mathematical peculiarity; it is the physically meaningful case of an *evanescent* wave. In this case, the wave propagates horizontally but decays exponentially in the vertical direction. Such waves will be first encountered in Sect. 8.5 in connection with reflection at the water–sediment interface.

In the following section, losses due to energy dissipation are considered, and are shown to lead to complex values for the wavenumber, k_p. All of the relations of this section hold without modification, requiring only the understanding that k_p is complex.

8.3 Losses and Energy Flux

In Chs. 9 and 10, specific acoustic loss mechanisms will be discussed. Even without knowledge of specific mechanisms, it is possible to obtain useful results regarding losses. Losses can be accounted for, at least approximately, by replacing either the bulk modulus, K_b, or the bulk density, ρ, by complex, frequency-dependent effective parameters. The complex bulk modulus can be broken into the components

$$K_b = K_b' - iK_b'' , \tag{8.15}$$

where both K_b' and K_b'' are real and positive. The imaginary part of the modulus must be negative in order that loss leads to decaying rather than growing wave amplitude. To understand this, note that a plane wave traveling in the positive x-direction has spatial dependence $\exp(ik_p x)$ and will decay if k_p has a positive imaginary part, that is,

$$k_p = k_p' + ik_p'' , \tag{8.16}$$

where both k_p' and k_p'' are real and positive. Because of the reciprocal relation between wavenumber and sound speed (8.10), the sound speed must have a *negative* imaginary part. Finally, as sound speed is proportional to the square root of the modulus (8.6), the modulus must also have a negative imaginary part. Similarly, (8.6) requires that complex density must have a *positive* imaginary part,

$$\rho = \rho' + i\rho'' , \tag{8.17}$$

where both ρ' and ρ'' are real and positive. Complex moduli are encountered in the literature quite often (e.g., [Morse and Ingard 1968, Ch. 6],

[Stoll and Kan 1981, Chotiros 1995a]), but complex densities are also used (e.g., [Morse and Ingard 1968, Williams 2001b]). The use of one or the other (or both) is not a matter of taste, but a matter of physics. As will be seen, each results from different loss mechanisms and each has distinct physical consequences. In either case, (8.6) yields a complex sound speed.

It is common, but not essential, to assume that the complex compressional speed is independent of frequency. In this case, both k_p' and k_p'' increase linearly with frequency, and the ratio of the imaginary and real parts,

$$\delta_p = \frac{k_p''}{k_p'} , \tag{8.18}$$

is independent of frequency. This assumption translates into proportionality of attenuation with frequency. While this assumption has some support in experimental data ([Hamilton and Bachman 1982, Buckingham 2000]), it can at best be an approximation owing to fundamental constraints imposed by causality. These constraints, expressed by the Kramers–Kronig relations, are discussed in Appendix I. As will be seen in Ch. 10, Biot theory predicts that attenuation will have a nonlinear frequency dependence. In any case, it should not generally be assumed that δ_p is independent of frequency.

Whether or not the assumption of frequency independence is made, an attenuation coefficient, α_p, measured in dB m^{-1} is commonly used. The relation between k_p'' and α_p can be obtained by equating the factors expressing the decrease in squared pressure magnitude for a plane wave propagating a distance r:

$$e^{-2k_p'' r} = 10^{-0.1\alpha_p r} . \tag{8.19}$$

This yields

$$\alpha_p = \frac{20 k_p''}{\ln(10)} . \tag{8.20}$$

Since wave speed is complex in the lossy case, it is necessary to make a connection with the measured speed, which is real by definition. The measured speed is usually the *phase speed*,

$$c_{pphase} = \frac{\omega}{k_p'} . \tag{8.21}$$

Because of the inverse relationship between wavenumber and wave speed, the phase speed is not simply the real part of the complex speed, rather

$$\frac{1}{c_{pphase}} = \mathrm{Re}\left\{\frac{1}{c_p}\right\} . \tag{8.22}$$

The notation c_{pphase} is rather awkward, and the simpler form, V_p, is used for compressional wave phase speed in Ch. 5. It is convenient to define the complex speed ratio

$$a_p = \frac{c_p}{c_w} , \tag{8.23}$$

where c_w is the sound speed in water. It is more common to use a real parameter for the speed ratio, represented here by the symbol ν_p and in earlier chapters by V_pR:

$$\nu_p = \frac{c_{pphase}}{c_w} . \tag{8.24}$$

The relation between these two ratios can be found by using (8.10), (8.16), (8.18), (8.21), (8.23), and (8.24) to obtain

$$a_p = \frac{\nu_p}{1 + i\delta_p} . \tag{8.25}$$

The attenuation coefficient can be written in terms of the dimensionless loss parameter as

$$\alpha_p = \frac{40\pi f \delta_p}{\nu_p c_w \ln(10)} . \tag{8.26}$$

There are other commonly used dimensionless measures of attenuation. The loss (in dB) per wavelength, $\alpha_{p\lambda}$, is obtained by multiplying (8.26) by the wavelength in sediment, c_{pphase}/f, with the result

$$\alpha_{p\lambda} = \frac{40\pi\delta_p}{\ln(10)} . \tag{8.27}$$

The "quality factor," Q_p, is

$$Q_p = \frac{1}{2\delta_p} , \tag{8.28}$$

and the "logarithmic decrement," Δ_p, is

$$\Delta_p = 2\pi\delta_p . \tag{8.29}$$

The above definitions of loss or attenuation parameters are applicable to the other wave types that will be encountered later in this monograph, with the subscripting simply changed to denote shear waves and Biot waves.

The energy flux density of an acoustic wave is called the intensity. The intensity is the average power per unit area carried by an acoustic wave and can be computed from the product of pressure and velocity. In order to understand the expression for intensity, one must appreciate the fact that complex numbers P and $\mathbf{v}$ are being used to represent pressure and velocity, but real, time-varying pressure and velocity are required to compute power. The real time functions are obtained by taking the real part of the product of the complex number (P or $\mathbf{v}$) and $\exp(-i\omega t)$ (see Appendix E). Instantaneous power per unit area is the product $\mathrm{Re}\{P \exp(-i\omega t)\}\mathrm{Re}\{\mathbf{v} \exp(-i\omega t)\}$, where Re denotes the real part. Averaging over one cycle to obtain average power per unit area, one obtains the well-known expression for acoustic intensity, $\mathbf{I}$ [Morse and Ingard 1968, Ch. 6], [Medwin and Clay 1998, Ch. 2], [Kinsler et al. 1999, Ch. 5]:

$$\mathbf{I} = \frac{1}{2}\text{Re}\{P^*\mathbf{v}\} . \tag{8.30}$$

Note that $\mathbf{v} = -i\omega\mathbf{u}$ and that $\mathbf{I}$ is a vector whose direction is the direction of power flow. For a plane wave, the connection between displacement and pressure (8.11) can be used to obtain the magnitude of the intensity:

$$I = \frac{|P|^2}{2}\text{Re}\{\frac{1}{\rho c_p}\} . \tag{8.31}$$

The product ρc_p is the *acoustic impedance* of the fluid medium, a complex quantity in lossy media.

8.4 Boundary Conditions

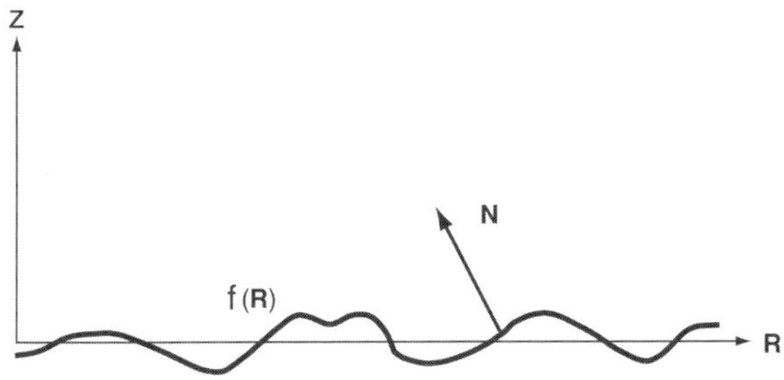

Fig. 8.1. Two-dimensional representation of an arbitrary boundary, $z = f(\mathbf{R})$, between two dissimilar media.

The boundary conditions of interest are the conditions for matching solutions of the equation of motion (8.5) across a boundary such as that depicted in Fig. 8.1. This figure shows two dissimilar media separated by an arbitrarily shaped boundary specified by the equation

$$z = f(\mathbf{R}) , \tag{8.32}$$

where $\mathbf{R} = (x,\ y)$ is a two-dimensional vector comprised of the horizontal coordinates. One boundary condition is continuity of pressure, which can be expressed as

$$P|_{z=f(\mathbf{R})^+} = P|_{z=f(\mathbf{R})^-} . \tag{8.33}$$

This continuity condition can be inferred from (8.1) which states that the acceleration is proportional to the gradient of pressure (volume force). A step change in pressure across the boundary would yield an infinite gradient and infinite acceleration, an impossibility. The other boundary condition is continuity of displacement normal to the boundary

$$\mathbf{N} \cdot \mathbf{u}|_{z=f(\mathbf{R})^+} = \mathbf{N} \cdot \mathbf{u}|_{z=f(\mathbf{R})^-} , \tag{8.34}$$

where

$$\mathbf{N} = \mathbf{e}_z - \nabla f(\mathbf{R}) \tag{8.35}$$

is a vector normal to the boundary (Fig. 8.1), and $\mathbf{e}_z$ is the unit-normal vector in the z-direction. This boundary condition is also implied by the equations of motion. In this case, (8.5) can be divided by the position-dependent bulk density, ρ, and integrated with respect to the z-coordinate. For convenience, the coordinate system will be tilted so that the z-axis is normal to the boundary, and the point of interest on the boundary will be placed at $z = 0$. If density undergoes a step change across this boundary, the z-derivative portion of the divergence may appear to be infinite at the boundary (it's not), but the integral can be performed to give

$$\frac{\partial}{\partial x}\left[\frac{\partial P}{\partial x}|_{z=0}\int_{z=0^-}^{z=0^+}\frac{dz}{\rho(\mathbf{r})}\right]+\frac{\partial}{\partial y}\left[\frac{\partial P}{\partial y}|_{z=0}\int_{z=0^-}^{z=0^+}\frac{dz}{\rho(\mathbf{r})}\right]+$$

$$\frac{1}{\rho_w}\frac{\partial P}{\partial z}|_{z=0^+} - \frac{1}{\rho}\frac{\partial P}{\partial z}|_{z=0^-} = -\omega^2 P\int_{z=0^-}^{z=0^+}\frac{dz}{\rho(\mathbf{r})c_p^2(\mathbf{r})} . \tag{8.36}$$

Here, ρ with no argument denotes the density immediately below the boundary. The partial derivatives of pressure with respect to x and y have been taken outside their respective integrals because continuity of pressure across the boundary insures that these derivatives have no abrupt z-dependence at the boundary. Because all the integrands in (8.36) are finite, the integrals vanish as the end points move toward the boundary. This leads to the continuity condition

$$\frac{1}{\rho_w}\frac{\partial P}{\partial z}|_{z=0^+} = \frac{1}{\rho}\frac{\partial P}{\partial z}|_{z=0^-} . \tag{8.37}$$

Finally, (8.4) shows that the left and right sides of (8.37) are proportional to the z-component (normal component) of displacement, yielding continuity of the normal component of displacement.

8.5 Reflection and Transmission

Figure 8.2 illustrates the geometry for a plane wave incident at grazing angle θ_w on a perfectly flat seafloor. The water–sediment interface is assumed to pass through the origin of the coordinate system, so that the equation specifying the interface is simply $z = 0$. Although a perfectly flat seafloor is an idealized situation, it illustrates several important features of acoustic interaction with the seafloor and provides the simplest example of a model derived from fluid theory. A more detailed account of the reflection problem can be found in [Brekhovskikh and Godin 1990].

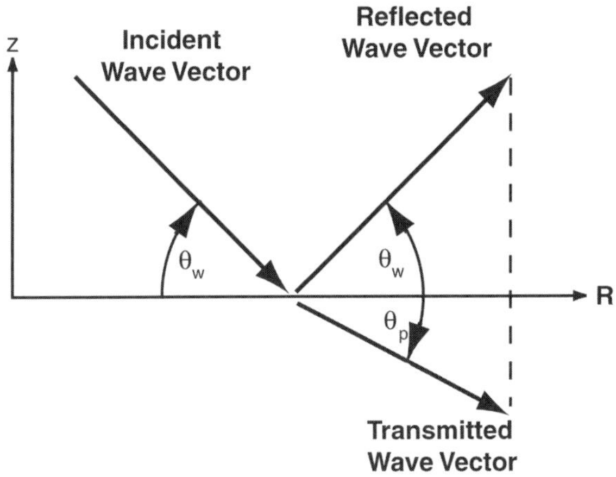

Fig. 8.2. Two-dimensional representation of a planar boundary, $z = 0$, between two dissimilar fluids. The arrows indicate the wave vectors for an incident plane wave and the resulting reflected and transmitted plane waves. All three have the same horizontal components.

The pressure in the water at an arbitrary point, $\mathbf{r} = (\mathbf{R}, z), z > 0$ is

$$P_w(\mathbf{r}) = e^{i\mathbf{K}\cdot\mathbf{R} - ik_w\beta_w(K)z} + V_{ww}e^{i\mathbf{K}\cdot\mathbf{R}+ik_w\beta_w(K)z} , \tag{8.38}$$

and the pressure within the sediment $(z < 0)$ is

$$P_p(\mathbf{r}) = V_{wp}e^{i\mathbf{K}\cdot\mathbf{R}-ik_p\beta_p(K)z} . \tag{8.39}$$

The subscripts w and p denote the acoustic wave in water and the compressional wave in sediment, respectively. The pressure field in the water contains two terms, a unit-amplitude, downgoing incident wave and a reflected, upgoing wave. The complex amplitude of this wave is denoted V_{ww} and is known as the *reflection coefficient* (Ch. 2). The wave in the sediment has complex amplitude, V_{wp}, the *pressure transmission coefficient*. Without loss of generality, the x-axis is assumed to be aligned with the vertical plane of incidence, that is, the incident wave vector is parallel to the x–z plane:

$$\mathbf{K} = k_w\mathbf{e}_x \cos\theta_w , \tag{8.40}$$

where $k_w = \omega/c_w$ is the wavenumber in water. In this case, the function

$$\beta_w(K) = \sqrt{1 - \frac{K^2}{k_w^2}} \tag{8.41}$$

is the sine of the incident grazing angle:

$$\beta_w(K) = \sin\theta_w . \tag{8.42}$$

In (8.38) and (8.39) the dependence of all fields on the horizontal coordinates is the same, namely, $\exp(i\mathbf{K}\cdot\mathbf{R})$. If this were not true, the continuity conditions of Sect. 8.4 could never be satisfied at all points along the $z=0$ boundary. This fact leads to Snell's law as follows. Equality of $\mathbf{K}\cdot\mathbf{R}$ in the water and in the sediment requires that K be the same in the water and in the sediment:

$$K = k_w \cos\theta_w = k_p \cos\theta_p \ , \tag{8.43}$$

or

$$\frac{c_w}{\cos\theta_w} = \frac{c_p}{\cos\theta_p} \ . \tag{8.44}$$

This is Snell's law. This law can be applied to lossy seafloors simply by asserting that the expression

$$\beta_p(K) = \sin\theta_p \ , \tag{8.45}$$

analogous to (8.42), defines the grazing angle in the sediment. Then

$$\sin\theta_p = \sqrt{1 - a_p^2 \cos^2\theta_w} \ . \tag{8.46}$$

The grazing angle θ_p is a complex number and cannot be interpreted in the usual geometric sense, although its real part will still approximately indicate the wave propagation direction when the imaginary part is much smaller than the real part.

Imposing the continuity of pressure condition (8.33) by equating (8.38) and (8.39) for $z=0$ yields

$$1 + V_{ww} = V_{wp} \ . \tag{8.47}$$

Similarly, (8.37) yields

$$\frac{k_w\beta_w(K)}{\rho_w}(1 - V_{ww}) = \frac{k_p\beta_p(K)}{\rho}V_{wp} \ . \tag{8.48}$$

Equations (8.47) and (8.48) can be solved for the pressure reflection coefficient:

$$V_{ww} = \frac{z_{wp} - 1}{z_{wp} + 1} \ , \tag{8.49}$$

with

$$z_{wp} = \frac{z_p}{z_w} \ , \tag{8.50}$$

where

$$z_w = \frac{\rho_w c_w}{\sin\theta_w} \tag{8.51}$$

and

$$z_p = \frac{\rho c_p}{\sin\theta_p} \tag{8.52}$$

are angle-dependent acoustic impedances for the two media with $\sin\theta_p$ being complex. The reflection coefficient depends only on dimensionless ratios of density and sound speed, as can be seen by writing

$$z_{wp} = \frac{a_\rho a_p \sin\theta_w}{\sin\theta_p} , \tag{8.53}$$

where the density ratio is defined as

$$a_\rho = \frac{\rho}{\rho_w} , \tag{8.54}$$

and with a_p given by (8.23).

Using (8.47), the pressure transmission coefficient, V_{wp}, is

$$V_{wp} = \frac{2z_{wp}}{z_{wp} + 1} . \tag{8.55}$$

Both the transmission and reflection coefficient are dimensionless, being ratios of pressures. Some authors define transmission and reflection coefficients in terms of displacement or in terms of potentials for displacement or velocity. In these cases, (8.49) still holds, but (8.55) must be altered.

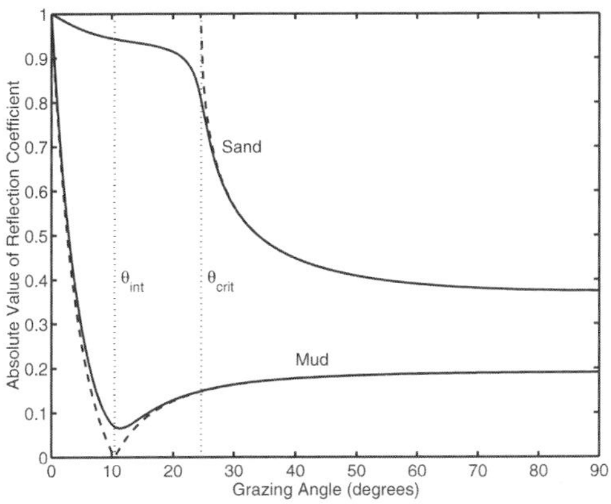

Fig. 8.3. Flat-interface reflection coefficient for representative sand and mud seafloors. The acoustic parameters for the sand seafloor are $\nu_p = 1.1$, $a_\rho = 2.0$, and the parameters for the mud seafloor are $\nu_p = 0.98$, $a_\rho = 1.5$. For each of these two types, the lossless case ($\delta_p = 0$) is shown as a dashed line and the lossy case ($\delta_p = 0.01$) is shown as a solid line. The critical and intromission angles are indicated by vertical dotted lines.

The reflection coefficient reveals a great deal about the acoustic properties of the sediment. Figure 8.3 shows two reflection coefficient examples, one for a sand seafloor and the other for a mud (muds are comprised of silt- and clay-sized particles) seafloor. For each example, both lossless and lossy cases are shown. The sand seafloor is an example of a "fast" seafloor for which sediment sound speed is greater than the water sound speed. In this case, Snell's law dictates that the grazing angle in the sediment be smaller than the incident grazing angle in the water ($\theta_p < \theta_w$). This statement only applies strictly to the lossless case, as θ_p is complex in the lossy case. In fact, in the lossless case, there exists a critical grazing angle, θ_{crit}, for which $\theta_p = 0$, that is, the refracted acoustic energy travels horizontally. For incident grazing angles smaller than the critical angle, refraction will prevent acoustic propagation into seafloor. This results in total reflection back into the water, resulting in unit magnitude for the reflection coefficient. Figure 8.3 shows this behavior for the lossless sand case. The critical angle for the lossless case can be found by taking c_p to be real, setting the left-hand side of (8.46) equal to zero, and solving for $\theta_w = \theta_{crit}$. The result is

$$\theta_{crit} = \cos^{-1}(1/\nu_p) . \tag{8.56}$$

In the sand example of Fig. 8.3, $\theta_{crit} = 24.6°$. Note that in the lossy case the reflection coefficient has magnitude somewhat less than unity for grazing angles smaller than the critical angle, but decreases rapidly for angles greater than the critical angle. Thus, while the critical angle may not be perfectly defined in the lossy case, the value given by (8.56) for the lossless case is nonetheless useful. In optics, the critical angle is also known as the "angle of total internal reflection."

The mud examples in Fig. 8.3 are "slow" seafloors having sediment sound speed smaller than the water sound speed. In this case, refraction is in the downward direction and no critical angle exists. An interesting phenomenon is evident in Fig. 8.3, however, which shows that, in the lossless case, the reflection coefficient vanishes at one particular angle. This is the "angle of intromission," for which the incident acoustic energy passes into the seafloor without any reflection. The angle of intromission can be found by setting $z_{wp} = 1$ for the lossless case and solving for the resultant angle, $\theta_w = \theta_{int}$:

$$\theta_{int} = \cos^{-1}\left(\sqrt{\frac{a_\rho^2 - 1/\nu_p^2}{a_\rho^2 - 1}}\right) . \tag{8.57}$$

For the mud examples in Fig. 8.3, the angle of intromission is about 10.5°.

In order to see that there is a physical difference between complex sound speed and complex density, consider the normalized impedance for vertical incidence:

$$z_{wp}|_{\theta_w = \pi/2} = a_\rho a_p . \tag{8.58}$$

If sound speed is complex, a_p has a *negative* imaginary part, while if density is complex, a_ρ has a *positive* imaginary part. Thus, each contributes oppositely to the phase of the boundary impedance and each will contribute oppositely to the phase of the reflection coefficient. As noted in Sect. 11.3, this may have implications for tests of the validity of the various wave theories (fluid, elastic, Biot, and Buckingham).

The fluid reflection model is idealized and should not be expected to faithfully model all acoustic seafloor interactions. Real seafloors are rough and heterogeneous and may exhibit layering or other forms of vertical gradients in physical properties. These problems are discussed in Chs. 13 and 14.

8.6 Nonplane Waves, Flat Seafloors

Acoustic sources have finite size and launch spherically diverging waves. A useful idealization of this situation is a point source having no spatial extension. More complicated sources can be represented by superpositions of point sources. This section will outline the *wavenumber integration* formalism for point sources, often encountered in low-frequency underwater acoustics, but also forming the basis for many treatments of high-frequency seafloor scattering and penetration.

The solution of a wave equation, including all applicable boundary effects, for a point source of unit amplitude is called the *Green's function.* For a fluid, the equation to be solved for the Green's function is [Jensen et al. 1994, Eq. 2.81]

$$\rho(\mathbf{r})\nabla\cdot(\frac{1}{\rho(\mathbf{r})}\nabla G(\mathbf{r},\mathbf{r}')) + \frac{\omega^2}{c_p^2(\mathbf{r})}G(\mathbf{r},\mathbf{r}') = -4\pi\delta(\mathbf{r}-\mathbf{r}') \ . \tag{8.59}$$

In this equation, $\delta(\mathbf{r}-\mathbf{r}') = \delta(x-x')\delta(y-y')\delta(z-z')$ is a three-dimensional delta function specifying a point source at the position $\mathbf{r}'$. The factor -4π is used for convenience, so that the pressure field in water having uniform density and wavenumber, k_w, is

$$G_0(\mathbf{r},\mathbf{r}') = \frac{e^{ik_w|\mathbf{r}-\mathbf{r}'|}}{|\mathbf{r}-\mathbf{r}'|} \ . \tag{8.60}$$

Many authors do not include the factor 4π in the delta-function source term with the result that it appears in the denominator of the Green's function [Jensen et al. 1994, Eq. 2.50]. With the convention adopted here, inspection of (8.60) shows that the pressure has a magnitude of unity at a distance of one unit of length. In the terminology of [Pierce 1989, p. 160], this is the case of a point source with "unit monopole amplitude." The field $G_0(\mathbf{r},\mathbf{r}')$ will be referred to as the "free-space Green's function." If it is expanded as a Fourier integral of plane waves, then it will be possible to use the previously derived plane wave reflection and transmission coefficients to calculate the field of a point source placed in the water over the seafloor. The required expansion can be found in several references [Zipfel and DeSanto 1972], [Voronovich 1994,

Ch. 2], [Jensen et al. 1994, Ch. 4], [Brekhovskikh and Godin 1999, Ch. 1], and is

$$G_0(\mathbf{r},\ \mathbf{r}') = \frac{i}{2\pi k_w} \int \frac{e^{i\mathbf{K}\cdot(\mathbf{R}-\mathbf{R}')+ik_w\beta_w(K)|z-z'|}}{\beta_w(K)} d^2K \ . \tag{8.61}$$

The integration is over the two variables k_x and k_y (recall $\mathbf{K} = (k_x,\ k_y)$), with limits $\pm\infty$. Thus, the magnitude, K, of the two-dimensional wave vector, $\mathbf{K}$, becomes larger than the wavenumber, k_w, over part of the domain of integration. As noted in connection with (8.14), this is the domain of evanescent waves where $\beta_w(K)$ becomes imaginary. Thus, the expansion of the free-space Green's function not only includes freely propagating plane waves, but also contains evanescent waves that decay exponentially as the field point moves vertically away from the source.

If cylindrical integration coordinates are used in (8.61), the angular integral can be performed, leaving

$$G_0(\mathbf{r},\ \mathbf{r}') = \frac{i}{k_w} \int_0^\infty J_0(K|\mathbf{R}-\mathbf{R}'|) e^{ik_w\beta_w(K)|z-z'|} \frac{KdK}{\beta_w(K)} \ , \tag{8.62}$$

where J_0 is the zeroth-order Bessel function of the first kind.

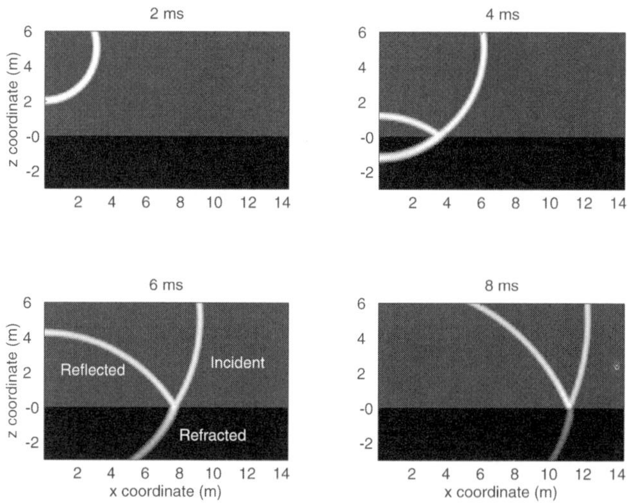

Fig. 8.4. Time evolution of field due to a 20-kHz pulsed point source situated 5 m above a flat, homogeneous sandy seafloor. The acoustic parameters for the seafloor are $\nu_p = 1.16$, $a_\rho = 2.0$, $\delta_p = 0.01$. The time evolution of the reflected and refracted waves is clear, but the lateral and evanescent waves are not seen owing to the particular geometry and loss used.

To obtain an expression for the field due to a monopole at some arbitrary position $\mathbf{r}'$ over a flat, homogeneous seafloo, 8.4, it is only necessary to take (8.61) as the plane-wave expansion o source field and apply the plane wave reflection coefficient to each Fourier component, finally integrating over all plane waves:

$$G_{ww}(\mathbf{r},\ \mathbf{r}') =$$

$$G_0(\mathbf{r},\ \mathbf{r}') + \frac{i}{k_w}\int_0^\infty V_{ww}(\mathbf{K})J_0(K|\mathbf{R}-\mathbf{R}'|)e^{ik_w\beta_w(K)(z+z')}\frac{KdK}{\beta_w(K)}\ . \tag{8.63}$$

In (8.63), the subscript ww is used on the Green's function to indicate that both the source and field point are in the water, that is, $z,\ z' > 0$. The first term, $G_0(\mathbf{r}, \mathbf{r}')$, represents the spherical wave emanating from the source, and is evaluated using (8.60). The integral term represents the reflected wave, which may exhibit rather complicated behavior. The factor $\exp(ik_w\beta_w(K)z')$ in (8.63) is the phase shift in moving vertically from the source to the interface at $z = 0$, and is obtained by setting $z = 0$ in (8.61). The reflection coefficient is generalized to include both evanescent and propagating waves. This is simply a matter of using (8.42) and (8.45) in (8.53),

$$z_{wp}(K) = \frac{a_\rho a_p \beta_w(K)}{\beta_p(K)}\ , \tag{8.64}$$

so that the reflection coefficient (8.49) is defined for all K (thus the reflection coefficient is denoted $V_{ww}(\mathbf{K})$ in (8.63)).

The field within the sediment due to a source in the water is readily found by using the same approach, with the transmission coefficient, V_{wp}, substituting for the reflection coefficient and with the appropriate phase factor for propagation in the sediment used in place of the factor $\exp[ik_w\beta_w(K)z]$:

$$G_{wp}(\mathbf{r}, \mathbf{r}') =$$

$$\frac{i}{k_w}\int_0^\infty V_{wp}(\mathbf{K})J_0(K|\mathbf{R}-\mathbf{R}'|)e^{ik_w\beta_w(K)z' - ik_p\beta_p(K)z}\frac{KdK}{\beta_w(K)} \tag{8.65}$$

Note that, in this expression, $z' > 0$ and $z < 0$. Equations (8.63) and (8.65) were used to compute the fields pictured in Fig. 8.4. This computation assumed as an exciting waveform a Gaussian-shaped pure-tone pulse having a 5-kHz 3-dB full bandwidth centered at 20 kHz. The single-frequency Green's functions of (8.63) and (8.65) were computed at a few hundred equally spaced frequencies covering the band of interest, multiplied by the Fourier transform of the exciting waveform, and the result was inverse transformed. The main features seen in Fig. 8.4 are the field reflected back into the water and the field refracted into the sediment. In fact, there are other interesting effects not obvious in the figure. One effect of importance at high frequencies is the "evanescent" wave, the exponentially decaying vestige of the incident wave

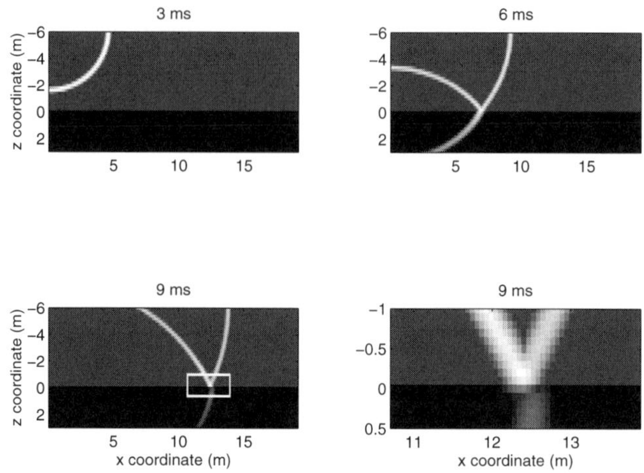

Fig. 8.5. Time evolution of field due to a 20-kHz pulsed point source situated 5 m above a sandy seafloor showing detail for incidence below the critical angle. The refracted wave is relatively weak in the lower two panels, having traveled through the sediment at small grazing angle over a large distance. The enhanced field strength very near the interface is due to evanescent penetration of the incident field. This field is slightly retarded compared to the refracted field, which has traveled through the sediment at higher speed than the waterborne incident wave. The acoustic parameters for the seafloor are $\nu_p = 1.2$, $a_\rho = 2.0$, $\delta_p = 0.01$.

in water that occurs when the grazing angle is smaller than the critical angle. This phenomenon is visible in Fig. 8.5, for which the sound speed ratio, ν_p, was increased to 1.2 in order to enhance the visibility of the evanescent wave. While not dramatic in appearance, the evanescent wave can be important in sonars designed to detect buried objects (see Chs. 1 and 15).

A phenomenon of some importance at low frequencies is the "lateral wave," resulting from reradiation into the water of energy traveling horizontally in the sediment. At high frequencies, attenuation in the sediment makes the horizontally traveling refracted wave weak, hence the lateral wave is weak, even when the source is placed very near the interface, as in Fig. 8.6, and both attenuation and sound-speed ratio are assigned unrealistic values (too low and too high, respectively). Generally speaking, the lateral wave is unimportant at high frequencies, but it may be an issue in close-range measurements of reflection (Sect. 11.1).

An important property of the Green's function can be found by considering the case of propagation from a point, $\mathbf{r}$, within the sediment to a point, $\mathbf{r}'$, in the water. By simply interchanging indices in (8.65), one obtains

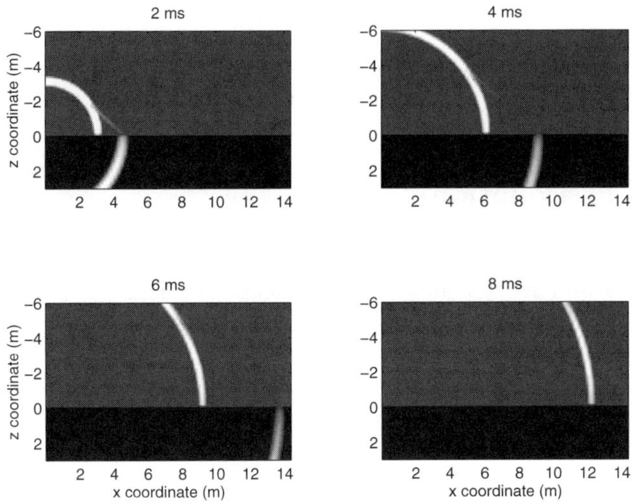

Fig. 8.6. Time evolution of field due to a 20-kHz pulsed point source situated 0.05 m above a sandy seafloor showing the lateral wave, the weak feature propagating in the water away from the boundary. The lateral wave precedes the incident wave, having originated from the rapidly traveling refracted wave. The acoustic parameters for the seafloor are $\nu_p = 1.5,\ a_\rho = 2.0,\ \delta_p = 0.0005$.

$$G_{pw}(\mathbf{r}', \mathbf{r}) =$$

$$\frac{i}{k_p} \int_0^\infty V_{pw}(\mathbf{K}) J_0(K|\mathbf{R} - \mathbf{R}'|) e^{ik_w \beta_w(K) z' - ik\beta_p(K) z} \frac{K dK}{\beta_p(K)} . \tag{8.66}$$

By a similar interchange of indices, one can show that the sediment-to-water pressure transmission coefficient is related to the water-to-sediment pressure transmission coefficient as follows:

$$V_{pw}(\mathbf{K}) = V_{wp}(\mathbf{K}) \frac{k_p \beta_p(K)}{a_\rho k_w \beta_w(K)} . \tag{8.67}$$

Upon substitution in (8.66), one finds

$$G_{pw}(\mathbf{r}', \mathbf{r}) = \frac{1}{a_\rho} G_{wp}(\mathbf{r}, \mathbf{r}') . \tag{8.68}$$

This is an expression of the fundamental principle of *reciprocity*, and a general statement can be derived for fluid media having arbitrary position dependence of sound speed and density. Using (8.59) and Green's theorem, it can be shown that, in general,

$$\frac{G(\mathbf{r}',\mathbf{r})}{\rho(\mathbf{r}')} = \frac{G(\mathbf{r},\mathbf{r}')}{\rho(\mathbf{r})} . \tag{8.69}$$

This is a generalization of the expression encountered for homogeneous media in which the density factors are absent [Pierce 1989, p. 199].

8.7 Wood's Equation

Sediments have a certain degree of rigidity, that is, they can support static shear stresses and are not fluids in the strict sense. This rigidity results from contact between sediment grains, and, if the forces acting at these contacts are ignored, a fluid approximation results. To characterize acoustic behavior in this approximation, one must develop equations for the effective density and bulk modulus of a volume of sediment that is large enough to contain many grains, yet much much smaller than the acoustic wavelength. It is easy to see that the bulk density, ρ, is

$$\rho = \beta\rho_w + (1-\beta)\rho_g , \tag{8.70}$$

where β is the fractional porosity, ρ_w is the density of the pore water, and ρ_g is the density of the sediment grains. With pressure, P, applied, the fractional volume change of the pore water is P/K_w, where K_w is the bulk modulus of the water, and the fractional volume change of the grains is P/K_g, where K_g is the bulk modulus of the grains. Overall, the fractional volume change of a small quantity of saturated sediment will be $P\beta/K_w + P(1-\beta)/K_g$, so, dividing this by pressure, the bulk modulus, K_b, of the sediment is given by [Wood 1964]:

$$1/K_b = \beta/K_w + (1-\beta)/K_g . \tag{8.71}$$

Sound speed can be determined using (8.70) and (8.71) in (8.6). The result will be denoted c_0 to distinguish it from other theoretical values for compressional wave speed:

$$c_0 = \frac{1}{\sqrt{[\beta/K_w + (1-\beta)/K_g][\beta\rho_w + (1-\beta)\rho_g]}} . \tag{8.72}$$

This equation is referred to as Wood's equation, but it will be convenient in later developments to refer to (8.70)–(8.72) collectively as *Wood's equations*. Wood's approach usually gives a rather poor prediction of sediment sound speed [Hamilton 1971a, Hamilton and Bachman 1982], as it ignores much of the dynamics of the grain-pore water system. First, contact between grains produces forces not accounted in Wood's treatment. These forces are included in Gassmann's equations (Sect. 10.1.1). Second, inertial and viscous forces are neglected, yet become important at high frequencies. Wood's equations are useful, however, in the context of more ambitious sediment acoustic theories [Buckingham 1997, Buckingham 1998, Buckingham 2000,

Williams 2001b], and in interpretation of laboratory measurements in which grain-contact and density-differential effects have been deliberately reduced [Richardson et al. 2002a].

8.8 Research Issues

Sediment propagation theory is an active area of research, but fluid theory is a special case. Consequently, discussion of research issues is deferred to Ch. 10.

9 Elastic Theories

Elasticity theory incorporates shear forces not included in fluid theory. Elastic media support two types of waves: compressional waves analogous to the acoustic waves of fluid theory and shear waves. Both types of waves are attenuated due to energy loss, and the term "viscoelasticity" is often used to designate the relevant theory [Hamilton 1971a]. In this monograph, elasticity theory includes viscoelastic effects. In unconsolidated sediments such as clay, silt, and sand, shear waves travel at considerably lower speeds than the water sound speed (Sect. 5.2), and elasticity theory provides rather small corrections to models based on the fluid approximation. The fluid approximation fails entirely for rocky seafloors such as sandstone, granite, and basalt. In these materials, the shear wave speed is often greater than the water sound speed, and the theory of elasticity should be used. This chapter outlines elements of elasticity theory and notation needed for later discussion of modeling. Shear wave speed and attenuation are the two additional parameters required by elasticity theory compared to fluid theory. In situ and laboratory techniques commonly used to measure these parameters are presented in Sect. 5.2. Empirical regressions that can be used to predict shear speed attenuation in near-surface sediments are also given in Sect. 5.2 along with information on vertical gradients.

Because an elastic medium can support shear stresses, pressure is not an appropriate field for describing wave motion. In an elastic medium, deformation is defined in terms of the components of *strain*, which involve spatial derivatives of the three components of the displacement vector. There are nine such derivatives, organized in the *strain tensor*. Similarly, the force per unit volume is expressed in terms of the *stress tensor* which also has nine components. Hooke's law, expressing the assumed linear relation between stress and strain, can then be written as a tensor relation with a grand total of 81 elastic constants, or moduli. Fortunately, physical constraints and symmetry considerations reduce many of the moduli to zero and provide relationships that leave a relatively small number of independent moduli. In this monograph, only *isotropic* elastic media will be considered, in which case there are only two independent moduli and for which tensor notation can be largely avoided. The stresses and strains of immediate interest are "dynamic," and in the linear regime (with stress less than 10^{-5}). Larger static and dynamic strains are of geotechnical and seismic interest. In particular, large static strain resulting from overburden is responsible for shear wave speed gradients (Sect. 5.2). Higher amplitude dynamic strains are typically discussed in the geotechnical literature [Mitchell 1993].

The following development of the theory of waves in elastic media can be supplemented by a number of references, including [Landau and Lifshitz 1970, Brekhovskikh and Godin 1990].

9.1 Equations of Motion

The volume force equation for fluids (8.1) must be generalized for elastic media. This generalization involves replacing the pressure with the stress tensor, σ_{ij}:

$$f_i = \sum_j \partial_j \sigma_{ij} \ , \tag{9.1}$$

where the indices i, j represent the coordinates $(x,\ y,\ z)$, and $\partial_j = \frac{\partial}{\partial x}, \frac{\partial}{\partial y}$, or $\frac{\partial}{\partial z}$. In an isotropic elastic medium, the stress tensor can be written in terms of the displacement, $\mathbf{u}$, as follows:

$$\sigma_{ij} = \lambda \delta_{ij} \nabla \cdot \mathbf{u} + \mu(\partial_i u_j + \partial_j u_i) \ , \tag{9.2}$$

where δ_{ij} is the Kronecker delta function, and the coefficients λ and μ are the Lamé parameters. Combining this constitutive relation with the volume force equation (9.1), Newton's second law (8.2) yields the following equation of motion:

$$\omega^2 \rho u_i + \partial_i(\lambda \nabla \cdot \mathbf{u}) + \sum_j \partial_j \left[\mu(\partial_i u_j + \partial_j u_i)\right] = 0 \ , \tag{9.3}$$

where the density, ρ, and the parameters λ and μ can be functions of position. Note that λ occupies a place in the equation of motion analogous to the bulk modulus in the fluid case (8.7). The parameter μ multiplies the *strain tensor*, $(\partial_i u_j + \partial_j u_i)/2$. This parameter characterizes the medium's ability to support shear forces and, when $\mu = 0$, (9.3) reduces to the fluid equation of motion. Another special case of interest is the medium for which the Lamé parameters are independent of position. In this case, (9.3) can be written in vector form:

$$\omega^2 \rho \mathbf{u} + (\lambda + 2\mu) \nabla(\nabla \cdot \mathbf{u}) - \mu \nabla \times \nabla \times \mathbf{u} = 0 \ . \tag{9.4}$$

It is convenient to express the displacement field in terms of scalar and vector potentials [Brekhovskikh and Lysanov 1991]:

$$\mathbf{u}(\mathbf{r}) = \nabla \phi(\mathbf{r}) + \nabla \times \boldsymbol{\psi}(\mathbf{r}) \ . \tag{9.5}$$

The scalar potential, ϕ, describes compressional waves, and the vector potential, $\boldsymbol{\psi}$, describes shear waves. Without loss of generality, one can require

$$\nabla \cdot \boldsymbol{\psi}(\mathbf{r}) = 0 \ . \tag{9.6}$$

In homogeneous media, for which ρ, λ, and μ are independent of position, (9.4) reduces to two uncoupled Helmholtz equations in the two potentials,

$$\nabla^2 \phi + \frac{\omega^2}{c_p^2} \phi = 0 \tag{9.7}$$

and

$$\nabla^2 \boldsymbol{\psi} + \frac{\omega^2}{c_t^2} \boldsymbol{\psi} = 0 \ . \tag{9.8}$$

In these equations, the compressional and shear waves speeds are

$$c_p = \sqrt{\frac{\lambda + 2\mu}{\rho}} \tag{9.9}$$

and

$$c_t = \sqrt{\frac{\mu}{\rho}} \ , \tag{9.10}$$

respectively, where the subscript p denotes the compressional wave and the subscript t denotes the shear wave. While the subscript "s" might be more logical and consistent with seismic notation, it will be reserved for other purposes, and "t" (for "transverse") will be used to denote shear waves. Note that compressional waves have "irrotational" displacement, $\nabla \times \mathbf{u} = 0$, since $\mathbf{u} = \nabla \phi$. In contrast, shear wave displacement is "solenoidal," having $\nabla \cdot \mathbf{u} = 0$, since $\mathbf{u} = \nabla \times \boldsymbol{\psi}$. This means that shear waves do not produce any time-varying change in the volume of the (homogeneous, isotropic) medium.

9.2 Plane Waves

As in the fluid case, plane-wave solutions exist for the displacement field if the medium is homogeneous, that is, if ρ, λ, and μ are independent of position. Compressional plane waves have potential of the form

$$\phi(\mathbf{r}) = \phi_0 e^{i\mathbf{k}_p \cdot \mathbf{r}} \ , \tag{9.11}$$

where

$$k_p = \frac{\omega}{c_p} \ . \tag{9.12}$$

Taking the gradient, the corresponding displacement is

$$\mathbf{u}(\mathbf{r}) = i\mathbf{k}_p \phi_0 e^{i\mathbf{k}_p \cdot \mathbf{r}} \ . \tag{9.13}$$

Thus, displacement for a compressional wave is parallel to the direction of propagation. A shear plane wave has vector potential of the form

$$\boldsymbol{\psi}(\mathbf{r}) = \boldsymbol{\psi}_0 e^{i\mathbf{k}_t \cdot \mathbf{r}} , \tag{9.14}$$

where

$$k_t = \frac{\omega}{c_t} . \tag{9.15}$$

Taking the curl, the displacement is

$$\mathbf{u}(\mathbf{r}) = i\mathbf{k}_t \times \boldsymbol{\psi}_0 e^{i\mathbf{k}_t \cdot \mathbf{r}} . \tag{9.16}$$

This shows that displacement for a shear wave is perpendicular to the direction of propagation. As the complex vector amplitude $\boldsymbol{\psi}_0$ is arbitrary, the vector $i\mathbf{k}_t \times \boldsymbol{\psi}_0$ in (9.16) could be replaced by any complex vector that is perpendicular to the direction of propagation. The direction of this vector will be said to define the "polarization" of the shear wave, somewhat analogous to the polarization of an electromagnetic wave.

As in the fluid case, stratification of the seafloor gives special significance to the vertical (z) direction, and it is useful to differentiate between those plane waves whose propagation direction is tilted downward and those whose propagation direction is tilted upward:

$$\mathbf{k}_q^{\pm}(\mathbf{K}) = (\mathbf{K}, \pm k_q \beta_q(K)) , \qquad q = p,\ t. \tag{9.17}$$

The wave equations obeyed by the potentials determine the relation between the transverse and vertical components of the wave vectors. This gives the following complex sines for the compressional and shear waves:

$$\beta_q(K) = \sqrt{1 - K^2/k_q^2} . \tag{9.18}$$

The following unit vectors specifying the directions of propagation are also useful:

$$\mathbf{e}_q^{\pm}(\mathbf{K}) = \mathbf{k}_q^{\pm}(\mathbf{K})/k_q . \tag{9.19}$$

There are two plane-wave shear polarizations to consider, both having particle displacement normal to the direction of propagation. The shear wave having particle displacement in the direction

$$\mathbf{e}_h^{\pm}(\mathbf{K}) = \mathbf{k}_t^{\pm}(\mathbf{K}) \times \mathbf{e}_z/K = (K_y/K, -K_x/K, 0) \tag{9.20}$$

will be referred to as "horizontally polarized" as the particle displacement is in a horizontal plane. In this case, the polarization vector does not depend on whether propagation is upward or downward, but the superscripts $\pm$ are used for consistency with the vertical polarization case. A plane shear wave having transverse wave vector, $\mathbf{K}$, and particle displacement in the direction of the unit vector

$$\mathbf{e}_v^{\pm}(\mathbf{K}) = \mathbf{e}_h^{\pm}(\mathbf{K}) \times \mathbf{e}_t^{\pm}(\mathbf{K}) = \mp \mathbf{K}\beta_t/K + \mathbf{e}_z K/k_t \tag{9.21}$$

will be referred to as "vertically polarized," as the displacement is in a vertical plane. In this case, the polarization vector depends upon whether the wave propagation direction is tilted upward or downward. Scattering by interface roughness and inhomogeneities within the seabed cause conversion between the wave types defined above. For example, while incidence of a plane compressional wave in the water on a flat interface only gives rise to a vertically polarized shear wave, scattering by interface roughness causes conversion of compressional energy in the water to both the vertical and horizontal shear polarizations. Scattering by heterogeneity of the seabed causes conversion between all three wave types.

Figure 9.1 summarizes the conventions defined here for compressional and shear plane waves as to propagation directions and directions of displacement. Note that the horizontal polarization vector satisfies the relationship

$$\mathbf{e}_h^{\pm}(\mathbf{K}) = \mathbf{e}_t^{\pm}(\mathbf{K}) \times \mathbf{e}_v^{\pm}(\mathbf{K}) \ . \tag{9.22}$$

9.3 Losses

As in the fluid case, the simplest method of dealing with loss is to assume that the density and elastic moduli (the Lamé parameters) have small imaginary parts. If the complex moduli and the (possibly complex) density are inserted in the expressions for compressional and shear wave speeds (9.9) and (9.10), the resulting complex wave speeds will have small, negative imaginary parts, and the corresponding wave vectors will have small, positive imaginary parts.

As in the fluid case, it is convenient to define complex speed ratios.

$$a_q = \frac{c_q}{c_w} = \frac{\nu_q}{1 + i\delta_q}, \quad q = p,\ t \ . \tag{9.23}$$

The real parameters, ν_q, are the sediment/water phase speed ratios, and the real parameters, δ_q, are the loss parameters, which can be used to obtain the attenuation coefficients, α_q, for the compressional and shear waves:

$$\alpha_q = \frac{40\pi f \delta_q}{\nu_q c_w \ln(10)} \ . \tag{9.24}$$

The dimensionless attenuation parameters defined for the fluid case in Sect. 8.3 can be used here with the obvious change in subscripting.

9.4 Boundary Conditions

Consider the problem of matching wave solutions between water and an elastic seafloor. As in Fig. 8.1, the media are separated by a boundary given by the equation

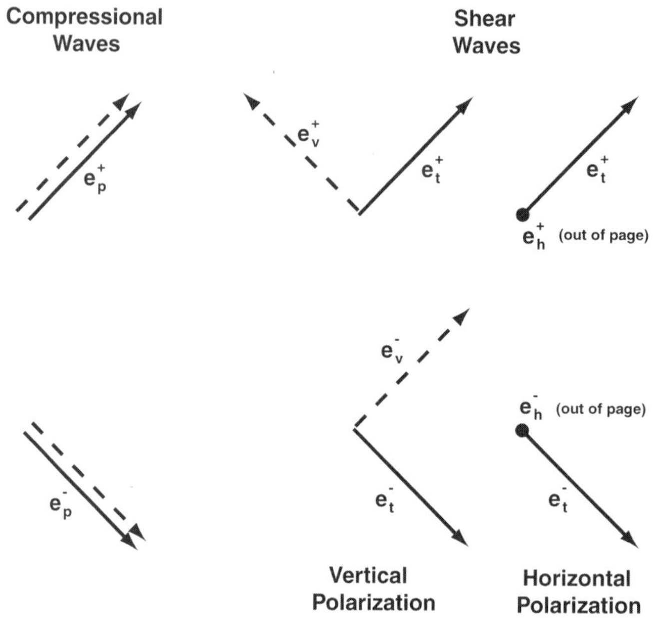

Fig. 9.1. The conventions used for unit vectors defining elastic plane-wave propagation directions (arrows with solid lines) and polarizations (direction of particle motion, arrows with dashed lines). This is a side view in the vertical plane defined by the wave vectors. Upgoing and downgoing compressional waves have directions given by the unit vectors $\mathbf{e}_p^+$ and $\mathbf{e}_p^-$, respectively, and particle motion is parallel to the propagation direction. Upgoing and downgoing shear waves have directions given by the unit vectors $\mathbf{e}_t^+$ and $\mathbf{e}_t^-$, respectively, and particle motion is in the directions $\mathbf{e}_v^{\pm}$ and $\mathbf{e}_h^{\pm}$ for vertical and horizontal polarization, respectively.

$$z = f(\mathbf{R}) \, , \tag{9.25}$$

where $\mathbf{R} = (x, \ y)$. As in the pure fluid case, one boundary condition is continuity of the normal component of displacement

$$\mathbf{N} \cdot \mathbf{u}|_{z=f(\mathbf{R})^-} = \mathbf{N} \cdot \mathbf{u}|_{z=f(\mathbf{R})^+} \, , \tag{9.26}$$

where the normal vector $\mathbf{N}$ is defined in (8.35). The other boundary condition is continuity of normal and transverse *tractions* for the stress tensor,

$$\sum_j N_j \sigma_{ij}|_{z=f(\mathbf{R})^-} = \sum_j N_j \sigma_{ij}^w|_{z=f(\mathbf{R})^+} \, , \tag{9.27}$$

where the superscript, w, denotes the fluid medium (water) and no superscript is used for the elastic medium (seafloor material). As with the pressure continuity condition, this condition guarantees that the volume force will be finite across the boundary. The traction boundary condition (9.27) is actually three equations, so there are a total of four conditions that must be satisfied in matching fields across the boundary.

9.5 Reflection and Transmission

When a plane acoustic wave is incident from water onto an elastic medium, an acoustic wave is reflected, and compressional and shear waves are transmitted into the elastic medium as shown in Fig. 9.2.

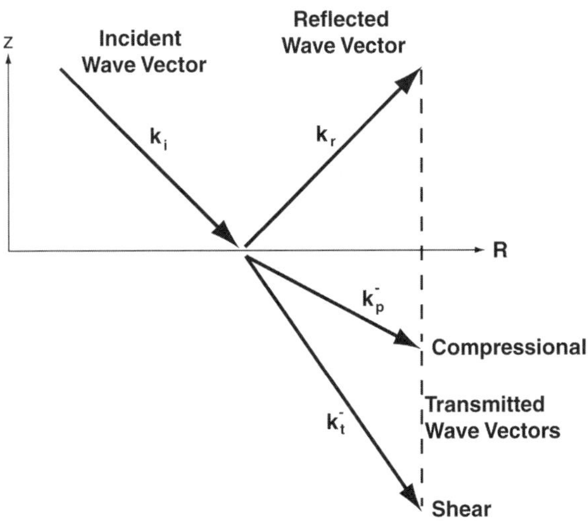

Fig. 9.2. Two-dimensional representation of a planar boundary, $z = 0$, between water and an elastic medium. The arrows indicate the wave vectors for an incident plane wave and the resulting reflected wave and the two transmitted waves, one a compressional wave and the other a shear wave. All four wave vectors have the same horizontal component, $\mathbf{K}_i$.

The scalar displacement potential in the water at an arbitrary point, $\mathbf{r} = (\mathbf{R}, z),\ z > 0$, is

$$\phi_w(\mathbf{r}) = e^{i\mathbf{k}_i\cdot\mathbf{r}} + V_{ww}e^{i\mathbf{k}_r\cdot\mathbf{r}} , \tag{9.28}$$

where

$$\mathbf{k}_i = \mathbf{K}_i - \mathbf{e}_z k_w \beta_w(K_i) \tag{9.29}$$

and

$$\mathbf{k}_r = \mathbf{K}_i + \mathbf{e}_z k_w \beta_w(K) \ . \tag{9.30}$$

The scalar potential representing the compressional wave in the sediment is

$$\phi(\mathbf{r}) = V_{wp} e^{i\mathbf{k}_p^- \cdot \mathbf{r}} \ , \tag{9.31}$$

and the vector potential representing the shear wave in the sediment is

$$\psi(\mathbf{r}) = -\mathbf{e}_h^- V_{wt} e^{i\mathbf{k}_t^- \cdot \mathbf{r}} \ . \tag{9.32}$$

The transmitted shear wave is vertically polarized, but the horizontal polarization vector is used in the vector potential, because the curl operation will then give displacement in the vertical plane. As in the treatment of reflection from a fluid–fluid interface, the x-axis is assumed to be aligned with the vertical plane of incidence so that

$$\mathbf{K} = k_w \mathbf{e}_x \cos\theta_w \ ,. \tag{9.33}$$

Imposition of the four boundary conditions is sufficient to determine the four reflection and transmission coefficients. One of these, the transmission coefficient for the horizontally polarized shear wave, vanishes. The reflection coefficient is

$$V_{ww}(\theta_w) = \frac{z_e(\theta_w) - 1}{z_e(\theta_w) + 1} \ , \tag{9.34}$$

where the normalized acoustic impedance, $z_e(\theta_w)$, is

$$z_e(\theta_w) = z_{wp} \cos^2 2\theta_t + z_{wt} \sin^2 2\theta_t \ . \tag{9.35}$$

Here,

$$z_{wp} = \frac{a_\rho a_p \sin\theta_w}{\sin\theta_p} \tag{9.36}$$

is the normalized boundary impedance for compressional waves, identical to (8.53), and

$$z_{wt} = \frac{a_\rho a_t \sin\theta_w}{\sin\theta_t} \tag{9.37}$$

is the equivalent impedance for shear waves. The variables, θ_p and θ_t, are the complex grazing angles of the compressional and shear waves in the seafloor material, obtained from Snell's law:

$$\cos\theta_q = a_q \cos\theta_w \ , \qquad q = p,\ t \ . \tag{9.38}$$

A trigonometric identity can be used to obtain

$$\cos 2\theta_t = 2a_t^2 \cos^2\theta_w - 1 \ . \tag{9.39}$$

Equation (9.35) also requires

$$\sin^2 2\theta_t = 1 - \cos^2 2\theta_t \tag{9.40}$$

as an input. In later applications of (9.35)–(9.40), the grazing angle in water, θ_w, is sometimes replaced by the scattered grazing angle in water, θ_s. This will require double subscripting of θ_t, either θ_{ti} or θ_{ts}. The reflection coefficient is used to define the bottom loss, L (2.11).

The transmission coefficients for the potentials of compressional- and shear-wave particle displacements are [Brekhovskikh and Godin 1990]

$$V_{wp}(\theta_w) = -\frac{a_p \cos 2\theta_t \sin\theta_w}{\sin\theta_p}[1 - V_{ww}(\theta_w)] \tag{9.41}$$

and

$$V_{wt}(\theta_w) = -\frac{a_t \sin 2\theta_t \sin\theta_w}{\sin\theta_t}[1 - V_{ww}(\theta_w)] \; . \tag{9.42}$$

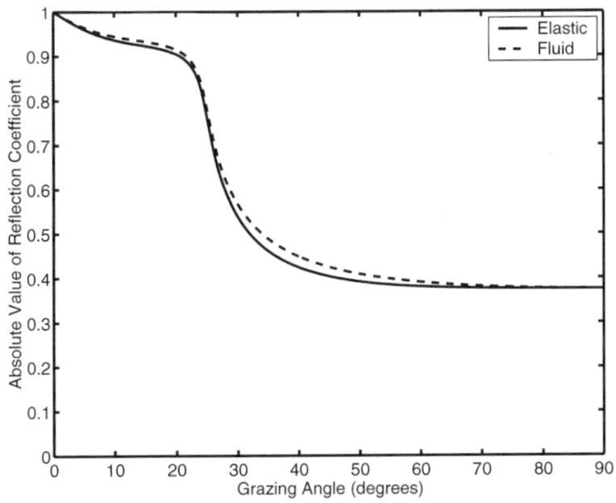

Fig. 9.3. Flat-interface reflection coefficient for a sand seafloor, comparing the fluid and elastic models. The acoustic parameters are $c_w = 1500$ m s^{-1}, $\nu_p = 1.1$, $a_\rho = 2.0$, $\delta_p = 0.01$, $\nu_t = 0.167$, $\delta_t = 0.01$. The shear speed is 250 m s^{-1}, a rather large value chosen so that the difference between the fluid and elastic models is clearly visible.

Figure 9.3 shows the magnitude of the reflection coefficient for a sand seafloor with and without inclusion of elastic effects. The calculation for the fluid case uses the same values of ν_p, δ_p, and a_ρ as for the elastic case. The shear speed is taken to be 250 m s^{-1}; for a more realistic value of 100 m s^{-1} the difference between the two models is negligible.

While elastic theory provides only a modest correction to fluid theory for unconsolidated sediments having shear speeds of a few hundred m s^{-1} or less, it yields much different results for rock seafloors, where the shear speed is typically greater than the water sound speed. Figure 9.4 shows that the difference between the two models can be quite dramatic. The fluid model predicts very low loss for grazing angles below the compressional wave critical angle, 66.4°. The elastic model predicts large loss in the vicinity of the shear wave critical angle, 39.7°. This loss is mainly due to transmission of energy into the seafloor and will be important when considering volume scattering in rock in Ch. 14.

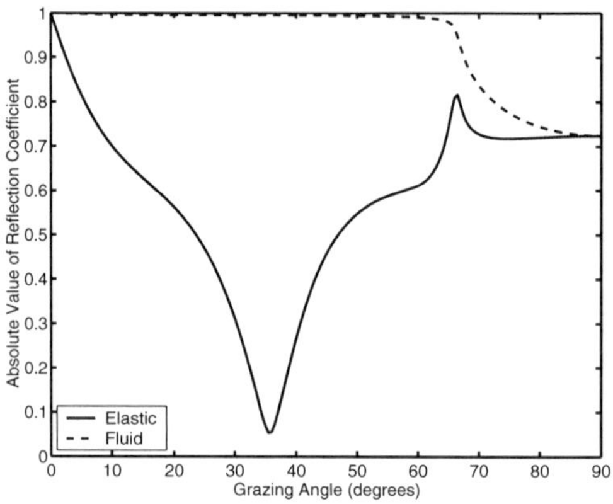

Fig. 9.4. Flat-interface reflection coefficient for a rock seafloor, comparing the fluid and elastic models. The material parameters are $\nu_p = 2.5$, $a_\rho = 2.5$, $\delta_p = 0.02$, $\nu_t = 1.3$, $\delta_t = 0.1$.

9.6 Nonplane Waves

The fields produced by a point source in the water column above a flat, homogeneous elastic seafloor can be computed using the approach given in Ch. 8 for fluid theory. The wavenumber integral expression for the field in water is unchanged from (8.63), except that the elastic expression for the reflection coefficient is to be substituted for the fluid expression. Similarly, the compressional and shear waves in the seafloor are computed using the appropriate transmission coefficients and wave vector z-components. Like the fluid case, lateral and evanescent waves are seen, but new effects emerge. This can be anticipated by examining the integrand of the wavenumber integration, specifically either the reflection or transmission coefficients, which are evaluated for all real values of wave vector, including those whose magnitude is greater than the wavenumber in water, the evanescent domain. Figure 9.5 shows a large peak corresponding to the Scholte (or Stonely) wave [Brekhovskikh and Godin 1990]. Unlike the lateral and evanescent waves, the Scholte wave is a true surface wave, decaying exponentially in both upward and downward directions. Although the reflection coefficient magnitude shows a peak value greater than unity at the Scholte wavenumber, this does not violate energy conservation, as it occurs in the evanescent domain, where the wave does not propagate away from the interface and therefore cannot transport energy away from the interface. The reflection coefficient shows small features near the compressional and shear wavenumbers. These wavenumbers correspond to incident grazing angles near the corresponding critical angles. The reflection coefficient nearly vanishes near the "Rayleigh" wavenumber [Brekhovskikh and Godin 1990, p. 111]. This is due to the existence of the "leaky Rayleigh wave," which is not a true surface wave since it is not a freely propagating mode in the lossless case. This does not mean that this wave is unimportant. The near vanishing of the reflection coefficient due to this wave renders the interface nearly transparent acoustically for angles slightly smaller than the shear critical angle and will have important consequences in scattering.

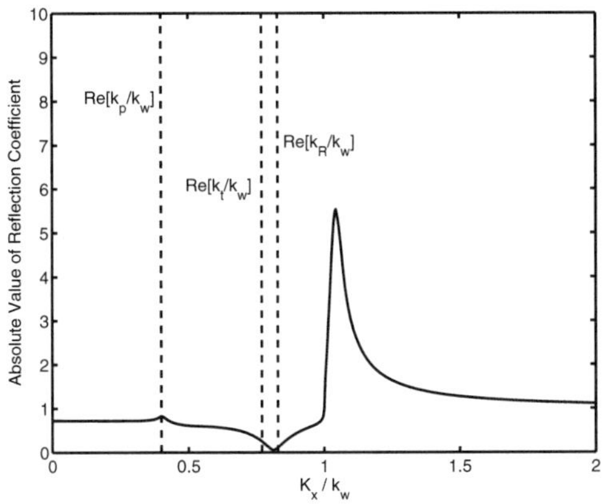

Fig. 9.5. Magnitude of flat-interface reflection coefficient for a rock seafloor, plotted as a function of horizontal wavenumber normalized by the wavenumber in water. Points on the abscissa greater than unity correspond to evanescent waves. The large peak just inside the evanescent domain is due to the Scholte wave. In order of increasing wavenumber, the three vertical dashed lines correspond to the real parts of the compressional, shear, and Rayleigh wavenumbers. The material parameters are $\nu_p = 2.5,\ a_\rho = 2.5,\ \delta_p = 0.02,\ \nu_t = 1.3,\ \delta_t = 0.1$.

9.7 Gassmann's Equations

Wood's approach for saturated sediments (Sect. 8.7) has been generalized by Gassmann [Gassmann 1951, Berryman 1999] to take account of the elastic properties of the "frame" of solid particles in fluid-saturated porous materials. One of Gassmann's equations gives the bulk modulus of a fluid-saturated, porous sediment [Anderson and Hampton 1980a]

$$K_b = K_g \frac{K_f + Q'}{K_g + Q'} , \tag{9.43}$$

where

$$Q' = K_w \frac{K_g - K_f}{\beta(K_g - K_w)} , \tag{9.44}$$

where β is the fractional porosity, K_g is the bulk modulus of the individual sediment grains, K_w is the bulk modulus of the pore water, and K_f is the bulk modulus of the frame, drained of water. The second Gassmann equation states that the shear modulus of the saturated medium is the same as that of the frame. In terms of the Lamé parameters, the sediment bulk modulus is $K_b = \lambda + 2\mu/3$. The modulus $\lambda + 2\mu = K_b + 4\mu/3$ is convenient, as it gives the compressional wave speed via (9.9).

Gassmann's equations are appropriate when one can neglect the differential motion of the fluid and frame, typically when permeability is low, as in mud [Stoll and Bautista 1998] and rock. The effects of movement of the pore fluid with respect to the frame are incorporated in Biot theory (Ch. 10).

9.8 Buckingham's Theory

Elastic theory as presented above treats attenuation empirically, without providing a physical explanation for the loss mechanism. This is in distinction to Biot theory, which, as will be seen, ascribes at least a portion of acoustic energy loss to viscous movement of water in the pores between sediment particles. One problem with the purely empirical approach is that one might inadvertently assign impossible frequency dependencies to sound speed and attenuation, that is, dependencies that violate causality (Appendix I). Buckingham [Buckingham 1997, Buckingham 1998, Buckingham 2000, Buckingham 2005] has developed a theory that postulates viscous-like forces between sediment grains. For the purposes of this monograph, Buckingham's theory will be divided into two parts, a microscopic part and an effective medium part. At the microscopic level, sediment grains come into contact, but lack bonding. Thus the dynamic frame bulk and shear moduli are zero. Instead, two types of shearing, translational and torsional, are assumed to occur at grain contacts and act as stress relaxation mechanisms. This shearing, referred to as a "stick-slip" mechanism by Buckingham, results from the applied strain and the microroughness of the grains (Sect. 3.4). The three parameters characterizing this process are the strain-hardening index, n, the compressional rigidity coefficient, γ_p, and the shear rigidity coefficient, γ_t. The values of these parameters cannot be derived from microscopic properties in the present state of the theory. Instead, they are inferred from a relatively few field measurements of sound speed and attenuation, shear wave speed, and sediment porosity and mean grain size. In [Buckingham 2005] values of these parameters so obtained are assumed to be universal and are used to predict the dependence of compressional and shear wave speeds and attenuations on frequency, depth in sediment, mean grain size, and porosity.

The microscopic stress relaxation mechanism may only be applicable to sands. It seems doubtful that this relaxation mechanism can be applied to muddy sediments where electrostatic repulsive and attractive forces and adhesion of organic matter control particle-to-particle interactions, and the flexure of clay particles may provide a dissipation mechanism. In addition, carbonate sediments often exhibit interparticulate bonding, or cementation, which should lead to significant values of frame bulk and shear moduli. Although quite intriguing, it is obvious that additional research is needed to fully define the physics of particle interactions required for the Buckingham theory.

Whether or not the microscopic part of Buckingham's theory has universal validity, the effective medium part can be examined independently, considering the parameters needed in Buckingham's wave equation to be given. Even in this restricted view, the theory makes testable predictions regarding propagation, reflection, and scattering. This theory has the formally attractive property of having causality "built in" at the outset.

Using arguments motivated by the microscopic picture, Buckingham arrives at the following equation of motion :

$$\rho \frac{\partial \mathbf{v}}{\partial t} = -\nabla p + \lambda_p \nabla \left\{ \nabla \cdot [h_p(t) \otimes \mathbf{v}] \right\} + \frac{4}{3} \eta_t \nabla \left\{ \nabla \cdot [h_t(t) \otimes \mathbf{v}] \right\}$$
$$-\eta_t \nabla \times \nabla \times [h_t(t) \otimes \mathbf{v}] = 0 \;, \tag{9.45}$$

where $\mathbf{v}$ is the particle velocity

$$\mathbf{v} = \frac{\partial \mathbf{u}}{\partial t} \;, \tag{9.46}$$

λ_p and η_t are "stress relaxation coefficients," and ρ is the density given by Wood's equation (8.70). The relation between pressure, p, and displacement, $\mathbf{u}$, is

$$p = -K_b \nabla \cdot \mathbf{u} \;, \tag{9.47}$$

where K_b is the (real) bulk modulus given by Wood's equation (8.71). Equation (9.45) is only applicable to homogeneous media and is written in the time domain as opposed to the frequency domain representation used previously for fluid and elastic theories. The notational convention used previously in this monograph has been abused slightly in that $\mathbf{u}$ now represents real, time-dependent displacement rather than complex displacement amplitude for time-harmonic excitation. Lowercase p is used to denote real, time-dependent pressure as opposed to uppercase P, used for complex pressure amplitude. This temporary excursion into the time domain is helpful because the essence of the theory is "memory" in the dissipation terms, expressed as temporal convolutions of "material impulse response functions" (MIRFs), $h_p(t)$ and $h_t(t)$ with the velocity, e.g.,

$$h(t) \otimes \mathbf{v}(t) = \int_{-\infty}^{\infty} h(t')\mathbf{v}(t-t')dt' \ . \tag{9.48}$$

The MIRFs are

$$h_p(t) = t_p^{-1}(1 + \frac{t}{t_p})^{-n} \tag{9.49}$$

and

$$h_t(t) = t_t^{-1}(1 + \frac{t}{t_t})^{-m} \ , \tag{9.50}$$

where these functions vanish for negative t, and the characteristic times t_p and t_t are very short compared to the inverse of the acoustic frequency. The parameters m and n are the "strain-hardening indices" and are not integers as the notation might suggest. Buckingham's equation reduces to the linearized Navier–Stokes equation for acoustic waves in a viscous fluid [Kinsler et al. 1999, Ch. 8] when the MIRFs are set equal to delta functions. In this case, λ_p becomes the "bulk viscosity" and η_t becomes the "shear viscosity."

If the equation of motion is expressed in the frequency domain, with complex displacements, etc., and $\exp(-i\omega t)$ time dependence, the convolutions become products

$$\omega^2 \rho \mathbf{u} + K_b \nabla(\nabla \cdot \mathbf{u}) - i\omega[\lambda_p H_p(\omega) + \frac{4}{3}\eta_t H_t(\omega)]\nabla(\nabla \cdot \mathbf{u})$$

$$+ i\omega\eta_t H_t(\omega) \nabla \times \nabla \times \mathbf{u} = 0 \ , \tag{9.51}$$

where

$$H_p(\omega) = \int_{-\infty}^{\infty} h_p(t) e^{i\omega t} dt \tag{9.52}$$

and

$$H_t(\omega) = \int_{-\infty}^{\infty} h_t(t) e^{i\omega t} dt \ . \tag{9.53}$$

Equation (9.51) is in the form of the elastic equation of motion (9.4) with complex Lamé parameters

$$\lambda + 2\mu = K_b - i\omega[\lambda_p H_p(\omega) + \frac{4}{3}\eta_t H_t(\omega)] \ , \tag{9.54}$$

$$\mu = -i\omega\eta_t H_t(\omega) \ . \tag{9.55}$$

From this it follows that there will be compressional and shear waves with complex speeds

$$c_p = \sqrt{\frac{\lambda + 2\mu}{\rho}} \tag{9.56}$$

and

$$c_t = \sqrt{\frac{\mu}{\rho}} \; . \tag{9.57}$$

Buckingham gives the following approximations for $H_p(\omega)$ and $H_s(\omega)$:

$$H_p(\omega) = \left(\frac{i}{\omega t_p}\right)^{1-n} \Gamma(1-n) \tag{9.58}$$

$$H_t(\omega) = \left(\frac{i}{\omega t_t}\right)^{1-m} \Gamma(1-m) \; , \tag{9.59}$$

where Γ is the gamma function. Specializing to the case $m = n$, the complex speeds are

$$c_p = c_0 \sqrt{1 + \frac{3\gamma_p + 4\gamma_t}{3\rho c_0^2}(-i\omega T)^n} \tag{9.60}$$

and

$$c_t = c_0 \sqrt{\frac{\gamma_t}{\rho c_0^2}}(-i\omega T)^{n/2} \; , \tag{9.61}$$

where

$$c_0 = \sqrt{\frac{K_b}{\rho}} \tag{9.62}$$

is the (real) wave speed given by Wood's equation (8.72). The parameters γ_p and γ_t are called the compressional and shear relaxation coefficients and are given by

$$\gamma_p = \frac{\lambda_p}{t_p}\left(\frac{t_p}{T}\right)^n \Gamma(1-n) \tag{9.63}$$

and

$$\gamma_t = \frac{\eta_t}{t_t}\left(\frac{t_t}{T}\right)^n \Gamma(1-n) \; . \tag{9.64}$$

The parameter T has the units of time and may be assigned any convenient value (e.g., $T = 1$ s), as changes in T will be compensated by changes in γ_p and γ_t so that the velocities remain unchanged. As in the elastic case, complex ratios of compressional and shear speeds (9.23) can be defined along with attenuation coefficients (9.24). Buckingham notes the simple result, obtained from (9.61):

$$\delta_t = \tan(n\pi/4) \; . \tag{9.65}$$

Thus, the shear wave attenuation sets the value of n via the dimensionless loss parameter, δ_t. Typically $n << 1$. The parameter γ_t can be related to measurable quantities through (9.61) in the form

$$\gamma_t = \frac{\rho c_{tphase}^2 \cos^2(n\pi/4)}{(\omega T)^n} , \tag{9.66}$$

with the shear wave phase velocity given by

$$\frac{1}{c_{tphase}} = \mathrm{Re}\left\{\frac{1}{c_t}\right\} . \tag{9.67}$$

Equation (9.66) can be evaluated at any frequency for which phase velocity is known. If the theory is reasonably accurate, γ_t should be essentially independent of frequency. The remaining parameter, γ_p, can be determined by forcing agreement between measured compressional wave phase speed at some convenient frequency and phase speed as determined from (9.60). Again, the relation between complex speed and phase speed is

$$\frac{1}{c_{pphase}} = \mathrm{Re}\left\{\frac{1}{c_p}\right\} . \tag{9.68}$$

Thus, Buckingham's theory predicts the frequency dependence of compressional and shear phase speeds and attenuations given measurements at a single frequency of the two phase speeds and the shear attenuation. Such measurements are available for an experiment conducted in Florida [Williams et al. 2002a], and result in the frequency dependencies shown in Fig. 9.6. Speeds increase slowly with frequency and attenuations increase approximately linearly with frequency. When phase speed changes with frequency, propagating pulses will become distorted owing to propagation at different speeds of the frequency components making up the pulse. This effect is known as "dispersion." Dispersion must be accompanied by frequency-dependent attenuation, and all frequency dependencies must satisfy the *Kramers–Kronig* relations (Appendix I). Dispersion and attenuation in Buckingham's theory satisfy these relations.

Buckingham's theory is compared with frequency-dependent sound speed and attenuation data and Biot theory in Ch. 10. It is concluded that each theory seems to explain some, but not all, features of propagation in the sand sediment of SAX99. In addition to tests of frequency dependence, the wave-theoretic part (as opposed to the microscopic part) of Buckingham's theory can be tested through measurements of reflection and scattering. At a fixed frequency, Buckingham's wave theory is equivalent to the elastic theory discussed in previous sections of this chapter, or, if shear effects are negligible, to the fluid theory of Ch. 8. Thus any measurements that discriminate between elastic or fluid theory and Biot theory can equally be said to discriminate between Buckingham's theory and Biot theory. As an example, for sandy seafloors, Biot theory tends to predict somewhat lower levels of reflection (Sect. 10.1.5) and scattering (Sect. 13.2.3) than fluid or elastic theory.

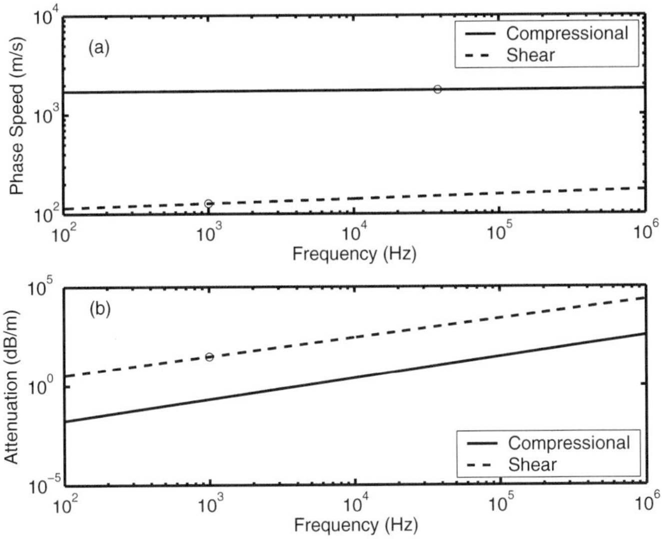

Fig. 9.6. Phase speeds (a) and attenuations (b) versus frequency using Buckinghams' theory [Buckingham 2000] with inputs from [Williams et al. 2002a]. These inputs were determined from compressional wave speed measurement at 38 kHz and shear wave speed and attenuation measurement at 1 kHz. These points are shown as circles on the plots. The parameters are $c_0 = 1653.4$ m s^{-1}, $\rho = 2074$ kg/m^3, $\gamma_p = 2.48 \times 10^8$ Pa, $\gamma_t = 1.517 \times 10^7$ Pa, and $T = 1$ s.

9.9 Research Issues

Alternative approaches to compressional and shear wave propagation in sediments (such as Buckingham's theory) are of current interest, but discussion of research issues will be deferred to Ch. 10.

10 Poroelastic Theories

Sediments are porous, with the possibility that the fluid and granular phases will vibrate differently in response to acoustic excitation. Poroelastic theory (or Biot theory) treats both porosity and elasticity. As will be seen, poroelastic effects are noticeable in sand, but are presumably less important in softer sediments where Gassmann's equations may be adequate. While this theory has had some success in application to sediments, it appears likely that a complete understanding of the acoustics of sediments may require a modification or generalization of Biot theory. Some proposed modifications are mentioned later in this chapter and in Ch. 11.

The application of Biot theory as described in this chapter requires values of 13 physical parameters. These include the physical properties of the pore water (dynamic viscosity, bulk modulus, and density), sediment particles (bulk modulus and density), and the sediment frame (permeability, tortuosity, porosity, pore size parameter, and the real and imaginary parts of the bulk and shear moduli). Methods to predict physical properties of pore water are described in Sect. 4.7 and Appendix B. Methods to measure sediment particle properties and typical values are given in Sect. 4.5. Techniques to measure or predict permeability, porosity, tortuosity, and pore size are given in Sects. 4.3 and 4.4. As noted in Sect. 10.3, accurate determination of several of the Biot parameters is difficult, making rigorous tests of the theory nearly impossible given the current state of measurement techniques as described in Chs. 3 and 4. Given this situation, knowledge of the microscopic structure of sediments (Ch. 3) can be a guide in deciding which Biot parameters are truly important for various sediment types. For example, permeability is unlikely to be important in clays. One approach to the problem of determining the Biot parameters has been to use geoacoustic measurements (Ch. 5) to constrain unmeasured parameters. The success of this approach would be evident if the resulting Biot predictions were found to be valid outside the range of frequencies of the original geoacoustic measurements.

10.1 Biot Theory

Biot theory accounts for the fact that, in a two-phase system such as a sediment composed of grains and water (Fig. 10.1), these two components may move with different displacements when waves are excited. In this theory, the sediment grains constitute an elastic "frame" coupled to the pore fluid. One result of this coupling is that the acoustic wave of the fluid (with longitudinal displacement) and the compressional and shear waves of the frame become mixed together into three hybrid wave types, two having longitudinal displacement and one being a shear wave with transverse displacement (and two different polarizations). Of the two longitudinal waves, one has speed similar to the water sound speed (the "fast" wave) and one has speed an order of magnitude lower (the "slow" wave). As indicated in Fig. 10.1, during passage of a longitudinal wave, the grain and fluid displacement may differ. Rather complete treatments of the subject are available [Biot 1956a, Biot 1956b, Biot 1962a, Biot 1962b, Deresiewicz and Skalak 1963, Stoll 1974, Stoll 1989, Pierce et al. 2005], but the discussion in this chapter is necessarily limited, giving results with some explanation but little derivation. It will use notation and a development similar to that of [Stoll and Kan 1981].

Fig. 10.1. Simulated section through a sandy sediment, with grains indicated as white and pore fluid as black. The arrows indicate that longitudinal wave motion may result in different average displacements for grains and fluid; they may even move in opposite directions as in this illustration. The displacements suggested by the arrows are greatly exaggerated.

10.1.1 Biot Constitutive Relations

In Biot theory, constitutive relations connecting stress and strain are necessarily more complicated than for elastic media, where two moduli suffice for the isotropic case. In an isotropic, elastic porous medium, two additional moduli are required. The constitutive relation for stress is

$$\sigma_{ij} = [(H - 2\mu)\nabla \cdot \mathbf{u} - C\nabla \cdot \mathbf{w}]\delta_{ij} + \mu(\partial_i u_j + \partial_j u_i) , \tag{10.1}$$

where

$$\mathbf{w} = \beta(\mathbf{u} - \mathbf{U}) \tag{10.2}$$

is the *relative displacement*, a measure of the displacement of the pore fluid relative to the frame. The displacement field of the frame is denoted $\mathbf{u}$, and the displacement field of the pore fluid is denoted $\mathbf{U}$. These are spatial averages over volumes large compared to the grain size, yet small compared to the wavelengths of the three wave types that will be encountered. It must be realized that averaging involves division not by the averaging volume, but by the fraction of this volume that is occupied by each constituent. Thus, the average is not "diluted" by the presence of the other constituent. The definition of $\mathbf{w}$ is such that the integral of its normal component over an area gives the negative of the volume of fluid transported relative to the frame. Comparing (10.1) with the corresponding elastic relation (9.2), one sees that $H-2\mu$ takes the place of the Lamé parameter, λ. The other Lamé parameter, μ, is the shear modulus of the frame. An additional term not found in the elastic case appears in (10.1) with an additional modulus, C, that couples the fluid and frame motions. Expression (10.1) represents the total stress acting on a small volume of both frame and fluid. A second constitutive relation involves the pressure, P_f, of the pore fluid:

$$P_f = M\nabla \cdot \mathbf{w} - C\nabla \cdot \mathbf{u} . \tag{10.3}$$

This relation introduces a fourth complex, frequency-dependent modulus, M.

These four moduli can be related to parameters of the constituent media following a procedure given by [Stoll 1974]. The relationships are

$$H = (K_g - K_f)^2/(D - K_f) + K_f + 4\mu/3 , \tag{10.4}$$

$$C = K_g(K_g - K_f)/(D - K_f) , \tag{10.5}$$

$$M = K_g^2/(D - K_f) , \tag{10.6}$$

$$D = K_g[1 + \beta(K_g/K_w - 1)] , \tag{10.7}$$

where K_g is the bulk modulus of the individual sediment grains, K_w is the bulk modulus of the pore water, and K_f is the bulk modulus of the frame. While K_g and K_w are usually taken to be real and frequency independent, the frame bulk modulus, like the frame shear modulus, μ, is usually assumed to be complex. The two complex frame moduli account for a portion of the loss in porous materials. These moduli are usually assumed to be frequency independent, although this entails a slight violation of causality (Appendix I), being equivalent to the fluid assumption of complex, frequency-independent compressional wave speed. Equation (10.4) is equivalent to Gassmann's equation (9.43).

10.1.2 Biot Equations of Motion

The volume force due to stress is given by (9.1), and Newton's second law applied to a small volume is

$$\sum_j \partial_j \sigma_{ij} = -\omega^2(\rho u_i - \rho_w w_i) \ , \tag{10.8}$$

where

$$\rho = \beta\rho_w + (1-\beta)\rho_g \tag{10.9}$$

is Wood's expression for density (8.70). The pore water density is denoted ρ_w, and ρ_g is the density of the material comprising the sediment grains. In (10.8), the factor $-\omega^2$ corresponds to a double time derivative. Note that the right-hand side of (10.8) can be written in the more physically obvious form $-\omega^2[(1-\beta)\rho_g u_i + \beta\rho_w U_i]$.

A second equation of motion involves the fluid pressure gradient

$$-\partial_i P_f = -\omega^2 \rho_w (u_i - \frac{\alpha}{\beta} w_i) + i\omega\frac{\bar{\eta}}{\kappa} w_i \ . \tag{10.10}$$

The double time-derivative terms represent inertia, and the single time derivative ($i\omega$) term represents viscous loss. The real parameter, α, is called the "tortuosity" and models the increase in inertia due the complicated pore structure as compared to flow in a straight line. If $\alpha = 1$, the inertial terms combine to form $\rho_w \ddot{U}_i$, as expected for the ordinary fluid equation of motion (8.4). The viscous term is a modified form of Darcy's law for fluid flow through a porous medium. The version of this law expressed in (4.13) can be put in the form

$$-\nabla P_f = \frac{\eta}{\kappa}\mathbf{V} \ , \tag{10.11}$$

where η is the dynamic viscosity, replacing the symbol μ used in (4.13). The permeability of the porous medium is denoted κ. This equation is analogous to the point form of Ohm's law for electrical conduction, with pressure analogous to electrostatic potential, fluid velocity ($\mathbf{V}$) analogous to current density, and η/κ analogous to resistivity. Darcy's law in this form has greater predictive power than Ohm's law in that the analog of resistivity is a ratio of the viscosity, an intrinsic property of the fluid, and permeability, an intrinsic property of the porous medium. Thus, this law makes predictions for the effect of changing the type of fluid while keeping the porous medium the same and vice-versa. Darcy's law only applies strictly for steady flow. For oscillatory flow, as frequency is increased, the effects of inertia will ultimately be felt. Flow will decrease and will be shifted in phase. In Biot theory, this is modeled by replacing the dynamic viscosity by a complex, frequency-dependent value,

$$\bar{\eta} = F\eta \tag{10.12}$$

A common form for the complex correction factor, F, is obtained by considering oscillatory flow in straight tubes of circular cross section having radius a:

$$F(\epsilon) = \frac{\frac{\epsilon}{4}T(\epsilon)}{1 - \frac{2i}{\epsilon}T(\epsilon)} , \tag{10.13}$$

with

$$T = \frac{-\sqrt{i}J_1(\epsilon\sqrt{i})}{J_0(\epsilon\sqrt{i})} , \tag{10.14}$$

where $J_0(\epsilon\sqrt{i})$ and $J_1(\epsilon\sqrt{i})$ are cylindrical Bessel functions, and

$$\epsilon = a\sqrt{\frac{\omega\rho_w}{\eta}} . \tag{10.15}$$

The parameter a is called the "pore size." Since the pores are not actually circular cylinders, pore size is not a simple radius. The pore size can be considered to be a material property characteristic of the medium, but some authors [Johnson et al. 1987, Turgut 2000, Williams et al. 2002a, ?] use the relation

$$\kappa = \frac{a^2\beta}{8\alpha} \tag{10.16}$$

that is obtained by treating the pores as tubes of circular cross section and radius a. Note that a single length scale is assigned to the pore size, although [Hovem 1980, Yamamoto and Turgut 1988] have argued that multiple scales are involved.

Figure 10.2 shows a typical dependence of the correction factor, F, on frequency. In this example, the correction is slight for frequencies below 1 kHz, but becomes significant above 10 kHz. The viscous loss term is a large contributor to attenuation in Biot theory as applied to sandy sediments, as will be seen. The resulting frequency-dependent attenuation is consistent with causality as expressed by the Kramers–Kronig relations (Appendix I) as it is derived from a dynamic physical model rather than being imposed ad hoc.

Combining the constitutive relations with the force–acceleration equations, the following poroelastic equations of motion are obtained:

$$\partial_i \left[(H - 2\mu)\nabla \cdot \mathbf{u} - C\nabla \cdot \mathbf{w}\right] + \sum_j \partial_j \left[\mu(\partial_i u_j + \partial_j u_i)\right] = -\omega^2 \rho u_i + \omega^2 \rho_w w_i , \tag{10.17}$$

$$\partial_i \left[C\nabla \cdot \mathbf{u} - M\nabla \cdot \mathbf{w}\right] = -\omega^2 \rho_w (u_i - \frac{\alpha}{\beta} w_i) + i\omega \frac{\bar{\eta}}{\kappa} w_i . \tag{10.18}$$

These equations [Biot 1962b] apply to heterogeneous media, but are usually specialized to the homogeneous case in which the various parameters are constant with respect to position. It is important, however, to have the more general equations at hand when considering boundary conditions and scattering by heterogeneity.

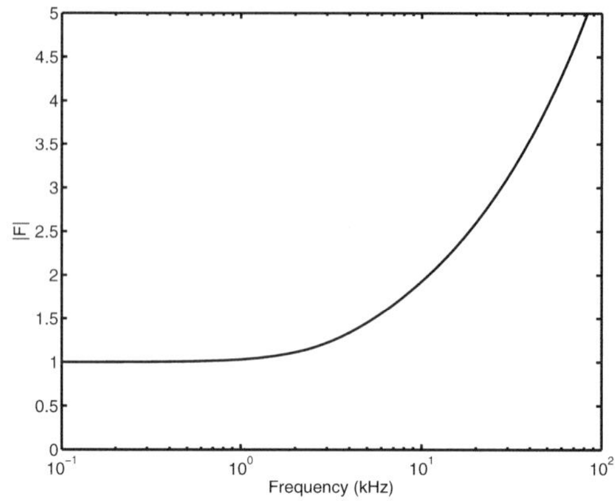

Fig. 10.2. The correction factor (10.13) as a function of frequency. The parameter values of Table 10.1 have been inserted in (10.13)–(10.16).

Specializing for the moment to the homogeneous medium, solutions that correspond to pure longitudinal waves (analogous to compressional waves) and shear waves can be found. These solutions are most readily obtained in terms of the potentials $\phi_s, \phi_f, \boldsymbol{\psi}_s, \boldsymbol{\psi}_f$:

$$\mathbf{u} = \nabla \phi_s + \nabla \times \boldsymbol{\psi}_s \, , \tag{10.19}$$

$$\mathbf{w} = \nabla \phi_f + \nabla \times \boldsymbol{\psi}_f \, . \tag{10.20}$$

In terms of these potentials, Biot's equations for homogeneous porous media become

$$H\nabla^2 \phi_s - C\nabla^2 \phi_f = -\omega^2 \rho \phi_s + \omega^2 \rho_w \phi_f \, , \tag{10.21}$$

$$C\nabla^2 \phi_s - M\nabla^2 \phi_f = -\omega^2 \rho_w \phi_s + \frac{\omega^2 \alpha \rho_w}{\beta} \phi_f + \frac{i\omega F \eta}{\kappa} \phi_f \, , \tag{10.22}$$

$$\mu \nabla^2 \boldsymbol{\psi}_s = -\omega^2 \rho \boldsymbol{\psi}_s + \omega^2 \rho_w \boldsymbol{\psi}_f \, , \tag{10.23}$$

and

$$\frac{-i\omega F \eta}{\kappa} \boldsymbol{\psi}_f = -\omega^2 \rho_w \boldsymbol{\psi}_s + \frac{\omega^2 \alpha \rho_w}{\beta} \boldsymbol{\psi}_f \, . \tag{10.24}$$

10.1.3 Plane Waves

Consider an infinite, homogeneous porous medium, and generalize the elastic case to include an additional longitudinal wave (the slow wave). The various plane waves have spatial dependence

$$\exp(i\mathbf{k}_q^{\pm} \cdot \mathbf{r}) \, , \quad q = 1, \; 2, \; t,$$

with the subscript q denoting the wave vectors for fast longitudinal waves (1), slow longitudinal waves (2), and shear waves (t), respectively. As before, the superscripts + and − denote the direction of the wave propagation, tilted upward and downward, respectively.

The three-dimensional wave vectors can be written in terms of the horizontal and vertical components,

$$\mathbf{k}_q^{\pm}(\mathbf{K}) = (\mathbf{K}, \pm k_q \beta_q(K)) , \tag{10.25}$$

with

$$\beta_q(K) = \sqrt{1 - K^2/k_q^2} . \tag{10.26}$$

The complex speed ratios

$$a_q = c_q/c_w \tag{10.27}$$

will prove useful in simplifying later expressions.

The following unit vectors specify the directions of propagation:

$$\mathbf{e}_q^{\pm}(\mathbf{K}) = \mathbf{k}_q^{\pm}(\mathbf{K})/k_q . \tag{10.28}$$

The frequency dependencies of the three wavenumbers are found by inserting plane wave solutions into Biot's equations for the infinite medium. Consider longitudinal waves of the form

$$\phi_s = \exp(i\mathbf{k}_q \cdot \mathbf{r}) , \quad q = 1,\ 2 , \tag{10.29}$$

$$\phi_f = \gamma_q \exp(i\mathbf{k}_q \cdot \mathbf{r}) , \tag{10.30}$$

where $q = 1,\ 2$ correspond to the fast and slow waves, respectively, and where the $\pm$ superscripts have been dropped as superfluous. When the plane-wave solutions, (10.29) and (10.30) are inserted in Biot's equations of motion, two homogeneous, simultaneous linear equations result. Setting the corresponding determinant to zero, one obtains

$$(Hk_q^2 - \rho\omega^2)(m\omega^2 - Mk_q^2 + i\frac{F\eta\omega}{\kappa}) + (Ck_q^2 - \rho_w\omega^2)^2 = 0 , \tag{10.31}$$

with $m = \alpha\rho_w/\beta$, to be solved for the fast and slow wavenumbers. The ratio of the two potentials, Φ_f/Φ_s, follows from the homogeneous equations as

$$\gamma_q = \frac{\rho c_q^2 - H}{\rho_w c_q^2 - C} . \tag{10.32}$$

Recalling the definitions of the two scalar potentials (10.19) and (10.20), the displacement of the fluid is proportional to the displacement of the frame according to the following relation:

$$\mathbf{U} = (1 - \frac{\gamma_q}{\beta})\mathbf{u} , \quad q = 1,\ 2 . \tag{10.33}$$

The analogous relations for shear waves are

$$\boldsymbol{\psi}_s = \mathbf{e}_h(\mathbf{K}) \exp(i\mathbf{k}_t \cdot \mathbf{r}) , \tag{10.34}$$

$$\boldsymbol{\psi}_f = \gamma_t \mathbf{e}_h(\mathbf{K}) \exp(i\mathbf{k}_t \cdot \mathbf{r}) , \tag{10.35}$$

$$(\mu k_t^2 - \rho\omega^2)(m\omega^2 + i\frac{F\eta\omega}{\kappa}) + \rho_w^2\omega^4 = 0 , \tag{10.36}$$

$$\gamma_t = \frac{\rho}{\rho_w} - \frac{\mu}{\rho_w c_t^2} , \tag{10.37}$$

$$\mathbf{U} = (1 - \frac{\gamma_t}{\beta})\mathbf{u} . \tag{10.38}$$

As in the elastic case, there are two plane-wave shear polarizations to consider having displacements in the directions of the "horizontal" and "vertical" unit vectors, $\mathbf{e}_h^{\pm}(\mathbf{K})$ and $\mathbf{e}_v^{\pm}(\mathbf{K})$. Again, only the vertically polarized wave is excited at the plane boundary, but the corresponding vector potential has direction parallel to $\mathbf{e}_h^{\pm}(\mathbf{K})$. For the fast and slow waves, displacement is parallel to the unit vectors specifying propagation direction, $\mathbf{e}_1^{\pm}(\mathbf{K})$ and $\mathbf{e}_2^{\pm}(\mathbf{K})$.

The dependence of wavenumber on frequency is found by solving (10.31) for longitudinal (fast and slow) waves and (10.36) for shear waves. For the longitudinal waves

$$k_q = \frac{-b \pm \sqrt{b^2 - 4ac}}{2a} , \quad q = 1,\ 2. \tag{10.39}$$

The (+) sign corresponds to the fast wave ($q = 1$) and the (−) sign corresponds to the slow wave ($q = 2$). The parameters appearing in (10.39) are

$$a = C_H^2 C_M^2 - C_C^4 , \tag{10.40}$$

$$b = -C_H^2 \omega \Omega + 2C_C^2 \omega^2 - \frac{\rho}{\rho_w} C_M^2 \omega^2 , \tag{10.41}$$

$$c = \frac{\rho}{\rho_w} \Omega \omega^3 - \omega^4 , \tag{10.42}$$

$$C_H^2 = \frac{H}{\rho_w} , \tag{10.43}$$

$$C_M^2 = \frac{M}{\rho_w} , \tag{10.44}$$

$$C_C^2 = \frac{C}{\rho_w} , \tag{10.45}$$

$$\Omega = \omega \frac{\alpha}{\beta} + \frac{iF\eta}{\rho_w \kappa} . \tag{10.46}$$

The wavenumber for the shear wave is

$$k_t = \omega c_t \ , \tag{10.47}$$

where

$$c_t = \sqrt{\frac{\rho - \omega\rho_w/\Omega}{\mu}} \ . \tag{10.48}$$

The complex speed ratios and associated parameters are

$$a_q = \frac{c_q}{c_w} = \frac{\nu_q}{1 + i\delta_q}, \quad q = 1,\ 2,\ t \ . \tag{10.49}$$

The real parameters, ν_q, are the sediment/water phase speed ratios, and the real parameters δ_q are the loss parameters, from which attenuation in dB/unit length can be obtained:

$$\alpha_q = \frac{40\pi f \delta_q}{\nu_q c_w \ln(10)} \ . \tag{10.50}$$

The dimensionless attenuation parameters defined for the fluid case in Sect. 8.3 can be used here with a simple change in subscripting.

Phase speed is the ratio of angular frequency divided by the real part of the wavenumber, thus

$$\frac{1}{c_{qphase}} = \mathrm{Re}\left\{\frac{1}{c_q}\right\} \ . \tag{10.51}$$

Figure 10.3 shows typical frequency dependence for phase speed and loss parameter for the fast and slow Biot waves. The fast wave phase speed depends weakly, but significantly, on frequency, while the slow wave speed is a strong function of frequency. The loss parameters for both waves are strong functions of frequency, but losses are extremely high for the slow wave. As frequency is reduced, the slow wave loss parameter approaches unity, which is characteristic of solutions of the diffusion equation. Thus, the slow wave at low frequencies is not really a wave in the usual sense.

Figure 10.4 presents another view of the fast-wave speed and attenuation for the same parameters as used in plotting Fig. 10.3. The vertical dashed line is placed at Biot's characteristic frequency [Biot 1956a, Biot 1956b],

$$f_c = \frac{\beta^2 \eta}{2\pi\rho_w\kappa} \ . \tag{10.52}$$

For frequencies much smaller than f_c, the attenuation increases as the second power of frequency, while for frequencies much greater than f_c, the attenuation increases as the square root of frequency. This behavior is evident in the lower panel of Fig. 10.4. Additional attenuation results from the use of complex frame moduli, but this is insignificant for the parameters used here (Table 10.1), with frame moduli small compared to the bulk modulus of water.

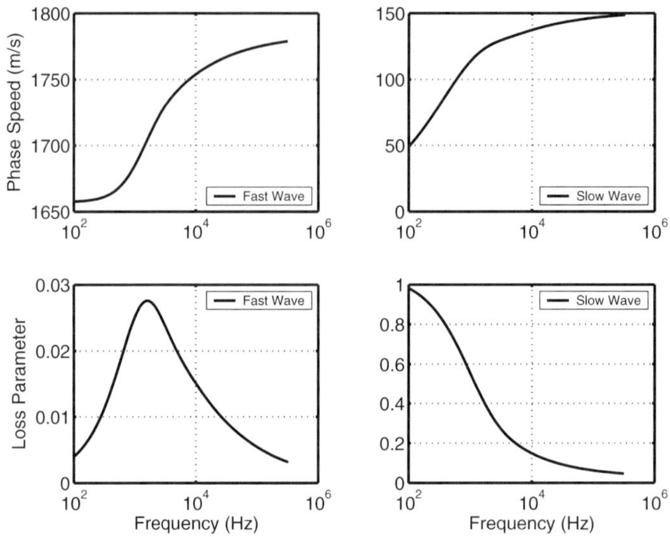

Fig. 10.3. Phase speeds and loss parameters for the fast and slow Biot waves as functions of frequency. The parameter values of Table 10.1 have been used.

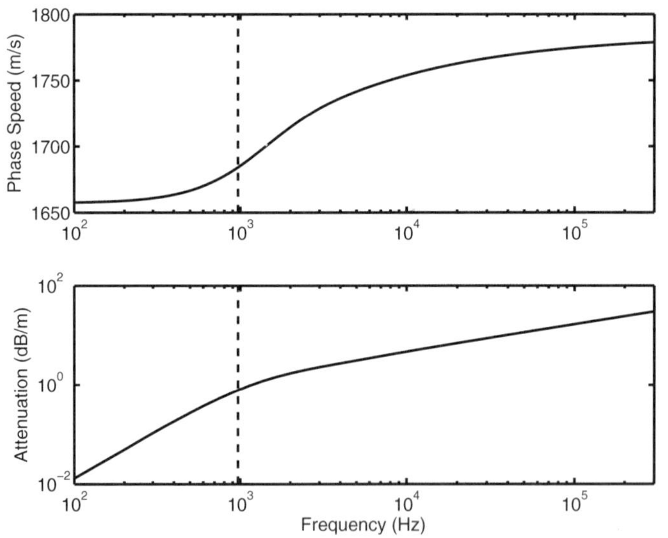

Fig. 10.4. Fast-wave phase speed and attenuation for the parameters of Fig. 10.3. The dashed vertical line is placed at Biot's characteristic frequency, f_c.

10.1.4 Boundary Conditions

The boundary conditions applying to the interface between two dissimilar porous media are more involved than those for fluid or elastic media. Although the main interest in this monograph is the sediment–water boundary, it is useful to begin with the more general case of two porous media in contact at an interface defined by $z = f(\mathbf{R})$, with normal vector, $\mathbf{N}$, given by (8.35). The boundary conditions can be inferred from the equations of motion for the heterogeneous medium by considering a thin transition layer in which the material properties change smoothly from one medium to the next. This is analogous to the development in the fluid case (Sect. 8.4). Letting the thickness of this layer approach zero, the following field quantities must be continuous across the interface [Gurevich and Schoenberg 1999]:

$$\sum_j N_j \sigma_{ij} = \text{Tractions} , \tag{10.53}$$

$$P_f = \text{Fluid Pressure} , \tag{10.54}$$

$$\mathbf{u} = \text{Frame Displacement} , \tag{10.55}$$

$$\mathbf{N} \cdot \mathbf{w} = \text{Normal Component, Relative Displacement} . \tag{10.56}$$

These conditions are general enough to apply to interfaces between a porous medium and a second medium that may be porous, elastic, or fluid. Note there are a total of eight continuity conditions, sufficient to determine the eight unknown wave amplitudes (two shear waves, two longitudinal waves in each medium) for any given incident wave. For the fluid case, it is useful to replace the continuity condition on $\mathbf{N} \cdot \mathbf{w}$ by the equivalent condition of continuity of fluid flow

$$\mathbf{N} \cdot (\mathbf{u} - \mathbf{w}) = \mathbf{N} \cdot [(1 - \beta)\mathbf{u} + \beta\mathbf{U})] . \tag{10.57}$$

This condition is obtained by subtracting (10.56) from the normal component of (10.55) and requires that the average displacement (frame plus fluid) in the poroelastic medium be equal to the displacement of the fluid medium. This formulation avoids the problem that the frame displacement, $\mathbf{u}$, is undefined in the fluid because, if $\beta = 1$, the frame displacement disappears from consideration in (10.57). To summarize the situation for the boundary between a fluid and a porous medium, continuity of tractions, fluid pressure, and fluid flow must be enforced. This is a total of five conditions, sufficient to determine the five unknown wave amplitudes. Continuity of fluid flow is the "open-pore" boundary condition, used by [Stoll and Kan 1981], and discussed in the literature [Deresiewicz and Skalak 1963]. The "closed-pore" boundary condition [Gurevich and Schoenberg 1999] is not applicable to the problems of interest in this monograph.

If the boundary $z = f(\mathbf{R})$ separates a homogeneous porous medium from a homogeneous fluid (water), the boundary conditions can be expressed in terms of the potentials introduced earlier. These conditions are continuity of tractions

$$\{[(H-2\mu)\nabla^2\phi_s - C\nabla^2\phi_f]\mathbf{N} + 2\mu(\mathbf{N}\cdot\nabla)(\nabla\phi_s)\}\,|_{z=f(\mathbf{R})^-}$$

$$+\{\mu(\mathbf{N}\cdot\nabla)(\nabla\times\boldsymbol{\psi}_s) + \mu\nabla(\nabla\times\boldsymbol{\psi}_s\cdot\mathbf{N})\}\,|_{z=f(\mathbf{R})^-} = K_w\nabla^2\phi_w\mathbf{N}|_{z=f(\mathbf{R})^+}\ , \tag{10.58}$$

continuity of fluid pressure

$$(M\nabla^2\phi_f - C\nabla^2\phi_s)|_{z=f(\mathbf{R})^-} = -K_w\nabla^2\phi_w|_{z=f(\mathbf{R})^+}\ , \tag{10.59}$$

and continuity of fluid flow

$$\mathbf{N}\cdot(\nabla\phi_s - \nabla\phi_f + \nabla\times\boldsymbol{\psi}_s - \nabla\times\boldsymbol{\psi}_f)|_{z=f(\mathbf{R})^-} = \mathbf{N}\cdot\nabla\phi_w|_{z=f(\mathbf{R})^+}\ . \tag{10.60}$$

The subscript w denotes the upper medium (water). In the last term on the left-hand side of (10.58) the gradient operator acts on the curl but not on the interface normal.

In the problem of reflection of a plane acoustic wave from a flat interface between water and a porous seafloor [Stoll and Kan 1981], there are only four boundary conditions, as required to determine four unknowns: the compressional field in the water and three fields in the seafloor material (fast- and slow-wave fields and the shear field having vertical polarization). Scattering by the rough interface causes excitation of the other possible shear polarization (horizontal), so the additional boundary condition is needed to determine this additional unknown.

10.1.5 Reflection and Transmission

When a plane acoustic wave is incident from water onto a poroelastic medium, an acoustic wave is reflected, and longitudinal (fast and slow) and shear waves are transmitted into the poroelastic medium.

The scalar displacement potential in the water is

$$\phi_w(\mathbf{r}) = e^{i\mathbf{k}_i\cdot\mathbf{r}} + V_{ww}e^{i\mathbf{k}_r\cdot\mathbf{r}}\ , \tag{10.61}$$

with all variables having the same meaning as in Sect. 9.5.

The scalar potentials representing the two longitudinal waves in the sediment are

$$\phi_{sq}(\mathbf{r}) = V_{wq}e^{i\mathbf{k}_q^-\cdot\mathbf{r}}\ , \quad q=1,\ 2\ , \tag{10.62}$$

$$\phi_{fq}(\mathbf{r}) = \gamma_q V_{wq}e^{i\mathbf{k}_q^-\cdot\mathbf{r}}\ , \quad q=1,\ 2\ . \tag{10.63}$$

The subscripts s and f identify the potentials for frame displacement (10.19) and differential fluid-frame displacement (10.20), and the subscript q identifies the wave type. Similarly, the vector potentials representing the shear wave in the sediment are

$$\boldsymbol{\psi}_s(\mathbf{r}) = \mathbf{e}_h^- V_{wt} e^{i\mathbf{k}_t^- \cdot \mathbf{r}} \tag{10.64}$$

and

$$\boldsymbol{\psi}_f(\mathbf{r}) = \mathbf{e}_h^- \gamma_t V_{wt} e^{i\mathbf{k}_t^- \cdot \mathbf{r}} \ . \tag{10.65}$$

The ratios of the pore fluid and frame potentials (10.33) and (10.38) have already been determined in the free-space solution of the Biot equations and are incorporated in (10.63) and (10.65), so there are a total of four unknown coefficients. Imposition of the four boundary conditions gives the following simultaneous equations:

$$AV = D \ , \tag{10.66}$$

where

$$V = \begin{bmatrix} V_{ww}^s \\ V_{w1}^s \\ V_{w2}^s \\ V_{wt}^s \end{bmatrix} , \tag{10.67}$$

$$D = \begin{bmatrix} k_{iz} \\ \rho_w \omega^2 \\ \rho_w \omega^2 \\ 0 \end{bmatrix} , \tag{10.68}$$

$$A = \begin{bmatrix} k_{iz} & k_{1z}(1-\gamma_1) & k_{2z}(1-\gamma_2) & -K_x(1-\gamma_t) \\ -\rho_w\omega^2 & k_1^2(H-\gamma_1 C) - 2\mu K_x^2 & k_2^2(H-\gamma_2 C) - 2\mu K_x^2 & -2\mu k_{tz} K_x \\ -\rho_w\omega^2 & k_1^2(C-\gamma_1 M) & k_2^2(C-\gamma_2 M) & 0 \\ 0 & 2K_x k_{1z} & 2K_x k_{2z} & k_{tz}^2 - K_x^2 \end{bmatrix} . \tag{10.69}$$

While these equations can be solved analytically, it is easy to employ a numerical solution. Using the parameters of Table 10.1, the bottom loss ($-20\log_{10}|V_{ww}|$) obtained in this fashion is as shown in Fig. 10.5. For comparison, this figure also shows bottom loss computed in the fluid approximation using complex sound speed equal to the complex fast wave speed and density equal to the total density (10.9). This choice is made as sound speed, attenuation, and total density are often measured more-or-less directly and used in bottom-loss calculations. As Fig. 10.5 makes evident, Biot effects lead to a significant increase in bottom loss, about 2 dB, near normal incidence in this case. This is true even though the slow wave is excited to an insignificant degree in comparison to the fast wave. The ratio of the transmission coefficients for these two waves provides a measure of their relative levels of excitation. The magnitude of this ratio, $|V_{w2}/V_{w1}|$, in the present example is 0.0056 at normal incidence and about 0.0001 at small grazing angles. Given the insignificance of the slow wave, it is natural to ask why the fluid model does not match the Biot model better. The answer lies in the differential movement of the pore fluid and frame for the fast wave. This behavior cannot be mimicked by the simplest fluid model, but it *can* be approximated by a fluid model in which density is replaced by a reduced, or *effective*, value (Sect. 10.2). This difference in reflection has been used as a method of discriminating between fluid and Biot models and constraining Biot parameters (Sect. 11.2.2).

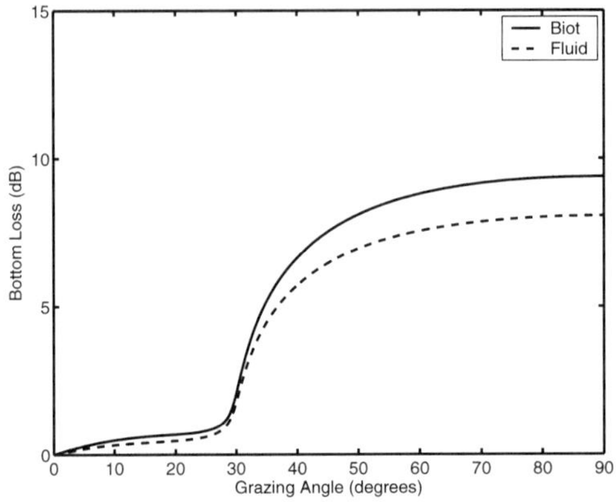

Fig. 10.5. Comparison of bottom loss using Biot and fluid theory. The Biot parameters are taken from Table 10.1, and the frequency is 20 kHz.

10.2 Effective Density Fluid Approximation

Williams has developed a fluid approximation to Biot theory [Williams 2001a]. Noting that the frame moduli are small compared to the grain and fluid moduli (see Table 10.1), the frame moduli are set to zero. This reduces the number of independent moduli to one and provides an effective fluid modulus

$$K_{eff} = H = C = M = (\frac{1-\beta}{K_g} + \frac{\beta}{K_w})^{-1} . \tag{10.70}$$

Note that the effective modulus is the same as (8.71) used in obtaining Wood's equation. The equations of motion (10.17) and (10.18) become

$$\nabla[K_{eff}\nabla \cdot (\mathbf{u} - \mathbf{w})] = -\omega^2(\rho\mathbf{u} - \rho_w\mathbf{w}) , \tag{10.71}$$

$$\nabla[K_{eff}\nabla \cdot (\mathbf{u} - \mathbf{w})] = -\omega^2(\rho_w\mathbf{u} - \rho\mathbf{w}) + i\frac{\omega F\eta}{\kappa}\mathbf{w} . \tag{10.72}$$

Defining the effective displacement

$$\mathbf{u}_{eff} = \mathbf{u} - \mathbf{w} = (1-\beta)\mathbf{u} + \beta\mathbf{U} , \tag{10.73}$$

the variable $\mathbf{w}$ can be eliminated, leaving

$$\nabla[K_{eff}\nabla\cdot\mathbf{u}_{eff}] = -\omega^2\rho_{eff}\mathbf{u}_{eff} \ , \tag{10.74}$$

where the effective density is

$$\rho_{eff} = \frac{\rho\tilde{\rho} - \rho_w^2}{\tilde{\rho} + \rho - 2\rho_w} \ , \tag{10.75}$$

$$\tilde{\rho} = \frac{\alpha\rho_w}{\beta} + \frac{iF\eta}{\kappa\omega} \ . \tag{10.76}$$

Equation (10.74) is the same form as the fluid equation of motion, (8.7). Equations (10.71) and (10.72) can be combined to show that the two displacements have a fixed proportionality,

$$\mathbf{w} = \gamma_1\mathbf{u} \ , \tag{10.77}$$

with

$$\gamma_1 = \frac{\rho_w - \rho}{\tilde{\rho} - \rho_w} \ . \tag{10.78}$$

It can be shown that (10.32) does reduce to (10.78) for the case at hand.

The effective equation of motion can be written in the equivalent form

$$\rho_{eff}\nabla\cdot(\frac{1}{\rho_{eff}}\nabla P_{eff}) + \frac{\omega^2}{c_1^2}P_{eff} = 0 \ , \tag{10.79}$$

where

$$P_{eff} = -K_{eff}\nabla\cdot\mathbf{u}_{eff} \tag{10.80}$$

and

$$c_1 = \sqrt{\frac{K_{eff}}{\rho_{eff}}} \ . \tag{10.81}$$

This completes the derivation of the effective density fluid approximation, with constitutive relation and equation of motion for the inhomogeneous medium in the same forms as those of the fluid case. Using the arguments of Ch. 8, it follows that equivalent fluid boundary conditions must be used, that is, continuity of normal effective displacement and continuity of effective pressure. This is a proof of the boundary conditions assumed in [Williams 2001a]. It follows that the effective density approximation can be used in the fluid interface roughness scattering models of Ch. 13 by the simple replacement of density by effective density and use of the measured sound speed as the compressional wave speed. In addition, because the effective density approximation has been extended to the case of heterogeneous sediments, the fluid volume scattering models of Ch. 14 can be extended to the poroelastic case by means of the same replacements.

In many practical applications, the complex speed, c_1, is known, as phase speed and attenuation are two of the most commonly measured geoacoustic parameters. In this case only one more parameter must be determined in order to implement the effective density approximation. One may either determine effective density from (10.75) or effective modulus from (10.70). As the latter only requires readily available parameters (porosity, real grain and fluid moduli), this method might be preferred in typical circumstances. The effective density is complex and frequency dependent. Figure 10.6 shows the real and imaginary parts of the effective density as functions of frequency for the Biot parameters given in Table 10.1. The real part is smaller than the true density and the imaginary part is positive as required for the $\exp(-i\omega t)$ convention used here. At low frequencies, the pore fluid and frame move together and the effective density is equal to the true density (10.9). At high frequencies, the real part of the effective density is lower than the true density because the frame moves with lower amplitude than the fluid.

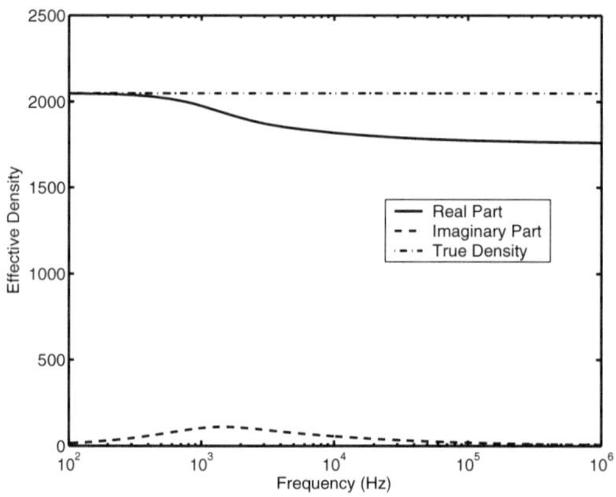

Fig. 10.6. Real and imaginary parts of the effective density as functions of frequency for the Biot parameters given in Table 10.1.

In Sect. 10.3, it will be seen that the effective density fluid approximation closely matches the sound speed and attenuation behavior of Biot theory. It has also been shown [Williams 2001a] that this approximation essentially replicates Biot results for flat-interface reflection and rough-interface scattering. Figure 10.7 compares the reflection coefficient magnitude at 20 kHz using full Biot theory and the effective density fluid approximation. This example uses the parameters of Table 10.1 and shows that the approximation is quite accurate under these circumstances. An example of the usefulness of this approximation for scattering models is given in Ch. 13.

A simple version of the effective density approximation results if one assumes the effective density is real (as a further approximation), takes the pore water properties to be identical to the overlying water, and identifies the fast wave as the compressional wave with real phase speed, c_{pphase}. The effective density definition can be written as $c_{pphase} = \sqrt{K_{eff}/\rho_{eff}}$. This is identical to the effective density definition given by [Nolle et al. 1963], although these authors evidently did not envision its application to situations in which the fluid and grains had separate, differing displacements. Then the Wood expression for bulk modulus, (10.70), gives

$$\frac{\rho_{eff}}{\rho_w} = \frac{1}{\nu_p^2[\beta + (1-\beta)K_w/K_g]} . \tag{10.82}$$

Using the values of moduli in Table 10.1, $K_g/K_w = 13.4$. Equation (10.82) can be evaluated using measured sound speed and porosity and substituted for the density ratio in fluid models. This provides a readily implemented means of estimating the magnitude of poroelastic behavior in sands. This procedure requires measured values of compressional wave speed and porosity, and must assume a value for the grain bulk modulus. Table 4.7 gives values for the bulk modulus of typical constituents of shallow-water sediments. In order to bound the uncertainty associated with the simple effective density procedure, the range of bulk modulus for the two most common constituents, quartz and calcite, will be used. Figure 10.8 shows bottom loss computed for bulk modulus values from 32 GPa (highest bottom loss) to 75 GPa (lowest bottom loss). The 32 GPa value is slightly below the range given in Table 4.7, and is taken from Table 10.1. It can be seen that the simple EDFM prediction is rather insensitive to the choice of grain bulk modulus, with an uncertainty in bottom loss at normal incidence of about ± 0.5 dB for a 2:1 range of grain bulk modulus.

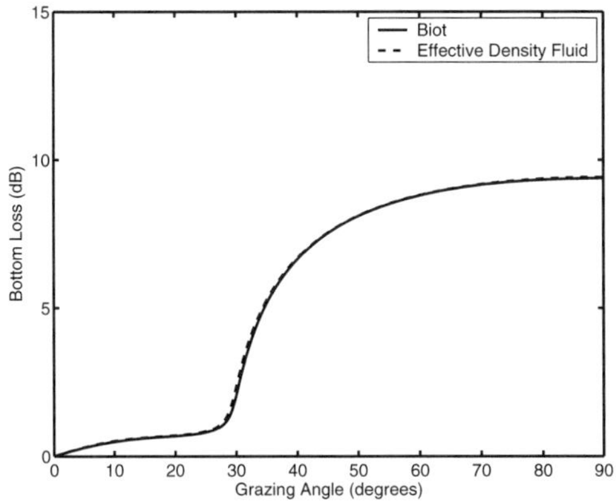

Fig. 10.7. Comparison of bottom loss at 20 kHz for Biot theory and the effective density fluid approximation using the parameters given in Table 10.1.

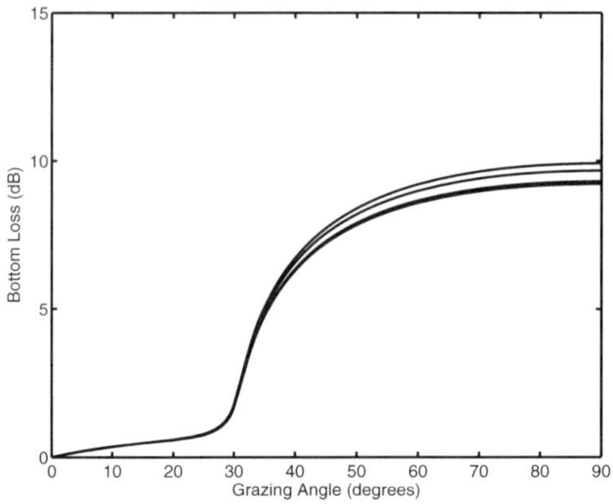

Fig. 10.8. Bottom loss computed using the simple EDFM procedure with various values of grain bulk modulus, $K_g = 32$, 40, 64, and 75 GPa. The water bulk modulus is taken as 2.395 GPa. The sound speed ratio, ν_p, is 1.16, and the porosity, β, is 0.40. The curves follow the order of the inverse of the grain bulk modulus, with the highest bottom loss corresponding to the lowest modulus, etc.

Expression (10.82) can be used to show that the effective density will be nearly equal to the bulk density in muds. In muds, Wood's equation gives a reasonable approximation to the sound speed. Inserting the corresponding expression for the sound speed ratio in (10.82) gives $\rho_{eff} = \rho$.

10.3 Experimental Tests of Biot Theory

Considering the great deal of attention Biot theory has received in connection with seafloor acoustics, relatively little has been published regarding tests of its validity. The principal feature of Biot theory at high frequencies is the frequency dependence predicted for sound speed and attenuation, but tests of the theory are also possible using reflection and scattering measurements. Reflection data are compared with Biot predictions in Ch. 11 and similar comparisons with scattering data are given in Sect. 13.2.4. The main concern of this section is the frequency dependence of sound speed and attenuation. Laboratory measurements in the frequency range 20–300 kHz using water-saturated glass beads [Hovem and Ingram 1979] showed "good agreement with theory." The field measurements of [Turgut and Yamamoto 1990] show frequency dependence that is presented by the authors as support for Biot theory. More indirectly, [Maguer et al. 2000b] used measurements of the critical angle to obtain the frequency dependence of sound speed. Field measurements supporting Biot theory were made by [Simpson et al. 2003] at a sandy site over the frequency range 3–80 kHz. The measurement system used an omnidirectional source in the water and a receiver that could be set at various depths in the sediment. The resulting data showed dispersion and frequency-dependent attenuation consistent with the Kramers–Kronig relations (Appendix I). The results appear consistent with Biot theory, but attenuation is essentially a linear function of frequency over the measured range, therefore consistent with Buckingham's theory (Sect. 9.8).

One problem with experimental tests of Biot theory is the large number of parameters (13) that must be determined if completely rigorous tests with no curve fitting are to be performed. Five of the Biot parameters, those specifying the properties of pore fluid and sediment grains, can be assigned handbook values, leaving 8 to be measured or inferred by some means. For example, [Badiey et al. 1998] constrain some of the Biot parameters using geophysical and geotechnical information. A compilation of the Biot parameters used by several investigators is given in [Chotiros 1995a], but many of these parameters are inferred rather than measured. A parameter set in which most have been measured is given in Table 10.1. These data are from the SAX99 experiment [Williams et al. 2002a], with most parameters being measured either in situ or in the laboratory, while a few were adjusted to improve fits to the data shown in Figs. 10.9 and 10.10. The pore size parameter in the table was computed using (10.16).

Table 10.1. Biot parameters from [Williams et al. 2002a].

Parameter	Symbol	Units	Value
Bulk Modulus of Grains	K_g	Pa	3.2×10^{10}
Permeability	κ	m^2	2.5×10^{-11}
Tortuosity	α	Dimensionless	1.35
Porosity	β	Dimensionless	0.385
Dynamic Viscosity of Water	η	kg m^{-1} s^{-1}	0.00105
Mass Density of Grains	ρ_g	kg m^{-3}	2690
Bulk Modulus of Water	K_w	Pa	2.395×10^9
Mass Density of Water	ρ_w	kg m^{-3}	1023
Shear Modulus of Frame	μ	Pa	$(2.92 - i0.18) \times 10^7$
Bulk Modulus of Frame	K_f	Pa	$(4.36 - i0.208) \times 10^7$
Pore Size	a	m	2.65×10^{-5}

Figure 10.9 compares the measured frequency dependence of sediment sound speed (phase speed) with various theories: fluid, Buckingham, Biot, and effective density. The observed dispersion is matched best by the Biot and effective density theories. Figure 10.10 makes the same comparison of measured attenuation with theoretical predictions. Here, Buckingham's theory provides a better fit, showing that there are fundamental issues yet to be addressed in the theory of sediment acoustics.

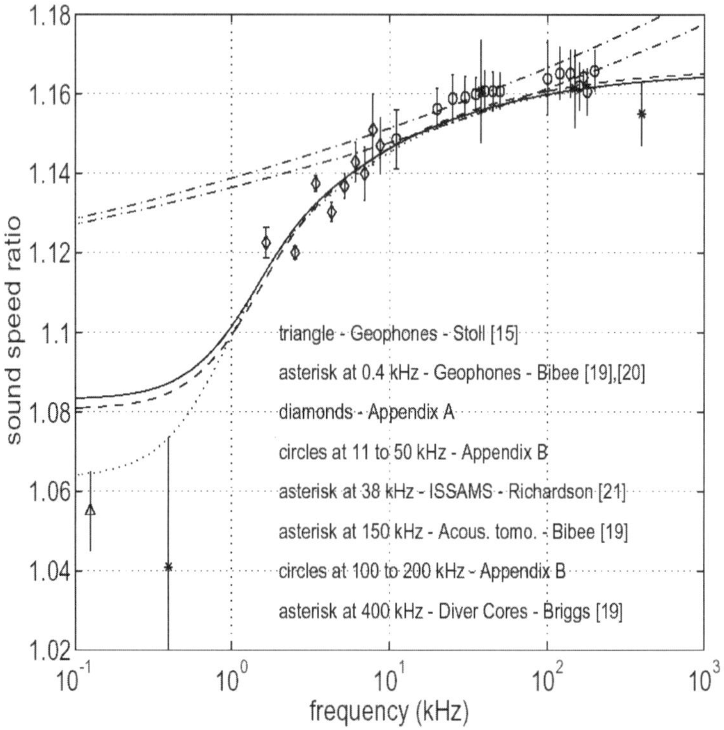

Fig. 10.9. Comparison of sediment phase speed measurements from the SAX99 experiment with various theories. The figure is from [Williams et al. 2002a], and the references are from that article. The Biot sound speed prediction based on measured parameter values is shown as a solid line while the dotted line results if the adjusted values of Table 10.1 are used. The dashed line is the effective density fluid approximation result (Sect. 10.2) using measured parameters. The upper dash-dot curve is the prediction of Buckingham's model using the parameters given in Sect. 9.8, and the lower dash-dot curve is the Buckingham result with the same parameters, except $n = 0.085$.

For the Biot parameter set used here, the slow wave phase speed at 20 kHz is 141 m s^{-1}, to be compared with the fast wave speed, 1763 m s^{-1}. At 20 kHz, the fast wave attenuation coefficient is 7.0 dB m^{-1}, while that of the slow wave is 852 dB m^{-1}. This extreme attenuation is one reason that many authors discount the significance of the slow wave. At 20 kHz, the ratio of fluid to frame displacement (10.33) is about 1.8 for the fast wave and about -1.6 for the slow wave. Thus, the fluid and frame move in the same direction for the fast wave but in opposition for the slow wave. Even if the slow wave is unobservable, the properties of the fast wave are significantly

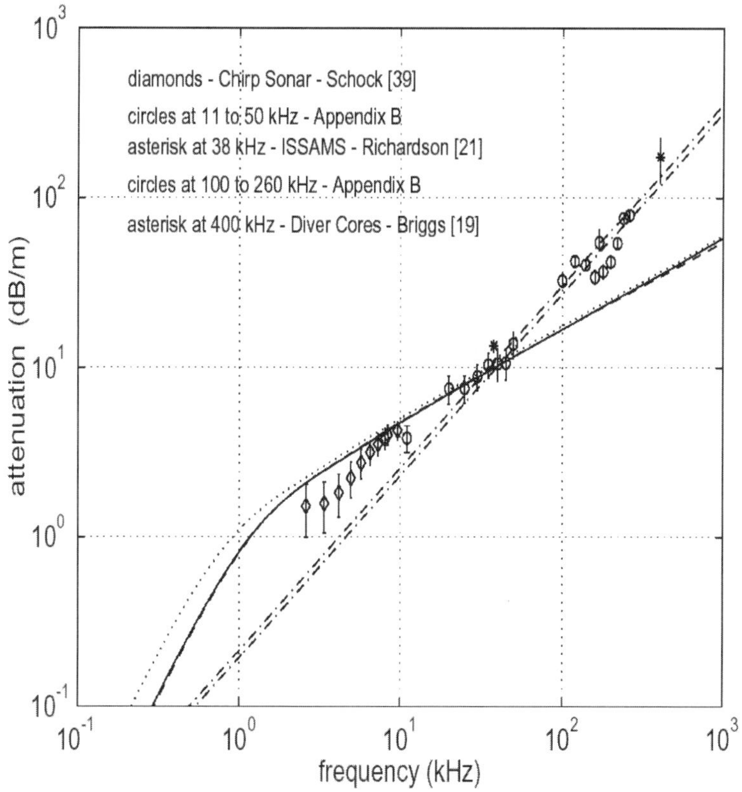

Fig. 10.10. Comparison of sediment attenuation measurements from the SAX99 experiment with various theories. The figure is from [Williams et al. 2002a], and the references are from that article. The various curves have the same significance as in Fig. 10.9, and the upper dot-dash curve (Buckingham theory) employs the same parameters as the upper curve in Fig. 10.9.

different than those of the ordinary compressional wave. This difference is embodied in the effective density and provides a basis for additional experimental tests beyond observations of frequency-dependent wave speed and attenuation (Sects. 10.1.5 and 13.2.4, and Ch. 11).

10.4 Theoretical Alternatives

Biot theory as presented in this chapter is not the only existing formulation, but is a commonly used form based largely on the work of Stoll [Stoll 1974, Stoll and Kan 1981, Stoll 1989]. This approach is sometimes referred to as "Biot–Stoll" theory. A review of alternatives to this theory is beyond the scope of this monograph, but several examples will be cited to illustrate possible directions of future research. One experimental fact motivating alternative versions of Biot theory is the same as that motivating Buckingham's theory: the tendency of attenuation to increase linearly with frequency at high frequencies in such a manner as to exceed the attenuation predicted by Biot theory (Fig. 10.10). Some of this "excess" attenuation has been attributed to scattering from larger shells or other forms of heterogeneity [Richardson and Briggs 1996, Williams et al. 2002a]. However, measured high-frequency attenuation data at 400 kHz, collected during SAX04 [Briggs et al. 2005b], where shells and other forms of local heterogeneity are nearly absent, were also much higher (92 dB m^{-1}) than predicted by Biot theory. Excess attenuation might be accommodated within the Biot–Stoll formalism by increasing the magnitude of the imaginary part of the frame moduli, but this would cause a problem at low frequencies [Pierce et al. 2005], where a quadratic frequency dependence is indicated by the data. These authors suggest a modification to Biot theory that would preserve the quadratic dependence at low frequencies while giving linear dependence at high frequencies. The frequency dependence due to viscous losses in Biot theory would also be modified by the proposals of [Hovem 1980, Yamamoto and Turgut 1988] to incorporate multiple pore sizes. In another approach to this problem, [Chotiros and Isakson 2004] argue that fluid flow between grains ("squirt flow") alters the high-frequency behavior of the frame moduli. The above approaches are motivated by the application of Biot theory to sands, but [Leurer 1997] has presented a squirt-flow model for fine-grained sediments in which fluid flows between water layers formed by swelling of clays (Sect. 3.2.1) and the pore space. This process leads to an effective (complex) frequency-dependent grain modulus. Another alternative is discussed by [Hickey and Sabatier 1997] who note that Biot's original formulation allows an additional free parameter, the "coefficient of fluid content." This alternative was examined by [Chotiros 2002b] who also considered the hypothesis that intraparticulate porosity (water trapped in grains) and mobile solid material in the pore fluid may require alterations to Biot theory.

10.5 Research Issues

Experimental efforts to validate theories for high-frequency acoustics in marine sediments have been partly successful, but have highlighted the need for better data. Future work should employ a combination of field and laboratory acoustic measurements. As will be seen in Ch. 11, reflection measurements are an important adjunct to sound speed and attenuation measurements. As suggested in Chs. 4 and 5, better and more accurate physical and geoacoustic measurement techniques are needed, and the same can be said for reflection, scattering, and penetration measurements. While the primary interest in this monograph is high frequencies, it appears that measurements are needed over a wide range of frequencies (roughly 100 Hz to 500 kHz) to understand the physics of sound propagation in sediments. Efforts are underway to develop alternatives to the present theories, in some cases through generalization and, in others, through less ambitious modification. One may say that there is a search for a unified theory capable of treating a wide range of sediment types. However, because of the fundamental difference in the nature of mud versus sand sediments, it may be unrealistic to expect a single theory to apply to both.

11 Reflection

In spite of its seeming simplicity, reflection by the seafloor is a difficult topic, both experimentally and theoretically. The importance of reflection is threefold: (1) it is an essential component of propagation in shallow water, (2) it is used to infer acoustic impedance and other properties of the seafloor, and (3) it provides rather direct tests of theories for wave propagation in the seafloor. In particular, reflection measurements have been advocated as a means of measuring the poroelastic parameters of sediments [Chotiros 1995b, Drevet et al. 1999, Chotiros 2002b, Isakson and Nielsen 2006]

Measurement of reflection by the seafloor has a long history, and a review of much of the early work is given by [Zhitkovski and Lysanov 1967]. Most reflection measurements have been made at frequencies below 5 kHz and will not be considered here owing to likely complication by layering and gradients. Seafloor reflection at vertical incidence has become an important means of classifying sediments [Parrott et al. 1980, Pace and Langhorne 1993, Panda et al. 1994, Schock 2004], but this large subject is beyond the scope of this monograph. Of high-frequency measurements, relatively few are accompanied by sufficient characterization of the seafloor to be of interest in comparisons between theory and models. Although [Mackenzie 1960] is primarily concerned with low-frequency reflection, some data were obtained at 16 kHz along with grain size information. This reference is also useful in giving a detailed explanation of the fluid reflection coefficient (8.49). References to newer data sources will be given as the discussion proceeds.

Other parts of this monograph can be consulted for background on the reflection problem. For example, Sect. 2.3.1 gives a brief outline of reflection, and Sects. 8.5, 9.5, 9.6, and 10.1.5 provide examples of theoretical predictions for reflection in fluid, elastic, and poroelastic theories. The effects of nonplanarity of the incident field are discussed in Sect. 8.6, and measurement issues are treated in Sects. G.1 and G.2.2. The effects of roughness (Ch. 6) on scattering are discussed in Sect. 13.1 where the incoherent reflection coefficient is defined.

This chapter begins with a discussion of the considerable difficulties of measuring reflection, and the remainder focuses on comparisons of reflection data with theory and models as well as application of the geoacoustic regressions of Ch. 5. Reflection measurement results will be given in terms of "bottom loss," in decibels, as defined in (2.11).

11.1 Reflection Measurement Issues

Reflection measurements require care to avoid errors resulting from several sources: interface roughness, volume scattering, layering, gradients in sediment properties, and near-field effects. As in any acoustic measurement, uncertainties in transducer calibration and in geometry must also be considered.

Error due to interface roughness falls into two categories: fluctuations in signal level and scattering away from the specular direction. The antidote to fluctuations is averaging over multiple reflections from different portions of the interface [Chotiros 1995b]. In Sect. 13.1 and Appendix L, the coherent and incoherent reflection coefficients, relevant to reflection from rough surfaces, are discussed. By averaging the squared pressure, an estimate of the sum of the squared coherent and incoherent reflection coefficients is obtained. This sum, in turn, is approximately equal to the squared magnitude of the flat-interface reflection coefficient. It must be emphasized that other types of average, e.g., average envelope or average signal level in dB, will be biased, that is, will not lead to a good approximation to the true reflection coefficient. The mean-square-average procedure is most successful if the scattered energy falls within a narrow cone centered on the specular direction. That is, the angular separation from the specular direction, measured in radians, is much less than unity. The bistatic scattering models of Ch. 13 provide a means of estimating this angular spread in terms of roughness statistics. If this scattering angle criterion is met, the magnitude of the flat-interface reflection coefficient can be obtained, provided the source and receiver transducers are not so directional as to exclude a portion of this energy. The scattering inversion algorithm of Sect. G.2.2 provides a method of compensating for transducer directivity in the case of normal-incidence measurements. The discussion of this algorithm also makes the point that the echo elongation caused by scattering can be a problem if it is comparable to or greater than the source pulse length. Here again, the algorithm provides a means of compensation

Volume scattering occurs in all sediments, and usually at a level that is comparable to, or even greater than, the level of scattering due to interface roughness. For grazing incidence, one can estimate this level using the models of Ch. 14. For normal incidence, one can hope to exploit the fact that volume scattering returns are somewhat delayed relative to interface roughness returns. In the best of cases, there is still some overlap, and the algorithm of Sect. G.2.2 can be used to attempt a separation.

Gradients (Sect. 5.1.5) and layering in sediment properties pose obstacles to interpretation of reflection data. Bulk density and sound speed can be expected to diminish as the interface is approached from below. That is, there is often a transition layer in which bulk density and sound speed change from at-depth values to smaller values at the interface. Transition layers can have strong effects on reflection, yet may not be evident in low-resolution core data. The grain-size relations of [Mourad and Jackson 1989] are a crude attempt to account for such gradients, and these authors present a sample calculation which shows that a density transition layer of 3-cm thickness can cause a 6-dB increase in bottom loss at normal incidence at 20 kHz. Scaling this result, a similar transition layer of 3-mm thickness would cause a 6-dB increase in loss at 200 kHz. A more extensive treatment of this issue is provided by [Lyons and Orsi 1998], who employ CT-measured density profiles in their calculations (see Sect. 7.5.5). These profiles were fitted by the analytic function (7.36) with the parameter a assigned values in the approximate range 10–200 m^{-1}. Note that the value $a = 200\ \mathrm{m}^{-1}$ corresponds to a transition layer thickness of 5 mm. In some cases, large reductions in reflection coefficient (10 dB and greater) compared to the gradient-free case are predicted. Positive gradients in sound speed (increasing downward) could have strong effects on grazing angle dependence near the critical angle owing to upward refraction of the wave in the sediment. The resulting interference with the wave initially reflected from the interface can cause oscillations of the reflection coefficient with respect to grazing angle. These issues require attention and present challenges in geoacoustic measurement.

The strategy for measurement of the reflection coefficient outlined in Sect. G.1 assumes that the plane-wave reflection coefficient will result from the measurement even though the incident and reflected waves are spherically diverging. This approximation is valid if the propagation distances are many wavelengths and if the lateral wave (Sect. 8.6) can be ignored. Spherical wave effects in reflection measurements have been examined by [Camin and Isakson 2006]. While innocuous in many experiments, these effects can enter in short-range laboratory measurements. As an example, Fig. 11.1 compares the plane-wave bottom loss with values of a simulated measurement situation. The simulations employed wavenumber integration (8.63) to obtain the reflected pressure for an incident spherical wave due to a point acoustic source. The simulations assumed a water sound speed of 1530 $\mathrm{m\ s}^{-1}$ and a sand sediment with geoacoustic parameters $a_\rho = 2.0$, $\nu_p = 1.16$, and $\delta_p = 0.01$. Both source and receiver were assumed to be 10 cm above the interface, with horizontal separation varied in order to vary the grazing angle. The 200-kHz example shows significant departures from the plane-wave case, while the 500-kHz example is better represented by the plane-wave case. The oscillations occurring for angles less than the critical angle (30.5°) are due to interference of the reflected and lateral waves [Chin-Bing et al. 1982]. These results can be scaled for application to other frequencies. For example, these curves would apply to measurements at 20 and 50 kHz if the transducers were 1 m above the seafloor.

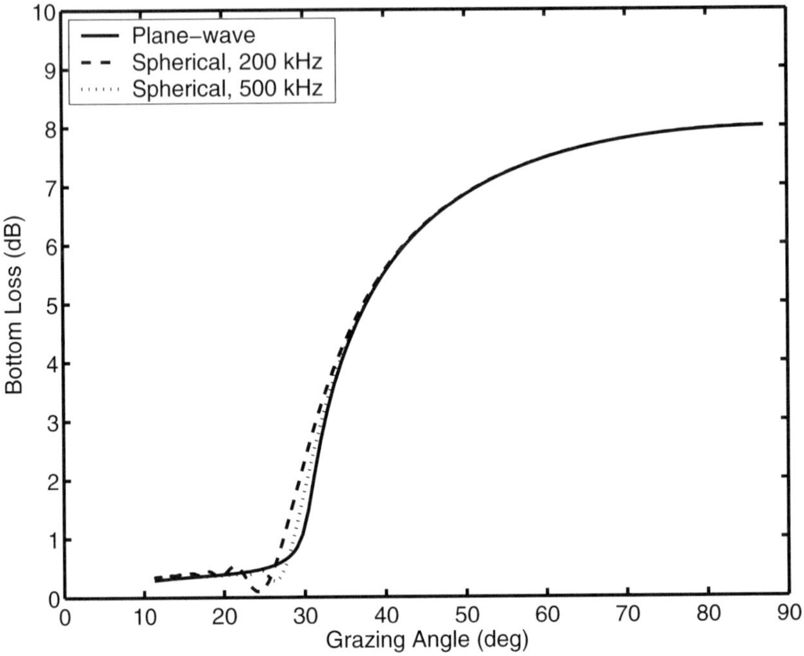

Fig. 11.1. Simulation of spherical-wave effects in measurement of the reflection coefficient. The source and receiver are 10 cm above the sediment–water interface, and the geoacoustic parameters are appropriate to sand. Wavenumber integration is used to obtain the curves labeled "Spherical."

11.2 Reflection Data Compared with Theory and Models

Reflection measurement results will be divided into field data and data obtained in the laboratory. Laboratory measurements allow greater control of geometry and sediment parameters, and are preferred for tests of theory and models. The natural environment is more complex, and field measurements are required to verify that model simplifications have not obscured the true behavior of the seafloor.

11.2.1 Field Measurements of Reflection

Normal-incidence reflection data obtained at 16 kHz were reported by [Mackenzie 1960], and a rather large normal-incidence reflection data set was acquired by [Parrott et al. 1980] who employed a broadband impulsive source and associated their acoustic data with grain size analyses from an earlier survey.

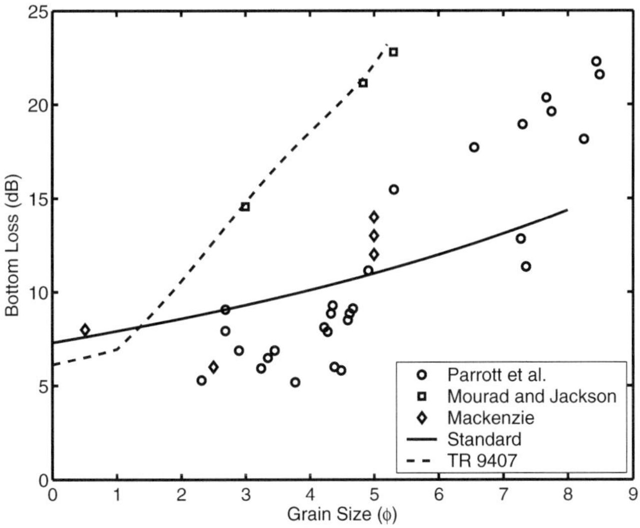

Fig. 11.2. Bottom loss at normal incidence versus grain size as reported by [Mackenzie 1960, Parrott et al. 1980, Mourad and Jackson 1989]. The "standard" theoretical curve uses grain-size regressions given in this monograph while the "TR 9407" curve uses the ad hoc relations of [APL-UW TR 9407]. Both curves employ the fluid approximation to the seafloor reflection coefficient.

More recently, [Mourad and Jackson 1989] presented data in the frequency range 20–30 kHz. These data sets are shown in Fig. 11.2 along with theoretical curves using the fluid reflection coefficient and two different sets of grain-size–geoacoustic relations. The "standard" relations are the regressions presented in Table 5.4 and in Eq. (5.15). The "TR 9407" relations are those first given by [Mourad and Jackson 1989] and expanded in a technical report [APL-UW TR 9407] to cover a greater range of grain sizes. These relations are

$$\begin{aligned} a_\rho &= 0.007797M_z^2 - 0.17057M_z + 2.3139 \;, && -1 \le M_z < 1 \;, \\ &= -0.0165406M_z^3 + 0.2290201M_z^2 - 1.1069031M_z + 3.0455 \;, && 1 \le M_z < 5.3 \;, \\ &= -0.0012973M_z + 1.1565 \;, && 5.3 \le M_z \le 9 \;, \quad (11.1) \end{aligned}$$

$$\begin{aligned} \nu_p &= 0.002709M_z^2 - 0.056452M_z + 1.2778 \;, && -1 \le M_z < 1 \;, \\ &= -0.0014881M_z^3 + 0.0213937M_z^2 - 0.1382798M_z + 1.3425 \;, && 1 \le M_z < 5.3 \;, \end{aligned}$$

$$= -0.0024324M_z + 1.0091\ , \qquad 5.3 \le M_z \le 9\ , \quad (11.2)$$

$$\begin{aligned} k &= 0.4566\ , & -1 \le M_z < 0\ , \\ &= 0.0245M_z + 0.4556\ , & 0 \le M_z < 2.6\ , \\ &= 0.1245M_z + 0.1978\ , & 2.6 \le M_z < 4.5\ , \\ &= 0.20098M_z^2 - 2.5228M_z + 8.0399\ , & 4.5 \le M_z < 6.0\ , \\ &= 0.0117M_z^2 - 0.2041M_z + 0.9431\ , & 6.0 \le M_z < 9.5\ , \\ &= 0.0601\ , & 9.5 \le M_z\ . \quad (11.3) \end{aligned}$$

Expressions (11.1) and (11.2) for the density and sound speed ratios are adjustments of relations given by [Hamilton and Bachman 1982]. These adjustments are based on limited sets of acoustic and geoacoustic data, therefore, (11.1) and (11.2) should not be regarded as geoacoustic regressions. Rather, they are empirical expressions intended to improve acoustic model-data fits. Expression (11.3) for the attenuation factor, k (dB m^{-1} kHz^{-1}), is taken directly from [Hamilton 1972] and is therefore a geoacoustic regression. Section 5.1.4 compares Hamilton's results with more recent data.

As noted in Sect. 9.5, inclusion of shear would have a negligible effect on the predicted bottom loss. Even though the data span a considerable range in frequency, the theoretical reflection coefficient has no frequency dependence given the assumed lack of layering and with the further assumptions of frequency-independent sound speed and linear dependence of attenuation on frequency.

The standard regressions fit the data of [Mackenzie 1960, Parrott et al. 1980] reasonably given the scatter of both the acoustic data and the geoacoustic data on which the regressions are based. The good fit of the "TR 9407" relations to the data of [Mourad and Jackson 1989] is simply due to the fact that these authors used this data set in developing these ad hoc relations. These relations cannot be considered to be predictors of geoacoustic properties. Rather, they were intended to make up for shortcomings in reflection and backscattering predictions as compared to data available at the time. In fact, these relations perform well when used to predict more recent data [Briggs et al. 2002c]. Compared to geoacoustic regression relations (Ch. 5), these relations tend to give lower values of sound speed and density ratios for fine-grained sediments. This may indicate that, at many of the shallow water sites included in the acoustic data sets, bioturbation has increased the porosity of the upper few centimeters of sediment [Mourad and Jackson 1989], but this supposition is not supported by the measurements presented in Ch. 5. As will be seen, the measurement sites considered here seem to fall into two classes: those with loss in rough agreement of predictions based on standard geoacoustic regressions and those with anomalously high loss.

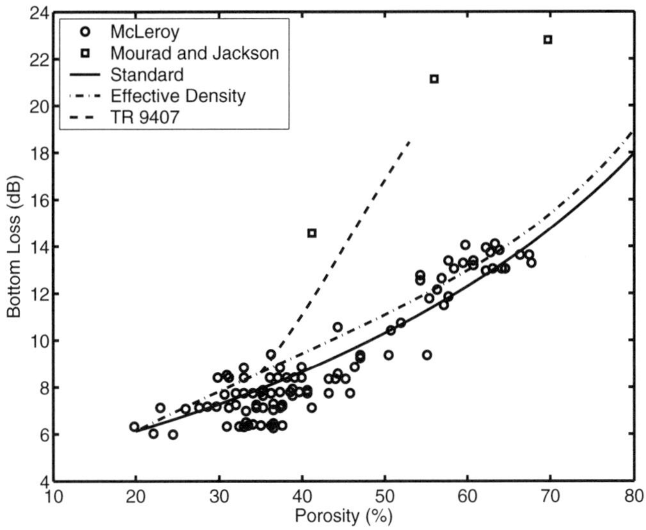

Fig. 11.3. Bottom loss at normal incidence versus porosity as reported by [McLeroy 1972, Mourad and Jackson 1989]. The "standard" theoretical curve uses grain-size regressions given in this monograph while the "TR 9407" curve uses the ad hoc relations of [APL-UW TR 9407] and a porosity–grain-size relationship given in that reference. Both curves employ the fluid approximation to the seafloor reflection coefficient.

A rather large set of normal-incidence reflection data obtained at 12 kHz and accompanied by grain-size analysis has been presented by [McLeroy 1972]. Figure 11.3 compares their data and those of [Mourad and Jackson 1989] with theoretical curves based on the fluid approximation. The "standard" and "TR 9407" curves have the same significance as in Fig. 11.2. The data of [McLeroy 1972] agree rather well with predictions based on geoacoustic regressions and fall into the "standard" loss class.

One of the earliest reflection data sets was obtained by [Liebermann 1948] who used an interference technique to measure the reflection coefficient at 24 kHz at a grazing angle of 9°. Rather than measuring the reflected pulse amplitude directly, interference between the direct and reflected paths was observed by slowly changing the depth of the receiving transducer. It appears this technique gave results that were significantly biased toward higher loss values owing to seafloor roughness. This can be understood by considering a case of very low loss. In this case, the reflected and direct amplitude are nearly equal, and one expects very deep interference nulls. Interface roughness will alter the phase and amplitude of the reflected signal, however, "washing out" these nulls. Thus, very low loss values will not be measured, even if the bottom loss is, in fact, low.

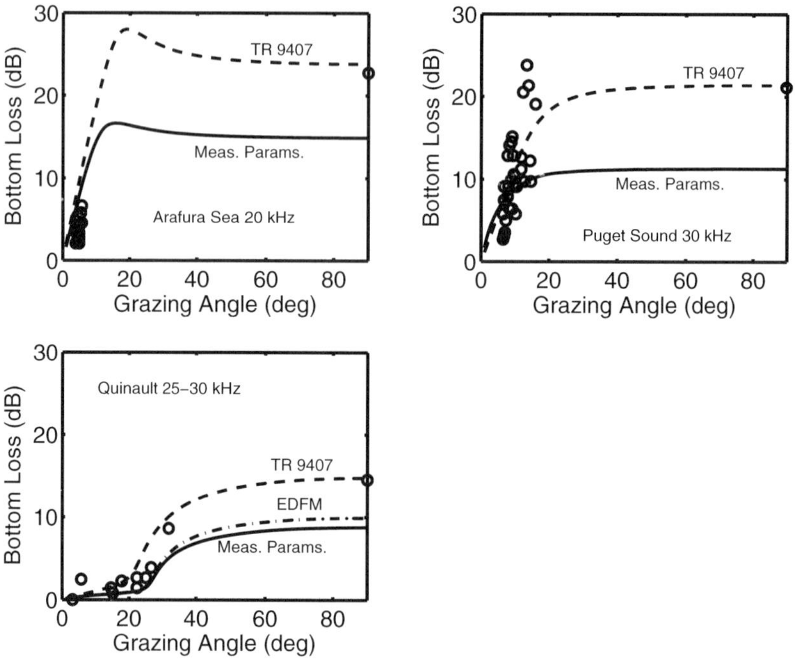

Fig. 11.4. Bottom loss versus grazing angle as reported by [Mourad and Jackson 1989]. The "TR 9407" curves have the same significance as in Fig. 11.2, and the curves labeled "Meas. Params." use the following values: Arafura, $a_\rho = 1.456$, $\nu_p = 0.988$, $\delta_p = 0.024$; Puget Sound, $a_\rho = 1.74$, $\nu_p = 1.006$, $\delta_p = 0.01$; Quinault, $a_\rho = 1.92$, $\nu_p = 1.117$, $\delta_p = 0.014$. The EDFM curve was computed using the Quinault parameters given above with a_ρ replaced by (10.82) with $\beta = 0.418$ and $K_g/K_w = 13.4$.

The data of [Mourad and Jackson 1989] include small grazing angles as well as normal incidence and are shown in Fig. 11.4. The loss curves labeled "Meas. Params." used measured values for density ratio, sound speed ratio, and loss parameter. For Arafura and Quinault, these were taken from Table 5.1. For Puget Sound, the values reported by [Mourad and Jackson 1989] were used. This data set exhibits anomalously high loss and was used in determining the "TR 9407" relations. Note that the simple effective density approximation (10.82) underestimates bottom loss at the sandy Quinault site.

A broadband reflection data set was obtained by [McConnell and DeProspo 1994]. These data are averages over measurements at 1, 3, and 5 kHz and fall below the frequency range of other data in this chapter, but are presented as another example of high-loss behavior. As seen in Fig. 11.5,

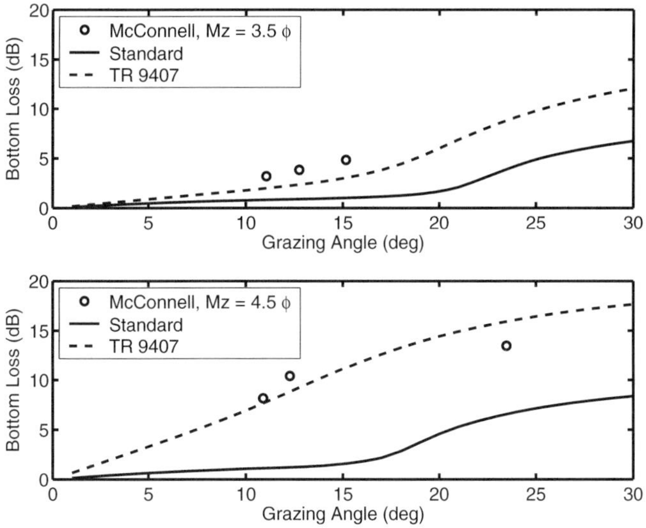

Fig. 11.5. Bottom loss versus grazing angle as reported by [McConnell and DeProspo 1994]. The "standard" and "TR 9407" curves have the same significance as in Fig. 11.2.

these reflection data are matched better by the "TR 9407" curves than by the standard regressions.

11.2.2 Laboratory Measurements of Reflection

Laboratory measurements of reflection offer better control of geometry and interface roughness than field measurements, and have been used both for testing wave theories and for inversion to obtain geoacoustic parameters. Most of these measurement have employed sandy sediments. Figure 11.6 presents measurements by [Nolle et al. 1963, Drevet et al. 1999, Worley 2004]. The data of [Drevet et al. 1999] have a small, downward-pointing bump just below the critical angle that may be due to spherical wave effects. Figure 11.1 was computed for parameters similar to those of this measurement, and shows similar behavior in the 500-kHz case. The data from the three measurements are compared in Fig. 11.6 with various curves computed using the parameters given in Table 11.1. These curves include the fluid approximation, the effective density fluid approximation (EDFM), and the "TR 9407" model. For the EDFM, (10.82) was used to obtain the effective density for substitution in the fluid model with all other parameters unchanged. The "TR 9407" model employed grain size from Table 11.1 in the grain-size relations of [APL-UW TR 9407]. Table 11.1 gives parameters reported by the cited references except, in the case of [Drevet et al. 1999],

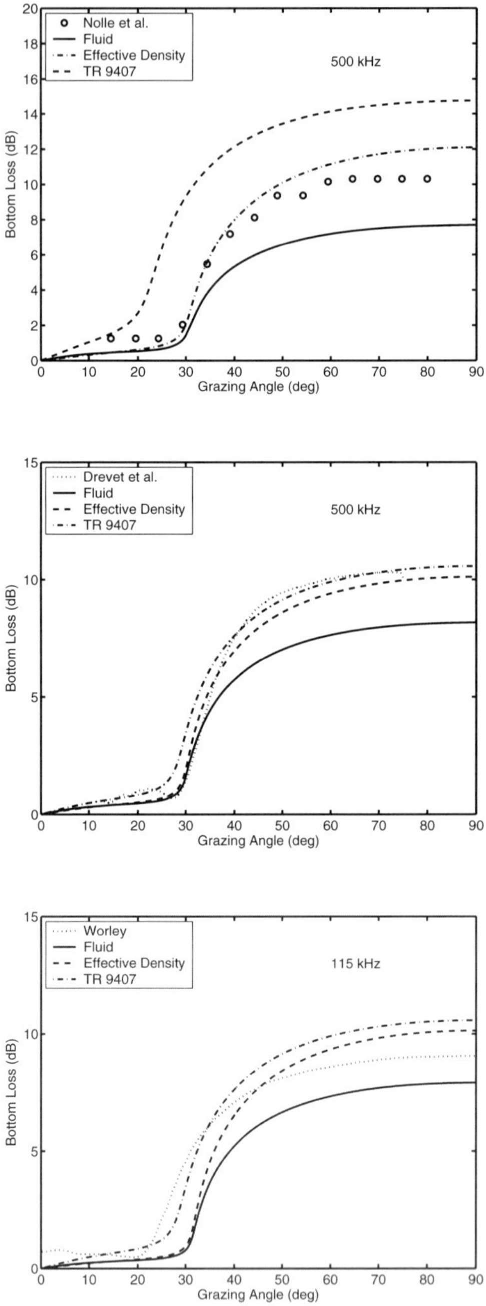

Fig. 11.6. Bottom loss versus grazing angle for laboratory measurements on sand reported by: (top) [Nolle et al. 1963], (middle) [Drevet et al. 1999], (bottom) [Worley 2004]. The curve labeled "Fluid" is computed in the fluid approximation using values given in Table 11.1. This table also gives the parameters used to compute the effective density fluid model and "TR 9407" curves.

the sound speed ratio was determined by fitting the reflection loss data near the critical angle, and, for [Nolle et al. 1963], the loss parameter was taken to match that reported by [Mourad and Jackson 1989] for the Quinault site, which had similar grain size.

In all three examples of Fig. 11.6, the fluid approximation yields bottom loss significantly lower than the laboratory data. The "TR 9407" curve overestimates loss for [Nolle et al. 1963], and, overall, the EDFM curves come closest to matching the three data sets. This is an indication that Biot effects are significant for sandy seafloors. The last column of Table 11.1 gives the ratio of effective density to true bulk density. These values appear reasonable when compared to the value obtained by inserting the Biot parameter set of Table 10.1 in the more complete EDFM equation (10.75). For a frequency of 250 kHz, $\rho_{eff}/\rho = 0.657 + \mathrm{i}\ 0.0046$.

The data sets pictured in Fig. 11.6 have been used for geoacoustic inversion by several investigators [Drevet et al. 1999, Chotiros 2002b, Worley 2004]. Reasonable fits are obtained in most cases using Biot theory, but [Chotiros 2002b] advocates modifications to standard Biot theory in order to improve the fits. Two different versions are proposed, "Composite materials: The possibility that the frame may contain fluid and that the pore fluid may contain loose grains" and "Independent coefficient of fluid content: The possibility that porosity may change with pore fluid pressure." These proposals and efforts by others to modify [Yamamoto and Turgut 1988, Chotiros and Isakson 2004] or replace [Buckingham 2000, Buckingham 2005] Biot theory show that this subject is not a closed book.

Table 11.1. Parameters associated with curves in Fig. 11.6.

Reference	Mean Grain Size M_z (ϕ)	Porosity β	Density Ratio a_ρ	Sound Speed Ratio ν_p	Loss Parameter δ_p	ρ_{eff}/ρ
[Nolle et al. 1963]	3.0	0.48	2.07	1.16	0.014	0.69
[Drevet et al. 1999]	2.0	0.41	1.973	1.155	0.0114	0.84
[Worley 2004]	2.0	0.403	2.00	1.173	0.01	0.81

Laboratory reflection experiments were carried out by [Yargus 2003] for frequencies in the neighborhood of 250 kHz in Ottawa sand having mean grain size $M_z = 1.84\ \phi$. Using normal-incidence reflection for various boundary conditions, he was able to estimate some, but not all, of the Biot parameters. The effective density was found to be consistent with the value resulting from the parameter set of [Williams et al. 2002a] and comparable to the values in Table 11.1 for [Drevet et al. 1999] and [Worley 2004].

One interesting possibility for future laboratory experiments is measurement of the phase of the reflection coefficient. This may provide a means of distinguishing between Biot theory and the fluid, elastic, and Buckingham alternatives. As noted in Sect. 8.5, the phase of the reflection coefficient at normal incidence is opposite if losses are incorporated in the density rather than in the moduli. With typical Biot parameters, the complex effective density dominates the losses due to complex frame moduli. Thus, with $\exp(-i\omega t)$ time dependence, the imaginary part of the reflection coefficient at normal incidence will be positive if poroelastic behavior is dominant and negative if either fluid, elastic, or Buckingham theory is correct. Unfortunately, this phase is very small, being comparable (when measured in radians) to the loss parameter, δ_p.

11.3 Research Issues

Reflection measurements offer one of the most direct means of testing theories for wave propagation in sediments. As noted at several points in this monograph, poroelastic effects reduce the reflection coefficient relative to the values expected on the basis of fluid or elastic theory. In this context, Buckingham's theory would fall in the class of fluid or elastic theories and would predict lower reflection loss than Biot theory, given the same acoustic wave speed and bulk density. Ambiguities remain in the interpretation of present data sets, so future work must be undertaken with increased effort at characterization of the medium. Both laboratory and field measurements are useful, and, in both, roughness statistics should be determined. To combat roughness effects, averaging over multiple echos from statistically equivalent seafloor patches is required. Model calculations using measured roughness statistics can be used to estimate bias due to scattering. Contamination of reflection data by scattering from within the sediment volume deserves more attention than it has received to date. Gradients due to transition layers immediately below the interface must be accounted for in the interpretation of measurements, unless the wavelength is much greater than the layer thickness. This point is illustrated by the occasional success of the ad-hoc "TR 9407" geoacoustic model in fitting reflection data. This model yields smaller-than-normal sound speed and bulk density for a given mean grain size. At higher frequencies (100–500 kHz), measurement of these gradients with the required millimeter resolution is challenging. Computed tomography (CT, Sect. 7.1) can measure density profiles with sufficient resolution if sediment samples are undisturbed, but no means is available to measure sound speed with similar resolution. It is unfortunate that the effects of a transition layer, of roughness, and those of poroelasticity are similar: all cause a reduction in reflection coefficient compared to the simplest fluid treatment. Discrimination between these effects will require extreme care, and may only be possible in the laboratory.

12 Seafloor Scattering Experiments

This chapter will review published data on high-frequency scattering by the seafloor, pointing out regularities (and lack of regularity as well) in the data without attempting physical explanations. Discussion of physical mechanisms will be reserved for Chs. 13 and 14, where data and scattering models are compared. Attention will be restricted to measurements of scattering strength (defined in Sect. 2.3.2). Reviews of seafloor sound scattering can be found in [Bunchuk and Zhitkovskii 1980, McCammon 1993, Chotiros 2002a].

A universal problem with older seafloor scattering data is lack of physical characterization, or at least lack of sufficient characterization to constrain modern scattering models. Only since the 1980s have experimenters devoted sufficient resources to the problem of characterization, and progress has been aided by the development of new geoacoustic measurement techniques. Even though older scattering data are unaccompanied by sufficient seafloor characterization, they are useful in revealing trends such as angular dependence, frequency dependence, and correlation of scattering level with seafloor type.

One can place seafloor scattering measurements in a hierarchy based on the degree of control of experimental conditions. Reverberation data have been gathered during naval fleet exercises in areas of operational interest. This type of data gathering is often aimed at the development of "through-the-sensor" capability [Brown and Barlett 2005] to assess the environment in tactical situations. Such measurements may suffer from poor calibration, errors in knowledge of measurement geometry, and lack of ground-truth, but provide scattering strength data over a wide range of conditions. Similarly, acoustic survey efforts have produced large data sets on seafloor scattering, generally with reasonable calibration and knowledge of geometry. Better experimental control is possible if acoustic platforms are deployed on the seafloor allowing acoustic measurements of improved accuracy and permitting physical and geoacoustic characterization of the area surrounding the acoustic platform. Such improved control comes at the price of smaller statistical samples. These types of experiment are typical of the present scientific effort and are the dominant subject of this chapter and the model-data comparisons of Chs. 13, 14, and 15. An even greater degree of experimental control is possible with seafloor-deployed systems if well-characterized manipulations of the seafloor are carried out, e.g., [Richardson et al. 2001b].

Laboratory experiments offer even greater control and rigor in model-data comparisons, but may not mimic the natural world accurately. Here one may use artificial sediments (e.g., small glass beads) or reconstituted natural sediments. The ultimate degree of control can be obtained in laboratory measurements using model seafloors, e.g., plastic machined to have specified roughness [Mellema 1999, Soukup et al. 2005].

Many different experimental setups have been used in field measurements of seafloor scattering. One example is shown in Fig. 1.1, in which a rotating sonar mounted on a tripod provided backscattering data, while also serving as a source for bistatic measurements that employed a receiving array deployed from a nearby vessel. Other investigators have used towed sonar platforms (e.g., [D. Jackson et al. 1986b]), fixed seafloor-mounted platforms (e.g., [Stanic and Kennedy 1992]), diver-movable tripods [Williams et al. 2002b], platforms moving on rails deployed on the seafloor [Maguer et al. 2000a, Williams et al. 2005] (see Figs. 1.2 and 12.1), and autonomous underwater vehicles.

Fig. 12.1. Transducers used by investigators at the NATO Undersea Research Centre for seafloor acoustics experiments. The transmitter is a parametric source.

In interpreting experimental results, it is important to realize that acoustic scattering measurements are challenging, and errors can be difficult to identify and quantify. For example, backscattering strength at small grazing angles is determined from the echo signal arising from ranges much greater than the transducer height above the seafloor. In shallow water, such ranges may be affected by multipath propagation or by unwanted scattering by the sea surface. Such effects can be assessed by modeling of the experimental arrangement or by acoustic tests of the direction of arrival, but this was not the usual practice in older experiments. Other factors in measurement of seafloor scattering strength are discussed in Sect. G.2.

12.1 Monostatic Experiments

Monostatic experiments usually provide measurements of the backscattering strength of the seafloor, but may also provide information on signal envelope statistics or spatial or temporal correlations (Ch. 16). The aim of the present section is to draw broad conclusions regarding the dependence of backscattering strength on grazing angle, acoustic frequency, and seafloor type. More detailed conclusions will be reached in Chs. 13 and 14.

12.1.1 Grazing Angle Dependence

Figure 12.2 displays backscattering strength data as a function of grazing angle for "mud" seafloors (mean grain size $M_z \geq 5\ \phi$). The same presentation is given in Fig. 12.3 for sand ($5\ \phi < M_z \leq 0\ \phi$), in Fig. 12.4 for gravel ($M_z \leq 0\ \phi$), and in Fig. 12.5 for rock. The older (pre-1985) data have been edited to remove data at smaller grazing angles, which may be contaminated by multipath reverberation and therefore biased upward. Section 2.3.2 gives an argument that, in most cases, scattering strength must decrease rapidly as grazing angle approaches zero. More recent data have typically been subjected to scrutiny to detect the presence of unwanted scattered energy [Boehme et al. 1985, D. Jackson et al. 1986b] and have been found to exhibit this rapid decrease at small angles.

Lambert's law has been fitted to each of the data sets in Figs 12.2–12.5, with the fit restricted to grazing angles less than 60°. The Lambert parameter $10 \log_{10} \mu$ was computed by fitting each broad data grouping (mud, sand, gravel, and rock) and used in plotting a Lambert-law curve for comparison with the data. The data of Figs. 12.2 and 12.3 for mud and sand seafloors have a grazing angle dependence that follows Lambert's law reasonably well for grazing angles less than about 60°, but the data rise above Lambert's law for steep grazing angles, showing that scattering is not entirely diffuse, but increases sharply near the specular direction. The data sets for scattering by gravel, Fig. 12.4, and rock, Fig. 12.5, are very sparse, but show clearly that seafloors of these types can be strong scatterers of sound. In fact, the extremely high scattering levels for the rock seafloor of [APL-UW TR 9407] exceed the Lambert scattering limit (see Eq. (2.23) and Fig. 2.7) and are suggestive of backscattering enhancement (Sect. 2.3.2).

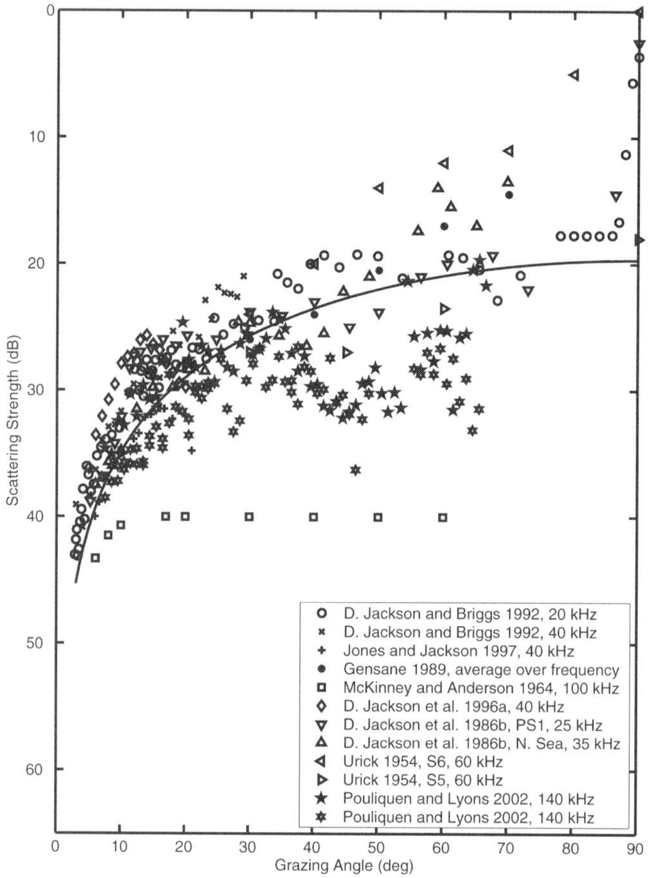

Fig. 12.2. Backscatter data from selected sites having mud seafloors. The solid curve is Lambert's law (2.22) using $10\log_{10}\mu = -19.7$.

The best-fit Lambert's law curve has $10\log_{10}\mu = -19.7$ dB for the mud data and $10\log_{10}\mu = -20.2$ dB for the sand data. The spread in the range of scattering strengths for mud and sand seafloors is very large, about ±5–8 dB. This large spread shows that sediment name is not a good predictor of

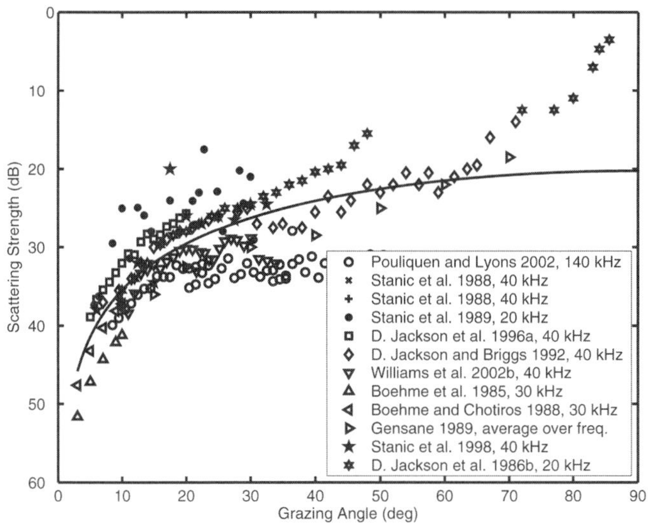

Fig. 12.3. Backscatter data from selected sites having sand seafloors. The solid curve is Lambert's law (2.22) using $10\log_{10}\mu = -20.2$.

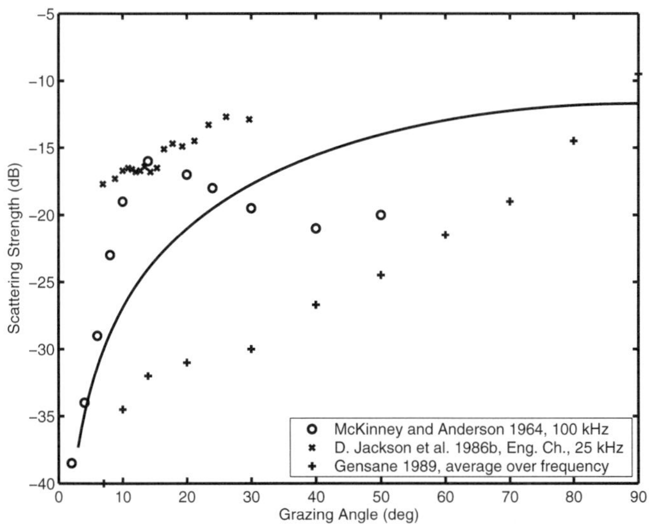

Fig. 12.4. Backscatter data from selected sites having gravel seafloors. The solid curve is Lambert's law (2.22) using $10\log_{10}\mu = -11.7$.

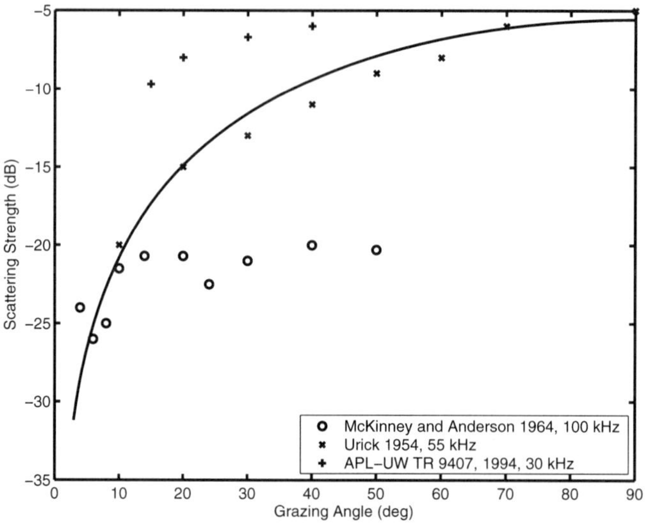

Fig. 12.5. Backscatter data from selected sites having rock seafloors. The solid curve is Lambert's law (2.22) using $10 \log_{10} \mu = -5.6$.

scattering strength. Since average scattering strengths for sand are comparable to those for mud, it can also be said that scattering strength is not a good predictor of sediment type. Equivalently, there are large errors in predicting scattering strength in terms of grain size and vice versa [Chotiros 2002a]. Even so, grain size is often used in lieu of any better seafloor descriptor [Mourad and Jackson 1989, APL-UW TR 9407].

While measurement of scattering strength may allow coarse classification into "hard" and "soft" seafloors (e.g., gravel and rock versus mud and sand), finer classification requires additional information. Measurement of seafloor reflection at normal incidence provides values for acoustic impedance. Since impedance is well correlated with several sediment properties (Sect. 5.1.8), reflection coefficient measurement in concert with backscatter measurement offers the possibility of improved acoustic seafloor classification.

It is difficult to determine errors in measurements of scattering strength, and many authors have avoided this issue, while some give rough error estimates or provide "error bars" showing the observed spread in measured values. Errors arise from insufficient statistical sample size (Sect. G.2), errors in calibration of transducer source level, receiving sensitivity, and directivity (Appendix F), errors in system gain, patchiness of the seafloor (see Ch. 16), and other causes. When all these errors are taken into account, it is unlikely that the overall accuracy of the most careful measurement of scattering strength will be better than about 2 dB, and larger errors are to be expected in most cases. Because scattering strength shows a rapid angular variation near vertical incidence, the approximation used to obtain the simple sonar equation (G.13) may fail, and the experimentalist may have to resort to a more complicated means to extract scattering strength from echo data. The problem is exacerbated by the fact that the angular resolution corresponding to a given pulse length is poor near normal incidence. An approach to this problem is given in Sect. G.2.2.

12.1.2 Frequency Dependence

Figures 12.6–12.9 show the frequency dependence of published backscattering strength data for mud, sand, gravel, and rock at selected grazing angles. The mud data show, on average, a slight increase in scattering strength with increasing frequency, roughly 3 dB for each decade increase in frequency. Most of the sand sites show a definite increase of scattering strength with frequency, in the neighborhood of 6–8 dB/decade. The gravel and rock data are too sparse and span too small a frequency range to allow definite conclusions.

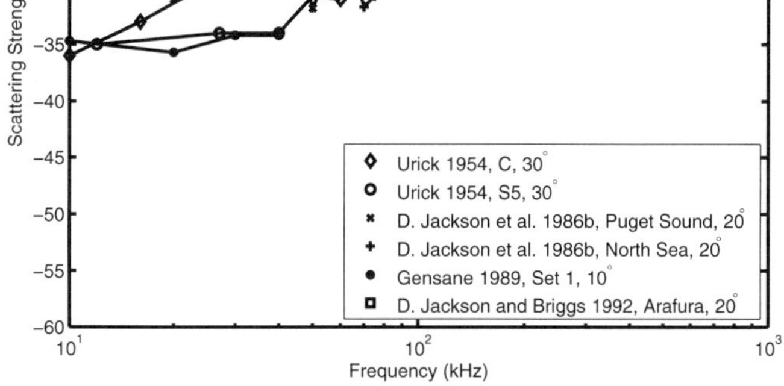

Fig. 12.6. Frequency dependence of backscatter strength at fixed grazing angle for sites having mud seafloors. The straight line has a slope of 3 dB/decade.

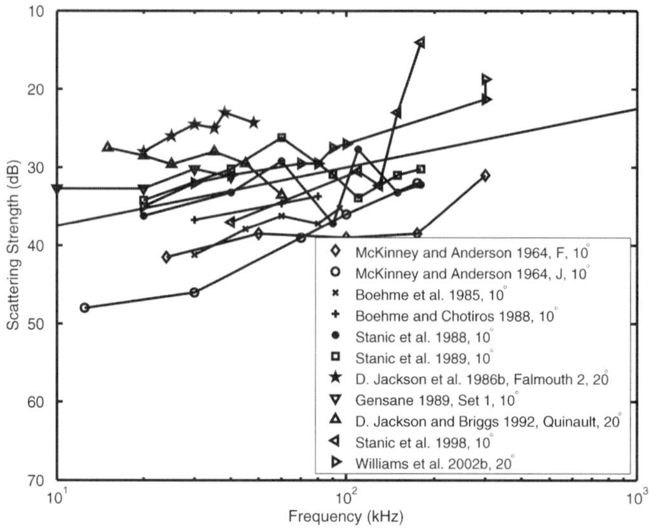

Fig. 12.7. Frequency dependence of backscatter strength at fixed grazing angle for sites having sand seafloors. The straight line has a slope of 7.5 dB/decade.

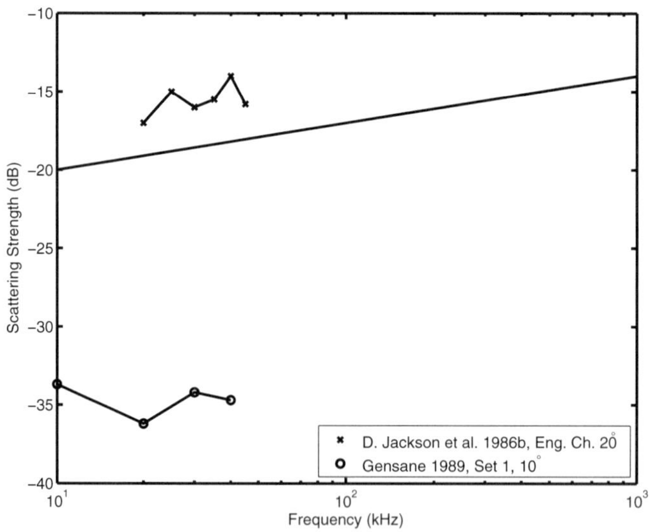

Fig. 12.8. Frequency dependence of backscatter strength at fixed grazing angle for sites having gravel seafloors. The straight line has a slope of 3 dB/decade.

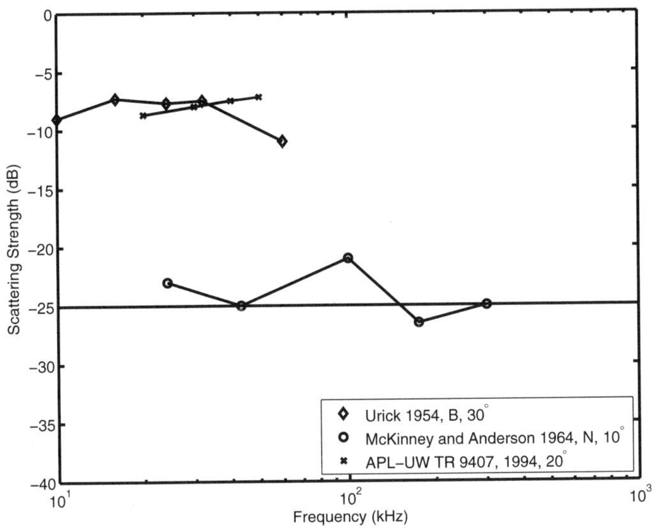

Fig. 12.9. Frequency dependence of backscatter strength at fixed grazing angle for sites having rock seafloors. The straight line has a slope of 0 dB/decade.

12.2 Bistatic Experiments

Bistatic seafloor scattering data are much more difficult to acquire than monostatic data, largely because two separate platforms are required (one for the source, one for the receiver). The positions of these platforms must be known accurately, as must the pointing directions of the source and receiver. Further complication arises from the fact that the ensonified region is not sharply defined in the angular sense. The ensonified region is defined by the product of the source and receiver directivity patterns and the intersection of this product with an elliptical band defined by the source pulse.

An early observation of bistatic scattering at 22 kHz was made by [Urick 1960] at a shallow-water (10–20 m) site near Panama City, Florida. Because the directional source and directional receiver were separated by a distance (3 km) much greater than the water depth, multipath propagation prevents a direct use of these data in testing scattering models. The data are interesting, however, in showing little dependence of scattering strength on the bistatic angle (Sect. 2.3.2). While Urick's data probe the azimuthal dependence of bistatic scattering, the measurements of [Schmidt 1971] were intended to examine the grazing angle dependence. These data were obtained over the frequency band 3.2–6.3 kHz. Unfortunately, it appears that these measurements were compromised by surface-reflected multipaths to a greater extent than appreciated by the author, as can be seen by considering the simple model for these effects given by [Mourad and Jackson 1993]. Bistatic scattering near the backward direction was studied by [Stanic et al. 1991, Stanic et al. 1993] over the frequency range 20–180 kHz. The measurements showed little variation in scattering strength for small horizontal and vertical angles with respect to the backscattering direction. Thus, these measurements showed no appreciable backscattering enhancement, consistent with expectations for sandy seafloors having moderate scattering strength (Sect. 2.3.2).

The discussion of more recent bistatic scattering data, which allow more direct comparison with model predictions, will be deferred to Ch. 13.

12.3 Research Issues

While there have been many high-frequency seafloor scattering experiments, there are still large gaps to be filled. It is important that future measurements be accompanied by detailed geoacoustic measurements characterizing roughness, discrete scatterers, gradients, and heterogeneity. With regard to backscatter measurements, most previous experiments have had limited coverage of either grazing angle or frequency (or both), and this has hampered model-data comparisons. Temporal and spatial variations in the seafloor lead to corresponding variations in acoustic, physical, and geoacoustic data Thus, care should be taken in model-data comparisons that all data sets are concurrent in space and time. There is a relative lack of data for muddy and rocky seafloors and corresponding weaknesses in model tests. Finally, further effort on bistatic measurements is warranted, and this may entail development of new methods of coping with geometric uncertainties. Laboratory measurements may be especially helpful in allowing controlled experiments, but the problem of scaling laboratory measurements to the natural world must be appreciated.

13 Roughness Scattering Models

This chapter deals with models that predict the scattering cross section (Sect. 2.3.2) in terms of statistical information on seafloor roughness. Most research in this area has assumed that seafloor relief can be described by a continuous function of the two horizontal coordinates, that is, roughness is the type of microtopography discussed in Ch. 6. This work is the primary subject of this chapter, but recent work on scattering due to discrete roughness features, such as shell fragments lying on the sediment–water interface, will be treated briefly.

As a starting point for the discussion, Fig. 13.1 shows model curves for backscattering strength for a sandy seafloor at 30 kHz. The curves were computed using one of the roughness scattering approximations to be discussed in this chapter and a volume scattering approximation to be discussed in Ch. 14. Several features evident in Fig. 13.1 are seen in other acoustic scattering models and in data for sand seafloors. First, roughness scattering shows a peak near vertical incidence. This is the remnant of the vertical reflection that would occur if the seafloor were flat. Second, roughness scattering has another peak near the critical grazing angle (about 25° in this example), and decreases as grazing angle becomes greater than the critical angle. This is because the interface becomes more "transparent" acoustically for angles greater than the critical angle. At the same time, and for the same reason, volume scattering increases for angles greater than the critical angle. For soft sediments such as mud (sediments comprised primarily of silt- and clay-sized particles, Table 4.1), the compressional wave speed is often less than the water sound speed (Figs. 5.2 and 5.7), so there is no critical angle. In such cases, volume scattering tends to dominate roughness scattering except for angles near vertical. In the other extreme, for "hard" sediments, such as coarse sand, one expects roughness scattering to dominate volume scattering, although this may not always be the case. In fact, it is not clear that a separation into roughness and volume scattering components is always possible or desirable, particularly for seafloors composed of gravel-sized grains or for seafloors whose volume heterogeneity is partly due to vertical undulations of a stratified sediment (see [Ivakin 1998a] and Sect. M.1.2). Nevertheless, this separation is made in nearly all published models and will be used here.

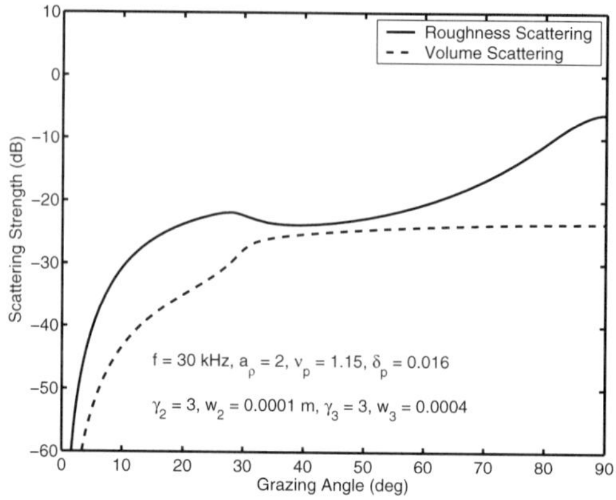

Fig. 13.1. Model curves for backscattering strength for seafloor showing the separate contributions of scattering by roughness and volume heterogeneity. The parameters listed were chosen arbitrarily, but are typical of a sand seafloor. These parameters have been defined in previous chapters and will be encountered again in this chapter and in Ch. 14. The roughness scattering contribution was computed in the fluid small-slope approximation, and the volume scattering contribution was computed in the fluid perturbation approximation.

In order to obtain numerically tractable models, investigators invariably introduce approximations and idealizations in addition to those already contained in the underlying theory. Thus, for example, there can be a variety of fluid models for scattering by seafloor roughness. In order to make the main results of scattering models for the seafloor readily accessible, this chapter and the following chapter (on scattering by volume heterogeneity) will only give the essential assumptions and final expressions for each model along with a discussion of the physics involved and a comparison with data. In addition, the accuracy expected for several of the approximations is discussed. The derivation of each model is given in the appendices, and the necessary statistical characterizations of roughness are defined in Sects. 6.1 and D.1. For deeper treatments of wave scattering, the reader can consult some of the books and review articles on this subject, including [Bass and Fuks 1979, Ogilvy 1991, Voronovich 1994, Ishimaru 1997, Elfouhaily and Guérin 2004].

13.1 General Properties of Roughness-Scattering Models

This section takes the unusual approach of beginning with final mathematical expressions for common scattering models, deferring explanation until later in the chapter and placing derivations in the appendices. This choice was made to allow a classification of the various models prior to detailed discussion. This classification is based on striking, but not widely appreciated, mathematical similarities in the disparate models. This approach also makes the models more accessible to the user, but may require some patience on the part of those whose main interest is in the physics of the models.

The two most widely used approximations for scattering by seafloor roughness are the small-roughness perturbation method (sometimes known as Rayleigh–Rice perturbation theory) and the Kirchhoff approximation (also known as the tangent-plane approximation). Each has its own separate domain of validity, with perturbation theory tending to be most accurate for scattering at wide angles relative to the specular (flat-interface reflection) direction [Thorsos and Jackson 1989] and the Kirchhoff approximation being better for scattering near the specular direction [Thorsos 1990].

The object of primary interest here is the bistatic scattering cross section, defined in Ch. 2. It is a function of the angles defining the direction from the source to the scattering region and from the scattering region to the receiver. These directions, in turn, can be specified in terms of wave vectors as follows:

$$\mathbf{k}_i = k_w(\mathbf{e}_x \cos\theta_i \cos\phi_i + \mathbf{e}_y \cos\theta_i \sin\phi_i - \mathbf{e}_z \sin\theta_i) \;, \tag{13.1}$$

$$\mathbf{k}_s = k_w(\mathbf{e}_x \cos\theta_s \cos\phi_s + \mathbf{e}_y \cos\theta_s \sin\phi_s + \mathbf{e}_z \sin\theta_s) \;. \tag{13.2}$$

Here, k_w is the wavenumber in water, the incident direction is specified by $\mathbf{k}_i$, and the direction toward the receiver is specified by $\mathbf{k}_s$. The angles used here are defined in Fig. 2.4, where it is noted that one may set $\phi_i = 0$ without loss of generality if the seafloor is isotropic in the statistical sense. All three components of the wave vectors cannot be specified independently, as the magnitude must be equal to k_w. Thus, the following horizontal components of the wave vectors are sufficient to define the incident and scattered directions:

$$\mathbf{K}_i = k_w(\mathbf{e}_x \cos\theta_i \cos\phi_i + \mathbf{e}_y \cos\theta_i \sin\phi_i) \;, \tag{13.3}$$

$$\mathbf{K}_s = k_w(\mathbf{e}_x \cos\theta_s \cos\phi_s + \mathbf{e}_y \cos\theta_s \sin\phi_s) \;. \tag{13.4}$$

It is convenient to define the following wave vector differences:

$$\mathbf{\Delta k} = \mathbf{k}_s - \mathbf{k}_i \;, \tag{13.5}$$

$$\mathbf{\Delta K} = \mathbf{K}_s - \mathbf{K}_i \;, \tag{13.6}$$

$$\Delta k_z = k_{sz} - k_{iz} \;. \tag{13.7}$$

Note

$$\Delta k^2 = \Delta K^2 + \Delta k_z^2 \;, \tag{13.8}$$

where the magnitude of a difference vector is denoted without boldface.

The small-roughness perturbation, Kirchhoff, and small-slope approximations yield equations whose form does not depend on the particular wave theory employed. This assertion is true in general for the perturbation and

Kirchhoff approximations, but is only true for the small-slope approximation when the seafloor is homogeneous (i.e., without gradients or layering). Given these general forms, one has only to fill in a few blanks to obtain the complete result within a given wave theory. Consequently, it is convenient to make the major division of roughness scattering models along the lines of the scattering approximation being used, rather than the wave theory employed. The basic equations for each of these three approximations will be given next, without derivation and with little explanation. This is done in order to expose the formal similarities and differences in these approximations. Each approximation will then be discussed in some detail in separate sections of this chapter, and appendices provide derivations.

The bistatic scattering cross section in the small-roughness perturbation approximation is

$$\sigma = k_w^4 |A_{ww}|^2 W(\boldsymbol{\Delta}\mathbf{K}) \ , \tag{13.9}$$

where $W(\boldsymbol{\Delta}\mathbf{K})$ is the roughness spectrum defined in Sect. 6.1. In the Kirchhoff approximation:

$$\sigma = \frac{|V_{ww}(\theta_{is})|^2}{8\pi}[\frac{\Delta k^2}{\Delta K \Delta k_z}]^2 I_K \ , \tag{13.10}$$

and in the small-slope approximation:

$$\sigma = \frac{k_w^4 |A_{ww}|^2}{2\pi \Delta K^2 \Delta k_z^2} I_K \ . \tag{13.11}$$

It is useful to note the similarities between these equations and to identify the essential factors that must be supplied for each different choice of wave theory. The factor A_{ww} depends on choice of wave theory, but is common to both the perturbation and small-slope approximations. The "Kirchhoff integral"

$$I_K = \frac{\Delta K^2}{2\pi} \int e^{-i\boldsymbol{\Delta}\mathbf{K}\cdot\mathbf{R}}[e^{-\frac{1}{2}\Delta k_z^2 S(\mathbf{R})} - e^{-\Delta k_z^2 h^2}]d^2 R \tag{13.12}$$

does not depend on the choice of wave theory and is common to both the Kirchhoff and small-slope approximations. Roughness statistics enter this integral through the "structure function," $S(\mathbf{R})$, defined in Sect. D.1. The reflection coefficient, V_{ww}, appearing in the Kirchhoff expression depends on the choice of wave theory, and is evaluated at the grazing angle

$$\theta_{is} = \sin^{-1}(\frac{\Delta k}{2k_w}) \tag{13.13}$$

that corresponds to specular reflection from the source to the receiver with the rough surface tilted in such a way as to provide such a reflection (Thorsos, private communication as cited in [D. Jackson 1994]).

In summary, wave theory must be applied to obtain A_{ww} for the small-roughness perturbation and small-slope approximations or, correspondingly, V_{ww} for the Kirchhoff approximation. Once these theoretical tasks are complete, only numerical computation remains, including evaluation of (13.12).

The scattering cross section describes the angular dependence of the incoherent field. In general, one must also consider the coherent field, which can be described in terms of the coherent reflection coefficient, V_{wwc}, discussed in Ch. 2. The second term in the integrand of (13.12) represents subtraction of the coherently reflected energy from the total energy leaving the interface such that the remainder represents the incoherently scattered energy. For both the Kirchhoff and small-slope approximations, the coherent reflection coefficient is the flat-interface reflection coefficient, $V_{ww}(\theta)$, reduced by an exponential factor:

$$V_{wwc}(\theta) = V_{ww}(\theta) e^{-2k_w^2 h^2 \sin^2 \theta_i} \ . \tag{13.14}$$

This result is widely used and is often associated with [Eckart 1953], who derived the analogous result for scattering by the sea surface. One must be careful in applying (13.14) in acoustic simulations. This expression is sometimes used as the total reflection coefficient at the seafloor and scattering is ignored. This approach assumes that the incoherently scattered energy is "lost" and can be ignored in computations of the field at ranges beyond that at which the reflection occurred. This may be a good approximation if the scattering occurs predominantly at wide angles with respect to the specular direction, but if this is not the case, the scattered field may be an important component of the down-range field.

To this point, the three roughness scattering approximations are general enough to treat the case of anisotropic roughness, e.g., seafloors with directional ripples. In the examples to follow, isotropy will be assumed for the sake of simplicity. The spectrum (6.8) is an example of isotropy, as it depends only on the magnitude, K, of its two-dimensional vector argument, $\mathbf{K}$. Likewise, in the isotropic case, the structure function will only depend on the magnitude, R, of its argument, $\mathbf{R}$. In this case the Kirchhoff integral (13.12) can be reduced to a one-dimensional integral by using polar integration coordinates and integrating over the angular coordinate with the result

$$I_K = \int_0^\infty J_0(u) [e^{-\frac{1}{2}\Delta k_z^2 S(\frac{u}{\Delta K})} - e^{-\Delta k_z^2 h^2}] u du \ , \tag{13.15}$$

where J_0 is the zeroth-order cylindrical Bessel function of the first kind. For the power-law case (6.8), h is infinite, and (D.11) can be used to put (13.15) in the form

$$I_K = \int_0^\infty J_0(u) e^{-q u^{2\alpha}} u du \ , \tag{13.16}$$

where

$$q = \frac{1}{2} C_h^2 \Delta k_z^2 \Delta K^{-2\alpha} \ . \tag{13.17}$$

The parameters C_h and α are given in terms of the power-law spectrum parameters in (D.12) and (D.13). Evaluation of this integral requires special consideration numerically [Drumheller and Gragg 2001]. Carefully chosen, more general forms of the roughness spectrum can also lead to tractable numerical integrals [Wurmser 2005].

A comparison of the three roughness-scattering approximations defined previously is made in Fig. 13.2. Fluid theory is used to compute backscattering strength, and the parameters are appropriate to a medium sand seafloor. For these parameters, the small-slope approximation is accurate over a wide range of angles (though perhaps not at the smallest grazing angles). The Kirchhoff approximation matches the small-slope approximation near vertical incidence, but gives excessively large scattering strengths for moderate and small angles. In fact, the increasing trend as grazing angle approaches zero is an indication of the violation of energy conservation (see Ch. 2). Perturbation theory behaves in a fashion complementary to the Kirchhoff approximation, matching the small-slope approximation at small-to-moderate angles and giving overlarge scattering strengths near vertical incidence. This behavior, too, violates energy conservation and is in part an artifact due to the singularity in the assumed roughness spectrum.

The complementary behavior of the Kirchhoff and perturbation approximations has led numerous investigators to combine the two in the "composite-roughness" approximation discussed in some detail in Sect. 13.5. Most of this work employed a version of the Kirchhoff approximation in which a high-frequency limit is taken. The high-frequency Kirchhoff approximation is discussed in in Sect. 13.3 and Appendix L.

With the advent of the small-slope approximation, motivation for use of the composite-roughness and Kirchhoff approximations is reduced. The small-slope approximation provides a single expression that covers all angles and is likely to be at least as accurate as either the Kirchhoff or perturbation approximations. The relationship between the three scattering approximations can be clarified by considering various limiting cases. First, if the acoustic wavelength is much smaller than the RMS roughness, h, the exponential factor containing the structure function in the Kirchhoff integral (13.12) can be expanded to second order in $k_w h$. Using the definition of the structure function in terms of the covariance (D.6) and the Fourier transform relation between the covariance and the spectrum, (6.5), it can be shown that the small-slope cross section (13.11) reduces to the perturbation result (13.9). This is not surprising, as the small-slope approximation was constructed to reduce to the perturbation approximation in the small-roughness limit.

As is well known, the Kirchhoff and small-slope approximations for the bistatic scattering cross section coincide near the specular direction. As shown in Appendix L, this is a result of a general identity (L.9) linking the reflection coefficient and the first-order perturbation coefficient appearing in the

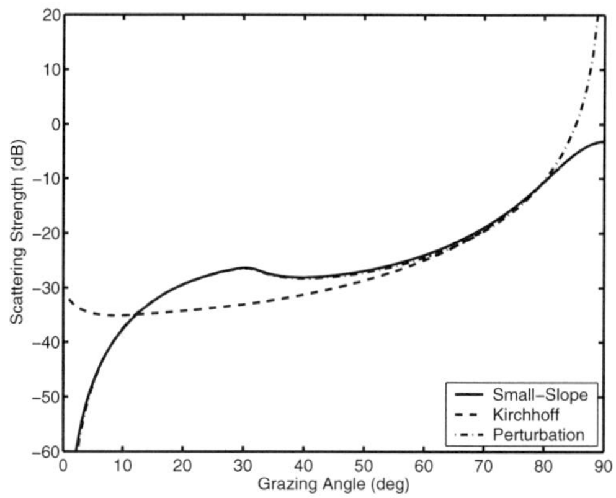

Fig. 13.2. Comparison of backscattering strengths computed in the small-slope, Kirchhoff, and small-roughness perturbation approximations using fluid theory. The acoustic frequency is 30 kHz, and the input parameters are typical of a medium sand seafloor: $a_\rho = 1.845$, $\nu_p = 1.1782$, $\delta_p = 0.01624$, $\gamma_2 = 3.25$, and $w_2 = 0.000141$ $\text{m}^{0.75}$. In this case, the curves for the small-slope and perturbation approximations overlap for grazing angles smaller than about 70°, and those for the Kirchhoff and small-slope approximations overlap for angles larger than this.

small-slope cross section (13.11). Appendix L defines an incoherent reflection coefficient that is useful whenever the bistatic scattering cross section is peaked near the specular direction. This reflection coefficient gives a simple means of applying the sonar equation to estimate the scattered intensity near the specular direction. Because of the near equality of the Kirchhoff and small-slope approximations in the vicinity of the specular direction, this incoherent reflection coefficient is the same in both cases and equal to

$$|V_{wwi}(\theta_i)|^2 = |V_{ww}(\theta_i)|^2 - |V_{wwc}(\theta_i)|^2 . \tag{13.18}$$

This expresses a simple and physically reasonable property. If the interface roughness does not spread the scattered energy far from the specular direction, the combined coherent and incoherent fields will yield the same intensity as that reflected by a flat interface [McDonald and Spindel 1971]. In an optical analogy, one may imagine sunlight reflected from a flat mirror. If the surface of the mirror is deformed slightly, one would not expect the average intensity of the reflected sunlight to change. This incoherent reflection coefficient gives the RMS incoherently scattered pressure near the specular direction if it is inserted into the usual expression for flat-interface reflection.

This property of rough-interface scattering can be exploited to allow measurement of the flat-interface reflection coefficient, even if the interface is moderately rough. In such applications, the source and receiver must be nondirectional, or at least substantially less directional than the near-specular peak of the bistatic scattering cross section. Also, the derivation leading to (13.18) assumes that the incident wave is a continuous sine wave. If a pulsed source is used, the pulse length must be great enough that the scattered intensity can reach the steady-state value that CW excitation would provide. This situation is discussed in detail for the case of vertical-incidence measurements in Sect. G.2.2.

13.2 Small-Roughness Perturbation Approximation

Small-roughness perturbation theory is widely employed in modeling acoustic scattering by the seafloor and provides a basis for the composite-roughness and small-slope approximations. Furthermore, perturbation theory yields an intuitive picture of the physics of scattering. The roughness spectrum, $W(\mathbf{\Delta K})$, appearing in the small-roughness cross section (13.9), is central to this picture. The spectrum is evaluated at the so-called "Bragg wave vector," the difference of the scattered and incident horizontal wave vectors. A useful and physically instructive mnemonic for the Bragg wave vector follows from rearranging (13.6) to read $\mathbf{K}_s = \mathbf{K}_i + \mathbf{\Delta K}$. If, as in quantum mechanics, one interprets wave vector as momentum, this relation expresses conservation of momentum, with the momentum of the scattered wave being equal to the incident momentum plus the momentum gained by interaction with the interface. If the roughness spectrum has a large peak at some wave vector, $\mathbf{K}_0$, there will be strong scattering in the two directions corresponding to $\mathbf{K}_s = \mathbf{K}_i \pm \mathbf{K}_0$, the "Bragg condition," a term originating in X-ray diffraction. If there is such a peak in the spectrum, the seafloor will have a strongly rippled appearance, with the ripples perpendicular to the direction of $\mathbf{K_0}$ and having wavelength $\lambda_0 = 2\pi/K_0$. Two scattering directions exist, because the spectrum is symmetric in the sense $W_2(\mathbf{K}) = W_2(-\mathbf{K})$, so the existence of a peak at $\mathbf{K}_0$ requires a peak at $-\mathbf{K}_0$. The Helmholtz equation requires that the z-component of the scattered wave vectors satisfy

$$k_{sz} = \sqrt{k_w^2 - K_s^2} \, , \tag{13.19}$$

so that all wave vectors have the same length, k_w, as indicated by the dashed semicircles in Fig. 13.3. Note also that $\mathbf{K}_s = \mathbf{K}_i$ corresponds to the specular direction, so the two scattered waves have directions above and below the specular direction in the two-dimensional example of Fig. 13.3(a). If K_s is sufficiently large, the z-component of the scattered wave vector will be imaginary, that is, the wave will be evanescent and will not contribute significantly to the scattered field in the water column (Fig. 13.3(b)).

If the roughness of the seafloor is random and nondirectional (isotropic), the spectrum will be spread over a wide range of wave vectors, allowing the Bragg condition to be satisfied at all angles. Most spectra have a peak at or near the origin (in two-dimensional K-space), giving rise to a peak in

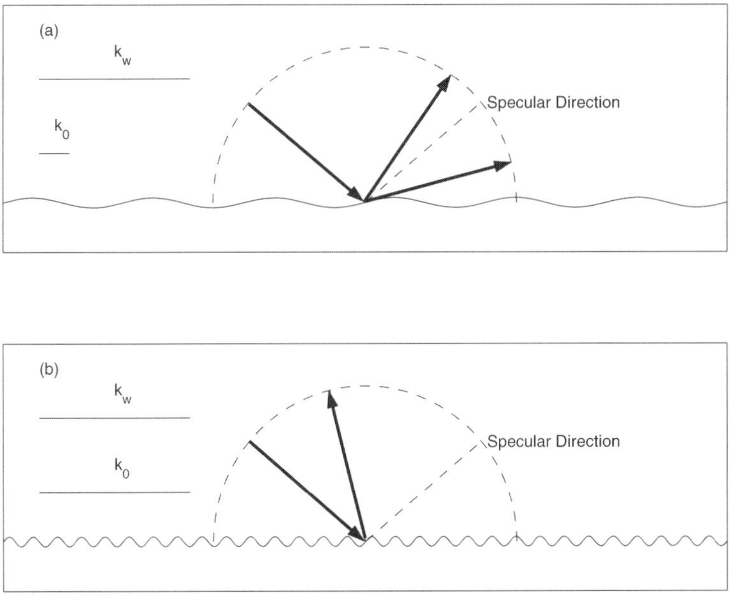

Fig. 13.3. A two-dimensional example illustrating Bragg scattering by sinusoidal ripple fields. In panel (a), the ripple wavelength is longer than the acoustic wavelength, with the result that there are two scattering directions, both near the specular direction. In panel (b), the ripple wavelength is equal to the acoustic wavelength, and one of the scattered waves has become evanescent and the other is in the backward direction. The arrows represent the wave vectors of plane waves and do not imply interaction at a single point on the interface. The interaction occurs over the entire interface.

scattering near the specular direction. In the special case of backscattering, the Bragg condition, obtained from (13.6), takes on the simple form

$$\Delta K = 2k_w \cos\theta_i \ , \tag{13.20}$$

where ΔK is the Bragg wavenumber, k_w is the acoustic wavenumber, and $\theta_i = \theta_s$ is the grazing angle. At small angles, this amounts to the statement that the wavelength of the roughness features responsible for scattering is one-half the acoustic wavelength. An experiment to demonstrate "Bragg scattering" from artificially created ripples is reported in [Richardson et al. 2001b] and discussed briefly in Sect. 6.4.

As noted earlier, the small-roughness perturbation approximation has been applied to a variety of seafloor models, including the fluid, elastic, and poroelastic cases. Each of these cases will be treated in order, with the main results contained in expressions for the factor A_{ww} appearing in the scattering cross section. The discussion of perturbation theory concludes with comparisons of model predictions and data.

13.2.1 Fluid Model

The fluid perturbation model was among the first physics-based seafloor scattering models to be developed. Expressions for the case of a lossless fluid sediment were given by [Kuo 1964], and these are easily generalized to the case of finite attenuation [D. Jackson and Briggs 1992]. This model assumes the sediment is homogeneous, without layering or gradients of any form. Several investigators have employed this model [Crowther 1983, Ivakin 1983, Stanic et al. 1989, Mourad and Jackson 1989, D. Jackson and Briggs 1992, Stanic et al. 1993, Kuo 1995, D. Jackson et al. 1996a, Williams and Jackson 1997, Williams and Jackson 1998]. Others have extended it to allow for gradients and layering in the seafloor [Ivakin 1994b, Lyons et al. 1994, Moe and Jackson 1994].

A convenient expression for the factor A_{ww} needed in the expression for the scattering cross section (13.9) can be found in [D. Jackson 1994] and derivations are given in Sects. K.1 and M.1.2:

$$A_{ww} = \frac{1}{2}[1 + V_{ww}(\theta_i)][1 + V_{ww}(\theta_s)]G \; , \tag{13.21}$$

where

$$G = (1 - 1/a_\rho)[\cos\theta_i \cos\theta_s \cos\phi_s - B] - 1 + \frac{1}{a_p^2 a_\rho} \; , \tag{13.22}$$

$$B = \frac{\sin\theta_{pi} \sin\theta_{ps}}{a_p^2 a_\rho} \; , \tag{13.23}$$

$$\sin\theta_{pi} = \sqrt{1 - a_p^2 \cos^2\theta_i} \; , \tag{13.24}$$

$$\sin\theta_{ps} = \sqrt{1 - a_p^2 \cos^2\theta_s} \; . \tag{13.25}$$

As defined in Ch. 8, a_p is the complex ratio of sediment compressional wave speed to water sound speed, a_ρ is the sediment–water density ratio, and $V_{ww}(\theta)$ is the flat-interface reflection coefficient evaluated at the grazing angle θ. Expression (13.23) is only applicable to the homogeneous fluid case. A more general expression applicable to horizontally stratified seafloors [Moe and Jackson 1994] can be obtained by replacing (13.23) with the following:

$$B = \sin\theta_i \sin\theta_s \frac{[1 - V_{ww}(\theta_i)][1 - V_{ww}(\theta_s)]}{[1 + V_{ww}(\theta_i)][1 + V_{ww}(\theta_s)]} a_\rho \; . \tag{13.26}$$

Expressions (13.21), (13.22), (13.26) can be applied to seafloors with gradients and layering simply by using the appropriate reflection coefficient. In such cases a_p and a_ρ appearing in (13.22) and (13.26) are to be evaluated immediately below the sediment–water interface. It is assumed that gradients and layering are such that these parameters are substantially constant over the range of vertical coordinate of the interface relief function. If this is not true, the approach of [Ivakin 1998a] may be useful (see Sect. M.1.2). The seafloor below the rough interface is assumed not to have any variations in its properties with respect to the horizontal coordinates, and it must be fluid-like in the region immediately below the rough interface. Apart from those restrictions, the lower portion of the seafloor may consist of layers of any type, fluid, elastic, or poroelastic. The effect of these lower layers is embodied in the reflection coefficient, $V_{ww}(\theta)$. Any attempt to model an elastic or poroelastic upper layer by letting the thickness of the upper fluid layer vanish will fail, as the boundary conditions at the rough interface will not be satisfied.

Figure 13.4 illustrates the small-roughness perturbation model for two different seafloor types, a "fast" sand seafloor having compressional wave speed greater than the water sound speed ($\nu_p > 1$) and a "slow" mud seafloor having $\nu_p < 1$. All other parameters are taken to be the same for both cases except that the sand seafloor has greater density ($a_\rho = 2.0$) than the mud seafloor ($a_\rho = 1.5$). The sediment is assumed to be homogeneous, and the roughness spectrum has the power-law form of (6.8). A maximum in scattering strength near the critical angle is evident for the sand case and absent for the mud case. This feature becomes less sharp-edged as attenuation increases (δ_p increases). The mud scattering strength is lower than the sand scattering strength, owing to the reduced acoustic contrast in this case. As will be seen, this deficit in scattering strength is often made up by increased volume scattering in muddy sediments. The curve for mud shows no feature at the angle of intromission (about 10.5°). The scattering strength curves for both cases become arbitrarily large as the grazing angle approaches 90°. As mentioned earlier, this artifact is due to the singularity in the idealized roughness spectrum at $K = 0$.

The model represented by (13.21), (13.22), and (13.26) has only one rough interface (the water–sediment interface), and all buried interfaces are required to be smooth. Models with rough buried interfaces have been developed by [McDaniel 1992, Lyons et al. 1994, Ivakin 1994b, Tang 1996a].

13.2.2 Elastic Model

As shown in Ch. 9, shear effects on the flat-interface reflection coefficient only become significant when the shear wave speed becomes substantial in comparison to the water sound speed. The same will be found true in the small-roughness scattering approximation, where the inclusion of shear effects is only a slight refinement of the fluid model for sands with shear speeds on the order of 100 m s^{-1}, but shear effects are dominant

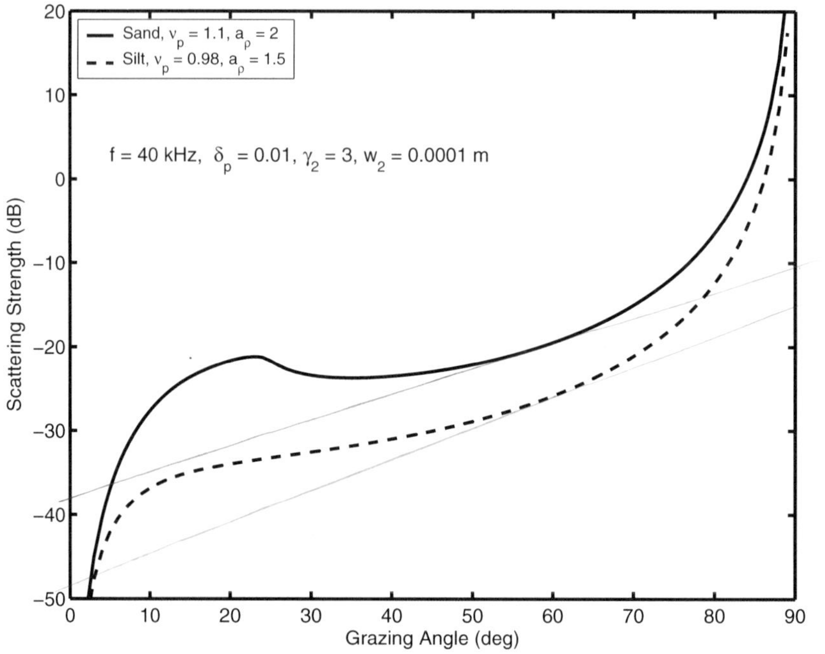

Fig. 13.4. Comparison of small-roughness perturbation model curves for backscattering strength for sand and mud seafloors differing only in their sound speed and density ratios. Note that the critical angle feature evident in the curve for sand is absent in the mud case, where the compressional wave speed is assumed to be smaller than the water sound speed. Model parameters common to both cases are $\gamma_2 = 3.0$, $w_2 = 0.0001$ m, $\delta_p = 0.01$.

in rock, where shear speeds of 2–4 km s^{-1} are seen (Sect. 5.6). Perturbation theory has been applied to the elastic seafloor scattering problem by several authors [Lapin 1964, Lapin 1966, Kuperman and Schmidt 1986, Dacol and Berman 1988, Kuo 1992, Essen 1994, D. Jackson and Ivakin 1998], using various approaches and notation, but the formalisms are basically equivalent. The results to be presented here are valid only for homogeneous seafloors, that is, seafloors with no spatial dependence in their geoacoustic parameters.

To minimize the notational complexity, a compact formalism is developed in Appendix K. The results can be expressed in terms of the expression for the factor A appearing in the scattering cross section (13.9):

$$A_{ww} = \frac{1}{2}(D_1[1 + V_{ww}(\theta_s)][1 + V_{ww}(\theta_i)] + D_2[1 - V_{ww}(\theta_s)][1 + V_{ww}(\theta_i)]$$

$$+D_3[1+V_{ww}(\theta_s)][1-V_{ww}(\theta_i)]+D_4[1-V_{ww}(\theta_s)][1-V_{ww}(\theta_i)]) \ , \quad (13.27)$$

where $V_{ww}(\theta)$ is the usual flat-interface reflection coefficient, and the variables D_n are

$$D_1 = -1 + S + \frac{1}{a_p^2 a_\rho \cos 2\theta_{ts} \cos 2\theta_{ti}}$$

$$-\frac{a_t^2[(a_t^{-2} - 2\cos^2\theta_s - 2\cos^2\theta_i + 2S)S + 2\cos^2\theta_s\cos^2\theta_i]}{a_\rho \cos 2\theta_{ts}\cos 2\theta_{ti}} \ , \quad (13.28)$$

$$D_2 = -\frac{4a_t^3 \sin\theta_s \sin\theta_{ts}}{\cos 2\theta_{ts}\cos 2\theta_{ti}}[a_p^{-2}\sin^2\theta_{pi}\cos^2\theta_s + (\cos^2\theta_i - S)S] \ , \quad (13.29)$$

$$D_3 = -\frac{4a_t^3 \sin\theta_{ti} \sin\theta_i}{\cos 2\theta_{ts}\cos 2\theta_{ti}}[a_p^{-2}\sin^2\theta_{ps}\cos^2\theta_i + (\cos^2\theta_s - S)S] \ , \quad (13.30)$$

$$D_4 = \frac{2a_t^4 a_\rho \sin\theta_s \sin\theta_i \sin\theta_{ts}\sin\theta_{ti}}{\cos 2\theta_{ts}\cos 2\theta_{ti}} \times$$

$$[2(a_t^{-2} - 2S)S - 4\cos^2\theta_s\cos^2\theta_i(1 - 2a_t^2 a_p^{-2})] - (a_\rho - 1)\sin\theta_s\sin\theta_i \ . \quad (13.31)$$

The variable S is

$$S = \cos\theta_s \cos\theta_i \cos\phi_s \ , \quad (13.32)$$

and the following complex cosines are found from Snell's law

$$\cos\theta_p = a_p \cos\theta \ , \quad (13.33)$$

$$\cos\theta_t = a_t \cos\theta \ , \quad (13.34)$$

where a_t is the complex shear speed/water sound speed ratio, the angle θ may be either θ_i or θ_s, and the sines are obtained using $\sqrt{1-\cos^2\theta}$. The cosines of the form $\cos 2\theta_t$ can either be found using a trigonometric identity or by using (9.39).

One advantage of the form given above for the cross section is that reciprocity is manifest, that is, the cross section is obviously invariant under the interchange of θ_i and θ_s, as it must be. This invariance is present, but hidden, in most other published expressions, except in the small-slope formalism of [Wurmser 1996].

Figure 13.5 compares scattering strengths at 30 kHz computed using elastic perturbation theory with those computed using fluid theory. All parameters are equal for the two cases, except that the complex shear speed is effectively zero for the fluid case. Note that, for the sand example, there is only a slight difference between the two models, and this difference has been exaggerated by assuming a rather high shear speed (250 m s^{-1}). The rock example shows major differences between the two models, indicating that the fluid model has failed for this case in which the shear speed is high (ν_t = 1.3). As for the reflection loss case, Fig. 9.4, the largest departure from the fluid model occurs near the shear critical angle. In fact, the dip in scattering cross section appears near the so-called "Rayleigh angle" [Berman 1991, Yang and Broschat 1994]. Geoacoustic parameters for rock can be found in Sect. 5.6 and in [Hamilton 1978, Hamilton 1979b].

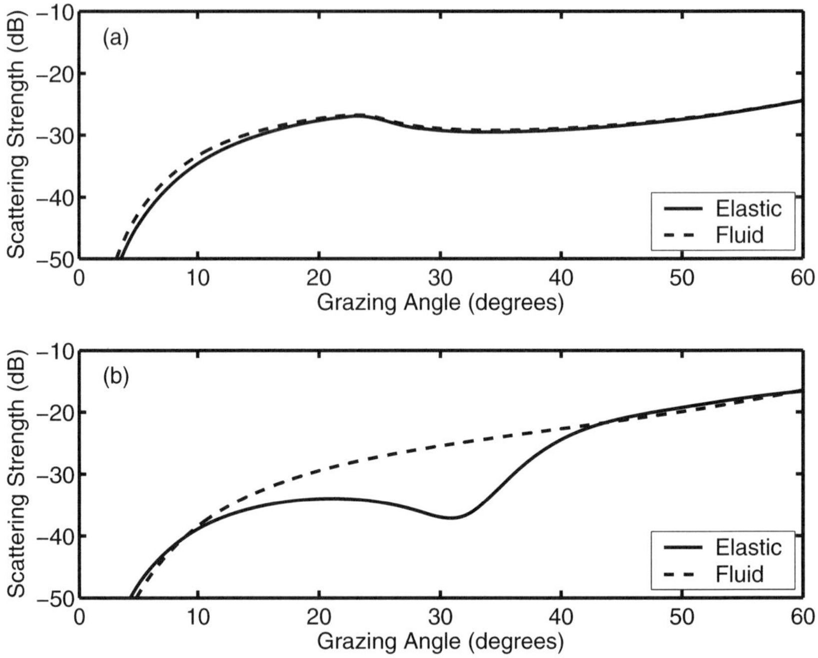

Fig. 13.5. Comparison of fluid and elastic perturbation theory models for rough-interface backscattering, with parameters appropriate for sand (a) and rock (b). For the sand example, the parameters of Fig. 9.3 are used, and for the rock example, the parameters of Fig. 9.4 are used. Both the sand and rock examples have the same spectral parameters, $w_2 = 0.000141$ m$^{0.75}$, $\gamma_2 = 3.25$, and the same acoustic frequency, 30 kHz.

13.2.3 Poroelastic Model

This section presents expressions for first-order perturbation theory for the poroelastic case. As with the elastic case presented previously, the results to be given apply only to homogeneous seafloors. The expressions listed below are taken from [Williams et al. 2001a], and an alternative derivation is sketched in Appendix K. The development here employs the matrix method of Ivakin [D. Jackson and Ivakin 1998] in order to reduce algebraic complexity.

The factor A_{ww} appearing in the general expression for the scattering cross section in perturbation theory (13.9) is the first element of the column matrix

$$A_{all}(\mathbf{K}_s, \mathbf{K}_i) = \begin{bmatrix} A_{ww}(\mathbf{K}_s, \mathbf{K}_i) \\ A_{w1}(\mathbf{K}_s, \mathbf{K}_i) \\ A_{w2}(\mathbf{K}_s, \mathbf{K}_i) \\ A_{wv}(\mathbf{K}_s, \mathbf{K}_i) \\ A_{wh}(\mathbf{K}_s, \mathbf{K}_i) \end{bmatrix} . \tag{13.35}$$

As noted in Appendix K, the elements of $A_{all}(\mathbf{K}_s, \mathbf{K}_i)$ determine, to first order in roughness amplitude, the plane-wave spectra ("T-matrices") of waves scattered into the water and into the seafloor in the form of fast (1), slow (2), and shear waves. Two shear wave polarizations must be considered, vertical (v) and horizontal (h). These wave types are discussed in Chs. 9 and 10. The matrix A_{all} is computed as follows:

$$A_{all}(\mathbf{K}_s, \mathbf{K}_i) = Y_1(\mathbf{K}_s)[P^{(3)}(\mathbf{K}_s)]^{-1} B(\mathbf{K}_s, \mathbf{K}_i) V(\mathbf{K}_i) . \tag{13.36}$$

In this equation, $V(\mathbf{K}_i)$ is a six-row column vector comprised of the five reflection and transmission coefficients discussed in Ch. 10 and supplemented with unity in the last row,

$$V(\mathbf{K}_i) = \begin{bmatrix} V_{ww} \\ V_{w1} \\ V_{w2} \\ V_{wt} \\ 0 \\ 1 \end{bmatrix} . \tag{13.37}$$

Note the transmission coefficient for horizontally polarized shear waves is zero. The matrix $B(\mathbf{K}_s, \mathbf{K}_i)$ is

$$B(\mathbf{K}_s, \mathbf{K}_i) =$$

$$k_w^{-1}[(K_{s1}-K_{i1})E^{(1)}(\mathbf{K}_i)+(K_{s2}-K_{i2})E^{(2)}(\mathbf{K}_i)]-E^{(3)}(\mathbf{K}_i)Y_2(\mathbf{K}_i) . \tag{13.38}$$

The matrices Y_1 and Y_2 are

$$Y_1(\mathbf{K}) = \begin{bmatrix} \nu_w(K) & 0 & 0 & 0 & 0 \\ 0 & \nu_1(K) & 0 & 0 & 0 \\ 0 & 0 & \nu_2(K) & 0 & 0 \\ 0 & 0 & 0 & \nu_t(K) & 0 \\ 0 & 0 & 0 & 0 & \nu_t(K) \end{bmatrix} , \tag{13.39}$$

$$Y_2(\mathbf{K}) = \begin{bmatrix} \nu_w(K) & 0 & 0 & 0 & 0 & 0 \\ 0 & -\nu_1(K) & 0 & 0 & 0 & 0 \\ 0 & 0 & -\nu_2(K) & 0 & 0 & 0 \\ 0 & 0 & 0 & -\nu_t(K) & 0 & 0 \\ 0 & 0 & 0 & 0 & -\nu_t(K) & 0 \\ 0 & 0 & 0 & 0 & 0 & -\nu_w(K) \end{bmatrix} , \tag{13.40}$$

where

$$\nu_q = \beta_q(K)/a_q \ , \qquad q = w,\ 1,\ 2,\ t\ . \tag{13.41}$$

The factors $\beta_q(K)$ and a_q are, respectively, the complex sines and speed ratios for all wave types of interest, generalized slightly to include $a_w = 1$. The subscripts 1 and 2 on Y_1 and Y_2 do not relate to orders of perturbation or fast and slow waves; they merely serve to distinguish between the two matrices. There are three matrices $E^{(n)}(\mathbf{K})$, defined for each of the coordinate indices, $n = 1,\ 2,\ 3$ corresponding to $x,\ y\ , z$:

$$E^{(n)}(\mathbf{K}) = \left[\begin{array}{ccccc|c} P^{(n)}_{w1}(\mathbf{K}) & P^{(n)}_{11}(\mathbf{K}) & P^{(n)}_{21}(\mathbf{K}) & P^{(n)}_{v1}(\mathbf{K}) & P^{(n)}_{h1}(\mathbf{K}) & Q^{(n)}_{1}(\mathbf{K}) \\ P^{(n)}_{w2}(\mathbf{K}) & P^{(n)}_{12}(\mathbf{K}) & P^{(n)}_{22}(\mathbf{K}) & P^{(n)}_{v2}(\mathbf{K}) & P^{(n)}_{h2}(\mathbf{K}) & Q^{(n)}_{2}(\mathbf{K}) \\ P^{(n)}_{w3}(\mathbf{K}) & P^{(n)}_{13}(\mathbf{K}) & P^{(n)}_{23}(\mathbf{K}) & P^{(n)}_{v3}(\mathbf{K}) & P^{(n)}_{h3}(\mathbf{K}) & Q^{(n)}_{3}(\mathbf{K}) \\ P^{(n)}_{w4}(\mathbf{K}) & P^{(n)}_{14}(\mathbf{K}) & P^{(n)}_{24}(\mathbf{K}) & P^{(n)}_{v4}(\mathbf{K}) & P^{(n)}_{h4}(\mathbf{K}) & Q^{(n)}_{4}(\mathbf{K}) \\ P^{(n)}_{w5}(\mathbf{K}) & P^{(n)}_{15}(\mathbf{K}) & P^{(n)}_{25}(\mathbf{K}) & P^{(n)}_{v5}(\mathbf{K}) & P^{(n)}_{h5}(\mathbf{K}) & Q^{(n)}_{5}(\mathbf{K}) \end{array}\right] , \tag{13.42}$$

where the vertical bars in Eq.(13.42) separate it into the 5×5 matrices, $P^{(n)}(\mathbf{K})$, and the column matrices, $Q^{(n)}(\mathbf{K})$. The five rows of $E^{(n)}(\mathbf{K})$ correspond to the five boundary conditions: the first five columns correspond to the five wave types: reflected and scattered pressure waves in water, fast and slow waves in sediment, and vertically and horizontally polarized shear waves in sediment. The sixth column corresponds to the incident wave in water.

The elements of $P^{(n)}(\mathbf{K})$ and $Q^{(n)}(\mathbf{K})$ are

$$P^{(n)}_{wm}(\mathbf{K}) = K_w k_w \delta_{mn} \ , \quad m = 1,2,3, \tag{13.43}$$

$$P^{(n)}_{w4}(\mathbf{K}) = -K_w k_w \delta_{3n} \ , \tag{13.44}$$

$$P^{(n)}_{w5}(\mathbf{K}) = -e^{+}_{wn}(\mathbf{K}) \ , \tag{13.45}$$

$$P^{(n)}_{1m}(\mathbf{K}) = -k_1^2 k_w^{-1}[(H - 2\mu - \gamma_1 C)\delta_{mn} + 2\mu e^{-}_{1m}(\mathbf{K}) e^{-}_{1n}(\mathbf{K})] \ , \quad m = 1,2,3, \tag{13.46}$$

$$P^{(n)}_{14}(\mathbf{K}) = -k_1^2 k_w^{-1}(M\gamma_1 - C)\delta_{n3} \ , \tag{13.47}$$

$$P^{(n)}_{15}(\mathbf{K}) = k_1 k_w^{-1}(1 - \gamma_1) e^{-}_{1n}(\mathbf{K}) \ , \tag{13.48}$$

$$P^{(n)}_{2m}(\mathbf{K}) = -k_2^2 k_w^{-1}[(H - 2\mu - \gamma_2 C)\delta_{mn} + 2\mu e^{-}_{2m}(\mathbf{K}) e^{-}_{2n}(\mathbf{K})] \ , \quad m = 1,2,3, \tag{13.49}$$

$$P^{(n)}_{24}(\mathbf{K}) = -k_2^2 k_w^{-1}(M\gamma_2 - C)\delta_{n3} \ , \tag{13.50}$$

$$P^{(n)}_{25}(\mathbf{K}) = k_2 k_w^{-1}(1 - \gamma_2) e^{-}_{2n}(\mathbf{K}) \ , \tag{13.51}$$

$$P_{vm}^{(n)}(\mathbf{K}) = \mu k_t^2 k_w^{-1} [e_{tm}^-(\mathbf{K}) e_{vn}^-(\mathbf{K}) + e_{tn}^-(\mathbf{K}) e_{vm}^-(\mathbf{K})] \;, \quad m = 1, 2, 3, \tag{13.52}$$

$$P_{v4}^{(n)}(\mathbf{K}) = 0 \;, \tag{13.53}$$

$$P_{v5}^{(n)}(\mathbf{K}) = -k_t k_w^{-1} (1 - \gamma_t) e_{vn}^-(\mathbf{K}) \;, \tag{13.54}$$

$$P_{hm}^{(n)}(\mathbf{K}) = -\mu k_t^2 k_w^{-1} [e_{tm}^-(\mathbf{K}) e_{hn}^-(\mathbf{K}) + e_{tn}^-(\mathbf{K}) e_{hm}^-(\mathbf{K})] \;, \quad m = 1, 2, 3, \tag{13.55}$$

$$P_{h4}^{(n)}(\mathbf{K}) = 0 \;, \tag{13.56}$$

$$P_{h5}^{(n)}(\mathbf{K}) = k_t k_w^{-1} (1 - \gamma_t) e_{hn}^-(\mathbf{K}) \;, \tag{13.57}$$

$$Q_m^{(n)}(\mathbf{K}) = K_w k_w \delta_{mn} \;, \quad m = 1, 2, 3, \tag{13.58}$$

$$Q_4^{(n)}(\mathbf{K}) = -K_w k_w \delta_{n3} \;, \tag{13.59}$$

$$Q_5^{(n)}(\mathbf{K}) = -e_{wn}^-(\mathbf{K}) \;. \tag{13.60}$$

In these expressions, δ_{mn} is the Kronecker delta ($\delta_{mn} = 1$ if $m = n$, $\delta_{mn} = 0$ otherwise), and K_w is the bulk modulus of the upper fluid (water) taken equal to the pore fluid modulus. The unit vectors defined in Sect. 10.1.3 appear here in the form $e_{\eta n}^{\pm}$, with $n = x, y, z$ indicating the three spatial components and with $\eta = w, 1, 2, t, h, v$. The unit vectors $\mathbf{e}_w^{\pm}$ are the obvious extension of the propagation unit vectors for waves in the sediment to the case of waves in the water. The coefficients γ_1, γ_2, and γ_t are defined in Sect. 10.1.3.

The small-roughness Biot model is compared with the fluid and effective density fluid approximation in Fig. 13.6. Note that the effective density approximation matches the Biot results very well and therefore provides a simpler alternative for scattering strength modeling. The complex sound speed and complex effective density for this approximation were computed using (10.81) and (10.75), which were then inserted in the fluid model of Ch. 8. The fluid approximation yields generally higher scattering strengths than the Biot model, but this difference diminishes as frequency takes on low values where the fluid and frame tend to move together. The lower two frequencies in Fig. 13.6 are only included for the sake of illustration. These frequencies are most likely below the band of applicability of the model, as penetration of the seafloor at such low frequencies would be so deep as to expose layering and gradients not included in the model.

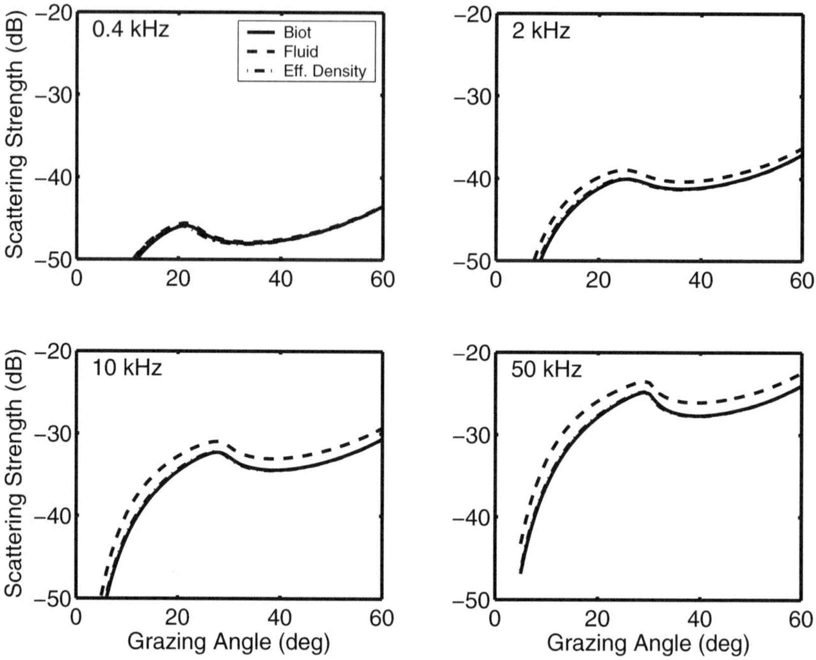

Fig. 13.6. Comparison of bottom backscattering strength computed in the small-roughness perturbation approximation with the sediment treated using Biot, fluid, and effective density fluid theory. The Biot parameters are taken from Table 10.1, and the roughness parameters are $w_2 = 0.0000433 \text{ m}^{0.96}$, $\gamma_2 = 3.04$.

13.2.4 Experimental Tests of Small-Roughness Models

The small-roughness, fluid model has been compared with scattering data for various seafloor types, but the clearest tests of the model come from experiments on sand seafloors, where volume scattering can be assumed to be relatively small. Figure 13.7 compares the model with data from six experiments conducted at sandy sites. The model input parameters are given in Table 13.1. These parameters differ slightly in some cases from those of the original references, as they have been obtained by reprocessing (Kevin Briggs, personal communication) the original data in a manner appropriate to the frequency (35–40 kHz) of these model-data comparisons. Thus, the sound speed and density ratios are averages over the upper 2 cm of the core samples rather than averages over the entire length of the cores. The fit between the model and data appears to be reasonably good, with the greatest model-data difference occurring for the two intermediate grain size cases represented in panels (c) and (d) of Fig. 13.7. A detailed error analysis has been performed for the SAX99 data [Williams et al. 2002b], with the conclusion that the model-data difference is significant and likely due to neglect of poroelastic effects, to be discussed shortly. The similar model-data differences seen in panel (d) [Stanic et al. 1988] may be due to the same cause. In any event, the fluid small-roughness model appears to perform satisfactorily for the examples of Fig. 13.7 in the neighborhood of 40 kHz. One significant point of agreement between model and data is the scattering strength maximum occurring near the critical angle, visible in (c), (d), and (e). No maximum is predicted or seen for the finer grain sediment of (f).

Table 13.1. Geoacoustic input parameters used in model-data comparisons of Figs. 13.7 and 13.9.

Site	Grain Size (ϕ)	a_ρ	ν_p	δ_p	γ_2	w_2
Jacksonville	0.51	1.956	1.087	0.049	2.47	0.00000455
Panama City 98	0.62	1.98	1.113	0.0324	3.12	0.00014
Panama City 93	0.68	1.948	1.097	0.032	3.12	0.0001475
SAX99	1.28	2.00	1.16	0.010	3.04	0.0000433
Falmouth 2	2.07	1.96	1.16	0.004	3.0	0.000035
Panama City 84	2.63	1.93	1.115	0.0185	1.89	0.000000356
Quinault	2.95	1.94	1.113	0.0115	3.92	0.0041
Key West	7.00	1.499	1.003	0.0153	3.29	0.000464

Bistatic measurements offer the possibility of stringent model tests, but are difficult to carry out and may be subject to larger errors than backscatter measurements. A bistatic measurement was conducted by [Day and Yamamoto 1999] at frequencies of 3.75, 7.5, and 15 kHz at a sandy site on the Florida Atlantic coast. The authors indicate that roughness scattering was likely the dominant contributor, but lack of measured roughness parameters prevented a definitive test of scattering models. Reasonable agreement between bistatic data and the perturbation model was found by [Choi et al. 2001] at 240 kHz, but the model parameters were assigned values from the literature and not measured. Williams [Williams and Jackson 1997, Williams and Jackson 1998] has conducted bistatic scattering measurements at well-characterized sites using the apparatus depicted in Fig. 1.1. Figure 13.8 compares the fluid model with bistatic scattering data. The model employs the Kirchhoff approximation for angles near the specular direction and small-roughness perturbation theory elsewhere, with a smooth interpolation between the two approximations. The model-data agreement is consistent with the measurement errors. It should be noted that no account is taken of error in the model inputs, only in the acoustic measurements.

A further test of the small-roughness model is in the frequency dependence it predicts for backscattering strength. If the frequency dependence

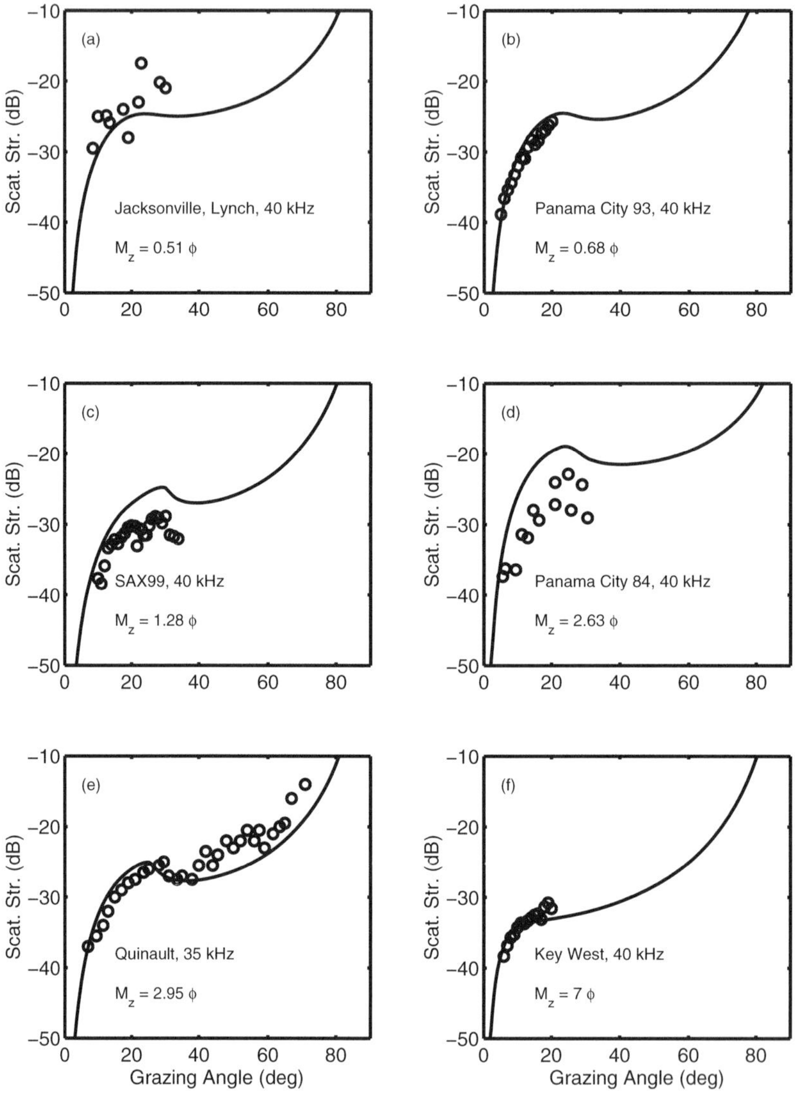

Fig. 13.7. Comparison of fluid small-roughness perturbation model with data from six experiments conducted at sandy sites. The examples are given in order of increasing grain size, and the primary acoustic and geoacoustic data sources are: (a) [Stanic et al. 1989], (b) [D. Jackson et al. 1996a], (c) [Williams et al. 2002b], (d) [Stanic et al. 1988], (e) [D. Jackson and Briggs 1992], and (f) [Briggs et al. 2002a].

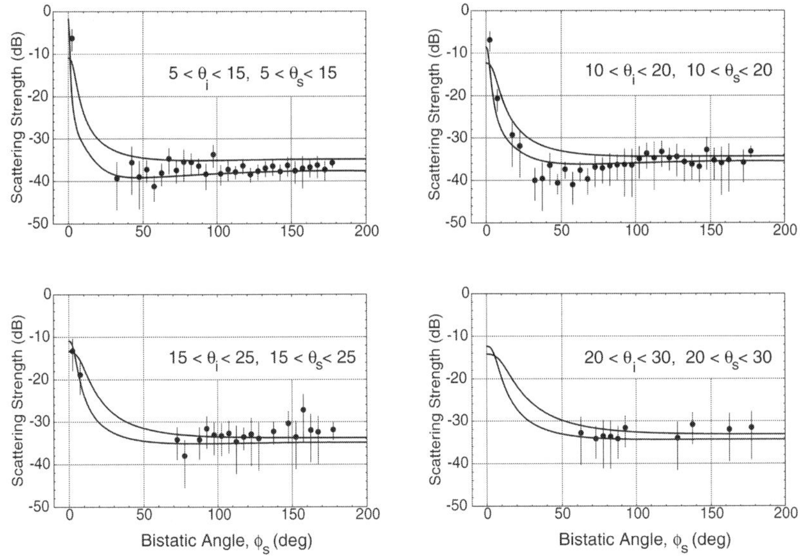

Fig. 13.8. Comparison of fluid model with bistatic scattering data [Williams and Jackson 1997]. The model uses interpolation between the Kirchhoff approximation near the specular direction and perturbation theory for angles away from the specular direction.

of the sediment sound speed and loss parameter (δ_p) can be neglected, the factor A_{ww} will have negligible frequency dependence, so that the product $k_1^4 W_2(\mathbf{\Delta K})$ governs the frequency dependence. Both k_1 and $\mathbf{\Delta K}$ are proportional to frequency, f, so (13.9) in combination with (6.8) yields

$$\sigma = \text{constant} f^{4-\gamma_2} \ . \tag{13.61}$$

With typical spectral exponents in the range $3 < \gamma_2 < 4$, scattering strength should increase with frequency at a negligible rate for $\gamma_2 = 4$ to rates of 10 dB/decade for $\gamma_2 = 3$. Figure 13.9 compares the frequency dependence predicted by the fluid model with data from six sandy sites for which roughness scattering should be dominant over sediment volume scattering. The slopes of the model curves match the slopes of the data reasonably well up to 100 kHz. Given likely errors of 2–3 dB for most of these data sets, the mismatch of the model to the measured absolute scattering strengths is satisfactory except for a tendency for the model to overpredict scattering at the Panama City 98, SAX99, and Panama City 84 sites. As noted earlier, part of this mismatch could be due to poroelastic effects. The data from Jacksonville and Panama City 84 fall below the model for frequencies above 100 kHz, indicating that the model has failed at these higher frequencies. The apparent success of the model in predicting the relative lack of frequency dependence of the Quinault data is tempered by the fact that the roughness was highly

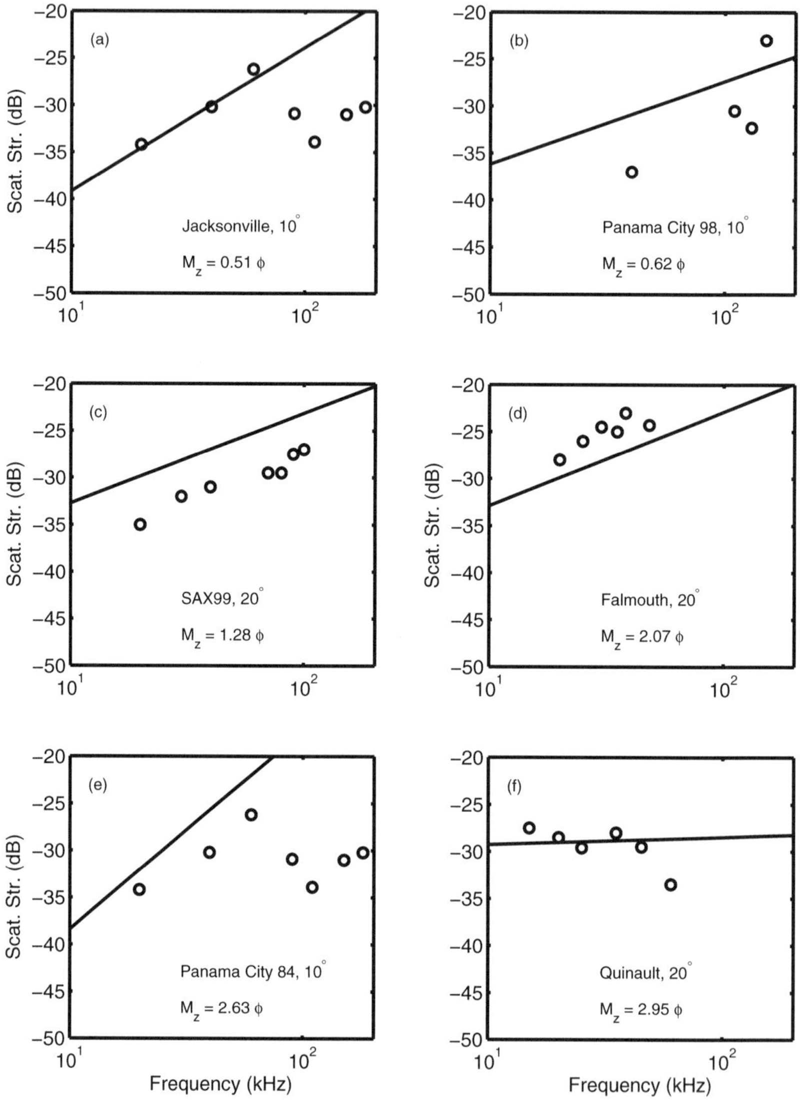

Fig. 13.9. Comparison of frequency dependence of fluid small-roughness fluid perturbation model with data from six experiments conducted at sandy sites. The grazing angle is fixed at either 10° or 20°. The examples are given in order of increasing grain size, and the primary acoustic and geoacoustic data sources are: (a) [Stanic et al. 1989], (b) [Stanic et al. 1998], (c) [Williams et al. 2002b], (d) [D. Jackson et al. 1986b], (e) [Stanic et al. 1988], and (f) [D. Jackson and Briggs 1992].

directional [D. Jackson and Briggs 1992]. The isotropy implied by the spectral parameters in Table 13.1 is not realistic, and a more detailed comparison of model and data using an anisotropic spectrum is in order. Perturbation theory was found to give a reasonable fit to backscatter data at 140 kHz from a silty-sand site ("Tellaro") [Pouliquen and Lyons 2002]. These authors note that vertical gradients in sediment properties had a significant effect on roughness scattering.

The Biot model has been compared with backscattering strength data by [Williams et al. 2002b]. Figure 13.10 shows that, within the uncertainties of acoustic measurement and the uncertainties in model inputs, the Biot model provides a better fit to the data than the fluid model. The effective density approximation would give essentially the same result as the Biot model and might be preferred on the grounds of simplicity. Arguments are given in [Williams et al. 2002b] that, for the measurement site in question, roughness scattering should dominate sediment volume scattering, at least for frequencies below 100 kHz, so the comparison of the small-roughness Biot model with data is appropriate. Figure 13.11, taken from [Williams et al. 2002b], compares the frequency dependence of the Biot model with data at a fixed grazing angle of 20°. The agreement is satisfactory for frequencies of 100 kHz or lower, but substantial disagreement is seen at 300 kHz. The falloff in the model curves at frequencies above 150 kHz is caused by a change in slope of the roughness spectrum at the corresponding Bragg wavenumber. The difference between the model and data at 300 kHz is tentatively ascribed to volume scattering of unknown type.

13.3 Kirchhoff Approximation

The Kirchhoff approximation is used in electromagnetic scattering [Beckmann and Spizzichino 1963, Ogilvy 1991, Ishimaru 1997] and was applied to acoustic scattering by the sea surface at a rather early date [Eckart 1953]. It is widely used in seafloor scattering, where its main utility is near the specular direction (vertical incidence in the case of backscattering) [D. Jackson et al. 1986a, de Moustier 1986]. The Kirchhoff approach has been applied to elastic seafloors by [Dacol 1990], and approximations to the Kirchhoff integral (13.16) have been employed in monostatic [Mourad and Jackson 1989] and bistatic scattering models [Williams and Jackson 1998]. The Kirchhoff approximation has been applied to inverse problems, either for extracting roughness parameters from vertical-incidence data [Matsumoto et al. 1993, Michalopoulou et al. 1994, Premus and Alexandrou 1994, Talukdar et al. 1995, Michalopoulou and Alexandrou 1996] or sediment acoustic parameters [Chotiros 1994, Chotiros 1995b]. In all these applications, expressions for the scattering cross section in the Kirchhoff approximation have been employed with the implication that average echo properties will be used. Pouliquen

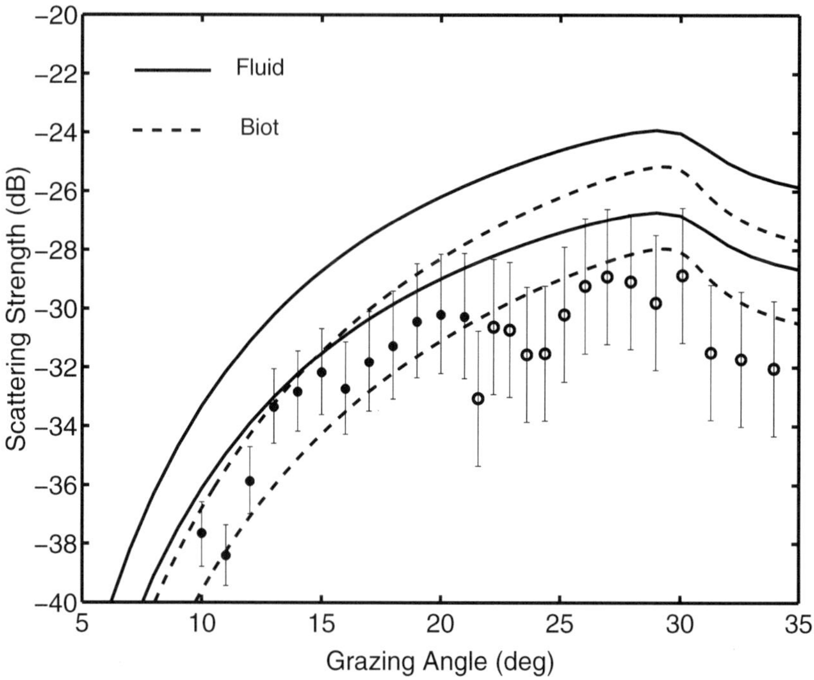

Fig. 13.10. Comparison of measured bottom backscattering strength at 40 kHz with small-roughness perturbation fluid model (solid curves) and Biot model (dashed curves). The Biot parameters are those of Table 10.1, and the nominal roughness parameters are $w_2 = 0.0000433$ m$^{0.96}$, $\gamma_2 = 3.04$. The upper and lower curves in each case give the 95% uncertainty bounds for the models based on the uncertainty bounds for the roughness data. The vertical error bars give the corresponding 95% uncertainty bounds for the acoustic measurements. The data denoted by asterisks were taken at the same site as the roughness measurements, whereas the data denoted by open cicles were obtained using a different apparatus at a site removed by a few hundred meters. The slight offset between these data sets may be due to either acoustic calibration error, difference in roughness at the two sites, or a combination of these two. From [Williams et al. 2002b].

and collaborators [Pouliquen et al. 1999, Bergem et al. 1999] use the Kirchhoff approximation without the formal averaging that yields the scattering cross section. Instead, the Kirchhoff approximation is used to obtain an approximate boundary condition at the rough interface so that the scattered field can be computed for individual roughness realizations. The formalism is developed in the time domain and allows generation of synthetic time series for seafloor reverberation.

The roughness scattering cross section in the Kirchhoff approximation is derived in Sect. L.1. In applying the Kirchhoff approximation, the general

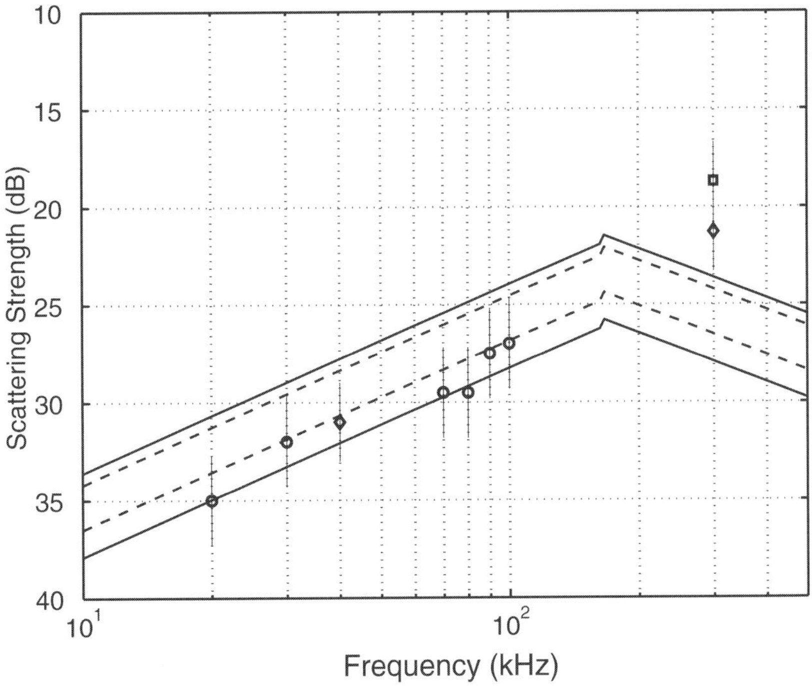

Fig. 13.11. Comparison of measured bottom backscattering strength versus frequency with the small-roughness Biot model for a grazing angle of 20°. All parameters are as in Fig. 13.10, except the roughness spectrum departs from the simple power law at high wavenumbers in keeping with measurement. The solid lines denote 95% error bounds based on roughness spectrum uncertainty at spatial frequencies corresponding to an acoustic frequency of 300 kHz, and the dashed lines are bounds for 80 kHz. The acoustic data were obtained at three locations within a few hundred meters of each other, and these locations are identified by different symbols. The roughness data were taken at the site denoted by diamonds. From [Williams et al. 2002b].

expression (13.10) is used with reflection coefficient, V_{ww}, obtained for the boundary conditions of interest, and evaluated at the angle corresponding to specular reflection from an appropriately tilted flat surface (13.13). If the sediment is treated as a fluid, or if the effective density approximation to Biot theory is used, the reflection coefficient is obtained from (8.49). If the shear wave properties of the sediment are to be considered, or if Buckingham's theory is employed, (9.34) can be used. If the sediment is modeled using poroelastic theory, the reflection coefficient can be obtained by solving (10.66). The final step in applying the Kirchhoff approximation is to evaluate the Kirchhoff integral for the angles of interest. If the seafloor roughness statistics are not isotropic, the general expression (13.12) must be used. If isotropy can be assumed, (13.15) can be used, and the further simplification (13.16) results if the roughness spectrum obeys the power law (6.8). It should be remembered that the Kirchhoff integral expressions given here are based on the assumption that the seafloor relief function, $f(\mathbf{R})$, is a two-dimensional Gaussian random process. This assumption is not central to the Kirchhoff approximation, but is made in the interest of tractability. The literature contains examples of scattering formalisms that do not assume Gaussian relief statistics [Beckmann 1973, Henyey and Wurmser 1990, Guérin 2002].

The Kirchhoff approximation predicts interesting frequency dependence for backscattering near vertical incidence. An example is shown in Fig. 13.12. In this case, an isotropic power-law roughness spectrum is assumed (6.8) with exponent $\gamma_2 = 3.25$. As frequency increases, the scattering strength becomes less sharply peaked in the vertical direction. This is as expected based on the simple argument that power-law spectra having $\gamma_2 < 4$ represent seafloor roughness that becomes relatively greater when viewed on smaller and smaller scales. Thus, as frequency increases, energy is scattered over a wider range of angles, leading to the behavior seen in Fig. 13.12.

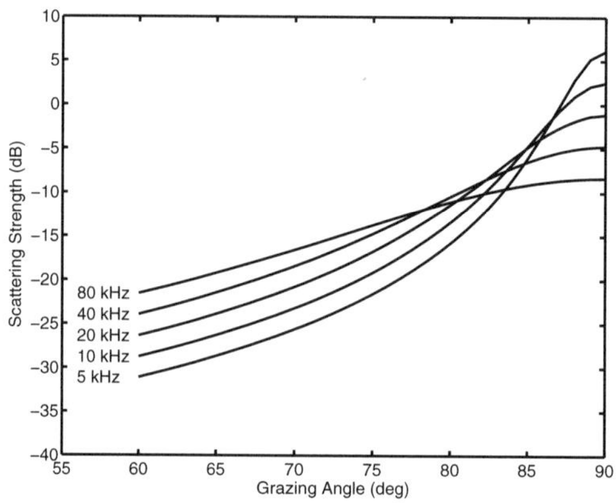

Fig. 13.12. Near-vertical backscattering strength in the Kirchhoff approximation at several frequencies. A medium sand seafloor is assumed with all parameters as in Fig. 13.2.

The widening and lowering of the backscattering strength curve as frequency increases also occurs as the roughness (as determined by w_2) increases or as the spectral exponent, γ_2, decreases. Thus, the angular and frequency dependence of near-vertical backscattering can be used to determine, or at least constrain, the parameters of the roughness spectrum [Demoustier and Alexandrou 1991, Matsumoto et al. 1993, Dziak et al. 1993, Michalopoulou et al. 1994, Premus and Alexandrou 1994, Talukdar et al. 1995, Michalopoulou and Alexandrou 1996].

13.3.1 High-Frequency Kirchhoff Approximation

A simplified form of the Kirchhoff formalism is often used, based on a high-frequency limit. This further approximation is variously known as the high-frequency Kirchhoff approximation, the geometric optics approximation, the facet model, or the broken mirror model. As shown in Appendix L, (13.10) reduces in the high-frequency limit to

$$\sigma = \frac{1}{4}|V_{ww}(\theta_{is})|^2 \frac{\Delta k^4}{\Delta k_z^4} p_s(\mathbf{s}) \ , \tag{13.62}$$

where θ_{is} is the specular angle given by (13.13) and

$$p_s(\mathbf{s}) = \frac{1}{2\pi||\mathbf{B}_s||^{1/2}} e^{-\frac{1}{2}\mathbf{s}^T\mathbf{B}_s^{-1}\mathbf{s}} \tag{13.63}$$

is the bivariate Gaussian probability density function for interface slope,

$$\mathbf{s} = \begin{bmatrix} \Delta K_x/\Delta k_z \\ \Delta K_y/\Delta k_z \end{bmatrix} \ . \tag{13.64}$$

This column matrix contains the x- and y-components of slope corresponding to specular reflection, and

$$\mathbf{B}_s = \begin{pmatrix} < (\frac{\partial f}{\partial x})^2 > & < \frac{\partial f}{\partial x}\frac{\partial f}{\partial y} > \\ < \frac{\partial f}{\partial x}\frac{\partial f}{\partial y} > & < (\frac{\partial f}{\partial y})^2 > \end{pmatrix} \tag{13.65}$$

is the covariance matrix for slope, with determinant $||\mathbf{B}_s||$.

The notation used in (13.62) and (13.63) requires a good deal of explanation. The relevant geometry is illustrated in Fig. 13.13. The entire geometry is specified once the incoming and outgoing directions for which the bistatic cross section is desired, are given. These two directions are defined as the directions of the incident and scattered wave vectors, $\mathbf{k}_i$ and $\mathbf{k}_s$. The difference wave vector defined earlier, $\mathbf{\Delta k} = \mathbf{k}_s - \mathbf{k}_i$, plays a key role. This vector is normal to the hypothetical plane surface that would give specular reflection between the source and receiver. From Fig. 13.13, it follows that the slope of the hypothetical surface is

$$\mathbf{s} = \frac{\mathbf{\Delta K}}{\Delta k_z} \ , \tag{13.66}$$

where $\mathbf{\Delta K}$ and Δk_z are the horizontal and vertical components of $\mathbf{\Delta k}$, as defined in (13.6) and (13.7). To summarize, knowing the incident and scattered directions, one can determine $\mathbf{s}$ and evaluate (13.63) from which the cross section follows (13.62).

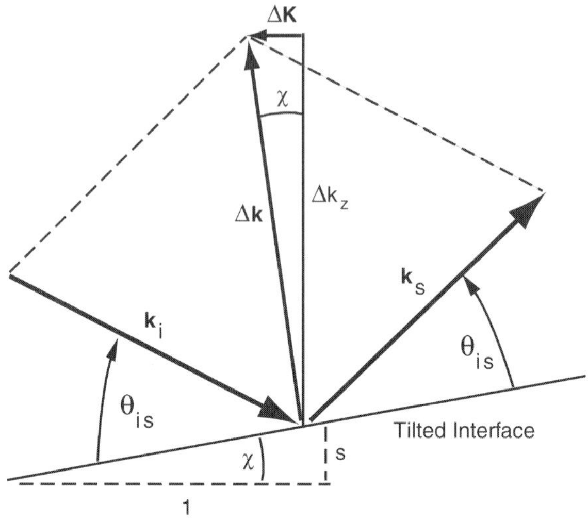

Fig. 13.13. Diagram illustrating the geometry relevant to the high-frequency Kirchhoff approximation.

The high-frequency Kirchhoff approximation has an appealing physical interpretation that is suggested by Fig. 13.13. In this approximation, scattering is due to specular reflection off those portions of the interface that happen to be tipped in just the proper way [Barrick 1968]. This is analogous to sunlight glinting off rippled water. For a given scattering direction, the strength of the scattering is determined by the probability density function for slope, evaluated at the slope giving specular reflection.

Equation (13.62) can be applied to bistatic scattering from seafloors with anisotropic roughness statistics (e.g., directional sand ripples). It has a conservation property similar to that inherent in the Kirchhoff and small-slope approximations. If the near-specular lobe of the scattering cross section has a small angular width ($<< 1$ radian), the incoherent reflection coefficient will be nearly equal to the flat-interface reflection coefficient. Thus, with a nondirective source and nondirective receiver, the received RMS pressure level for a sufficiently long transmitted pulse will be nearly equal to that received if the surface is flat.

In the isotropic case, the slope covariance matrix is equal to the unit matrix multiplied by a constant, and (13.62) becomes

$$\sigma = \frac{|V_{ww}(\theta_{is})|^2}{8\pi\sigma_s^2} \frac{\Delta k^4}{\Delta k_z^4} e^{-\frac{\Delta K^2}{2\sigma_s^2 \Delta k_z^2}} , \tag{13.67}$$

where

$$\sigma_s^2 =< (\frac{\partial f}{\partial x})^2 >=< (\frac{\partial f}{\partial y})^2 > \tag{13.68}$$

is the RMS slope for a 1D track. In terms of the (isotropic) spectrum,

$$\sigma_s^2 = \pi \int K^3 W(K) dK . \tag{13.69}$$

The high-frequency Kirchhoff approximation is most commonly applied to backscatter from a seafloor with isotropic roughness statistics, for which (13.67) takes the form

$$\sigma = \frac{|V_{ww}(90°)|^2}{8\pi\sigma_s^2 \sin^4\theta_i} e^{-\frac{\cot^2\theta_i}{2\sigma_s^2}} . \tag{13.70}$$

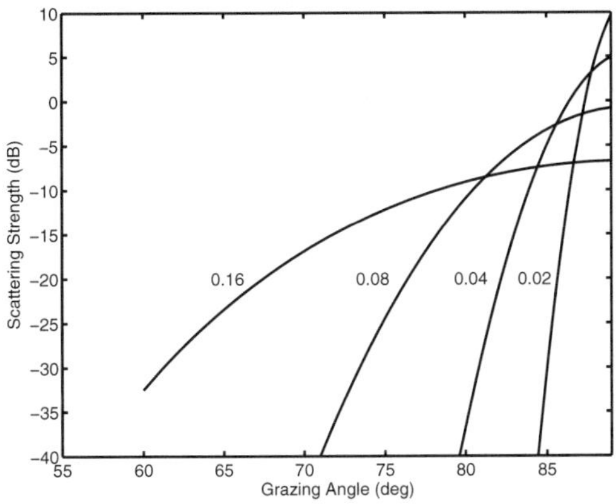

Fig. 13.14. Backscattering strength in the high-frequency Kirchhoff approximation for the indicated RMS slopes. A medium sand seafloor is assumed with all parameters as in Fig. 13.2, except the roughness is characterized by RMS slope rather than the spectral parameters w_2 and γ_2.

Backscattering strength curves computed using (13.70) are shown in Fig. 13.14. The curves are labeled with the values of RMS slope, σ_s. These results differ from the full Kirchhoff approximation (Fig. 13.12) in two significant ways. First, scattering strength is independent of frequency, and second, the high-frequency approximation yields backscattering strength curves that are concave downward without the inflection seen in the curves of Fig. 13.12. Both of these properties are due to the neglect of diffraction in the high-frequency approximation, which includes only reflection from specular points.

The slope covariance matrix can be obtained from the roughness spectrum as follows:

$$B_{smn} = \int K_m K_n W(\mathbf{K}) d^2K \ , \tag{13.71}$$

where $m,\ n = x,\ y$. This integral diverges except for spectra that fall off more rapidly than K^{-4} at large wavenumbers, which excludes many spectra of interest. In the examples given in Table 6.1, the "spectral slope" is the rate of falloff of the 1D spectrum. Adding one to obtain the falloff rate of the 2D spectrum, only four entries in the table fall off more rapidly than K^{-4}. Mean-square slope, then, is undefined unless a high-frequency cutoff is introduced, either on physical or mathematical grounds. This exposes the main weakness of the high-frequency Kirchhoff approximation as applied to scattering by the seafloor. If a cutoff is invoked at a wavenumber moderately greater than the upper limit of measurement (typically, this would correspond to wavelengths on the order of 1 mm or less), the resulting RMS slopes would be very large, approaching unity. These large slopes are due to short-wavelength roughness features that are not of the gently undulating sort that the Kirchhoff approximation can deal with. It can be reasonably argued that such features can be neglected if their wavelength is much shorter than the acoustic wavelength. In this view, a frequency-dependent cutoff might be imposed on the integral (13.71), resulting in frequency-dependent slope. Though ad hoc, this approach finds some support in the composite-roughness approximation of Sect. 13.5.

13.3.2 Experimental Tests of the Kirchhoff Approximation

Comparison of the Kirchhoff approximation with data is hampered by the lack of accurate near-specular measurements accompanied by roughness data. One example in which an experimental test has been made is shown in Fig. 13.8. In this case, the Kirchhoff approximation has been used near the specular direction and gives reasonable agreement with the data. A comparison of measured near-vertical backscattering strength with the Kirchhoff approximation is made in [D. Jackson et al. 1986a], and Fig. 13.15 shows these model-data comparisons near vertical incidence. Model-data differences as large as 10 dB occur in this comparison, but these differences may be caused by errors in measurement of the near-vertical scattering strength, as its angular width is small. Appendix G discusses some of the problems that may be encountered in field measurements. In controlled laboratory measurements, [Thorne and Pace 1984] have found good agreement between measurements of near-vertical scattering and Kirchhoff predictions, while at smaller grazing angles, [Thorne et al. 1988] find that model and data are in "moderate accord."

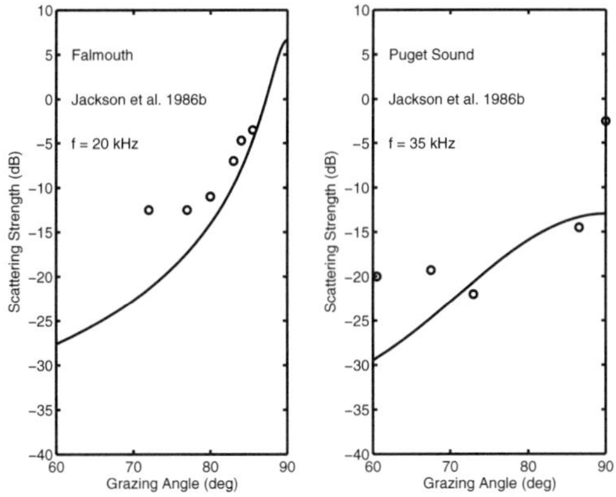

Fig. 13.15. Comparison of the Kirchhoff approximation with the data of [D. Jackson et al. 1986a].

13.4 Small-Slope Approximation

The small-slope approximation was originally developed by [Voronovich 1985] for scattering by the sea surface, but, as noted in Sect. 13.1, can be readily applied to other boundary conditions provided the lower medium (the seafloor in the present case) is homogeneous, without layering or gradients. The small-slope approximation is a systematic expansion [Thorsos and Broschat 1995] that can be carried to higher orders in order to achieve greater accuracy [Berman 1991, Broschat and Thorsos 1997]. In this monograph, the lowest-order version of the small-slope approximation is used, in the interest of simplicity. Even this version is an improvement over the more traditional small-roughness perturbation and Kirchhoff approximations, as it provides complete coverage of incident and scattering angles with a single approximation and with no loss in accuracy (Sect. 13.7) compared to the traditional approximations. With regard to seafloor scattering, there has been a substantial effort in applying the small-slope approximation to the fluid-elastic boundary (which includes the fluid–fluid boundary as a special case) [Berman 1991, Yang and Broschat 1994, Wurmser 1996, Gragg et al. 2001], and the fluid–poroelastic case has also been considered [Yang et al. 2002].

In applying the lowest-order small-slope approximation to a given boundary, one must first obtain the factor, A_{ww}, appearing in the scattering cross section (13.11). This is the same factor that must be evaluated in order to apply small-roughness perturbation theory, and can be obtained using (13.21)–(13.23) for the fluid–fluid boundary, (13.27) for the fluid–elastic boundary, or (13.36) for the fluid–poroelastic boundary. In comparing the predictions of the small-slope approximation with data, one is restricted to well-characterized seafloors for which density, sound speed, roughness, etc., have been measured. This effectively reduces the set of available measurements to a subset of those made on mud and sand seafloors, as these are the only types that have been characterized sufficiently in the geoacoustic sense. In these cases, the small-slope approximation produces essentially the same results as the perturbation and Kirchhoff approximations when each is applied to the angular region where it is the most accurate. This is evident in Fig. 13.2 and in model–model comparisons made in [APL-UW TM 2-00]. Consequently there is no need to consider model–data comparisons for the small-slope approximation, the discussions regarding the perturbation approximation in Sect. 13.2 and the Kirchhoff approximation in Sect. 13.3 suffice. Although no experiments have yet been conducted on well-characterized gravel or rock seafloors, [Soukup and Gragg 2003] find that fits of the small-slope model to scattering data for rock seafloors yield plausible model parameters.

13.5 Composite-Roughness Approximation

The success of the composite-roughness approximation in modeling scattering by the sea surface [Kur'yanov 1963, Bachmann 1973, Galybin 1976, McDaniel and Gorman 1982] led to subsequent application to seafloor scattering [D. Jackson et al. 1986a, Mourad and Jackson 1989, Caruthers and Novarini 1993]. Although this approximation can be derived from first principles [Brown 1978, Brown 1980, McDaniel and Gorman 1983], the discussion here will be heuristic, exploiting the fact that the composite-roughness approximation has a physically intuitive interpretation. The essence of this approximation is a division of seafloor roughness into large- and small-scale parts as illustrated in Fig. 13.16. In terms of the spectrum of seafloor roughness, it is assumed that the small-scale (shorter-wavelength) roughness causes scattering, while the large-scale (longer-wavelength) roughness simply tilts the seafloor, altering the grazing angle and causing acoustic shadowing. In this approach, scattering near vertical incidence is not treated. Application of the composite-roughness near vertical incidence will be discussed near the end of this section.

The division into large and small scales is made at the "cutoff" wavenumber, K_c. The choice of the cutoff might be clear in some cases, for example, if the seafloor roughness consists of a ripple field of relatively long wavelength with random, shorter-wavelength roughness superimposed. If the ripple wavelength is much larger than the acoustic Bragg wavelength, then the cutoff wavenumber should be chosen somewhere between the ripple wavenumber and the Bragg wavenumber. For backscattering, the Bragg wavenumber is given by (13.20). Ideally, the scattering cross section computed in the composite-roughness approximation will be insensitive to the choice of cutoff wavenumber within the limits set by the boundaries of the large and small scales. The division of the roughness spectrum into large- and small-scale parts is illustrated in Fig. 13.17. In this example, a power-law spec-

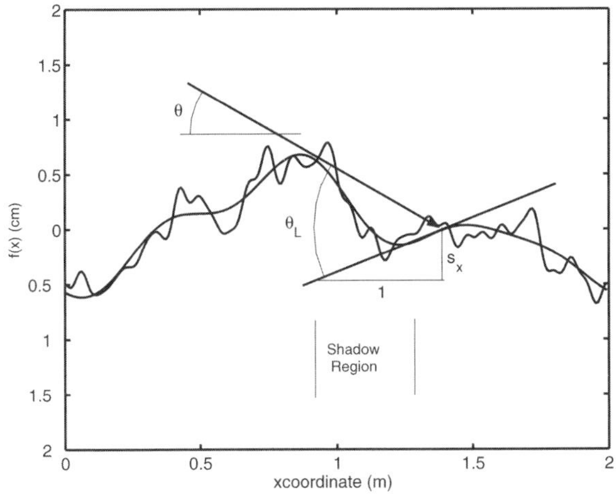

Fig. 13.16. Sketch (in two dimensions for simplicity) illustrating the main ideas of the composite-roughness approximation. The large-scale roughness is shown as a smoothly undulating curve, and the sum of the large- and small-scale roughness is the more rapidly varying curve. The slanting arrow shows a ray incident at grazing angle θ, tangent to a crest of the large-scale surface. This defines the right-hand boundary of a shadow region. The ray strikes the large-scale surface at a local grazing angle θ_L, where the slope of the large-scale surface is s_x.

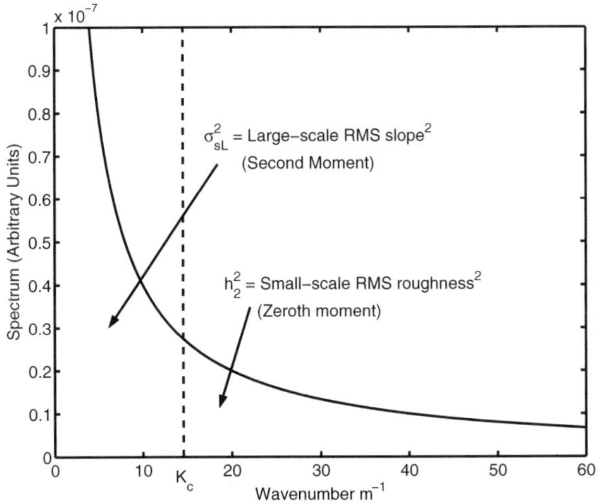

Fig. 13.17. Division of a power-law roughness spectrum into large- and small-scale parts. The vertical dashed line marks the cutoff wavenumber at which this division is made.

trum (6.8) is assumed, and the choice of the cutoff is not obvious owing to the lack of any particular feature in the spectrum. Several authors, including [Bachmann 1973, Brown 1978, Bass and Fuks 1979], recommend setting the cutoff such that the roughness of the small-scale surface is sufficiently small that perturbation theory is accurate (see Sect. 13.7). In particular, [D. Jackson et al. 1986a] use the criterion

$$k_w h_s = \frac{1}{2} . \tag{13.72}$$

Here, h_s is the RMS roughness of the small-scale surface, obtained by integrating the roughness spectrum over wave vectors whose magnitude is greater than the cutoff wavenumber:

$$h_s^2 = \int_{K>K_c} W(\mathbf{K}) d^2K . \tag{13.73}$$

For the pure power-law spectrum of (6.8),

$$h_s^2 = \frac{2\pi w_2}{(\gamma_2 - 2) K_c^{\gamma_2 - 2}} . \tag{13.74}$$

This can be inserted in (13.72) to obtain the cutoff wavenumber,

$$K_c = \left(\frac{8\pi w_2 k_w^2}{\gamma_2 - 2} \right)^{1/(\gamma_2 - 2)} . \tag{13.75}$$

The composite-roughness approximation can only be used if the cutoff wavenumber is less than the Bragg wavenumber. This restricts application to grazing angles less than the maximum found by setting $\Delta K = K_c$ and solving (13.20) for θ. Knowing the cutoff wavenumber, the RMS slope, σ_{sL}, of the large-scale surface can be found in terms of the second moment of the spectrum

$$\sigma_{sL}^2 = \frac{1}{2} \int_{K<K_c} K^2 W(\mathbf{K}) d^2K . \tag{13.76}$$

This expression is an adaptation of (13.85), derived in Sect. 13.7. For the power-law spectrum,

$$\sigma_{sL}^2 = \frac{\pi w_2 K_c^{4-\gamma_2}}{4 - \gamma_2} . \tag{13.77}$$

If the RMS slope of the large-scale surface is small, that is, if

$$\sigma_{sL} < 0.1 , \tag{13.78}$$

and if the grazing angle is not too large ($\theta < 70°$), then the local grazing angle pictured in Fig. 13.16 can be approximated as

$$\theta_L = \theta + s_x , \tag{13.79}$$

where s_x is the local value of large-scale slope in the vertical plane containing the line connecting the source and scattering region. This slope is assumed to be Gaussian-distributed with standard deviation σ_{sL}, and the scattering cross section is obtained as an average over large-scale slope as follows:

$$\sigma_{cr}(\theta) = \frac{R(\theta,\ \sigma_{sL})}{\sqrt{2\pi}\sigma_{sL}} \int_{-\theta}^{\infty} \sigma(\theta + s_x) \exp(-\frac{s_x^2}{2\sigma_{sL}^2}) ds_x \ . \tag{13.80}$$

In this expression, $\sigma(\theta)$ is the backscattering cross section computed in an approximation suited to small roughness. First-order perturbation theory is typically used, as it provides an analytic expression for insertion into the above integrand. The lower limit of the integral is taken to be the negative of the grazing angle, as more negative slopes must be shadowed. This shadowing correction is not sufficient, as crests will cast shadows on some portions of the surface regardless of slope. An additional shadowing correction is provided by the factor [Wagner 1967]

$$R(\theta,\ \sigma_{sL}) = \frac{1 - e^{-2Q}}{2Q} \ , \tag{13.81}$$

where

$$Q = \frac{1}{4t}\{\pi^{1/2}e^{-t^2} - t[1 - \mathrm{erf}(t)]\} \ , \tag{13.82}$$

with erf being the error function, and

$$t = \frac{\tan\theta}{\sqrt{2}\sigma_{sL}} \ . \tag{13.83}$$

The composite-roughness approximation is compared with the small-roughness perturbation model in Fig. 13.18. Each panel of the figure employs the parameters of the corresponding panel of Fig. 13.7. In most of the cases illustrated, the corrections imposed by the composite-roughness approximation are small, except in panels (b) and (f) where the large-scale RMS slopes are large, violating condition (13.78). In these two cases, the composite-roughness corrections are suspect and do not improve the fit of the model to data, as can be seen by comparing Figs. 13.7 and 13.18. The composite-roughness approximation could not be employed for panel (d), for which the spectral exponent was $\gamma_2 = 1.89$, causing the integral for small-scale RMS roughness (13.73) to diverge. A similar problem was noted by [D. Jackson and Briggs 1992] where an exponent slightly less than 4 caused the RMS slope to be unrealistically large. These problems stem from extrapolation of the power-law spectrum to regions outside its domain of validity rather than from weakness of the composite-roughness approximation. The composite-roughness approximation as developed here has an obvious weakness, however. The cutoff wavenumber is dependent on a rather arbitrary criterion (13.72), and different, but equally plausible, criteria would give different cutoffs. This, in turn, would result in substantially different values

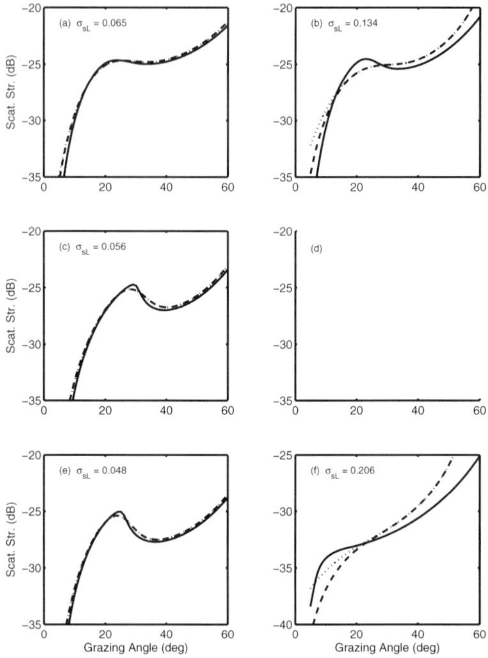

Fig. 13.18. Composite roughness corrections to the perturbation model results of Fig. 13.7. The uncorrected model is shown as a solid line. The composite-roughness result with both slope and shadowing corrections is shown as a dashed line, and the result with slope but no shadowing correction is shown as a dotted line. The composite roughness approximation could not be applied to the Panama City 84 case (panel d), as the spectral exponent is less than 2.

for the RMS large-scale slope, the primary determinant of the composite-roughness corrections. Thus, for power-law spectra, the composite-roughness approximation is sensitive to the choice of cutoff, and no unique criterion is available for fixing the cutoff. One possible choice is to set the cutoff so as to provide an approximate fit to a more accurate approximation, e.g., the Kirchhoff approximation near normal incidence [Novarini and Caruthers 1994] (see Fig. 13.19).

In the approach outlined above, the large-scale part of the surface only modulates the grazing angle and does not cause scattering. Especially near vertical incidence, where the Bragg wavelength becomes large, large-scale roughness may cause scattering through diffraction [McDaniel 1986]. One approach [McDaniel and Gorman 1982] ascribes all scattering near vertical incidence to the large-scale roughness, employing the high-frequency Kirchhoff approximation. The spectrum is divided as usual into large- and small-scale parts, and the RMS slope of the large-scale part is used in (13.70). Following [Novarini and Caruthers 1994], the reflection coefficient in (13.70) is replaced by the coherent reflection coefficient (13.14) evaluated using the RMS roughness, h_s, of the small-scale surface. With the criterion (13.72), the result is simply division of the flat-interface reflection coefficient by $\sqrt{e}$ (where e is the base of the natural logarithms). The physical picture is that near-vertical scattering is due to specular reflection off roughened facets. The roughness of the facets scatters a portion of the incident acoustic energy at wide angles relative to the vertical direction, and this energy is accounted for by (13.80). As in the development given above for smaller grazing angles, the choice of cutoff wavenumber is dependent on acoustic frequency, leading to a frequency-dependent RMS slope for the large-scale surface. This, in turn, imparts a frequency dependence to the near-vertical backscattering strength.

Examples of the composite-roughness approximation in the form defined above applied to near-specular seafloor scattering can be found in [Novarini and Caruthers 1994] who use (13.72) to divide the spectrum. These authors use a version of (13.76) without the factor 1/2, so their large-scale mean-square slope is twice that given by (13.77). Figure 13.19 shows comparisons between backscattering strength computed in the full Kirchhoff approximation and the high-frequency Kirchhoff approximation. In the latter case, the large-scale slope is obtained using (13.77) and the coherent reflection coefficient is equal to the flat-interface coefficient divided by $\sqrt{e}$. The correspondence between the two approximations is reasonable, but the high-frequency approximation is consistently lower.

The high-frequency approximation yields scattering strength curves near vertical incidence that lack the inflection seen in the full Kirchhoff approximation. This problem is less severe in the case of scattering by the sea surface, where spectral exponents are relatively large ($\gamma_2 \approx 4$), and the Kirchhoff approximation yields curves with little inflection.

Comparisons of data with the composite roughness model at smaller grazing angles [D. Jackson et al. 1986a, Mourad and Jackson 1989] show agreement within experimental errors, but the corrections in these examples are small compared to the model-data differences. Together with the examples of Fig. 13.18, this suggests the composite-roughness approximation may be most useful where its corrections are small. Comparisons with exact calculations in the case of scattering by the sea surface (Sect. 13.7) tend to support this conclusion. If this is the case, the composite-roughness approximation does little to extend the domain of seafloor scattering models, at least when the roughness spectrum can be approximated as a power law. One motivation for use of the composite-roughness approximation is numerical economy. Present computing speeds are sufficient to allow use of the more accurate Kirchhoff and small-slope approximations. If, however, the interface roughness shows a clear separation of scales, say, prominent ripples with wave-

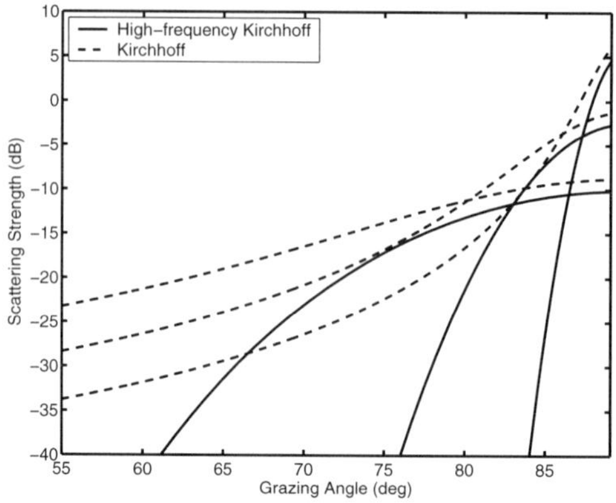

Fig. 13.19. Comparison of the high-frequency Kirchhoff approximation with the full Kirchhoff approximation at 20 kHz. The high-frequency Kirchhoff curves are computed using a scheme similar to that of [Novarini and Caruthers 1994], with the reflection coefficient replaced by the coherent reflection coefficient. This is found using (13.14) with the mean-square roughness of the small-scale surface. A medium sand seafloor is assumed with all parameters as in Fig. 13.2, except w_2 is assigned the values 0.0000422, 0.0001406, and 0.0004218. The corresponding RMS slopes, σ_s, provided by the cutoff prescription are 0.0227, 0.0595, and 0.1432. In order to identify the value of w_2 for each curve, note that the width of the scattering strength peak increases as w_2 increases.

length much greater than the acoustic wavelength with superimposed small, random roughness, the composite-roughness approximation may be useful. The lowest-order small-slope approximation does not incorporate the effects of large-scale tilting and shadowing.

13.6 Scattering by Discrete Features

The models described previously in this chapter ignore the possibility of scattering by discrete features, such as rocks, shell fragments, or live animals (e.g., sand dollars), lying on, or partially embedded in, the sediment–water interface. More deeply buried scatterers are treated in Sect. 14.1.7. A model for scattering by rock facets has been applied to low-frequency reverberation data by [Dorfman and Dyer 1999]. A simple model for scattering by spheres on an otherwise flat interface has been presented by [Stanton 2000]. He assumes that the scatterers are large enough that the geometric scattering approximation can be employed (see Sect. 14.1.7). Further, the influence of the water–sediment interface is neglected, so that the scatterers can be treated as a thin layer suspended in water. It is shown that the seafloor backscattering cross section at normal incidence in this case is the product of two dimensionless factors. One factor is the backscattering cross section of an individual scatterer, normalized by division by the cross-sectional area of the scatterer. This factor is expected to be essentially independent of frequency and scatterer size, but dependent on the composition of the scatterer. The second factor is the "packing factor," the fraction of seafloor area covered by scatterers. To account approximately for shadowing and other multiple scattering effects expected at small grazing angles, the normal-incidence cross section is multiplied by the Lambert factor (Sect. 2.3.2), $\sin^2 \theta$, where θ is the grazing angle. Another single scattering model is given by [Williams et al. 2001b], who include geometric reflection by the seafloor. This model was found to agree reasonably well with experimental data at 40 kHz from a sandy site at which glass spheres (marbles of radius 1.75 cm) had been randomly distributed over the seafloor. The results of both [Stanton 2000] and [Williams et al. 2001b] suggest that scattering by discrete targets on the seafloor can be important in some cases and therefore is deserving of further theoretical and experimental study.

13.7 Mathematical Accuracy of Approximations

In considering the validity of the roughness scattering models discussed here, it is important to distinguish between *physical* and *mathematical* accuracy. Physical accuracy has to do with the geoacoustic model employed: Does it faithfully model the important properties of the rough interface or does it gloss over essential factors? This difficult question can only be answered by measurement, and one may hope that if the model and data are in reasonable agreement, the model is accurate in the both the mathematical and physical senses. The two accuracy issues may become confused in some cases. There have doubtless been instances in which a poor geoacoustic model has been combined with a bad mathematical approximation such that the two errors tend to cancel and yield good agreement with data. The question of mathematical accuracy is more straightforward: Given a geoacoustic model specifying sediment roughness statistics and sediment bulk acoustic properties, do the approximations used in the scattering model fit the exact solution of the given acoustic theory? That is, once a geoacoustic model and acoustic theory have been settled on, there exists an exact solution which can usually be obtained only with considerable computational effort. Approximations (perturbation theory, Kirchhoff, etc.) are used to give results with reasonable computational effort. The question of mathematical accuracy is completely different than that of physical accuracy and can be answered using computations rather than measurements.

As the various published criteria for mathematical accuracy of scattering approximations depend on such roughness parameters as RMS roughness, RMS slope, and correlation length, it is useful to connect these parameters to the statistical roughness measures employed earlier in this monograph.

First, the RMS roughness is the square root of the mean-square roughness, h^2, which is equal to the relief covariance evaluated at zero lag (D.2). Equation (13.65) gives the slope covariance as a matrix involving derivatives of the relief covariance. The trace of this matrix provides a convenient scalar measure of mean-square slope:

$$< (\frac{\partial f}{\partial x})^2 > + < (\frac{\partial f}{\partial y})^2 >= -[\frac{\partial^2 B}{\partial x^2} + \frac{\partial^2 B}{\partial y^2}]_{\mathbf{R}=\mathbf{0}} \ . \tag{13.84}$$

In the case of isotropic roughness, it is simpler to define, as before, the mean-square slope, $\sigma_s^2 =< (\frac{\partial f}{\partial x})^2 >=< (\frac{\partial f}{\partial y})^2 >$,

$$\sigma_s^2 = -\frac{\partial^2 B}{\partial R^2}|_{\mathbf{R}=\mathbf{0}} \ . \tag{13.85}$$

The correlation length of the random function, $f(\mathbf{R})$, can be defined in any of several ways, but for the Gaussian-shaped covariance function,

$$B(\mathbf{R}) = h^2 e^{-\frac{R^2}{L^2}} \ , \tag{13.86}$$

the parameter L is usually designated the correlation length. Because the Fourier transform of a Gaussian is a Gaussian, the Gaussian relief covariance yields a Gaussian-shaped spectrum. A general correlation length definition that yields the correct value for the Gaussian-shaped covariance can be obtained by expanding the Gaussian in a power series,

$$B(\mathbf{R}) = h^2[1 - \frac{R^2}{L^2} + \cdots] \ . \tag{13.87}$$

Restricting attention to the isotropic case, the corresponding general definition of correlation length can be obtained by comparing the expansion of the general covariance with (13.87) to obtain

$$L = \frac{\sqrt{2}h}{\sqrt{-\frac{\partial^2 B}{\partial R^2}|_{\mathbf{R}=\mathbf{0}}}} \ . \tag{13.88}$$

This definition is inapplicable to 2D relief having a power-law spectrum with exponent, γ_2, less than 4, or to 1D relief with power-law spectrum exponent, γ_1 less than 3. For these cases, the second derivative of the relief covariance is infinite at zero lag. Section 13.5 shows how an RMS slope can be defined in these cases by truncating the spectrum at high wavenumbers. This issue does not arise with the Gaussian covariance function or for power-law spectra having sufficiently rapid falloff at high wavenumbers. Using (13.85) in (13.88),

$$L = \frac{\sqrt{2}h}{\sigma_s} \ . \tag{13.89}$$

According to this definition, the correlation length is not an additional measure of roughness, but can be obtained from the ratio of RMS roughness and RMS slope. As suggested, however, this definition is not universal, merely convenient for the following discussion of mathematical accuracy of roughness scattering approximations.

Most of the studies of the mathematical accuracy of the small-roughness perturbation and Kirchhoff approximations have employed exact numerical calculations for the pressure-release boundary condition. In order to make these computations tractable, two-dimensional geometry has been assumed in every case. Application of these results to the three-dimensional case is generally not considered problematic, but requires care in choosing the one-dimensional counterpart of the two-dimensional spectrum [Thorsos 1990].

Some of the accuracy studies employing the Gaussian-shaped covariance and spectrum will be summarized here, although there are questions as to their applicability to the seafloor scattering problem, not only due to the inappropriate boundary condition, but also due to the single-scale nature of the roughness. The term "single-scale" means that roughness features tend to have similar wavelengths owing to the fact that the Gaussian spectrum is finite at zero wavenumber and falls off rapidly with increasing wavenumber. The idealized power-law spectrum often used to model seafloor roughness is infinite at zero wavenumber and has a relatively slow fall off with increasing wavenumber. As such, one cannot identify any particular wavelength for the roughness features, and the random relief function can be characterized as "multiscale." Of course, if the seafloor has prominent ripples, with well-defined wavelength this picture is not quite correct, but even with ripples present, there remains a multiscale background. Fortunately, studies of accuracy with multiscale rough surfaces are available, and some of the key results of this work will be summarized later.

The Gaussian relief covariance is completely characterized by RMS roughness and correlation length, and the two dimensionless parameters, $k_w h$ and $k_w L$, are commonly used to delineate regions of accuracy. The RMS slope, σ_s, is also used. For the most part, it is found that the Kirchhoff and perturbation approximations are complementary, being accurate in disjoint regions of $k_w h$–$k_w L$ space. The use of dimensionless parameters allows scaling of results so that frequencies and roughness parameters other than those used in the numerical calculations can be treated. Such scaling is perfectly valid in the mathematical world considered here, but questionable in the case of experimental data owing to the complexity of the real seafloor.

Figure 13.20 summarizes work on the validity of perturbation theory and the Kirchhoff approximation. The perturbation results are taken from [Thorsos and Jackson 1989] who considered the Gaussian covariance for a pressure-release interface. Exact numerical calculations and higher-order perturbation theory were used, and regions of validity in $k_w h$–$k_w L$ space where theory was accurate to within 1 dB were mapped out. The cross section computed in lowest-order perturbation theory was denoted $\sigma^{(2)}$, because it is proportional to the second power of the RMS roughness. In the next order of approximation, the cross section is denoted $\sigma^{(4)}$, as it includes terms proportional to the fourth power of RMS roughness (Appendix K). Considering lowest-order perturbation theory, the region of accuracy is bounded approximately by $k_w h < 0.6$. The downturn of the bounding curve on the left side restricts validity to moderately small slopes. This can be seen by rewriting (13.89) to obtain $\sigma_s = \sqrt{2}h/L$. Perturbation theory is also inaccurate at large correlation lengths such that the Gaussian spectrum has fallen to very small values at the Bragg wavenumber. In this case the scattering cross section is dominated by higher-order terms in the perturbation series. This difficulty is not expected for multi-scale roughness. Results on the accuracy of higher-order perturbation theory can be found in [Thorsos 1990, Berman 1991, Thorsos and Jackson 1991, Broschat and Thorsos 1997].

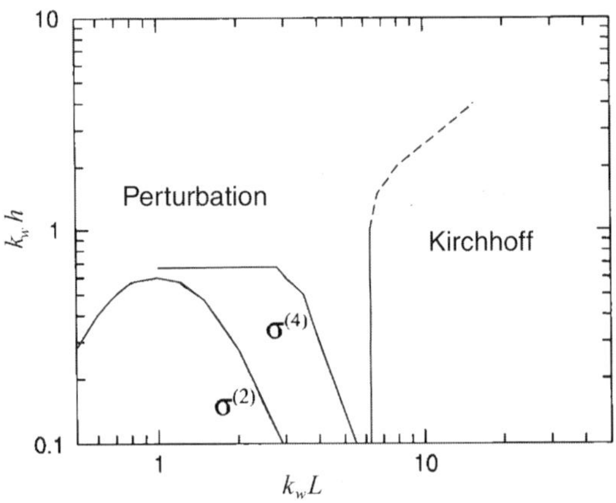

Fig. 13.20. Contours for 1-dB accuracy of the bistatic scattering cross section for one-dimensional random interfaces having the Dirichlet (pressure-release) boundary condition and Gaussian covariance. The contour labeled $\sigma^{(2)}$ applies to lowest-order perturbation theory, the form discussed in this chapter. The contour labled $\sigma^{(4)}$ applies to the next higher order of perturbation approximation (K.13). The two forms of perturbation theory are accurate to 1 dB or better over all bistatic angles for the regions below the contours. The Kirchhoff approximation is accurate to 1 dB to the right of its contour, with angles near grazing excluded. Adapted from [Kaczkowski and Thorsos 1994].

Similar studies have been made of the mathematical accuracy of the Kirchhoff approximation for nonpenetrable rough interfaces with Gaussian covariance [Thorsos 1988, Thorsos and Jackson 1991, Broschat and Thorsos 1997]. The Kirchhoff approximation is accurate for "gently undulating" rough interfaces. In the older literature [McDaniel and Gorman 1983,

D. Jackson et al. 1986a], the accuracy criterion was given in terms of the curvature of the interface, but [Chen and Fung 1988, Thorsos 1988] find that accuracy depends on correlation length rather than curvature. As indicated in Fig. 13.20, the Kirchhoff approximation is accurate to within 1 dB for bistatic scattering from pressure-release surfaces with Gaussian-shaped roughness spectra if

$$k_w L > 6 \ . \tag{13.90}$$

This condition must be supplemented by the condition that the incident and scattered grazing angles be larger than $2\sigma_s$, that is,

$$\theta_i, \ \theta_s > 2\sigma_s \ . \tag{13.91}$$

If these angles are not larger than twice the RMS slope, shadowing and multiple scattering effects impair the accuracy of the Kirchhoff approximation. Thorsos finds the Kirchhoff approximation can be accurate for $k_w L < 6$ provided attention is restricted to scattering near the specular direction. Thus, the Kirchhoff approximation is quite robust for near-forward scattering, which includes near-vertical backscattering.

Figure 13.20 shows that there is a substantial gap in $k_w h$–$k_w L$ space where neither lowest-order perturbation theory nor the Kirchhoff approximation is accurate. This gap can be filled in by using further terms in the perturbation series or by using iterations of the Kirchhoff approximation. Neither approach has found widespread use owing to the analytical complexity and numerical effort involved. Instead, attention has focused on new approximation schemes, the most widely used being the small-slope method. Because this approximation reduces to the perturbation and Kirchhoff approximations in the appropriate limits, one might hope that it would fill in some of the gap in $k_w h$–$k_w L$ space. In fact, [Broschat and Thorsos 1997] find that this is the case for higher-order small-slope approximations, but it is not evident that the lowest-order small-slope approximation considered here dramatically outperforms perturbation theory or the Kirchhoff approximation in the gap region. Even so, the lowest-order small-slope approximation combines the best features of perturbation theory and the Kirchhoff approximation, and provides a convenient approach to scattering by multiscale roughness.

Studies have also been made of the mathematical accuracy of scattering models when applied to surfaces having multiscale roughness. Some of these studies are listed in Table 13.2 which gives the boundary conditions and parameters characterizing the rough interface. The idealized power-law spectrum was not used in these studies owing to the singularities noted earlier. For studies of scattering by the sea surface, the "Pierson-Moskowitz" spectrum in the one-dimensional form

$$W_1(K_x) = \frac{\alpha}{4|K_x|^3} e^{-\frac{\beta g^2}{K_x^2 U^4}} \tag{13.92}$$

Table 13.2. Parameters for tests of mathematical accuracy in two dimensions of scattering models applied to multiscale rough interfaces. The parameter γ_1 gives the asymptotic power behavior as wavenumber becomes large. The parameters given here are a subset of those used in the references and are chosen as the largest roughness case giving reasonable accuracy.

Reference	Boundary	Freq. (Hz)	Spectrum	γ_1	$k_w h$
[Thorsos 1990]	Pressure-Release	200	Pierson-Moskowitz, $U = 20$ m s^{-1}	3	1.79
[Berman 1991]	Pressure-Release	150	Pierson-Moskowitz, $U = 20$ m s^{-1}	3	1.34
[Berman 1991]	Elastic (Rock)	150	von Karmen $L_c = 50$ m	4	2.1
[Thorsos et al. 2000a]	Fluid (Sand)	20000	von Karmen $L_c = 0.227$ m	2	1.0

has been used. The exponential factor suppresses the spectrum as $K_x \to 0$, eliminating the singularity at large scales. For large K_x, the inverse third-power term dominates, and is equivalent to $\gamma_2 = 4$ in the two-dimensional power-law spectrum. As such, the RMS slope is infinite and cannot be defined. In (13.92), $\alpha = 8.1 \times 10^{-3}$, $\beta = 0.74$, $g = 9.81$ m^2/s, and U is the wind speed in m s^{-1} at an altitude of 19.5 m. For studies of scattering by the seafloor, the "von Karman" spectrum has been used in the one-dimensional form

$$W_1(K_x) = \frac{w_1}{[K_x^2 + L_c^{-2}]^{\frac{\gamma_1}{2}}} . \tag{13.93}$$

Here, the exponent γ_1 is related to the power-law exponent in two dimensions by $\gamma_2 = \gamma_1 + 1$. The parameter L_c is a low-frequency cutoff that causes the spectrum to approach a finite value as $K_x \to 0$.

In order to interpret the multiscale results, it is necessary to express the mean-square roughness and slope in terms of the relief spectrum. Equation (6.5) can be used to show that

$$h^2 = \int W(\mathbf{K}) d^2K , \tag{13.94}$$

$$< (\frac{\partial f}{\partial x})^2 > + < (\frac{\partial f}{\partial y})^2 >= \int K^2 W(\mathbf{K}) d^2K . \tag{13.95}$$

These are two-dimensional expressions. For the one-dimensional spectra given above, the double integrals are replaced by single integrals over K_x:

$$h^2 = \int W_1(K_x) dK_x , \tag{13.96}$$

$$\sigma_s^2 = \int K_x^2 W_1(K_x) dK_x \; . \tag{13.97}$$

With reference to Table 13.2, [Thorsos 1990] found that the Kirchhoff approximation was accurate for near-specular bistatic scattering from a pressure-release surface with Pierson-Moskowitz spectrum for rather large RMS roughness ($k_w h = 1.79$). Accuracy worsened for small incident grazing angles ($\leq 15°$), and was always poor at wide angles from specular, including the backscattering direction. Interestingly, perturbation theory was found to be quite accurate for this large-roughness case for bistatic angles not too close to the specular direction. Note that the Kirchhoff approximation and the perturbation approximation are performing in complementary fashion with respect to angular coverage.

Tests of the composite-roughness approximation are also presented in [Thorsos 1990] for sea-surface backscattering at 200 kHz with a Pierson-Moskowitz spectrum. Two choices of cutoff were used, $K_c = k_w/2$ and $K_c = k_w$. Wind speeds of 10 m s^{-1} and 20 m s^{-1} were assumed, corresponding to $k_w h = 0.45$ and $k_w h = 1.79$, respectively, with large-scale RMS slopes ranging from 0.072 to 0.117. For grazing angles of 15° and 20°, the composite-roughness corrections due to slope averaging were small, yielding increases in scattering strength from 1 to 3 dB. These corrections improved the accuracy of the perturbation approximation to 0.5 dB or less. At a grazing angle of 10° and wind speed of 20 m s^{-1}, composite roughness slope averaging increased scattering strength 5–6 dB, where a 3-dB correction was actually desired. Shadowing corrections reduced this discrepancy by only 0.5 dB.

The mathematical accuracy of the perturbation, Kirchhoff, and small-slope approximations has been tested by [Berman 1991]. For a pressure-release surface having Pierson-Moskowitz spectrum such that $k_w h = 1.34$, the Kirchhoff approximation performed well near the specular direction, while the perturbation approximation was accurate over virtually the entire range of scattering directions. One of the most severe tests in [Berman 1991] used an elastic boundary condition appropriate to a rock seafloor and von Karman spectrum with $k_w h = 2.1$. The Kirchhoff approximation was accurate to within about 2 dB for scattered grazing angles greater than 30° for an incident grazing angle of 57°. The small-slope approximation was tested at second order (in this nomenclature, the small-slope approximation as used in this monograph is first-order). Curiously, the small-slope approximation did not perform as well as the Kirchhoff approximation, being accurate only within about 20° of the specular direction.

In another example [Thorsos et al. 2000a], perturbation theory compared well with exact calculations for a fluid seafloor with realistic von Karman spectrum with $k_w h = 1.0$ and parameters appropriate to a sandy bottom. The incident grazing angle was 20°, and 1-dB accuracy was observed over the entire range of bistatic scattering angle (Figure 15.12 of this monograph).

The picture regarding accuracy of the lowest-order small-slope approximation as applied to multiscale surfaces is not especially clear. Given the similarities with perturbation theory and the Kirchhoff approximation, however, it is reasonable to suppose that, if these are accurate, the lowest-order small-slope approximation will be accurate as well.

13.8 Research Issues

Roughness scattering models based on the perturbation, Kirchhoff, and small-slope approximations have been successfully applied to seafloors of moderate roughness and more-or-less simple composition. It is common, however, to encounter roughness conditions where these single-scattering approximations fail. Higher-order small-slope approximations may be useful here if they can be put into numerically tractable form. Rough, rocky seafloors are an especially challenging area for future research, both because of the lack of theoretical tools and the difficulty of geoacoustic characterization. The problem of scattering by discrete objects on the seafloor, such as shell fragments, has received relatively little attention and deserves further study.

14 Sediment Volume Scattering

Sediment heterogeneity can produce appreciable levels of sound scattering, particularly in soft sediments for which transmission into and out of the seafloor is facilitated by the small contrast of sediment density and sound speed compared to water. Heterogeneity, as demonstrated in Ch. 7, can take many forms: layering due to widespread erosional and deposition processes, continuous spatial fluctuations in acoustic properties due to the activity of benthic animals [Jones and Jackson 1997], buried shell fragments [Ivakin 2005, Lyons 2005], gas bubbles [Tang et al. 1994, Boyle and Chotiros 1995a, Lyons et al. 1996], and the inherent granularity of the sediment itself [Williams et al. 1988, Chotiros 2002a, Greenlaw et al. 2004]. It is generally agreed that sediment volume scattering is an important contributor to seafloor reverberation [Ivakin 1981, Ivakin and Lysanov 1981a, Ivakin and Lysanov 1981b, D. Jackson et al. 1986a, Ivakin 1990, Hines 1990, D. Jackson and Briggs 1992, Gensane 1993, Lyons et al. 1994, Pace 1994, Boyle and Chotiros 1995b, Yamamoto 1996, Hines 1996, Pouliquen et al. 2000b, Pouliquen and Lyons 2002].

In the previous chapter, it was possible to organize the discussion of roughness scattering on the basis of general expressions that were independent of the wave theory employed. Similar general expressions do not exist for the volume scattering models to be considered, because volume interaction depends directly on the particular wave theory employed. For example, including shear waves in the theory provides an additional scattering mechanism compared to a fluid model having only compressional waves. Because of this, volume scattering models will be broadly categorized according to the wave theory employed: fluid or elastic. Studies of volume scattering in poroelastic media are at a very early stage [Yelton and Chotiros 1995, Gurevich et al. 1998] and will not be discussed here. The effective density approximation (Sect. 10.2) offers one promising approach, allowing application of fluid models to porous sandy sediments.

As in Ch. 7, heterogeneity will be divided into random and nonrandom parts. Nonrandom heterogeneity includes the effects of gradients and layering. Specifically, the nonrandom part of a geoacoustic parameter is its average value, assumed here to be a function of depth, but not of horizontal position. The random part of a geoacoustic parameter is its difference, at a given position, with the average, nonrandom part. This division is motivated by the perturbation approach that is widely used in modeling sediment volume scattering.

14.1 Fluid Sediment Volume Scattering Models

Many of the published models for high-frequency sediment volume scattering employ a fluid model for the sediment. The simplest of these can be cast in a common form, such that they differ only in the value of a volume scattering parameter. More general cases [Ivakin 1998a] are treated in Appendix M. The simplest fluid models assume that the average complex sound speed and average (possibly complex) density of the sediment are constant with respect to depth. Complex sound speed and complex density are discussed in Sect. 8.3. This is a reasonable assumption at high frequencies and low grazing angles, where acoustic penetration depths are relatively small, but may be a poor approximation at steep grazing angles or at frequencies below roughly 10 kHz, where deeper penetration may expose sediment layers. It may also be a poor approximation in some cases at frequencies above 100 kHz, because sediments may exhibit strong gradients in physical properties in the first few millimeters, the transition between water and sediment [Lyons and Orsi 1998, Pouliquen and Lyons 2002, Tang et al. 2002].

14.1.1 Basic Volume Scattering Model

The basic fluid sediment volume scattering model depicted in Fig. 14.1 was developed by [Stockhausen 1963]. The key assumptions are that (1) volume scattering is sufficiently weak that the scattered acoustic field in the sediment is much smaller than the incident field, (2) sediment volume scattering can be characterized by a cross section that increases linearly with the ensonified volume, (3) attenuation in the sediment limits penetration to depths that are much smaller than the distances from the source and receiver to the scattering region, and (4) the scattering effects of the sediment–water interface are negligible. More will be said later about the weak scattering assumption. The second assumption may seem self-evident to those familiar with treatments of volume scattering in the ocean water column. In this case, scattering by marine organisms is nearly always characterized by a volume scattering strength, the decibel equivalent of a scattering cross section per unit solid angle per unit volume. In fact, this assumption may not always be valid in the case of sediment volume scattering [Jones and Jackson 1997, Briggs et al. 2002a], where scattering may be limited to distances within a wavelength or so of the water–sediment interface, so that the incident field may not be readily approximated by a plane wave and may vary significantly in strength over short distances. The approach to be used here takes partial account of the nonplane nature of the incident field within the sediment, but does not completely account for the proximity of the interface. The assumption of small penetration depth is generally a good one at high frequencies and allows treatment of sediment volume scattering in terms of an effective interface scattering strength, the decibel equivalent of scattering cross section per unit solid angle per unit area. That is, sediment volume scattering at high frequencies can be treated as an interface process and can be characterized both experimentally and theoretically in the same terms as scattering by seafloor roughness. Clearly there is a potential conflict between the requirement of small penetration depth and the assumption that the interface is not too close to the scatterers. The assumption that the interface is flat is introduced for simplicity and cannot be strictly true. These issues will be discussed later in this chapter.

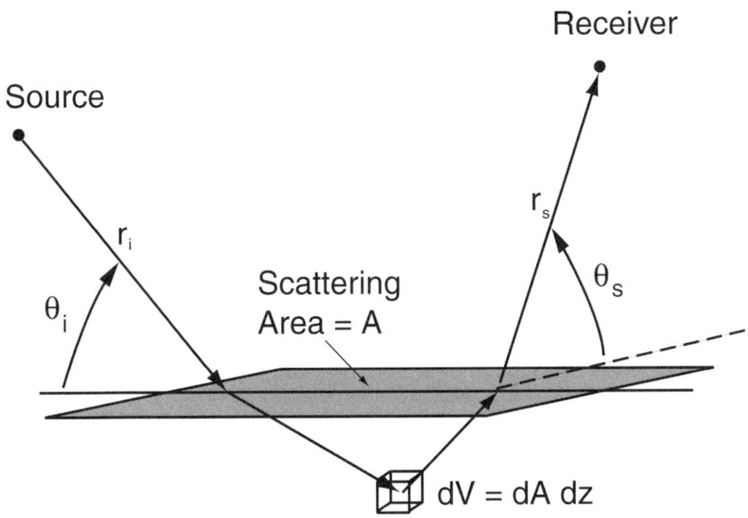

Fig. 14.1. The geometry relevant to the simplest fluid sediment volume scattering models. The figure is not to scale in that the ranges $r_i = |\mathbf{r}_i|$ and $r_s = |\mathbf{r}_s|$ are assumed large compared to the depth of penetration.

Referring to Fig. 14.1, the mean-square pressure at the receiver, located at $\mathbf{r}_s$, due to a unit point source (Sect. 8.6), located at $\mathbf{r}_i$, is

$$< |P(\mathbf{r}_s)|^2 >= \int |G_{wp}(\mathbf{r},\ \mathbf{r}_i)|^2 |G_{pw}(\mathbf{r}_s,\ \mathbf{r})|^2 \sigma_v(\mathbf{r}) dV \ . \tag{14.1}$$

The Green's functions, G_{wp} and G_{pw}, are defined in Sect. 8.6. The function $G_{wp}(\mathbf{r},\ \mathbf{r}_i)$ is the Green's function giving the field in the sediment at $\mathbf{r}$ due to a unit point source in the water at $\mathbf{r}_i$. Similarly, $G_{pw}(\mathbf{r}_s,\ \mathbf{r})$ is the Green's function giving the field in the water at $\mathbf{r}_s$ due to a unit point source in the sediment at $\mathbf{r}$. This expression is a modest improvement over Stockhausen, who used a ray method that predicts zero volume scattering for incident or scattered grazing angles smaller than the critical angle. By employing the Green's functions, subcritical scattering, though usually small, will not be ignored. The volume scattering level is expressed in terms of the volume scattering cross section, $\sigma_v(\mathbf{r})$, which has the units $\mathrm{m}^2/\mathrm{m}^3 = \mathrm{m}^{-1}$. The scattering patch has area A, and it will be assumed that this area is small enough that all quantities in the integrand can be treated as constants with regard to the horizontal position variables, x and y, which are taken to be zero at the center of the scattering patch. The squared Green's function factors give the propagation gain (< 1) from the source to the infinitesimal scattering volume, dV, and from this volume to the receiver. It will be assumed that the incident field can be approximated by a plane wave at the water–sediment interface, so that

$$|G_{wp}(\mathbf{r},\ \mathbf{r}_i)|^2 \simeq \frac{|V_{wp}(\theta_i)|^2}{r_i^2} e^{2\mathrm{Im}[k_p \beta_p(K_i)]z} \ . \tag{14.2}$$

Here, k_p is the complex wavenumber in the sediment, and $\beta_p(K_i)$ is the complex sine of the refracted angle in the sediment. The factor $1/r_i^2$ gives the spreading loss from the source to the scattering region, the plane-wave transmission coefficient, $V_{wp}(\theta_i)$, gives the pressure change across the interface (assumed to be flat, see (8.55)), the exponential factor gives the subsequent decay of the wave amplitude in the sediment, and Im denotes the imaginary part. This decay is governed by the imaginary part of the z-component of the wave vector, which, in turn, depends on $K_i = k_w \cos\theta_i$. Note that, even though the incident field is assumed to approximate a plane wave in the water, it is *not* plane in the sediment owing to the decay factor. The negative of the coordinate, z, is the distance of the infinitesimal volume below the interface. It is assumed that the distance, $-z$, and the linear dimensions of the scattering patch are much smaller than the source range, r_i, so that the incident plane-wave assumption is valid over the entire scattering volume. If the field point range, r_s, is similarly large, the reciprocity relation (8.69) can be used to express the other squared Green's function as

$$|G_{pw}(\mathbf{r_s},\ \mathbf{r})|^2 \simeq \frac{|V_{wp}(\theta_s)|^2}{|a_\rho|^2 r_s^2} e^{2\mathrm{Im}[k_p \beta_p(K_s)]z} \ , \tag{14.3}$$

where a_ρ is the sediment/water density ratio and $K_s = k_w \cos\theta_s$. Since a unit source is assumed, the squared incident pressure is

$$|P_i|^2 = \frac{1}{r_i^2} \ . \tag{14.4}$$

Using these results, (14.1) can be rewritten as

$$< |P(\mathbf{r}_s)|^2 >=$$

$$\frac{|P_i|^2 A |V_{wp}(\theta_i)|^2 |V_{wp}(\theta_s)|^2}{|a_\rho|^2 r_s^2} \int_{-\infty}^{0} \sigma_v(z) e^{2(\mathrm{Im}[k_p \beta_p(K_i) + k_p \beta_p(K_s)])z} dz \; , \quad (14.5)$$

where the volume scattering strength has been assumed to depend on the vertical coordinate only. Comparison with (2.15) shows that, with the assumptions given above, sediment volume scattering cannot be distinguished from interface roughness scattering; that is, the scattered intensity has the same dependence on range and ensonified area. Thus, one can identify the equivalent interface bistatic scattering cross section

$$\sigma = \frac{|V_{wp}(\theta_i)|^2 |V_{wp}(\theta_s)|^2}{|a_\rho|^2} \int_{-\infty}^{0} \sigma_v(z) e^{2k_w \mathrm{Im}[P(\theta_i) + P(\theta_s)]z} dz \; , \quad (14.6)$$

where

$$P(\theta) = \sqrt{a_p^{-2} - \cos^2\theta} \; , \quad (14.7)$$

with a_p being the complex compressional wave speed/water sound speed ratio. If the volume scattering strength is independent of depth, the integral can be performed, leaving

$$\sigma = \frac{|V_{wp}(\theta_i)|^2 |V_{wp}(\theta_s)|^2 \sigma_v}{2k_w |a_\rho|^2 \mathrm{Im}[P(\theta_i) + P(\theta_s)]} \; . \quad (14.8)$$

While the assumptions and approximations that lead to this result may seem restrictive, they are satisfied reasonably well in the majority of cases encountered in practice. The assumptions regarding range and size of the ensonified region guarantee that the plane-wave approximation is satisfied. There are subtle points in this connection that are discussed in Appendix J. It must be realized that if the plane-wave assumption fails, then the concept of an interface scattering cross section loses its usefulness. As noted in Appendix J, there are unduly pessimistic accounts in the literature that might lead one to discard the interface scattering strength even when it has utility.

Equation (14.8) forms the basis for several sediment volume scattering models, which differ only in the assumptions used to obtain the volume scattering cross section, σ_v [Stewart and Chotiros 1992]. The simplest approach is to treat σ_v empirically, that is, as a quantity that must be obtained by fitting data [D. Jackson et al. 1986a] rather than calculated on the basis of theory. In [Mourad and Jackson 1989], the following dimensionless parameter is used to quantify sediment volume scattering:

$$\sigma_2 = \frac{\sigma_v}{\alpha_p} \; , \quad (14.9)$$

where α_p is the attenuation in dB m^{-1}, given by (8.26). Typical values of σ_2 range from 0.0001 to 0.004. Values larger than 0.004 may indicate failure of the weak scattering assumption, that is, they may signal the onset of multiple scattering [D. Jackson et al. 1986a]. If σ_2 is taken to be frequency independent, then (14.8) will yield a scattering cross section that is nearly frequency independent. This can be seen by noting that if attenuation increases approximately linearly with frequency, then, according to (14.9), σ_v will increase linearly with frequency, and this dependence is canceled by the factor k_w in the denominator of (14.8). Finally, the transmission coefficients and $P(\theta)$ have only weak frequency dependence because they depend only on the real sound speed ratio, ν_p, and the loss parameter, δ_p.

Examples of backscattering strength computed using (14.8) are shown in Fig. 14.2. Of the three examples, one is a "slow" bottom with sound speed ratio 0.98 and no critical angle. The other two examples have sound speed ratios greater than unity and critical angles of about 18° and 25°. For grazing angles smaller than the critical angle, volume scattering is suppressed owing to the relatively small depth of acoustic penetration into the sediment.

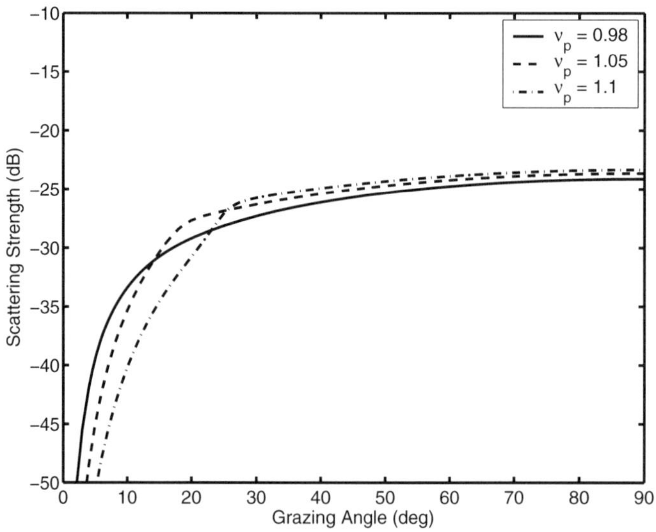

Fig. 14.2. Backscattering strength curves computed using (14.8) for a frequency of 30 kHz with $c_w = 1500$ m s^{-1}, $a_\rho = 1.5$, $\delta_p = 0.015$, $\sigma_2 = 0.002$, and the indicated values of sound speed ratio, ν_p.

14.1.2 Roughness–Volume Interaction

The separation of scattering into separate, independent roughness and volume components is somewhat artificial, as noted in Ch. 13. At the very least, the rough interface will alter the field entering the sediment from that expected for a flat interface. A rather straightforward approach to this problem is offered by [Ivakin and Lysanov 1981b], who employ an approximation analogous to the composite-roughness model. That is, account is taken of the larger-scale slope of the seafloor by averaging the volume scattering cross section over the expected distribution of local grazing angles. This approach is readily implemented numerically via (13.80) as in [D. Jackson et al. 1986a, Mourad and Jackson 1989], but neglects diffraction by smaller-scale components of roughness. More difficult is the question of correlation between roughness and volume scattering, which would be expected if the vertical undulations of sediment layers follow to some degree the undulations of the sediment–water interface. In such cases, the division between roughness and volume scattering is not clear, and more general approaches may be required (Sect. M.1.2).

14.1.3 Small-Perturbation Fluid Approximation

Fluctuations in sediment mass density and sound speed are always present owing to biological, biogeochemical, and hydrodynamic activity. These fluctuations result in smooth changes in sediment acoustic parameters as opposed to the abrupt changes due to the presence of discrete scatterers such as shell inclusions or bubbles. Smooth heterogeneities are usually weak enough that the method of small perturbations can be applied [Jones 1999]. This is a single scattering approximation in which it is assumed that the scattered field is weaker than the incident field and is consistent with the assumptions used in deriving (14.8). The perturbation approach is widely used [Ivakin and Lysanov 1981a, Hines 1990, Tang 1991, Lyons et al. 1994, D. Jackson 1994, Yamamoto 1996, Pouliquen and Lyons 2002] and appears to be a satisfactory solution to at least part of the problem of sediment volume scattering. It is often referred to as the "Born approximation."

The expression for the bistatic volume scattering cross section (14.8) requires a means of obtaining the parameter σ_v. Two different theoretical approaches have been used. The simplest is to neglect the presence of the water–sediment interface and find the plane-wave scattering amplitude for a localized heterogeneous region. This approach cannot treat the subcritical case, where the x- and y-components of the incident and scattered wave vectors are real, but the z-components are complex. The volume scattering cross section for this more general case is given by several authors, and details are provided in Appendix M. The central result is

$$\sigma_v = \frac{\pi |k_p|^4}{2} \times$$

$$\left[W_{\kappa\kappa}(\mathbf{\Delta k}_p) + 2\mathrm{Re}\{ \frac{\mathbf{k}_{ps} \cdot \mathbf{k}_{pi}\, k_p^{*2}}{|k_p|^4} W_{\rho\kappa}(\mathbf{\Delta k}_p)\} + \frac{|\mathbf{k}_{ps} \cdot \mathbf{k}_{pi}|^2}{|k_p|^4} W_{\rho\rho}(\mathbf{\Delta k}_p) \right] . \tag{14.10}$$

In this expression, k_p is the compressional wavenumber, and the various vectors subscripted p are to be defined later. The function $W_{\kappa\kappa}$ is the three-dimensional spectrum for the fluctuations in normalized sediment compressibility. The normalization consists in dividing the fluctuating part of the compressibility (which is the inverse of the bulk modulus) by the mean compressibility. Similarly, $W_{\rho\rho}$ is the spectrum for normalized density fluctuations, and $W_{\rho\kappa}$ is the cross-spectrum that expresses the correlation between these two random variables. These spectra are discussed in Sects. 7.2, 14.1.4, and D.2. Careful consideration of the bistatic angular dependence implied by the factor $\mathbf{k}_{ps} \cdot \mathbf{k}_{pi}$ would show that density fluctuations yield a dipole scattering pattern in the sediment, with maxima along the direction of the refracted incident wave. Compressibility fluctuations yield an omnidirectional (monopole) scattering pattern. The term involving the cross-spectrum is the product of monopole and dipole patterns. This simple division into monopole and dipole terms motivates the use of density and compressibility as the pair of random variables of interest. Other choices can be made; for example, practical considerations suggest the use of density and sound speed, as these are the most commonly measured pair. Equations connecting one pair to the other will be developed later in this section.

The incident and scattered directions are defined by the in-water wave vectors as in Ch. 13:

$$\mathbf{k}_i = k_w(\mathbf{e}_x \cos\theta_i \cos\phi_i + \mathbf{e}_y \cos\theta_i \sin\phi_i - \mathbf{e}_z \sin\theta_i) , \tag{14.11}$$

$$\mathbf{k}_s = k_w(\mathbf{e}_x \cos\theta_s \cos\phi_s + \mathbf{e}_y \cos\theta_s \sin\phi_s + \mathbf{e}_z \sin\theta_s) . \tag{14.12}$$

Using Snell's law, the corresponding wave vectors for the incident and scattered compressional waves in the sediment are

$$\mathbf{k}_{pi} = \mathbf{K}_i - \mathbf{e}_z k_w P(\cos\theta_i) , \tag{14.13}$$

$$\mathbf{k}_{ps} = \mathbf{K}_s + \mathbf{e}_z k_w P(\cos\theta_s) , \tag{14.14}$$

where, as usual, the uppercase, boldface symbols represent the horizontal components, which are the same in-water and in-sediment:

$$\mathbf{K}_i = k_w(\mathbf{e}_x \cos\theta_i \cos\phi_i + \mathbf{e}_y \cos\theta_i \sin\phi_i) , \tag{14.15}$$

$$\mathbf{K}_s = k_w(\mathbf{e}_x \cos\theta_s \cos\phi_s + \mathbf{e}_y \cos\theta_s \sin\phi_s) . \tag{14.16}$$

Note that

$$\mathbf{k}_{pi} \cdot \mathbf{k}_{pi} = \mathbf{k}_{ps} \cdot \mathbf{k}_{ps} = k_p^2 , \tag{14.17}$$

with the complex compressional wavenumber

$$k_p = \frac{k_w(1 + i\delta_p)}{\nu_p} . \tag{14.18}$$

The spectra in (14.10) are evaluated at the Bragg wave vector, $\mathbf{\Delta k}_p$. Unlike the small-roughness perturbation case, where the relevant Bragg wave vector is the difference of two-dimensional scattered (14.16) and incident (14.15) wave vectors, in the case of volume scattering, the Bragg wave vector is three-dimensional.

$$\mathbf{\Delta k}_p = \mathrm{Re}\{\mathbf{k}_{ps} - \mathbf{k}_{pi}\} . \tag{14.19}$$

The fact that the Bragg wave vector is the real part of the wave vector difference is derived in Appendix M. If the complex difference were required instead, one would be faced with the issue of analytic continuation of the spectrum to complex arguments. As will be seen, this issue cannot be avoided when treating volume scattering in elastic or layered media.

14.1.4 Statistical Description of Continuous Fluid Heterogeneity

In the fluid small-perturbation model, spectra describing fluctuations in density and compressibility are needed. These spectra are discussed in detail in Ch. 7 and Sect. D.2, consequently, only a few key definitions will be given here. Normalized fluctuations are defined for density,

$$\gamma_\rho = \frac{\delta\rho}{<\rho>} , \tag{14.20}$$

and compressibility,

$$\gamma_\kappa = \frac{\delta\kappa}{<\kappa>} . \tag{14.21}$$

The relevant covariances are

$$B_{\rho\rho}(\mathbf{r}) = <\gamma_\rho(\mathbf{r} + \mathbf{r}_0)\gamma_\rho(\mathbf{r}_0)> , \tag{14.22}$$

$$B_{\kappa\kappa}(\mathbf{r}) = <\gamma_\kappa(\mathbf{r} + \mathbf{r}_0)\gamma_\kappa(\mathbf{r}_0)> , \tag{14.23}$$

and

$$B_{\rho\kappa}(\mathbf{r}) = <\gamma_\rho(\mathbf{r} + \mathbf{r}_0)\gamma_\kappa(\mathbf{r}_0)> . \tag{14.24}$$

These are related to the spectra by three-dimensional Fourier transforms

$$W_{\alpha\beta}(\mathbf{k}) = \frac{1}{(2\pi)^3} \int B_{\alpha\beta}(\mathbf{r}) e^{-i\mathbf{k}\cdot\mathbf{r}} d^3r , \tag{14.25}$$

where $\alpha\beta$ corresponds to all pairs of ρ and κ. Since the covariances are dimensionless, the spectra have dimensions (length)^3. Although compressibility is a convenient variable in discussing sediment volume scattering (because it yields simple monopole scattering) it is not commonly measured in sediments, but can be inferred from density and sound speed measurements. Sections 7.3 and D.2.2 treat this topic.

In order to avoid the problem of determining all three of the fluctuation spectra, various authors have made simplifying assumptions, as discussed in Sect. D.2.2. Some neglect the correlation between sound speed and density [Lyons et al. 1994], but, looking at the relation between density and sound speed (Ch. 4), it is evident that there *must* be some correlation between the two. In fact, it is common [Hines 1990, Yamamoto 1996, Hines 1996, Ivakin 2001, Pouliquen and Lyons 2002] to assume that the two fluctuations are perfectly correlated, that is, density and sound speed fluctuations have the same ratio at every point in the sediment. Equivalently, [D. Jackson et al. 1996a, Williams and Jackson 1998, Briggs et al. 2002a] have assumed that density and compressibility fluctuations are perfectly correlated. The scatter seen in plots of measured sound speed as a function of density, such as in Ch. 4, suggests that the correlation cannot be perfect. In this case it might seem that all points would lie on a single curved line. This picture is misleading, however, as such plots combine data from many different sites. The present issue is whether or not, for a given site, density and sound speed have a deterministic relationship. To resolve this issue, density and sound speed measurements could be made on single sediment samples in consolidation (for mud samples) and packing experiments (for sand samples). If the resulting density–sound speed plot shows little scatter, then the correlation is high. This interesting issue has yet to be resolved. In the interim, a general strategy is to assume an adjustable degree of correlation such that the compressibility spectrum and density–compressibility cross-spectrum have the same functional dependence as the density spectrum, but differing strengths:

$$W_{\kappa\kappa} = \mu^2 W_{\rho\rho} \ , \tag{14.26}$$

$$W_{\rho\kappa} = \rho_{\rho\kappa} \mu W_{\rho\rho} \ . \tag{14.27}$$

Without loss of generality, it is assumed that $\mu \geq 0$. The parameter $\rho_{\rho\kappa}$ is the density–compressibility correlation coefficient and is restricted to the range $-1 \leq \rho_{\rho\kappa} \leq 1$.

Spectra for density fluctuations have been given by [Yefimov et al. 1988, Tang and Orsi 2000a,Tang and Orsi 2000b, Tang et al. 2002, Pouliquen and Lyons 2002], but both density (sometimes porosity) and sound speed spectra can be found in [Lyons et al. 1994, Briggs 1994, Yamamoto 1995, D. Jackson et al. 1996a, Briggs et al. 2002a]. Determination of cross spectra such as $W_{\rho\kappa}$ without the simplifying assumptions given above is difficult and is a research topic rather than a routine undertaking.

Section 7.2 gives expressions for model spectra that have appeared in the literature. Computations in this chapter will employ the simplest spectral form, the pure power law. Repeating (7.16) for convenience,

$$W_{\alpha\beta}(\mathbf{k}) = \frac{w_{3\alpha\beta}}{k^{\gamma_3}} . \tag{14.28}$$

As noted in Sect. 7.2, the case $\gamma_3 = 3$ is self-similar, and the parameter $w_{3\alpha\beta}$ is dimensionless. It has been noted that $\gamma_3 = 3$ gives a nearly frequency-independent scattering cross section [Ivakin 1981, D. Jackson and Briggs 1992]. This can be seen by counting wavenumber factors: k^{-1} in (14.8), k^4 in (14.10), and k^{-3} in the spectra. There will still be a weak frequency dependence due to dispersion, that is, due to the frequency dependence of ν_p and δ_p. As Fig. 12.6 suggests, interface scattering from soft sediments sometimes is substantially frequency independent, indicating that the self-similar spectrum may be a reasonable approximation in some cases.

The correlation between density and compressibility has a strong effect on the directionality of scattering [Ivakin 1998b, Jones and Jackson 2000, Ivakin 2001, Jones and Jackson 2001]. As noted above, density fluctuations scatter with a dipole radiation pattern with axis along the direction given by Snell's law for the incident wave in the sediment. Compressibility fluctuations scatter with an omnidirectional pattern. If the correlation is zero, the net pattern is the sum of the intensities of these two. With a nonzero correlation, however, there will be interference between these two patterns. For example, if $\mu = 1$ and $\rho_{\rho\kappa} = -1$, the dipole and monopole patterns cancel in the forward direction, with substantial backscattering. This situation is illustrated in the upper panel of Fig. 14.3. This is the case in which normalized sound speed fluctuations are much smaller than normalized density fluctuations ($\eta << 1$). This may happen for soft sediments where the sound speed versus density curve has a stationary point (Ch. 4). The lower panel of Fig. 14.3 shows that if the correlation vanishes, forward scattering is substantial. While the difference between the two cases illustrated in the figure is dramatic, this difference will be somewhat obscured by roughness scattering, which invariably has a strong peak in the specular direction.

14.1.5 Experimental Tests of the Small-Perturbation Fluid Model

In one of the earliest tests of the perturbation approximation for sediment volume scattering [Lyons et al. 1994], the model compared favorably with measured backscattering data at 6.5 kHz, falling within the ±5-dB data spread over the measured grazing angle range of 12°–57°. The seafloor consisted of a silt-clay layer overlying sand. The density and compressibility spectra were estimated from fits of the exponential covariance to core data, and cross-correlation was neglected.

Perturbation model predictions have been compared with data from a soft-sediment site in the Straits of Florida [Rogers and Yamamoto 1999]. The spectrum of sound speed was estimated from acoustic tomographic measurements at the same frequencies used in the backscattering measurements (7.5 and 15 kHz). These measurements had resolution of approximately 10 cm, so extrapolation of the measured spectrum was necessary to reach the smaller

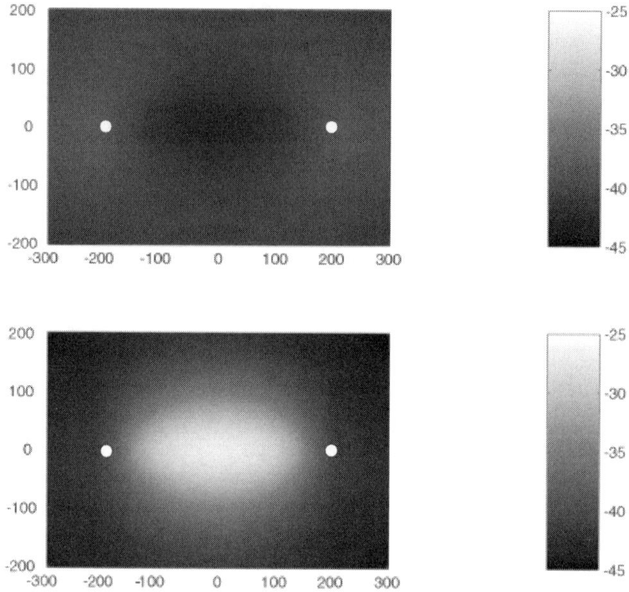

Fig. 14.3. Bistatic scattering strength (dB) for a mud seafloor with scattering strength projected onto the seafloor at the position responsible for the scattering. The source and receiver are located at $x = \pm$ 200 m, $y = 0$ (filled white circles), and both are positioned 50 m above the seafloor. Scattering strength is computed using (14.10) for a frequency of 35 kHz with $c_w = 1500$ m s^{-1}, $a_\rho = 1.5$, $\nu_p = 0.98$, and $\delta_p = 0.01$. Roughness scattering is ignored. Power-law spectra of the form (14.28) are used with $\gamma_3 = 3.0$, $w_{3\rho\rho} = 0.0005$, and $\mu = 1$. The two panels differ only in the assumed density–compressibility correlation, $\rho_{\rho\kappa} = -1$ in the upper panel and $\rho_{\rho\kappa} = 0$ in the lower panel.

scales set by the Bragg wavenumber (2.5-cm resolution needed at 15 kHz). The spectral exponent estimates ranged from $\gamma_3 = 2.3$ to $\gamma_3 = 2.8$. The perturbation model was in general agreement with the backscattering data, but with differences as large as 10 dB for some grazing angles.

Figure 14.4 compares higher-frequency backscattering data obtained by several different investigators with the perturbation model. These model-data comparisons are the same, except for details, as those given in the primary references, with geoacoustic parameters given in Table 14.1. The model-data agreement is satisfactory, but many of the comparisons are not as rigorous as one might wish. In particular, the volume heterogeneity data for the Arafura, San Francisco, and Orcas sites were obtained from small-diameter, short core samples and were analyzed in the vertical coordinate only. Thus, the aspect ratio was not determined and isotropy was assumed. The heterogeneity parameters for these sites are subject to the uncertainties discussed in Sect. 7.4. In particular, bias in these parameters is expected due to averaging over core sections. In addition, bias may occur due to disturbance intrinsic to core collection and subsequent manipulations, including sectioning (see Sects. 4.3 and 7.6). The Arafura site had a large concentration of buried shell fragments, calling into question the use of the perturbation approach. In all cases, the density–compressibility correlation was not measured, but assigned the commonly used value −1. In summary, the model-data comparisons are clouded by uncertainties in the heterogeneity parameters, but offer support of the perturbation model for sediment volume scattering.

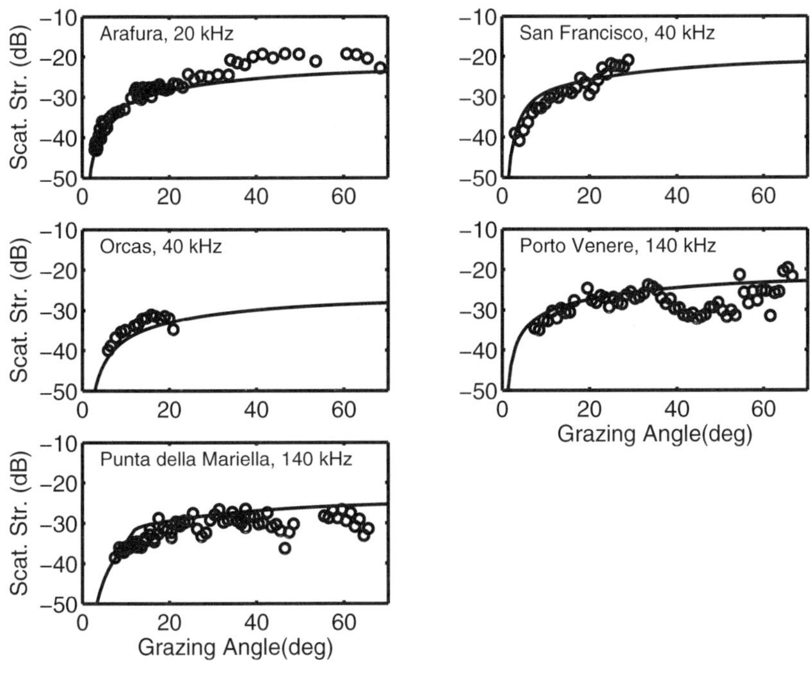

Fig. 14.4. Comparison of the fluid volume scattering perturbation model with backscattering strength measured at five sites having soft sediments. The primary acoustic and geoacoustic data sources are: Arafura – [D. Jackson and Briggs 1992], San Francisco – [D. Jackson and Briggs 1992], Orcas – [Jones and Jackson 1997], Porto Venere – [Pouliquen and Lyons 2002], Punta della Mariella – [Pouliquen and Lyons 2002]. The volume heterogeneity parameters for the Arafura and San Francisco sites were taken from [Yamamoto 1996].

Table 14.1. Geoacoustic input parameters used in model-data comparisons of Fig. 14.4. The citations for the parameter values are given in Fig. 14.4. The units of w_3 are $\mathrm{m}^{3-\gamma_3}$. The density–compressibility correlation coefficient, $\rho_{\rho\kappa}$, is taken to be −1 for all sites, and the aspect ratio, Λ, is taken to be 1.

Site	Grain Size (ϕ)	a_ρ	ν_p	δ_p	γ_3	w_3	L_c (m)	μ
Arafura	5.6	1.39	0.988	0.0055	3.45	1.32×10^{-3}	∞	1.389
San Francisco	6.4	1.41	1.002	0.0104	3.63	1.37×10^{-2}	∞	1.200
Orcas	8.1	1.41	0.977	0.0019	4.00	6.44×10^{-3}	0.032	0.934
Porto Venere	9.1	1.46	0.993	0.0011	7.60	5.13×10^{9}	0.0014	1.667
P. della Mariella	7.2	1.66	1.002	0.0028	8.00	3.20×10^{11}	0.0012	1.667

14.1.6 Mathematical Accuracy of the Small-Perturbation Fluid Model

Jones has used exact two-dimensional calculations to test the accuracy of the perturbation approximation for volume scattering in fluid sediments [Jones 1999, Jones and Jackson 2001]. The covariance for density fluctuations was assumed to be exponential with a spectrum that is the 2D equivalent of (7.17):

$$W_{\rho\rho}(\mathbf{k}) = \frac{\hat{\sigma}_\rho^2/(2\pi L_c)}{(k^2 + L_c^{-2})^{3/2}} \,. \tag{14.29}$$

The parameter $\hat{\sigma}_\rho^2$ is the normalized variance of the density fluctuations (Sect. 7.2). The other spectra are assumed proportional to $W_{\rho\rho}$ as in (14.26) and (14.27), with unit-magnitude, negative correlation between density and compressibility fluctuations, $\rho_{\rho\kappa} = -1$. Figure 14.5 shows a comparison of exact and perturbation calculations using parameters typical of a sandy sediment. The computations were performed with the low-frequency cutoff length, L_c, equal to the acoustic wavelength in water. Because of this, there is no need to specify the absolute acoustic frequency. The figure shows two different perturbation calculations: one is the full perturbation method with no additional approximations, and the other neglects the windowing effects discussed in Appendix M (these are referred to as "half-space effects" in [Jones and Jackson 2001]). These effects are neglected in most volume scattering models, including those discussed in this chapter.

The exact calculations displayed in Fig. 14.5 show that windowing effects are significant near the specular direction, but that the "ordinary" perturbation model is quite accurate in other directions. While these computations do not exhaustively explore the parameter space of the perturbation model, they indicate that, with typical sediment parameters, the model is comfortably within its realm of accuracy.

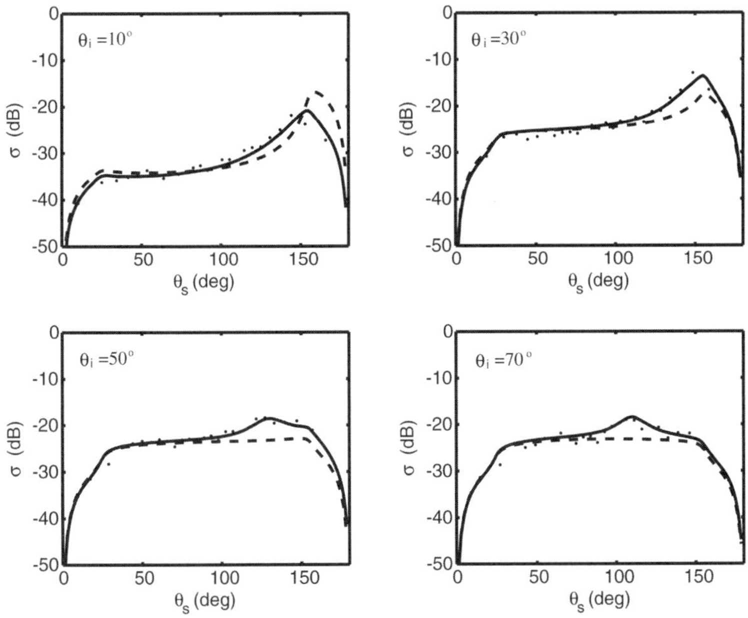

Fig. 14.5. Comparison of exact and approximate calculations of the bistatic scattering cross section due to volume heterogeneity. These are two-dimensional calculations with $a_\rho = 2.0$, $\nu_p = 1.10$, $\delta_p = 0.00183$, $\mu = 2$, and $\hat{\sigma}_\rho = 0.1$. The incident grazing angle is indicated in each panel, and the abscissa is the scattered grazing angle. The dots are the result of exact Monte Carlo calculations with 50 realizations. The solid curves are the results of the full perturbation model, while the dashed curves show the more commonly used perturbation model in which windowing effects due to layer interfaces are neglected. From [Jones and Jackson 2001].

14.1.7 Scattering by Discrete Heterogeneity

Buried gas bubbles and shell fragments are discrete forms of heterogeneity that cannot be modeled using perturbation theory in the form developed above. The scattering model presented in this section is perhaps the simplest available in the literature, and might be considered a tool for approximate calculations and a starting point for more sophisticated models. Thus, the single scattering assumption inherent in (14.8) will be employed, and a simple model will be used to obtain the volume scattering cross section, $\sigma_v(z)$, appearing in (14.6). The volume scattering cross section depends on the number of discrete scatterers per unit volume and the probability density function for scatterer size, a. Both of these are incorporated in the "size distribution," $N(a)$, the number of scatterers per unit volume per unit size. At this stage, "size" has not been defined clearly; that definition will depend on the physical situation of interest. For example, in the case of spherical bubbles, the "size" will be chosen to be the bubble radius. If the differential scattering cross section, $\sigma_d(a)$, for individual discrete scatterers of size a is known, the desired volume scattering cross section follows as an integral over all sizes [Boyle and Chotiros 1995a]:

$$\sigma_v = \int_0^\infty N(a)\sigma_d(a)da \; . \tag{14.30}$$

It should be noted that this simple model assumes that scattering is isotropic so that knowledge of the angular orientation of each scatterer is not required. If both the size distribution, $N(a)$, and the individual scattering cross section, $\sigma_d(a)$, are known, (14.30) can be evaluated and inserted in (14.6) or, if the volume scattering strength is independent of depth, in (14.8) to obtain the desired seafloor scattering cross section.

Scattering by Buried Gas Bubbles

Methane is a common product of biochemical reactions occurring in sediments that are rich in organic material [Kaplan 1974, Martens et al. 1998]. Section 3.3 discusses the biochemical processes responsible for methane bubble formation and presents CT and X-ray images of bubbles in sediments. The acoustic effects of bubbles have been discussed by, e.g., [Anderson and Hampton 1980a, Anderson and Hampton 1980b, Lyons et al. 1996, Wilkens and Richardson 1998, Anderson et al. 1998, Medwin and Clay 1998, Ch. 8]. Bubble scattering is a difficult topic that is still under development. Rather than attempting to fully document current models in this evolving research area, a brief summary will be given at the end of this section. The differential scattering cross section for individual, spherical bubbles of radius a is [Medwin and Clay 1998, Ch. 8]

$$\sigma_d(a) = \frac{a^2}{[(\omega_0/\omega)^2 - 1]^2 + \delta^2} \; . \tag{14.31}$$

The differential scattering cross section has dimensions (length)^2. Scattering is omnidirectional, hence there are no angular variables in (14.31), as assumed in (14.30). This expression applies to bubbles in an infinite fluid medium, and so does not include any water–sediment interface effects. When bubbles are large with respect to sediment grain or pore size (so-called Type III bubbles, Sect. 3.3) such as typically found in organic-rich muddy sediment, the resonant frequency of spherical bubbles is [Anderson and Hampton 1980a, Lyons et al. 1996, Gardner and Sills 2001]

$$\omega_0 = \frac{1}{a}\sqrt{\frac{3\gamma P_0}{A\rho} + \frac{4\mu'}{\rho}} \; , \tag{14.32}$$

where

$$A = (1 + B^2) \left[1 + \frac{3(\gamma - 1)}{X} \left(\frac{\sinh X - \sin X}{\cosh X - \cos X} \right) \right] , \tag{14.33}$$

$$B = 3(\gamma - 1) \left[\frac{X(\sinh X + \sin X) - 2(\cosh X - \cos X)}{X^2(\cosh X - \cos X) + 3X(\gamma - 1)(\sinh X - \sin X)} \right] , \tag{14.34}$$

$$X = a(2\omega_0 \rho_0 S_p / C_0)^{1/2} , \tag{14.35}$$

and μ' is the real part of the sediment shear modulus. The larger bubbles seen in the images of Sect. 3.3 are nonspherical, but [Lyons et al. 1996] define an equivalent spherical radius (the radius of the sphere having the same volume as the bubble) for such bubbles and use (14.31)–(14.35). The parameters characterizing the gas are: S_p, the specific heat at constant pressure; γ, the specific heat at constant pressure divided by the specific heat at constant temperature; ρ_0, the mass density; and C_0, the thermal conductivity. Although propagation in the scattering model is treated in the fluid approximation, it is essential to include the effect of shear on the bubble resonance, even for soft sediments [Wilkens and Richardson 1998]. Losses are described by the dimensionless parameter, δ (not to be confused with the loss parameters δ_p and δ_t), which is made up of three terms:

$$\delta = \delta_r + \delta_h + \delta_s . \tag{14.36}$$

Radiation by the bubble is responsible for the part of damping given by

$$\delta_r = a\mathrm{Re}\left\{k_p\right\} , \tag{14.37}$$

damping due to loss of heat to the surrounding water is described by

$$\delta_h = B(\omega_0/\omega)^2 , \tag{14.38}$$

and shear losses contribute the term

$$\delta_s = \frac{4\mu''}{\rho\omega^2 a^2} , \tag{14.39}$$

where μ'' is the negative of the imaginary part of the sediment shear modulus. The bubble resonance and loss equations above are the versions given by [Lyons et al. 1996] who corrected some of the equations given by [Anderson and Hampton 1980a]. The parameter a is the static equilibrium bubble radius, so-called because the radius oscillates when the bubble is excited acoustically. The acoustic radian frequency is denoted ω, and P_0 is the ambient, hydrostatic pressure

$$P_0 = D\rho_w g . \tag{14.40}$$

Here, D is the water depth, ρ_w is the density of of the overlying seawater, and g is the acceleration of gravity. Pressure is evaluated at the sediment–water interface on the assumption that the bubbles of interest are near this interface. As before, ρ is the sediment mass density, and k_p is the compressional wavenumber in the sediment. The gas density is proportional to total pressure (atmospheric + hydrostatic), so

$$\rho_0 = (1 + \frac{P_0}{P_A})\rho_0|_{D=0} , \tag{14.41}$$

where atmospheric pressure, P_A, is approximately 101.3 kPa.

The expression for resonant frequency (14.32) is a transcendental equation, as the factor A depends on resonant frequency through the variable X. A simple iterative method of solution starts with $A = 1$, computes ω_0, substitutes this in (14.35) to find an improved value for A, substitutes this in (14.32), and repeats this process until convergence is achieved.

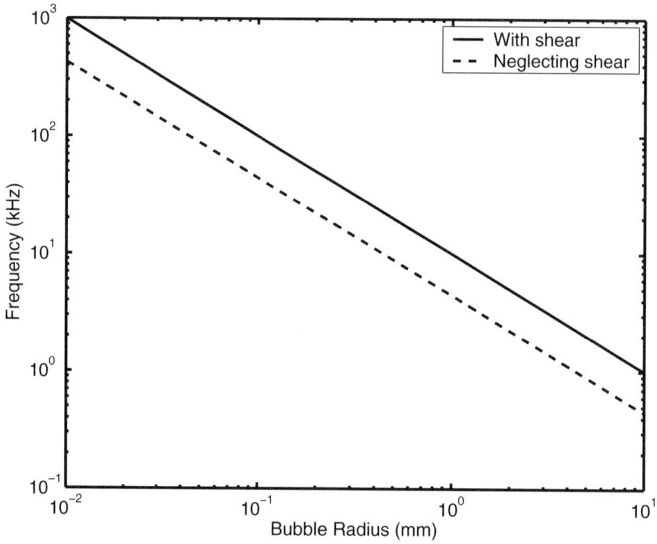

Fig. 14.6. Single-bubble resonant frequency as a function of bubble radius, a. The solid curve is computed using the solution of (14.32)–(14.35), and the shear-free case is shown as the dashed line. The hydrostatic pressure is 2.5 atmospheres (corresponding to a depth of about 25 m), and the gas and sediment parameters are given in Table 14.2.

To illustrate the predictions of the bubble resonance model, the parameters of Table 14.2 will be used. The sediment in this case is very soft, having low bulk density and low shear phase speed (28.4 m s^{-1}). Figure 14.6 shows the dependence of resonant frequency on bubble radius. It is interesting that shear increases the resonant frequency appreciably, when shear has little influence on roughness scattering in sands and muds (Sect. 13.2.2).

Table 14.2. Parameters (SI units) used in computations for Figs. 14.6–14.8. The gas parameters are appropriate to methane.

Parameter	Symbol	Value
Gas specific heat (const. press.)	S_p	2256 J (kg °C)$^{-1}$
Gas specific heat ratio	γ	1.31
Gas thermal conductivity	C_0	0.034 J (s m °C)$^{-1}$
Gas mass density (1 atm)	$\rho_0\|_{D=0}$	0.717 kg m^{-3}
Sediment mass density	ρ	1250 kg m^{-3}
Real part, sed. shear modulus	μ'	10^6 Pa
−Imag. part, sed. shear modulus	μ''	10^5 Pa
Water sound speed	c_w	1500 m s^{-1}
Sed. sound speed ratio	ν_p	0.991

As shown in Fig. 14.7, all three loss mechanisms contribute significantly to the loss factor, δ. The loss factor increases with frequency, but is much smaller than unity for frequencies less than 1 MHz .

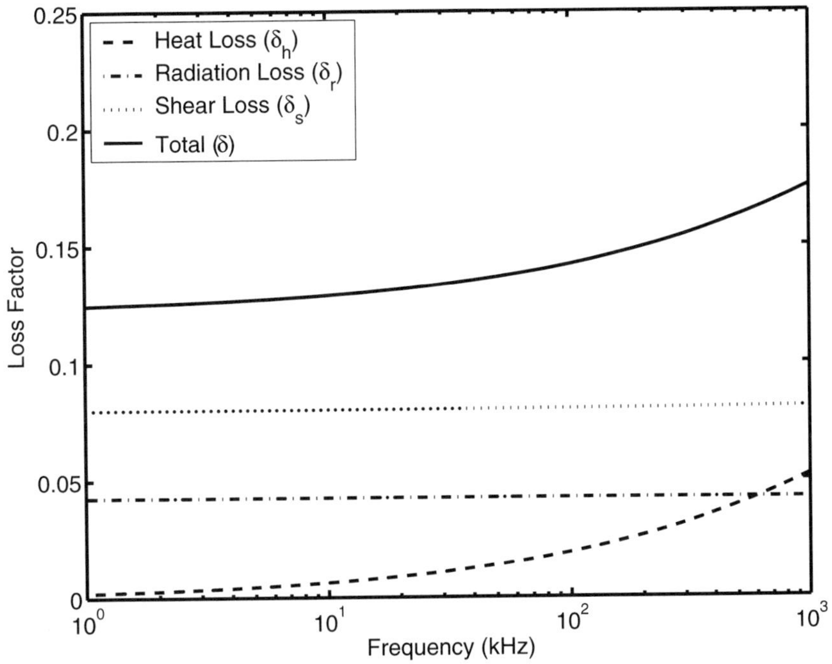

Fig. 14.7. Single-bubble loss factors as functions of resonant frequency. All parameters are as in Fig. 14.6.

Figure 14.8 shows the frequency dependence of the scattering cross section (expressed in dB as target strength) for a single methane bubble in sediment. The resonance is rather narrow, and the peak target strength is much larger than that for geometric backscattering, $20\log_{10}(a/2) = -66.0$ dB m^{-2}.

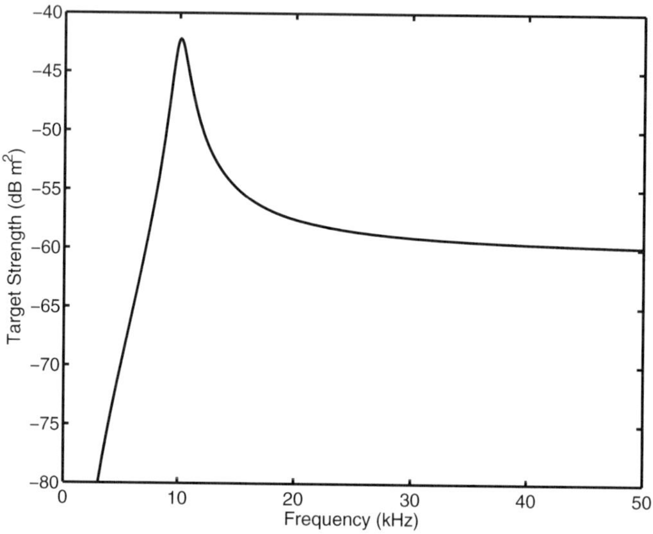

Fig. 14.8. Single-bubble target strength, 10 $\log_{10}$ $\sigma(a)$, for a bubble of 1-mm radius as a function of frequency, computed using (14.32)–(14.39). Parameters are as in Fig. 14.6.

A convenient approximation to the integral (14.30) was employed by [Boyle and Chotiros 1995a] following [Medwin 1977], who treats the expression for $\sigma(a)$ (14.31) as proportional to a delta function in a. This is justified if the resonance has a high quality factor (the quality factor is denoted Q, where $Q = 1/(2\delta)$), that is, if it is narrow in the frequency domain. The percentage width in a will be approximately equal to the percentage width in ω. This approximation reduces (14.30) to the form

$$\sigma_v = \frac{\pi a^3 N(a)}{2\delta} , \tag{14.42}$$

where the bubble radius is to be evaluated at the resonance value

$$a = \frac{1}{\omega_0}\sqrt{\frac{3\gamma P_0}{A\rho} + \frac{4\mu'}{\rho}} . \tag{14.43}$$

An example of bubble size distribution data is shown in Fig. 14.9 [Anderson et al. 1998]. These data were obtained at Eckernförde Bay in Germany by

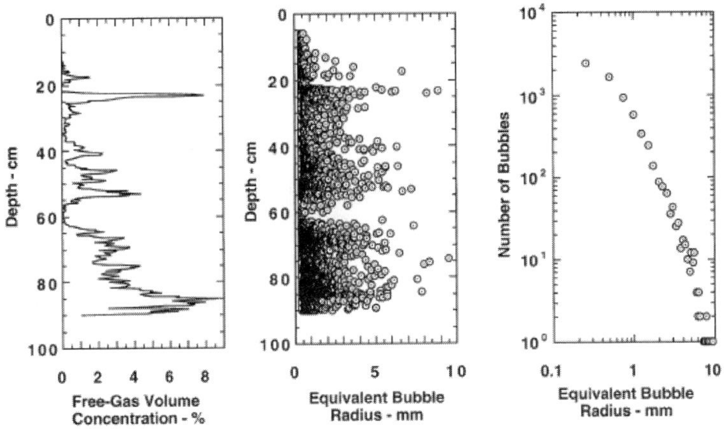

Fig. 14.9. Bubble size distribution data from Eckernförde Bay in Germany [Anderson et al. 1998].

means of CT scanning. The core volume was approximately 0.01 m^3, so the approximate size distribution, $N(a)$ (in mm^{-1} m^{-3}), can be obtained by dividing the data from the right-hand panel of Fig. 14.9 by this volume.

The bubble scattering model defined above has the same angular dependence as the empirical model (Fig. 14.2). The frequency dependence is governed by the bubble size distribution and is discussed by [Boyle and Chotiros 1995a]. They conclude that concentrations of gas of order 10^{-5} by volume could cause backscattering at observed levels. Unfortunately, for frequencies above 10 kHz, resonant bubble sizes are not resolved by present methods. For sandy sediments, [Boyle and Chotiros 1995b] have argued that the size distribution can be constrained by knowledge of the grain size distribution. Because sands are not as rich in organic material as finer sediments, methane is unlikely to form in concentrations sufficient to produce bubbles. Oxygen produced by photosynthesis in shallow water has been suggested as a possible source of surficial bubbles in sand [Greenlaw et al. 2004]. In a laboratory experiment covering the frequency range from 150 kHz to 1 MHz, [Holliday et al. 2004] showed that such bubbles did form and were important scatterers of sound, with the strongest scattering occurring at the lower frequencies.

Layering of sedimentary bubble populations is commonly observed and will affect both the frequency and angular dependence of scattering [Tang et al. 1994, Boyle and Chotiros 1995a]. Bubbles in sediments are usually nonspherical, particularly when sizes exceed about 1 mm (Sect. 3.3). Bubble scattering models that account, at least approximately, for nonsphericity can be found in [Lyons et al. 1996, Tang 1996b, Chu et al. 1997, Anderson et al. 1998]. These models also allow for depth dependence of the bubble size distribution, and have been tested against acoustic data from Eckernförde Bay in Germany. Bubble sizes were determined using X-ray computed tomography to a resolution of 1 mm (Sect. 3.3). Reasonable agreement between models and data was obtained, but strong conclusions cannot be drawn owing to assumptions needed to fill gaps in the characterization of the bubble population. The resonance model outlined above was used by [Lyons et al. 1996, Anderson et al. 1998], who treat multiple scattering. At a frequency above resonance (40 kHz), a geometric optics approximation has been used by [Tang 1996b, Chu et al. 1997]. These authors assume single scattering, but [Chu et al. 1997] note that multiple scattering effects must be significant.

Scattering by Buried Shell

While shell fragments are irregular in shape and have nonzero shear modulus, they are treated as fluid spheres in the work to be outlined here. The scattering cross section of a fluid sphere is discussed in [Morse and Ingard 1968, Ch. 8] and [Medwin and Clay 1998, Ch. 7]. For low frequencies such that the wavelength is much smaller than the sphere circumference, the scattering cross section increases as the fourth power of frequency. In this "Rayleigh" scattering regime, the backscattering cross section is [Morse and Ingard 1968, Ch. 8]

$$\sigma_r = (k_p')^4 a^6 \left[\frac{e-1}{3e} + \frac{(h-1)}{2h+1} \right]^2 . \tag{14.44}$$

In this expression, a is the sphere radius, k_p' is the real part of the compressional wavenumber in the sediment,

$$e = \frac{K_d}{K_b'} \tag{14.45}$$

is the specific modulus of the fluid sphere, and

$$h = \frac{\rho_g}{\rho} \tag{14.46}$$

is the specific density. In (14.45), K_d is the bulk modulus of the sphere, and K_b' is the real part of the bulk modulus of the sediment. Since shell pieces have substantial rigidity, there is ambiguity in the choice of K_d. In the following, K_d will be taken to be the modulus $\lambda + 2\mu$ (Ch. 9) that determines the compressional wave speed.

At higher frequencies, the scattering cross section approaches a frequency-independent value, the "geometric" cross section. For an impenetrable sphere, the geometric backscattering cross section is equal to $a^2/4$. This result follows from the knowledge that a large sphere scatters the acoustic power passing through its cross sectional area, πa^2, omnidirectionally over the full solid angle, 4π. The backscattering cross section for a fluid sphere has been approximated over the full frequency range by [Ivakin 2004] in terms of the expression

$$\sigma_d = \frac{\sigma_r}{1 + \sigma_r/(V_d^2 a^2/4)} . \tag{14.47}$$

Here,

$$V_d = \frac{\sqrt{eh} - 1}{\sqrt{eh} + 1} \tag{14.48}$$

is the reflection coefficient for a flat interface between the sediment and the fluid material of the sphere (Sect. 8.5). At high frequencies, the individual backscattering cross section (14.47) approaches $|V_d|^2 a^2/4$, which is the geometric cross section reduced by the squared magnitude of the normal-incidence reflection coefficient. Expression (14.47) is illustrated in Fig. 14.10. This smooth fit to the backscattering cross section of a fluid sphere is a reasonable approximation considering the uncertainties inherent in taking the particles to be fluid and spherical. The exact cross section for the fluid sphere [Medwin and Clay 1998, Ch. 7] exhibits oscillations as a function of $k_p'a$ for $k_p'a > 1$, but follows the Rayleigh curve at lower frequencies. There has been research on the issue of random scatterer shape [Thorne et al. 1995, Palmer 1996] and on the effects of elasticity [Stanton et al. 2000, Stanton 2000].

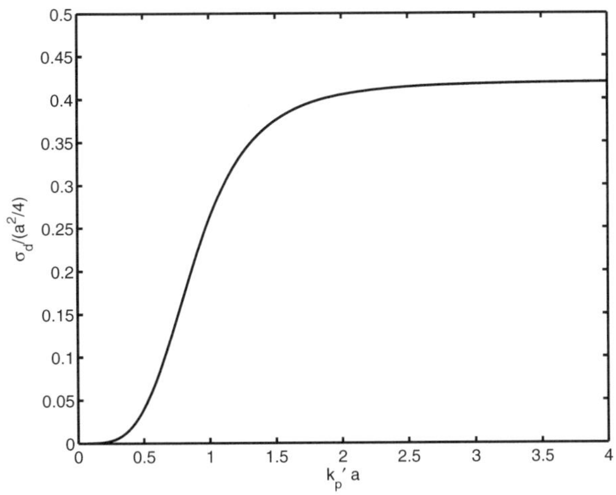

Fig. 14.10. Approximate backscattering cross section (14.47) for a fluid sphere, having specific modulus $e = 15.7$ and specific density $h = 1.4$. The cross section is normalized by division by $a^2/4$, the scattering cross section for a large, impenetrable sphere. The abscissa is the product of wavenumber and radius, and should be thought of as a dimensionless frequency variable.

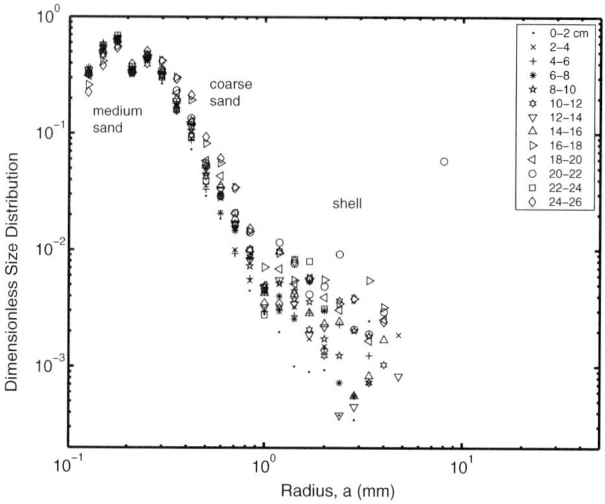

Fig. 14.11. Size distributions for shell fragments at various depths at the SAX99 site as provided by Briggs and converted to a dimensionless distribution in terms of an equivalent spherical radius, a, in [Ivakin 2005].

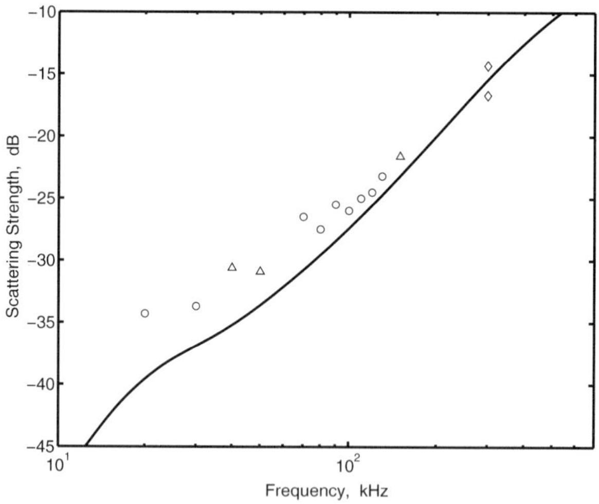

Fig. 14.12. Comparison of backscattering strength prediction of the discrete scattering model of [Ivakin 2005] with data from [Williams et al. 2002b] for a grazing angle of 35°.

The single scattering approximation of (14.6) and (14.30) has been applied to scattering by buried shell fragments at the SAX99 site in [Ivakin 2004, Ivakin 2005]. The shell size distribution at various depths is shown in Fig. 14.11. The distribution in this figure is dimensionless and can be converted to $N(a)$ by multiplication by $3/(4\pi a^4)$. The radius, a, is the equivalent spherical radius correponding to the particle volume. The predicted seafloor scattering strength for a grazing angle of 35° is compared with backscattering data in Fig. 14.12. The model curve lies below the data at the lower frequencies, a desirable result, since roughness scattering (Sect. 13.2.4) has been shown to account for this portion of the data. When the roughness scattering model and the present discrete volume scattering model are combined, agreement with data is reasonable over the entire frequency range [Ivakin 2005].

The single scattering, fluid-sphere approximation was applied to a site near Panama City, Florida by [Lyons 2005]. The sediment contained numerous buried shell pieces, and the size distribution was determined using CT. The size distribution is shown in Fig. 7.8 where the radius is chosen so as to yield sphere volumes approximately equal to the individual particle volumes. The smooth curve is a fit to the size distribution used in computing backscattering strength. The units of this distribution are mixed: it is the number of particles per unit radius (mm) per unit volume (m^3). Using this distribution in (14.30) with the individual cross section given by the Rayleigh expression (14.44), Lyons found that volume scattering would not compete with roughness scattering at the Panama City site at 20 kHz. At 100 kHz, however, and for grazing angles above the critical angle, volume scattering is predicted to be comparable to roughness scattering. In these calculations, the specific modulus and density of Fig. 14.10 were used. The accuracy of the single scattering approximation in this case was verified by [Lyons 2005] through use of a multiple scattering approximation. He concluded that multiple scattering effects would become important for shell concentrations larger than 1% by volume; this is greater than the concentration at the Panama City site.

14.2 Elastic Sediment Volume Scattering

Acoustic scattering due to volume heterogeneity in rocky seafloors has received little attention, yet is potentially significant. The small-perturbation method has been applied to scattering by continuous heterogeneity in elastic seafloors, with rather surprising but untested results (Sect. 14.2.3).

14.2.1 Small Perturbation Elastic Approximation

The small-perturbation method for volume scattering in elastic seafloors has been developed by [Ivakin 1990]. The presentation here follows that given in [D. Jackson and Ivakin 1998, Ivakin and Jackson 1998] who give examples for sand and rock seafloors. When an elastic model is used for the seafloor, one must consider heterogeneity in shear wave speed as well as in the density and compressional wave speed. Scattering can convert compressional wave energy into shear wave energy and vice versa. As illustrated by Fig. 14.13, there are several such scattering mechanisms. For example, one mechanism is scattering of compressional waves by density fluctuations, with conversion to shear waves. These shear waves can be converted to waterborne compressional waves on encountering the interface. If the shear waves are polarized such that the particle motion is in a vertical plane, conversion is possible even for the flat interface. As a result, this scattering mechanism is evident in a first-order perturbation treatment. Scattering to the other possible polarization (with particle motion in a horizontal plane) will only be evident at higher order, as such waves suffer total internal reflection at a horizontal, flat interface.

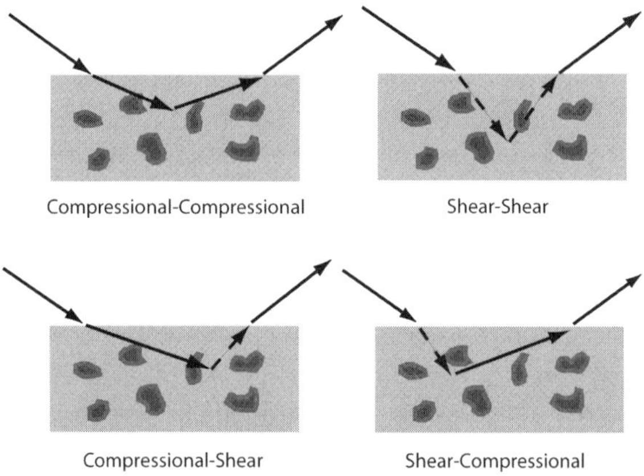

Fig. 14.13. Volume scattering mechanisms contributing in first order for an inhomogeneous, elastic seabed. In addition to the four conversion processes indicated, there are three types of heterogeneity to consider: density, compressional wave speed, and shear wave speed fluctuations.

The equivalent interface scattering cross section due to volume heterogeneity is derived in Sect. M.2 and is of the form

$$\sigma = -\frac{\pi k_w^4 a_\rho^2}{2} \mathrm{Im} \left(\sum_{\eta,\alpha,\eta',\beta} d_{\eta\alpha} d^*_{\eta'\beta} \frac{W_{\alpha\beta}\left((\mathbf{\Delta k}_\eta + \mathbf{\Delta k}^*_{\eta'})/2\right)}{(\Delta k_{\eta z} - \Delta k^*_{\eta' z})} \right) , \tag{14.49}$$

where $W_{\alpha\beta}$ is a matrix of cross-spectra for volume heterogeneities analogous to (14.25). The elastic case is discussed briefly in Sect. 14.2.2. The α and β sums run over the three types of heterogeneity: density, compressional wave speed, and shear wave speed ($\alpha, \beta = \rho, p, t$). Thus, following [D. Jackson and Ivakin 1998], spectra for wave speed heterogeneity are employed as opposed to the choice of spectra for modulus heterogeneity in the fluid development above. As noted earlier, the choice is arbitrary and is made primarily for convenience. The fluctuations in density and wave speeds may be correlated, thus the $\alpha \neq \beta$ cross-terms may be important. Note that the arguments of the cross-spectra are complex; the Fourier transform relation between the spectra and the corresponding (real) covariances defines the continuation of the spectra into the complex domain. The sum in (14.49) is purely imaginary, so the operator, Im, for the imaginary part is only needed for numerical reasons. The η and η' sums run over the four types of wave conversion caused by volume scattering: (1) compressional to compressional ($\eta = pp$), (2) shear to compressional ($\eta = pt$), (3) compressional to shear ($\eta = tp$), and (4) shear to shear ($\eta = tt$). Symbolically,

$$\eta = qq' ,$$

where q and q' run over the two types of waves ($q, q' = p, t$). As noted earlier, volume heterogeneities cause conversion to both shear polarizations, but upgoing shear waves with horizontal particle motion suffer total reflection at the interface and do not contribute to the scattering cross section in first order. The following wave vector differences appear in (14.49):

$$\mathbf{\Delta k}_\eta = \mathbf{k}_q^+(\mathbf{K}_s) - \mathbf{k}_{q'}^-(\mathbf{K}_i) \ . \tag{14.50}$$

The upgoing and downgoing wave vectors, $\mathbf{k}_q^+(\mathbf{K})$ and $\mathbf{k}_q^-(\mathbf{K})$, are defined in (9.17). Expression (14.50) gives the changes in wave vector for the corresponding conversions, and these wave vector differences appear as arguments in the cross-spectra in (14.49). The coefficients $d_{\eta\alpha}$ are

$$d_{\eta\alpha} = w_\eta D_{\eta\alpha} \ , \tag{14.51}$$

where

$$w_{qq'} = V_{wq}(\mathbf{K}_s) V_{wq'}(\mathbf{K}_i) a_q^{-1} a_{q'}^{-1}, \ q, q' = p, t \ . \tag{14.52}$$

The transmission coefficients, V_{wq}, are defined in Ch. 9, where angular arguments are used. The use of transverse wave vector arguments instead is straightforward. The coefficients $D_{\eta\alpha}$ are elements of the 3-column matrix

$$D = (D_\rho | D_p | D_t) \ , \tag{14.53}$$

where $D_\rho = D'_\rho + D_t/2$ and

$$D'_\rho = \begin{pmatrix} 1 - b_{pp} \\ b_{pv} \\ -b_{vp} \\ b_{vv} \end{pmatrix} , \tag{14.54}$$

$$D_p = \begin{pmatrix} 2 \\ 0 \\ 0 \\ 0 \end{pmatrix} , \tag{14.55}$$

$$D_t = 2 \begin{pmatrix} 2g^2 b_{pp}^2 - 2g^2 \\ -2g b_{pt} b_{pv} \\ 2g b_{tp} b_{vp} \\ -b_{vt} b_{tv} - b_{vv} b_{tt} \end{pmatrix}, \tag{14.56}$$

with $g = a_t/a_p = c_t/c_p$ and

$$b_{qq'} = \mathbf{e}_q^+(\mathbf{K}_s) \cdot \mathbf{e}_{q'}^-(\mathbf{K}_i), \tag{14.57}$$

where $q, q' = p, t, v$ and the unit vectors $\mathbf{e}_q^{\pm}(\mathbf{K})$ are defined in Ch. 9. The subscripts p and t denote unit vectors in the direction of propagation of compressional and shear waves, while the subscript v denotes the polarization vector of the vertically polarized shear wave.

14.2.2 Statistical Description of Elastic Heterogeneity

The statistics of volume heterogeneities in the elastic case appear in (14.49) as a matrix of spatial (3-dimensional) cross-spectra, $W_{\alpha\beta}(\mathbf{k})$. These are Fourier transforms of cross-correlation functions for the normalized fluctuations $\gamma_\rho = \delta\rho/ < \rho >, \gamma_p = \delta c_p/ < c_p >, \gamma_t = \delta c_t/ < c_t >$. These spectra provide a sufficient description of heterogeneity in the small-perturbation approximation. The form defined for the fluid case (7.14) is useful here, with the indices expanded to include shear $(\alpha, \beta = \rho, p, t)$. As this form is analytic over the domain of interest, it automatically gives the analytic continuation required to define the spectra for complex arguments.

14.2.3 Numerical Examples

Several numerical examples are provided in [Ivakin and Jackson 1998], where it is noted that shear has a negligible effect on volume scattering in sand but is important in rock. Figure 14.14 compares backscattering due to volume heterogeneity and interface roughness for parameters appropriate to rock (see Sect. 5.6 and [Hamilton 1978, Hamilton 1979b]). For lack of information regarding heterogeneity, the cross-spectra have been set to zero and the spectra for the three fluctuation types have been given equal strength. It is noteworthy that appreciable volume scattering is predicted in spite of the relatively low value assigned the spectral strengths, $w_{3\alpha\beta}$. Also, one might have supposed that a seafloor composed of rock would not be acoustically penetrable, but this is not the case, as noted in Ch. 9. Penetration is large near the shear critical angle (39.7° in this case), where volume scattering increases, roughness scattering decreases, and reflection loss increases, owing to proximity to the Rayleigh angle (Sect. 13.2.2).

A comparison of bistatic scattering due to heterogeneity and roughness is given in Fig. 14.15, for the same conditions as in Fig. 14.14. Surprisingly, volume scattering is predicted to dominate roughness scattering near the specular direction.

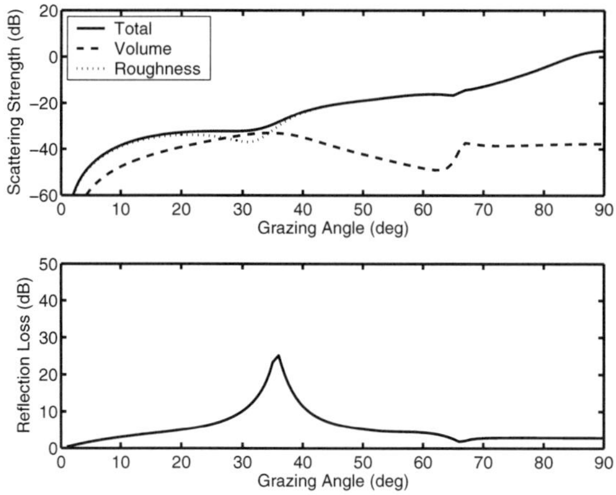

Fig. 14.14. Upper: Backscattering due to volume fluctuations compared to interface roughness scattering. Lower: Flat-interface reflection loss. The model parameters are appropriate to rock and are: $a_\rho = 2.5$, $\nu_p = 2.5$, $\delta_p = 0.02$, $\nu_t = 1.3$, $\delta_t = 0.1$, $w_2 = 0.000141 \text{ m}^{0.75}$, $\gamma_2 = 3.25$, $w_{3\rho\rho} = w_{3pp} = w_{3tt} = 6 \times 10^{-6}$, $\gamma_3 = 3.0$. All cross-spectra are set to zero. The frequency is 30 kHz.

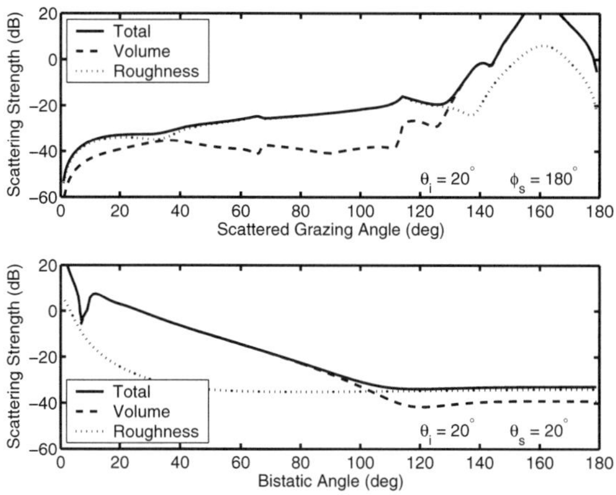

Fig. 14.15. Upper: Bistatic scattering due to volume fluctuations for rock compared to interface roughness scattering. The incident grazing angle is 20° and the bistatic angle is fixed at 180° so that the x-axis, the scattered grazing angle, covers all directions in the plane of incidence, including the backward (20°) and specular (160°) directions. Lower: Bistatic scattering as a function of bistatic angle, with incident and scattered grazing angles set to 20°. The specular direction corresponds to a bistatic angle of 0° and the backscatter direction to 180°. All other parameters are as in Fig. 14.14.

In summary, perturbation theory predicts strong and interesting effects in volume scattering in rock seafloors. Laboratory experiments may be the best means of testing the predictions of theory, as characterization of the heterogeneity of rock in situ is a difficult undertaking.

14.3 Research Issues

While considerable effort has been devoted to the development and testing of volume scattering models based on perturbation theory, the experimental tests of these models have not been particularly convincing owing to uncertainties in the characterization of heterogeneity. These models assume that heterogeneity is continuous spatially and do not incorporate discrete scatterers such as shell fragments. While models for discrete scatterers may not be difficult to formulate, experimental tests will be challenging, again because of difficulties in characterization. Further work is also needed in the "gray area" where scattering cannot be clearly divided into interface and volume components. Examples include worm tubes, shell deposits on the interface, and thin layers having rough boundaries.

15 Acoustic Penetration of the Seafloor

While scattering of sound by the seafloor into the water column is a central problem in high-frequency ocean acoustics, acoustic penetration into the sediment is a subject of increasing importance. Acoustic penetration makes possible sonar detection and identification of buried mines, pipelines, cables, and archeological objects. One might expect the penetrating field strength to be very low for "fast" sediments, whose compressional wave speed exceeds the water sound speed. For these sediments, compressional waves tend to be refracted upward and out of the sediment, with small penetration depths compared to "slow" sediments such as silt and clay. Both theoretically [Jensen and Schmidt 1987], and in controlled measurements [Williams et al. 1989, Altenberg et al. 1991], the penetrating field has been found to agree reasonably well with expectations based on fluid sediment theory. That is, penetration is substantial if the incident grazing angle is larger than the critical angle and is slight otherwise, being due to the evanescent wave (Sect. 8.6). Earlier literature (e.g., [Muir et al. 1979]) had noted apparent strong subcritical penetration that was later explained as due to use of a parametric source having narrow beamwidth [Wingham et al. 1986, Jensen and Schmidt 1987].

From these early results, there developed an expectation that it would be difficult for a high-frequency sonar to detect objects buried in sand, for which the critical grazing angle is usually about 30°. This assumes the sonar is attempting detection at ranges much greater than the water depth. In such cases, the grazing angle is smaller than the critical angle, and subcritical penetration would be required for detection of buried targets. It was assumed that the only practical means of detecting targets buried in sand would involve the evanescent wave (Sects. 8.2 and 8.6), whose penetration depth is on the order of the acoustic wavelength. This would require operation at frequencies below 10 kHz. Contrary to these expectations, it was found that targets buried in sand (e.g., [Chotiros and Boyle 1992]) were often detected when they supposedly should have been invisible. A recent example of subcritical detection was reported by [Piper et al. 2002] and illustrated in Fig. 15.1. In addition to unexpected target detections, anomalous subcritical penetration was seen in measurements on sandy seafloors (e.g., [Boyle and Chotiros 1992, Lopes 1996]).

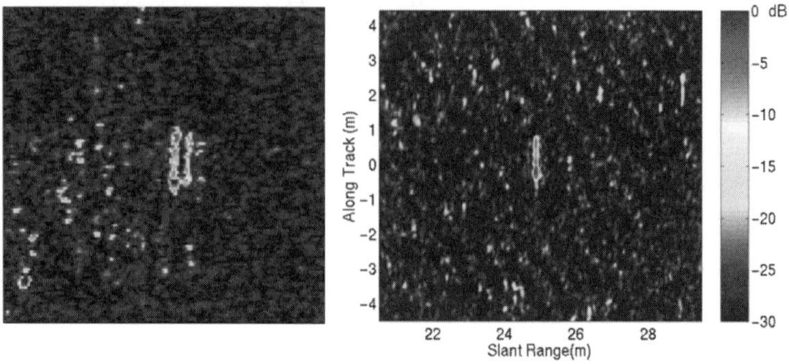

Fig. 15.1. Images of cylindrical targets buried in a sandy sediment. These images were obtained using a synthetic-aperture sonar operating at 20 kHz [Piper et al. 2002]. In the left image the top of the target is 50 cm below the interface, and the grazing angle is 3.3°–4.3°. In the right image the top of the target is 15 cm below the interface, and the grazing angle is 8°–10°. A ripple field having mean wavelength of about 70 cm was present at the site. (see second color insert)

As indicated in Fig. 15.2, various mechanisms have been proposed to explain anomalous penetration. The two that have received the most attention are refraction of a Biot slow wave (which has no critical angle and always refracts downward), and scattering by interface roughness. The Biot slow wave mechanism [Chotiros 1995a] can provide significant penetrating energy if the Biot parameters of the medium depart considerably from those employed by most investigators, yielding a slow wave speed in the neighborhood of 1000 m s^{-1}, considerably faster than the expectations of most investigators. This possibility has been the subject of some debate [Stoll 1998, Chotiros 1998] and of experimental investigation, discussed below. Rough-interface models have been used to show that subcritical data can be explained by interface scattering [Moe 1996, Thorsos et al. 2000a, Pouliquen et al. 2000a, Maguer et al. 2000a, D. Jackson et al. 2002]. In particular, [Thorsos et al. 2000a] show that downward scattering by interface roughness can give rise to incoherent energy fronts that could be misinterpreted as waves traveling at speeds in the neighborhood of 1000 m s^{-1}.

An experiment (SAX99) was designed to differentiate between the candidate subcritical penetration mechanisms [Thorsos et al. 2001a, Thorsos et al. 2001b]. This and other experiments in the field [Maguer et al. 2000a, Simpson et al. 2003] and in the laboratory [Mellema99,Simpson and Houston 2000, Lim et al. 2001] have given strong evidence that subcritical penetration at high frequencies is due to interface scattering. Also, these experiments failed to show the presence of a "fast" Biot slow wave, that is, one having speed near 1000 m s^{-1}.

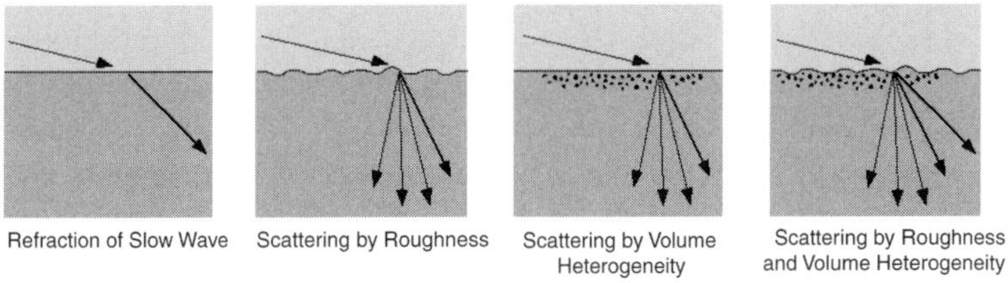

Fig. 15.2. Possible mechanisms for anomalous subcritical penetration.

The various aspects of acoustic penetration are illustrated by the data of [Maguer et al. 2000b] shown in Fig. 15.3. These data were obtained using the apparatus of Fig. 1.2, which allowed coverage of both a wide range of grazing angles and acoustic frequencies. In the figure, the largest grazing angle (31.5°) is slightly greater than the critical angle, and the penetrating field is large at all frequencies. For the lower grazing angle (16°, well below the critical angle), the penetrating field at the lowest frequencies is due to evanescent penetration, for which the penetrating field strength drops rapidly as frequency increases. Eventually, however, this rapid decline ceases, and pressure levels in the sediment are much greater than flat-interface, fluid theory would predict. This anomalous penetration was ascribed to scattering by seafloor roughness, which was substantial at this site (Fig. 15.4).

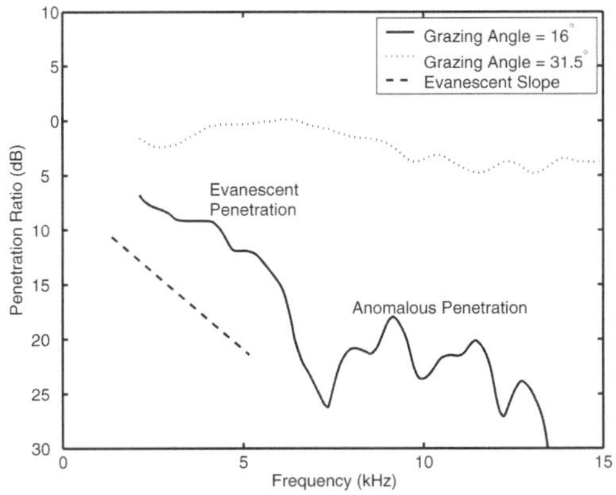

Fig. 15.3. Measured penetrating field strength at a depth of 25 cm, figure adapted from [Maguer et al. 2000b]. The "Penetration Ratio" is the ratio of the pressure magnitude at 25-cm depth and the incident pressure magnitude at the interface. The straight, dashed line labeled "evanescent slope" gives the rate of falloff with increasing frequency for evanescent penetration. The vertical position of this line is arbitrary.

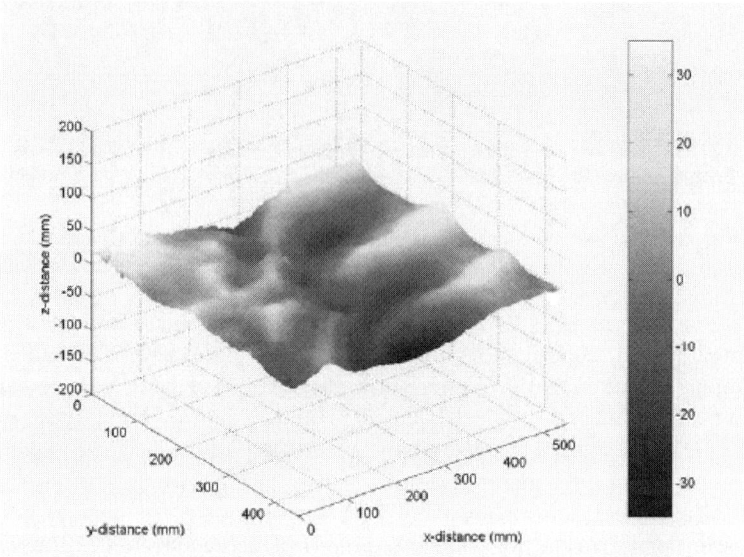

Fig. 15.4. Interface relief as measured by digital stereophotography in the experiment of [Maguer et al. 2000b].

Models for acoustic penetration due to seafloor roughness can be divided into two basic types: those that provide individual realizations of the acoustic field in the seafloor (given a realization of the rough seafloor) and those that provide formal averages (Sect. 2.3.2) of the mean-square acoustic field. The models providing individual realizations are suited to Monte Carlo simulations and have been implemented in both the small-roughness perturbation [Maguer et al. 2000a, Thorsos et al. 2000a] and Kirchhoff [Pouliquen et al. 2000a, Maguer et al. 2000a] approximations. This monograph will primarily treat the formally averaged approach, as it is most similar to the bulk of the scattering models already presented. As noted in Sect. L.1, the formally averaged Kirchhoff approximation employs an additional assumption not needed in the single-realization approach. In the strict Kirchhoff approximation, the reflection coefficient (or transmission coefficient if penetration is of interest) varies with position owing to the varying local slope of the interface. In the formally averaged approach, it is necessary to use a constant value, chosen somewhat arbitrarily. This is not a serious problem if the incident grazing angle is greater than the critical angle, but this domain is of little interest in the penetration problem. As a consequence, the following discussion is limited to the perturbation model.

15.1 Roughness Perturbation Model

Moe [Moe 1996, Moe and Jackson 1998] developed a first-order, formally averaged, perturbative model for the acoustic penetration problem. This model yields predictions for the mean-square pressure in the sediment and has been compared to experimental data by [Thorsos et al. 2000a], who also examine a first-order single-realization model.

The starting point for the perturbation model is the T-matrix giving the spectrum of plane waves scattered into the seafloor due to interface roughness. Appendix K defines the perturbation series for arbitrary orders, but here, only the first two terms are needed. In perturbation theory, the order of a term determines its dependence on the roughness amplitude. Thus, the zeroth-order term is independent of roughness, the first-order term increases linearly with roughness amplitude, and so on. It will be assumed that a unit-amplitude, single-frequency, plane wave is incident on the seafloor.

The zeroth-order term is the solution to the flat-interface problem, and can be written without reference to the T-matrix. From Sect. 8.5, noting that the pressure transmission coefficient is equal to one plus the reflection coefficient,

$$P_0(\mathbf{R},\ z) = [1 + V_{ww}(\mathbf{K}_i)]e^{i\mathbf{K}_i\cdot\mathbf{R} - ik_p\beta_p(K_i)z} \ , \tag{15.1}$$

where V_{ww} is the flat-interface reflection coefficient, $\mathbf{K}_i$ is the 2D vector composed of the horizontal components of the incident wave vector, and $\beta_p(K_i)$ is the complex sine of the angle in the sediment, as given by Snell's law. The pressure field expressed by (15.1) will be evanescent for subcritical grazing angles but will be large for grazing angles greater than the critical angle. Thus, this term describes both evanescent and refractive acoustic penetration.

The first-order term for the scattered pressure in the sediment at $\mathbf{r} = [\mathbf{R},\ z],\ \ z < 0$, is

$$P_s(\mathbf{R},\ z) = \int T_{wp}^{(1)}(\mathbf{K}_s,\ \mathbf{K}_i)e^{i\ \mathbf{K}_s\cdot\mathbf{R} - ik_p\beta_p(K_s)z}d^2K_s \ , \tag{15.2}$$

where $T_{wp}^{(1)}(\mathbf{K}_s,\ \mathbf{K}_i)$ is the first-order T-matrix given by (K.45). It can be put in the form

$$T_{wp}^{(1)}(\mathbf{K}_s,\ \mathbf{K}_i) = \frac{ik_w}{\beta_w(K_s)}A_{wp}(\mathbf{K}_s,\ \mathbf{K}_i)F(\mathbf{K}_s - \mathbf{K}_i) \ , \tag{15.3}$$

with

$$A_{wp}(\mathbf{K}_s,\ \mathbf{K}_i) = \frac{1}{2}[1 + V_{ww}(\mathbf{K}_i)][1 + V_{ww}(\mathbf{K}_s)] \times$$
$$\left\{(1 - 1/a_\rho)\left[\frac{\mathbf{K}_s\cdot\mathbf{K}_i}{k_w^2} + \frac{\beta_p(K_i)\beta_w(K_s)}{a_p}\right] - 1 + \frac{1}{a_p^2 a_\rho}\right\} \ . \tag{15.4}$$

In the expressions above, k_w is the wavenumber in water, a_ρ is the sediment/water density ratio, and a_p is the complex sound-speed ratio. The development to this point has not introduced formal averaging. Before doing so, a deterministic problem for which averaging is unnecessary will be considered: sinusoidal ripple.

15.1.1 Sinusoidal Ripple

It is instructive to begin with the simplest relevant scenario, diffraction of a unit-amplitude incident plane wave by a sinusoidal interface. In first-order perturbation theory (Sect. 13.2), scattering by sinusoidal ripple having transverse (horizontal) wave vector $\mathbf{K}_0$ results in a change in the transverse part of the three-dimensional acoustic wave vector from the incident value $\mathbf{K}_i$ to either of two values $\mathbf{K}_s = \mathbf{K}_i \pm \mathbf{K}_0$. The three-dimensional wave vector in the sediment has a vertical component $-k_p\beta_p(K_s)$. If this vertical component has a real part much larger in magnitude than its imaginary part, the scattered wave will not be evanescent and will propagate in the sediment. This is an example of subcritical penetration. The grazing angle of the downward-propagating acoustic wave can be most simply obtained in the case for which loss is neglected, and the incident acoustic energy propagates in a direction normal to the strike of the ripple. In this case, the problem becomes two-dimensional as illustrated by Fig. 15.5. The grazing angle, θ_{ps}, of the downward-propagating acoustic wave has a cosine equal to the ratio, $K_s/k_p|_{\delta=0}$, giving

$$\cos\theta_{ps} = \nu_p(\cos\theta_{wi} - \frac{\lambda_w}{\lambda_0}) \ . \tag{15.5}$$

In (15.5), θ_{wi}, ν_p, λ_w, and λ_0 are, respectively, the incident grazing angle, sediment/water sound-speed ratio, acoustic wavelength in water, and ripple wavelength. Note that (15.5) yields a real angle only if the condition

$$\lambda_w > \lambda_0(\cos\theta_{wi} - 1/\nu_p) \tag{15.6}$$

is satisfied. This is a high-frequency cutoff condition: penetration will be slight for frequencies greater than the value for which the acoustic wavelength in water is equal to $\lambda_0(\cos\theta_{wi} - 1/\nu_p)$.

The ripple relief function will be taken to be of the form

$$f(\mathbf{R}) = \sqrt{2}h_0\cos(\mathbf{K}_0\cdot\mathbf{R} + \theta_0) \ , \tag{15.7}$$

where $\mathbf{R} = x\mathbf{e}_x + y\mathbf{e}_y$ and

$$\mathbf{K}_0 = \frac{2\pi}{\lambda_0}(\cos\phi_0\mathbf{e}_x + \sin\phi_0\mathbf{e}_y \ ). \tag{15.8}$$

In (15.7), h_0 is the RMS ripple amplitude, ϕ_0 is the ripple "propagation" angle relative to the x-direction, and θ_0 is the phase of the ripple relative to ripple having a peak at the origin. The Fourier transform of the ripple relief

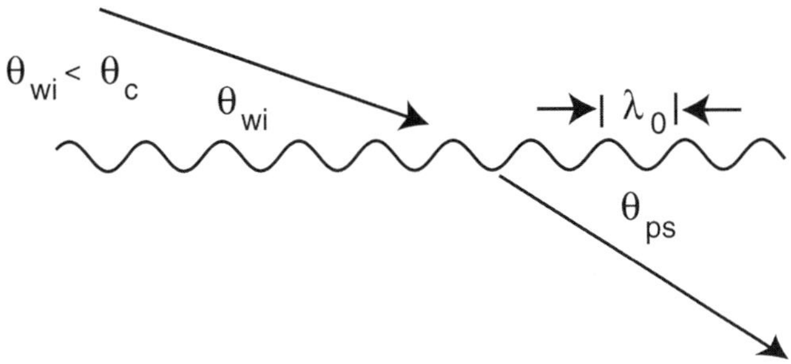

Fig. 15.5. Illustration of downward diffraction by sinusoidal ripple when the propagation direction of the incident acoustic field is normal to the strike (line parallel to crests) of the ripple.

function (15.7) consists of two delta functions with the result that (15.2)–(15.4) yield

$$P_-(\mathbf{R},\ z) = \frac{k_w h_0}{\sqrt{2}\beta_w(K_s)} A_{wp}(\mathbf{K}_s,\ \mathbf{K}_i) e^{-i\theta_0 + i\mathbf{K}_s\cdot\mathbf{R} - ik_2\beta_p(K_s)z}|_{\mathbf{K}_s=\mathbf{K}_i-\mathbf{K}_0} \ . \tag{15.9}$$

This is one of two diffracted waves, this one having transverse wave vector equal to $\mathbf{K}_s = \mathbf{K}_i - \mathbf{K}_0$. There will also be a diffracted contribution, $P_+(\mathbf{R},\ z)$, having transverse wave vector given by $\mathbf{K}_s = \mathbf{K}_i + \mathbf{K}_0$. This contribution can be found by substituting $\mathbf{K}_s = \mathbf{K}_i + \mathbf{K}_0$ in the equations given above, but will generally be small except near the interface, as this wave is evanescent under the conditions of interest here. The total penetrating field up to first order in perturbation theory is

$$P(\mathbf{R},\ z) = P_0(\mathbf{R},\ z) + P_-(\mathbf{R},\ z) + P_+(\mathbf{R},\ z) \ , \tag{15.10}$$

where $P_0(\mathbf{R},\ z)$ is given by (15.1).

Although both the zeroth-order and first-order fields have magnitudes that are constant with respect to the horizontal coordinates, interference between these fields gives rise to the position dependence of penetrating field magnitude seen in Fig. 15.6. In this figure, field magnitude is expressed as "loss," defined as $-20\log_{10}|P(\mathbf{R},\ z)|$, with a unit-amplitude incident plane wave.

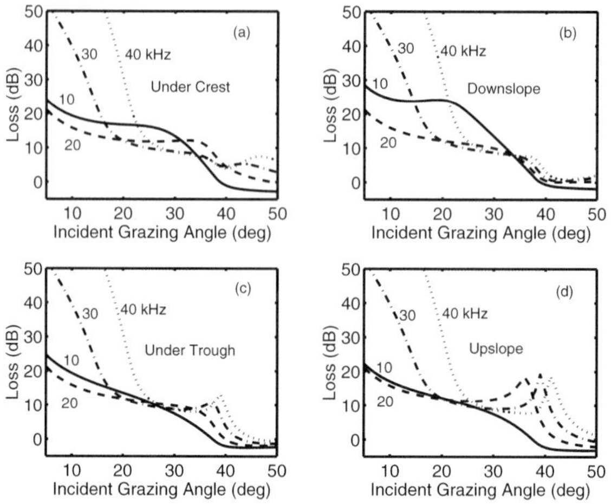

Fig. 15.6. Signal loss due to penetration computed using simple diffraction theory with ripple parameters $h_0 = 0.01$ m, $\lambda_0 = 0.48$ m. The one-way loss is computed at a depth of 17 cm with the incident wave propagating normal to the ripple crests. Loss is position dependent due to interference between the zeroth- and first-order terms. The receiver positions assumed are, moving in increments of one-quarter ripple wavelength (12 cm) away from source: (a) under a ripple crest, (b) under the maximum downslope, (c) under a trough, (d) under the maximum upslope. From [D. Jackson et al. 2002].

15.1.2 Formally Averaged Perturbation Model

Acoustic penetration due to random interface roughness has been treated using formal averaging (Sect. 2.3.2) by Moe [Moe 1996, Moe and Jackson 1998]. A bistatic scattering cross section for scattering *into* the seafloor is defined in analogy with (K.9):

$$\sigma_{wp}(\mathbf{K}_s, \mathbf{K}_i) = \frac{k_w^4}{\nu_p^2} |A_{wp}(\mathbf{K}_s, \mathbf{K}_i)|^2 W(\mathbf{K}_s - \mathbf{K}_i) \ . \tag{15.11}$$

With reference to Fig. 15.7, the incident and scattered horizontal wave vectors are evaluated in terms of the source–receiver geometry as follows:

$$\mathbf{K}_i = k_w(\mathbf{R} - \mathbf{R}_i)/|\mathbf{r} - \mathbf{r}_i| \tag{15.12}$$

and

$$\mathbf{K}_s = k_w(\mathbf{R}_s - \mathbf{R})/|\mathbf{r}_s - \mathbf{r}| \ , \tag{15.13}$$

with $\mathbf{r}_i = (\mathbf{R}_i, \ z_i)$ and $\mathbf{r}_s = (\mathbf{R}_s, \ z_s)$.

In direct analogy with the situation for bistatic scattering into the water column, (J.15), the mean-square scattered penetrating pressure is computed as follows:

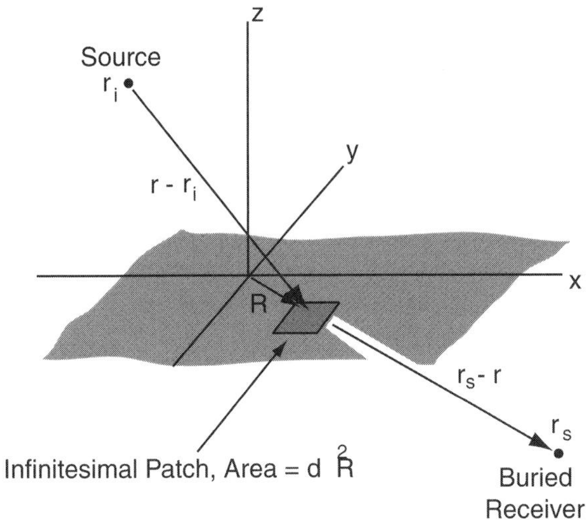

Fig. 15.7. Geometry for integration over mean scattering interface to obtain the mean-square scattered pressure in the sediment. The vector $\mathbf{r} = [\mathbf{R},\ 0]$ lies on the mean plane ($z = 0$) and *not* on the rough interface. The same is true of the area element, d^2R.

$$< |P_s(\mathbf{r}_s)|^2 >= \int_{z=0} \sigma_{wp}(\mathbf{K}_s, \mathbf{K}_i)|P_i(\mathbf{r})|^2|\mathbf{r}_s - \mathbf{r}|^{-2}e^{-2k_p''|\mathbf{r}_s-\mathbf{r}|}d^2R\ . \quad (15.14)$$

This integration assumes that the far-field criteria defined in Sect. J.2 are satisfied by the penetrating field. These criteria require the interface to be in the far field of the source, the mean-square scattered field to have a rather broad angular dependence, and the receiver to be at least a few wavelengths from the interface.

While (15.14) applies to single frequency (continuous wave) transmissions, it can be adapted to pulsed-wave situations [Moe and Jackson 1998]. Assuming the source is omnidirectional, (15.14) is replaced by

$$< |P_s(\mathbf{r}_s,\ t)|^2 >= s_0^2 \int_{z=0} \frac{\sigma_{wp}(\mathbf{K}_s, \mathbf{K}_i)|u(t - t_d)|^2 e^{-2k_p''|\mathbf{r}_s-\mathbf{r}|}}{|\mathbf{r} - \mathbf{r}_i|^2|\mathbf{r}_s - \mathbf{r}|^2} d^2R\ . \quad (15.15)$$

In this expression, $u(t)$ is the complex envelope of the source waveform, having peak magnitude unity. The source level corresponding to this peak is given in terms of the parameter s_0 by (F.7). The squared transmit envelope is evaluated with time delay, t_d, which depends on the integration coordinates, $\mathbf{R}$:

$$t_d = \frac{1}{c_w}(|\mathbf{r} - \mathbf{r}_i| + |\mathbf{r}_s - \mathbf{r}|/\nu_p)\ . \quad (15.16)$$

The sediment attenuation factor is evaluated using the imaginary part of the compressional wavenumber, k_p'', at the center frequency. The numerical implementation of (15.15) is straightforward. The integration domain is gridded, and the time delay, t_d, is evaluated for each grid point. For each grid point, the squared transmit envelope is weighted, delayed, and added to the time series for the mean-square pressure. An example of the application of (15.15) is given in Fig. 15.8 for several depths in the sediment. In this example, the incident grazing angle is about 11°, significantly smaller than the critical angle (about 30°). For this geometry, the zeroth-order penetrating field is evanescent and entirely negligible. The penetrating field is due to scattering by interface roughness, assumed here to be isotropic. If the pulse is sufficiently

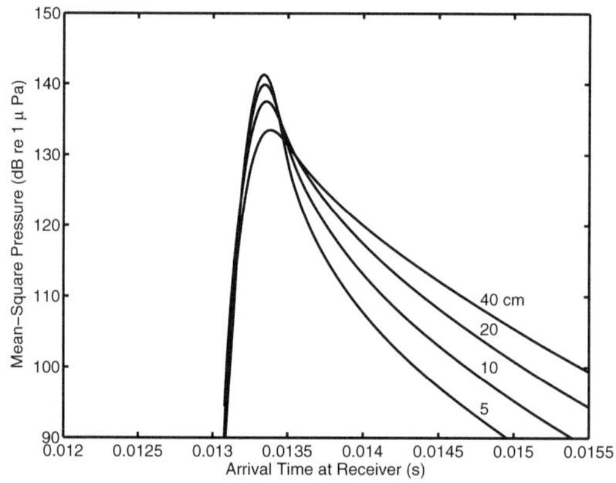

Fig. 15.8. An example of subcritical penetration due to interface scattering. Time series for mean-square pressure have been computed using (15.15) at points in the sediment having depths indicated by the curve labels. These points are on a vertical line, and the source is at a horizontal distance of 20 m and 4 m above the seafloor. The sediment geoacoustic and roughness parameters are taken from Table 13.1 for the sandy SAX99 site. The roughness spectrum was of the form (D.7) with $K_0 =$ 10 m^{-1}. The water speed is 1530 m s^{-1}, and the transmitted pulse has a Gaussian envelope, $u(t) = \exp(-t^2/t_s^2)$, with $t_s = 100$ μs and source level 190 dB re 1 μPa @ 1 m.

broadband that frequency-dependent attenuation may cause pulse distortion, the strategy given by [Moe and Jackson 1998, D. Jackson et al. 2002] can be used.

15.1.3 Experimental Tests of the Penetration Model

The formally averaged perturbation model compared favorably [Thorsos et al. 2000a] to the experimental data of [Chotiros 1995a]. This test was qualitative in nature, and did not consider absolute pressure levels. Most of the quantitative model-data comparisons for acoustic penetration have

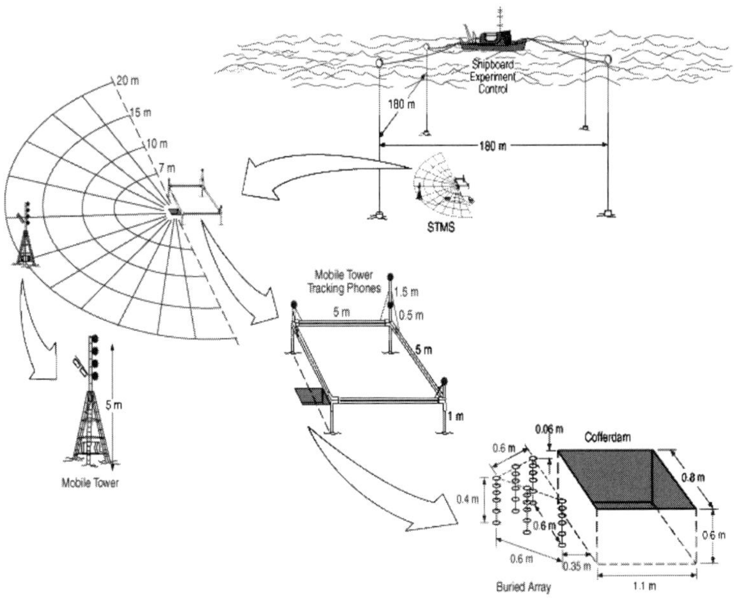

Fig. 15.9. Apparatus used to measure acoustic penetration into the seafloor during the SAX99 experiment.

used the individual-realization approach rather than the formal averaging approach developed here. Nonetheless, tests of the perturbation approximation based on single realizations are certainly relevant to the validity of the averaging approach. The perturbation prediction was found to be consistent with measured penetrating field levels for a rippled sandy seafloor by [Maguer et al. 2000a], subject to the uncertainties stemming from the unavoidable difference of the actual seafloor relief and that generated for input to the model. The SAX99 experiment (Fig. 15.9) included measurements of penetration for another rippled, sandy seafloor [D. Jackson et al. 2002, Chotiros et al. 2002a]. The predictions of the simple sinusoidal ripple model for the diffraction angle (15.5) compared well with data from SAX99, as seen in Fig. 15.10.

The measured levels of the penetrating field were also found to be in substantial agreement with the perturbation model, as shown in Fig. 15.11. In this comparison, the acoustic frequency was 30 kHz and the individual-realization model was used Monte Carlo fashion to generate ten realizations for comparison with the data.

Further tests of the first-order penetration model have been undertaken by [Lopes et al. 2002, Lopes et al. 2003, Nesbitt and Lopes 2004, Lopes et al. 2005]. Carefully controlled backscattering measurements were made using buried targets in a test pool, motivated in part by the apparent failure of the perturbation model to explain the strong return seen in Fig. 15.1 for

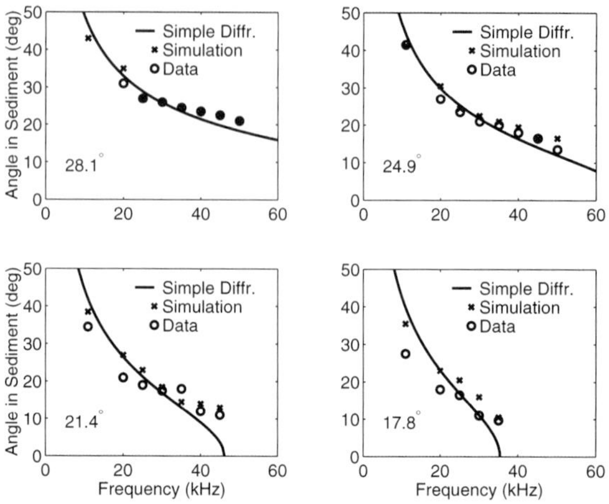

Fig. 15.10. Comparison of measured and modeled diffraction angles in sediment for incident grazing angles noted in each panel. The predictions of Eq. (15.5) are indicated by the solid lines, and the symbols "x" show the predictions of a more detailed model. From [D. Jackson et al. 2002], where geometric and geoacoustic parameters are given.

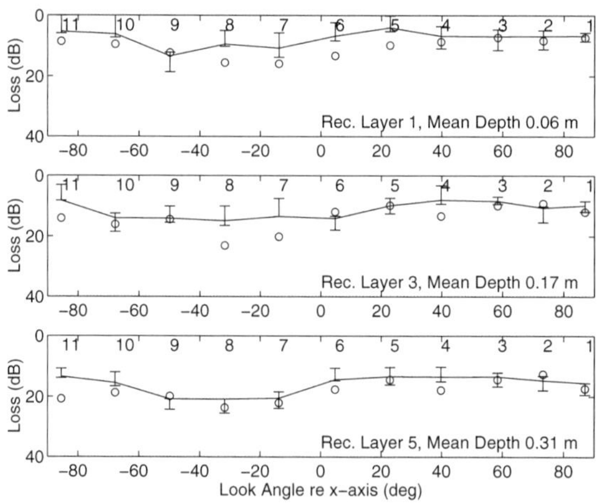

Fig. 15.11. Comparison of measured and modeled field levels, expressed as loss relative to the incident pressure level at the seafloor. The nominal grazing angle was 20°, to be compared with the critical angle of about 30°. The horizontal axis gives the azimuthal coordinate of the movable source with numbers indicating each location. The data are represented by open circles and are averages over three hydrophones with mean depths as indicated in each panel. The solid line indicates the perturbation prediction for an arbitrarily chosen realization, and the error bars show the extremes over ten realizations. From [D. Jackson et al. 2002]

the target 50 cm below the interface. In the test pool experiments, ripples were formed by scraping the sandy bottom with a machined rake, and the resulting ripple field was measured using IMP2 [Tang 2004]. Both cylindrical and spherical targets were employed, the latter filled with silicone oil to increase target strength. Modeling employed single-realization perturbation theory, rather than the formally averaged approach. To test the possibility that failure of the first-order perturbation approximation might explain the difficulty with the data of Fig. 15.1, the perturbation approximation was carried to higher order (Sect. K) [Lim and Sammelmann 2004]. In the tests of [Lopes et al. 2005], the first-order perturbation approximation performed satisfactorily for frequencies between 15 kHz and 40 kHz with ripple having wavelength 75 cm and RMS amplitude 1.7 cm. When the ripple RMS amplitude was increased to 2.5 cm at the same wavelength, the second-order approximation was found to improve model-data agreement noticeably. While these tests gave convincing validation of the first-order roughness penetration model, they did not explain the modeling difficulties with the field data of Fig. 15.1, which may have to do with uncertainties in the environment, target burial, or target physics.

15.2 Mathematical Accuracy

Exact two-dimensional calculations [Thorsos et al. 2000a] have shown that the perturbation approximation applied to subcritical penetration has a wider range of validity than might be expected. Figure 15.12 shows the results of an exact calculation using the two-dimensional roughness spectrum (13.93) and parameters appropriate to a sand seafloor. The scattering strength plotted in the figure correspond to the 2D decibel equivalents of σ (cross section for scattering into the water) and σ_{wp} (cross section for scattering into the sediment). As discussed in Sect. 13.7, the perturbation model is accurate to within 1 dB over all scattered grazing angles for scattering into the water. It is interesting that accuracy is still better for scattering into the sediment. As explained in [Thorsos et al. 2000a], this is due to the fact that the sound speed contrast between the water and sediment is not large ($\nu_p = 1.13$).

15.3 Research Issues

The roughness penetration model has been validated in several controlled experiments combining acoustic and physical measurements. Nevertheless, there remain unexplained detections of deeply buried targets in synthetic-aperture sonar exercises. While these detections may be due to experimental uncertainty, further research is needed to clarify the situation. Higher-order perturbation theory has been developed to accommodate larger ripple amplitudes, and the small-slope approximation should be useful here as

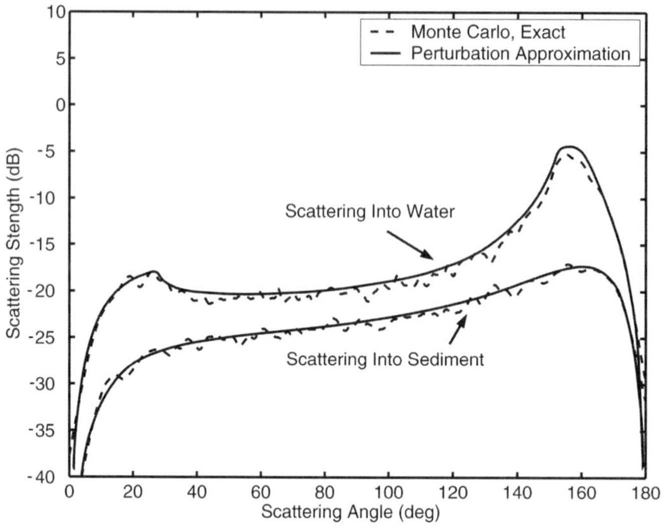

Fig. 15.12. Comparison of perturbation predictions with exact two-dimensional Monte Carlo calculations employing 50 rough-interface realizations. The roughness spectrum has parameters $w_1 = 0.0002$ m, $L_c = 0.227$ m, the water sound speed is 1500 m s^{-1}, and the acoustic frequency is 20 kHz, corresponding to $k_w h = 1.0$, where h is the RMS relief. The sediment parameters are $\nu_p = 1.13$, $a_\rho = 2.0$, and $\delta_p = 0.0155$. The incident grazing angle is 20°, and the "scattering angle" is defined such that 20° is the backscatter direction and 160° is the specular direction. Adapted from [Thorsos et al. 2000a].

well. Penetration experiments can help discriminate between sediment acoustic propagation theories by determining the frequency dependence of wave speed and attenuation and by measuring particle motion (wave polarization) [Osler and Lyons 2004]. Thus, there is ample practical and fundamental motivation for future penetration experiments.

16 Backscatter Statistics

The scattering cross section is proportional to the second moment of received pressure and determines the average bottom reverberation level for a given seafloor. In sonar target detection, it is useful to understand the fluctuations about this average, as strong fluctuations may be interpreted as targets. These fluctuations are usually described in terms of the "probability of false alarm," which is the probability that the received pressure envelope will exceed some preset threshold. It is often assumed that complex scattered pressure is a Gaussian random process, and, in this case, the envelope obeys Rayleigh statistics. While Rayleigh statistics provide a good approximation in many cases, departures from Rayleigh statistics for large envelope values are of interest in setting detection thresholds. Figure 16.1 illustrates the threshold problem using simulated reverberation time series. The upper panel of the figure is a time series for an envelope obeying Rayleigh statistics and the lower panel is an envelope times series obeying Weibull statistics (Sect. 16.3.2). Both series are normalized to have unit mean squares. In these examples, a detection threshold set at the value $\sqrt{10}$ (10 dB above the RMS envelope) produces zero threshold crossings in the displayed time interval for the Rayleigh series while the same threshold produces numerous crossings for the Weibull series. If the sonar designer wishes to set the threshold to achieve a specified rate of false alarms, it is useful to know the statistics obeyed by the reverberation envelope.

This chapter gives a brief account of some of the models developed for the statistics of the backscattered envelope. These models are different in spirit than fluctuation models for electromagnetic and acoustic propagation [Uscinski 1978, Flatté et al. 1979, Ishimaru 1997], which usually begin with the assumption of spatial stationarity of statistics of the medium and then employ approximate forward-scattering theory to obtain statistical measures of the field. In the propagation problem, strong fluctuations are usually caused by focusing due to fluctuations in the acoustic index of refraction (these fluctuations are largely due to fluctuations in water temperature and salinity). Even a medium with Gaussian fluctuations may cause non-Gaussian fluctuations in the field, owing to the nonlinear relation between properties of the medium and the resulting scattered field. It is possible that application of a properly chosen scattering approximation [Jakeman 1982] may lead to useful

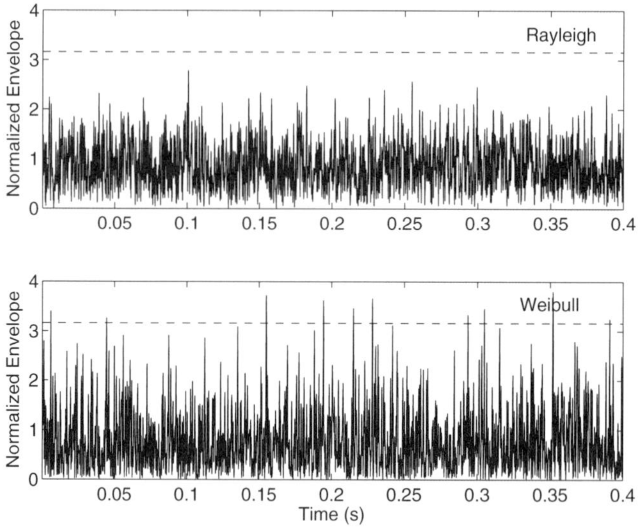

Fig. 16.1. Simulated reverberation envelopes obeying Rayleigh statistics (upper panel) and Weibull statistics (lower panel). The Weibull parameter is $\beta = 1.3$ (Sect. 16.3.2). The RMS envelope in both cases is unity, and the horizontal dashed line represents a detection threshold set 10 dB above the RMS envelope.

results regarding fluctuation statistics for relatively featureless seafloors, but investigators have for the most part assumed that envelope fluctuations result from seafloor patchiness. An example of patchiness is shown in Fig. 16.2, in which patches of mud are found on a rippled sand seafloor.

As automated target detection involves many decisions per ping, very low probabilities of false alarm are of interest. In the Weibull example of Fig. 16.1, there are 4000 samples in the time domain, and 18 of these exceed the detection threshold. In this case the "probability of false alarm" (PFA) is about $18/4000 = 0.0045$. The PFA for the Rayleigh process illustrated in the figure is evidently so low that no threshold crossings are seen. To estimate the PFA in this case, considerably more data would be needed. Although this is an artificial example, it shows that measurements of envelope statistics can be difficult, as it is necessary to build envelope histograms from many thousands of data points, necessitating repeated pinging over a statistically uniform area of seafloor. If a towed system is used, depth changes and platform instability may introduce spurious fluctuations. While fixed platforms offer stability, the available scattering area is limited. If the sonar scans this area by rotating, anisotropy may introduce additional variability not associated with patchiness. In some experiments, a fixed sonar position and orientation have been used, and investigators have depended on ping-to-ping fluctuations to obtain a large statistical sam-

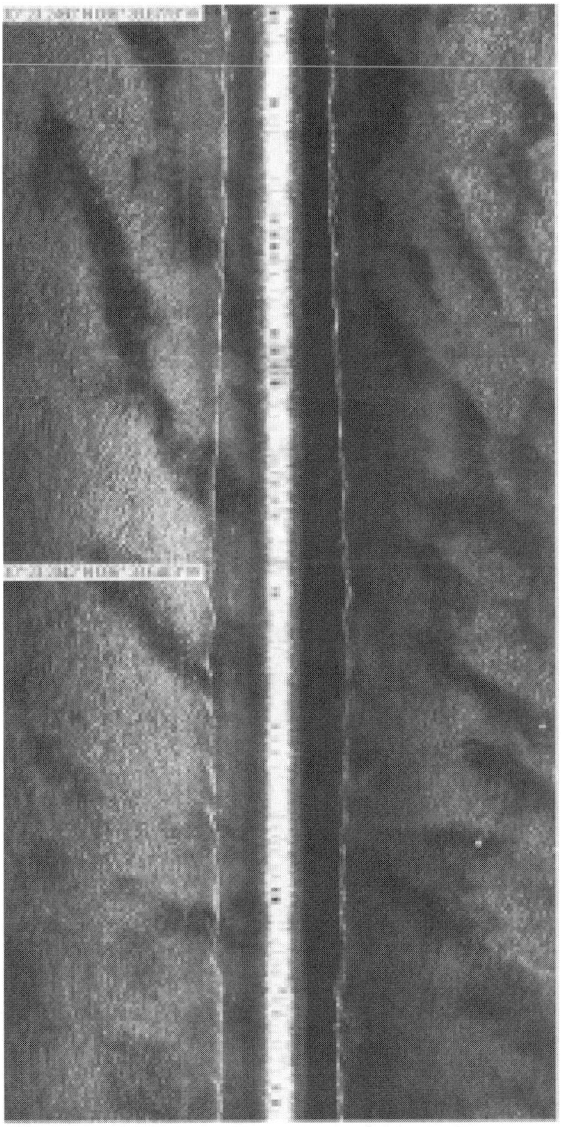

Fig. 16.2. Side-scan sonar (300 kHz) image of the SAX04 experimental site after the passage of Hurricane Ivan. Ripples, with wavelengths of ∼75 cm, are evident. Mud deposits are seen as dark patches of 10–50 m extent [Richardson et al. 2005].

ple [Stanic and Kennedy 1992, Stanic and Kennedy 1993, Trevorrow 2004]. These envelope fluctuations are induced by fluctuations in the water column and may not correspond to those obtained with a sonar moving across the seafloor.

Departures from Rayleigh statistics have been observed in high-frequency field measurements by several investigators [Crowther 1980, Wilson and Powell 1983, Chotiros et al. 1985, Gensane 1989, Lyons et al. 1997, Lyons and Abraham 1999, Lyons et al. 2002a], in laboratory measurements [Becker 2004], and in low-frequency field measurements [Dorfman and Dyer 1999]. Various models have been developed for envelope statistics, some purely empirical and some with at least a partial basis in physics. This chapter will discuss several of these models and will define and discuss the associated envelope probability density functions.

16.1 General Statistical Properties of Backscattered Field

The statistics of the complex backscattered pressure field time series, $P(t)$ (see Appendix E), are of primary interest in this chapter. Here, P with no time argument displayed will denote a complex sample at some instant of time of the echo times series. Most, but not all, of the statistical questions of interest have to do with an ensemble of values of P, obtained at the same round-trip travel time, t, from separate pings on different parts of the seafloor. The measurement geometry and the statistical nature of the seafloor itself will be assumed to be the same for all pings in the ensemble. Breaking P into its real and imaginary parts,

$$P = U + iV \ , \tag{16.1}$$

one may expect that both U and V are sums of terms arising from numerous scatterers. Since the phase of the pressure from each scatterer depends strongly on range, the phase of each scatterer should be completely random and independent of the phases of the other scatterers. Because the phase is random, the contributions of a given scatterer to U and V should be independent in the statistical sense. The mean of each will be zero, as the sums have both positive and negative terms. Both U and V should have the same variance, σ^2. Thus, one expects

$$< U >=< V >= 0 \ , \tag{16.2}$$

$$< U^2 >=< V^2 >= \sigma^2 \ , \tag{16.3}$$

$$< UV >= 0 \ . \tag{16.4}$$

From these conditions it follows that

$$< P >= 0 \; , \tag{16.5}$$

$$< PP^* >= 2\sigma^2 \; , \tag{16.6}$$

$$< P^2 >= 0 \; . \tag{16.7}$$

The object of primary interest is the probability density function (PDF) for the envelope, $p_x(x)$, where the envelope is

$$x = |P| = \sqrt{U^2 + V^2} \; . \tag{16.8}$$

The conditions given above require the phase of P to be uniformly distributed over the interval $[-\pi, \; \pi]$. This is expected for backscatter, but is not a physical law, merely a plausible assumption inherent in many of the models to be discussed. Note that for forward scattering near the specular direction, the phase may not be entirely random if the coherent reflection coefficient (see Sect. 2.3.2) is not zero.

In terms of the envelope PDF, the probability of false alarm introduced earlier is

$$P_{fa}(x) = \int_x^\infty p_x(x')dx' \; . \tag{16.9}$$

Examples of measured and modeled false-alarm probabilities are shown in Fig. 16.3.

The probability of false alarm can be used to examine the behavior of the PDF for large excursions of the envelope from the mean, the situation of interest in sonar performance prediction. Another measure of fluctuation is the "scintillation index," which is the normalized variance of the squared envelope, equivalently, the normalized variance of the intensity:

$$\sigma_I^2 = \frac{< x^4 > - < x^2 >^2}{< x^2 >^2} \; . \tag{16.10}$$

Random processes with $\sigma_I^2 << 1$ can be said to have low levels of fluctuation, and the opposite is true if $\sigma_I^2 >> 1$. It will be seen that a process with Gaussian statistics has a scintillation index of unity.

16.2 Gaussian Statistics

If the backscattered pressure results from a large number of scatterers, the real and imaginary parts of the received pressure will each be given by a sum with many terms, each term having random amplitude, with both positive and negative values. In this case, if the scatterers have comparable amplitude and are not "bunched" together into a few small clusters, the Central Limit Theorem states that both U and V should obey Gaussian statistics, at least approximately. This argument can be made in many cases of interest and

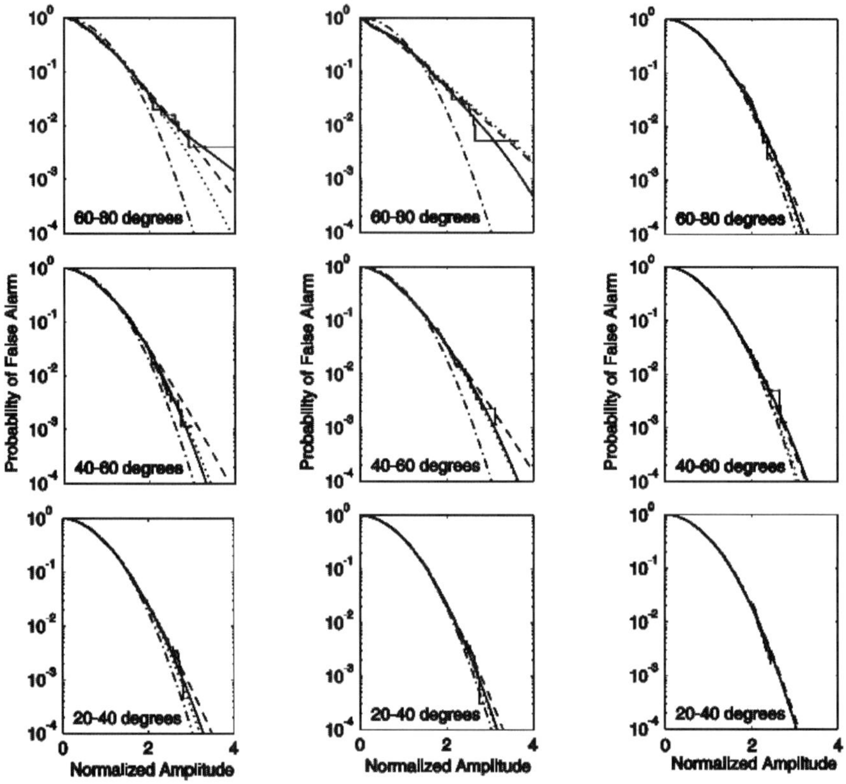

Fig. 16.3. Probability of false alarm versus normalized envelope (mean-square envelope, $< x^2 > = 1$) at 80 kHz for three different sites. The data displayed in the left column are from a shell-covered site, those in the center column are from a sand site, and those in the right column are from a mud site. The data are indicated by the stepped lines and various fits are shown: (dashed-dotted lines, Rayleigh), (dotted lines, Weibull), (dashed lines, K-distribution), (solid lines, Rayleigh-mixture). The grazing angle range is shown in each panel, and the increasing departure from Rayleigh statistics as angle increases is due to the reduction in area of the ensonified region. From [Lyons and Abraham 1999].

accounts for the fact that the statistics of backscattered pressure are often approximately Gaussian. More generally, the complex pressure time series, $P(t)$, is often assumed to be a Gaussian random process [Bendat and Piersol 2000, Sect. 5.3]. This is a stronger statement than the assertion that the statistics of the pressure at any one round-trip time are Gaussian and has profound implications for a number of statistical issues.

It can be shown that if the complex pressure is Gaussian, with independent, identically distributed real and imaginary parts, the squared envelope, $y = x^2$, will have an exponential PDF,

$$p_y(y) = \frac{1}{< x^2 >} e^{-y/<x^2>} \ , \quad y \geq 0 \ , \qquad (16.11)$$
$$= 0 \ , \qquad y < 0 \ ,$$

and the envelope will have a Rayleigh PDF,

$$p_x(x) = \frac{2x}{< x^2 >} e^{-x^2/<x^2>} \ , \quad x \geq 0 \ , \qquad (16.12)$$
$$= 0 \ , \qquad x < 0 \ ,$$

where

$$< x^2 >=< y >= 2\sigma^2 \ . \qquad (16.13)$$

The probability of false alarm is

$$P_{fa}(x) = e^{-x^2/<x^2>} \ . \qquad (16.14)$$

Rayleigh envelope statistics can be considered the "typical" behavior of reverberation, at least until one considers large excursions of the envelope, $x > \sqrt{< x^2 >}$. The level of fluctuation for the Rayleigh case is conveniently stated in terms of the natural logarithm of the squared envelope,

$$b = \ln(y) \ . \qquad (16.15)$$

Using (16.11), it can be shown that [Dyer 1970]

$$< b >= \ln(< x^2 >) \ - \ \gamma \ , \qquad (16.16)$$

$$\sigma_b = \sqrt{< b^2 > - < b >^2} = \frac{\pi}{\sqrt{6}} \ , \qquad (16.17)$$

where $\gamma = 0.5772157$ is Euler's constant. The standard deviation of the natural logarithm of the squared envelope, σ_b, can be converted to decibels through multiplication by $10\log_{10}(e)$, which gives a standard deviation of 5.57 dB. Thus, reverberation pressures are expected to fluctuate a good deal, even in the absence of seafloor patchiness. To obtain the scintillation index (16.10), one needs the fourth moment of the envelope. Equation (16.11) can be used to find $< x^4 >=< y^2 >= 2 < x^2 >^2$. From this, the scintillation index for reverberation obeying Rayleigh envelope statistics is unity.

Low-order moments of the backscattered pressure field time series, $P(t)$, are also of interest, and the Gaussian case provides a tractable example that serves as a baseline with which to compare more complicated behavior. The covariance of the field,

$$< P(t)P^*(t+\tau) >= B_{PP}(t, \ \tau) \ , \qquad (16.18)$$

will be assumed to be independent of the variable t, that is, $P(t)$ will be taken as stationary. This can only be an approximation, as sonar echoes invariably show temporal variations in average intensity, but the approximation can be improved by applying time-varying gain to the backscattered signal. In this case, the covariance is a real, even function of the lag, τ. Expression (16.18) can be considered a generalization of (16.6), and (16.7) generalizes to

$$< P(t)P(t+\tau) >= 0 \ . \tag{16.19}$$

By differentiating (16.18) with respect to τ and setting $\tau = 0$, one can obtain the result

$$< P(t)\frac{P^*(t)}{dt} >= 0 \ , \tag{16.20}$$

as any finite-bandwidth random process has $\partial B_{PP}/\partial t = 0$ for $\tau = 0$.

One important general property of Gaussian random processes is that the first moment (assumed zero here) and covariance suffice to determine *all* statistical measures. This is so because the covariance completely determine the multivariate Gaussian PDF for samples taken at different times within a ping. Equivalently, if the process is stationary, the power spectrum contains all necessary statistical information. The reverberation power spectrum can be found as the Fourier transform of the covariance,

$$W_{PP}(\omega) = \frac{1}{2\pi}\int_{-\infty}^{\infty} B_{PP}(\tau)e^{-i\omega\tau}d\tau \ . \tag{16.21}$$

If the scattering cross section has only a weak dependence on frequency, say, varying by less than 1 dB over the bandwidth of the transmitted signal, then the spectrum of reverberation should be approximately proportional to the spectrum of the transmitted signal. This, in turn, is proportional to the Fourier transform of the autocorrelation of the transmitted signal. An inverse transform then gives [Plemons et al. 1972]

$$B_{PP}(\tau) =< |P|^2 > \rho_s(\tau) \tag{16.22}$$

where

$$\rho_s(\tau) = \frac{\int_{-\infty}^{\infty} s(t)s^*(t+\tau)dt}{\int_{-\infty}^{\infty} |s(t)|^2 dt} \tag{16.23}$$

is the normalized autocorrelation of the transmitted signal, represented in baseband as $s(t)$ (Appendix E). It should be noted that Gaussian statistics were not used in obtaining (16.22), so this expression should be approximately true for any scattering process that does not unduly "color" the backscattered spectrum. As a counterexample, extreme frequency dependence could result from layering of the seafloor. The rapid dependence of scattering strength on grazing angle seen in models for layered seafloors [Ivakin 1986, Ivakin 1989, Mourad and Jackson 1993, Essen 1994, Moe and Jackson 1994] implies a rapid dependence on frequency as well. In such cases, the reverberation spectrum may exhibit deep nulls at certain frequencies, and this can void the assumptions leading to (16.22). It is easy to see why this should be so, as the reverberation will "repeat" itself due to reflections, causing the covariance to peak at lags corresponding to the times between reflections. Such a situation is unlikely to exist at high frequencies owing to the extreme regularity that would be demanded of the layering.

For a pure-tone (constant frequency), unshaded (rectangular envelope) signal, the autocorrelation is a triangular function of lag having width twice the transmitted pulse length. In this case, the reverberation autocorrelation vanishes for lags greater than the pulse length. While this result is based on the idealization that the echo signal is a stationary random process, it provides a basis for approximate statements regarding temporal correlation of reverberation in general. Suppose, for example, that one wishes to know the minimum temporal sampling interval that will provide statistically independent samples of the envelope. For a Gaussian random process, independence and lack of correlation are synonymous, as the covariance determines all statistics. Equation (16.22) shows that the samples should be spaced by at least the pulse time resolution in order to be independent. The pulse time resolution is given by the width of the autocorrelation (16.23), and is comparable to the inverse of the bandwidth of the transmitted signal. This resolution is about equal to the pulse length for pure-tone signals and can be much shorter for frequency-modulated or coded signals intended for matched-filter processing [Medwin and Clay 1998, Sect. 6.5].

It must be remembered that the temporal covariance is defined in terms of a hypothetical average over an infinite ensemble. An estimate of the covariance based on a finite number of echoes will not vanish for lags greater than the pulse length, but these nonzero values may be statistical noise that reduces as the number of echoes in the sample increases.

As a useful example of the analytical convenience attending stationary Gaussian random processes, consider the temporal covariance of intensity (more precisely, the temporal covariance of the squared envelope),

$$< y(t)y(t+\tau) > - < y(t) >^2 = < P(t)P^*(t)P(t+\tau)P^*(t+\tau) > - < |P|^2 >^2 \ . \tag{16.24}$$

This is a fourth moment of the complex pressure , but a useful result is readily obtained, as higher moments of Gaussian random processes can be found in terms of the first and second moments. If a, b, c, and d are real, jointly-Gaussian random variables with zero mean [Bendat and Piersol 2000, Ch. 3],

$$< abcd > = < ab >< cd > + < ac >< bd > + < ad >< bc > \ . \tag{16.25}$$

For a complex stationary Gaussian random process, this, together with (16.18), (16.19), (16.22), and (16.24), gives [McDaniel 1990]

$$< y(t)y(t+\tau) > - < y(t) >^2 = < |P|^2 >^2 \rho_s^2(\tau) \ . \tag{16.26}$$

16.3 Non-Gaussian Statistics

Most models for non-Gaussian reverberation assume that the seafloor is patchy, with localized regions of high scattering strength as in Fig. 16.4. The ensonified region pictured in the figure is discussed in Sect. G.2.1, and has range extent determined by the pulse time resolution and angular extent determined by the sonar azimuthal directivity. If the ensonified area encompasses many patches of comparable strength, the Central Limit Theorem comes into play, and the complex pressure is Gaussian with Rayleigh envelope statistics. Non-Gaussian behavior emerges as the number of patches in the ensonified region becomes smaller.

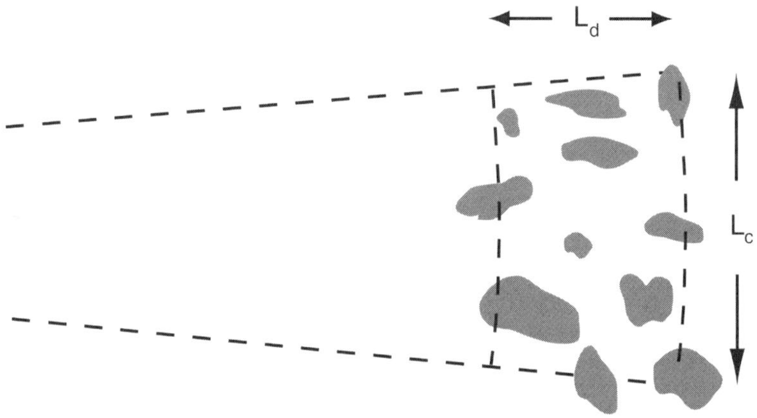

Fig. 16.4. Cartoon showing the elements of models for non-Rayleigh envelope statistics. Patches having elevated scattering cross section give rise to non-Rayleigh behavior if the number of such patches within the ensonified area (dimensions $L_c \times L_d$) is not large.

In considering alternatives to the Rayleigh PDF, there is a trade-off between flexibility of the model and its complexity (number of adjustable parameters). The Rayleigh PDF has only one parameter, the mean-square envelope. If the mean-square envelope is set equal to the measured value, the Rayleigh PDF has no freely adjustable parameters. In fitting non-Rayleigh envelope statistics, the PDFs to be tested should have two or more parameters. The probabilities of false alarm for several such PDFs are compared with the Rayleigh case in Fig. 16.5.

The idealizations used in the development of the models of Chs. 13 and 14 remove much of the complexity needed to understand non-Rayleigh behavior, so the envelope statistics models tend to be either purely empirical or heuristic rather than physical. Most models incorporate the main feature seen in data and illustrated in Fig. 16.4: as the ensonified area is reduced, departures from Rayleigh statistics become more pronounced.

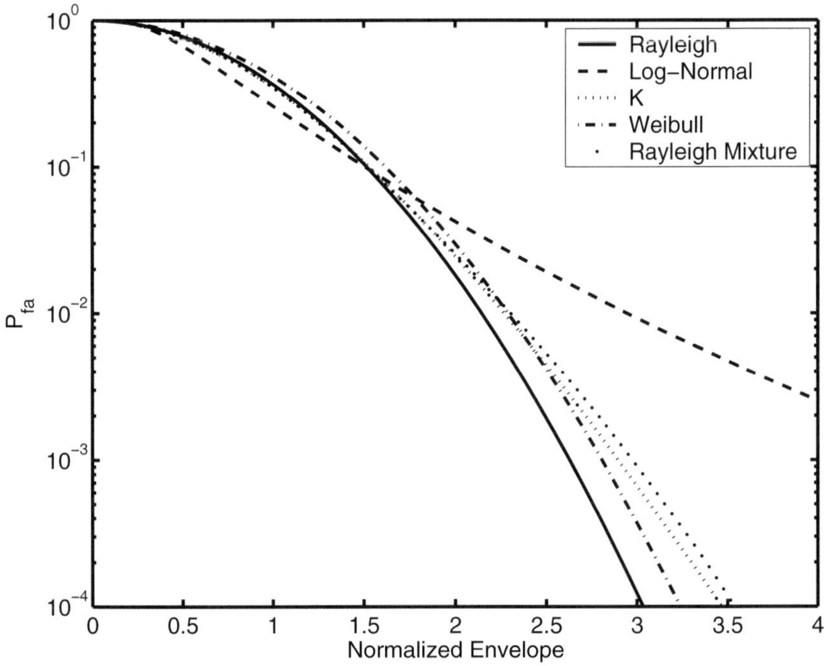

Fig. 16.5. Probability of false alarm versus normalized envelope for several different probability density functions. The mean-square envelope for all curves is unity, and the other parameters used are explained in the main text.

16.3.1 Lognormal PDF

The lognormal PDF is equivalent to a Gaussian PDF for the logarithm of the envelope, $d = \ln x$:

$$p_d(d) = \frac{1}{\sqrt{2\pi\sigma_d^2}} e^{-\frac{(d-\bar{d})^2}{2\sigma_d^2}} , \quad -\infty < b < \infty . \tag{16.27}$$

As one normally takes logarithms of dimensionless quantities, the envelope, x, can be assumed to have been normalized by division by a reference parameter having the same units. Changing variables from the log-envelope to the envelope,

$$p_x(x) = \frac{1}{x\sqrt{2\pi\sigma_d^2}} e^{-\frac{(\ln x-\bar{d})^2}{2\sigma_d^2}} , \quad 0 < x < \infty . \tag{16.28}$$

This is the lognormal PDF [Skolnik 2001, Ch. 7], having two adjustable parameters, $\bar{d}$, the mean of the log-envelope, and σ_d^2, its variance. The mean-square envelope is

$$< x^2 >= e^{2(\bar{d}+\sigma_d^2)} . \tag{16.29}$$

If the mean-square envelope is set to unity, one free parameter remains. This freedom was exercised in [Chotiros et al. 1985] by setting the standard deviation of the log-envelope (expressed in dB) to the value 5.57 dB, as for Rayleigh statistics. This same constraint is used in the lognormal curve of Fig. 16.5. In [Chotiros et al. 1985], envelope statistics data are compared with the Rayleigh and lognormal PDFs. The data were obtained with a fixed sonar position and direction for two different 3-dB full beamwidths, 2.8° and 21°. For the wider beamwidth, Rayleigh statistics provided a good fit to the probability of false alarm, but, for the narrower beamwidth, the low-probability tail ($P_{fa} < 0.03$) was better fit by lognormal statistics, while Rayleigh statistics provided a better fit to the higher-probability portion. The acoustic frequency was 30 kHz, the pulse length was 0.25 ms, and the data were taken at ranges from 80 to 101 m with the source-receiver at a height of 4.5 m above the seafloor. The sediment was a fine sand with mean grain size $M_z = 3.5\phi$, density ratio $a_\rho = 1.9$, compressional speed ratio $\nu_p = 1.10$, and attenuation $\alpha_p = 14.1$ dB/m. The departure from Rayleigh statistics for the narrower beamwidth supports the expectation that smaller ensonified areas may yield non-Gaussian statistics.

Envelope statistics data from several seafloor types were compared to the Rayleigh, lognormal, Weibull, chi-squared, and Rice PDFs by [Gensane 1989]. The lognormal PDF was found to provide the best overall fit, but the author cautions that the fit is suspect as the data failed statistical tests for stationarity.

16.3.2 Weibull PDF

The Weibull PDF is a simple extension of the Rayleigh PDF that adds one more parameter. Where the squared envelope has an exponential PDF in the Rayleigh case (16.11), Weibull statistics give an exponential PDF for the envelope raised to the power β. From this the PDF for the envelope follows as [Skolnik 2001, Ch. 7]

$$p_x(x) = \alpha\beta x^{\beta-1} e^{-\alpha x^\beta} \ , \quad x \geq 0 . \tag{16.30}$$

The mean-square envelope is

$$< x^2 >= \alpha^{-2/\beta}\Gamma(2/\beta + 1), \tag{16.31}$$

where Γ is the gamma function, and the probability of false alarm is

$$P_{fa}(x) = e^{-\alpha x^\beta} . \tag{16.32}$$

Note that for $\beta = 2$, Weibull statistics become Rayleigh. Figure 16.5 shows a plot of the Weibull probability of false alarm for $< x^2 >= 1$ and $\beta = 1.5$. Examples of fits of the Weibull PDF to envelope statistics can be found in [Lyons and Abraham 1999] (also see Fig. 16.3 of this monograph).

16.3.3 K Distribution

The K distribution (more properly, the K PDF) for the envelope is [Lyons and Abraham 1999]

$$p_x(x) = \frac{4}{\sqrt{\alpha}\Gamma(\nu)} \left(\frac{x}{\sqrt{\alpha}}\right)^{\nu} K_{\nu-1}(2x/\sqrt{\alpha}) \ , \tag{16.33}$$

where Γ is the gamma function, and K_ν is the modified Bessel function of the second kind (also known as the modified Bessel function of the third kind, or the Basset function). The mean-square envelope is

$$< x^2 >= \alpha\nu, \tag{16.34}$$

and the probability of false alarm is

$$P_{fa}(x) = \frac{2}{\Gamma(\nu)} \left(\frac{x}{\sqrt{\alpha}}\right)^{\nu} K_{\nu}(2x/\sqrt{\alpha}) \ . \tag{16.35}$$

Figure 16.5 shows a plot of the probability of false alarm for the K distribution with $< x^2 >= 1$ and $\nu = 10$. Examples of fits of the K distribution to envelope statistics can be found in [Lyons and Abraham 1999] (also see Fig. 16.3 of this monograph).

The K distribution approaches the Rayleigh distribution as the parameter ν becomes large compared to unity. An interpretation of this parameter is given by [Abraham and Lyons 2002b], who associate ν with the number of patches in the ensonified region (ν need not be an integer). These authors give illustrative derivations of the K distribution, and one, in particular, is physically suggestive. It is shown that K-distributed reverberation will result if the ensonified region contains a number, 2ν, of Gaussian scattering patches, each having random area, exponentially distributed. In this picture, the scattering patches are surrounded by a non-scattering background. The particulars of this picture may not be relevant, though the association of ν with the number of patches seems significant. In [Abraham and Lyons 2002a, Abraham and Lyons 2004] low-frequency (450–700 Hz) reverberation data are fitted by the K distribution. The expected increase in ν was observed as pulse resolution increased (bandwidth increased) up to a point, but actually decreased as resolution approached its highest values. The authors develop a model accounting for the large aspect ratio (L_c/L_d) of the ensonified region at high resolution and find that this model can be fitted to the bandwidth dependence of the envelope statistics data.

16.3.4 Rayleigh Mixture PDF

Non-Rayleigh statistics can be modeled in terms of two or more Rayleigh processes having different second moments, with each assigned a probability of occurrence [Stewart et al. 1994, Lyons and Abraham 1999]. Then

$$p_x(x) = \sum_{m=1}^{M} P_m \frac{2x}{\sigma_m^2} e^{-x^2/\sigma_m^2} \ , \quad x \geq 0 \ , \tag{16.36}$$

with

$$\sum_{m=1}^{M} P_m = 1 \ . \tag{16.37}$$

The Rayleigh mixture model assumes that the seafloor consists of M different types of patch, with the mth patch type having a scattering cross section proportional to σ_m^2 and occurring with probability P_m. These probabilities give the fraction of seafloor occupied by each type of patch. The patches are larger than the ensonified region so that there is never more than one patch type ensonified for any given ping. It is important to realize that (16.36) is not the PDF for the sum of Rayleigh random variables, as the PDF for a sum of independent random variables is obtained by convolution of the individual PDFs. Nor can (16.36) be regarded as the PDF for the envelope of the sum of Gaussian-distributed pressures. Such a sum would be Gaussian and its envelope would have a Rayleigh PDF.

The mean-square envelope is

$$< x^2 >= \sum_{m=1}^{M} P_m \sigma_m^2 \ , \tag{16.38}$$

and the probability of false alarm is

$$P_{fa}(x) = \sum_{m=1}^{M} P_m e^{-x^2/\sigma_m^2} \ . \tag{16.39}$$

Although this PDF provides an unlimited number of fitting parameters, [Lyons and Abraham 1999] obtain good fits with only two terms (see Fig. 16.3). Figure 16.5 shows a plot of the probability of false alarm for a two-component Rayleigh mixture PDF with $P_1 = 0.75$, $P_2 = 0.25$, $\sigma_2^2/\sigma_1^2 = 2$.

16.3.5 Crowther's Model

An early seafloor fluctuation model was presented by [Crowther 1980] who assumed that scattering strength was spatially dependent with two kinds of patches, one with higher scattering cross section than the other. The observed scattering cross section is then a spatial average over the ensonified region. As a result, the scattering cross section is a random variable with statistics dependent on the dimensions of the ensonified region. The joint PDF for the squared envelope and cross section is

$$p_{ys}(y, \ s) = \frac{1}{s} e^{-\frac{y}{s}} p_s(s) \ , \tag{16.40}$$

where s is proportional to the random scattering cross section. The proportionality is chosen for convenience such that $< s >= 1$, giving mean-square envelope unity. The probability of false alarm is

$$P_{fa}(x) = \int_0^\infty e^{-\frac{x^2}{s}} p_s(s) ds \ . \tag{16.41}$$

Crowther obtains an expression for $P_{fa}(s)$ using a Markov process argument in one dimension. There are inconsistencies in the expressions of [Crowther 1980] which have been corrected here.

The ensonified region is represented by a straight line of length X, and this region is divided into two types of patches, one having a relatively low backscattering cross section and the other having cross section larger by a factor g. The fraction of the ensonified line that is occupied by patches of the stronger scattering type is denoted by the random variable β. The probability that any arbitrarily chosen point on the line is in a high-scattering region is denoted p. Thus, $1 - p$ is the probability that an arbitrarily chosen point is in a low-scattering region. It follows that

$$< \beta >= p \ . \tag{16.42}$$

The relation between the scattering cross section and β is

$$s = \frac{1 - \beta + g\beta}{q} \ , \tag{16.43}$$

where q is a normalizing constant chosen so that $< s >= 1$:

$$q = 1 + (g - 1)p \ . \tag{16.44}$$

It is convenient to first derive the PDF, $p_\beta(\beta)$, from which the PDF for s follows as

$$p_s(s) = \frac{q}{g - 1} p_\beta(\beta) \ . \tag{16.45}$$

The random variables are limited in range, with $0 < \beta < 1$, $1/q < s < g/q$.

The Markov argument uses probabilities for transition from one type of scattering region to another as one moves along the ensonified line. For example, the probability that a transition from low scattering (arbitrarily designated state 0) to high scattering (state 1) will occur in a short interval δx is

$$p\delta x/L \ ,$$

and the probability of a transition from state 1 to state 0 is

$$(1 - p)\delta x/L \ .$$

The parameter L is a characteristic length determining the size of the patches. The PDF for β can be found by considering the various possibilities for the division of the ensonified line into patches: (1) only one subpatch, either state 1 or state 0, (2) two patches, state 1 followed by state 0 and vice versa, etc. For example, the probability that the entire patch is in state 0 is

$$P_0 = (1-p)[1-p\frac{\delta x}{L}]^{X/\delta x} ,$$

where the interval $0 < x < X$ has been subdivided into small subintervals of equal length, δx. For each subinterval, $1 - p\delta x/L$ is the probability that a transition to state 1 is *not* made. Taking the natural logarithm of P_0, $\ln(1+a) \to a$ as $a \to 0$ can be used to simplify the expression. Finally, exponentiating the result gives

$$P_0 = (1-p)e^{-\nu p} , \tag{16.46}$$

where

$$\nu = X/L . \tag{16.47}$$

Similarly, the probability that the entire patch is in state 1 is

$$P_1 = pe^{-\nu(1-p)} . \tag{16.48}$$

The probability that the interval $(0,\ X)$ is comprised of two patches, the first in state 1 having length βX followed by a patch in state 0 having length $(1-\beta)X$, is

$$pe^{-(1-p)\nu\beta}\nu(1-p)e^{-\nu p(1-\beta)}\delta\beta .$$

The transition occurs in a small length interval $X\delta\beta$ with probability $\nu(1-p)\delta\beta$. This case contributes a term

$$p_{10}(\beta) = \nu p(1-p)e^{-\nu(1-p)\beta-\nu p(1-\beta)} \tag{16.49}$$

to the PDF for β. Terms involving more than one transition require integration over all possible lengths of those patches whose length is not dictated by the requirement that the sum of the lengths of all patches in state 1 must be equal to βX. As an example, five (simple) integrations are required to obtain

$$p_{1010101}(\beta) = \frac{1}{2!\ 3!}(p\beta)^3[(1-p)(1-\beta)]^2\nu^5 p_{10}(\beta) . \tag{16.50}$$

The PDF for β can represented as a power series in the variable

$$\mu = 2\nu\sqrt{p(1-p)\beta(1-\beta)} . \tag{16.51}$$

The terms in this series can be segregated into series for two Bessel functions, resulting in

$$p_\beta(\beta) = P_0\delta(\beta)+P_1\delta(1-\beta)+2\{I_0(\mu)+\nu[p\beta+(1-p)(1-\beta)]I_1(\mu)/\mu\}p_{10}(\beta) , \tag{16.52}$$

where I_0 and I_1 are modified Bessel functions of the first kind. Although derived for the 1D case, this expression can be applied to the 2D case with the parameter ν being a measure of the number of strong scattering patches in the ensonified area.

Using (16.41), (16.45), and (16.52), the probability of false alarm with unit mean-square envelope is

$$P_{fa}(x) = (1-p)e^{-qx^2-\nu p} + pe^{-qx^2/g-\nu(1-p)} + \frac{2\nu p(1-p)q}{g-1}\int_{1/q}^{g/q}$$

$$\times e^{-x^2/s}\{I_0(\mu)+\nu[p\beta+(1-p)(1-\beta)]I_1(\mu)/\mu\}e^{-\nu(1-p)\beta-\nu p(1-\beta)}ds\,. \quad (16.53)$$

This expression has three free parameters, g, the ratio of backscattering cross sections (strong/weak) for the two seafloor types, p, the fraction of the seafloor occupied by strong scattering patches, and ν, a measure of the number of strong scattering patches in the ensonified region. Expression (16.53) approaches the Rayleigh false-alarm probability as $g \to 1$, $p \to 0$, $p \to 1$, or $\nu \to \infty$. Figure 16.6 compares the Crowther and Rayleigh false-alarm probabilities for a single choice of these parameters.

In [Crowther 1980], this model was fitted to data obtained at several sites at frequencies of 1.8, 4.1, and 8.1 kHz with satisfactory results. This model has sufficient free parameters that a good fit is not surprising, but the variation of model parameters with site, range, frequency, and beamwidth was deemed plausible and encouraging. Expressions are also given in [Crowther 1980] for the temporal covariance of the squared envelope, a non-Gaussian generalization of (16.26). Comparison of this covariance with data was used to infer the characteristic patch size, L.

16.3.6 McDaniel's Models

Two models have been developed [McDaniel 1990, McDaniel 1996] for fluctuations in backscattering due to seafloor roughness. Both models are based on first-order perturbation theory, but the later model has a stronger physical basis. Even though the physics of the earlier model is questionable, this model might be useful for empirical fits to envelope statistics.

The freely adjustable parameters of the model [McDaniel 1990] are the "number of degrees of freedom," n_c and n_d, in the cross-range and down-range directions. The envelope PDF in this model is

$$p_x(x) = \frac{2x}{\alpha}\int_0^\infty p_{n_d}(\frac{x^2}{\alpha v})p_{n_c}(v)\frac{dv}{v}\,, \quad (16.54)$$

where

$$p_n(u) = \frac{u^{\frac{n}{2}-1}e^{-u}}{\Gamma(n/s)} \quad (16.55)$$

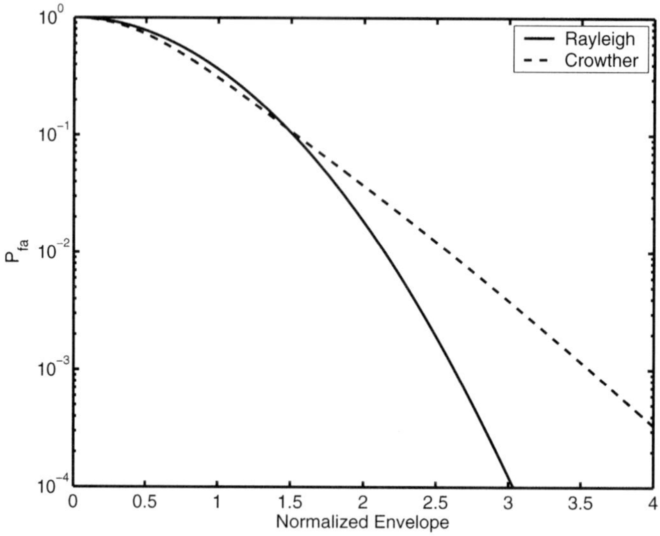

Fig. 16.6. Probability of false alarm versus normalized envelope for the model of [Crowther 1980] compared to that for Rayleigh statistics. The mean-square envelope for both curves is unity, $\nu = 8$, $g = 10$, and $p = 0.1$.

is the chi-square PDF for n degrees of freedom, and α has the same significance as in the K distribution. The mean-square envelope is

$$< x^2 >= \frac{\alpha n_c n_d}{4} \, , \tag{16.56}$$

and the probability of false alarm is

$$P_{fa}(x) = \frac{1}{\Gamma(n_d/2)\Gamma(n_c/2)} \int_0^\infty v^{\frac{n_c}{2}-1} e^{-v} [\Gamma(n_d/2) - \gamma(n_d/2, \, \frac{x^2}{\alpha v})] dv \, , \tag{16.57}$$

where

$$\gamma(n, \, u) = \int_0^u v^{n-1} e^{-v} dv \tag{16.58}$$

is the incomplete gamma function. In spite of appearances, (16.54) and (16.57) are symmetric with respect to the parameters n_c and n_d. If either parameter is set to the value 2, the integral in (16.57) can be written in terms of the modified Bessel function of the second kind, yielding the K distribution result (16.35) with $\nu = n_c/2$ or $n_d/2$, using whichever parameter is not fixed at the value 2. False-alarm probabilities computed in this model are sensitive to slight departures of either n_c or n_d from the value 2 when the other parameter takes on large values. While this model provides a generalization of the K distribution, the author notes in a later publication [McDaniel 1996] that it does not result from a rigorous application of rough-surface scattering theory. Thus, this model is best treated as a potentially useful PDF with the significance of parameters n_c and n_d to be determined.

A physically based model has been developed by [McDaniel 1996]. Small-roughness perturbation theory is used to obtain expressions for the scintillation index and the temporal covariance of intensity. Gaussian statistics are assumed for the interface relief function, $f(\mathbf{R})$, and it is shown that stationarity yields a scintillation index of unity except at very short ranges. This, in turn, suggests that the statistics of the complex pressure are Gaussian. To obtain non-Gaussian behavior, nonstationarity, or patchiness, is introduced by allowing the RMS roughness to vary with position, giving a roughness spectrum that also varies with position. In perturbation theory, this imparts the same position dependence to the backscattering cross section at fixed grazing angle, $\sigma(\mathbf{R})$. The following discussion departs slightly from [McDaniel 1996] in using the cross section rather than the spectrum as the spatially dependent variable. This is done in the expectation that the results so obtained may be applicable when small-roughness perturbation is not, e.g., for large roughness or for sediment volume scattering. The sonar will "see" a spatial average, s, of the position-dependent scattering cross section:

$$s = \int \Omega(\mathbf{R})\sigma(\mathbf{R})d^2R \; . \tag{16.59}$$

In (16.59), single-frequency operation is assumed, so that pulse length effects do not alter the ensonified region. The directivity of the source and receiver are represented by the function $\Omega(\mathbf{R})$, and the grazing angle dependence of the scattering cross section is assumed to be negligible over the ensonified region defined by $\Omega(\mathbf{R})$. Without loss of generality it will be assumed that

$$\int \Omega(\mathbf{R})d^2R = 1 \; . \tag{16.60}$$

Non-Gaussian behavior will be caused by fluctuations in s, but an ensemble of echos taken from different regions having the *same* value of s will be assumed to have Gaussian statistics and an exponential PDF for the squared envelope, y. In this case, the joint PDF for s and y is given by (16.40), presuming y is scaled appropriately. It follows that $< y >=< s >$ and $< y^2 >= 2 < s^2 >$. This gives a scintillation index

$$\sigma_I^2 = \frac{2 < s^2 >}{< s >^2} - 1 \; . \tag{16.61}$$

The first and second moments of s can be evaluated using (16.59) and are

$$< s >=< \sigma > \tag{16.62}$$

and

$$< s^2 >= \int\int \Omega(\mathbf{R}_1)\Omega(\mathbf{R}_2) < \sigma(\mathbf{R}_1)\sigma(\mathbf{R}_2) > d^2R_1 d^2R_2 \ . \tag{16.63}$$

The spatial correlation of the backscattering cross section will be assumed to be of the form

$$< \sigma(\mathbf{R}_1)\sigma(\mathbf{R}_2) >=< \sigma >^2 [1 + \gamma C_\sigma(\mathbf{R}_1 - \mathbf{R}_2)] \ . \tag{16.64}$$

In this expression, γ is the normalized variance of fluctuations in the backscattering cross section (variance divided by the squared mean), and $C_\sigma(\mathbf{R}_1 - \mathbf{R}_2)$ is the normalized spatial covariance of these fluctuations ($C_\sigma(\mathbf{0}) = 1$). Note that the fluctuations that represent nonstationarity are themselves being represented by a stationary random process. This potentially confusing point will be revisited later in this section.

Using (16.61)–(16.64), the scintillation index is

$$\sigma_I^2 = 1 + 2\gamma \int\int \Omega(\mathbf{R}_1)\Omega(\mathbf{R}_2) C_\sigma(\mathbf{R}_1 - \mathbf{R}_2) d^2R_1 d^2R_2 \ . \tag{16.65}$$

In order to obtain numerical results, it is necessary to assume specific forms for the ensonification function, $\Omega(\mathbf{R})$, and the normalized covariance. The following Gaussian forms allow analytical evaluation of (16.65):

$$\Omega(\mathbf{R}) = \frac{1}{\pi L_c L_d} e^{-\left(\frac{x^2}{L_d^2} + \frac{y^2}{L_c^2}\right)} \ , \tag{16.66}$$

$$C_\sigma(\mathbf{R}) = e^{-\frac{R^2}{L^2}} \ . \tag{16.67}$$

The result is

$$\sigma_I^2 = 1 + \frac{2\gamma}{\sqrt{(1 + L_c^2/L^2)(1 + L_d^2/L^2)}} \ . \tag{16.68}$$

This equation has three freely adjustable parameters: L_c/L, L_d/L, and γ. McDaniel also considers pulsed operation for which the ensonified region has a down-range dimension, L_d, that is determined by the pulse length. For a transmitted pulse having the Gaussian baseband (App. E) representation $\exp(-t^2/t_s^2)$, (16.68) still holds, with the replacement

$$L_d = \frac{c_w t_s}{2\cos\theta} \ , \tag{16.69}$$

and with L_c determined by the range and the azimuthal widths of the source and receiver directivity patterns. In (16.69), c_w is the water sound speed, and θ is the grazing angle.

This model has the expected dependence of statistics on the size of the ensonified region, that is, the scintillation index rises above the value unity as the region becomes smaller. Figure 16.7 shows this dependence for a range of the two parameters determining the size of the ensonified region. A fit to the

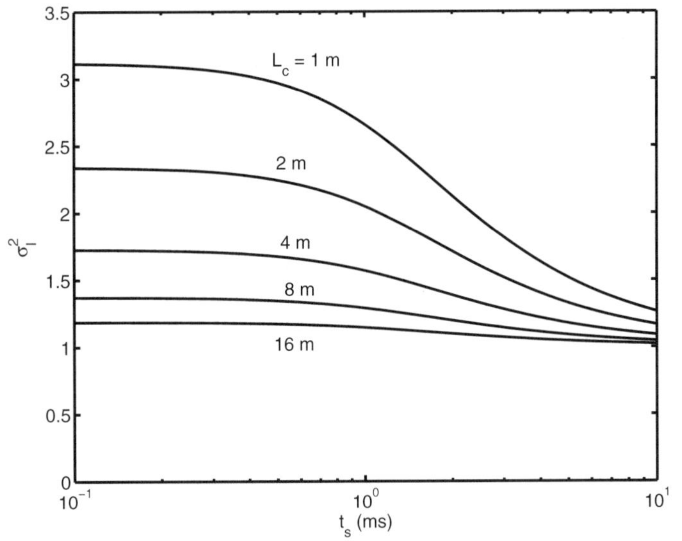

Fig. 16.7. Scintillation index as a function of the Gaussian pulse length parameter, t_s, for the model of [McDaniel 1996]. The curves are labeled by values of cross-range dimension, L_c. Other parameters are: water sound speed $c_w = 1500$ m s^{-1}, correlation length $L = 1$ m, normalized variance of scattering cross section fluctuations $\gamma = 1.5$, and grazing angle $\theta = 30°$.

data of [Gensane 1989] in [McDaniel 1996]) produced plausible values for the correlation length ($L = 1$ m) and for the normalized variance of scattering cross section ($\gamma = 1.5$). The time-domain intensity covariance is also obtained in [McDaniel 1996], providing a non-Gaussian counterpart to (16.26).

This model has several properties worthy of note. As mentioned earlier, [McDaniel 1996] finds that, in perturbation theory, seafloor roughness obeying stationary, Gaussian statistics can give rise to non-Gaussian scattered pressure statistics. Although this behavior only occurs at impractically short ranges, it is of formal interest. One might suppose that, as the scattered pressure in perturbation theory is a linear functional of the relief function, $f(\mathbf{R})$, the pressure must also be Gaussian. This is because any sum of Gaussian random variables is also Gaussian. However, the relief function is a real Gaussian random process and the pressure, P, is complex. The linear relationship between $f(\mathbf{R})$ and P given by Eq. (3) of [McDaniel 1996] does not enforce the condition (16.7). As a result, P need not be a complex Gaussian random variable of the type that yields Rayleigh envelope statistics.

There is another circumstance under which stationary, Gaussian roughness statistics might give rise to non-Gaussian scattering. If the RMS roughness of the seafloor is such as to violate the conditions for validity of first-order perturbation theory (Sect. 13.7), there is a possibility of significant departure from Gaussian statistics, even for relatively featureless, non patchy seafloors. For example, for sand ripples having RMS height comparable to the acoustic wavelength, first-order perturbation theory would fail completely and this leaves open the possibility of non-Gausssian behavior.

The primary source of non-Gaussian behavior in the model of [McDaniel 1996] is nonstationarity of seafloor roughness. There is a potential for confusion in this terminology, as the seafloor statistics in this model are actually stationary. This must be so because the statistical description of patchiness embodied in (16.64) is not tied to any particular location $(x,\ y)$. Thus, this model, and most conceivable models for patchiness will actually use stationary statistics! This seems contrary to the notion of patchiness, but this is only a question of semantics. In [McDaniel 1996], two types of averages are taken, the first over the nonstationary Gaussian randomness of roughness with the roughness covariance having a dependence on position. At this stage, roughness statistics are nonstationary. This model employs a second stage of averaging, however, in which this spatial dependence is assumed to be random, and all dependence of average quantities on the coordinates x and y is lost. Thus, the statistics of the model are stationary, as asserted at the beginning of this paragraph, even though the model incorporates patchiness.

16.4 Research Issues

Although backscatter statistics have been an object of study for a number of years, several problems have slowed progress. A key problem is the lack of appropriate environmental measurements. Another problem is that most current models either are purely empirical or are obtained from heuristic physical arguments. In addition, the acoustic measurements are difficult because large ensembles of echo data must be gathered without significant variations due to unwanted causes such as platform instability and slow environmental changes. If progress is to be made, investigators must find methods of isolating mechanisms causing non-Gaussian behavior, perhaps in part by means of laboratory measurements. Theory can be a guide here, and it should not be assumed that non-Rayleigh envelope statistics only result from complicated environmental scenarios. For example, stationary, Gaussian roughness can produce non-Gaussian scattering due to elevated scattering from steep slopes.

Future field measurements should begin with relatively uncomplicated seafloor types, and methods of quantifying patchiness should be developed and applied. Such efforts, coupled with the development of physical models, should lead to a better understanding of reverberation statistics.

A List of Symbols

The following list of mathematical symbols includes most, but not all of those used in the main text and appendices. The arguments of functions are not shown in this list, as different arguments are used in different contexts. In many cases, the same symbol is used for more than one parameter, but only if the parameter appears in well-separated portions of the monograph. In other cases, the *same* parameter is represented by two different symbols, adhering to differences in acoustic and geoacoustic notation. Definitions are not given here for some of the symbols that appear only once, in which case the reader should be able to find the definition in nearby text.

$\mathbf{A}$	A diagonal matrix with elements taken from $A_\alpha^\pm$
A	Area
Å	Angstroms (10^{-10} m)
A_{all}	Generalization of A_{ww} to the poroelastic case
A_{wp}	Analogous to A_{ww}, but for scattering into sediment
A_{ww}	Function appearing in perturbation and small-slope approximations
$A_\alpha^\pm$	A transform used in perturbation theory
a	Pore size, radius, coefficient in modified Archie's law, parameter in bulk density profile function
a_p	Complex ratio of compressional wave speed to water sound speed
a_q	Complex ratio of wave speed to water sound speed for wave type q
a_t	Complex ratio of shear wave speed to water sound speed
a_α	Amplitude of zeroth-order compressional wave
a_ρ	Ratio of sediment and water mass densities
$\mathbf{B}_s$	Interface slope covariance matrix
B	Interface relief covariance, matrix in poroelastic roughness scattering
B_{PP}	Covariance of random, complex, pressure time series
B_α	A wave vector function used in perturbation theory
$B_{\alpha\beta}$	Volume fluctuation covariance
$\tilde{B}_{\alpha\beta}$	Estimated volume fluctuation covariance
b	Azimuthally averaged directivity
b_i	Source complex directivity function in wave vector space
b_r	Receiver complex directivity function
b_x	Source complex directivity function

C Modulus in Biot theory, a factor in the T-matrix second moment
C_h Structure constant for interface relief
C_0 Gas thermal conductivity
c_p Speed (complex) for compressional wave
c_q Speed (complex) for wave type q
c_{qphase} Phase speed (real) for wave type q
c_t Speed (complex) for shear wave
c_w Sound speed in water
c_0 Wood's value for compressional wave speed, reference wave speed
$\mathbf{D}$ Two-dimensional difference coordinate, a matrix appearing in perturbation theory
D Matrix appearing in model for volume scattering in elastic media, magnitude of $\mathbf{D}$, water depth
D_b Bioturbation mixing coefficient (diffusivity)
D_{mn} Elements of matrix, $\mathbf{D}$
D_r Receiver directivity function in dB, relative density of sediment
D_x Source directivity function in dB
D_1 A function appearing in volume perturbation theory
D_2 A function appearing in volume perturbation theory
d Transducer aperture size, cross-range, sediment grain diameter, dimension of ensonified region, inner diameter of core liner, depth below water–sediment interface
d_i Range from source to scattering patch
d_n Thickness of nth seafloor layer
d_s Range from scattering patch to receiver
d_0 Wave orbital diameter at the seafloor
$\mathbf{E}$ A diagonal matrix appearing in perturbation theory
$E^{(n)}$ Rectangular matrices appearing in poroelastic scattering theory
$\mathbf{e}_h^{\pm}$ Horizontal polarization vector for shear wave
$\mathbf{e}_q^{\pm}$ Unit vector in direction of propagation for wave type q
$\mathbf{e}_v^{\pm}$ Vertical polarization vector for shear wave
$\mathbf{e}_x$ Unit vector in x-direction (similarly for y and z)
e Sediment void ratio (not to be confused with base of natural logarithms), specific modulus
$\mathbf{F}$ Fourier transform of $\mathbf{f}$
$\mathbf{F}_t$ Proportional to Fourier transform of $\nabla \times \mathbf{f}$
F Flow correction factor in Biot theory, Fourier transform of interface relief function (f), formation factor, spatial frequency
F_n Fourier transform of f^n
F_p Proportional to Fourier transform of $\nabla \cdot \mathbf{f}$
$\mathbf{f}$ Force per unit volume, first-order vector source term
f Frequency, seafloor interface relief function
f_c Biot's characteristic frequency
f_i Components of $\mathbf{f}$
G_{pw} Green's function, fluid sediment to water
G_{wp} Green's function, water to fluid sediment
G_{ww} Green's function, water to water
G_α Green's functions for unbounded, homogeneous elastic medium
G_0 Green's function for unbounded, homogeneous fluid

g Acceleration of gravity, dipole Green's function, parameter in Crowther's false-alarm model
H Modulus in Biot theory, transfer function, height, hydraulic tortuosity
H_p Fourier transform of material impulse response function
H_t Fourier transform of material impulse response function
H' Real part of transfer function
H'' Imaginary part of transfer function
h RMS interface roughness, impulse response, specific density
h_p Material impulse response function
h_t Material impulse response function
h_0 Ripple RMS relief, ripple peak-to-trough amplitude
h' Inverse Fourier transform of H'
h'' Inverse Fourier transform of iH''
$\mathbf{I}$ Intensity
I Magnitude of $\mathbf{I}$, electric current
I_K Kirchhoff integral
IOI Index of impedance
$\mathbf{J}$ Electrical current density
$\mathbf{K}$ Two-dimensional horizontal wave vector
$\mathbf{K}_i$ Two-dimensional horizontal incident wave vector
$\mathbf{K}_s$ Two-dimensional horizontal scattered wave vector
$\mathbf{K}_0$ Two-dimensional horizontal wave vector for ripple
K Magnitude of $\mathbf{K}$, hydraulic conductivity
K_G Graphic kurtosis
K_b Bulk modulus
K_b' Real part of bulk modulus
K_b'' Negative of imaginary part of bulk modulus
K_{eff} Effective modulus
K_f Bulk modulus of (drained) sediment frame
K_g Bulk modulus of sediment grain material
K_i Magnitude of $\mathbf{K}_i$
K_s Magnitude of $\mathbf{K}_s$
K_w Bulk modulus of water
K_x x-component of $\mathbf{K}$ (similarly for K_y)
K_0 Magnitude of $\mathbf{K}_0$, coefficient of earth pressure, spectral cutoff wave number
$\mathbf{k}$ Three-dimensional wave vector
$\mathbf{k}_i$ Incident three-dimensional wave vector
$\mathbf{k}_p$ Three-dimensional compressional wave vector
$\mathbf{k}_{pi}$ Incident, down-going compressional wave vector
$\mathbf{k}_{pi}^{(\alpha)}$ Incident, up- and down-going compressional wave vectors
$\mathbf{k}_{ps}$ Scattered, down-going compressional wave vector
$\mathbf{k}_{ps}^{(\alpha)}$ Scattered, up- and down-going compressional wave vectors
$\mathbf{k}_q^{\pm}$ Up-going and down-going wave vectors for wave type q
$\mathbf{k}_r$ Reflected three-dimensional wave vector
$\mathbf{k}_s$ Scattered three-dimensional wave vector
$\mathbf{k}_t$ Three-dimensional shear wave vector
k Wavenumber, attenuation factor (dB m^{-1} kHz^{-1})
k_b Thermal conductivity

k_p	Compressional wavenumber
k_p'	Real part of compressional wavenumber
k_p''	Imaginary part of compressional wavenumber
k_q	Wavenumber for wave type q
k_s	Shear wave attenuation factor, sediment thermal conductivity
k_t	Shear wavenumber
k_w	Acoustic wavenumber in water
L	Bottom loss in dB, correlation length of interface roughness, length parameter in Crowther's false-alarm model, length
L_b	Depth of mixing by benthic fauna
L_c	Low-frequency cutoff length in spectrum, correlation length, cross-range dimension of ensonified region
L_d	Down-range dimension of ensonified region
L_f	Correlation length of surface field
$\mathbf{M}$	A matrix appearing in perturbation theory
M	Modulus in Biot theory, mass of sediment sample
M_d	Graphic median grain size
M_z	Graphic mean grain size
m	Exponent in Archie's law, strain-hardening index
$\mathbf{N}$	Normal to rough interface (not a unit vector)
N	Bubble size distribution
N_j	Components of $\mathbf{N}$
$\mathbf{n}$	Unit normal to rough interface
n	Refractive index, strain-hardening index
n'	Real part of n
n''	Imaginary part of n
P	Acoustic pressure field (complex amplitude), absolute water pressure (including atmospheric pressure), normalized vertical wave vector component
P_A	Atmospheric pressure
P_{eff}	Effective pressure
P_f	Pore fluid pressure
P_{fa}	Probability of false alarm
P_i	Incident pressure (complex amplitude)
P_m	Probability associated with mth component in Rayleigh-mixture PDF
P_r	Reflected pressure (complex amplitude)
P_s	Scattered pressure (complex amplitude)
P_{surf}	Effective surface field
P_0	Hydrostatic pressure, zeroth-order acoustic pressure field, amplitude of plane pressure wave, vertical component of overburden pressure
P_1	First-order acoustic pressure field
P_+	Upward-diffracted acoustic pressure field
P_-	Downward-diffracted acoustic pressure field
$P^{(n)}$	Square matrices appearing in poroelastic scattering theory
PL	Propagation loss
p	Time-domain pressure field, a function appearing in the volume perturbation approximation, probability, probability parameter in Crowther's false-alarm model
p_d	Probability density for natural logarithm of (real) envelope

p_{ref} Reference pressure (usually 1 μPa)
p_s Probability density for scattering cross section
p_x Probability density for (real) envelope
p_y Probability density for squared envelope
p_{ys} Joint probability density for squared envelope and scattering cross section
p_β Probability density appearing in Crowther's false-alarm model
$\mathbf{Q}$ Heat flux, a matrix appearing in small-roughness perturbation theory
$Q^{(n)}$ Column matrices appearing in poroelastic scattering theory
Q_p Quality factor for compressional wave
Q_q Quality factor for wave type q
$\mathbf{R}$ Two-dimensional horizontal position vector
$\mathbf{R}'$ Two-dimensional horizontal position vector
$\mathbf{R}_i$ Horizontal coordinates of source
$\mathbf{R}_s$ Horizontal coordinates of receiver
R Magnitude of $\mathbf{R}$
R' Magnitude of $\mathbf{R}'$
RS Receiver sensitivity
$\mathbf{r}$ Three-dimensional position vector
$\mathbf{r}'$ Three-dimensional position vector
$\mathbf{r}_i$ Position vector for source
$\mathbf{r}_s$ Position vector for receiver
r Magnitude of $\mathbf{r}$, range
r_{ref} Reference length (usually 1 m)
r_s Range from scattering patch (at origin) to receiver
r_t Range corresponding to travel time, t
r_1 Range from source to reflection point
r_2 Range from reflection point to receiver
r^2 Coefficient of determination
S Structure function for interface relief, first-order scalar source term, salinity, degree of sediment saturation
S_b Bottom scattering strength
Sk_I Inclusive graphic skewness
S_p Specific heat at constant pressure, pore water salinity
SL Source level
$\mathbf{s}$ Interface slope (vector or column matrix), a column matrix
s Scattering cross section, complex envelope of transmitted signal
s_r Receiving sensitivity as a voltage/pressure ratio
s_0 RMS source pressure times range
$\mathbf{T}$ A column matrix representing the components of the T-matrix
$\mathbf{T}^{(m)}$ mth-order term in expansion of $\mathbf{T}$
T Temperature
T_{wh} T-matrix for scattering from water into horizontally polarized shear wave
T_{wp} T-matrix for scattering from water into sediment compressional wave
T_{wv} T-matrix for scattering from water into vertically polarized shear wave
T_{ww} T-matrix for scattering from water back into water
T_{wws} Scattered (incoherent) part of T_{ww}
$T_{wp}^{(n)}$ nth-order contribution to T_{wp}
$T_{ww}^{(n)}$ nth-order contribution to T_{ww}

t	Time
t_d	Propagation delay time
t_p	Characteristic time of material impulse response function
t_t	Characteristic time of material impulse response function
t_0	Acoustic travel time from source to receiver via scattering patch
$\mathbf{U}$	Displacement
U	Wind speed, real part of complex pressure, horizontal velocity
U_c	Coherently reflected power
U_i	Incident power, components of $\mathbf{U}$
U_s	Mean scattered power
$\mathbf{u}$	Displacement field
$\mathbf{u}_{eff}$	Effective displacement
$\mathbf{u}_0$	Zeroth-order displacement field
$\mathbf{u}_1$	First-order displacement field
u	Complex envelope of source waveform
u_i	Components of $\mathbf{u}$
$\mathbf{V}$	Hydraulic flow rate
V	Volume, sediment volume, voltage, imaginary part of complex pressure
V_p	Phase speed (real) of compressional wave
V_{pw}	Transmission coefficient, fluid sediment to water
V_r	RMS receiver output voltage
V_s	Phase speed (real) of shear wave, volume of sediment solid particles
V_{sed}	Sediment volume
V_w	Water volume
V_{wq}	Transmission coefficient, water to wave type q
V_{ww}	Reflection coefficient for flat interface
V_{wwc}	Coherent reflection coefficient (complex)
V_{wwi}	Incoherent reflection coefficient
$V_{ww}^{(2)}$	Reflection coefficient to second order for rough interface
V_pR	Sediment–water sound-speed ratio
$\mathbf{v}$	Velocity
v	Function giving depth dependence of zeroth-order compressional wave
W	Two-dimensional roughness spectrum, weight of a sediment sample
W_{PP}	Power spectral density of random, complex, pressure time series
W_s	Weight of solid material in a sediment sample
W_w	Water weight of a sediment sample
$W_{z\alpha\beta}$	One-dimensional volume fluctuation cross spectrum
$W_{\alpha\beta}$	Three-dimensional volume fluctuation cross spectrum
W_1	One-dimensional roughness spectrum
$\mathbf{w}$	Relative displacement
w	Horizontal range extent of ensonified region in backscattering, water content of sediment
w_i	Components of $\mathbf{w}$
w_η	A factor incorporating elastic transmission coefficients
w_1	Interface roughness one-dimensional spectral strength
w_2	Interface roughness two-dimensional spectral strength
$w_{3\alpha\beta}$	Volume fluctuation cross-spectral strength
x	x-coordinate, envelope of signal
Y_1	A diagonal matrix used in the poroelastic roughness scattering model
Y_2	A diagonal matrix used in the poroelastic roughness scattering model

y y-coordinate, volume-averaging function, squared envelope (x^2)
z z-coordinate, depth below water–sediment interface
z_c Mean of z_1 and z_2
z_e Normalized impedance for elastic boundary
z_i z-coordinate of source
z_n z-coordinate of upper interface of $(n+1)$th seafloor layer
z_p Boundary impedance for compressional wave
z_s z-coordinate of receiver
z_w Boundary impedance for acoustic wave in water
z_{wp} Normalized boundary impedance for compressional wave
z_1 Integration z-coordinate
z_2 Integration z-coordinate

α Tortuosity, structure function exponent, any spatially fluctuating sediment physical parameter, constant in Pierson-Moskowitz spectrum, parameter in the Weibull and K PDFs, attenuation in seawater, exponent of structure function
$\bar{\alpha}$ Mean value of fluctuating parameter α
α_p Attenuation, dB/ unit length for compressional wave
α_q Attenuation, dB/ unit length for wave type q
$\alpha_{q\lambda}$ Attenuation, dB/wavelength for wave type q
α_s Attenuation, dB/ unit length for shear wave
α_t Attenuation, dB/ unit length for shear wave
α^2 Sediment thermal diffusivity
β Fractional porosity, constant in Pierson-Moskowitz spectrum, parameter in the Weibull PDF, random variable in Crowther's false-alarm model
β_q Complex sine of propagation grazing angle for wave type q
γ Specific heat ratio for a gas
γ_p Compressional relaxation coefficient in Buckingham's theory, normalized compressional wave speed fluctuation
γ_q Displacement ratio for wave type q
γ_t Shear relaxation coefficient in Buckingham's theory, normalized shear wave speed fluctuation
γ_1 Relief (or volume) one-dimensional spectral exponent
γ_2 Relief two-dimensional spectral exponent
γ_3 Volume fluctuation three-dimensional spectral exponent
γ_α Normalized fluctuation of parameter α
$\tilde{\gamma}_\alpha$ Measured value of γ_α
γ_κ Normalized compressibility fluctuation
γ_ρ Normalized density fluctuation
$\Delta\mathbf{K}$ Two-dimensional horizontal wave vector difference
ΔK Magnitude of $\Delta\mathbf{K}$
ΔK_x x component of $\Delta\mathbf{K}$
ΔK_y y component of $\Delta\mathbf{K}$
Δ_z Length of core section
$\Delta\mathbf{k}$ Three-dimensional wave vector difference
$\Delta\mathbf{k}_p$ Three-dimensional, real compressional wave vector difference
$\Delta\mathbf{k}_\eta$ General three-dimensional wave vector difference
Δk Magnitude of $\Delta\mathbf{k}$
Δk_z Difference in vertical wave vector components
Δ_p Logarithmic decrement for compressional wave
Δ_q Logarithmic decrement for wave type q
Δ_s Logarithmic decrement for shear wave
$\Delta\theta$ Angular spread
δ Bubble damping parameter
δ_h Bubble damping due to heat loss
δ_p Loss parameter for compressional wave
δ_q Loss parameter for wave type q
δ_r Bubble damping due to radiation
δ_s Bubble damping due to shear losses
δ_t Loss parameter for shear wave
δ_κ Compressibility fluctuation

δ_ρ Density fluctuation
δ_1 Frequency-independent loss parameter
η Dynamic viscosity, compressional speed/density fluctuation factor, index denoting two types of waves in elastic medium, porosity in percent, anorbital ripple height
$\bar{\eta}$ Effective viscosity in Biot theory
η_t Stress relaxation coefficient
θ Grazing angle
θ_{crit} Critical grazing angle
θ_i Incident grazing angle
θ_{int} Intromission angle
θ_{is} Local grazing angle for specular reflection
θ_p Grazing angle of compressional wave
θ_{ps} Scattered grazing angle of compressional wave
θ_s Scattered grazing angle
θ_t Grazing angle of shear wave
θ_w Grazing angle in water
θ_{wi} Incident grazing angle in water
θ_0 Ripple phase angle, transducer depression angle
κ Intrinsic coefficient of permeability, compressibility
κ_0 Unperturbed sediment compressibility
Λ Horizontal/vertical fluctuation aspect ratio, wavelength
λ Wavelength, Lamé parameter
λ_p Stress relaxation coefficient
λ_w Wavelength in water
λ_0 Ripple wavelength, unperturbed Lamé parameter
μ Lambert parameter, Lamé parameter (shear modulus), compressibility/density fluctuation factor, dynamic (or absolute) viscosity
μ' Real part of shear modulus
μ'' Negative of imaginary part of shear modulus
μ_0 Unperturbed Lamé parameter
ν Shape parameter in the K distribution, parameter in Crowther false-alarm probability, kinematic viscosity
ν_p Ratio of wave phase speed (real) to water sound speed for compressional wave
ν_q Ratio of wave phase speed (real) to water sound speed, wave type q
ν_t Ratio of wave phase speed (real) to water sound speed for shear wave
ν_α Negative of z-component of compressional wave vector
ρ Bulk density of sediment, effective density, seawater density
$\bar{\rho}$ Mean bulk density of sediment
ρ' Real part of complex effective density
ρ'' Imaginary part of complex effective mass density
ρ_e Electrical resistivity
ρ_{eff} Effective density
ρ_{es} Sediment electrical resistivity
ρ_{ew} Water electrical resistivity
ρ_f Mass density of pore fluid
ρ_g Mass density of sediment particle
ρ_s Normalized autocorrelation of transmitted signal, parameter in bulk density profile function, mass density of sediment particle

ρ_{sw} Mass density of sediment pore water
ρ_w Mass density of water
ρ_0 Mass density of gas, unperturbed sediment mass density, parameter in bulk density profile function
$\rho_{\rho p}$ Density–compressional speed correlation coefficient
$\rho_{\rho\kappa}$ Density–compressibility correlation coefficient
σ Interface scattering cross section (per unit area per unit solid angle), standard deviation
σ_I Inclusive graphic standard deviation (sorting)
σ_I^2 Scintillation index
σ_d Standard deviation of natural logarithm of envelope, scattering cross section of discrete object
σ_{ij} Components of stress tensor
σ_m Set of parameters in the Rayleigh-mixture PDF
σ_r Rayleigh scattering cross section of fluid sphere
σ_s RMS interface slope
σ_v Volume scattering cross-section
σ_{wp} Cross section for scattering into the sediment
σ_2 Volume scattering parameter
σ_0' Mean effective stress
$\sigma^{(2)}$ Roughness scattering cross section to second order
$\sigma^{(4)}$ Roughness scattering cross section to fourth order
$\hat{\sigma}_\alpha$ Normalized standard deviation of parameter α
τ Pulse length, time lag, horizontal shear stress in water
Φ Spectrum of reflected and scattered plane waves, two-dimensional spatial frequency roughness spectrum
Φ_i Spectrum of incident plane waves
Φ_0 A function appearing in the small-slope approximation
Φ_1 One-dimensional spatial frequency roughness spectrum
ϕ Scalar potential, azimuthal angle, time-dependent phase angle, sediment grain size in base-2 logarithmic units, autoregressive coefficient
ϕ_N Value of the grain diameter (in ϕ units) at cumulative frequency N (%)
ϕ_f Scalar potential for relative displacement
ϕ_i Incident azimuthal angle
ϕ_s Scattered azimuthal angle, scalar potential for frame displacement
ϕ_w Scalar potential for displacement in water
ϕ_0 Ripple azimuthal direction, amplitude of plane compressional wave
ϕ_1 One-dimensional spectral intercept, first-order displacement scalar potential
χ Angle of incidence (measured from vertical)
Ψ Angular width of ensonified region
$\boldsymbol{\psi}$ Vector potential
$\boldsymbol{\psi}_f$ Vector potential for relative displacement
$\boldsymbol{\psi}_s$ Vector potential for frame displacement
$\boldsymbol{\psi}_0$ Amplitude of plane shear wave
$\boldsymbol{\psi}_1$ First-order displacement vector potential
Ω Term used in expressions in Biot theory, function defining ensonification of seafloor

ω Angular frequency
ω_c Center frequency
ω_0 Bubble resonant frequency
ω_1 Reference frequency

B Calculation of Properties of Seawater

Determinations of the values of the physical properties of seawater (density, sound speed, attenuation, and viscosity) are often required as part of ocean acoustic experiments. Sound speed and density of the water overlying the seafloor is required in models for reflection (Chs. 8–11), scattering (Chs. 13 and 14), and penetration (Ch. 15). Sound speed and attenuation of seawater are required in interpretation of most acoustic measurement data. Sound not only travels through the water column prior to and after interacting with the seafloor, but values of pore water sound speed, density and viscosity are required for many seafloor propagation and scattering models. Pore water density is required in determinations of sediment bulk density (Sect. 4.3) and is a parameter in all propagation theories (Ch. 8–10). Pore water resistivity is an important variable in measurements of porosity using resistivity techniques (Sect. 4.3) and viscosity is required in Biot theory (Ch. 10). In nearly every case, it is assumed that pore water properties can be calculated using the formulas developed for seawater.

Seawater physical properties are rarely measured directly but instead calculated using empirical relationships involving temperature, conductivity or salinity, and pressure or depth. These relationships are obtained from regressions based on numerous, mostly laboratory, measurements over the range of temperature, salinity, and pressure expected in the world's oceans. The measurements are usually made under conditions where particulate matter (organic and inorganic) is absent (i.e., pure filtered seawater diluted with distilled water). Many of the relationships for seawater properties require several steps such as calculation of values of properties for freshwater at atmospheric pressure and then making corrections for salinity or depth. Some empirical relationships have evolved through time (especially those for sound speed) with corrections added to earlier empirical regressions as new data became available. Recommendations by the Joint Panel on Oceanographic Tables and Standards (JPOTS) culminated in 1981 in a new definition of practical salinity (PSS 78) and an equation of state for seawater (EOS 80), establishing precise and reproducible standards for salinity, conductivity, density, specific volume and density volume anomalies, pressure–depth conversions, freezing point, specific heat, adiabatic lapse rate, potential temperature, and sound speed [Fofonoff and Millard 1983, Fofonoff 1985]. The most recent versions of the equation of state for seawater are based on the Gibbs thermodynamic potential function [Feistel 2003]. All are based on best-fit polynomial regressions derived from carefully collected data. Unless otherwise indicated, temperature, T, is in °C, and salinity, S, is dimensionless. A variety of units are employed in the literature for pressure, P, which may be expressed in bars, kg cm^{-2}, or water depth. In many cases the algorithms used to calculate seawater properties far exceed the accuracy requirements of high-frequency acoustic modeling, but the required computations are not difficult.

B.1 Seawater Density

Seawater density has been a fundamental property required by dynamical oceanographers for over a century. Knudsen's Hydrographic Tables were developed to calculate in situ seawater density (actually specific gravity) from temperature, salinity, and depth (pressure) [Knudsen 1901, Eckman 1908, Hesselberg and Sverdrup 1914, Matthews 1932, Cox et al. 1970]. These tables are based on the relationship between density and salinity, the heat expansion of seawater at atmospheric pressure, and the compressibility of seawater all of which were determined through careful laboratory measurements [Knudsen et al. 1902, Forch et al. 1902, Eckman 1908]. More direct techniques including the use of sound speed measurements to determine seawater compressibility and vibrating flow densimeters to measure density were used to develop the current international standard for density [Millero et al 1080, Millero and Poisson 1981, Fofonoff and Millard 1983]. Seawater density (ρ, kg m^{-3}) is a function of the density of seawater at atmospheric pressure ($\rho(S,T,0)$), pressure (P, bars above 1 atmosphere), and the secant bulk modulus of water, K,

$$\rho(S,T,P) = \frac{\rho(S,T,0)}{1 - P/K(S,T,P)} , \tag{B.1}$$

where $\rho(S,T,0)$ is given by

$$\begin{aligned}\rho(S,T,0) = \rho(0,T,0) + S(0.824493 - 4.0899 \times 10^{-3}T + 7.6438 \times 10^{-5}T^2 \\ -8.2467 \times 10^{-7}T^3 + 5.3875 \times 10^{-9}T^4) + S^{1.5}(-5.72466 \times 10^{-3} \\ +1.0227 \times 10^{-4}T - 1.6546 \times 10^{-6}T^2) + 4.8314 \times 10^{-4}S^2 ,\end{aligned} \tag{B.2}$$

T is the temperature (°C), and S is in PSU (practical salinity units, Sect. B.6) determined from electrical conductivity of seawater. The density of pure water $\rho(0,T,0)$ is given by

$$\begin{aligned}\rho(0,T,0) = 999.842594 + 6.793952 \times 10^{-2}T - 9.095290 \times 10^{-3}T^2 \\ +1.001685 \times 10^{-4}T^3 - 1.120083 \times 10^{-6}T^4 + 6.536332 \times 10^{-9}T^5 ,\end{aligned} \tag{B.3}$$

and the secant bulk modulus ($K(S,T,P)$) is given by

$$K(S,T,P) =$$

$$\begin{aligned}
&19652.21 + 148.4206T - 2.327105T^2 + 1.360477\times10^{-2}T^3 - 5.155288\times10^{-5}T^4 \\
&+S(54.6746 - 0.603459T + 1.09987\times10^{-2}T^2 - 6.167\times10^{-5}T^3) \\
&+S^{1.5}(7.944\times10^{-2} + 1.6483\times10^{-2}T - 5.3009\times10^{-4}T^2) \\
&+P[3.239908 + 1.43713\times10^{-3}T + 1.16092\times10^{-4}T^2 - 5.77905\times10^{-7}T^3 \\
&+S(2.2838\times10^{-3} - 1.0981\times10^{-5}T - 1.6078\times10^{-6}T^2) \\
&+S^{1.5}(1.91075\times10^{-4})] + P^2[8.50935\times10^{-5} - 6.12293\times10^{-6}T \\
&+5.2787\times10^{-8}T^2 + S(-9.9348\times10^{-7} + 2.0816\times10^{-8}T + 9.1697\times10^{-10}T^2)]\,.
\end{aligned} \tag{B.4}$$

This polynomial regression was adopted by the UNESCO Joint Panel on Oceanographic Tables and Standards in 1980 and is referred to as the international equation of state for seawater (for brevity, EOS 80). The algorithm is valid for a temperature range of -2 to 40°C, salinity between 0 and 42 ppt (parts per thousand), and pressure from 0 to 1000 bars, with an accuracy of 0.05 kg m^{-3} for all the world's oceans [Fofonoff 1985]. The accuracy primarily arises from the smallness of variations in the composition of dissolved salts in the world's oceans [Millero 2000] and the use of conductivity techniques to measure practical salinity. The range of density over full ocean depths is about 7% (1072 to 1000 kg m^{-3}) but in coastal waters ($>$ 100 m) the range is 3% (1032 to 1000 kg m^{-3}) which is primarily a function of salinity.

Several more computationally efficient equations of state have been developed for dynamical oceanographic applications where repeated density calculations are required [Bryan and Cox 1972, Friedrich and Levitus 1972, Mellor 1991, Levitus and Isayev 1992, Jackett and McDougall 1995, Wright 1997, Brydon et al. 1999]. All of these algorithms are adequate for the high-frequency applications considered in this monograph. More recent equations of state are derived from the generalized Gibbs thermodynamic potential function (or specific free enthalpy) of seawater [Feistel and Hagen 1995, McDougall et al. 2003, Feistel 2003]. This allows calculation of all thermodynamic equilibrium properties in a self-consistent manner (heat capacity, thermal expansion, density, compressibility or sound speed, free enthalpy) from practical salinity, absolute temperature (ITS-90 standard), and applied pressure. Feistel's free-enthalpy Gibbs function requires 101 coefficients to calculate all thermodynamic equilibrium properties over the full range of oceanic and estuarine conditions (-2 to 40°C, 0–42 PSU, 0–1000 bars). The difference between the density calculated for continental shelf conditions ($<$ 100 m) using the Feistel thermodynamic approach and the EOS 80 algorithms is less than 0.01 kg m^{-3} which is insignificant for acoustic modeling applications.

B.2 Sound Speed in Seawater

The first strong motivation to develop an accurate relationship between sound speed and temperature, salinity, and pressure (depth) in seawater was based on the requirements of echo sounding during the 1920s. Sound speed profiles were required to determine accurate seafloor depths based on two-way acoustic travel time. The first comprehensive sound speed tables [Matthews 1927, Matthews 1939, Kuwahara 1939] were based on analysis of specific volumes rather than actual sound speed measurements. The development of new underwater naval weapons systems required more accurate predictions of sound speed gradients in oceanic waters. The first tables based on measured sound speeds at atmospheric pressure for the range of temperature and salinity found in oceanic water were produced by [Del Grosso 1952, Greenspan and Tschiegg 1959]. The first accurate laboratory measurements of sound speed in seawater covering the range of temperature, salinity, and depth found in the world's oceans were made by [Wilson 1960a] who first determined sound speed over the range of temperature (−4 to 30°C), salinity (33–37 ppt), and pressure (1.033–1000 kg cm^{-2}) based on a total of 581 measurements. Later measurements included salinities of 10, 20, and 30 ppt [Wilson 1960b]. All measurements were made on filtered seawater using a 5-MHz velocimeter similar to that developed by [Greenspan and Tschiegg 1959]. Different salinities were obtained by diluting oceanic seawater with distilled water. Measurements (96) were also made on distilled water samples covering a range of temperature (0–90°C) and pressure (1–965 kg cm^{-2}) exceeding that of oceanic conditions [Wilson 1959]. The standard deviations of the polynomial equations developed by Wilson were 0.22 m s^{-1} for the first data set (over the range of temperature −4 to 30°C, salinity 33 to 37 ppt, and absolute pressure 1.033 to 1000 kg cm^{-2}) and 0.30 m s^{-1} for the combined data set which included the measurements made on distilled water. Deviations up to 2–4 m s^{-1} were noted from the sound speeds tabulated by [Matthews 1939] and [Kuwahara 1939]. Wilson's results [Wilson 1959, Wilson 1960a, Wilson 1960b] have been reanalyzed by numerous authors [Leroy 1969, Frye and Pugh 1971, Anderson 1971, Chen and Millero 1976] by correcting errors, limiting the data to the oceanic regime, and constructing simpler predictive regressions.

A new set of laboratory measurements based on an ultrasonic interferometer were made by [Del Grosso and Mader 1972a] covering the salinity (30–41 ppt), temperature (0–40°C), and depth (0–1000 kg cm^{-2}) ranges typical of open ocean waters. Del Grosso [Del Grosso 1974] combined these data with 148 sound speed measurements made in distilled water at atmospheric pressure [Del Grosso and Mader 1972b] to develop a new equation describing sound speed in most naturally occurring waters. Judging from a comparison of deviations in predicted sound speeds between the Del Grosso [Del Grosso 1974] and previous predictive relationships, these regressions should not be extended outside the range of data used to generate them. Differences were generally less than 0.5 m s^{-1} for oceanic waters but greater for lower salinity waters common in coastal regions and at pressures greater than 500 kg cm^{-2} (> 5000 m).

Chen and Millero [Chen and Millero 1977] determine values of sound speed measured over a wide range of temperature (0–40°C), salinity (5–40 ppt.) and applied pressure (0–1000 bars). Dilution or evaporation of filtered Copenhagen water was used to construct waters of different salinities. Differences in time delays between the sample water (306 samples) and distilled water at the same pressure and temperature were measured, and system calibration was based on sound speed in distilled water, calculated from the 19-term equation given by [Chen and Millero 1976]. Their regression was based on sound speed at 1-atmosphere measured by [Del Grosso and Mader 1972b] and the pressure effects determined by [Wilson 1959]. Millero and Kubinski [Millero and Kubinski 1975] had previously determined a relationship between sound speed at atmospheric pressure and temperature and salinity using similar techniques. These two equations [Millero and Kubinski 1975, Chen and Millero 1976] were combined with the measured sound speed differences and a new sound speed equation was proposed by [Chen and Millero 1977]. This polynomial regression was adopted by the UNESCO Joint Panel on Oceanographic Tables and Standard in 1980 [Fofonoff and Millard 1983] and is referred to as the international equation of state for seawater (EOS 80).

Recent long-range, low-frequency acoustic pulse propagation measurements suggest that the sound speeds predicted by the international standard algorithm derived by [Chen and Millero 1977] are too high, especially at depths below 1000 meters and for temperatures between 0–14°C [Spiesberger and Metzger 1991a, Spiesberger and Metzger 1991b, Spiesberger 1993, Dushaw et al. 1993, Meinen and Watts 1997], and that the algorithm presented by [Del Grosso 1974] produces more accurate results for these cold, deep-ocean depths. Corrections to the Chen and Millero algorithm by [Millero and Li 1994] improve the estimates of sound speed in the deep sea but are still less accurate than the Del Grosso predictions. It has been demonstrated [Dushaw et al. 1993] that the Del Grosso algorithm deteriorates for pure water under pressure and outside of the salinity range (33–38 ppt) of his original measurements [Del Grosso and Mader 1972a]. Differences between the Del Grosso and Chen and Millero algorithms are less than 0.3 m s^{-1} for most realistic conditions in bays, estuaries, and on the continental shelf. If the corrections to the Chen and Millero algorithm suggested by Millero and Li are implemented, the differences are less than 0.12 m s^{-1} for water-column properties encountered during high-frequency experiments. The Del Grosso algorithm has only 19 coefficients compared to the 41 coefficients of the Chen and Millero algorithm and is presented below:

$$c_w = 1402.392 + 5.01109398873T - 5.50946843172 \times 10^{-2}T^2$$

$$+2.2153596924 \times 10^{-4}T^3 + 1.32952290781S + 1.28955756844 \times 10^{-4}S^2$$
$$+0.156059257041P + 2.44998688441 \times 10^{-5}P^2 - 8.83392332513 \times 10^{-9}P^3$$
$$-1.27562783426 \times 10^{-2}TS + 6.35191613389 \times 10^{-3}TP$$
$$+2.65484716608 \times 10^{-8}T^2P^2 - 1.59349479045 \times 10^{-6}TP^2$$
$$+5.22116437235 \times 10^{-10}TP^3 - 4.38031096213 \times 10^{-7}T^3P$$
$$-1.61674495909 \times 10^{-9}S^2P^2 + 9.6840315641 \times 10^{-5}T^2S$$
$$+4.85639620015 \times 10^{-6}TS^2P - 3.40597039004 \times 10^{-4}TSP\ , \qquad \text{(B.5)}$$

where S is salinity (ppt), T is temperature (°C), and P is absolute pressure (i.e., includes atmospheric pressure) (kg m^{-2}).

More recent equations of state derived from the generalized Gibbs thermodynamic potential function (or specific free enthalpy) of seawater have been used to predict sound speed in seawater [Feistel and Hagen 1995, McDougall et al. 1903, Feistel 2003]. Sound speed differences between the Del Grosso algorithm (B.5) and the thermodynamic approach tend to be less than 0.1 m s^{-1} for typical shallow-water conditions ($<$ 100 m) and do not justify the use of 101 coefficients needed for sound speed calculation.

B.3 Seawater Absorption and Attenuation

Attenuation (α, dB m^{-1}) in filtered seawater arises from a combination of two different processes: viscous dissipation of energy, and chemical relaxation of magnesium sulfate and boric acid. Magnesium sulfate ($MgSO_4$) and boric acid (H_3BO_3) molality (concentrations) in standard seawater (35 ppt) are 0.00561 and 0.00037 mole per kilogram of water, respectively. This is only 0.84% and 0.055% of the total molality of seawater but relaxation processes from these two constituents dominate compressional wave attenuation in seawater below 200 kHz.

Francois and Garrison [Francois and Garrison 1982a, Francois and Garrison 1982b] reviewed previous laboratory and at-sea attenuation measurements and provided an empirical algorithm (B.6)–(B.12) to calculate compressional wave attenuation (α, dB m^{-1}) from salinity (S, ppt), temperature (T, °C), depth (z, meters), and seawater pH, where f is the acoustic frequency (kHz), c is the calculated sound speed, and f_1 and f_2 are, respectively, the relaxation frequencies (kHz) for magnesium sulfate and boric acid. The first two terms in (B.6) represent chemical relaxation processes (attenuation) due to boric acid and magnesium sulfate and the last term represents the viscous dissipation in pure water. In this empirical formulation, also used by [Medwin and Clay 1998] and [Richards 1998] as well as many others, the dynamic coefficients of shear and bulk viscosity for freshwater are incorporated into the terms A_3 and P_3. Attenuation (α, dB m^{-1}) is

$$\alpha = 10^{-3}\left(\frac{A_1P_1f_1f^2}{f^2+f_1^2}+\frac{A_2P_2f_2f^2}{f^2+f_2^2}+A_3P_3f^2\right) , \tag{B.6}$$

where

$$A_1 = \frac{8.86}{c}10^{(0.78pH-5)} , \quad A_2 = 21.44\frac{S}{c}(1+0.025T) , \tag{B.7}$$

$$P_1 = 1 , \quad P_2 = 1 - 1.37\times10^{-4}z + 6.2\times10^{-9}z^2 ,$$

$$P_3 = 1 - 3.83\times10^{-5}z + 4.9\times10^{-10}z^2 , \tag{B.8}$$

$$f_1 = 2.8\left(\frac{S}{35}\right)^{0.5}\times10^{[4-1245/(273+T)]} , \quad f_2 = \frac{8.17\times10^{[8-1990/(273+T)]}}{1+0.0018(S-35)} , \tag{B.9}$$

$$c = 1412 + 3.21T + 1.19S + 0.0167z , \tag{B.10}$$

and, for $T \leq 20°\mathrm{C}$,

$$A_3 = 4.937\times10^{-4} - 2.59\times10^{-5}T + 9.11\times10^{-7}T^2 - 1.50\times10^{-8}T^3 , \tag{B.11}$$

and, for $T \geq 20°\mathrm{C}$,

$$A_3 = 3.964\times10^{-4} - 1.146\times10^{-5}T + 1.45\times10^{-7}T^2 - 6.5\times10^{-10}T^3 . \tag{B.12}$$

The terms P_1, P_2, and P_3 include the pressure effects on absorption by boric acid and magnesium sulfate and viscosity, respectively. The terms A_1, A_2, and A_3 include the effects of temperature and viscosity on absorption. Several other empirical algorithms have been proposed to calculate sound absorption in seawater. The [Francois and Garrison 1982b] and [Fischer and Simmons 1977] algorithms are the most comprehensive and most often cited. A simplified version of the Francois–Garrision algorithm was proposed by [Ainslie and McColm 1998].

Based on the Francois–Garrision sound attenuation algorithm, attenuation below 1 kHz is primarily a result of boric acid relaxation; between 1 and 5 kHz the importance of boric acid relaxation decreases; and between 5 and 100 kHz magnesium sulfate absorption dominates (Fig. B.1). Between 100 and 500 kHz the importance of magnesium sulfate absorption decreases, and above 500 kHz absorption is primarily due to water viscosity alone. Attenuation is 1–3 orders of magnitude higher for salinities 5–40 ppt. than for freshwater alone (Fig. B.2). Over the frequency range where magnesium sulfate relaxation dominates seawater attenuation, attenuation increases with temperature by an order of magnitude from 0 to 40°C but decreases with increasing temperature where absorption by viscosity dominates (B.3). The pH of seawater in open ocean conditions generally ranges between 7.8 and 8.3, and increasing acidity increases the boric acid relaxation contribution to attenuation but has no apparent effect on absorption by magnesium sulfate or water which dominates attenuation of the frequency band of interest to high-frequency acoustics.

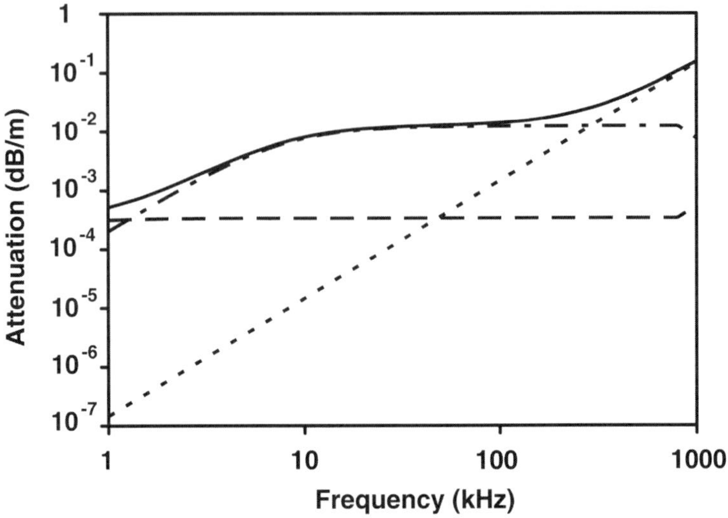

Fig. B.1. Attenuation versus frequency for seawater based on the algorithm of [Francois and Garrison 1982b] calculated for 25°C, 35 ppt, pH 8.0, and atmospheric pressure. Attenuation due to boric acid relaxation is represented by the dashed line, magnesium sulfate relaxation by the dash-dot line, freshwater attenuation by the dotted line, and total attenuation by the solid line.

The Francois–Garrision sound absorption algorithm is based on field and laboratory measurements made for open ocean conditions (S = 30–40 ppt) and may not be appropriate for lower-salinity, shallow-water estuarine conditions depicted by the lower-attenuation curves in Fig. B.2. These empirical relationships were determined for filtered or pure seawater. Many high-frequency acoustic applications occur in estuarine and coastal regions where considerable suspended organic and inorganic matter is present in the water column. The presence of solid particles in suspension gives rise to two additional attenuation mechanisms: viscous absorption and scattering. These attenuation mechanisms are of the same order of magnitude as magnesium sulfate absorption at acoustic frequencies of 1 to 100 kHz even at relatively low concentrations of suspended particulate matter (0.1 kg m^{-2}) [Richards et al. 1996] and may dominate at higher concentrations (1.0 kg m^{-2}) [Richards 1998].

B.4 Seawater Viscosity

The viscosity of seawater appears as a parameter in Biot theory and is also needed in determinations of sediment permeability. Viscosity is the measure of a fluid's internal, or molecular, resistance to shear stress. The thicker or

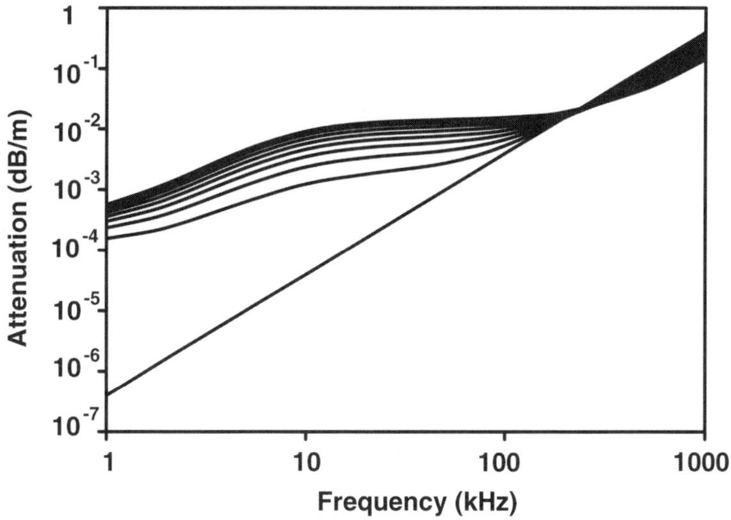

Fig. B.2. Attenuation versus frequency over the salinity range of 0–40 ppt, for 25°C, pH 8, and atmospheric pressure. Freshwater absorption is indicated by the linear curve, and successive increases in salinity in 5-ppt increments give successively higher attenuation primarily over the frequencies below 200 kHz.

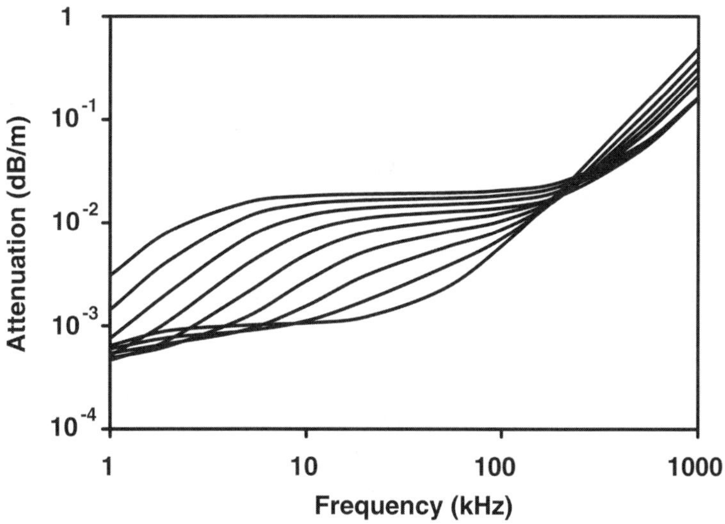

Fig. B.3. Attenuation versus frequency over the temperature range of 0–40° C (5°C increments). All values of attenuation are calculated for 35 ppt, pH 8, and atmospheric pressure. Attenuation increases with temperature at frequencies below 200 kHz and decreases with temperature at frequencies above 500 kHz where absorption is dominated by viscosity.

more consistent the fluid, the more viscous is the fluid. Water is fortunately a Newtonian fluid (i.e., at a given temperature and pressure, viscosity is independent of shearing stress or the duration of the shearing). Kinematic viscosity, ν, is a molecular property of water that is defined for laminar flow by

$$\tau = \rho\nu\frac{dU}{dz} , \tag{B.13}$$

where τ is the horizontal shear stress, ρ is the water density, U is the horizontal velocity, and z is the vertical coordinate. Absolute, or dynamic, viscosity, μ, is related to kinematic viscosity as follows:

$$\mu = \rho\nu . \tag{B.14}$$

Absolute viscosity is denoted η in older texts and in Ch. 10, where the symbol μ is used for the shear modulus. Viscosity can be measured using Poiseuille flow (laminar, steady-state flow) through a tube,

$$q = \frac{\pi p r^4}{8L\mu} , \tag{B.15}$$

where L is the length of the tube, r is the radius, p is the difference in pressure at the ends, and q is the volume of flow per second.

Absolute viscosity is typically reported in poise (g $\text{cm}^{-1}\text{s}^{-1}$) or centipoise (cP) whereas kinematic viscosity is given in stokes ($\text{cm}^2\ \text{s}^{-1}$) with fluid density, ρ, in g cm^{-3}. Seawater viscosity decreases with temperature, increases with salinity, and only slightly increases with pressure. Kinematic viscosity is used for most fluid flow applications in oceanography and seafloor dynamics. Both kinematic and absolute viscosities are used in acoustic propagation theory and care must be exercised with units and whether values of kinematic or absolute viscosity are required. For example, [Stoll 1989] uses an absolute viscosity of (0.01 dyne s cm^{-2} or poise) as input to the Biot model whereas [Urick and Ament 1949] use kinematic viscosity as part of a theory for propagation in composite media. Stoll's value of viscosity is appropriate for laboratory conditions for seawater at 23°C and freshwater at 20°C.

A variety of techniques have been used to measure fluid viscosity. These include measurement of the fluid flow through a tube, discharge of a known volume of fluid through an orifice or capillary, the torque of rotating viscometers immersed in a fluid, fall rates of calibrated spheres through a fluid, rise time of bubbles through a fluid, or damping of fluid vibration [Miller 1996]. In most cases the measurement techniques are calibrated with fluids of known viscosity. For most acoustic applications, viscosity can be predicted from seawater temperature, salinity and pressure.

Values of kinematic viscosity of distilled water and seawater at atmospheric pressure have been tabulated by [Myers et al. 1969] and are presented in Fig. B.4. Interpolation between these curves can be used to estimate seawater viscosity between 0 and 35 ppt and −2 to 35°C.

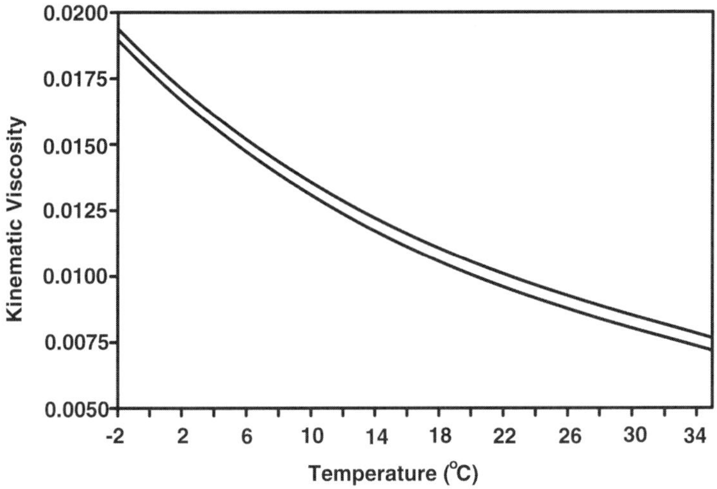

Fig. B.4. Kinematic viscosity (cm^2 s$^{-1)}$) for freshwater (lower curve) and seawater (35 ppt) at atmospheric pressure.

An empirical relationship for absolute viscosity (μ, Pa s) is presented by [Richards 1998]. Viscosity in this relationship is dependent on temperature (T, °C), salinity (S, ppt), and pressure (P, dbar) and based on laboratory measurements summarized by [Matthäus 1972],

$$\begin{aligned}\mu = 0.1(1.79 \times 10^{-2} - 6.1299 \times 10^{-4}T + 1.4467 \times 10^{-5}T^2\\ -1.6826 \times 10^{-7}T^3) + P(-1.8266 \times 10^{-7} + 1.3817 \times 10^{-8}T\\ -2.6363 \times 10^{-10}T^2) + P^2(9.8972 \times 10^{-12} - 6.3255 \times 10^{-13}T\\ +1.2116 \times 10^{-14}T^2) + S(2.4727 \times 10^{-5} + 4.8429 \times 10^{-7}T\\ -4.7172 \times 10^{-8}T^2 + 7.5986 \times 10^{-10}T^3) .\end{aligned} \tag{B.16}$$

The viscosity of distilled water decreases with temperature from 1.80 cP at 0°C to 0.72 cP at 35°C. The viscosity of seawater (35 ppt) over the same temperature range is about 2–7% higher than the viscosity of distilled water. The increase in viscosity with pressure over full ocean depths is less than 0.5% and can be neglected for acoustic applications in continental shelf depths (< 100 m).

B.5 Electrical Conductivity of Seawater

The electrical conductivity of seawater can either be measured in the laboratory from water bottle samples or in situ using a variety of CTD (Conductivity Temperature Depth) profilers and is generally used to calculate practical salinity, S (Sect. B.6). Laboratory bench galvanic or inductive salinometers are calibrated with Standard Seawater (SSW) and, if carefully maintained, have reported accuracy of 0.002 derived salinity (conductivity, 0.0002 S m^{-1}) [Müller 1999]. The Practical Salinity Scale, PSS78, has no units but is often considered to be equivalent to salinity measured as parts per thousand (ppt). Salinity profiles determined from well-calibrated CTDs in combination with water samples may have accuracy near that of laboratory salinometers but a more common accuracy, especially for self-recoding CTDs, is near 0.02 PSU (conductivity, 0.002 S m^{-1}). In situ electrical conductivity is derived from the conductivity of reference salinity samples, in situ temperature, and in situ pressure by numerical back iteration of the international equations of state given by [Fofonoff and Millard 1983]. Seawater temperature, pressure, and conductivity provide the three thermodynamic parameters needed to determine the state of seawater.

Table B.1. Seawater conductivity (S m^{-1}) at atmospheric pressure calculated from the algorithms given by [Fofonoff 1985].

	S (PSU)								
T (°C)	0	5	10	15	20	25	30	35	40
5	0.4806	0.5566	0.6364	0.7201	0.8064	0.8960	0.9875	1.0814	1.1767
10	0.9171	1.0613	1.2123	1.3702	1.5337	1.7020	1.8753	2.0526	2.2324
15	1.3363	1.5436	1.7620	1.9899	2.2259	2.4693	2.7190	2.9744	3.2340
20	1.7415	2.0110	2.2938	2.5886	2.8941	3.2087	3.5318	3.8618	4.1978
25	2.1371	2.4663	2.8109	3.1705	3.5426	3.9262	4.3202	4.7223	5.1317
30	2.5242	2.9104	3.3160	3.7374	4.1742	4.6257	5.0875	5.5595	6.0393
35	2.9036	3.3456	3.8090	4.2914	4.7913	5.3067	5.8350	6.3749	6.9242
40	3.2756	3.7717	4.2918	4.8338	5.3947	5.9728	6.5658	7.1714	7.7880

The inverse of conductivity, resistivity, ρ_{ew}, of sediment pore water is used to convert measured sediment resistivity, ρ_{es}, to a dimensionless formation factor, $F = \rho_{es}/\rho_{ew}$ (Sect. 4.4.2), which is dependent on sediment porosity, pore shape and connectivity, and particle size and shape. In most cases, the resistivity of the overlying seawater and sediment are measured under the same conditions (S, T, P), often at the same time, and then used to calculate a formation factor without calibrations (see Sect. 4.4.2). Several authors, however, have derived laboratory values of formation factor by comparing measured sediment resistivity to calculated pore water resistivity. This is especially important when pore water is not available in sufficient quantities from core collected sediments or when reference measurements of seawater or salt solutions are used to calibrate resistivity probes. In either case pore water resistivity must be calculated from temperature and salinity and if measured in situ from pressure. The definitions of salinity and temperature have changed as instrument precision increased complicating the conductivity to salinity–chlorinity relationships. However, given the accuracy of sediment resistivity measurements interpolation of tabular values of seawater resistivity is probably of sufficient accuracy to calculate sediment formation factors which are rarely reported to more than two decimal places and are probably accurate to ±1%. The tabular values of seawater conductivity were calculated using the methods proposed by [Fofonoff 1985] and adopted as part of the Unesco algorithms for computation of fundamental properties of seawater [Fofonoff and Millard 1983]. The conductivity of seawater at 15°C and 35 PSU salinity (4.29140 S m^{-1}) is used to convert the calculated conductivity ratio to absolute electrical conductivity in Table B.1 [Fofonoff 1985]. Figure B.5 demonstrates the almost linear relationships between seawater conductivity and both temperature and salinity justifying linear interpolation of seawater conductivity from Table B.1. The effects of pressure account for less than 0.1% increase with depth over the 100-m continental shelf and can be ignored for determination of sediment formation factor. For more precise applications the algorithms of [Fofonoff and Millard 1983] can be used.

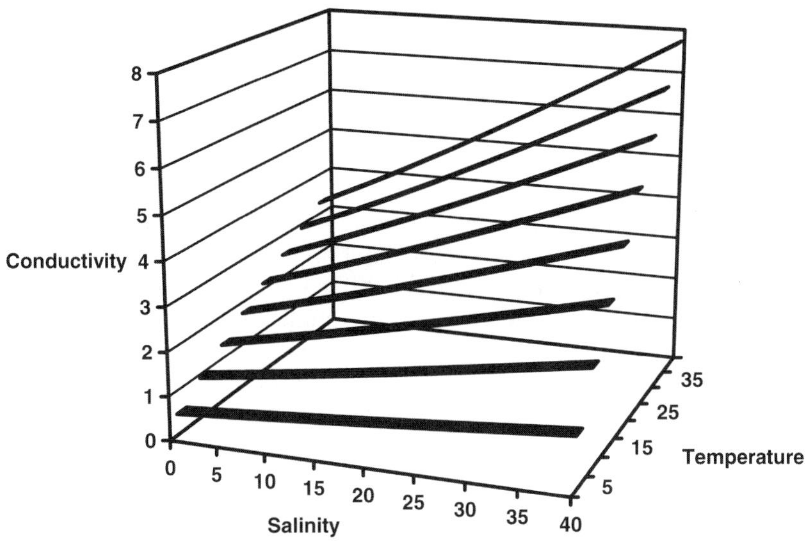

Fig. B.5. Relationship between seawater conductivity (S m^{-1}), salinity (PSU), and temperature (°C) at atmospheric pressure. The apparent linear relationship justifies linear interpolation to obtain estimates of seawater conductivity from temperature and salinity.

B.6 Practical Salinity

Prior to the advent of conductivity bridges, salinity, S, was determined from chlorinity (Cl) by titration with silver nitrate or as total halide concentration. The assumption was that the ratios of the major inorganic compounds in seawater were nearly constant in all ocean waters and measurement of one conservative constituent allowed the calculation of salinity (S = 0.030 + 1.8050Cl). Salinity was defined as the mass of dissolved salts in a given mass of water in parts per thousand (ppt). This approach, the Knudsen Tables, was used by oceanographers for over 60 years. The excessive variability between chlorinity (and thus calculated salinity) and density lead the redefinition of salinity in terms of electrical conductivity [Cox et al. 1967]. Based on worldwide measurements of conductivity and salinity, [Cox et al. 1970] found salinity (S, ppt) could be calculated from a conductivity ratio R_{15} which is the ratio of measured conductivity at 15°C to that of standard seawater of 35 ppt measured at 15°C with both measured at atmospheric pressure,

$$S = -0.08996 + 28.29720R_{15} + 12.80832R_{15}^2 - 10.67869R_{15}^3 + 5.98624R_{15}^4 - 1.32311R_{15}^5 \ . \tag{B.17}$$

The relationship between salinity and chlorinity was redefined as (S = 1.80655Cl), eliminating the slight offset caused by the lack of consistency in the ionic composition of seawater. Minor differences in the relative concentrations of inorganic constituents in samples of standard seawater led to use of KCl solutions as the conductivity standard for seawater [Millero et al. 1977]. In 1978 the Practical Salinity Scale (PSS 78) was proposed as part of the new equation of state for seawater [Fofonoff and Millard 1983]. Salinity (S with no units, but often referred to as PSU) is expressed in terms of an electrical conductivity ratio (RT) of a given sample $C(S, T, 0)$ to the conductivity of a KCl solution with a salinity of 35 ppt and at a temperature of 15°C. The conductivity of the reference KCl solution was calibrated with a single batch of standard seawater (SSW P79). Practical salinity can be determined from conductivity, temperature, and pressure [Fofonoff 1985]. Detailed descriptions and algorithms to calculate the Practical Salinity ratio from conductivity, temperature, and pressure and the inverse to calculate the resistivity from temperature, salinity, and pressure are given in [Fofonoff 1985]. The inverse calculation seems a bit circular as the salinity was probably determined from conductivity but does allow conductivity to be predicted at different conditions of temperature and pressure. In addition to conductivity, which was discussed in the previous section, salinity can also be determined by laboratory measurements of density (±0.002), refractive index (±0.05), or sound speed (±0.03) [Millero 1996, p. 47]. The best precision laboratory measurement of salinity is conductivity (±0.001).

The practical salinity scale assumes a constant relative composition of major ions in seawater. This has been shown not to be true for different water masses, for the deep sea, and especially for estuarine waters where mixing of seawater and river water may not yield the same conductivity as mixing seawater with pure water [Millero 2000]. The differences in true salinity (based on a complete chemical analysis of seawater) and salinity measured by conductivity (PSS) may be as high as 0.04 increasing with dilution of seawater with river water. If the density of estuarine water is estimated from salinity derived from conductivity measurements, these differences could result in $\pm 50 \times 10^{-3}$ kg m^{-3} errors. Differences between the ratio of the major inorganic constituents between pore water and the overlying seawater are largely unknown, but calcium (Ca^{2+}) and magnesium (Mg^{2+}) are both dissolved and precipitated in pore waters in carbonate sediments, sulfate (SO_4^{2-}) produced in anoxic pore waters and potassium (K^+) and Mg^{2+} concentrations altered during diagenetic alteration of clay minerals. These effects are briefly discussed in Ch. 3.

C Coefficient of Variation of Sediment Parameters

Table C.1. Coefficients of variation for sediment parameters: sound speed (V_p), compressional wave attenuation (α_p), mean grain size (M_z), porosity (β), bulk density (ρ), and particle (or grain) density (ρ_g). Data are listed in order of increasing grain size, with siliciclastic sediments first and carbonate sediments second. Coefficients of variation are not given for sound speed ratio, porosity in percent (η), and the attenuation parameter (k), as these are the same as for the quantity from which they are derived.

Siliciclastic							
Site	M_z	CV	CV	CV	CV	CV	CV
		V_p	α_p	M_z	β	ρ	ρ_g
SG98-6	0.08	1.41	14.99	366.91	4.06	1.62	-
SG98-1	0.84	1.94	30.32	9.49	2.88	0.94	0.18
MVCO/crse	0.85	1.05	30.92	22.90	2.07	0.52	0.19
PCII	0.85	1.69	33.18	16.71	8.49	2.87	0.39
KB/lyn	0.90	2.01	27.53	75.76	6.03	2.06	0.59
Misby/crse	0.95	1.76	44.53	11.67	-	-	-
PC93	0.98	1.26	28.00	39.44	4.55	1.53	1.33
PE00	1.21	0.90	23.56	5.53	3.46	1.17	0.91
SAX99	1.27	0.72	31.44	13.82	2.04	0.79	0.96
PE99	1.28	0.61	32.26	18.86	2.56	1.17	1.09
KB/bar	1.33	1.05	39.36	17.42	4.14	1.27	0.24
Hood Canal	1.34	1.07	41.80	17.95	4.43	1.35	0.49
Charl/crse	1.44	1.05	26.45	19.33	3.90	1.25	-
SAX04	1.48	0.54	30.00	10.70	1.84	0.59	1.51
TBay/crse	1.49	2.72	23.97	79.66	18.60	6.90	1.09
SG98-10	1.65	2.41	79.21	77.44	22.33	7.54	1.51
IRB	1.77	1.85	83.22	37.83	4.00	2.06	2.62
NS	1.87	1.91	47.72	22.39	8.51	3.26	1.36
Charl/fine	1.97	0.97	44.05	11.12	4.75	1.54	-
JDF2	2.0	1.49	34.03	6.91	6.29	2.03	0.23
MonPt	2.04	1.06	16.62	3.62	3.24	0.96	-

Table C.1. Continued

Siliciclastic							
Site	M_z	CV	CV	CV	CV	CV	CV
		V_p	α_p	M_z	β	ρ	ρ_g
NoSea/C1	2.09	0.43	29.04	3.81	4.03	1.13	0.09
SG98-8	2.14	2.43	61.20	23.81	4.51	1.39	1.19
ATB/B14	2.15	0.29	48.35	3.52	2.07	0.67	-
PCB99	2.24	0.56	40.37	4.34	2.46	0.89	2.89
JDF5	2.31	2.28	24.83	7.34	9.50	3.73	-
PCB I, II	2.34	1.62	41.49	4.65	6.45	1.98	0.31
ATB/G40	2.52	1.15	32.71	3.38	4.54	1.63	0.39
Duck	2.53	0.34	31.37	2.87	1.92	1.07	1.48
PC84	2.61	0.91	17.53	4.67	3.25	1.02	-
LTB	2.62	1.76	46.03	8.11	7.91	3.03	0.66
MVCO/fine	2.62	0.64	40.48	6.80	3.84	1.27	0.70
TBay/fine	2.89	0.69	24.19	3.12	3.52	1.18	0.51
JDF6	2.94	3.81	29.18	41.12	15.66	6.57	0.52
Quinault	2.94	1.23	42.91	3.53	5.19	1.78	-
Tirrenia	3.72	0.30	7.20	5.96	2.05	0.81	0.22
Misby/fine	3.77	1.37	20.77	12.92	-	-	-
VAzzura	4.03	1.22	21.28	7.36	5.83	2.19	-
JDF1	4.37	0.92	14.10	9.70	4.98	2.78	-
ER/inshore	4.39	1.82	25.04	16.26	6.04	2.33	0.76
Monasteroli	5.12	2.65	15.58	27.62	6.76	1.90	-
Arafura	5.24	0.38	34.67	14.87	5.80	4.60	-
Tellaro	6.08	4.32	40.28	36.19	9.60	4.36	-
RussRiver	6.35	0.55	14.94	7.05	6.17	4.33	0.97
EelRiver/C70	6.46	1.44	12.05	5.72	8.86	4.10	0.40
ATB/scour	6.65	3.50	75.57	49.11	20.76	18.11	4.46
JDF4	6.93	0.29	20.31	3.34	3.72	3.52	1.16
EelRiver/S60	7.24	1.97	32.01	9.99	8.97	4.80	0.62
EelRiver/U70	7.39	1.53	24.02	8.85	9.81	5.72	0.99
LISound	7.64	0.80	-	11.43	6.36	5.65	-
EelRiver/S70/S80	7.86	0.90	26.51	5.52	6.72	4.07	0.56
Orcas	8.08	0.36	15.07	6.94	5.62	4.58	0.52
CLBight	8.10	0.13	41.48	3.01	1.82	2.44	1.80
EelRiver/O70	8.22	1.29	34.80	9.56	6.27	3.81	0.95
JDF7	8.50	0.21	21.88	2.70	3.46	3.86	0.78
EelRiver/I70	8.53	1.27	22.82	5.48	6.62	3.59	0.62
STeresa	8.78	1.53	83.53	10.22	12.99	9.49	-
Viareggio	8.98	0.59	51.22	2.23	7.45	4.75	-
EelRiver/L70	8.98	1.36	33.59	7.68	7.13	4.20	0.90
Portovenere	9.45	0.38	35.42	1.73	6.25	4.56	-
Eck93	9.88	0.25	22.27	3.52	1.47	1.61	3.84
Diga	10.05	0.24	88.90	0.80	2.45	1.90	-
SABay	10.94	0.13	24.86	3.81	2.33	2.42	3.92

Table C.1. Continued

Carbonate							
Site	M_z	CV	CV	CV	CV	CV	CV
		V_p	α_p	M_z	β	ρ	ρ_g
LFK/crse	0.54	1.33	35.65	31.90	3.84	1.44	0.27
Hawaii/crse	0.74	0.92	28.75	13.94	5.44	2.00	0.26
Dtortugas/crse	1.12	0.54	23.39	10.93	1.61	0.46	0.51
RebShoal	1.26	1.59	37.61	11.51	5.64	2.12	0.59
SG98-3/ooid sand	1.55	0.96	29.82	3.44	1.78	0.53	0.27
SG98-2	1.57	1.52	28.35	6.12	4.30	1.96	0.42
SG98-3/skel. sand	1.78	1.69	28.10	3.93	4.36	1.74	0.20
Hawaii-2	2.33	0.46	32.58	14.06	5.08	1.95	0.73
Hawaii-4	3.88	0.30	5.99	3.24	3.29	1.89	0.45
LFK/fine	5.40	1.18	14.01	17.71	7.34	4.20	0.76
SG98-5	5.85	0.53	13.86	2.79	6.52	4.12	-
MarqKeys	6.15	0.73	37.15	15.66	8.90	5.57	0.83
Dtortugas/fine	6.58	0.64	14.10	5.84	7.51	4.72	1.79
Hawaii/mud	8.67	0.37	10.04	5.50	3.14	3.42	1.48

D Roughness and Volume Statistics

Most of the scattering models discussed in this monograph require power spectra for roughness or heterogeneity as inputs. Often, these are not available in the form required by the model in question, and some sort of conversion is required. A commonly encountered example is conversion of 1D roughness spectra to the 2D form needed for roughness scattering models. Technical issues such as these are the subject of this appendix.

D.1 Interface Roughness

This section expands the statistical description of seafloor roughness given in Sect. 6.1. The roughness scattering models considered in Chs. 13 and 15 assume the seafloor relief function, $f(\mathbf{R})$, (8.32) is a zero-mean, stationary random process with spatial covariance that depends only on the difference of the coordinates of the two points in question:

$$B(\mathbf{R}) = < f(\mathbf{R}_0 + \mathbf{R}) f(\mathbf{R}_0) > \ . \tag{D.1}$$

In the Kirchhoff and small-slope approximations (Sects. 13.3, 13.4, and Appendix L), the additional assumption is made that $f(\mathbf{R})$ is a Gaussian random process. In this case, the covariance suffices to completely define the statistics of interface roughness.

If the covariance depends only on the magnitude, R, of the lag vector, the roughness is said to be *isotropic*. This will not be the case if the seafloor relief includes directional ripples. The mean-square roughness, h^2, is equal to the covariance at zero lag:

$$h^2 = B(\mathbf{0}) \ . \tag{D.2}$$

The Fourier transform of the covariance is the roughness spectrum (the Wiener–Khintchine theorem), used in the perturbation approximation:

$$W(\mathbf{K}) = \frac{1}{(2\pi)^2} \int B(\mathbf{R}) e^{-i\mathbf{K}\cdot\mathbf{R}} d^2 R \ . \tag{D.3}$$

From this it follows that the mean-square roughness is the integral of the spectrum over all $\mathbf{K}$:

$$h^2 = \int W(\mathbf{K}) d^2K \ . \tag{D.4}$$

The small-slope and Kirchhoff approximations require the structure function, $S(\mathbf{R})$,

$$S(\mathbf{R}) = < [f(\mathbf{R}_0 + \mathbf{R}) - f(\mathbf{R}_0)]^2 > \ . \tag{D.5}$$

This gives the mean-square height difference of two points on the seafloor separated horizontally by the vector $\mathbf{R}$. Carrying out the square operation in (D.5) and using (D.1), the structure can be related to the covariance as follows:

$$S(\mathbf{R}) = 2[B(\mathbf{0}) - B(\mathbf{R})] \ . \tag{D.6}$$

D.1.1 Specific Roughness Spectral Forms

The "von Karman" spectral form is sometimes used in scattering models:

$$W(\mathbf{K}) = \frac{w_2}{(K^2 + K_0^2)^{\gamma_2/2}} \ . \tag{D.7}$$

The mean-square roughness in this case is

$$h^2 = \frac{2\pi w_2}{(\gamma_2 - 2) K_0^{(\gamma_2 - 2)}} \ . \tag{D.8}$$

The von Karman spectrum approaches a constant value as wavenumber approaches zero, and for large wavenumbers ($K >> K_0$) approaches zero as an inverse power of wavenumber. The parameter K_0 is sometimes referred to as the "cutoff" wavenumber, and its inverse defines the "outer scale," determining the largest horizontal scale of importance. In most of the examples presented in this monograph, the roughness spectrum will be assumed to obey a simple power law,

$$W(\mathbf{K}) = \frac{w_2}{K^{\gamma_2}} \ . \tag{D.9}$$

This is the von Karman spectrum with the outer scale set to infinity. The parameter w_2 will be called the "spectral strength," and the parameter γ_2 will be called the "spectral exponent." This spectrum depends only on the magnitude, K, of the wave vector, that is, the seafloor roughness is assumed to be isotropic. This idealized spectrum is infinite at $\mathbf{K} = \mathbf{0}$, and this non-physical behavior must be remembered when interpreting model results. One resulting difficulty is that the covariance cannot be defined, as the RMS roughness is infinite. In such cases the following transform [Ishimaru 1997, Appendix B] can be used to obtain the structure function from the spectrum without the covariance as an intermediary:

$$S(\mathbf{R}) = 2 \int (1 - \cos \mathbf{K} \cdot \mathbf{R}) W(\mathbf{K}) d^2K \ . \tag{D.10}$$

For the power-law spectrum (D.9), this transform yields a structure function that is also power-behaved [D. Jackson et al. 1986a]:

$$S(\mathbf{R}) = C_h^2 R^{2\alpha} \ , \tag{D.11}$$

where the square of the "structure constant" is related to the parameters of the power-law spectrum through

$$C_h^2 = \frac{2\pi w_2 \Gamma(2-\alpha) 2^{-2\alpha}}{\alpha(1-\alpha)\Gamma(1+\alpha)} \ , \tag{D.12}$$

where Γ is the gamma function, and the exponent is related to the spectral exponent, γ_2, as follows:

$$\alpha = \frac{\gamma_2}{2} - 1 \ . \tag{D.13}$$

In addition to its utility in the Kirchhoff and small-slope approximations, the structure function reveals an important property of roughness that has (approximately) a power-law spectrum. It shows that the RMS roughness must increase as the size of the patch of seafloor over which it is measured increases. For example, if the spectral exponent is $\gamma_2 = 3$, then the structure function is $S(\mathbf{R}) = C_h^2 R$. Thus, the RMS height difference of two points on the seafloor increases as the square root of their separation. In general, for the normal range of spectral exponent ($2 < \gamma_2 < 4$), the relief of a patch of seafloor increases with the size of the patch. This behavior persists until the patch size approaches the outer scale at which the spectrum "rolls over," departing from the pure power law.

D.1.2 Conversion of Roughness Spectra

Conversion of 1D measured spectra to the 2D form needed for scattering models is based on the 1D Fourier transform relation between the covariance measured along a 1D track and the 1D spectrum:

$$W_1(K_x) = \frac{1}{2\pi} \int B(x,\ 0) e^{-iK_x x} dx \ . \tag{D.14}$$

Setting $y = 0$ ($\mathbf{R} = (x,\ 0)$) in the inverse Fourier transform gives the covariance in terms of the spectrum,

$$B(x,\ 0) = \int\int W(K_x,\ K_y) e^{iK_x x} dK_x dK_y \ . \tag{D.15}$$

Finally, the 1D spectrum can be obtained from the 2D spectrum as follows:

$$W_1(K_x) = \int W(K_x,\ K_y) dK_y \ . \tag{D.16}$$

The 1D spectrum is the "marginal spectrum" obtained by integrating the 2D spectrum over the wave vector component orthogonal to the 1D measurement track. Applied to the von Karman spectrum, (D.16) yields

$$W_1(K) = \frac{w_1}{(K^2 + K_0^2)^{\gamma_1/2}} , \tag{D.17}$$

with

$$\gamma_2 = \gamma_1 + 1 \tag{D.18}$$

and

$$w_2 = w_1 \frac{\Gamma(\frac{\gamma_2}{2})}{\sqrt{\pi}\Gamma(\frac{\gamma_2 - 2}{2})} . \tag{D.19}$$

In general, the inverse procedure cannot be carried out, as there is no unique 2D spectrum corresponding to a given 1D spectrum. If isotropy is assumed, however, the relation

$$W(\mathbf{K}) = \frac{1}{\pi} \int_0^\infty \left[W_1(K) - W_1(\sqrt{K^2 - q^2}) \right] q^{-2} dq \tag{D.20}$$

can be used [Henyey 1991, Williams et al. 2002b].

In addition to conversion of the dimensionality of the spectrum, it is sometimes necessary to convert between differing spectral definitions. For example, [Briggs 1989] presents 1D spectra in terms of spatial frequency, F, where

$$K = 2\pi F . \tag{D.21}$$

The 1D spectrum in terms of spatial frequency will be denoted $\Phi_1(F)$, and is related to the wavenumber spectrum as follows:

$$\Phi_1(F) = 2\pi W_1(2\pi F) . \tag{D.22}$$

With this definition, the integral of $\Phi(F)$ over all F is equal to the mean-square roughness, h^2. Briggs' results are summarized by fits to the pure power-law spectrum

$$\Phi_1(F) = \frac{\phi_1}{F^{\gamma_1 - 1}} , \tag{D.23}$$

where

$$\phi_1 = \frac{w_1}{(2\pi)^{\gamma_1 - 1}} . \tag{D.24}$$

This expression allows conversion of Briggs' spectra to 1D wavenumber spectra, and (D.18) and (D.19) provide the final link to the 2D spectrum required by the models presented in this monograph.

D.1.3 Units of Roughness Spectra

The units or dimensions of roughness spectra deserve some comment. Usually, the term "units" is applied to specific measures, meters, kilograms, etc., while the term "dimensions" is generic: length, mass, etc. The term "dimension" will be used in the generic sense with the warning that it must not be confused with the use of "dimension" in the previous section, where one thinks of 1D and 2D spectra. Considering (6.4), it can be seen that the dimensions of covariance are $(\text{length})^2$. Then, from (6.5), the dimensions of the 2D spectrum, $W(\mathbf{K})$, are $(\text{length})^4$. The dimensions of the spectral strength, w_2, follow from (D.9) as $(\text{length})^{4-\gamma_2}$. This rather awkward dimension can be simplified by introducing a dimensional constant, h_0, in (D.9) so that $W(\mathbf{K}) = w_2/(h_0 K)^{\gamma_2}$, e.g., [Mourad and Jackson 1993, Williams et al. 2001a]. In this approach, w_2 has dimensions $(\text{length})^4$, independent of the value of γ_2. This approach invites error when changing units (as from cm to m) by suggesting that w_2 in cm^4 should be divided by 10^8 to obtain w_2 in m^4. The more awkward dimensioning of w_2 has the virtue of leading one directly to the proper scaling (w_2 in cm units should be divided by $100^{4-\gamma_2}$ to obtain w_2 in m units). Equations (D.19) and (D.24) show that w_1 and ϕ_1 have the same units as w_2.

D.2 Volume Heterogeneity

This section expands on the material of Sect. 7.2 and supports the use of heterogeneity spectra in volume scattering models (Sects. 14.1.4 and 14.2.2). Several technical issues arise in the use of spectra for volume heterogeneity in acoustic scattering models. Often, the available fluctuation data do not include all the geoacoustic parameters required by the model. For example, density fluctuation data may be available, but both density and compressibility spectra may be needed. As for the roughness case, it may be necessary to convert 1D measured spectra to higher dimensionality (3D). Finally, the finite resolution of many heterogeneity measurements can cause bias in spectral estimates. These issues are the subject of this section.

It will be assumed that, over the region of interest, the normalized fluctuations of the geoacoustic parameters, α and β (Sect. 7.2), are spatially stationary so that the general covariance depends only on lag, $\mathbf{r} = (x, y, z)$:

$$< \gamma_\alpha(\mathbf{r}_1)\gamma_\beta(\mathbf{r}_1 + \mathbf{r}) >= B_{\alpha\beta}(\mathbf{r}) \ . \tag{D.25}$$

The following symmetry property follows from (D.25):

$$B_{\alpha\beta}(\mathbf{r}) = B_{\beta\alpha}(-\mathbf{r}) \ , \tag{D.26}$$

hence the ordinary covariance is an even function of the lag:

$$B_{\alpha\alpha}(\mathbf{r}) = B_{\alpha\alpha}(-\mathbf{r}) \ . \tag{D.27}$$

A 3D Fourier transform can be performed to obtain the cross-power spectrum

$$W_{\alpha\beta}(\mathbf{k}) = \frac{1}{(2\pi)^3} \int B_{\alpha\beta}(\mathbf{r}) e^{-i\mathbf{k}\cdot\mathbf{r}} d^3r \ . \tag{D.28}$$

Here, $\mathbf{k} = (k_x, k_y, k_z)$ is the wave vector expressing spatial radian frequency. If the fluctuations are stationary only over a layer of finite thickness, then the covariance is not defined over the infinite range of lags appearing in the integration of (D.28). In such cases one may hope that the covariance can be fitted by some model that allows the necessary extrapolation. Setting $\alpha = \beta$ in (D.28), one has the ordinary spectrum. It is important to note that the spectral definition used here has a normalization such that the integral over all of wave vector space yields the second moment (covariance at zero lag) of the corresponding fluctuations. That is,

$$\int W_{\alpha\beta}(\mathbf{k}) d^3k = < \gamma_\alpha(\mathbf{r}) \gamma_\beta(\mathbf{r}) > \ . \tag{D.29}$$

Because of the assumed stationarity, the value of the coordinate vector, $\mathbf{r}$, on the right-hand side of (D.29) is immaterial. For the ordinary spectrum ($\alpha = \beta$), the integral over all $\mathbf{k}$ gives the variance of γ_α,

$$\hat{\sigma}_\alpha^2 = \int W_{\alpha\alpha}(\mathbf{k}) d^3k \ . \tag{D.30}$$

If the covariance is real, the Fourier transform relation (D.28) requires

$$W_{\alpha\beta}(\mathbf{k}) = W_{\alpha\beta}^*(-\mathbf{k}) \ . \tag{D.31}$$

The symmetry condition (D.26) can be used to obtain the following relation:

$$W_{\alpha\beta}(\mathbf{k}) = W_{\beta\alpha}^*(\mathbf{k}) \ . \tag{D.32}$$

Setting $\alpha = \beta$, these conditions require that the ordinary spectrum, $W_{\alpha\alpha}$, be both real and even, but the cross-spectra need not be real or even.

Other symmetries of the covariance and spectrum are also of interest. It is often assumed that fluctuation statistics are isotropic. In particular, this means that the covariance depends only on the magnitude, r, of the lag vector, $\mathbf{r}$, not its direction. It follows that the spectrum would then depend only on the wavenumber, k, which is the magnitude of the wave vector, $\mathbf{k}$. Isotropy is at best an approximate symmetry owing to the general dependence of sediment properties on depth. A weaker and more likely symmetry is transverse isotropy, in which the covariance is independent of the direction of the horizontal lag vector, $\mathbf{R}$, that is, dependent on $R = \sqrt{x^2 + y^2}$ but not on $\phi = \tan^{-1}(y/x)$. Note that transverse isotropy may hold even when the fluctuation statistics are not stationary in the vertical. In this case, a 3D spectrum cannot be defined.

D.2.1 Converting Dimensionality of Heterogeneity Spectra

The conversion from 3D spectra to 1D can be obtained by following the same logic used in Sect. D.1.2 for the rough interface spectra, the 1D spectrum is

$$W_{z\alpha\beta}(k_z) = \frac{1}{2\pi} \int B_{\alpha\alpha}(0,\ 0,\ z) e^{-ik_z r} dz \ . \tag{D.33}$$

Using the inverse of (D.28) and setting $x = y = 0$, one obtains

$$W_{z\alpha\beta}(k_z) = \int\int W_{\alpha\alpha}(k_x,\ k_y,\ k_z) dk_x dk_y \ . \tag{D.34}$$

This 1D spectrum in k_z is the marginal spectrum obtained by integrating the 3D spectrum over the orthogonal variables, k_x and k_y. For the case of the model spectrum (7.14), this yields

$$W_{z\alpha\alpha}(k_z) = \frac{2\pi w_{3\alpha\alpha}}{(\gamma_2 - 2)\Lambda^2 (k_z^2 + L_c^{-2})^{(\gamma_3 - 2)/2}} \ . \tag{D.35}$$

If the measured 1D spectrum can be approximated by this form, the parameters γ_3 and $w_{3\alpha\alpha}\Lambda^2$ can be determined. If isotropy can be assumed ($\Lambda = 1$), $w_{3\alpha\alpha}$ is then fully determined. More generally [Yefimov et al. 1988, Ishimaru 1997], if the 3D spectrum is isotropic (dependent only on the magnitude, k, of the wave vector, $\mathbf{k}$),

$$W_{\alpha\alpha}(\mathbf{k}) = \frac{-1}{2\pi k} \frac{dW_{z\alpha\alpha}(k)}{dk} \ . \tag{D.36}$$

It can readily be verified that (D.36) applied to (D.35) in the isotropic case yields (7.14).

D.2.2 Relations between Parameter Fluctuations

Section 7.3 gives "exact" relations between fluctuating parameters. The quotation marks are used because these relations are only exact in the limit as fluctuation amplitude vanishes. Even so they are undoubtedly more accurate than the conjectured relations to be presented in this section.

Because few simultaneous determinations of density and compressibility (or sound speed) fluctuations exist, various authors have made simplifying assumptions regarding the relevant spectra. Most of these assumptions are special cases of the following hypothesized relations:

$$W_{\kappa\kappa}(\mathbf{k}) = \mu^2 W_{\rho\rho}(\mathbf{k}) \ , \tag{D.37}$$

$$W_{\rho\kappa}(\mathbf{k}) = \rho_{\rho\kappa} \mu W_{\rho\rho}(\mathbf{k}) \ . \tag{D.38}$$

The parameter $\rho_{\rho\kappa}$ is the density–compressibility correlation coefficient and is restricted to the range $[-1\ 1]$. Without loss of generality, it is assumed that the compressibility–density normalized fluctuation ratio, μ, is positive.

As sound speed fluctuation is sometimes measured, it is useful to express the correlation $\rho_{\rho\kappa}$ in terms of density–compressional wave speed spectra [D. Jackson et al. 1996a]. If these spectra can be fit by forms

$$W_{pp} = \eta^2 W_{\rho\rho} \tag{D.39}$$

and

$$W_{\rho p}(\mathbf{k}) = \rho_{\rho p}\eta W_{\rho\rho}(\mathbf{k}) \ , \tag{D.40}$$

with $\eta > 0$, then the two spectra involving compressibility can be determined using

$$\mu = \sqrt{1 + 4\rho_{\rho p}\eta + 4\eta^2} \tag{D.41}$$

and

$$\rho_{\rho\kappa} = -\frac{2\rho_{\rho p}\eta + 1}{\mu} \ , \tag{D.42}$$

with the positive square root used in (D.41). It is sometimes assumed that compressibility and density fluctuations are perfectly anticorrelated, that is, $\rho_{\rho\kappa} = -1$. In this case, (D.41) and (D.42) require $\rho_{\rho p} = \pm 1$. Looking at the density–compressional wave speed regression (Fig. 5.2), one expects $\rho_{\rho p}$ to be negative for very fine sediments having density less than 1.4 g cm^{-3}, very small for sediments having density near this value, and positive in all other cases.

The values for the compressibility–density fluctuation ratio, μ, in Table 14.1 were determined by two different strategies. For the Arafura, San Francisco, and Orcas sites [D. Jackson et al. 1996a], the compressional wave speed–density fluctuation ratio, η, was found as the ratio of coefficients of variation for the two measured fluctuations, and μ was determined using (D.41) with $\rho_{\rho p} = 1$. No compressional wave speed fluctuation measurements were available for the other two sites, and [Pouliquen and Lyons 2002] assumed $\eta = 1/3$ based on data presented in [Yamamoto 1995, Yamamoto 1996] and further assumed $\rho_{\rho p} = 1$. None of these examples provides an actual measurement of correlation, but efforts in this direction are underway (Briggs and Tang, personal communication).

D.2.3 Bias Due to Finite Resolution

The effects of finite resolution on estimated covariances and spectra will be derived from (7.31) for the averaging functions appropriate to core sections (7.33) and spherical resisitivity probes (7.34).

Filtering of data as defined by (7.31) results in filtering of the covariance. Using (D.25) and (7.31), the covariance obtained from the filtered (measured) data is

$$\tilde{B}_{\alpha\alpha}(\mathbf{r}) = \int B_{\alpha\alpha}(\mathbf{r}')v(\mathbf{r}' - \mathbf{r})d^3r' \ , \tag{D.43}$$

where $B_{\alpha\alpha}$ is the true covariance, and $v(\mathbf{r})$ is the autocorrelation of $y(\mathbf{r})$:

$$v(\mathbf{r}) = \int y(\mathbf{r}')y(\mathbf{r}' - \mathbf{r})d^3r' \ . \tag{D.44}$$

Equation (D.43) expresses the bias in measurements of covariance due to finite resolution. By using Fourier transforms, the equivalent expression for the biased spectrum is found to be

$$\tilde{W}_{\alpha\alpha}(\mathbf{k}) = W_{\alpha\alpha}(\mathbf{k})V(\mathbf{k}) \ , \tag{D.45}$$

where

$$V(\mathbf{k}) = \int v(\mathbf{r})e^{i\mathbf{k}\cdot\mathbf{r}}d^3r \ , \tag{D.46}$$

which can be reexpressed as

$$V(\mathbf{k}) = \left[\int y(\mathbf{r})e^{i\mathbf{k}\cdot\mathbf{r}}d^3r\right]^2 \ . \tag{D.47}$$

For the hockey-puck averaging function appropriate to core sections,

$$V(\mathbf{k}) = b_T^2(K)b_z^2(k_z) \ , \tag{D.48}$$

with

$$b_T(K) = \frac{2}{(Ka)^2}\int_0^{Ka} J_0(u)u du \tag{D.49}$$

and

$$b_z(k_z) = \frac{\sin(k_z\Delta z/2)}{k_z\Delta z/2} \ , \tag{D.50}$$

where

$$K = \sqrt{k_x^2 + k_y^2} \tag{D.51}$$

and

$$k_z = \sqrt{k^2 - K^2} \ . \tag{D.52}$$

Note that the filtering function in the wave vector domain, $V(\mathbf{k})$, is not isotropic in this case. Rather, it is transversely isotropic owing to the cylindrical symmetry of the core section. As porosity and density data from core sections are usually employed to obtain 1D spectra, (D.34) can be used to obtain an expression for the biased 1D spectrum. If the true spectrum is either isotropic or transversely isotropic,

$$\tilde{W}_{z\alpha\alpha}(k_z) = 2\pi b_z^2(k_z)\int_0^{\infty} W_{\alpha\alpha}(K,\ k_z)\left[\frac{2J_1(Ka)}{Ka}\right]^2 K dK \ . \tag{D.53}$$

Considering the bias due to finite resolution in resistivity measurement of porosity, (7.34) can be inserted in (D.47). Then the biased 3D spectrum is obtained by multiplying the true spectrum by

$$V(\mathbf{k}) = \left[ka \int_{ka}^{\infty} \frac{\sin x}{x^3} dx \right]^2 . \tag{D.54}$$

E Complex Representation of Signals

Acoustic measurements are often made using sources that emit pressure waveforms that are confined to a rather narrow band of frequencies. In this case, it is convenient to use complex representation [Ziomek 1995, Ch. 8] for the time dependence of quantities such as pressures and voltages. Let $p(t)$ be the (real) time-dependent pressure at some arbitrary point in space. The equivalent complex signal will be denoted $P(t)$ where the connection between the two is

$$p(t) = \text{Re}\{P(t)e^{-i\omega_c t}\} , \tag{E.1}$$

where $\text{Re}(z)$ is the real part of the complex number, z. The parameter ω_c is the "center frequency" of the pressure signal (in radians/s). To understand the motivation for this widely used representation, consider the magnitude and phase of the complex $P(t)$:

$$P(t) = |P(t)|e^{i\phi(t)} . \tag{E.2}$$

Then

$$p(t) = |P(t)| \cos[\omega_c t + i\phi(t)] . \tag{E.3}$$

The real pressure signal has time-varying peak amplitude equal to the magnitude of the complex signal and time-varying phase equal to the phase of the complex signal. This time-varying amplitude is usually called the "envelope" because, as shown in Fig. E.1, it bounds the positive and negative excursions of the real signal. Thus, the complex representation encapsulates the essential properties of the real signal, its amplitude and phase. If $p(t)$ is a "narrowband" signal, with a frequency spectrum occupying a narrow range of frequencies near ω_c, the complex signal will be slowly varying in time. Note that, in the limiting case in which $p(t)$ is a pure sinusoid having frequency ω_c, $P(t)$ is independent of time. The slowly varying nature of the complex signal provides one of the strongest motivations for its use: when dealing with sampled data, a lower sampling rate will suffice for the complex signal as compared to the real signal.

Although the above discussion uses time-dependent pressure as an example, the complex representation is used for a variety of waveforms, voltages, displacements, electric fields, etc. The complex signal is also referred to by the terms "baseband signal" or "complex envelope," and its real and imaginary

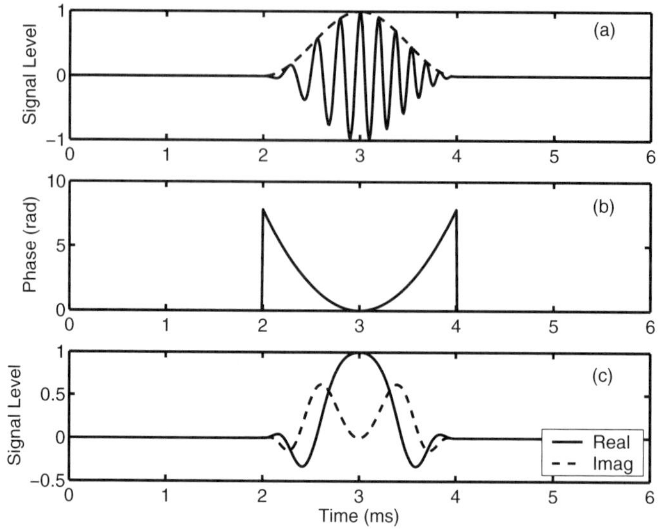

Fig. E.1. Illustration of complex representation of signals, using a frequency-modulated signal having cosine-shaped envelope as an example. (a) The real time-dependent signal (solid line) and its envelope (dashed line), (b) the phase, and (c) real and imaginary parts of the complex representation.

part are sometimes referred to as "quadrature components" or "in-phase and quadrature signals." While (E.1) shows how to obtain the real signal from the complex signal, the experimentalist is faced with the inverse process of obtaining $P(t)$ from $p(t)$ ("demodulation"). An algorithm for this purpose and additional information on complex envelopes can be found in [Ziomek 1995, Ch. 8].

In some applications, the "root-mean-square" (RMS) amplitude of a sinusoidally varying voltage, pressure, etc. is desired. This need arises particularly when considering power, either transmitted or dissipated. The RMS amplitude of a sinusoid having peak amplitude, $|A|$, is $|A|/\sqrt{2}$. The usual practice in acoustics is to measure pressure as RMS amplitude, so the $\sqrt{2}$ factor will be used at several points in this monograph.

Care is needed when time delays are introduced in complex time series. This is necessary in simulations of propagation and scattering, where the distance r, traveled by an acoustic wave at speed, c, translates into a time delay, $t_d = r/c$. Adapting (E.1),

$$p(t - t_d) = \text{Re}\{P(t - t_d)e^{i\omega_c t_d}e^{-i\omega_c t}\} . \tag{E.4}$$

Thus, the baseband representation for the delayed pressure is $P(t-t_d)e^{i\omega_c t_d}$, not simply $P(t - t_d)$.

F Acoustic Transducers

A variety of acoustic sources and receivers are used in high-frequency measurements, although most employ the piezoelectric effect to convert electrical to sound energy and vice versa. Many source and receiver transducers are "arrays" comprised of numerous transducer "elements." A detailed discussion of the physics of sound transduction is beyond the scope of this book; only definitions essential to the interpretation of experimental data will be given here. Figure F.1 shows the magnitude of the sound pressure field emitted by a simple acoustic source, a so-called circular piston, which imparts a uniform velocity to the fluid medium over the surface of the piston. This gives rise to a complicated sound field in the vicinity of the source, the "near field." At a greater distance, the field pattern becomes regular; this is the "far field" region in which the acoustic intensity (radiated power/unit area) obeys the inverse square law when absorption and refraction by the water are neglected. High-frequency seafloor acoustic measurements ordinarily are conducted with the source at a sufficient distance that the seafloor is in the far field region.

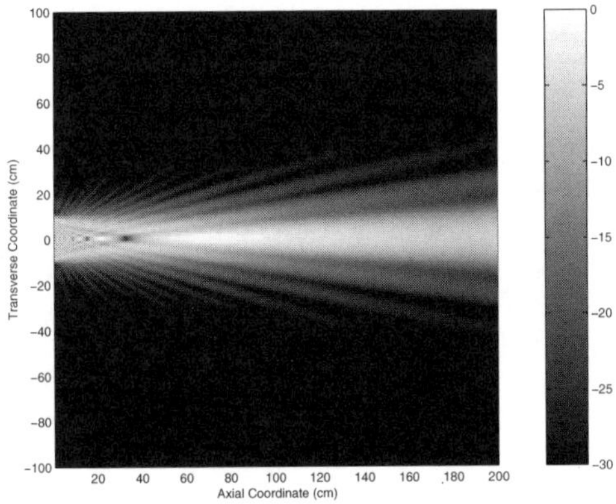

Fig. F.1. Computed pressure field for a circular piston transducer of radius 5 cm operating at 100 kHz. The gray scale is in dB relative to the maximum field value.

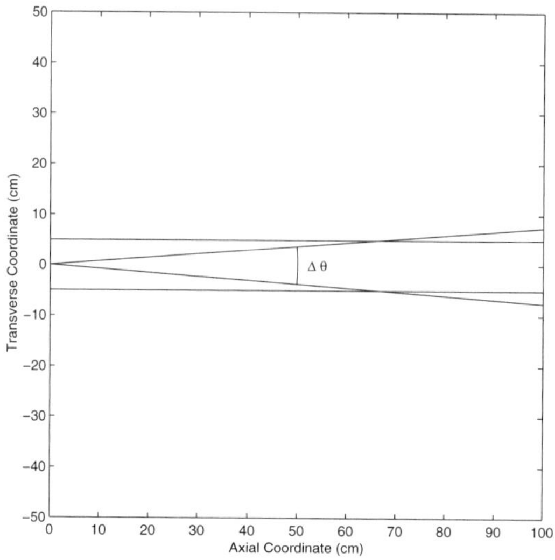

Fig. F.2. Rough estimate of far field region for a circular piston transducer of radius 5 cm operating at 100 kHz. The far field region lies at ranges greater than the range at which the diverging and nondiverging rays intersect (about 65 cm, compare Fig. F.1).

In the far field region, the radiated acoustic field can be at any point approximated as a plane wave having a well-defined direction of propagation (the direction from the transducer to the point in question). An approximate formula for the required minimum range, r, is

$$r > \frac{d^2}{\lambda} , \tag{F.1}$$

where d is the aperture size of the source and λ is the acoustic wavelength in water. Various forms of this criterion can be found, e.g., [Ziomek 1995, p. 413], [Medwin and Clay 1998, p. 143], but all yield similar values for the far-field range. As pictured in Fig. F.2, this condition can be derived quite simply from the fact that diffraction by an aperture of size d gives an angular spread $\Delta\theta$ (measured in radians):

$$\Delta\theta \approx \frac{\lambda}{d} . \tag{F.2}$$

The outgoing acoustic energy takes on well-defined directions of propagation at ranges, r, such that this angular spreading exceeds the size of the aperture

$$r\Delta\theta > d . \tag{F.3}$$

This, combined with (F.2), gives the far-field condition, (F.1). It might seem that the inequality symbol, >, in (F.3) is likely to underestimate the far-field range and should, perhaps, be replaced by >>. Numerical calculations in the literature [Lockwood and Willette 1973] and Fig. F.1 show that the criterion (F.1) is a reasonable and practical approximation. In some cases, the angular spread is caused not by diffraction, but by curvature of the transducer or by electronic beamforming. In these cases, the aperture size, d, is irrelevant, and it is best to estimate the far-field range by eliminating d to obtain

$$r > \frac{\lambda}{(\Delta\theta)^2} \, . \tag{F.4}$$

This formula is also applicable to endfire arrays [Ziomek 1995, Ch. 6] and parametric arrays [Westervelt 1963] where the assumed relation between physical aperture and angular spread does not apply.

The magnitude of the complex source pressure, P_i, in the far field is

$$|P_i| = \frac{\sqrt{2} s_0 |b_x(\theta, \ \phi)|}{r} \, , \tag{F.5}$$

where the function $b_x(\theta, \ \phi)$ is the source directivity, r, θ, and ϕ are spherical coordinates centered on the source, with θ measured from the horizontal rather than the vertical (see Fig. G.3). The subscript "x" on the directivity function is taken from an engineering abbreviation, "xmit" for "transmit." The parameter s_0 is related to the "source level" through (F.7).

While it would be logical to align the coordinate system with the transducer with, say, the z-axis chosen normal to the face of the transducer, it is more convenient in the applications of this monograph to use a system aligned with the seafloor. Thus, the directivity function defined here changes as the transducer depression (tilt) angle changes. In (F.5), refraction and absorption are neglected for simplicity. The subscript "i" is used because this pressure field will be the incident field for a seafloor reflection or scattering measurement. The factor of $\sqrt{2}$ in (F.5) is the ratio of the peak amplitude of a sinusoid to its RMS value. As explained in Appendix E, the magnitude of the complex signal is equal to the peak value of the corresponding real sinusoidal signal. The directivity function, $b_x(\theta, \ \phi)$, is complex in general, with a maximum magnitude of unity in the direction of the "maximum response axis" (MRA). It is convenient to express source directivity in decibels:

$$D_x(\theta, \ \phi) = 20 \log_{10}[|b_x(\theta, \ \phi)|] \, . \tag{F.6}$$

This gives the decibel level of radiated power in the far field relative to the maximum response axis, which may not be the direction specified by $\theta = 0$, $\phi = 0$. The units of D_x are simply "dB" because $b_x(\theta, \ \phi)$ is a dimensionless pressure ratio with the numerator equal to the pressure magnitude in the direction specified by the angular arguments and with the denominator equal

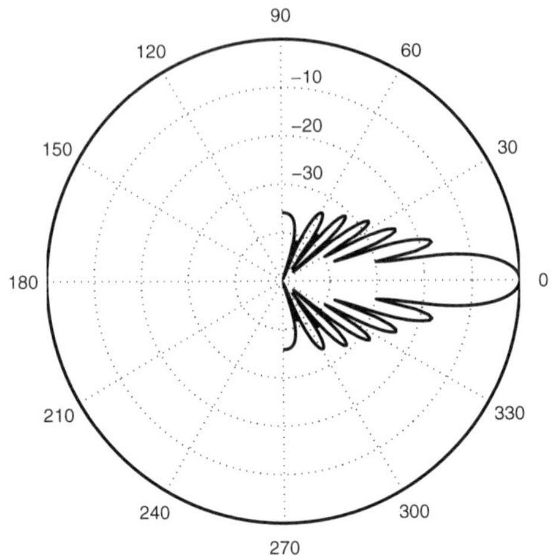

Fig. F.3. Computed directivity in dB for a circular piston transducer of radius 5 cm operating at 100 kHz. The angular variable (in degrees) is measured from the normal to the face of the piston.

to the pressure magnitude on the MRA. Figure F.3 shows the theoretical pattern in dB for the circular piston of Fig. F.1.

The parameter s_0 in (F.5), when divided by range, gives the far-field RMS pressure. The "source level" is defined as

$$SL = 20 \log_{10} \left[\frac{s_0}{p_{ref} r_{ref}} \right] , \tag{F.7}$$

where, conventionally, $p_{ref} = 1$ μPa and $r_{ref} = 1$ m. Dimensional analysis of (F.7) shows that s_0 has dimensions of pressure times length, hence source level would logically be given in units dB μPa m, but, instead, the usual convention in underwater acoustics will be employed in which source level is given in the rather cumbersome but descriptive terms "dB re 1 μPa @ 1 m."

Receiver transducers are also characterized by a complex directivity pattern, $b_r(\theta,\ \phi)$, which gives the angular dependence of the complex voltage at the transducer output terminals for an incident plane pressure wave incident at the direction specified by the angles θ and ϕ. The receiver directivity expressed in dB is

$$D_r(\theta,\ \phi) = 20 \log_{10}[|b_r(\theta,\ \phi)|] . \tag{F.8}$$

For a plane wave incident along the axis of maximum response, the "receiver sensitivity," RS, is defined as

$$RS = 20 \log_{10} \left[\frac{|V|}{|P|} \right] , \tag{F.9}$$

where $|V|$ is the magnitude of the voltage at the output terminals of the transducers for incident pressure having sinusoidal time dependence with magnitude $|P|$. As a ratio is involved, there is no issue as to whether RMS or peak amplitude is used, as long as both voltage and pressure are treated in the same way. The units of receiver sensitivity are dBV/μPa.

Reciprocity demands that, if a transducer is capable of both transmission and reception, its source and receiver directivities must be identical. This is the case for most transducers employing piezoelectric, electrodynamic, or magnetostrictive effects.

G Acoustic Measurements

Although the reflection coefficient and scattering cross section are defined in terms of incident plane waves, field experiments employ source fields that can be approximated as spherical waves. In the case of reflection, this presents no serious difficulty if the source and receiver are at least several wavelengths distant from the seafloor. For scattering, this condition must also be satisfied, but interpretation of the data is more involved, as will be seen.

G.1 Reflection Measurements

The reflection coefficient can be defined if the technical criteria discussed in Sect. 11.1 are met. In this section, it will be assumed that these criteria are satisfied, and the emphasis will be on inversion of the sonar equation to obtain the reflection coefficient. As indicated in Fig. G.1, the reflected wave will appear to originate from an image source and will have pressure magnitude

$$|P| = \sqrt{2}s_0|V_{ww}(\theta)||b_x(\theta,\ 0)|\frac{e^{-k''_w(r_1+r_2)}}{r_1+r_2} , \tag{G.1}$$

where the effective range is the sum of the two distances r_1 and r_2, that is, it is the range from the image source to the field point. As defined in Appendix F, the parameter s_0 determines the source level through relation (F.7), and $b_x(\theta_i,\ 0)$ determines the source directivity. The relation between the conventional attenuation coefficient in water and the imaginary part of the wavenumber in water, k''_w, is analogous to (8.20). In the present example, both the field point (the point at which reflected pressure is measured) and the source are assumed to lie in the x–z plane, where the azimuthal angle, ϕ, appearing in the directivity pattern vanishes. Taking the base-10 logarithm of (G.1), and using the relation between pressure and RMS voltage, V_r, at the terminals of the receiving transducer, one obtains a “sonar” equation in the form

$$20\log_{10}|V_r| = SL + RS - PL - L + D_x(\theta,\ 0) + D_r(\theta,\ 0) , \tag{G.2}$$

where SL, RS, D_x, and D_r are defined in Appendix F and are, respectively, the source level, receiver sensitivity, source directivity, and receiver directivity. The bottom loss (2.11) is denoted L. The propagation loss is

$$PL = 20\log_{10}\left(\frac{r_1 + r_2}{r_{ref}}\right) + \alpha_w(r_1 + r_2) \ . \tag{G.3}$$

Equation (G.2) forms the basis for most measurements of reflection coefficient. It should be noted that a continuous-wave source as assumed here is seldom used, because interfering reflection by the sea surface is best eliminated by time-gating the response to a pulsed source.

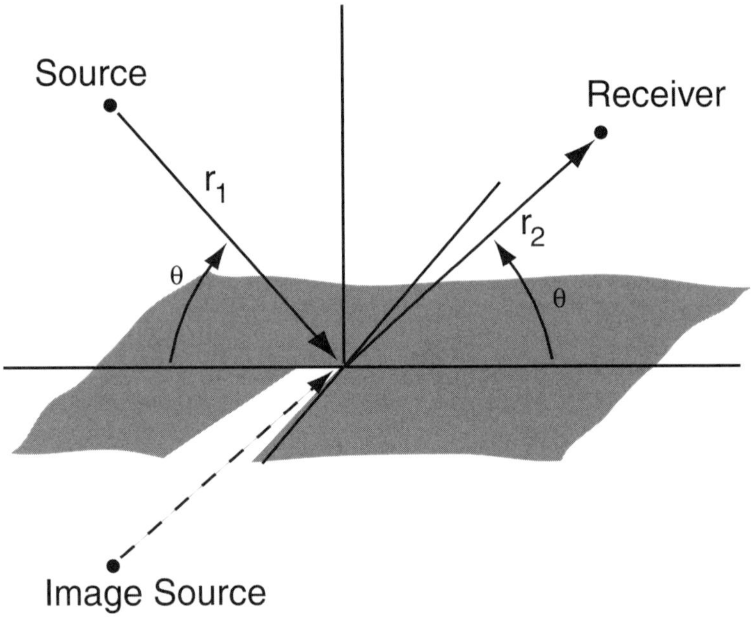

Fig. G.1. Geometry for reflection measurement. The pressure at the receiver location is the same as the pressure that would be produced by an image source, propagating through water, but reduced in amplitude by the reflection coefficient. The source, image source, and receiver locations lie in a vertical plane.

Measurements of pressure magnitude (envelope) yield only the magnitude of the reflection coefficient, not its phase angle. Phase angle measurements are difficult at high frequencies, because the distances from the source to the seafloor and from the seafloor to the receiving pressure transducer are likely to be hundreds or thousands of wavelengths, giving rise to large and variable phase shifts due to fluctuations in source and receiver locations and in the sound speed of the water. Small relative errors in determining propagation phase shift would result in large absolute errors in measured reflection coefficient phase. As noted in Sect. 11.3, however, laboratory measurements capable of measuring reflection phase would pay important dividends.

Technical issues that must be faced in reflection measurements are discussed in Sect. 11.1. One of the most important is the effect of scattering due to interface roughness, to be counteracted by averaging squared reflected pressure magnitude over a large number of measurements obtained at differing, but statistically similar, portions of the seafloor. An algorithm to invert such averaged data, compensating for roughness and sediment volume scattering, is given in Sect. G.2.2. For measurements conducted at ranges greater than a few tens of meters, refraction due to spatial dependence of sound speed in the water may alter the grazing angle, and it may be necessary to replace the spherical-spreading propagation loss (G.3) by one computed using ray tracing.

G.2 Scattering Measurements

The definition of scattering strength given in (2.15) and (2.17) is not immediately applicable to experimental data, as it pertains to an idealized experiment rather than a realistic measurement scenario. Consider the more realistic measurement situation depicted in Fig. G.2 in which a directional source transmits a pulse which is subsequently scattered by the seafloor and measured by a directional receiver. The generalization of (2.15) is simply an integration that sums the contributions of infinitesimal patches of area to obtain the mean-square voltage at the terminals of the receiver:

$$< |V_r(t)|^2 >= s_r^2 \int |P_i|^2 \sigma \frac{e^{-2k''_w |\mathbf{r}_s - \mathbf{r}|}}{|\mathbf{r}_s - \mathbf{r}|^2} |b_r(\theta_r,\ \phi_r)|^2 d^2R\ . \tag{G.4}$$

In (G.4), $RS = 20 \log_{10} s_r$ is the receiver sensitivity giving the voltage per unit pressure, and $b_r(\theta_r,\ \phi_r)$ is the receiver directivity pattern (see Appendix F).

The incoherent sum implied by (G.4) is justified in most cases of interest, as shown in Sect. J.2. As shown in Fig. G.2, the integration is over the plane defining the mean seafloor, $z = 0$, rather than over the rough water–seafloor interface. All the factors inside the integrand of (G.4) depend upon the integration coordinates, $\mathbf{R} = (x, y)$, with $d^2R = dxdy$. For example, the spreading and absorption losses due to propagation from the infinitesimal scattering patch to the receiver depend on the range $|\mathbf{r}_s - \mathbf{r}|$, where $\mathbf{r} = (\mathbf{R},\ 0)$. Likewise, incident source field magnitude, $|P_i|$, depends on $(x,\ y)$ through spreading and absorption loss factors as well as the source directivity pattern, $b_x(\theta_x,\ \phi_x)$. The angular variables in the directivities are denoted θ_x, ϕ_x, θ_r, ϕ_r to indicate that these angular coordinates are not simply related to the spherical coordinates aligned with the Cartesian system, $(x,\ y,\ z)$. In fact, determining the angular coordinates to be used in evaluating the source and receiver directivities as functions of the integration variables $(x,\ y)$ can be rather involved in the general bistatic case [Williams and Jackson 1998]. Finally, the scattering strength has angular dependence that translates into dependence on the integration coordinates, with the relationships given in Sect. J.1.

The integration limits in (G.4) depend on the length of the transmitted signal, and on the time, t, at which the receiver voltage is evaluated. The range–time relation is

$$|\mathbf{r} - \mathbf{r}_i| + |\mathbf{r}_s - \mathbf{r}| = c_w t_0(\mathbf{R}) \ , \tag{G.5}$$

where $t_0(\mathbf{R})$ is the time required for acoustic energy to travel the total distance from the source to the scattering patch and then to the receiver. This time varies with the position, $\mathbf{R}$, of the scattering patch.

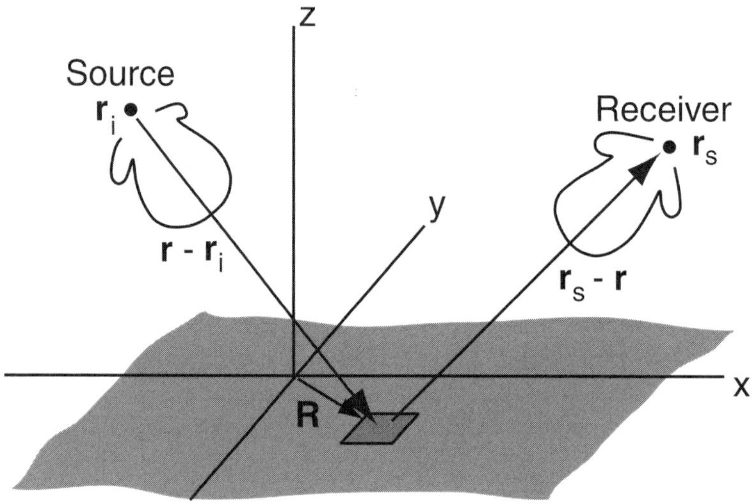

Fig. G.2. Geometry for computing the mean-square received voltage in a bistatic scattering scenario. Note that $\mathbf{r} = (\mathbf{R},\ 0)$.

In the simplest case, if the transmitted pulse has a rectangular envelope of length τ, the limits of the integration are set by constraining $(x,\ y)$ such that

$$t - \frac{\tau}{2} < t_0(\mathbf{R}) < t - \frac{\tau}{2} , \tag{G.6}$$

where t is the time at which one wishes to estimate the scattered mean-square pressure. The corresponding limits on the integration can be shown to be two ellipses. These, in turn, are the intersections with the mean seafloor of the two ellipsoids of revolution defined by (G.5) and (G.6). These ellipsoids have the source and receiver locations as foci. If the transmitted pulse is not rectangular, (G.4) can be evaluated for an infinitesimally short pulse and the result can be convolved with the squared transmitted pulse envelope, as in [Williams and Jackson 1998]. A simple, but not necessarily efficient, numerical means of evaluating (G.4) is to divide the integration area into a grid sufficiently fine to accommodate the variability of all the factors in the integrand. The travel time from the source to the receiver is evaluated for each grid point and the integral is approximated as a sum over all grid points of the squared transmitted pulse envelope, appropriately delayed and weighted by the factors in the integrand and the area represented by one grid point.

Evaluation of (G.4) is simple for the commonly encountered case of backscattering, in which $\mathbf{r}_i = \mathbf{r}_s$. Combining (G.5) and (G.6),

$$r_t - \frac{c_w \tau}{4} < |\mathbf{r} - \mathbf{r}_i| < r_t - \frac{c_w \tau}{4} , \tag{G.7}$$

where

$$r_t = c_w t/2 \tag{G.8}$$

is the range associated with the time of interest, t. Inequality (G.7) defines the range limits of the *ensonified region*, the portion of the seafloor contributing to the backscattered pressure at time t.

G.2.1 Moderate-to-Small Grazing Angles

If the pulse length is short compared to the distances involved ($c_w\tau/2 << r_t$), then, as shown in Fig. G.3, the ensonified region is a thin annular strip on the seafloor having width $w = (c_w\tau)/(2\cos\theta)$. For convenience, the source–receiver location has been placed on the z-axis, in which case the time-dependent grazing angle is

$$\theta = \sin^{-1}(z_i/r_t) . \tag{G.9}$$

Because the strip defined by the pulse duration is narrow compared to the range, the integration can be reduced to a one-dimensional integral over a single azimuthal variable. That is, $d^2R = wr_t \cos\theta d\phi$, where $r_t \cos\theta$ is the horizontal range, and ϕ is the azimuthal coordinate. Note that the cosine factors appearing in the strip width, w, and horizontal range cancel. Using cylindrical coordinates, $(R,\ \phi)$, the integral over R simply leaves a multiplier

$(c_w\tau/2)r_t$, because range, $r_1 \approx r_t$, and grazing angle, θ, are assumed to be constant over the region of integration. Combining these results, (G.4) reduces to

$$< |V_r(t)|^2 >= s_r^2\sigma\frac{e^{-2k_w''r_t}}{r_t^2}\int |P_i|^2|b_r(\theta,\ \phi)|^2 d\phi\ . \tag{G.10}$$

The incident pressure can be expressed in terms of the source level, propagation loss, and source directivity (Appendix F), giving

$$< |V_r(t)|^2 > /2 = (s_r s_0)^2\sigma\frac{e^{-4k_w''r_t}}{r_t^3}\frac{c_w\tau}{2}|b_r(\theta,\ 0)b_x(\theta,\ 0)|^2\Psi\ , \tag{G.11}$$

where

$$\Psi = \frac{\int |b_r(\theta,\ \phi)b_x(\theta,\ \phi)|^2 d\phi}{|b_r(\theta,\ 0)b_x(\theta,\ 0)|^2} \tag{G.12}$$

is a measure of the angular width of the ensonified region. Division by a factor of 2 on the left-hand side of (G.11) follows from definitions of Appendix F and expresses the fact that $|V_r(t)|^2/2$ is the mean square of real voltage, while $|V_r(t)|^2$ is the mean square of the envelope (see Appendix H). If (G.11) is expressed in dB by taking $10\log_{10}$ of both sides, a form of the sonar equation results:

$$10\log_{10}[< |V_r(t)|^2 > /2] = SL + RS + S_b + D_x(\theta,0) + D_r(\theta,0)$$

$$-2\alpha_p r_t - 30\log_{10}(r_t/r_{ref}) + 10\log_{10}[c_w\tau\Psi/(2r_{ref})]\ . \tag{G.13}$$

In (G.13), S_b is the bottom scattering strength (2.17), specialized to the monostatic (backscattering) case.

Recalling the relation between range and time, $r_t = c_w t/2$, one can examine the time dependence of the mean-square receiver voltage for a pulsed transmission. Figure G.4 is a sketch of the type of time dependence that might be expected for the conditions depicted in Fig. G.3. At the earliest time, a narrow impulsive feature is seen. This is due to reflection from the seafloor and is often referred to as the "fathometer return." The fathometer return arrives via a vertical path and may be weak if the combined source–receiver directivity function is small in the vertical direction. Following the fathometer return, a somewhat broader feature is seen. This feature is expected if a prominent sidelobe is pointed toward the seafloor, as in Fig. G.3. The sidelobe feature is followed by a broad feature due to the main lobe. The main lobe return is useful for measuring backscattering strength by solving (G.13) for the bottom scattering strength. The sidelobe return is usually of no value owing to uncertainty in the directivity functions.

The integral (G.12) used to obtain the angular width of the ensonified region can be performed numerically, but often sufficient accuracy can be obtained using an analytic approximation. One such approximation [D. Jackson,

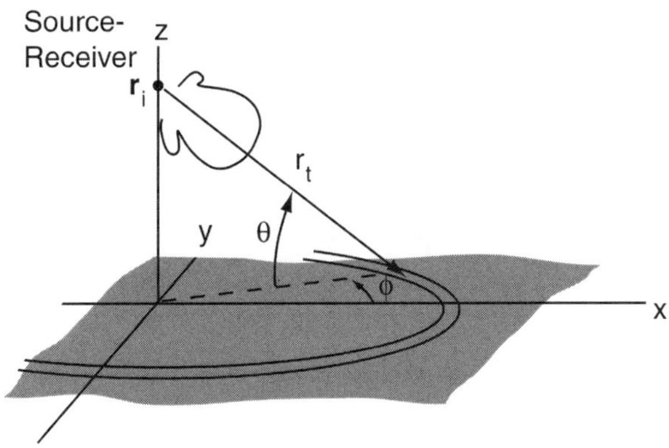

Fig. G.3. Geometry for computing the mean-square received voltage envelope in a backscattering, or monostatic, scenario.

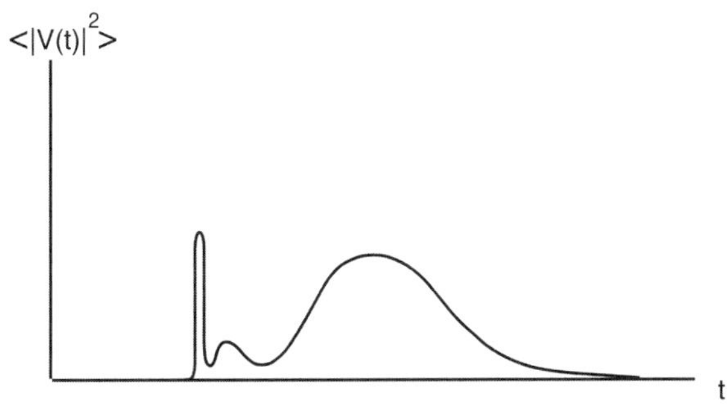

Fig. G.4. Sketch of time dependence of mean-square receiver output voltage for the monostatic configuration of Fig. G.3, showing the following features, in order of increasing time: fathometer return, sidelobe return, and main lobe return.

unpublished] assumes that the product of the squared source and receiver directivity functions can be factored into separate functions of a depression angle and an azimuthal angle. It is assumed that the source and receiver transducers are colocated, thus only the monostatic case is considered. As illustrated in Fig. G.5, a Cartesian coordinate system (x', y', z') is used with its origin on the transducer, with the x'-axis taken in the direction of the maximum response axis (MRA), the z'-axis pointing vertically downward when

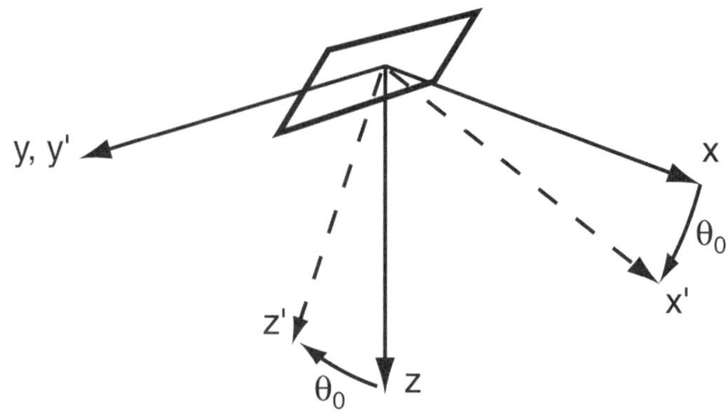

Fig. G.5. The two Cartesian coordinate systems used in deriving an analytic expression for the beam-pattern integral (G.12). The primed system is tied to the transducer (indicated by the tilted rectangle) and the unprimed system has its z-axis pointed vertically downward. The depression angle of the transducer is denoted θ_0.

the MRA is horizontal, and the y'-axis chosen to complete the right-handed coordinate system. In terms of these coordinates, the angles mentioned above are

$$\theta' = \tan^{-1}\frac{z'}{x'} \tag{G.14}$$

and

$$\phi' = \tan^{-1}\frac{y'}{x'} \; . \tag{G.15}$$

These angles are chosen because the directivity functions for rectangular transducers can be factorized into a product of separate functions of θ' and ϕ'. This factorization is only approximate in the general case, and improves as the angles become small compared to unity. Note that these are *not* spherical coordinates. The assumed factorization is

$$|b_r(\theta,\ \phi)b_x(\theta,\ \phi)|^2 = B_\theta(\theta')B_\phi(\phi') \; . \tag{G.16}$$

The convention $B_\phi(0) = 1$ will be adopted, so that the directivity factor needed in (G.11) can be expressed in the form

$$|b_r(\theta,\ 0)b_x(\theta,\ 0|^2 = B_\theta(\theta')] \; , \tag{G.17}$$

where it has been assumed without loss of generality that the vertical planes defined by $\phi' = 0$ and $\phi = 0$ are coincident. That is, the earth coordinate system is aligned in azimuth with the transducer coordinate system. Specifically, the earth coordinates $(x,\ y,\ z)$ are related to the transducer coordinates $(x',\ y',\ z')$ as follows:

$$x = x' \cos\theta_0 - z' \sin\theta_0 \ , \tag{G.18}$$

$$y = y' \ , \tag{G.19}$$

$$z = x' \sin\theta_0 + z' \cos\theta_0 \ . \tag{G.20}$$

The angle θ_0 is the transducer depression angle, measured positive downward. The angles θ and ϕ appearing in (G.12) are related to the earth Cartesian system as follows:

$$x = r \cos\theta \cos\phi \ , \tag{G.21}$$

$$y = r \cos\theta \sin\phi \ , \tag{G.22}$$

$$z = r \sin\theta \ , \tag{G.23}$$

where $r = \sqrt{x^2 + y^2 + z^2}$.

If the azimuthal directivity function, $B_\phi(\phi')$, is narrow (the horizontal beamwidth is of order 20° or less), the following approximate relations can be derived from the preceding expressions:

$$\theta' \approx \theta - \theta_0 \ , \tag{G.24}$$

$$\phi' \approx \frac{\cos\theta}{\cos(\theta - \theta_0)} \phi \ . \tag{G.25}$$

Relation (G.24) can be inserted into (G.26) to obtain one of the factors appearing in (G.11):

$$|b_r(\theta,\ 0) b_x(\theta,\ 0|^2 = B_\theta(\theta - \theta_0)] \ . \tag{G.26}$$

The function $B_\theta(\theta)$ is a vertical beam pattern, specifically, it is the product of the squared magnitudes of the source and receiving directivity functions. These are measured in a vertical plane through the MRA.

Finally, the effective angular width of the ensonified region can be approximated as

$$\Psi = \frac{\cos(\theta - \theta_0)}{\cos\theta} \int_{-\pi}^{\pi} B_\phi(\phi) d\phi \ . \tag{G.27}$$

The prime symbol on the dummy integration variable has been dropped as superfluous. An additional approximation can be made to allow analytic evaluation of (G.27). The function $B_\phi(\phi)$ is approximated by a Gaussian shape, which is the same as fitting a parabola in ϕ to the horizontal directivity in dB. After extending the integration limits to $\pm\infty$, the result is

$$\Psi = \frac{\cos(\theta - \theta_0)}{2\cos\theta} \sqrt{\frac{\pi}{\ln 2}} \Psi_3 \ , \tag{G.28}$$

where Ψ_3 is the 3-dB full horizontal width (in the transducer coordinate system) of the product of the source and receiver beam patterns. The angular factor, $\cos(\theta - \theta_0)/\cos\theta$, is nearly unity for small grazing and depression angles, but becomes important as the grazing angle becomes large. Some investigators have used $\Psi \approx \Psi_3$. Apart from the angular factor, this ignores the factor $\sqrt{\pi/\ln 2}/2$, an error of 0.3 dB. Another convenient approximation provides Ψ_3 in terms of the source and receiver horizontal 3-dB full widths, Ψ_x and Ψ_r.

$$\Psi_3 = \frac{1}{\sqrt{1/\Psi_x^2 + 1/\Psi_r^2}} \tag{G.29}$$

This result is based on Gaussian fits to the source and receiver horizontal directivities. As long as the narrower of these two directivities can be approximated by a Gaussian out to its 3-dB points, (G.29) provides a useful approximation.

It is implicit in the above discussion that the source and receiver transducers have the same maximum-response directions. If this is not the case, one need merely find the MRA of the product of the two directivity functions and proceed from there. In this case, however, (G.29) is not valid, and Ψ_3 should be determined from the 3-dB full width of the source–receiver directivity function product.

G.2.2 Near-Vertical Backscattering

Vertical-incidence backscattering measurements offer the possibility of separating the roughness and volume parts of the scattering cross section, as the volume return will be slightly delayed compared to the interface return. As will be seen, such measurements also offer the possibility of measuring the seafloor reflection coefficient (at normal incidence). Inversion of such data cannot be based on the simple sonar equation in the form (G.13) which lumps volume and roughness scattering together in a single interface scattering strength. By treating the volume component as one would treat volume scattering in the water column, a more detailed sonar equation approach can model the relevant time-domain effects. Further, the conventional sonar equation fails near vertical incidence as the scattering cross section and directivity vary too rapidly to be taken outside the integral (G.4), and the assumption that the ensonified area is a thin, annular strip is incorrect. More accurate time-domain models for backscattering near vertical incidence have been developed by [D. Jackson and Nesbitt 1988, Pouliquen 1992, Sternlicht and de Moustier 1997, Heald 2002, Sternlicht and de Moustier 2003a, Sternlicht and de Moustier 2003b]. Here the simplest such model [D. Jackson and Nesbitt 1988] will be defined as an aid to inversion of vertical incidence data. The mean-square received voltage envelope is taken as a sum of components due to roughness and volume scattering,

$$< |V_r(t)|^2 >=< |V_{rr}(t)|^2 > + < |V_{rv}(t)|^2 > \ , \tag{G.30}$$

where the subscripts rr denote "received, roughness," and the subscripts rv denote "received, volume."

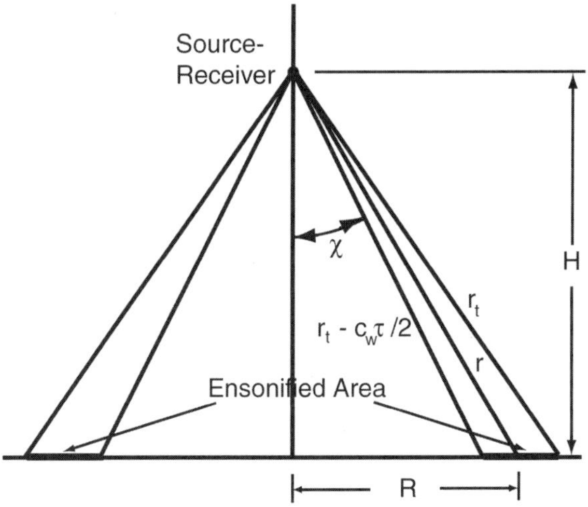

Fig. G.6. Geometry relevant to roughness backscattering near vertical incidence. This is a side view of the 3D geometry in which the ensonified area is seen edge-on. Here, the time, t, since the beginning of transmission is sufficient that the ensonified region is an annulus. The angle χ is exaggerated; the analysis is only valid if $\chi << 1$.

Figure G.6 illustrates the geometry of the interface scattering component of the model. It is assumed that the transmitted signal has a rectangular envelope of length τ. Placing the source–receiver at the origin for convenience, the backscattered signal received at time t (measured from the beginning of the transmitted pulse rather than the center as previously) is due to scatterers lying within the slant range interval

$$r_t - \frac{c_w \tau}{2} < r < r_t \ , \tag{G.31}$$

where r_t is defined in (G.8). Equation (G.4) can be written as

$$< |V_{rr}(t)|^2 > /2 = \frac{(s_0 s_r)^2 e^{-4k_w'' H}}{H^4} \int \sigma |b_r(\theta,\ \phi) b_x(\theta,\ \phi)|^2 d^2 R \ , \tag{G.32}$$

where H is the height of the source–receiver above the seafloor, and where it is assumed that scattering is confined to a rather narrow region immediately below the source–receiver. The source level ($20 \log_{10} s_0$) is measured in the vertical direction, which results in

$$|b_r(\frac{\pi}{2},\ 0) b_x(\frac{\pi}{2},\ 0)| = 1 \ . \tag{G.33}$$

It will be assumed that scattering is isotropic, so that σ does not depend on the azimuthal angle, ϕ. The integral over ϕ only involves the directivity, so the azimuthally averaged directivity is needed:

$$b(\theta) = \frac{1}{2\pi} \int_{-\pi}^{\pi} |b_r(\theta,\ \phi) b_x(\theta,\ \phi)|^2 d\phi \ . \tag{G.34}$$

This will be approximated by a Gaussian function

$$b(\theta) = e^{-\frac{(\pi/2-\theta)^2}{2\sigma_b^2}} \ . \tag{G.35}$$

A similar approximation will be used for the scattering cross section. The high-frequency Kirchhoff approximation (13.70) near vertical incidence can be written as

$$\sigma = \frac{|V_{ww}(90°)|^2}{8\pi\sigma_s^2} e^{-\frac{(\pi/2-\theta)^2}{2\sigma_s^2}} \ . \tag{G.36}$$

This should be regarded as a parameterization of the scattering cross section near vertical incidence, not as a use of the high-frequency Kirchhoff approximation. That is, the parameter σ_s should be considered to be a measure of the angular width of the backscattering cross section peak at vertical incidence, not as the RMS slope of the interface. As noted in Sect. 13.3.1, this form for the cross section has the physically reasonable property of yielding a scattered intensity equal to that of a flat interface with reflection coefficient $V_{ww}(90°)$, provided the scattered energy has a small angular spread. As such, it offers the possibility of obtaining the flat-interface reflection coefficient from backscatter data, even when the interface is moderately rough. Note that this formulation does not allow separation of the coherent and incoherent portions of the echo: they are lumped together in the scattering cross section. This is not a serious problem because, at high frequencies, the coherent component (13.14) is likely to be negligible. With these approximations, (G.32) can be written in the form

$$< |V_{rr}(t)|^2 > /2 = \frac{(s_0 s_r)^2 e^{-4k_w'' H} |V_{ww}(90°)|^2}{4H^4\sigma_s^2} \int_{R_1}^{R_2} e^{-\frac{(\pi/2-\theta)^2}{2\sigma_{sb}^2}} R dR \ , \tag{G.37}$$

where

$$\frac{1}{\sigma_{sb}^2} = \frac{1}{\sigma_s^2} + \frac{1}{\sigma_b^2} \ , \tag{G.38}$$

and R is the cylindrical radial coordinate defined in Fig. G.6 with limits determined by (G.31). Equation (G.37) is especially convenient, as it allows analytic integration [Heald 2002]. Defining the angle of incidence measured from the vertical

$$\chi = \pi/2 - \theta \ , \tag{G.39}$$

one can find the corresponding angular limits, which depend on elapsed time and pulse length. The elapsed time between the beginning of transmission and the leading edge of the seafloor echo is

$$t_0 = \frac{2H}{c_w} \ . \tag{G.40}$$

Before sufficient time has elapsed to allow a return from the seafloor, $t \le t_0$, $< |V_r(t)|^2 >= 0$. Next, for the time period within one pulse length of the first arrival, $t_0 < t \le t_0 + \tau$, the ensonified area is a circle

$$\chi_1^2 = 0 ,$$

$$\chi_2^2 = 2(t/t_0 - 1) . \tag{G.41}$$

Finally, for longer times, $t > t_0 + \tau$, the ensonified area is an annulus defined by

$$\chi_1^2 = 2[(t - \tau)/t_0 - 1] ,$$
$$\chi_2^2 = 2(t/t_0 - 1) . \tag{G.42}$$

If the integration variable in (G.37) is changed to χ^2, the integrand becomes a simple exponential, from which one obtains

$$< |V_{rr}(t)|^2 >= V_1^2 |V_{ww}(90°)|^2 \frac{g(t - t_0,\ T_{sb},\ \tau)}{1 + \sigma_s^2/\sigma_b^2} , \tag{G.43}$$

where

$$\begin{aligned} g(t,\ T,\ \tau) &= 0 , & t \le 0 , \\ &= 1 - e^{-t/T} , & 0 < t \le \tau , \\ &= e^{-(t-\tau)/T} - e^{-t/T} , & t > \tau , \end{aligned} \tag{G.44}$$

$$T_{sb} = \sigma_{sb}^2 t_0 , \tag{G.45}$$

and

$$V_1^2 = \frac{(s_0 s_r)^2 e^{-4k_w'' H}}{2H^2} \tag{G.46}$$

is the mean-square voltage envelope that would be measured if the interface were perfectly flat with reflection coefficient having magnitude unity. Figure G.7 shows an example computed using (G.43). The "normalized intensity" in the plot is the mean-square envelope (G.43) with the factor V_1^2 replaced by unity. The normalized intensity at first approaches an asymptote exponentially and then, after a time of one pulse length has elapsed, decays exponentially toward zero. The asymptote is less than that for pure reflection because the finite beamwidth excludes some of the scattered energy. This is expressed by the factor $1 + \sigma_s^2/\sigma_b^2$ in the denominator of (G.43). The rise time of the intensity provides a measure of the angular width, σ_s, of the backscattering cross section. Narrow widths give a rapid rise and vice versa.

For volume scattering, one must integrate over the ensonified volume, as pictured in Fig. G.8. To simplify the problem, the water–sediment interface is assumed to be flat, and ray bending is ignored. This is a reasonable approximation near vertical incidence. After integrating over the azimuthal angle, the integral for the mean-square envelope due to volume scattering becomes

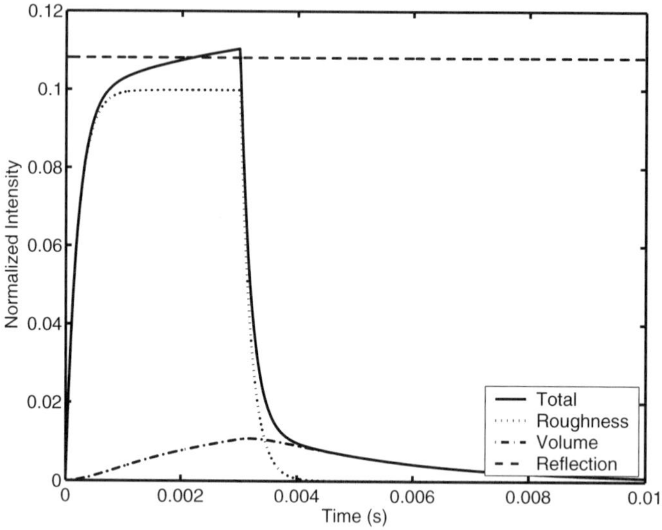

Fig. G.7. Simulated vertical incidence time series, showing the total intensity (normalized) and its roughness and volume components. The acoustic frequency is 20 kHz, the pulse length is 3 ms, the source–receiver height is $H = 10$ m, and the beamwidth parameter is $\sigma_b = 0.445$. The other parameters used in the simulation are $c_w = 1540$ m s^{-1}, $a_\rho = 2.0$, $a_p = 0.99$, $\delta_p = 0.01$, $\sigma_2 = 0.002$, and $\sigma_s = 0.13$.

$$< |V_{rv}(t)|^2 >= \frac{8\pi\sigma_v V_1^2 |V_{wp}(90^\circ)|^4}{a_\rho^2} \int_{r_1}^{r_2} \int_0^{\chi_2} b(\theta) e^{-4k_p''(r-H/\cos\chi)} \sin\chi d\chi dr \ , \tag{G.47}$$

where

$$V_{wp}(90^\circ) = 1 + V_{ww}(90^\circ) \ , \tag{G.48}$$

σ_v is the volume scattering strength, and the factor $|V_{wp}|^4/a_\rho^2$ accounts for round-trip transmission through the interface (see Sect. 14.1.1). It has been assumed that penetration of the seafloor is shallow compared to the source–receiver height, H. The exponential factor accounts for attenuation in the sediment, with k_p'' being the imaginary part of the wavenumber in the sediment, (8.16) and (8.18). The angular limit is

$$\chi_2 = \cos^{-1}(H/r) \ , \tag{G.49}$$

and the limits on the radial integration variable depend on time. For $t \le t_0$, the integral vanishes. For $t_0 < t \le t_0 + \tau$, the ensonified volume is a spherical section

$$r_1 = H \ , \tag{G.50}$$

$$r_2 = r_t \ . \tag{G.51}$$

For $t > t_0 + \tau$, the ensonified volume is a section of a spherical shell, as in Fig. G.8:

$$r_1 = r_t - c_w \tau/2 \ , \tag{G.52}$$

$$r_2 = r_t \ . \tag{G.53}$$

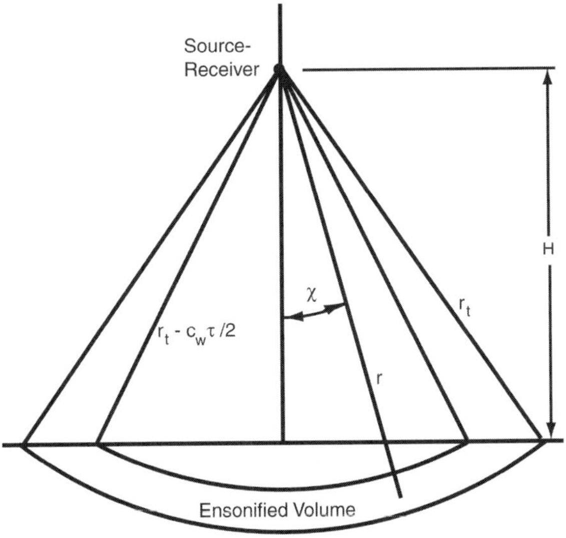

Fig. G.8. Geometry relevant to sediment volume backscattering near vertical incidence. This is a side view of the 3D geometry in which the ensonified volume is seen edge-on. Here, the time, t, since the beginning of transmission is sufficient that the ensonified volume is a truncated spherical shell. The angle χ is exaggerated; the analysis is only valid if $\chi << 1$.

Making a small-angle approximation ($\chi << 1$), the integrand can be written in terms of simple exponentials in χ^2 and r with the result

$$< |V_{rv}(t)|^2 >= \frac{8\pi\sigma_v V_1^2 |V_{wp}(90^\circ)|^4 H}{a_\rho^2(\sigma_p^{-2} - \sigma_b^{-2})} \left[\sigma_b^2 g(t - t_0, T_b, \tau) - \sigma_p^2 g(t - t_0, T_p, \tau)\right] \ , \tag{G.54}$$

with

$$T_b = \sigma_b^2 t_0 \tag{G.55}$$

and

$$T_p = \sigma_p^2 t_0 \ , \tag{G.56}$$

$$\sigma_p^2 = \frac{1}{4k_p'' H} \ . \tag{G.57}$$

Figure G.7 shows a time series computed using (G.54) with the factor V_1^2 set to unity as for the roughness example. The volume scattering contribution shows a more gradual rise and persists longer than the roughness contribution. This behavior offers the hope of separately determining the roughness and volume scattering strengths. In practice, the model is fitted to measured time series. The time series should be an average of the squared envelope over many pings (Sect. G.2.3), and this requires alignment of the leading edges of each return. It is best if the sonar is calibrated so that the parameters σ_b, s_0, and s_r are known. The height, H, can be determined from the average time of first arrival relative to the start of transmission, leaving four parameters to be determined by fitting data. A convenient choice of dimensionless parameters is $|V_{ww}(90°)|$, the reflection coefficient magnitude at vertical incidence, σ_s, the width parameter for roughness scattering, δ_p, the loss parameter, and σ_2, the normalized volume scattering strength (Sect. 14.1.1).

G.2.3 Statistical Error

The ensemble of measurements of the scattered field is necessarily finite, leading to statistical error in the estimated scattering cross section. A useful approximation to this error can be found by assuming the scattered field obeys Gaussian statistics (Sect. 16.2). In this case, the scattered intensity (power/unit area) for each measurement is exponentially distributed, with standard deviation equal to the mean. If there are N independent measurements, the standard deviation of the estimated average intensity will be equal to the theoretical mean divided by $\sqrt{N}$. It follows that the same is true of the estimated cross section, so one may approximate its standard deviation by the measured mean divided by $\sqrt{N}$. For example, if $N = 16$, the standard deviation is one-fourth of the mean, or roughly 1 dB.

H Measurement Units and Decibels

Acoustic variables such as pressure, scattering strength, reflection loss, and transducer directivity are usually given in decibels, abbreviated dB. While the term originated in connection with acoustics and was coined to honor Alexander Graham Bell, its application in acoustics has been inconsistent and unnecessarily obscure in comparison to application by electrical engineers. The underwater acoustics conventions will be used, but the electrical engineering equivalent will be noted as an option in some cases.

First, converting a quantity such as pressure to dB involves taking a base-10 logarithm, with the result dependent on the system of units employed. Hence it is crucial to understand the dimensions of the quantities concerned and the particular measurement units involved. In this book, SI units will be employed exclusively. Commentaries on the units relevant to underwater acoustics have been given by [Carey 1995] and [Hall 1995], and these authors also note some commonly encountered problems.

Decibels were originally used to quantify acoustic intensity, or power density, and the connection with power is the touchstone needed to avoid confusion. A student may ask: why is pressure converted to dB by multiplying the logarithm by 20 while scattering strength is converted using a multiplicative factor of 10? The answer is that acoustic power is proportional to the square of pressure while it is proportional to the first power of scattering strength. The question of units is the main pitfall in converting to dB. To a physicist, the argument of a function such as a logarithm, trigonometric function, or exponential must always be dimensionless. When forming dB, the argument of the logarithm is made dimensionless by dividing (figuratively) by a reference value of the quantity concerned. In underwater acoustics, the nearly universal reference value for pressure is the μPa (one millionth of a Nm^{-2}). Then the pressure envelope, $|P(t)|$, discussed in Appendix B has the dB equivalent

$$20 \log_{10}[\frac{|P(t)|}{\sqrt{2}p_{ref}}] \ . \tag{H.1}$$

The parameter p_{ref} is the reference pressure, and must be expressed in the same units as $P(t)$. If $|P(t)|$ has units of μPa, and $p_{ref} = 1$ μPa, then the units assigned to the dB quantity defined by (H.1) are to be given as "dB re 1 μPa." Engineering practice omits the "re 1," leaving "dB μPa." The factor $\sqrt{2}$ appears because one is concerned with time-averaged power, not peak power, and the average of the square of a sinusoid is one-half the square of its peak value (Appendix E). The factor is sometimes absorbed in the definition of the complex signal.

I Causality Constraints on Speed–Attenuation Relations

As noted in the main text, theories for propagation of sound in sediment should not violate the principle of causality which constrains the frequency dependence of wave speed and attenuation. These constraints are sometimes called "dispersion relations," because frequency dependence of wave speed causes wave packets to change shape, or disperse, as they travel through the medium in question. These fundamental constraints can be used to test candidate theories for sound propagation and attenuation in sediments as well as experimental data. Before considering the acoustic application of causality, it is helpful to consider the linear system point of view. Causality in this case is the simple requirement that the output must be zero for times earlier than the beginning of the input waveform. If the input is a unit-weight delta function at $t = 0$, the output is the "impulse response," $h(t)$. This function could represent, for example, the output voltage of a linear, time-invariant network in response to a driving impulse at the input. Or, it could represent a mechanical system, e.g., the displacement of a mass–spring combination in response to an impulsive force. Since an effect cannot precede its cause,

$$h(t) = 0 , \qquad t < 0 . \tag{I.1}$$

A number of interesting mathematical relationships follow from this seemingly trivial statement. Although there is a large literature on this topic spanning several decades, the results needed are scattered among several references. These results will be collected here along with brief, informal derivations.

Consider the Fourier transform, $H(\omega)$, of the impulse response, the so-called *transfer function*:

$$H(\omega) = \frac{1}{2\pi} \int_{-\infty}^{\infty} h(t) e^{i\omega t} dt . \tag{I.2}$$

If the impulse response is a real function of time,

$$H(-\omega) = H^*(\omega) . \tag{I.3}$$

Or, if the real and imaginary parts of the transfer function are defined

$$H(\omega) = H'(\omega) + iH''(\omega) , \tag{I.4}$$

then

$$H'(-\omega) = H'(\omega) \; , \tag{I.5}$$

$$H''(-\omega) = -H''(\omega) \; . \tag{I.6}$$

Causality provides a relationship between the real and imaginary parts of the transfer function. These relationships follow from consideration of the inverse transform

$$h(t) = \int_{-\infty}^{\infty} H(\omega) e^{-i\omega t} d\omega \; . \tag{I.7}$$

The integration path can be extended to form a closed path in the complex ω-plane, as shown in Fig. I.1. This path must follow the lower semicircle at

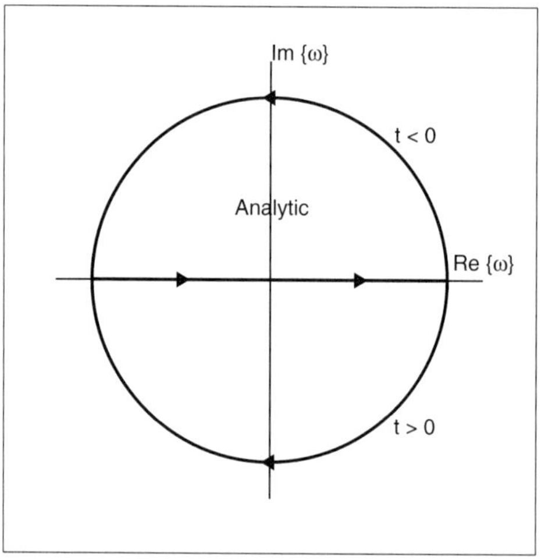

Fig. I.1. Integration contours for the inverse Fourier transform of a causal transfer function. The semicircular contours have infinite radius; the lower one is used for $t > 0$ and the upper one is used for $t < 0$.

infinity if $t > 0$ in order that the integrand vanish on this portion of the path. The upper semicircle must be followed if $t < 0$. The integral will vanish for all $t < 0$ only if the transfer function is analytic in the upper half plane. Then

$$\int_{C_1} \frac{H(\omega')d\omega'}{\omega' - \omega} = 0 \; , \tag{I.8}$$

where the integration is over the closed contour, C_1, shown in Fig. I.2. Taking the real and imaginary parts of (I.8),

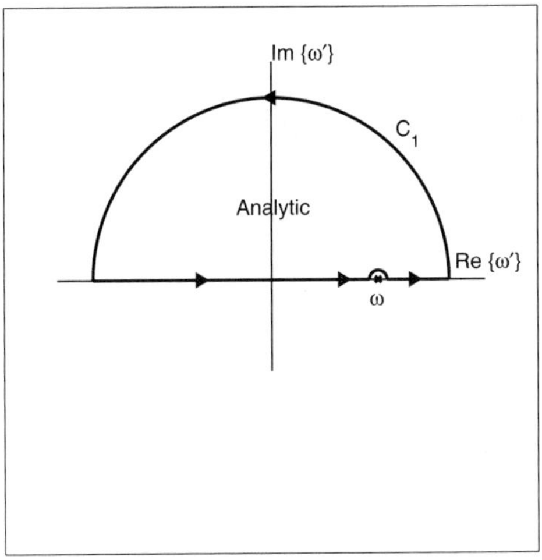

Fig. I.2. Integration contour for derivation of relationships between the real and imaginary parts of a causal transfer function. The smaller semicircle has vanishingly small radius, while the larger semicircle has infinite radius.

$$H'(\omega) = \frac{1}{\pi} P \int_{-\infty}^{\infty} \frac{H''(\omega')d\omega'}{\omega' - \omega} , \tag{I.9}$$

$$H''(\omega) = -\frac{1}{\pi} P \int_{-\infty}^{\infty} \frac{H'(\omega')d\omega'}{\omega' - \omega} . \tag{I.10}$$

In obtaining this result, the contribution from the part of the contour at infinity has been neglected. A sufficient condition for this is $|H(\omega)| \to 0$ as $|\omega| \to \infty$. The symbol P in front of the integrals stands for "Cauchy principal value," the limit as the integration is carried arbitrarily close to the pole at $\omega' = \omega$ while remaining on the real axis. That is, the integration excludes the portion of the real axis defined by the outer edges of the small semicircle pictured in Fig. I.2. Equations (I.9) and (I.10) are *Hilbert transforms.* Together, they constitute a Hilbert transform pair.

A common method for performing Hilbert transforms numerically involves Fourier transforms. It can be easily shown that the inverse transform of $H'(\omega)$ is an even, real function of t:

$$h'(t) = \int_{-\infty}^{\infty} H'(\omega)e^{-i\omega t} d\omega = \frac{1}{2} h(|t|) , \tag{I.11}$$

while the inverse transform of $iH''(\omega)$ is real and odd:

$$h''(t) = i \int_{-\infty}^{\infty} H''(\omega) e^{-i\omega t} d\omega = \frac{1}{2} \mathrm{sgn}(t) h(|t|) \ . \tag{I.12}$$

Here, sgn(t) is the sign function, equal to -1 for $t < 0$, $+1$ for $t > 0$, and zero for $t = 0$. The sum of the two time functions (I.11) and (I.12) is the impulse response, $h(t)$, as illustrated in Fig. I.3. Inserting $h(t) = h'(t) + h''(t)$ into

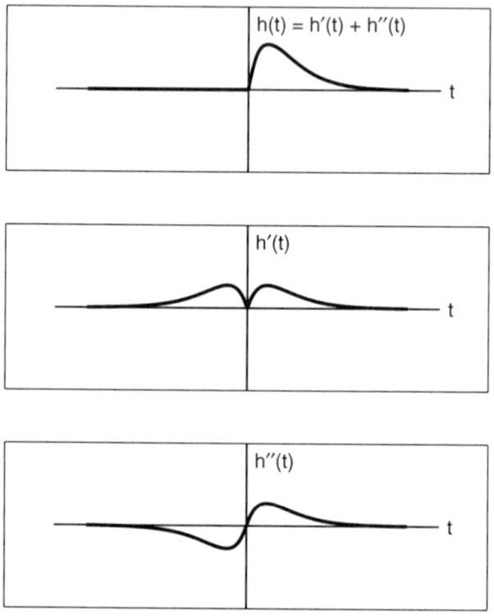

Fig. I.3. Decomposition of an impulse response into even and odd parts, relevant to numerical evaluation of Hilbert transforms.

(I.2),

$$H'(\omega) + iH''(\omega) = \frac{1}{\pi} \int_{-\infty}^{\infty} u(t) h'(t) e^{i\omega t} dt \ , \tag{I.13}$$

where $u(t)$ is the unit step, or Heaviside, function. Equation (I.13) provides a "recipe" for obtaining the Hilbert transform of the real (or imaginary) part of the transfer function. First, the inverse Fourier transform of $H'(\omega)$, (I.11), is evaluated to obtain $h'(t)$. The negative time part of $h'(t)$ is set to zero, corresponding to multiplication by $u(t)$, and the imaginary part of (I.13) is the desired $H''(\omega)$. In practice, of course, fast Fourier transforms are used in place of the continuous Fourier transforms employed here.

In treating causality constraints on wave speed and attenuation, the Green's function can be considered to be a transfer function. That is,

$$H(\omega) = \frac{e^{ik(\omega)r}}{r} \ . \tag{I.14}$$

The analyticity properties of the transfer function must be shared by the complex wavenumber, $k(\omega)$, that is, it must be analytic in the upper half of the complex frequency plane. Relationships are sought between the real and imaginary parts of the wavenumber, where

$$k(\omega) = k'(\omega) + ik''(\omega) \ . \tag{I.15}$$

The wavenumber may correspond to any sort of wave propagation, compressional, shear, etc. In discussing dispersion relations, most authors [Futterman 1962, Horton 1974, Horton 1981, Wingham 1985] employ a complex refractive index,

$$n(\omega) = \frac{c_0 k(\omega)}{\omega} \ , \tag{I.16}$$

where c_0 is a reference speed chosen on the basis of convenience. Corresponding to (I.3)–(I.6), one has

$$n(-\omega) = n^*(\omega) \ , \tag{I.17}$$

$$n(\omega) = n'(\omega) + in''(\omega) \ , \tag{I.18}$$

$$n'(-\omega) = n'(\omega) \ , \tag{I.19}$$

$$n''(-\omega) = -n''(\omega) \ . \tag{I.20}$$

The requirement that the wave amplitude must decay with distance gives the additional constraint

$$n''(\omega) \geq 0 \ , \qquad \omega > 0 \ . \tag{I.21}$$

If the transfer function, (I.14), is analytic in the upper half of the complex ω-plane, the refractive index must also be analytic there. Thus, one expects its real and imaginary parts to be related by a Hilbert transform pair identical to (I.9) and (I.10). The symmetry conditions (I.19) and (I.20) are often used to restrict the integral to positive frequencies:

$$n'(\omega) = \frac{2}{\pi} P \int_0^\infty \frac{n''(\omega')\omega' d\omega'}{\omega'^2 - \omega^2} \ , \tag{I.22}$$

$$n''(\omega) = -\frac{2}{\pi} P \int_0^\infty \frac{n'(\omega')\omega' d\omega'}{\omega'^2 - \omega^2} \ . \tag{I.23}$$

Equations (I.22) and (I.23) are known as the Kramers–Kronig relations, after the investigators who independently developed them. It must be remembered that these relationships were derived under the assumption that the magnitude of the transfer function (or refractive index in the present case) falls to zero as frequency increases without limit. The main interest, however, is in cases for which the refractive index has only a slight frequency dependence. For example, if attenuation increases linearly with frequency, the imaginary part of the index of refraction will be independent of frequency. The strategy in this case [Futterman 1962, Horton 1974,

Horton 1981, Wingham 1985] is to deal with the function $n(\omega)/(\omega - \omega_1)$, which is artificially forced to approach zero by division by the factor $(\omega - \omega_1)$. The frequency ω_1 is a free parameter that will be chosen later on the basis of convenience. Following the analyticity arguments made previously, one now begins with the statement

$$\int_{C_2} \frac{n(\omega')d\omega'}{(\omega' - \omega_1)(\omega' - \omega)} = 0 , \tag{I.24}$$

where the closed contour, C_2, is shown in Fig. I.4. Taking the real part of

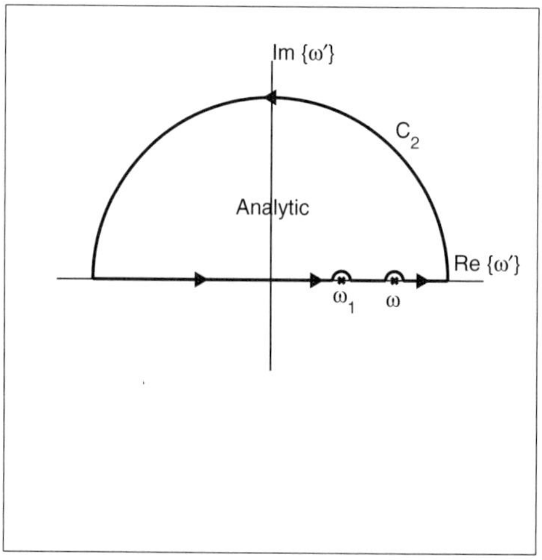

Fig. I.4. Integration contour for derivation of subtracted dispersion relations, applicable to complex refractive indices that do not approach zero as frequency increases.

(I.24) and invoking symmetry with respect to frequency, the end result is

$$n'(\omega) - n'(\omega_1) = \frac{2(\omega^2 - \omega_1^2)}{\pi} P \int_0^\infty \frac{n''(\omega')\omega' d\omega'}{(\omega'^2 - \omega_1^2)(\omega'^2 - \omega^2)} . \tag{I.25}$$

This is an example of a dispersion relation with one subtraction [Weinberg 1995], while (I.22) and (I.23) are dispersion relations with no subtraction. Equation (I.25) has been used to show that if attenuation increases linearly with frequency wave speed must increase logarithmically with frequency [Horton 1981].Specifically, choose the reference frequency, ω_1, such that the reference speed in (I.16) is $c_0 = \omega_1/k'(\omega_1)$. Then

$$n(\omega) = \frac{k(\omega)\omega_1}{k'(\omega_1)\omega} . \tag{I.26}$$

With this choice of reference frequency, the real part of the index of refraction is unity at $\omega = \omega_1$. Take the imaginary part of the index of refraction to be independent of frequency, specifically,

$$n''(\omega) = \delta_1 \;, \qquad\qquad \omega > 0 \;. \tag{I.27}$$

It should be remembered that $n''(\omega)$ is an odd function of frequency, thus equal to $-\delta_1$ for negative frequencies. The frequency-independent parameter, δ_1, is defined to be analogous to the dimensionless loss parameters introduced in the main text. With this choice, (I.25) can be evaluated to obtain

$$n'(\omega) = 1 - \frac{2\delta_1}{\pi} \ln\left|\frac{\omega}{\omega_1}\right| \;. \tag{I.28}$$

Wingham [Wingham 1985] has shown that the slight dispersion resulting from (I.28) agrees with laboratory data for sands for frequencies in the range 50–350 kHz. Similarly, Williams et al. [Williams et al. 2002a] find weak dispersion consistent with (I.28) for frequencies of several kHz and higher, but note greater dispersion occurs at lower frequencies (Ch. 10). This is consistent with results obtained earlier by [Turgut and Yamamoto 1990]. Note that the common assumption that wave speed is complex but frequency independent violates causality. This is because the assumption of constant complex sound speed results in attenuation that increases linearly with frequency, and this is inconsistent with the velocity dispersion demanded by causality. This violation leads to negligible error in most cases of interest [Thorsos et al. 2000a].

The dispersion relations discussed above involve infinite integrals, in principle requiring knowledge of either the real or imaginary part of a transfer function or index of refraction over all frequencies. In practice, data will be available over a finite frequency range. The obvious approach in this situation is to truncate the integration domain and hope that the resulting error is small. A more rigorous approach is provided by [Milton et al. 1997] who give bounds on dispersion for finite frequency-range data. Alternatively, [O'Donnell et al. 1981] provide "local" approximations that do not require integration.

It is sometimes possible to verify that the index of refraction obeys the causality constraint without resorting to dispersion relations. If one can show that the index is analytic in the upper half of the complex ω-plane, causality is assured. For example, [Williams et al. 2002a] add a term $s_0/(1 - i\omega\tau)$ to the index specified by (I.27) and (I.28). If $\tau > 0$, this term is analytic in the upper half of the complex ω-plane. Generally, if the index is obtained from a theory with a basis in physics, causality is satisfied. Thus, Biot theory yields a causal index as far as attenuation due to fluid flow is concerned, but causality can be violated if the complex moduli do not satisfy the causality constraint. This is the case when these are taken to be frequency-independent, but the resulting causality violation is insignificant for the parameters appropriate to sand. Considering the compressional wave in Buckingham's theory, (9.51) can be used to obtain

$$n^2 = \frac{1}{1 - i\omega[\lambda_p H_p(\omega) + \frac{4}{3}\eta_t H_t(\omega)]/K_b} \ . \tag{I.29}$$

Causality will be satisfied if the denominator of (I.29) is analytic in the upper half of the complex ω-plane and has no zeros there. For this, it is sufficient that $-i\omega H_p(\omega)$ and $-i\omega H_t(\omega)$ and their reciprocals be analytic in the upper half of the complex ω-plane. Such transfer functions are referred to as being "minimum-phase" [Papoulis 1962, pp. 204-206], and the most common examples are impedances and admittances. These, along with all other minimum-phase transfer functions, have real parts that are positive in the upper half of the complex ω-plane. In the approximation of (9.58) and (9.58), the minimum-phase requirement is clearly satisfied. Unpublished numerical calculations by one of the authors (DRJ) indicate that the minimum-phase requirement is met in general by (9.49) and (9.50). This, together with positivity of K_b, λ_p, and η_t, guarantees that the denominator of (I.29) can have no zeros in the upper half of the complex ω-plane, satisfying the causality constraint.

J The Scattering Cross Section and Plane Waves

This appendix develops expressions that are of both practical and theoretical use. An additional purpose is to derive criteria stating when it is acceptable to use the scattering cross section in sonar-equation calculations. Some of the scattering literature raises suspicions that the scattering cross section may be inapplicable in many common situations, particularly when the sonar is in the "near field" of the seafloor. In this view, the seafloor is regarded as being a sort of very large acoustic array. Much of the concern seen in the literature arises from misinterpretation of definitions, and this motivates a discussion of the finer points implied by the definitions given in Ch. 2.

The definition of scattering strength used in this monograph is inherently statistical, but this definition is not universal. Scattering strength is sometimes defined without averaging, in which case it becomes a random variable itself. There is nothing wrong with such a definition, as long as properties of the statistical average quantity are not imputed to the random one. The statistically- defined scattering cross section is invariant with respect to the parameters of the measurement system, and this invariance can be used to check experimental data. For example, if the scattering cross section is not independent of pulse length, this may indicate a flaw in the experiment or in data processing, as pulse length independence follows from very fundamental arguments [Henyey and Thorsos 1995]. Scattering strength should also be independent of geometrical parameters such as the height of the sonar transducer above the seafloor. If it is not, this may indicate that the echo signal does not originate from a nearly flat interface. It may, for example, originate from structures within the seafloor [D. Jackson et al. 1996a].

The first section of this appendix summarizes key definitions and results that simplify derivations of interface and volume scattering models and specify the manner in which the plane-wave scattering cross section can be used in the most general setting. The second section of this appendix is a derivation of the criteria that must be satisfied if the cross section is to be used.

J.1 General Expressions

The expressions to be given here are based on plane-wave expansions of the incident and scattered fields. As will be seen, the use of such Fourier expansions does not mean that the incident or scattered fields must be plane waves.

The essential element in these expansions is the "plane-wave T-matrix," $T_{ww}(\mathbf{K}_s,\ \mathbf{K}_i)$. This function is a decomposition of the scattered field into plane waves given that a plane wave of unit amplitude is incident. Specifically, express the incident pressure as a superposition of plane waves:

$$P_i(\mathbf{R}, z) = \int \Phi_i(\mathbf{K}_i) e^{i\mathbf{K}_i \cdot \mathbf{R} - ik_w \beta_w(K_i) z} d^2 K_i \ . \tag{J.1}$$

As usual, the subscript i is intended to denote "incident." The plane waves in the above integral have wave vectors given by (14.11). Recall $\beta_w(K) = \sqrt{1 - K^2/k_w^2}$.

The outgoing field in the water resulting from the field incident on the seafloor can also be expressed as a superposition of plane waves:

$$P(\mathbf{r}) = \int \Phi(\mathbf{K}_s) e^{i\mathbf{K}_s \cdot \mathbf{R} + ik_w \beta_w(K_s) z} d^2 K_s \ . \tag{J.2}$$

The incident and outgoing plane-wave amplitudes are related by the T-matrix:

$$\Phi(\mathbf{K}_s) = \int T_{ww}(\mathbf{K}_s, \mathbf{K}_i) \Phi_i(\mathbf{K}_i) d^2 K_i \ . \tag{J.3}$$

This is the definition of the T-matrix. The subscripts ww are employed following the convention used in the main text for reflection and transmission coefficients. In the present case, the T-matrix describes scattering of an incident acoustic wave in the water into another acoustic wave in the water. The subscripts wp and wt are used in this monograph for scattering from the water into sediment compressional and shear waves.

The T-matrix can be viewed as a generalized reflection coefficient for interfaces that are not uniform in the horizontal coordinates, that is, interfaces that may scatter sound away from the specular direction. For horizontally uniform seafloors, the Green's function can be found in terms of the reflection coefficient by means of (8.63). This can be generalized for the nonuniform seafloor to

$$G_{ww}(\mathbf{r}_s,\ \mathbf{r}_i) = G_0(\mathbf{r}_s, \mathbf{r}_i) +$$
$$\frac{i}{2\pi k_w} \int\int T_{ww}(\mathbf{K}_s,\ \mathbf{K}_i) e^{i(\mathbf{K}_s \cdot \mathbf{R}_s - \mathbf{K}_i \cdot \mathbf{R}_i) + ik_w[\beta_w(K_s) z_s + \beta_w(K_i) z_i]} \frac{d^2 K_i}{\beta_w(K_i)} d^2 K_s \ . \tag{J.4}$$

This expression can be found by using (8.61) to obtain the point-source plane-wave spectrum, $\Phi_i(\mathbf{K}_i)$, inserting this in (J.3), and then inserting the resulting expression for $\Phi(\mathbf{K}_s)$ in (J.2).

Reciprocity requires that $G_{ww}(\mathbf{r}_s,\ \mathbf{r}_i)$ be unchanged by an interchange of its two coordinate vector arguments. Using (J.4), this can readily be shown to require

$$\beta_w(K_s) T_{ww}(\mathbf{K}_s, \mathbf{K}_i) = \beta_w(K_i) T_{ww}(-\mathbf{K}_i, -\mathbf{K}_s) \ . \tag{J.5}$$

The treatment of the Kirchhoff approximations given in Appendix L is facilitated by the following exact expression for the T-matrix in terms of the surface field induced by a unit-amplitude incident plane wave:

$$T_{ww}(\mathbf{K}_s,\ \mathbf{K}_i) = \frac{-i}{8\pi^2 k_w \beta_w(K_s)}$$

$$\times \int [\nabla P(\mathbf{r},\ \mathbf{K}_i) + i\mathbf{k}_s P(\mathbf{r},\ \mathbf{K}_i)] \cdot \mathbf{n} e^{-i\mathbf{k}_s \cdot \mathbf{r}}|_{z=f(\mathbf{R})} dS \ . \tag{J.6}$$

This is a generalization to an arbitrary boundary condition of an expression for the T-matrix for a pressure-release (Dirichlet) interface [Winebrenner and Ishimaru 1985, Thorsos and Broschat 1995]. An equally general expression has been derived by [Berman and Dacol 1990]. Their expression has the advantage of obviously obeying reciprocity, but is somewhat more complicated than (J.6). The surface integral employs the upward-directed unit normal vector, $\mathbf{n}$, which is related to the previously introduced normal (8.35) by the following expressions:

$$\mathbf{n} = \frac{\mathbf{N}}{N} \ , \tag{J.7}$$

$$\mathbf{n} dS = \mathbf{N} d^2 R \ , \tag{J.8}$$

where $d^2R = dxdy$. In (J.6), $P(\mathbf{r},\ \mathbf{K}_i)$ is the total pressure field (incident plus scattered) at a point $\mathbf{r}$ in the water due to a unit-amplitude plane wave. The gradient in (J.6) acts *only* on $P(\mathbf{r},\ \mathbf{K}_i)$. The incident plane wave has three-dimensional wave vector

$$\mathbf{k}_i = \mathbf{K}_i - \mathbf{e}_z k_w \beta_w(K_i) \ , \tag{J.9}$$

and the scattered wave vector is

$$\mathbf{k}_s = \mathbf{K}_s + \mathbf{e}_z k_w \beta_w(K_s) \ . \tag{J.10}$$

In treating the statistics of scattering, the outgoing field is usually decomposed into coherent and incoherent parts. Correspondingly, the T-matrix will be expressed as follows:

$$T_{ww}(\mathbf{K}_s, \mathbf{K}_i) =< T_{ww}(\mathbf{K}_s, \mathbf{K}_i) > + T_{wws}(\mathbf{K}_s, \mathbf{K}_i) \ . \tag{J.11}$$

The mean, or first moment, of the T-matrix, $< T_{ww}(\mathbf{K}_s, \mathbf{K}_i) >$, describes the coherent reflection process. Useful constraints can be placed on the moments of the T-matrix if the statistics of the rough, heterogeneous seafloor are stationary in the two horizontal coordinates. That is, if all the probability density functions giving the spatial statistics of density, compressibility, roughness, etc., do not depend on the particular location being considered. This requires, for example, that the covariance functions for roughness, $B(\mathbf{R}_1,\ \mathbf{R}_2)$, and heterogeneity, $B_{\alpha\beta}(\mathbf{r}_1,\ \mathbf{r}_2)$, can be expressed as functions of $\mathbf{R}_1 - \mathbf{R}_2$. Practically speaking, this condition requires the random seafloor to be essentially the same over the region where measurements are made. If the stationarity condition is satisfied,

$$< T_{ww}(\mathbf{K}_s, \mathbf{K}_i) >= V_{wwc}(\mathbf{K}_i)\delta(\mathbf{K}_s - \mathbf{K}_i) \ , \tag{J.12}$$

where $V_{wwc}(\mathbf{K}_i)$ is the coherent reflection coefficient discussed in Chs. 2 and 13.

The incoherent, or scattered, T-matrix, $T_{wws}(\mathbf{K}_s, \mathbf{K}_i)$, is of primary interest. If the seafloor statistics are stationary in the sense defined above, the second moment of the scattered T-matrix has the following general form [Zipfel and DeSanto 1972]:

$$< T_{wws}(\mathbf{K}_s, \mathbf{K}_i) T^*_{wws}(\mathbf{K}'_s, \mathbf{K}'_i) >= C(\mathbf{K}_s, \mathbf{K}_i, \mathbf{K}'_i)\delta(\mathbf{K}_s - \mathbf{K}_i - \mathbf{K}'_s + \mathbf{K}'_i \ ) \ . \tag{J.13}$$

The following relationship between the second moment and the bistatic scattering cross section is provided by [Voronovich 1985, Thorsos and Jackson 1989]:

$$\sigma(\mathbf{K}_s \ , \mathbf{K}_i) = k_w^2 \beta_w^2(\mathbf{K}_s \ ) C(\mathbf{K}_s, \ \mathbf{K}_i, \ \mathbf{K}_i \ ) \ . \tag{J.14}$$

The connection between the wave vector arguments used here and more conventional angles is given in (14.15) and (14.16).

The utility of the expressions given above is mainly theoretical. While many derivations of scattering models given in the literature strive to mimic typical measurement conditions, with directive, non-plane-wave sources, etc., the mathematical apparatus defined above allows one to assume simple plane-wave ensonification. For this case, the spectrum of outgoing plane waves is the T-matrix, from which the scattering cross section can be immediately determined. The legitimacy of this procedure is established in the following section. A practical by-product of that discussion is the following expression for the mean-square scattered field:

$$< |P_s(\mathbf{r}_s)|^2 >= \int_{z=0} \sigma(\mathbf{K}_s, \mathbf{K}_i) |P_i(\mathbf{r})|^2 |\mathbf{r}_s - \mathbf{r}|^{-2} d^2 R \ , \tag{J.15}$$

where $\mathbf{r} = [\mathbf{R}, \ 0]$,

$$\mathbf{K}_i = k_w(\mathbf{R} - \mathbf{R}_i)/|\mathbf{r} - \mathbf{r}_i| \ , \tag{J.16}$$

and

$$\mathbf{K}_s = k_w(\mathbf{R}_s - \mathbf{R})/|\mathbf{r}_s - \mathbf{r}| \ . \tag{J.17}$$

The mean-square scattered pressure is obtained as an integral over the mean seafloor ($z' = 0$), with the angular arguments of the scattering cross section dependent on the integration coordinates. The geometry for this integration is shown in Fig. J.1. Expression (J.15) extends the applicability of the plane-wave scattering cross section to most situations of interest and obviates the need for more complicated methods, such as the Fresnel approximation [Horton and Melton 1970, McDonald and Spindel 1971, Thorne and Pace 1984, Pace et al. 1985]. Equation (J.15) is equivalent to (G.4), although the latter has integration limits restricted by the duration of the transmitted pulse and includes the effect of receiver directivity.

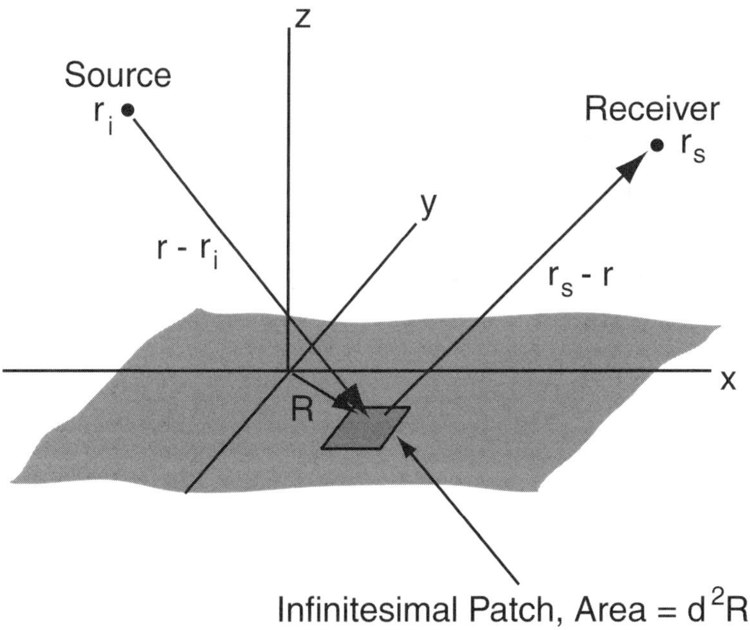

Fig. J.1. Geometry for integration over mean scattering interface to obtain the mean-square scattered pressure. The vector $\mathbf{r} = [\mathbf{R}, 0]$ lies on the mean plane ($z = 0$) and not on the rough interface. The same is true of the area element, d^2R.

J.2 Far-Field Issues

The definition of scattering strength assumes that the incident and scattered energy have well-defined directions, and this condition is sometimes misinterpreted. With regard to the incident field, this might be interpreted to mean that the incident direction must not vary significantly over the scattering (ensonified) region. With respect to the scattered energy, it is sometimes asserted that the receiver must be in the far field of the scattering region (also called the ensonified region) [Horton and Melton 1970]. Considering the "cross range" dimension, d, of this region (the dimension in the direction orthogonal to the line of sight between the sonar and the scattering region), it will grow linearly with range, r, $d = \Psi r$, where Ψ is the horizontal beamwidth (G.12). Then the far-field condition becomes $r < \lambda/\Psi^2$ which can never be satisfied at ranges of interest.

The purpose of this section is to derive rigorous criteria for the application of the scattering cross section to replace these heuristic conditions. One of the basic results is that given in [Winebrenner and Ishimaru 1986]: the length scale appearing in the far-field criterion is the correlation length of the random field at the interface (the "surface" field) rather than the dimension of the scattering region. As this correlation length is usually quite small, the far-field criterion can be satisfied in most cases of interest. A review and re-analysis of this problem has been given by [D. Jackson et al. 1997]. Their presentation employs, for the most part, expressions in wave vector space. The discussion in this appendix will employ coordinate-space expressions to a large degree, allowing more direct contact with the concepts of aperture size and correlation lengths.

The approach to be taken is to formulate the scattering problem in realistic but very general terms, with a directional source, and with no assumptions regarding the scattering mechanism other than horizontal stationarity. This will lead to the desired criteria and also (J.15) for computing the mean-square scattered field.

The first, and simplest, criterion for application of (J.15) is that the scattering region must be in the far field of the source. If this is not the case, there will not be a well-defined direction of incidence at each point on the scattering surface, that is, there will not be a definite $\mathbf{K}_i$, one of the arguments of the scattering cross section. In the T-matrix approach, all relevant source properties are contained in the plane-wave spectrum, $\Phi_i(\mathbf{K}_i)$. To relate this function to a far-field criterion, it will be written in the form

$$\Phi_i(\mathbf{K}_i) = b_i(\mathbf{K}_i)\, e^{-i\mathbf{K}_i\cdot\mathbf{R}_i + ikw\beta_w(K_i)z_i} \ . \tag{J.18}$$

The source is situated at the point $\mathbf{r}_i = [\mathbf{R}_i,\ z_i]$, and by explicitly factoring out the phase for propagation from this point to the origin, one is left with a factor, $b_i(\mathbf{K}_i)$, that is a slowly varying function of $\mathbf{K}_i$. Inserting this definition into (J.1), and setting $z = 0$ to obtain the incident field on the mean interface,

$$P_i(\mathbf{R}, 0) = \int b_i(\mathbf{K}_i) e^{i\mathbf{K}_i\cdot(\mathbf{R}-\mathbf{R}_i) + ik_w\beta_w(K_i)z_i} d^2K_i \ . \tag{J.19}$$

The incident field can be approximated in the far-field region by evaluating (J.19) in the stationary phase approximation. Expanding the exponent to second order in $\mathbf{K}_i$ about the stationary phase point and performing the integral analytically, the result is

$$P_i(\mathbf{R}, 0) = -2\pi i k_w\beta_w(K_i) b_i(\mathbf{K}_i) \frac{e^{ik_w|\mathbf{r}-\mathbf{r}_i|}}{|\mathbf{r}-\mathbf{r}_i|} \ , \tag{J.20}$$

where $\mathbf{r} = [\mathbf{R},\ 0]$, and $\mathbf{K}_i$ is to be evaluated at the stationary phase point

$$\mathbf{K}_i = k_w \frac{\mathbf{R}-\mathbf{R}_i}{|\mathbf{r}-\mathbf{r}_i|} \ . \tag{J.21}$$

Inspection of (J.20) shows that $b_i(\mathbf{K}_i)$ is a source directivity function, similar to $b_x(\theta,\ \phi)$ appearing in (F.5).

Anticipating that the correlation length of the field on the interface will be important in determining the far-field criterion, an effective surface field will be defined by setting $z = 0$ in (J.2):

$$P_{surf}\left(\mathbf{R}\right) = \int \Phi\left(\mathbf{K}_s\right) e^{i\mathbf{K}_s \cdot \mathbf{R}} d^2 K_s \ . \tag{J.22}$$

Although the surface field is usually taken to be the field *on* the rough interface, defining an equivalent surface field on the mean plane ($z = 0$) allows complete generality regarding details of the interface and the scattering mechanisms. Thus the present discussion applies to sediment volume scattering as well as to roughness scattering. Taking the inverse Fourier transform of (J.22) to obtain the plane-wave spectrum, $\Phi\left(\mathbf{K}_s\right)$, in terms of the effective surface field, and inserting the result in (J.2) yields

$$P_s(\mathbf{r}) = \frac{1}{2\pi} \int P_{surf}(\mathbf{R}') g(\mathbf{r},\ \mathbf{R}') d^2 R', \tag{J.23}$$

where

$$g(\mathbf{r},\ \mathbf{R}') = \frac{1}{2\pi} \int e^{i\mathbf{K}_s \cdot (\mathbf{R} - \mathbf{R}') + k_w \beta_w (K_s) z} d^2 K_s \ . \tag{J.24}$$

Comparison with (8.60) leads to

$$g(\mathbf{r},\ \mathbf{R}') = -\frac{\partial}{\partial z} G_0(\mathbf{r}, \mathbf{r}') \ , \tag{J.25}$$

with $z' = 0$. This is the field produced by a vertically oriented dipole situated on the mean interface at $[\mathbf{R}',\ 0]$. For distances greater than a few wavelengths from the interface,

$$g(\mathbf{r},\ \mathbf{R}') = \frac{-ik_w z e^{ik_w |\mathbf{r} - \mathbf{r}'|}}{|\mathbf{r} - \mathbf{r}'|^2} \ , \tag{J.26}$$

where

$$|\mathbf{r} - \mathbf{r}'| = \sqrt{r^2 + R'^2 - 2\mathbf{R} \cdot \mathbf{R}'} \ . \tag{J.27}$$

If (J.26) is inserted in (J.23), the resulting expression is valid at all ranges (near- and far-field) provided the shortest ranges exceed a few wavelengths. The integral is similar to that for the field radiated by a planar array, so it is logical to seek a far-field criterion. This criterion is obtained by expanding (J.27) in a power series,

$$|\mathbf{r} - \mathbf{r}'| = r - \frac{\mathbf{R} \cdot \mathbf{R}'}{r} + O(R'^2/r) \ . \tag{J.28}$$

The far-field, or Fraunhofer, approximation ignores terms that are second-order and higher in R'. This requires that the corresponding contribution to phase (the exponent) be small

$$\frac{k_w d^2}{r} << 1 \ . \tag{J.29}$$

This is equivalent to (F.1) where the "array size," d, is the range of the variable R' over which the surface field is appreciable, that is, d is the size of the scattering region. This is the conventional view of the far field and is relevant if one seeks to compute the scattered field in great detail. If, however, one merely seeks to compute the mean-square scattered field, this criterion is far too conservative, as will be seen.

The second moment of the scattered field is

$$< |P_s(\mathbf{r})|^2 >=$$

$$\frac{k_w^2 z^2}{(2\pi)^2} \int\int < P_{surf}(\mathbf{R}_1) P^*_{surf}(\mathbf{R}_2) > \frac{e^{ik_w(|\mathbf{r}-\mathbf{r}_1|-|\mathbf{r}-\mathbf{r}_2|)}}{|\mathbf{r}-\mathbf{r}_1|^2|\mathbf{r}-\mathbf{r}_2|^2} d^2R_1 d^2R_2 \ . \quad \text{(J.30)}$$

In this case, the expansion of the exponent involves

$$|\mathbf{r}-\mathbf{r}_1| - |\mathbf{r}-\mathbf{r}_2| = -\frac{(\mathbf{R}-\mathbf{R}')\cdot\mathbf{D}}{|\mathbf{r}-\mathbf{r}'|} + O(D^2/|\mathbf{r}-\mathbf{r}'|) \ , \quad \text{(J.31)}$$

where

$$\mathbf{R}' = \frac{1}{2}(\mathbf{R}_1 + \mathbf{R}_2) \quad \text{(J.32)}$$

and

$$\mathbf{D} = \mathbf{R}_1 - \mathbf{R}_2 \ . \quad \text{(J.33)}$$

Now the Fraunhofer approximation can be used if

$$\frac{k_w L_f^2}{|\mathbf{r}-\mathbf{r}'|} << 1 \ . \quad \text{(J.34)}$$

Here, L_f is the correlation length of the surface field, that is, the range of D for which the correlation of the surface field is appreciable. If this scale is small compared to the size of the entire scattering region, the far-field criterion (J.34) will be satisfied at much shorter ranges than the conventional criterion (J.29).

To proceed, an expression for the covariance of the effective surface field is needed. Using (J.2), (J.3), (J.11), and (J.13),

$$< P_{surf}(\mathbf{R}_1) P^*_{surf}(\mathbf{R}_2) >=$$

$$\int\int\int C(\mathbf{K},\ \mathbf{K}_1,\ \mathbf{K}_2)\phi_i(\mathbf{K}_1)\phi_i^*(\mathbf{K}_2) e^{i(\mathbf{K}_1-\mathbf{K}_2)\cdot\mathbf{R}_2} d^2K_1 d^2K_2 e^{i\mathbf{K}\cdot(\mathbf{R}_1-\mathbf{R}_2)} d^2K. \quad \text{(J.35)}$$

The integrals over $\mathbf{K}_1$ and $\mathbf{K}_2$ can be evaluated by stationary phase if the scattering surface is in the far field of the source and the T-matrix second moment, $C(\mathbf{K},\ \mathbf{K}_1,\ \mathbf{K}_2)$, is slowly varying in $\mathbf{K}_1$ and $\mathbf{K}_2$. Using (J.14) and (J.20), one obtains the result

$$< P_{surf}(\mathbf{R}_1) P^*_{surf}(\mathbf{R}_2) >=$$

$$|P_i(\mathbf{R}', 0)|^2 k_w^{-2} \int \sigma(\mathbf{K}_s, \mathbf{K}_i) e^{i\mathbf{K}_s\cdot\mathbf{D}} \beta_w^{-2}(K_i) d^2K_s. \qquad \text{(J.36)}$$

Equation (J.36) relates the spatial correlation of the effective surface field to the bistatic cross section through a Fourier transform. In obtaining (J.36), it has been assumed that the correlation length of the surface field is small compared to the width of the scattering region. As a consequence of the uncertainty relation for Fourier transforms, (J.36) implies that the correlation length of the surface field, L_f, is approximately equal to the inverse of the width of the cross section with respect to the variable $\mathbf{K}_s$. This allows two important conclusions. First, the condition set earlier on the slowness of the K-dependence of the second moment of the T-matrix is equivalent to the statement that the source is in the far field with respect to the correlation length of the effective surface field. Second, this correlation length is simply related to the angular width, $\Delta\theta$, of the narrowest feature of the scattering cross section: $L_f \approx \lambda/\Delta\theta$, analogous to (F.2) for arrays.

All the elements needed to derive the sonar equation relationship between the bistatic scattering cross section and the mean-square scattered field are now available. Changing integration variables in (J.30) to $\mathbf{R}'$ and $\mathbf{D}$, one uses (J.36) and the first-power expansion (J.31). The integration over $\mathbf{D}$ gives a delta function which enforces (J.17) and yields the desired sonar equation (J.15). Note that (J.16) is a result of the stationary phase approximation leading to (J.36). Equation (J.15) can be interpreted as application of the sonar equation to small scattering subpatches, with an incoherent summation of the resulting scattered mean-square fields.

To summarize, use of (J.15) is valid if the following conditions are met: (1) the scattering boundary is in the far field of the source, (2) the source and field points are both in the far field with respect to the correlation length, L_f, (3) the size of the ensonified region is much greater than L_f, and (4) the field point is at least a few wavelengths away from the boundary. This result is general, applying to both seafloor and sea-surface scattering as long as the statistical stationarity condition is satisfied.

This theoretical result can be explained using widely understood beamforming and diffraction concepts. As illustrated in [D. Jackson et al. 1997], the statistical averaging essential to the definition of the scattering cross section smooths out the angular "wiggles" that would appear in a single-ping measurement. That is, for a single transmission, the angular pattern of the scattered energy is very complex, and this angular dependence will not be independent of range until range exceeds the usual far-field value, d^2/λ, where d is the dimension of the ensonified region. This is the pessimistic criterion discussed earlier in this appendix. But the angular dependence of the mean-square field is usually broad and smooth, though not necessarily independent of range. Equation (J.15) will yield range dependence if the ranges are comparable to the size of the scattering region, as in [Pace et al. 1985].

While theory may provide estimates of the correlation length of the surface field, this is of little help to the experimentalist. Fortunately, this correlation scale is related to an experimentally accessible quantity, the angular width of the scattering strength function [Lysanov 1973]. This can be seen by considering the inverse of (J.36). Then (F.4) can be applied with $\Delta\theta$ taken to be angular width of the scattering cross section. The angular width is not unique, even in a given problem. The backscatter lobe may be very broad, indicating that a short correlation scale is relevant, while the forward lobe may be narrow, indicating a larger correlation scale. It follows that it may be necessary for the measurement transducers to be placed at a greater distance from the boundary in measurements of the forward part of the bistatic scattering cross section as compared to measurements in other directions.

It may be difficult to satisfy the far-field criteria in nonspecular directions, if the scattering cross section exhibits rapid variations with respect to incident and scattered angles. Such variations are expected if strong reflections occur due to seafloor stratification on scales much greater than the wavelength [Ivakin 1986, Ivakin 1989, Mourad and Jackson 1993, Essen 1994, Moe and Jackson 1994]. The depth of these reflections then appears as the field correlation length, requiring greater-than-usual ranges for satisfaction of the far-field criteria.

K Rough-Interface Perturbation Theory

The purpose of this appendix is to derive expressions given in Ch. 13 for the fluid, elastic, and poroelastic cases. While there are several excellent references on perturbation theory as applied to these cases, the notation and approach used is diverse. Here, a single approach applicable to all cases will be developed, making use of some of the general expressions given in Sect. J.1.

Perturbation theory employs a type of power series expansion which can be either an expansion of the scattered field or, more conveniently, the T-matrix:

$$T_{ww}(\mathbf{K}_s,\ \mathbf{K}_i) = \sum_{n=0}^{\infty} \frac{1}{n!} T_{ww}^{(n)}(\mathbf{K}_s,\ \mathbf{K}_i)\ . \tag{K.1}$$

The subscripts ww indicate that this T-matrix is the plane-wave expansion of the field scattered back into the water when a plane wave is incident from the water. In solving the scattering problem, it is necessary to include whatever fields exist in the sediment, e.g., for the fluid–fluid boundary, one must consider a similar expansion of T_{wp}, the T-matrix describing the compressional wave field in the sediment. The notation used in (K.1) is somewhat streamlined and requires explanation. The factor $1/n!$ is included in order to simplify later expressions. The first term in the expansion, $T_{ww}^{(0)}$, is independent of the interface relief function, $f(\mathbf{R})$. In fact, it is simply the solution of the flat-interface problem

$$T_{ww}^{(0)}(\mathbf{K}_s, \mathbf{K}_i) = V_{ww}(\mathbf{K}_i)\delta(\mathbf{K}_s - \mathbf{K}_i)\ , \tag{K.2}$$

where $V_{ww}(\mathbf{K}_i)$ is the flat-interface reflection coefficient discussed in the main text. The next term is the "first-order" approximation, proportional to the first power of interface roughness, even though this dependence is not shown explicitly. The first-order approximation is the most commonly employed form of perturbation theory, the so-called "Born approximation." In all cases (fluid, elastic, and poroelastic), first-order perturbation theory gives the following form for the incoherent, or scattered, part of the T-matrix:

$$T_{ww}^{(1)}(\mathbf{K}_s, \mathbf{K}_i) = \frac{ik_w}{\beta_w(K_s)} A_{ww}(\mathbf{K}_s, \mathbf{K}_i) F(\mathbf{K}_s - \mathbf{K}_i)\ , \tag{K.3}$$

where $F(\mathbf{K})$ is the Fourier transform of the interface relief function

$$F(\mathbf{K}) = \frac{1}{(2\pi)^2} \int e^{-i\mathbf{K}\cdot\mathbf{R}} f(\mathbf{R}) d^2R \ , \tag{K.4}$$

and $A_{ww}(\mathbf{K}_s, \mathbf{K}_i)$ is the factor appearing in the general expression for the small-roughness scattering cross section (13.9) and the small-slope scattering cross section (13.11). Equation (K.3) shows the sense in which $T_{ww}^{(1)}(\mathbf{K}_s, \mathbf{K}_i)$ is proportional to the first power of interface relief: it is proportional to the Fourier transform of the relief function. Similarly, the $n = 2$ term in (K.1) involves two powers of $F(\mathbf{K})$ in an integral expression that resembles a convolution.

As noted in Ch. 13 and Appendix L, there is a general relation between the zeroth- and first-order perturbation solutions,

$$A_{ww}(\mathbf{K}_i, \mathbf{K}_i) = -2\beta_w^2(K_i)V_{ww}(\mathbf{K}_i) \ . \tag{K.5}$$

Voronovich [Voronovich 1985] gives a proof of this relationship by considering the case of a flat interface displaced from $z = 0$. This result is true for all three theories considered in this monograph, provided the seafloor in homogeneous, without gradients or layering. This relationship is important in the small-slope approximation and also provides a useful test of numerical algorithms.

Returning to the first-order term, its first moment vanishes:

$$< T_{ww}^{(1)}(\mathbf{K}_s, \mathbf{K}_i) >= 0 \ , \tag{K.6}$$

because the interface relief function is assumed to have zero mean ($< f(\mathbf{R}) >= 0$). Considering the second moment, the connection between $A_{ww}(\mathbf{K}_s, \mathbf{K}_i)$ and the first approximation to the scattering cross section can be established by using (J.13) and (J.14):

$$< T_{ww}^{(1)}(\mathbf{K}_s, \mathbf{K}_i) T_{ww}^{(1)*}(\mathbf{K}'_s, \mathbf{K}'_i) >=$$

$$\frac{k_w^2}{\beta_w(K_s)\beta_w(K'_s)} A_{ww}(\mathbf{K}_s, \mathbf{K}_i) A_{ww}^*(\mathbf{K}'_s, \mathbf{K}'_i) < F(\mathbf{K}_s - \mathbf{K}_i) F^*(\mathbf{K}'_s - \mathbf{K}'_i) > \ . \tag{K.7}$$

Using (6.4) and (6.5),

$$< F(\mathbf{K}_1) F^*(\mathbf{K}_2) >= W(\mathbf{K}_1)\delta(\mathbf{K}_1 - \mathbf{K}_2) \ . \tag{K.8}$$

Combining (J.13), (J.14), (K.7), and (K.8), the key result is obtained:

$$\sigma^{(2)}(\mathbf{K}_s, \mathbf{K}_i) = k_w^4 |A_{ww}(\mathbf{K}_s, \mathbf{K}_i)|^2 W(\mathbf{\Delta K}) \ , \tag{K.9}$$

which is the same as (13.9), except the superscript (2) is used to indicate that this approximation involves the second power of the first-order approximation to the T-matrix. Hence, this approximation to the scattering cross section is actually second order in the interface roughness. Expression (K.9) applies for all boundary conditions (fluid–fluid, fluid–elastic, and fluid–poroelastic), even in the presence of gradients and layering, so long as the latter satisfy certain regularity conditions, to be discussed in Appendix M. In the present appendix, however, it will be assumed throughout that the sediment is homogeneous, with no gradients in density or complex sound speed.

In order to indicate how the higher-order terms can be used, it will be noted that the second-order approximation to the coherent field employs the expression

$$< T_{ww}^{(2)}(\mathbf{K}_s, \mathbf{K}_i) >= V_{ww}^{(2)}(\mathbf{K}_i)\delta(\mathbf{K}_s - \mathbf{K}_i) \; , \tag{K.10}$$

which can be taken as a definition of $V_{ww}^{(2)}(\mathbf{K}_i)$. Thus, the coherent reflection coefficient approximated to second order is $V_{ww}(\mathbf{K}_i) + V_{ww}^{(2)}(\mathbf{K}_i)$. If one is concerned with the reflected coherent power, the squared magnitude of the coherent reflection coefficient is of interest:

$$|V_{wwc}|^2 \approx |V_{ww}|^2 + |V_{ww}^{(2)}|^2 + 2\mathrm{Re}\left\{V_{ww}V_{ww}^{(2)*}\right\} \; . \tag{K.11}$$

One might be tempted to say that this expression is valid to fourth order in interface roughness, but this is not true owing to the omission of $< T_{ww}^{(4)} >$. It does, however, contain all second-order terms, so the squared coherent reflection coefficient is, to second order,

$$|V_{wwc}|^2 \approx |V_{ww}|^2 + 2\mathrm{Re}\left\{V_{ww}V_{ww}^{(2)*}\right\} \; . \tag{K.12}$$

In considering the division of power between the coherent and incoherent (scattered) fields, one can now consistently use (K.9) and (K.12) as they are both second order in interface roughness. If all fields, including those scattered into the sediment, are treated consistently to a given order of interface roughness, incident, coherently reflected, and scattered energy (power, actually) will balance perfectly. This may seem odd, as each term is an approximation, but one should think of the exact power balance equation and realize that the terms in a power series expansion of this equation must also balance.

The consistent application of higher orders to the scattering cross section requires some care. Using notation similar to that of [Thorsos and Jackson 1989], the cross section to fourth order is

$$\sigma_{ww}^{(4)} = \sigma_{ww}^{(2)} + \sigma_{ww}^{(22)} + \sigma_{ww}^{(13)} \; . \tag{K.13}$$

The superscript (22) indicates that this term comes from the second moment of $T_{ww}^{(2)}$, while the term having superscript (13) comes from the following moment: $< T_{ww}^{(1)}T_{ww}^{(3)*} >$.

K.1 Fluid–Fluid Boundary

Even though the focus of this appendix is first-order perturbation theory, the development will be arranged so that the nth-order case is clearly defined.

The derivation given here will be rather detailed in order to set a pattern to be used for the elastic and poroelastic cases.

Following the procedure outlined in the previous section, the incident field will be taken to be a plane wave, and the total pressure field in the water will be written in the form

$$P(\mathbf{r}) = e^{i\mathbf{K}_i\cdot\mathbf{R}-ik_w\beta_w(K_i)z} + \int T_{ww}(\mathbf{K}_s,\ \mathbf{K}_i)e^{i\mathbf{K}_s\cdot\mathbf{R}+ik_w\beta_w(K_s)z}d^2K_s\ . \quad \text{(K.14)}$$

Similarly, the pressure field in the sediment is

$$P(\mathbf{r}) = \int T_{wp}(\mathbf{K}_s,\ \mathbf{K}_i)e^{i\mathbf{K}_s\cdot\mathbf{R}-ik_p\beta_p(K_s)z}d^2K_s\ . \quad \text{(K.15)}$$

The T-matrices can be determined by imposing continuity of pressure and normal displacement on the boundary, $z = f(\mathbf{R})$ (see Ch. 8):

$$P|_{z=f(\mathbf{R})^+} = P|_{z=f(\mathbf{R})^-}\ , \quad \text{(K.16)}$$

$$a_\rho \mathbf{N}\cdot\nabla P|_{z=f(\mathbf{R})^+} = \mathbf{N}\cdot\nabla P|_{z=f(\mathbf{R})^-}\ , \quad \text{(K.17)}$$

where

$$\mathbf{N} = \mathbf{e}_z - \nabla f(\mathbf{R})\ . \quad \text{(K.18)}$$

In terms of the T-matrices, these conditions are

$$e^{i\mathbf{K}_i\cdot\mathbf{R}-ik_w\beta_w(K_i)f(\mathbf{R})} + \int T_{ww}(\mathbf{K}_s,\ \mathbf{K}_i)e^{i\mathbf{K}_s\cdot\mathbf{R}+ik_w\beta_w(K_s)f(\mathbf{R})}d^2K_s =$$

$$\int T_{wp}(\mathbf{K}_s,\ \mathbf{K}_i)e^{i\mathbf{K}_s\cdot\mathbf{R}-ik_p\beta_p(K_s)f(\mathbf{R})}d^2K_s \quad \text{(K.19)}$$

and

$$a_\rho[-ik_w\beta_w(K_i) - i\mathbf{K}_i\cdot\nabla f(\mathbf{R})]e^{i\mathbf{K}_i\cdot\mathbf{R}-ik_w\beta_w(K_i)f(\mathbf{R})} +$$

$$a_\rho\int[ik_w\beta_w(K_s) - i\mathbf{K}_s\cdot\nabla f(\mathbf{R})]T_{ww}(\mathbf{K}_s,\ \mathbf{K}_i)e^{i\mathbf{K}_s\cdot\mathbf{R}+ik_w\beta_w(K_s)f(\mathbf{R})}d^2K_s =$$

$$\int[-ik_p\beta_p(K_s) - i\mathbf{K}_s\cdot\nabla f(\mathbf{R})]T_{wp}(\mathbf{K}_s,\ \mathbf{K}_i)e^{i\mathbf{K}_s\cdot\mathbf{R}-ik_p\beta_p(K_s)f(\mathbf{R})}d^2K_s\ . \quad \text{(K.20)}$$

It is convenient to express these relations entirely in terms of wave vectors, removing the R-dependence by means of a Fourier transform. That is, operate on both (K.19) and (K.20) with

$$\int e^{-i\mathbf{K}'_s\cdot\mathbf{R}}d^2R\ .$$

The terms involving the gradient can be simplified by means of the following identity:

$$\int \nabla f(\mathbf{R}) e^{-i\mathbf{K}_1\cdot\mathbf{R} \pm i k_\alpha \beta_\alpha(\mathbf{K}_2) f(\mathbf{R})} d^2R =$$

$$\pm \frac{1}{i k_\alpha \beta_\alpha(\mathbf{K}_2)} \int [\nabla + i\mathbf{K_1}] e^{-i\mathbf{K}_1\cdot\mathbf{R} \pm i k_\alpha \beta_\alpha(\mathbf{K}_2) f(\mathbf{R})} d^2R \; . \tag{K.21}$$

The gradient term on the right-hand side of (K.21) integrates trivially to give the exponential evaluated at the boundary at infinity. Letting $f(\mathbf{R})$ vanish at infinity, this term can be discarded. The end result is replacement of $\nabla f(\mathbf{R})$ by $\pm\mathbf{K}_1/[k_\alpha\beta_\alpha(\mathbf{K}_2)]$. The continuity conditions can now be written in the forms

$$A_w^-(\mathbf{K}_s',\ \mathbf{K}_i) + \int A_w^+(\mathbf{K}_s',\ \mathbf{K}_s) T_{ww}(\mathbf{K}_s,\ \mathbf{K}_i) d^2K_s =$$

$$\int A_p^-(\mathbf{K}_s',\ \mathbf{K}_s) T_{wp}(\mathbf{K}_s,\ \mathbf{K}_i) d^2K_s \; , \tag{K.22}$$

$$a_\rho B_w(\mathbf{K}_s',\ \mathbf{K}_i) A_w^-(\mathbf{K}_s',\ \mathbf{K}_i)$$

$$-a_\rho \int B_w(\mathbf{K}_s',\ \mathbf{K}_s) A_w^+(\mathbf{K}_s',\ \mathbf{K}_s) T_{ww}(\mathbf{K}_s,\ \mathbf{K}_i) d^2K_s =$$

$$\int B_p(\mathbf{K}_s',\ \mathbf{K}_s) A_p^-(\mathbf{K}_s',\ \mathbf{K}_s) T_{wp}(\mathbf{K}_s,\ \mathbf{K}_i) d^2K_s \; , \tag{K.23}$$

where

$$B_\alpha(\mathbf{K}_1,\ \mathbf{K}_2) = \frac{1 - \mathbf{K}_1 \cdot \mathbf{K}_2/k_\alpha^2}{a_\alpha \beta_\alpha(\mathbf{K}_2)} \; , \tag{K.24}$$

and, using the definition given in [DeSanto 1979, D. Jackson et al. 1988],

$$A_\alpha^\pm(\mathbf{K}_1,\ \mathbf{K}_2) = \frac{1}{(2\pi)^2} \int e^{-i(\mathbf{K}_1 - \mathbf{K}_2)\cdot\mathbf{R} \pm i k_\alpha \beta_\alpha(\mathbf{K}_2) f(\mathbf{R})} d^2R \; . \tag{K.25}$$

Expressions (K.22) and (K.23) can be viewed as exact, though there is controversy associated with this position. The scattered field in the water as represented in (K.14) is comprised entirely of upgoing waves. One can readily imagine that, if the interface has deep valleys, waves will be scattered downward as well as upward. The assumption that only upgoing scattered waves need be considered is known as the "Rayleigh hypothesis." There is a large and contentious literature surrounding this topic, but [Voronovich 1994] presents a proof that he gave earlier in the Russian literature that, at least for the pressure release surface, the Rayleigh hypothesis is not germain to perturbation theory. A less general discussion is given in [D. Jackson et al. 1988]. These results were obtained by developing the perturbation series using a method that does not require the Rayleigh hypothesis and then showing that this series is identical to that obtained using the hypothesis. This identity seems self-evident when one considers a fixed interface relief function, $f(\mathbf{R})$, scaled by a real, positive factor, α. The perturbation series now becomes a simple power series in α. As α becomes sufficiently small, it is plausible that the series converges. If it does, it must be unique. As this argument makes no mention of the boundary conditions, it should apply generally. Thus, one may use the simpler Rayleigh approach in obtaining the perturbation series, with confidence that the result will be identical to one obtained with more difficulty by avoiding the Rayleigh hypothesis. Failure of the series to converge at large values of roughness signals a failure of perturbation theory that is unrelated to the Rayleigh hypothesis. It must be noted, however, that the Rayleigh hypothesis *is* sometimes relevant to nonperturbative methods.

The continuity conditions will be written in matrix notation to provide a general formalism for more complicated cases to be encountered later. Expressions (K.22) and (K.23) can be summarized in the form

$$\int \mathbf{D}(\mathbf{K}_s,\ \mathbf{K})\mathbf{A}(\mathbf{K}_s,\ \mathbf{K})\mathbf{T}(\mathbf{K},\ \mathbf{K}_i)d^2K = A_w^-(\mathbf{K}_s,\ \mathbf{K}_i)\mathbf{S}(\mathbf{K}_s,\ \mathbf{K}_i)\ , \quad \text{(K.26)}$$

where

$$\mathbf{D}(\mathbf{K}_s,\ \mathbf{K}) = \begin{bmatrix} -1 & 1 \\ a_\rho B_w(\mathbf{K}_s,\ \mathbf{K}) & a_\rho B_p(\mathbf{K}_s,\ \mathbf{K}) \end{bmatrix}, \quad \text{(K.27)}$$

$$\mathbf{A}(\mathbf{K}_s,\ \mathbf{K}) = \begin{bmatrix} A_w^+(\mathbf{K}_s,\ \mathbf{K}) & 0 \\ 0 & A_p^-(\mathbf{K}_s,\ \mathbf{K}) \end{bmatrix}, \quad \text{(K.28)}$$

$$\mathbf{T}(\mathbf{K},\ \mathbf{K}_i) = \begin{bmatrix} T_{ww}(\mathbf{K},\ \mathbf{K}_i) \\ T_{wp}(\mathbf{K},\ \mathbf{K}_i) \end{bmatrix}, \quad \text{(K.29)}$$

$$\mathbf{S}(\mathbf{K}_s,\ \mathbf{K}_i) = \begin{bmatrix} 1 \\ a_\rho B_w(\mathbf{K}_s,\ \mathbf{K}_i) \end{bmatrix}. \quad \text{(K.30)}$$

The next step in developing the perturbation series is to write the T-matrix as a series as in (K.1) with an analogous expansion for T_{wp}. It is necessary to expand $A_\alpha^\pm$ in powers of the interface relief function:

$$A_\alpha^\pm(\mathbf{K}_1,\ \mathbf{K}_2) = \sum_{n=0}^{\infty} \frac{1}{n!}[\pm i k_\alpha \beta_\alpha(\mathbf{K}_2)]^n F_n(\mathbf{K}_1 - \mathbf{K}_2)\ , \quad \text{(K.31)}$$

where

$$F_n(\mathbf{K}) = \frac{1}{(2\pi)^2}\int e^{-i\mathbf{K}\cdot\mathbf{R}} f^n(\mathbf{R})d^2R\ . \quad \text{(K.32)}$$

Note $F_0(\mathbf{K}) = \delta(\mathbf{K})$, and $F_1(\mathbf{K}) = F(\mathbf{K})$.

If these expansions are inserted into (K.26), a matrix equation relating the nth-power terms can be obtained:

$$[-i k_w \beta_w(K_i)]^n \mathbf{S}(\mathbf{K}_s,\ \mathbf{K}_i)F_n(\mathbf{K}_s - \mathbf{K}_i) =$$

$$\sum_{m=0}^{n} (i k_w)^{n-m}\binom{n}{m}\int \mathbf{D}(\mathbf{K}_s,\ \mathbf{K})\mathbf{E}^{n-m}(\mathbf{K})\mathbf{T}^{(m)}(\mathbf{K},\ \mathbf{K}_i)F_{n-m}(\mathbf{K}_s - \mathbf{K})d^2K\ , \quad \text{(K.33)}$$

where

$$\binom{n}{m} = \frac{n!}{m!(n-m)!} \tag{K.34}$$

are the binomial coefficients. The unknown mth-order T-matrices are represented by the column vector

$$\mathbf{T}^{(m)}(\mathbf{K}_s,\ \mathbf{K}_i) = \begin{bmatrix} T_{ww}^{(m)}(\mathbf{K}_s,\ \mathbf{K}_i) \\ T_{wp}^{(m)}(\mathbf{K}_s,\ \mathbf{K}_i) \end{bmatrix}, \tag{K.35}$$

and $\mathbf{E}(\mathbf{K})$ is the following diagonal matrix, to be raised to the power $n-m$:

$$\mathbf{E}(\mathbf{K}) = \begin{bmatrix} \beta_w(K) & 0 \\ 0 & -\beta_p(K)/a_p \end{bmatrix}. \tag{K.36}$$

The $m=n$ term on the right-hand side of (K.33) can be simplified by noting that $F_0(\mathbf{K}) = \delta(\mathbf{K})$. Isolating this term, one obtains

$$\mathbf{T}^{(n)}(\mathbf{K}_s,\ \mathbf{K}_i) = \mathbf{Q}(\mathbf{K}_s,\ \mathbf{K}_i)[-ik_w\beta_w(K_i)]^n F_n(\mathbf{K}_s - \mathbf{K}_i) -$$

$$\sum_{m=0}^{n-1} \binom{n}{m} \int \mathbf{M}(\mathbf{K}_s,\ \mathbf{K})\mathbf{E}^{n-m}(\mathbf{K})\mathbf{T}^{(m)}(\mathbf{K},\ \mathbf{K}_i)(ik_w)^{n-m} F_{n-m}(\mathbf{K}_s - \mathbf{K}) d^2K\,. \tag{K.37}$$

In this equation

$$\mathbf{Q}(\mathbf{K}_s,\ \mathbf{K}_i) = \mathbf{D}^{-1}(\mathbf{K}_s,\ \mathbf{K}_s)\mathbf{S}(\mathbf{K}_s,\ \mathbf{K}_i) \tag{K.38}$$

and

$$\mathbf{M}(\mathbf{K}_s,\ \mathbf{K}) = \mathbf{D}^{-1}(\mathbf{K}_s,\ \mathbf{K}_s)\mathbf{D}(\mathbf{K}_s,\ \mathbf{K})\,. \tag{K.39}$$

Equation (K.37) is a recursion relation from which the nth-order T-matrix can be found in terms of all lower orders. Setting n = 0,

$$\mathbf{T}^{(0)}(\mathbf{K}_s,\ \mathbf{K}_i) = \mathbf{Q}(\mathbf{K}_i,\ \mathbf{K}_i)\delta(\mathbf{K}_s - \mathbf{K}_i)\,. \tag{K.40}$$

With some algebra, the separated components of this two-element column matrix can be shown to be

$$T_{ww}^{(0)}(\mathbf{K}_s,\ \mathbf{K}_i) = V_{ww}(\mathbf{K}_i)\delta(\mathbf{K}_s - \mathbf{K}_i)\,, \tag{K.41}$$

$$T_{wp}^{(0)}(\mathbf{K}_s,\ \mathbf{K}_i) = V_{wp}(\mathbf{K}_i)\delta(\mathbf{K}_s - \mathbf{K}_i)\,. \tag{K.42}$$

As expected, the zeroth-order T-matrices contain the flat-interface reflection and transmission coefficients, independently derived in Ch. 8, multiplied by a delta function. Setting $n=1$ in (K.37)

$$\mathbf{T}^{(1)}(\mathbf{K}_s,\ \mathbf{K}_i) =$$

$$-ik_w[\beta_w(K_i)\mathbf{Q}(\mathbf{K}_s,\ \mathbf{K}_i) - \mathbf{M}(\mathbf{K}_s,\ \mathbf{K}_i)\mathbf{E}(\mathbf{K}_i)\mathbf{Q}(\mathbf{K}_i,\ \mathbf{K}_i)]F(\mathbf{K}_s - \mathbf{K}_i)\ . \tag{K.43}$$

Separating this expression into its two components and performing some algebraic manipulation,

$$T_{ww}^{(1)}(\mathbf{K}_s,\ \mathbf{K}_i) = \frac{ik_w}{2\beta_w(K_s)}[1 + V_{ww}(\mathbf{K}_i)][1 + V_{ww}(\mathbf{K}_s)] \times$$

$$\left\{(1 - 1/a_\rho)\left[\frac{\mathbf{K}_s \cdot \mathbf{K}_i}{k_w^2} - \frac{\beta_p(K_i)\beta_p(K_s)}{a_p^2 a_\rho}\right] - 1 + \frac{1}{a_p^2 a_\rho}\right\} F(\mathbf{K}_s - \mathbf{K}_i)\ , \tag{K.44}$$

$$T_{wp}^{(1)}(\mathbf{K}_s,\ \mathbf{K}_i) = \frac{ik_w}{2\beta_w(K_s)}[1 + V_{ww}(\mathbf{K}_i)][1 + V_{ww}(\mathbf{K}_s)] \times$$

$$\left\{(1 - 1/a_\rho)\left[\frac{\mathbf{K}_s \cdot \mathbf{K}_i}{k_w^2} + \frac{\beta_p(K_i)\beta_w(K_s)}{a_p}\right] - 1 + \frac{1}{a_p^2 a_\rho}\right\} F(\mathbf{K}_s - \mathbf{K}_i)\ . \tag{K.45}$$

Comparing (K.44) with (K.3), the factor A_{ww} can be recognized and is identical to (13.21)–(13.25) apart from the use of wave vector, rather than angular, arguments. A different derivation of these results is given in Sect. M.1.2. A more general approach is used in [Moe and Jackson 1994], in which the sediment is allowed to be vertically stratified, that is, its mean properties may have gradients, with the gradient vector pointing vertically. The result is given in Sect. 13.2.1. Expression (K.45) is useful in modeling acoustic penetration of the seafloor (Ch. 15), and an equivalent form is derived in [Moe and Jackson 1998].

K.2 Fluid–Elastic Boundary

A systematic development of rough-interface perturbation theory for fluid–elastic boundaries has been given by [Dacol and Berman 1988], and accounts are also found in [Essen 1994, D. Jackson and Ivakin 1998]. Since the goal of this appendix is to provide a common perturbation approach for all three wave theories of interest, a somewhat different analysis will be undertaken, even though the results are mathematically equivalent to those of the cited references. Chapter 9 summarizes the equations of motion and boundary conditions for fluid–elastic models. While these equations and boundary conditions are more complicated than those of the fluid–fluid case, the matrix formalism developed in the preceding section can be adopted to minimize the amount of additional analytical effort.

Following the approach in Ch. 9, the displacement field (9.5) will be decomposed into compressional and shear components described by scalar, $\phi(\mathbf{r})$, and vector, $\boldsymbol{\psi}(\mathbf{r})$, displacement potentials, respectively. The field in the water will be expanded in terms of the T-matrix as in (K.14), but with T-matrix relating scalar potential rather than pressure:

$$\phi_w(\mathbf{r}) = e^{i\mathbf{k}_w^-(\mathbf{K}_i)\cdot\mathbf{r}} + \int T_{ww}(\mathbf{K}_s,\ \mathbf{K}_i)e^{i\mathbf{k}_w^+(\mathbf{K}_s)\cdot\mathbf{r}}d^2K_s\ . \tag{K.46}$$

The scalar potential in the sediment is

$$\phi(\mathbf{r}) = \int T_{wp}(\mathbf{K}_s,\ \mathbf{K}_i)e^{i\mathbf{k}_p^-(\mathbf{K}_s)\cdot\mathbf{r}}d^2K_s\ . \tag{K.47}$$

A word of warning is necessary: By using scalar displacement potentials rather than pressure, T_{wp} as used here will not approach T_{wp} of the fluid case in the limit as shear effects become negligible. Thus, the notation used here is not optimum, but further subscripting of the T-matrix components would be unwieldy. The vector potential is

$$\boldsymbol{\psi}(\mathbf{r}) =$$

$$\int [-\mathbf{e}_h^-(\mathbf{K_s})T_{wv}(\mathbf{K}_s,\ \mathbf{K}_i) + \mathbf{e}_v^-(\mathbf{K_s})T_{wh}(\mathbf{K}_s,\ \mathbf{K}_i)]e^{i\mathbf{k}_t^-(\mathbf{K}_s)\cdot\mathbf{r}}d^2K_s\ . \tag{K.48}$$

Two T-matrix shear-wave components are needed, one, T_{wv}, describing vertically polarized shear waves and one, T_{wh}, describing horizontally polarized shear waves. The vertical and horizontal polarization unit vectors, $\mathbf{e}_v^-(\mathbf{K_s})$ and $\mathbf{e}_h^-(\mathbf{K_s})$, are defined in Ch. 9. Although the horizontal polarization vector multiplies the vertical T-matrix component and vice versa, the curl operation needed to obtain displacement gives the expected ordering. Using (9.5), the displacement field in the water is

$$\mathbf{u}_w(\mathbf{r}) = ik_w\mathbf{e}_w^-(\mathbf{K_i})e^{i\mathbf{k}_w^-(\mathbf{K}_i)\cdot\mathbf{r}} +$$

$$ik_w\int T_{ww}(\mathbf{K}_s,\ \mathbf{K}_i)\mathbf{e}_w^+(\mathbf{K_s})e^{i\mathbf{k}_w^+(\mathbf{K}_s)\cdot\mathbf{r}}d^2K_s\ . \tag{K.49}$$

The displacement field in the sediment due to compressional waves is

$$\mathbf{u}_p(\mathbf{r}) =$$

$$ik_p\int T_{wp}(\mathbf{K}_s,\ \mathbf{K}_i)\mathbf{e}_p^-(\mathbf{K_s})e^{i\mathbf{k}_p^-(\mathbf{K}_s)\cdot\mathbf{r}}d^2K_s\ , \tag{K.50}$$

and the displacement field due to shear waves is

$$\mathbf{u}_t(\mathbf{r}) =$$

$$ik_t\int [\mathbf{e}_v^-(\mathbf{K_s})T_{wv}(\mathbf{K}_s,\ \mathbf{K}_i) + \mathbf{e}_h^-(\mathbf{K_s})T_{wh}(\mathbf{K}_s,\ \mathbf{K}_i)]e^{i\mathbf{k}_t^-(\mathbf{K}_s)\cdot\mathbf{r}}d^2K_s\ . \tag{K.51}$$

Next, the continuity conditions for normal displacement (9.26) and tractions (9.27) can be imposed. Following the same steps as in the fluid case, a Fourier transform can be used to eliminate terms involving the gradient of the interface relief function so that the continuity conditions can be put in the form of (K.26). In this case, (K.26) is a set of four linear equations in the four unknown elements of the column vector,

$$\mathbf{T}(\mathbf{K},\ \mathbf{K}_i) = \begin{bmatrix} T_{ww}(\mathbf{K},\ \mathbf{K}_i) \\ T_{wp}(\mathbf{K},\ \mathbf{K}_i) \\ T_{wv}(\mathbf{K},\ \mathbf{K}_i) \\ T_{wh}(\mathbf{K},\ \mathbf{K}_i) \end{bmatrix} . \tag{K.52}$$

The order in which these equations is taken in forming the rows of the 4×4 matrix $\mathbf{D}(\mathbf{K}_s,\ \mathbf{K})$ and the column vector $\mathbf{S}(\mathbf{K}_s,\ \mathbf{K}_i)$ is arbitrary, but the following order will be used: x-traction, y-traction, z-traction, normal displacement. Thus, the row index of D_{mn} takes on the values $m = x,\ y,\ z,\ d$. The column index takes on the values $n = w,\ p,\ v,\ h$ corresponding to the four unknown T-matrix components. After a considerable amount of algebra, one finds

$$D_{jw} = \frac{K_{sj} - K_j}{k_w \beta_w(K)} , \qquad j = x,\ y , \tag{K.53}$$

$$D_{zw} = -1 , \tag{K.54}$$

$$D_{jp} = a_\rho[(1-2\frac{a_t^2}{a_p^2})\frac{K_{sj} - K_j}{k_w \beta_w(K)} - 2a_t^2 B_p(\mathbf{K}_s,\ \mathbf{K})K_{sj}/k_w] , \quad j = x,\ y , \tag{K.55}$$

$$D_{zp} = a_\rho[1 - \frac{2\mathbf{K}_s \cdot \mathbf{K}}{k_t^2}] , \tag{K.56}$$

$$D_{jv} = -a_\rho[1 - \frac{2\mathbf{K}_s \cdot \mathbf{K}}{k_t^2}]K_j/K , \quad j = x,\ y , \tag{K.57}$$

$$D_{zv} = -a_\rho[a_t^2 B_t(\mathbf{K}_s,\ \mathbf{K})K/k_w + \frac{\beta_t(K)\mathbf{K}_s \cdot \mathbf{K}}{k_t K}] , \tag{K.58}$$

$$D_{jh} = a_\rho[a_t^2 \frac{\mathbf{K}_s \cdot \mathbf{e}_h^-(\mathbf{K})K_j}{k_w^2 \beta_t(K)} - a_t B_t(\mathbf{K}_s,\ \mathbf{K})e_{hj}^-(\mathbf{K})] , \quad j = x,\ y \tag{K.59}$$

$$D_{zh} = -a_\rho a_t \frac{\mathbf{K}_s \cdot \mathbf{e}_h^-(\mathbf{K})}{k_w} , \tag{K.60}$$

$$D_{dw} = -B_w(\mathbf{K}_s,\ \mathbf{K}) , \tag{K.61}$$

$$D_{dp} = -B_p(\mathbf{K}_s,\ \mathbf{K}) , \tag{K.62}$$

$$D_{dv} = \frac{\mathbf{K}_s \cdot \mathbf{K}}{K k_w} , \tag{K.63}$$

$$D_{dh} = \frac{\mathbf{K}_s \cdot \mathbf{e}_h^-(\mathbf{K})}{\beta_t(K) k_w} . \tag{K.64}$$

The functions $B_\alpha(\mathbf{K}_s,\ \mathbf{K})$ are defined in (K.24), and the diagonal matrix $\mathbf{A}(\mathbf{K}_s,\ \mathbf{K})$ is

$$\mathbf{A}(\mathbf{K}_s,\ \mathbf{K}) = \begin{bmatrix} A_w^+(\mathbf{K}_s,\ \mathbf{K}) & 0 & 0 & 0 \\ 0 & A_p^-(\mathbf{K}_s,\ \mathbf{K}) & 0 & 0 \\ 0 & 0 & A_t^-(\mathbf{K}_s,\ \mathbf{K}) & 0 \\ 0 & 0 & 0 & A_t^-(\mathbf{K}_s,\ \mathbf{K}) \end{bmatrix} . \tag{K.65}$$

The column vector in the inhomogeneous term of (K.26) represents the contribution of the incident wave toward satisfaction of the continuity conditions and is given by

$$\mathbf{S}(\mathbf{K}_s,\ \mathbf{K}_i) = \begin{bmatrix} (K_{sx} - K_{ix})/[k_w \beta_w(K_i)] \\ (K_{sy} - K_{iy})/k_w \beta_w(K_i)] \\ 1 \\ B_w(\mathbf{K}_s,\ \mathbf{K}_i) \end{bmatrix} . \tag{K.66}$$

The nth-order T-matrix can now be found through a procedure formally identical to that used in the fluid case. Equation (K.37) applies, but with

$$\mathbf{T}^{(m)}(\mathbf{K},\ \mathbf{K}_i) = \begin{bmatrix} T_{ww}^{(m)}(\mathbf{K},\ \mathbf{K}_i) \\ T_{wp}^{(m)}(\mathbf{K},\ \mathbf{K}_i) \\ T_{wv}^{(m)}(\mathbf{K},\ \mathbf{K}_i) \\ T_{wh}^{(m)}(\mathbf{K},\ \mathbf{K}_i) \end{bmatrix} , \tag{K.67}$$

with $\mathbf{D}$ and $\mathbf{S}$ as defined above for the elastic case, and with

$$\mathbf{E}(\mathbf{K}) = \begin{bmatrix} \beta_w(K) & 0 & 0 & 0 \\ 0 & -\beta_p(K)/a_p & 0 & 0 \\ 0 & 0 & -\beta_t(K)/a_t & 0 \\ 0 & 0 & 0 & -\beta_t(K)/a_t \end{bmatrix} . \tag{K.68}$$

The approach used above is similar to that of [D. Jackson and Ivakin 1998], except that the matrix equation representing field continuity at the interface is arranged differently. The connection between the matrices used here and those in [D. Jackson and Ivakin 1998] is as follows:

$$\mathbf{D}(\mathbf{K}_s,\ \mathbf{K}) = k_w^{-1}[P^{(1)}(\mathbf{K})(K_{sx}-K_x)+P^{(2)}(\mathbf{K})(K_{sy}-K_y)]E^{-1}(\mathbf{K})-P^{(3)}(\mathbf{K}) , \tag{K.69}$$

$$\mathbf{S}(\mathbf{K}_s,\ \mathbf{K}) = [k_w\beta_w(K)]^{-1}[Q^{(1)}(\mathbf{K})(K_{sx}-K_x)+Q^{(2)}(\mathbf{K})(K_{sy}-K_y)]+Q^{(3)}(\mathbf{K}) . \tag{K.70}$$

To obtain the zeroth-order reflection and transmission coefficients, (K.40) can be used, with results identical to those given in Sect. 9.5. The first-order T-matrix can be found using (K.43). To make the connection with the scattering cross section, the first element of the first-order T-matrix, $T_{ww}^{(1)}$, can be compared with (K.3) to obtain the factor A_{ww}. It is possible to expand and rearrange the matrix expression for A_{ww} to obtain (13.27), although the required algebra is extensive.

K.3 Fluid–Poroelastic Boundary

The first-order rough-interface perturbation problem is developed in [Williams et al. 2001a]. The approach of this reference is used in Sect. 13.2.3 and also in the small-slope model of [Yang et al. 2002]. Consequently, no detailed derivation will be given here, but the connection with the general approach used in this appendix will be outlined. First, (K.69) and (K.70) can be used to find the matrices **D** and **S** by inserting the expressions of Sect. 13.2.3 for the matrices $P^{(n)}$ and $Q^{(n)}$. The nth-order equation (K.37) applies, with

$$\mathbf{E}(\mathbf{K}) = \begin{bmatrix} \beta_w(K) & 0 & 0 & 0 & 0 \\ 0 & -\beta_1(K)/a_1 & 0 & 0 & 0 \\ 0 & 0 & -\beta_2(K)/a_2 & 0 & 0 \\ 0 & 0 & 0 & -\beta_t(K)/a_t & 0 \\ 0 & 0 & 0 & 0 & -\beta_t(K)/a_t \end{bmatrix} . \tag{K.71}$$

The mth-order T-matrix is

$$\mathbf{T}^{(m)}(\mathbf{K},\ \mathbf{K}_i) = \begin{bmatrix} T_{ww}^{(m)}(\mathbf{K},\ \mathbf{K}_i) \\ T_{w1}^{(m)}(\mathbf{K},\ \mathbf{K}_i) \\ T_{w2}^{(m)}(\mathbf{K},\ \mathbf{K}_i) \\ T_{wv}^{(m)}(\mathbf{K},\ \mathbf{K}_i) \\ T_{wh}^{(m)}(\mathbf{K},\ \mathbf{K}_i) \end{bmatrix} . \tag{K.72}$$

As in Sect. 13.2.3, the fast Biot wave is denoted by a sub- or superscript "1" and the slow wave by a "2." The shear wave notation is identical to that of the preceding section. The zeroth-order reflection and transmission coefficients can be found using (K.40), and the first-order T-matrix can be found using (K.43). As usual, the scattering cross section follows from comparison of the first element of the first-order T-matrix, $T_{ww}^{(1)}$, with (K.3) to obtain the factor A_{ww}. The approach outlined in this appendix is an alternative to the equations presented in Sect. 13.2.3 and may be preferred if higher orders are of interest.

L The Kirchhoff and Small-Slope Approximations

This appendix sketches the derivations of Kirchhoff and small-slope approximations. In addition, limiting cases are considered: near-specular scattering and the high-frequency limit. The derivations are quite general, applicable for fluid, elastic, and poroelastic (Biot) theories. However, the presentation is abbreviated and intended mainly to establish notation and provide relations essential for the scattering models discussed in the main text. The main text also gives references that can be consulted for more detailed derivations.

L.1 Kirchhoff Approximation

The general T-matrix expression (J.6) provides a convenient starting point for developing the Kirchhoff approximation. The Kirchhoff approximation is often called the "tangent plane" approximation, as the surface field (J.1) and its normal derivative are approximated by the field that would exist if the rough interface were replaced at each point by a plane tangent to the true interface. Thus, the field at the surface is taken to be

$$P(\mathbf{r},\ \mathbf{K}_i) \approx [1 + V_{ww}(K_i,\ \mathbf{R})]e^{i\mathbf{k}_i\cdot\mathbf{r}}\ , \tag{L.1}$$

and the dot product of the gradient with the interface normal is approximated as

$$\mathbf{N}\cdot\nabla P(\mathbf{r},\ \mathbf{K}_i) \approx i\mathbf{N}\cdot\mathbf{k}_i[1 - V_{ww}(K_i,\ \mathbf{R})]e^{i\mathbf{k}_i\cdot\mathbf{r}}\ . \tag{L.2}$$

The reflection coefficient $V_{ww}(K_i,\ \mathbf{R})$ is allowed to be a function of the horizontal coordinates, $\mathbf{R}$, to account for the local tilting of the interface. With these approximations, (J.6) becomes

$$T_{ww}(\mathbf{K}_s,\ \mathbf{K}_i) = \frac{\Delta k^2}{8\pi^2 k_w \beta_w(\mathbf{K_s})\mathbf{\Delta k_z}} \int V_{ww}(K_i,\ \mathbf{R})e^{-i\mathbf{\Delta k}\cdot\mathbf{r}}|_{z=f(\mathbf{R})} d^2R\ . \tag{L.3}$$

Here, $\mathbf{r} = \mathbf{R} + \mathbf{e}_z f(\mathbf{R})$, and $\mathbf{\Delta k} = \mathbf{k}_s - \mathbf{k}_i$ with magnitude Δk and z-component $\Delta k_z = k_w[\beta_w(\mathbf{K_s}) + \beta_\mathbf{w}(\mathbf{K_i})]$ as in Sect. 13.1.

The first and second moments of the T-matrix, needed for the coherent reflection coefficient and the bistatic scattering cross section, can be readily obtained if the dependence of the reflection coefficient on horizontal position is neglected. In this case, V_{ww} is no longer a random variable, and the expectation values of interest are of the form

$$< e^{ix} >= e^{-\frac{1}{2}<x^2>} , \tag{L.4}$$

where x is a zero-mean, Gaussian, random variable. Expression (L.4) is the characteristic function of a Gaussian random variable. In applying this expression, it is assumed that the interface relief function, $f(\mathbf{R})$, is a Gaussian random process. The first moment of the T-matrix yields, through application of (J.12),

$$V_{wwc} = V_{ww} e^{-\frac{1}{2}\Delta k_z^2 h^2} , \tag{L.5}$$

where h is the RMS interface relief and $\Delta k_z = 2k_w\beta_w(\mathbf{K_i})$. The second moment of the T-matrix can be similarly evaluated and compared with (J.13) and (J.14) to obtain expression (13.10) for the bistatic cross section.

These derivations of the coherent reflection coefficient and bistatic scattering cross section employ an additional approximation beyond the traditional tangent plane approximation. The reflection coefficient has been assumed to be nonrandom and independent of horizontal coordinates. While this is true for the pressure-release surface (where the reflection coefficient is -1 for all angles), it is not true for the seafloor. Consequently, there remains the problem of making a reasonable choice for the argument of the reflection coefficient. As noted in Sect. 13.1, Thorsos (private communication) recommends evaluation of the reflection coefficient at the angle (13.13) that corresponds to specular reflection from the source to the receiver. This choice can be motivated by the high-frequency limit of the Kirchhoff approximation (Sect. 13.3) in which such specular reflections are responsible for *all* scattering. In the case of the coherent reflection coefficient, the angle provided by this recipe is simply the incident grazing angle. For the bistatic cross section, this angle is a function of both the incident and scattered directions.

L.2 Small-Slope Approximation

The lowest-order small-slope approximation assumes that the T-matrix can be written in a form similar to the Kirchhoff approximation [Voronovich 1985, Broschat and Thorsos 1997]:

$$T_{ww}(\mathbf{K}_s,\ \mathbf{K}_i) = \frac{\Phi_0(\mathbf{K}_s,\ \mathbf{K}_i)}{8\pi^2 k_w \beta_w(K_s)} \int e^{-i\boldsymbol{\Delta}\mathbf{k}\cdot\mathbf{r}}|_{z=f(\mathbf{R})} d^2R \ . \tag{L.6}$$

The function $\Phi_0(\mathbf{K}_s,\ \mathbf{K}_i)$ is determined by forcing consistency with the perturbation expansion carried out to first order. Realizing that the interface relief function appears in (L.6) through the coordinate vector $\mathbf{r}$ in the exponent, expansion to first order in $f(\mathbf{R})$ and comparison with the general zeroth- and first-order perturbation expressions (K.2) and (K.3) yields

$$\Phi_0(\mathbf{K}_i,\ \mathbf{K}_i) = 2k_w\beta_w(\mathbf{K}_i)V_{ww}(K_i) \tag{L.7}$$

and

$$\Phi_0(\mathbf{K}_s,\ \mathbf{K}_i) = -\frac{2k_w^2}{\Delta k_z}A_{ww}(\mathbf{K}_s,\mathbf{K}_i)\ . \tag{L.8}$$

Thus, the lowest-order small-slope approximation is completely determined once the first-order perturbation approximation has been employed to obtain $A_{ww}(\mathbf{K}_s,\mathbf{K}_i)$. This function depends on the wave theory employed and is given for the fluid, elastic, and poroelastic cases in Ch. 13. Further discussion is provided in Appendix K. Equations (L.7) and (L.8) taken together demand [Voronovich 1985]

$$A_{ww}(\mathbf{K}_i,\mathbf{K}_i) = -2\beta_w^2(\mathbf{K}_i)V_{ww}(\mathbf{K}_i)\ . \tag{L.9}$$

This is a fundamental relationship in perturbation theory. An important and well-known consequence of this relationship is equality of the Kirchhoff and small-slope T-matrices in the specular ($\mathbf{K}_s = \mathbf{K}_i$) direction, provided the reflection coefficient in (L.3) can be approximated by the position-independent value discussed above. This leads directly to equality of the coherent reflection coefficients and bistatic scattering cross sections (in the specular direction) for the two approximations.

L.3 The Incoherent Reflection Coefficient

If the scattering due to interface roughness is concentrated near the specular direction, it is possible to define a useful reflection coefficient for the incoherently scattered energy. First, consider the general equation (J.15) for the mean-square scattered pressure envelope at $\mathbf{r}_s$, with an omnidirectional point source situated at $\mathbf{r}_i$ and having source level $20\log_{10}s_0$. Accounting for the $\sqrt{2}$ ratio between envelope magnitude and RMS value,

$$<|P_s(\mathbf{r}_s)|^2>= 2s_0^2\int_{z'=0}\frac{\sigma(\mathbf{K}_s,\mathbf{K}_i)}{(d_id_s)^2}d^2R', \tag{L.10}$$

where

$$\mathbf{K}_i = k_w\frac{\mathbf{R}'-\mathbf{R}_i}{d_i}\ , \tag{L.11}$$

$$\mathbf{K}_s = k_w\frac{\mathbf{R}_s-\mathbf{R}'}{d_s}\ , \tag{L.12}$$

$$d_i = |\mathbf{r}'-\mathbf{r}_i|\ , \tag{L.13}$$

and

$$d_s = |\mathbf{r}_s-\mathbf{r}'|\ , \tag{L.14}$$

with $\mathbf{r}' = [\mathbf{R}',\ 0]$. Attenuation due to absorption in the water is neglected, but anisotropy in scattering is allowed. The cross section is a function of $\mathbf{K}_s$

and $\mathbf{K}_i$, consequently a change of integration variable will be made, using the Bragg wave vector

$$\Delta\mathbf{K} = \mathbf{K}_s - \mathbf{K}_i \; . \tag{L.15}$$

This choice is motivated by the fact that, near the specular direction, the cross section can be approximated as a function of $\Delta\mathbf{K}$. The elements of the Jacobian determinant for this change of variable are

$$\frac{\partial \Delta K_n}{\partial R'_m} =$$

$$k_w[-(\frac{1}{d_i}+\frac{1}{d_s})\delta_{mn} - \frac{1}{d_s^2}\frac{\partial d_s}{\partial R'_m}(R_{sn}-R'_n) + \frac{1}{d_i^2}\frac{\partial d_i}{\partial R'_m}(R'_n - R_{in}) \; , \quad m, \, n = x, \, y \; . \tag{L.16}$$

This will be approximated by assuming that the main contribution to the integral (L.10) comes near $\Delta\mathbf{K} = \mathbf{0}$, the specular direction, where

$$\frac{z_i}{d_i} = \frac{z_s}{d_s} \; , \tag{L.17}$$

$$\frac{\mathbf{R}' - \mathbf{R}_i}{d_i} = \frac{\mathbf{R}_s - \mathbf{R}'}{d_s} \; , \tag{L.18}$$

$$\nabla'(d_i + d_s) = \mathbf{0} \; . \tag{L.19}$$

Equation (L.19) is an expression of Fermat's principle.

These identities yield the following Jacobian:

$$d^2R' = \frac{d^2\Delta K}{[k_w(d_i + d_s)\sin\theta_i]^2} \; . \tag{L.20}$$

Here, $\sin\theta_i = z_i/d_i = \sin\theta_s = z_s/d_s$. With these results, (L.10) can be approximated as

$$< |P_s(\mathbf{r}_s)|^2 > = \frac{2s_0^2|V_{wwi}(\theta_i)|^2}{(d_i + d_s)^2} \; , \tag{L.21}$$

where

$$|V_{wwi}(\theta_i)|^2 = \frac{1}{k_w^2}\int \sigma d^2\Delta K \tag{L.22}$$

is the reflection coefficient for incoherently scattered energy. The key result is that the inverse-fourth-power spreading factor for each scattering subpatch has been converted to an inverse-square spreading factor for the entire seafloor. If (L.21) is compared to the equation for the peak pressure reflected from a flat surface (G.1), it can be seen that the two are of the same form, although the present result assumes a nondirectional source and neglects acoustic absorption in seawater. As the scattering cross section is often expressed in terms of angular variables, it may be preferable to express the incoherent reflection coefficient as

$$|V_{wwi}(\theta_i)|^2 = \frac{1}{\sin\theta_i} \int \sigma d^2 d\Omega_s \ , \tag{L.23}$$

using the solid-angle variable of Ch. 2.

Note that the energy conservation equation (2.20) can now be written as

$$|V_{wwi}|^2 + |V_{wwc}|^2 < 1 \ . \tag{L.24}$$

Near the specular direction, the bistatic scattering cross section in both the Kirchhoff (13.10) and small-slope (13.11) approximations can be written as

$$\sigma = \frac{[k_w |V_{ww}(\theta_i)| \sin\theta_i]^2}{(2\pi)^2} \int e^{-i\mathbf{\Delta K}\cdot\mathbf{R}} [e^{-\frac{1}{2}\Delta k_z^2 S(\mathbf{R})} - e^{-\Delta k_z^2 h^2}] d^2R \ . \tag{L.25}$$

Inserting this into (L.22), the integral over $\mathbf{\Delta K}$ gives rise to a delta function, $\delta(\mathbf{R})$, with the result

$$|V_{wwi}(\theta_i)|^2 = |V_{ww}(\theta_i)|^2 (1 - e^{-\Delta k_z^2 h^2}) \ , \tag{L.26}$$

or, recognizing the coherent reflection coefficient (L.5),

$$|V_{wwi}(\theta_i)|^2 = |V_{ww}(\theta_i)|^2 - |V_{wwc}(\theta_i)|^2 \ . \tag{L.27}$$

L.4 High-Frequency Limit of the Kirchhoff Approximation

To obtain the high-frequency limit, the Kirchhoff integral (13.12) is specialized to the large-roughness case for which the coherent reflection coefficient is negligible, making it possible to drop the second term in the integrand.

$$I_K = \frac{\Delta K^2}{2\pi} \int e^{-i\mathbf{\Delta K}\cdot\mathbf{R} - \frac{1}{2}\Delta k_z^2 S(\mathbf{R})} d^2R \ . \tag{L.28}$$

The high-frequency assumption enters through the statement that Δk_z in (L.28) is so large that the exponential $e^{-\frac{1}{2}\Delta k_z^2 S(\mathbf{R})}$ falls off with $\mathbf{R} = x\mathbf{e}_x + y\mathbf{e}_y$ so rapidly that only small values of R contribute. Thus, the structure function, $S(\mathbf{R})$, is expanded in a power series about $\mathbf{R} = \mathbf{0}$. The structure function is related to the roughness covariance, $B(\mathbf{R})$, by (D.5), so that

$$S(\mathbf{R}) \simeq -\frac{\partial^2 B}{\partial x^2} x^2 - 2\frac{\partial^2 B}{\partial x \partial y} xy - \frac{\partial^2 B}{\partial y^2} y^2 \ . \tag{L.29}$$

By taking partial derivatives of the roughness covariance in the form $B(\mathbf{R}_1 - \mathbf{R})$ and then setting $\mathbf{R}_1 = \mathbf{R}$, it can be shown that the second derivatives are simply connected to the mean-square slope of the rough interface as follows:

$$< (\frac{\partial f}{\partial x})^2 >= -\frac{\partial^2 B}{\partial x^2} , \tag{L.30}$$

$$< \frac{\partial f}{\partial x}\frac{\partial f}{\partial y} >= -\frac{\partial^2 B}{\partial x \partial y} , \tag{L.31}$$

$$< (\frac{\partial f}{\partial y})^2 >= -\frac{\partial^2 B}{\partial y^2} , \tag{L.32}$$

where the derivatives on the right-hand side of these equations are evaluated at $\mathbf{R} = 0$. It is convenient to define the slope covariance matrix

$$\mathbf{B}_s = -\begin{pmatrix} \frac{\partial^2 B}{\partial x^2} & \frac{\partial^2 B}{\partial x \partial y} \\ \frac{\partial^2 B}{\partial x \partial y} & \frac{\partial^2 B}{\partial y^2} \end{pmatrix} . \tag{L.33}$$

With this expansion, (L.28) becomes the two-dimensional Fourier transform of a Gaussian function, which is itself a Gaussian. Using (13.10), one obtains for the bistatic scattering cross section

$$\sigma = \frac{1}{4}|V_{ww}(\theta_{is})|^2 \frac{\Delta k^2}{\Delta k_z^2} p_s(\mathbf{S}) , \tag{L.34}$$

where $\mathbf{s}$ is the gradient that the surface relief function would have if it were oriented so as to give specular reflection from the source to the receiver, and $p_s(\mathbf{s})$ is the probability density function for the slope, evaluated at this gradient value:

$$p_s(\mathbf{s}) = \frac{1}{2\pi||\mathbf{B}_s||} e^{-\frac{1}{2}\mathbf{s}^T \mathbf{B}_s^{-1} \mathbf{s}} . \tag{L.35}$$

The variable $\mathbf{s}$ appearing in (L.35) is the column matrix comprised of the two slope components

$$\mathbf{s} = \begin{bmatrix} \Delta K_x / \Delta k_z \\ \Delta K_y / \Delta k_z \end{bmatrix} , \tag{L.36}$$

and $||\mathbf{B}_s||$is the determinant of the slope covariance matrix.

M Volume Perturbation Theory

Chapter 14 employs perturbation theory results for volume scattering without derivation. The purpose of this appendix is to provide the needed derivations as well as more ambitious models found in the literature.

M.1 Fluid Sediment

This section will develop volume perturbation theory for the fluid sediment based on Ivakin's unified approach [Ivakin 1998a]. This allows treatment of roughness and volume scattering on a common footing and provides avenues for obtaining additional results that would be difficult using standard methods.

The equation of motion for pressure, (8.5) can be written in the form

$$\nabla \cdot (\frac{1}{\rho} \nabla P) + \omega^2 \kappa P = 0 \; . \tag{M.1}$$

The compressibility and reciprocal density will be expanded to first order in the fluctuation strength,

$$\kappa \approx \kappa_0 (1 + \gamma_\kappa), \tag{M.2}$$

$$\frac{1}{\rho} \approx \frac{1}{\rho_0} (1 - \gamma_\rho), \tag{M.3}$$

where the normalized density fluctuations, γ_ρ, and compressibility fluctuations, γ_κ, are defined in (14.20) and (14.21). The pressure is also expanded to first order:

$$P \approx P_0 + P_1 \; . \tag{M.4}$$

Here, P_0 is the pressure (a function of three coordinates) obtained in the "zeroth-order" solution, that is, the solution of (M.1) when the fluctuations are set to zero. The zeroth-order equation is

$$\rho_0 \nabla \cdot (\frac{1}{\rho_0} \nabla P_0) + k_p^2 P_0 = 0 \; , \tag{M.5}$$

where

$$k_p = \frac{\omega}{c_p} , \tag{M.6}$$

with

$$c_p = \frac{1}{\sqrt{\rho_0 \kappa_0}} . \tag{M.7}$$

If the sediment is stratified, the zeroth-order sediment parameters, ρ_0 and κ_0, as well as the corresponding wave speed and wavenumber, c_p and k_p, will depend on the z-coordinate in the sediment. In any event, these functions are likely to have a step change at the sediment–water interface. The first-order correction to pressure due to fluctuations is denoted P_1 and is found as the solution of the first-order part of (M.1), which can be written in the form

$$\rho_0 \nabla \cdot (\frac{1}{\rho_0} \nabla P_1) + k_p^2 P_1 = -S(\mathbf{r}) , \tag{M.8}$$

where the source term is

$$S(\mathbf{r}) = -\rho_0 \nabla \cdot (\frac{1}{\rho_0} \gamma_\rho \nabla P_0) + k_p^2 \gamma_\kappa P_0 . \tag{M.9}$$

Equations (M.8) and (M.9) express the usual perturbation result that the first-order field obeys the same equation as the zeroth-order field, but with a source term that is first order in the fluctuations. One means of solving such an equation is to use the Green's function, $G_{pw}(\mathbf{r}, \mathbf{r}')$, the zeroth-order solution for the pressure in the water at $\mathbf{r}$ due to a unit point source in the sediment at $\mathbf{r}'$. Adapting (8.59) to the present situation,

$$\rho_0 \nabla \cdot [\frac{1}{\rho_0} \nabla G_{pw}(\mathbf{r}, \mathbf{r}')] + k_p^2 G_{pw}(\mathbf{r}, \mathbf{r}') = -4\pi\delta(\mathbf{r} - \mathbf{r}') . \tag{M.10}$$

Noting that the source, $S(\mathbf{r})$, can be viewed as a superposition of point sources, the first-order field in the water at $\mathbf{r}$ is

$$P_1(\mathbf{r}) = \frac{1}{4\pi} \int G_{pw}(\mathbf{r}, \mathbf{r}') S(\mathbf{r}') d^3 r' . \tag{M.11}$$

The required Green's function can be obtained by generalizing (8.65) to stratified sediments. It is easiest to begin with the Green's function for the field in the sediment due to a source in the water:

$$G_{wp}(\mathbf{r}, \mathbf{r}') = \frac{i}{2\pi k_w} \int \frac{e^{i\mathbf{K}\cdot(\mathbf{R}-\mathbf{R}') + ik_w\beta_w(K)z'} v(K,\ z)}{\beta_w(K)} d^2K , \quad z < 0,\ z' > 0. \tag{M.12}$$

This generalization consists in replacing the field in the homogeneous sediment due to a plane wave by a more general form. That is, $V_{wp}(K)\times \exp[-ik_p\beta_p(K)z]$ is replaced by $v(K,\ z)$. This function gives the z-dependence of the field in the sediment when a plane wave is incident from the water with horizontal wavenumber component, K. The plane wave in this case has amplitude unity (real, positive) at the origin. This function can be found using various computer codes, e.g., the OASES [Schmidt 1999] subroutine OASR. Continuity of pressure across the interface gives the condition

$$v(K,\ 0) = 1 + V_{ww}(K)\ , \tag{M.13}$$

where $V_{ww}(K)$ is the reflection coefficient for the stratified sediment. The desired Green's function can now be obtained using reciprocity (8.69) in the form

$$G_{pw}(\mathbf{r}, \mathbf{r}') = \frac{1}{a_\rho(z')} G_{wp}(\mathbf{r}', \mathbf{r})\ , \tag{M.14}$$

where the source at $\mathbf{r}'$ is in the sediment ($z' < 0$), and the density ratio depends on depth in the sediment,

$$a_\rho(z) = \frac{\rho_0(z)}{\rho_w}\ . \tag{M.15}$$

If the incident field is assumed to be a unit-amplitude plane wave, the first-order pressure can be expressed in terms of the first-order T-matrix as follows:

$$P_1(\mathbf{r}) = \int e^{i\mathbf{K}_s\cdot\mathbf{R}+ik_w\beta_w(K_s)z} T^{(1)}_{ww}(\mathbf{K}_s,\ \mathbf{K}_i) d^2K_s\ , \tag{M.16}$$

where $\mathbf{K}_i$ is the horizontal part of the wave vector for the incident plane wave. For the sake of completeness, the zeroth-order pressure is, adapting (8.38),

$$P_0(\mathbf{r}) = e^{i\mathbf{K}_i\cdot\mathbf{R}-ik_w\beta_w(K_i)z} + V_{ww}(K_i) e^{i\mathbf{K}_i\cdot\mathbf{R}+ik_w\beta_w(K_i)z}\ , \tag{M.17}$$

from which the zeroth-order T-matrix can be deduced,

$$T^{(0)}_{ww}(\mathbf{K}_s,\ \mathbf{K}_i) = V_{ww}(K_i)\delta(\mathbf{K}_s - \mathbf{K}_i)\ . \tag{M.18}$$

Comparing (M.11) and (M.16), the first-order T-matrix can be found after an interchange of orders of integration:

$$T^{(1)}_{ww}(\mathbf{K}_s,\ \mathbf{K}_i) =$$

$$\frac{i}{8\pi^2 k_w\beta_w(K)} \int e^{i\Delta\mathbf{K}\cdot\mathbf{R}} [D_1(\mathbf{K}_s, \mathbf{K}_i, z)\gamma_\kappa + D_2(\mathbf{K}_s, \mathbf{K}_i, z)\gamma_\rho)] \frac{d^3r}{a_\rho(z)}\ , \tag{M.19}$$

where

$$\Delta\mathbf{K} = \mathbf{K}_s - \mathbf{K}_i\ , \tag{M.20}$$

$$D_1(\mathbf{K}_s, \mathbf{K}_i, z) = k_p^2(z) v(K_s,\ z) v(K_i,\ z)\ , \tag{M.21}$$

$$D_2(\mathbf{K}_s, \mathbf{K}_i, z) = \mathbf{K}_s \cdot \mathbf{K}_i v(K_s,\ z) v(K_i,\ z) + \frac{\partial v(K_s,\ z)}{\partial z}\frac{\partial v(K_i,\ z)}{\partial z}\ . \tag{M.22}$$

In (M.19), γ_κ and γ_ρ are three-dimensional functions of the integration variable, $\mathbf{r}$, while ρ_0, k_p, and v depend only on the vertical component of $\mathbf{r}$, z. Expression (M.19) is the central result of this section. By making various assumptions as to the nature of the unperturbed and perturbed sediment parameters, all the roughness and volume scattering results of interest can be obtained, and the relation between roughness and volume scattering can be clarified [Ivakin 1998a].

M.1.1 Heterogeneous Fluid Sediment

The bistatic scattering cross section due to volume heterogeneity of a stratified fluid-like seafloor can be found by taking the second moment of (M.19) and using (J.13) and (J.14),

$$\sigma(\mathbf{K}_s,\ \mathbf{K}_i) = \frac{1}{(4\pi)^2}\int_{-\infty}^{0}\int_{-\infty}^{0}\int e^{-i\Delta\mathbf{K}\cdot\mathbf{R}} a_\rho^{-1}(z_1) a_\rho^{*-1}(z_2) \times$$
$$[D_1(\mathbf{K}_s,\mathbf{K}_i,z_1)D_1^*(\mathbf{K}_s,\mathbf{K}_i,z_2)B_{\kappa\kappa}(\mathbf{R},z_1,z_2) +$$
$$D_2(\mathbf{K}_s,\mathbf{K}_i,z_1)D_2^*(\mathbf{K}_s,\mathbf{K}_i,z_2)B_{\rho\rho}(\mathbf{R},z_1,z_2) +$$
$$D_1(\mathbf{K}_s,\mathbf{K}_i,z_1)D_2^*(\mathbf{K}_s,\mathbf{K}_i,z_2)B_{\rho\kappa}(\mathbf{R},z_2,z_1) +$$
$$D_2(\mathbf{K}_s,\mathbf{K}_i,z_1)D_1^*(\mathbf{K}_s,\mathbf{K}_i,z_2)B_{\rho\kappa}(\mathbf{R},z_1,z_2)]d^2R dz_1 dz_2\ . \tag{M.23}$$

Because the seafloor is stratified, the statistics of the fluctuations in density and compressibility may not be stationary with respect to the vertical direction. Consequently, (14.22), (14.23), and (14.24) have been generalized as follows:

$$B_{\rho\rho}(\mathbf{R},z_1,z_2) = < \gamma_\rho(\mathbf{R}+\mathbf{R}_0,z_1)\gamma_\rho(\mathbf{R}_0,z_2) > \ , \tag{M.24}$$

$$B_{\kappa\kappa}(\mathbf{R},z_1,z_2) = < \gamma_\kappa(\mathbf{R}+\mathbf{R}_0,z_1)\gamma_\kappa(\mathbf{R}_0,z_2) > \ , \tag{M.25}$$

and

$$B_{\rho\kappa}(\mathbf{R},z_1,z_2) = < \gamma_\rho(\mathbf{R}+\mathbf{R}_0,z_1)\gamma_\kappa(\mathbf{R}_0,z_2) > \ . \tag{M.26}$$

Transverse isotropy has been exploited in (M.23) in setting $B_{\rho\kappa}(\mathbf{R},z_2,z_1) = B_{\rho\kappa}(-\mathbf{R},z_2,z_1)$. Equation (M.23) is difficult numerically owing to the fourfold integration. Various investigators have used additional assumptions and approximations to obtain simpler models. Some of these models will be described in the remainder of this section.

Unstratified Sediment

If it is assumed that the unperturbed density and complex compressional wave speed are independent of the vertical coordinate, the vertical dependence of the unperturbed field is

$$v(K,z) = [1 + V_{ww}(K)]e^{-ik_p\beta_p(K)z} , \quad z < 0 . \tag{M.27}$$

Then

$$D_1(\mathbf{K}_s,\mathbf{K}_i,z) = k_p^2[1+V_{ww}(K_s)][1+V_{ww}(K_i)]e^{-ik_p[\beta_p(K_s)+\beta_p(K_i)]z} , \tag{M.28}$$

$$D_2(\mathbf{K}_s,\mathbf{K}_i,z) = \mathbf{k}_{ps}\cdot\mathbf{k}_{pi}[1 + V_{ww}(K_s)][1 + V_{ww}(K_i)]e^{-ik_p[\beta_p(K_s)+\beta_p(K_i)]z} , \tag{M.29}$$

with $\mathbf{k}_{pi}$ and $\mathbf{k}_{ps}$ as defined in (14.13) and (14.14). The scattering cross section takes the form

$$\sigma(\mathbf{K}_s,\ \mathbf{K}_i) = \frac{|1 + V_{ww}(K_s)|^2|1 + V_{ww}(K_i)|^2}{(4\pi)^2|a_\rho|^2}\int_{-\infty}^{0}\int_{-2z_c}^{2z_c}\int e^{-i\Delta\mathbf{k}_p\cdot\mathbf{r}} \times$$
$$e^{2(\alpha_s+\alpha_i)z_c}[|k_p|^4 B_{\kappa\kappa}(\mathbf{r}) + |\mathbf{k}_{ps}\cdot\mathbf{k}_{pi}|^2 B_{\rho\rho}(\mathbf{r}) +$$
$$2\mathrm{Re}\left\{\mathbf{k}_{ps}\cdot\mathbf{k}_{pi}\ k_p^{*2}\right\} B_{\rho\kappa}(\mathbf{r})]dR^2dzdz_c , \tag{M.30}$$

where $\Delta\mathbf{k}_p$ is the Bragg wave vector (14.19). Statistical stationarity in the vertical has been assumed so that the general covariance functions are replaced by (14.22), (14.23), and (14.24). In addition, it has been assumed that the covariances only depend upon the absolute value of the vertical lag. Other parameters and variables appearing in (M.30) are

$$\alpha_i = \mathrm{Im}\{k_p\beta_p(K_i)\} , \tag{M.31}$$

$$\alpha_s = \mathrm{Im}\{k_p\beta_p(K_s)\} , \tag{M.32}$$

$$\mathbf{r} = [\mathbf{R}, z] , \tag{M.33}$$

$$z = z_1 - z_2 , \tag{M.34}$$

$$z_c = \frac{1}{2}(z_1 + z_2) . \tag{M.35}$$

Equation (M.30) still involves four integrations. The main problem is that the z-integral is over finite limits. If these limits are extended to $\pm\infty$ as an approximation, the 3 integrations over $\mathbf{r}$ can be performed, replacing the covariances by the corresponding spectra (14.25). The validity of this approximation is discussed in [Jones and Jackson 2000, Jones and Jackson 2001], where it is noted that the finite limits on the z-integration produce a windowing effect, with the result that the true spectra should actually be replaced by modified, effective spectra. The main effect of this windowing occurs for scattering near the specular direction, as shown in Fig. 14.5. For example, windowing effects were shown to be unimportant in a particular bistatic model-data comparison for scattered grazing angles substantially removed from the specular angle [Briggs et al. 2002a]. Although the range of validity of the nonwindowing approximation is not specified explicitly in the literature, it depends on the correlation scale of the unmodified spectra, the attenuation coefficient, and the acoustic wavelength. If windowing is ignored, (M.30) is found to be identical to (14.8) with the volume scattering cross section, σ_v, given by (14.10).

Stratified Sediment with Homogeneous Layers

The nonwindowing approximation can also be used to improve the numerical tractability of the perturbation model for stratified seafloors. The problem is simplified if it is assumed that the unperturbed seafloor consists of homogeneous layers having horizontal boundaries, such that, for the nth layer,

$$z_{n-1} < z \leq z_n \ . \tag{M.36}$$

Taking $z_0 = 0$, the first layer has its upper interface at $z = 0$, the second at $z = z_1$, etc. Remembering that $z < 0$ in the seafloor, the thickness of the nth layer is of the form

$$d_n = z_{n-1} - z_n \ . \tag{M.37}$$

The z-dependence of pressure field due to an incident unit-amplitude plane wave is

$$v(K, z) = \sum_{\alpha=1}^{2} a_\alpha(K) e^{-i\nu_\alpha(K)(z-z_{n-1})} \ , \qquad z_{n-1} < z \leq z_n \ . \tag{M.38}$$

In each layer, there are assumed to be two compressional waves, one with wave vector tipped downward ($\alpha = 1$) and one with wave vector tipped upward ($\alpha = 2$). That is,

$$\nu_1(K) = k_p \beta_p(K) \ , \tag{M.39}$$

$$\nu_2(K) = -k_p \beta_p(K) \ . \tag{M.40}$$

It is convenient to define three-dimensional wave vectors analogous to those (14.13, 14.14) used in the unstratified case:

$$\mathbf{k}_{pi}^{(\alpha)} = \mathbf{K}_i - \mathbf{e}_z \nu_\alpha(K_i) \ , \tag{M.41}$$

$$\mathbf{k}_{ps}^{(\alpha)} = \mathbf{K}_s + \mathbf{e}_z \nu_\alpha(K_s) \ . \tag{M.42}$$

It should be realized that the parameters and functions $v(K, z)$, k_p, $\nu_\alpha(K)$, $\mathbf{k}_{pi}^{(\alpha)}$, and $\mathbf{k}_{ps}^{(\alpha)}$ are different for each layer and should bear a sub- or superscript, n. Attention will be restricted to the nth layer to minimize the possibility of confusion. With these definitions,

$$D_1(\mathbf{K}_s, \mathbf{K}_i, z) = k_p^2 \sum_{\alpha,\beta} a_\alpha(K_s) a_\beta(K_i) e^{-i(k_{psz}^{(\alpha)} - k_{piz}^{(\beta)})(z - z_{n-1})} , \qquad \text{(M.43)}$$

$$D_2(\mathbf{K}_s, \mathbf{K}_i, z) = \sum_{\alpha,\beta} \mathbf{k}_{ps}^{(\alpha)} \cdot \mathbf{k}_{pi}^{(\beta)} a_\alpha(K_s) a_\beta(K_i) e^{-i(k_{psz}^{(\alpha)} - k_{piz}^{(\beta)})(z - z_{n-1})} . \qquad \text{(M.44)}$$

If windowing due to the finite thickness of the layers is neglected, the contribution of the nth layer to the bistatic interface scattering cross section is

$$\sigma_n(\mathbf{K}_s, \mathbf{K}_i) =$$

$$\frac{\pi}{2|\alpha_{\rho n}|^2} \sum_{\alpha,\beta,\alpha',\beta'} \frac{1 - e^{-p(\alpha,\beta,\alpha',\beta')d_n}}{p(\alpha,\beta,\alpha',\beta')} a_\alpha(K_s) a_\beta(K_i) a_{\alpha'}^*(K_s) a_{\beta'}^*(K_i) \times$$

$$\{ \, |k_p|^4 W_{\kappa\kappa}(\mathbf{\Delta k}_p) + (\mathbf{k}_{ps}^\alpha \cdot \mathbf{k}_{pi}^\beta)(\mathbf{k}_{ps}^{\alpha'} \cdot \mathbf{k}_{pi}^{\beta'})^* W_{\rho\rho}(\mathbf{\Delta k}_p)$$

$$+ [(\mathbf{k}_{ps}^\alpha \cdot \mathbf{k}_{pi}^\beta) \, k_p^{*2} + (\mathbf{k}_{ps}^{\alpha'} \cdot \mathbf{k}_{pi}^{\beta'})^* \, k_p^2] W_{\rho\kappa}(\mathbf{\Delta k}_p) \, \} . \qquad \text{(M.45)}$$

In (M.45),

$$p(\alpha, \beta, \alpha', \beta') = -i(k_{psz}^{(\alpha)} - k_{piz}^{(\beta)} - k_{psz}^{(\alpha')*} + k_{piz}^{(\beta')*}) , \qquad \text{(M.46)}$$

$$\Delta \mathbf{k}_p = \mathbf{K}_s - \mathbf{K}_i + \frac{1}{2}\mathbf{e}_z(k_{psz}^{(\alpha)} - k_{piz}^{(\beta)} + k_{psz}^{(\alpha')*} - k_{piz}^{(\beta')*}) . \qquad \text{(M.47)}$$

The bistatic cross section is the sum over all layers of the terms given by (M.45). The deepest layer is semi-infinite and will have $a_2 = 0$. Although each term in the quadruple sum (over α, β, α', and β') is complex, the sum is real. An interesting feature of the layered case is that the variable, $\Delta \mathbf{k}_p$, that takes the place of the usual Bragg wave vector is complex. Thus, the spectrum is to be evaluated at complex arguments. The same situation is encountered in the unstratified, elastic case of Sects. 14.2.1 and M.2. Even though one normally thinks of the heterogeneity spectra as real-valued functions of real variables, the use of complex arguments is perfectly well defined through (14.25), simply requiring use of a complex wave vector in the exponent before integrating over $\mathbf{r}$. In practice, it is simpler to insert complex arguments into the spectra directly, e.g., in the widely used form (7.14).

M.1.2 Volume Treatment of Rough-Interface Scattering

Ivakin [Ivakin 1998a] has treated roughness scattering as a form of volume scattering by noting that the changes in sediment physical properties due to deformation of the interface can be written to first order in the seafloor relief function, $f(\mathbf{R})$, as

$$\gamma_\rho = -\frac{\rho_0'(z)}{\rho_0(z)} f(\mathbf{R}) , \qquad \text{(M.48)}$$

$$\gamma_\kappa = -\frac{\kappa_0'(z)}{\kappa_0(z)} f(\mathbf{R}) , \tag{M.49}$$

where the symbol $'$ denotes differentiation with respect to z. This formulation assumes that the surface relief lifts and depresses the sediment strata such that isosurfaces of constant density and compressibility are simply of the form $z = f(\mathbf{R}) + \text{const}$.

This formalism can be applied to cases in which the division between roughness and volume scattering is not clear-cut. The method will be illustrated by a simple case in which the sediment is homogeneous. For this case, $v(K, z)$ is given by (M.27). Step changes in density and compressibility give delta functions as follows

$$\gamma_\rho = -2\frac{a_\rho - 1}{a_\rho + 1} f(\mathbf{R})\delta(z) , \tag{M.50}$$

$$\gamma_\kappa = -2\frac{a_\rho a_p^2 - 1}{a_\rho a_p^2 + 1} f(\mathbf{R})\delta(z) . \tag{M.51}$$

When these expressions are inserted into (M.19), the delta functions allow the integrals to be performed analytically, with care needed in dealing with factors in the integrand that are discontinuous at $z = 0$. For these factors, integration with the delta function yields an average of the values above and below the interface. For example,

$$\int a_\rho^{-1}(z) \frac{\partial v(K_i,\ z)}{\partial z} \frac{\partial v(K,\ z)}{\partial z} \delta(z) dz =$$

$$\frac{1}{2}[1 + a_\rho(0^-)] a_\rho^{-2}(0^-) \frac{\partial v(K_i,\ 0^-)}{\partial z} \frac{\partial v(K,\ 0^-)}{\partial z} .$$

This result is obtained by noting that, while $\partial v(K,\ z)/\partial z$ is discontinuous at $z = 0$, when divided by density, the resulting quotient is continuous. These considerations allow the z-integral in (M.19) to be performed. The $\mathbf{R}$-integral yields the Fourier transform of the seafloor relief function, $F(\Delta\mathbf{K})$. The first-order T-matrix for roughness scattering is obtained in the form

$$T_{ww}^{(1)}(\mathbf{K}_s,\ \mathbf{K}_i) = \frac{ik_w}{\beta_w(K_s)} A_{ww}(\mathbf{K}_s,\ \mathbf{K}_i) F(\mathbf{K}_s - \mathbf{K}_i) . \tag{M.52}$$

This is identical to (K.3), and the factor, A_{ww}, is given by (13.21)–(13.25). As Appendix K shows that this completely determines the bistatic scattering cross section in lowest nontrivial order, the discussion can be considered complete.

M.2 Elastic Sediment

This section outlines the derivation of the results given in Sect. 14.2. The presentation follows, with some notational changes, [D. Jackson and Ivakin 1998,

Ivakin and Jackson 1998], who expanded on the earlier work of [Ivakin 1990]. Because the elastic problem is more difficult than the fluid problem, only the simplest case, the unstratified seafloor, will be considered.

Volume scattering is caused by fluctuation in the medium parameters, with density represented by $\rho_0 + \delta\rho$, and the Lamé parameters by $\lambda_0 + \delta\lambda$ and $\mu_0 + \delta\mu$. The zeroth-order parameters, ρ_0, λ_0, and μ_0, are assumed to be independent of depth in the sediment. The displacement field, approximated by the sum of zeroth- and first-order terms, will be represented by $\mathbf{u}_0 + \mathbf{u}_1$ with corresponding scalar and vector potentials, $\phi_0 + \phi_1$ and $\boldsymbol{\psi}_0 + \boldsymbol{\psi}_1$.

The zeroth-order potentials obey (9.7) and (9.8) with wave speeds given in terms of the unperturbed medium parameters,

$$c_p = \sqrt{\frac{\lambda_0 + 2\mu_0}{\rho_0}} \tag{M.53}$$

and

$$c_t = \sqrt{\frac{\mu_0}{\rho_0}} \; . \tag{M.54}$$

Using (9.31) and (9.32), the zeroth-order displacement field in the sediment is

$$\mathbf{u}_0 = i\mathbf{k}_p^-(\mathbf{K}_i)V_{wp}(K_i)e^{i\mathbf{k}_p^-(\mathbf{K}_i)\cdot\mathbf{r}} - i\mathbf{k}_t^-(\mathbf{K}_i) \times \mathbf{e}_h^- V_{wt}(K_i)e^{i\mathbf{k}_t^-(\mathbf{K}_i)\cdot\mathbf{r}} \; . \tag{M.55}$$

The first-order perturbation in the displacement field obeys the equation

$$\rho_0\omega^2\mathbf{u}_1 + \rho_0 c_p^2 \nabla(\nabla\cdot\mathbf{u}_1) - \rho_0 c_t^2 \nabla\times\nabla\times\mathbf{u}_1 = \mathbf{f} \; , \tag{M.56}$$

where $\mathbf{f}$ is a vector with components

$$f_i = -\delta\rho\omega^2 u_{0i} - \partial_i\left[\delta\lambda(\nabla\cdot\mathbf{u}_0)\right] - \sum_j \partial_j\left[\delta\mu(\partial_i u_{0j} + \partial_j u_{0i})\right], \quad i,j = x,y,z. \tag{M.57}$$

This corresponds to the following pair of first-order equations for the perturbed potentials:

$$(\nabla^2 + k_p^2)\nabla^2\phi_1 = \frac{1}{\rho_0 c_p^2}\nabla\cdot\mathbf{f} \; , \tag{M.58}$$

$$(\nabla^2 + k_t^2)\nabla^2\boldsymbol{\psi}_1 = -\frac{1}{\rho_0 c_t^2}\nabla\times\mathbf{f} \; . \tag{M.59}$$

Analogous to (M.11), Green's functions can be used as follows:

$$\nabla^2\phi_1(\mathbf{r}) = -\frac{1}{4\pi\rho_0 c_p^2}\int G_p(\mathbf{r}-\mathbf{r}')\nabla\cdot\mathbf{f}(\mathbf{r}')d^3r', \tag{M.60}$$

$$\nabla^2 \boldsymbol{\psi}_1(\mathbf{r}) = \frac{1}{4\pi\rho_0 c_t^2} \int G_t(\mathbf{r} - \mathbf{r}')\nabla \times \mathbf{f}(\mathbf{r}')d^3r' , \tag{M.61}$$

where $G_\alpha(\mathbf{r})\,(\alpha = p, t)$ are Green's functions for the unbounded elastic medium:

$$G_\alpha(\mathbf{r}) = \frac{e^{ik_\alpha r}}{r} , \quad \alpha = p, t . \tag{M.62}$$

These have plane-wave expansions, where the wave vectors are tilted upward toward the interface:

$$G_\alpha(\mathbf{r}) = \frac{i}{2\pi} \int \frac{d^2K}{k_\alpha \beta_\alpha(K)} e^{i\mathbf{k}_\alpha^+(\mathbf{K})\cdot\mathbf{r}} . \tag{M.63}$$

The fields represented by (M.60) and (M.61) do not obey the boundary condition at the interface ($z = 0$). Rather, they are the fields scattered by the heterogeneity. These fields propagate toward the interface, where they are partially reflected back into the sediment and partially transmitted into the water. These fields will be evaluated immediately below the interface, where it will be assumed that $\mathbf{f}$ and its first derivatives vanish. Then the region immediately below the interface is source-free, with the consequence that the Laplacians, ∇^2, in (M.60) and (M.61) can be replaced by $-k_p^2$ and $-k_t^2$. Next, (M.63) is substituted in (M.60) and (M.61). Windowing due to the seafloor–water interface is neglected, that is, the upper limit on the z-integrations is extended from 0 to $+\infty$. The result is

$$\phi_1(\mathbf{r}) = \int \frac{d^2K}{k_p \beta_p(K)} F_p(\mathbf{K}, \mathbf{K}_i) e^{i\mathbf{k}_p^+(\mathbf{K})\cdot\mathbf{r}} , \tag{M.64}$$

$$\boldsymbol{\psi}_1(\mathbf{r}) = \int \frac{d^2K}{k_t \beta_t(K)} \mathbf{F}_t(\mathbf{K}, \mathbf{K}_i) e^{i\mathbf{k}_t^+(\mathbf{K})\cdot\mathbf{r}} . \tag{M.65}$$

The functions $F_p(\mathbf{K}, \mathbf{K}_i)$ and $\mathbf{F}_t(\mathbf{K}, \mathbf{K}_i)$ are proportional to the three-dimensional Fourier transforms of $\nabla \cdot \mathbf{f}$ and $\nabla \times \mathbf{f}$, respectively. These can be obtained from the Fourier transform of $\mathbf{f}$,

$$\mathbf{F}(\mathbf{k}) = \frac{1}{(2\pi)^3} \int \mathbf{f}(\mathbf{r}) e^{-i\mathbf{k}\cdot\mathbf{r}} d^3r . \tag{M.66}$$

Thus,

$$F_p(\mathbf{K}, \mathbf{K}_i) = \frac{-\pi}{\rho_0 \omega^2} \mathbf{k}_p^+ \cdot \mathbf{F}(\mathbf{k}_p^+) , \tag{M.67}$$

$$\mathbf{F}_t(\mathbf{K}, \mathbf{K}_i) = \frac{\pi}{\rho_0 \omega^2} \mathbf{k}_t^+ \times \mathbf{F}(\mathbf{k}_p^+) . \tag{M.68}$$

The T-matrix giving the spectrum of scattered plane waves in the water can be obtained by considering the transmission through the interface of the scattered compressional and shear waves. Only the vertically polarized component of the shear wave is transmitted, so

$$T_{ww}^{(1)}(\mathbf{K}_s, \mathbf{K}_i) = V_{pw}(K_s) \frac{F_p(\mathbf{K}_s, \mathbf{K}_i)}{k_p \beta_p(K_s)} - V_{tw}(K_s) \frac{\mathbf{e}_h^+(\mathbf{K}_s) \cdot \mathbf{F}_t(\mathbf{K}_s, \mathbf{K}_i)}{k_t \beta_t(K_s)} , \tag{M.69}$$

where the $V_{\alpha w}$ are the transmission coefficients for conversion of the compressional and shear plane-wave potentials into the compressional potential in the water. These are related to the usual transmission coefficients (9.41) and (9.42) as follows:

$$V_{pw}(K) = \frac{a_\rho \beta_p(K)}{a_p \beta_w(K)} V_{wp}(K) , \tag{M.70}$$

$$V_{tw}(K) = -\frac{a_\rho \beta_t(K)}{a_t \beta_w(K)} V_{wt}(K) . \tag{M.71}$$

The Fourier transform, $\mathbf{F}(\mathbf{k})$, can be evaluated and is a linear combination of the Fourier transforms of the normalized fluctuations defined in Sect. 14.2.2. Using this in (M.69), the second moment of the T-matrix can be found in terms of the fluctuation spectra, $W_{\alpha\beta}$. Finally, (J.13) and (J.14) can be used to obtain the bistatic scattering cross section (14.49). The fluctuation spectra are evaluated for complex arguments, as in the layered fluid case of Sect. M.1.1.

References

[Abegg and Anderson 1997] F. Abegg and A. L. Anderson, "The acoustic turbid layer in muddy sediments of Eckernförde Bay, Western Baltic Sea: methane concentration, saturation and bubble characteristics," *Mar. Geol.* **137**, pp. 137-147.

[Abernethy 1965] S. H. Abernethy, "Improved equipment for a pulse method of sound velocity measurement in water, rock, and sediment," U.S. Navy Electronics Laboratory Technical Memorandum TM-824.

[Abraham and Lyons 2002a] D. A. Abraham and A. P. Lyons, "Reverberation envelope statistics and their dependence on sonar beamwidth and bandwidth," in *Impact of Littoral Environmental Variability on Acoustic Predictions and Sonar Performance*, N.G. Pace and F.B. Jensen (Eds.), Kluwer Academic Publishers, Netherlands, pp. 539-546.

[Abraham and Lyons 2002b] D. A. Abraham and A. P. Lyons, "Novel physical interpretations of K-distributed reverberation," *IEEE J. Ocean. Eng.* **27**, pp. 800-813.

[Abraham and Lyons 2004] D. A. Abraham and A. P. Lyons, "Reverberation envelope statistics and their dependence on sonar bandwidth and scattering patch size," *IEEE J. Ocean. Eng.* **29**, pp. 126-137.

[Agrawal and Pottsmith 2000] Y. C. Agrawal and H. C. Pottsmith, "Instruments for particle size and settling velocity observations in sediment transport," *Mar. Geol.* **168**, pp. 89-114.

[Ainslie and McColm 1998] M. A. Ainslie and J. G. McColm, "A simplified formula for viscous and chemical absorption in sea water," *J. Acoust. Soc. Am.* **103**, pp.1671-1672.

[Akal 1974] T. Akal, "Acoustical characteristics of the sea floor: Experimental techniques and some examples from the Mediterranean sea," in *Physics of Sound in Marine Sediments*, L. Hampton (Ed.), Plenum Press, New York, pp. 447-480.

[Akal and Hovem 1978] T. Akal and J. Hovem, "Two-dimensional space series analysis for sea-floor roughness," *Mar. Geotechnol.* **3**, pp. 171-182.

[Aki and Richards 2002] K. Aki and P. G. Richards, *Quantitative Seismology*, Second Edition, University Science Books, Sausalito, CA.

[Albert et al. 1998] D. B. Albert, C. S. Martens, and M. J. Alperin, "Biogeochemical processes controlling methane in gassy coastal sediments II. Groundwater flow control of acoustic turbidity in Eckernförde Bay sediments," *Cont. Shelf Res.* **18**, pp. 1771-1793.

[Altenberg et al. 1991] R. A. Altenberg, N. P. Chotiros, and C. M. Faulkner, "Plane-wave analysis of acoustic signals in a sandy sediment," *J. Acoust. Soc. Am.* **89**, pp. 165-170.

[Anandarajah and Lavoie 2002] A. Anandarajah and D. Lavoie, "Numerical simulation of the microstructure and compression behavior of Eckernförde Bay sediments," *Mar. Geol.* **182**, pp. 3-27.

[Anderson 1971] E. R. Anderson, "Sound speed in seawater as a function of realistic temperature-salinity-pressure domains," Ocean Sciences Department, Naval Undersea Research and Development Center TP 243.

[Anderson and Hampton 1980a] A. L. Anderson and L. D. Hampton, "Acoustics of gas-bearing sediments I. Background," *J. Acoust. Soc. Am.* **67**, pp.1865-1889.

[Anderson and Hampton 1980b] A. L. Anderson and L. D. Hampton, "Acoustics of gas-bearing sediments II. Measurements and models," *J. Acoust. Soc. Am.* **67**, pp.1890-1903.

[Anderson et al. 1998] A. L. Anderson, F. Abegg, J. A. Hawkins, M. E. Duncan, and A. P. Lyons, "Bubble populations and acoustic interaction with the gassy floor of Eckernförde Bay," *Cont. Shelf Res.* **18**, pp. 1807-1838.

[APL-UW TR 9407] *APL-UW High-Frequency Ocean Environmental Acoustic Models Handbook, Ch. IV, Bottom*, APL-UW TR 9407.

[APL-UW TM 2-00] D. R. Jackson *High-Frequency Bistatic Scattering Model for Elastic Seafloors*, APL-UW TM 2-00.

[Archie 1942] G. E. Archie, "The electrical resistivity log as an aid in determining some reservoir characteristics," *Petrol. Tech., No. 5, 1-8, Trans. AIME* **146**, pp. 54-61.

[Ardhuin et. al. 2003] F. Ardhuin, W. C. O'Reilly, T. H. C. Herbers, and P. F. Jessen, "Swell transformation across the continental shelf. Part I: Attenuation and directional broadening," *J. Phys. Oceanogr.* **33**, pp. 1921-1939.

[Bachman 1984] R. T. Bachman, "Intratest porosity in foraminifera," *J. Sediment. Petrol.* **54**, pp. 257-262.

[Bachman 1985] R. T. Bachman, "Acoustic and physical property relationships in marine sediment," *J. Acoust. Soc. Am.* **78**, pp. 616-621.

[Bachman 1989] R. T. Bachman, "Estimating velocity ratio in marine sediments," *J. Acoust. Soc. Am.* **86**, pp. 2029-2032.

[Bachmann 1973] W. Bachmann, "A theoretical model for the backscattering strength of a composite-roughness sea-surface," *J. Acoust. Soc. Am.* **54**, pp. 712-716.

[Badiey et al. 1998] M. Badiey, A. H-D. Cheng, and Y. Mu, "From geology to geoacoustics – Evaluation of Biot-Stoll sound speed and attenuation for shallow water acoustics," *J. Acoust. Soc. Am.* **103**, pp. 309-320.

[Bagnold 1956] R. A. Bagnold, "The flow of cohesionless grains in fluids," *Philos. Trans. R. Soc. London Ser. A* **249**, pp. 235-286.

[Barbagelata et al. 1991] A. Barbagelata, M. Richardson, B. Miaschi, E. Muzi, P. Guerrini, L. Troiano, and T. Akal, "ISSAMS: An in situ sediment acoustic measurement system," in *Shear Waves in Marine Sediments*, J. M. Hovem, M. D. Richardson, and R. D. Stoll (Eds.), Kluwer, Dordrecht, pp. 305-312.

[Barrick 1968] D. E. Barrick, "Rough surface scattering based on the specular point theory," *IEEE Trans. Antennas Propag.* **AP-16**, pp. 449-454.

[Bass and Fuks 1979] F. G. Bass and I. M. Fuks, *Wave Scattering from Statistically Rough Surfaces*, Pergamon, Oxford.

[Bathurst 1975] R. G. C. Bathurst, "Carbonate Sediments and Their Diagenesis," *Developments in Sedimentology 12*, Second Edition, Elsevier, Amsterdam.

[Bathurst 1993] R. G. C. Bathurst, "Microfabrics in carbonate diagenesis: a critical look at forty years of research," in *Carbonate Microfabric, Frontiers in Sedimentary Geology*, R. Rezak and D.L. Lavoie (Eds.), Springer-Verlag, New York, pp. 3-14.

[Bautista and Stoll 1995] E. O. Bautista and R. D. Stoll, "Remote determination of in situ sediment parameters using Love waves," *J. Acoust. Soc. Am.* **98**, pp. 1090-1096.

[Bayer and Niederdraenk 1993] G. Bayer and T. Niederdraenk, "Weak localization of acoustic waves in strongly scattering media," *Phys. Rev. Lett.* **70**, pp. 3884-3887.

[Becker 2004] K. M. Becker, "Effect of various surface-height distribution properties on acoustic backscattering statistics," *IEEE J. Ocean. Eng.* **29**, pp. 246-259.

[Beckmann and Spizzichino 1963] P. Beckmann and A. Spizzichino, *The Scattering of Electromagnetic Waves from Rough Surfaces* , Pergamon, Oxford, 1963, Reprinted 1987, Artech House, Norwood, MA.

[Beckmann 1973] P. Beckmann, "Scattering by non-Gaussian surfaces," *IEEE Trans. Antennas Propag.* **AP-21**, pp. 169-175.

[Bedford and Stern 1983] A. Bedford and M. Stern, "A model for wave propagation in gassy sediments," *J. Acoust. Soc. Am.* **73**, pp. 409-417.

[Bell 1979] D. W. Bell, "Shear Wave Propagation in Unconsolidated Fluid Saturated Porous Media," Technical report ARL-TR-79-31, Applied Research Laboratory, University of Texas at Austin.

[Bell and Shirley 1980] D. W. Bell and D. J. Shirley, "Temperature variation of the acoustic properties of laboratory sediments," *J. Acoust. Soc. Am.* **68**, pp. 227-231.

[Bendat and Piersol 2000] J. S. Bendat and A. G. Piersol, *Random Data: Analysis and Measurement Procedures*, Third Edition, Wiley-Interscience, New York.

[Bennell 1979] J. D. Bennell, "Acoustic Properties of Marine Sediments," PhD thesis, University College of North Wales, Menai Bridge, UK.

[Bennell and Taylor-Smith 1991] J. D. Bennell and D. Taylor-Smith, "A review of laboratory shear wave techniques and attenuation measurements with particular reference to the resonant column," in *Shear Waves in Marine Sediments*, J. M. Hovem, M. D. Richardson, and R. D. Stoll (Eds.), Kluwer, Dordrecht, pp. 83-93.

[Bennett and Lambert 1971] R. H. Bennett and D. N. Lambert, "Rapid and reliable technique for determining unit weight and porosity of deep-sea sediments," *Mar. Geol.* **13**, pp. 251-266.

[Bennett and Hulbert 1986] R. H. Bennett and M. H. Hulbert, *Clay Microstructure*, International Human Resources Development Cooperation, Boston.

[Bennett et al. 1989] R. H. Bennett, K. M. Fischer, D. L. Lavoie, W. R. Bryant, and R. Rezak, "Porometry and fabric of marine clay and carbonate sediments — determinants of permeability," *Mar. Geol.* **89**, pp. 127-152.

[Bennett et al. 1990a] R. H. Bennett, H. O. Li, D. N. Lambert, K. M. Fischer, D. J. Walter, C. E. Hickox, M. H. Hulbert, T. Yamamoto, and M. Badiey, "In situ porosity and permeability of selected carbonate sediment Great-Bahama-Bank. 1. Measurements," *Mar. Geotechnol.* **9**, pp. 1-28.

[Bennett et al. 1990b] R. H. Bennett, K. M. Fischer, H. Li, R. B. Baerwald, M. H. Hulbert, T. Yamamoto, and M. Badiey, "In situ porosity and permeability of selected carbonate sediment Great-Bahama-Bank. 2. Microfabric," *Mar. Geotechnol.* **9**, pp. 29-47.

[Bennett et al. 1991a] R. H. Bennett, H. Li, M. D. Richardson, P. Fleischer, D. N. Lambert, D. J. Walter, K. B. Briggs, C. R. Rein, W. B. Sawyer, F. S. Carnaggio, D. C. Young, and S. G. Tooma, "Geoacoustic and geological characterization of surficial marine sediments by in situ probe and remote sensing techniques," in *CRC Handbook of Geophysical Exploration at Sea, Second Edition Hydrocarbons*, R. A. Geyer (Ed.), CRC Press, Boca Raton, FL, pp. 295-350.

[Bennett et al. 1991b] R. H. Bennett, W. R. Bryant and M. H. Hulbert,"Microstructure of fine-grained sediments: From mud to shale," in *Frontiers in Sedimentary Geology*, Springer-Verlag, New York.

[Bennett et al. 1996] R. H. Bennett, M. H. Hulbert, M. M. Meyer, D. M. Lavoie, K. B. Briggs, D. L. Lavoie, R. J. Baerwald, and W. A. Chiou, "Fundamental response of pore-water pressure to microfabric and permeability characteristics: Eckernförde Bay," *Geo-Mar. Lett.* **16**, pp. 182-188.

[Bennett et al. 1999] R. H. Bennett, B. Ransom, M. Kastner, R. J. Baerwald, M. H. Hulbert, W. B. Sawyer, H. Olsen and M. W. Lambert, "Early diagenesis: impact of organic matter on mass physical properties and processes, California continental margin," *Mar. Geol.* **159**, pp. 7-34.

[Bentley and Nittrouer 2003] S. J. Bentley and C. A. Nittrouer, "Emplacement, modification, and preservation of event strata on a flood-dominated continental shelf: Eel Shelf, northern California," *Cont. Shelf Res.* **23**, pp. 1465-1493.

[Berge and Berryman 1995] P. A. Berge and J. G. Berryman, "Realizability of negative pore compressibility in poroelastic composites, *J. Appl. Mech.* **62**, pp. 1053-1062.

[Bergem et al. 1999] O. Bergem, E. Pouliquen, G. Canepa, and N. G. Pace, "Time-evolution modeling of seafloor scatter. II. Numerical and experimental evaluation," *J. Acoust. Soc. Am.* **105**, pp. 3142-3150.

[Bergmann and Spitzer 1946] P. G. Bergmann and L. Spitzer, *Physics of Sound in the Sea, Part II: Reverberation*, NRDC Division 6 Summary Technical Report **8**.

[Berkson and Matthews 1983] J. M. Berkson and J. E. Matthews, "Statistical properties of seafloor roughness," in *Acoustics of the Sea Bed*, N. G. Pace (Ed.), Oxford University Press, Oxford, pp. 215-223.

[Berman and Dacol 1990] D. H. Berman and D. A. Dacol, "Manifestly reciprocal scattering-amplitudes for rough interface scattering," *J. Acoust. Soc. Am.* **87**, pp. 2024-2032.

[Berman 1991] D. H. Berman, "Simulations of rough interface scattering," *J. Acoust. Soc. Am.* **89**, pp. 623-636.

[Berryman and Blair 1986] J. G. Berryman and S. C. Blair, "Use of digital image analysis to estimate fluid permeability of porous materials: application of 2-point correlation functions," *J. Appl. Phys.* **60**, pp. 1930-1938.

[Berryman 1999] J. G. Berryman, "Origin of Gassmann's equations," *Geophysics* **64**, pp. 1627-1629.

[Best et al. 1998] A. I. Best, J. A. Roberts, and M. L. Somers, "A new instrument for making in-situ acoustic and geotechnical measurements in seafloor sediments," *Underwater Technol.* **23**, pp. 123-131.

[Best and Gunn 1999] A. I. Best and D. E. Gunn, "Calibration of marine sediment core loggers for quantitative acoustic impedance studies," *Mar. Geol.* **160**, pp. 137-146.

[Best et al. 2001] A. I. Best, Q. J. Huggett, and A. J. K. Harris, "Comparison of in-situ and laboratory acoustic measurements on Lough Hyne marine sediments," *J. Acoust. Soc. Am.* **110**, pp. 695-709.

[Best et al. 2004] A. I. Best, M. D. J. Tuffin, and J. K. Dix, "Tidal height and frequency dependence of acoustic velocity and attenuation in shallow gassy marine sediments," *J. Geophys. Res. Solid Earth* **109**, BO8101.

[Bibee 1993] L. D. Bibee, "In situ measurements of seafloor shear-wave velocity and attenuation using seismic interface waves," in *Acoustic Classification and Mapping of the Seabed*, N.G. Pace and D.N. Langhorne (Eds.), Institute of Acoustics, University of Bath, UK, pp. 33-40.

[Biot 1956a] M. A. Biot, "Theory of propagation of elastic waves in a fluid-saturated porous solid. I. Low frequency range," *J. Acoust. Soc. Am.* **28**, pp. 168-178.

[Biot 1956b] M. A. Biot, "Theory of propagation of elastic waves in a fluid-saturated porous solid. II. Higher frequency range," *J. Acoust. Soc. Am.* **28**, pp. 179-191.

[Biot 1962a] M. A. Biot, "Mechanics of deformation and acoustic propagation in porous media," *J. Appl. Phys.* **33**, pp. 1482-1498.

[Biot 1962b] M. A. Biot, "Generalized theory of acoustic propagation in porous dissipative media," *J. Acoust. Soc. Am.* **34**, pp. 1254-1264.

[Blair et al. 1996] S. C. Blair, P. A. Berge, and J. G. Berryman, "Using two-point correlation functions to characterize microgeometry and estimate permeabilities of sandstones and porous glass," *J. Geophys. Res. Solid Earth* **101** (B9), pp. 20359-20375.

[Blott and Pye 2001] S. J. Blott and K. Pye, "GRADISTAT: A grain size distribution and statistics package for the analysis of unconsolidated sediments," *Earth Surface Processes and Landforms* **26**, pp. 1237-1248.

[Boehme et al. 1985] H. Boehme, N. P. Chotiros, L. D. Rolleigh, S. P. Pitt, A. L. Garcia, T. G. Goldsberry, and R. A. Lamb, "Acoustic backscattering at low grazing angles from the ocean bottom. I. Bottom backscattering strength," *J. Acoust. Soc. Am.* **77**, pp. 962-974.

[Boehme and Chotiros 1988] H. Boehme and N. P. Chotiros, "Acoustic backscattering at low grazing angles from the ocean bottom," *J. Acoust. Soc. Am.* **84**, pp. 1018-1029.

[Boudreau 1986] B. P. Boudreau, "Mathematics of tracer mixing in sediments. 2. Nonlocal mixing and biological conveyor-belt phenomena," *Am. J. Sci.* **286**, pp. 199-238.

[Boudreau and Imboden 1987] B. P. Boudreau and D. M. Imbodem, "Mathematics of tracer mixing. 3. The theory of nonlocal mixing within sediment," *Am. J. Sci.* **287**, pp. 693-719.

[Boudreau 1994] B. P. Boudreau, "Is burial velocity a master parameter for bioturbation?," *Geochim. Cosmochim. Acta* **58**, pp. 1243-1249.

[Boudreau 1998] B. P. Boudreau, "Mean mixed depth of sediment: The wherefore and the why," *Limnol. Oceanogr.* **43**, pp. 524-526.

[Boudreau and Bennett 1999] B. P. Boudreau and R. H Bennett, "New rheological and porosity equations for steady state compaction," *Am. J. Sci.* **299**, pp. 517-528.

[Boudreau et al. 2005] B.P. Boudreau, C. Algar, B. D. Johnson, I. Croudace, A. Reed, Y. Furukawa, K. M. Dorgan, P. A. Jumars, A. S. Grader, and B. S.

Gardiner, "Bubble growth and rise in soft sediments," *Geology* **33**, pp. 517-520.

[Bouma 1964] A. H. Bouma, "Notes on X-ray interpretation of marine sediments," *Mar. Geol.* **2**, pp. 278-309.

[Bourbié et al. 1987] T. Bourbié, O. Cossy, and B. Zinszer, *Acoustics of Porous Media*, Gulf Publishing Company, Houston TX.

[Bowles 1997] F. A. Bowles, "Observations on attenuation and shear-wave velocity in fine-grained, marine sediments," *J. Acoust. Soc. Am.* **101**, pp. 3385-3397.

[Boyce 1968] R. E. Boyce, "Electrical resistivity of modern marine sediments from the Bering Sea," *J. Geophys. Res.* **73**, pp. 4759-4766.

[Boyce 1973] R. E. Boyce, "Physical property methods," in Initial Reports of the Deep Sea Drilling Project, Appendix I **15**, pp. 1121-1124.

[Boyle and Chotiros 1992] F. A. Boyle and N. P. Chotiros, "Experimental detection of a slow acoustic wave in sediment at shallow grazing angles," *J. Acoust. Soc. Am.* **91**, pp. 2615-2619.

[Boyle and Chotiros 1995a] F. A. Boyle and N. P. Chotiros, "A model for acoustic backscatter from muddy sediments," *J. Acoust. Soc. Am.* **98**, pp. 525-530.

[Boyle and Chotiros 1995b] F. A. Boyle and N. P. Chotiros, "A model for high-frequency acoustic backscatter from gas bubbles in sandy sediments at shallow grazing angles," *J. Acoust. Soc. Am.* **98**, pp. 531-541.

[Boyle and Chotiros 1998] F. A. Boyle and N. P. Chotiros, "Nonlinear acoustic scattering from a gassy poroelastic seabed," *J. Acoust. Soc. Am.* **103**, pp. 1328-1336.

[Brekhovskikh and Godin 1990] L. M. Brekhovskikh and O. A. Godin, *Acoustics of Layered Media I, Plane and Quasi-Plane Waves*, Springer-Verlag, Berlin.

[Brekhovskikh and Lysanov 1991] L. M. Brekhovskikh and Yu. P. Lysanov, *Fundamentals of Ocean Acoustics*, Springer-Verlag, Berlin.

[Brekhovskikh and Godin 1999] L. M. Brekhovskikh and O. A. Godin, *Acoustics of Layered Media II, Point Sources and Bounded Beams*, Springer-Verlag, Berlin.

[Briggs et al. 1985] K. Briggs, M. D. Richardson, and D. K. Young, "Variability in geoacoustic and related properties of surface sediments from the Venezuela Basin, Caribbean Sea," *Mar. Geol.* **68**, pp. 73-106.

[Briggs et al. 1986] K. B. Briggs, P. Fleicher, R. I. Ray, W. B. Sawyer, D. K. Young, M. D. Richardson, and S. Stanic, "Environmental Support for a High Frequency Acoustic Bottom Backscatter Experiment off Charleston, South Carolina, 17-28 June 1983," NORDA Report 130.

[Briggs 1989] K. B. Briggs, "Microtopographical roughness of shallow-water continental shelves," *IEEE J. Ocean. Eng.* **14**, pp. 360-367.

[Briggs et al. 1989] K. B. Briggs, P. Fleischer, W. H. Jahn, R. I. Ray, W. B. Sawyer, and M. D. Richardson, "Investigation of High-Frequency Acoustic Backscattering Model Parameters: Environmental Data from the Arafura Sea," NORDA Report 197.

[Briggs 1994] K. B. Briggs, "High-frequency acoustic scattering from sediment interface roughness and volume inhomogeneities," Naval Research Laboratory, NRL/FR/7431–94-9617.

[Briggs and Percival 1997] K. B. Briggs and D. B. Percival, "Vertical porosity and velocity fluctuations in shallow-water surficial sediments and their use in modeling volume scattering," in *High Frequency Acoustics in Shallow Water*, N.G. Pace, E. Pouliquen, O. Bergem, and A.P. Lyons (Eds.), NATO SACLANT Undersea Res. Ctr., La Spezia, Italy, pp. 65-73.

[Briggs and Ray 1997] K. B. Briggs and R. I. Ray, "Seafloor roughness power spectra: trends and implications for high-frequency acoustic modeling," in *Shallow-Water Acoustics*, R. Zhang and J. Zhou (Eds.), China Ocean Press, Beijing, pp. 347-352.

[Briggs and Richardson 1997] K. B. Briggs and M. D. Richardson, "Small-scale fluctuations in acoustic and physical properties in surficial carbonate sediments and their relationships to bioturbation," *Geo-Mar. Lett.* **17**, pp. 306-315.

[Briggs et al. 1998] K. B. Briggs, P. D. Jackson, R. J. Holyer, R. C. flint, J. C. Sanidge, and D. K. Young, "Two-dimensional variability in porosity, density, and electrical resistivity of Eckernförde Bay sediment, *Cont. Shelf Res.* **18**, pp. 1939-1964.

[Briggs et al. 2001] K. B. Briggs, K. L. Williams, M. D. Richardson, and D. R. Jackson, "Effects of changing roughness on acoustic scattering: (1) natural changes," in *Proceedings of the Institute of Acoustics Conference*, T. G. Leighton, G. J. Heald, H. D. Griffiths, and G. Griffiths (Eds.), Southampton, United Kingdom.

[Briggs et al. 2002a] K. B. Briggs, K. L. Williams, D. R. Jackson, C. D. Jones, A. N. Ivakin, and T. H. Orsi, "Fine-scale sedimentary structure: implications for acoustic remote sensing," *Mar. Geol.* **182**, pp. 141-159.

[Briggs et al. 2002b] K. B. Briggs, D. Tang, and K. L. Williams, "Characterization of interface roughness of rippled sand off Fort Walton Beach, Florida," *IEEE J. Ocean. Eng.* **27**, pp. 505-514.

[Briggs et al. 2002c] K. B. Briggs, D. R. Jackson, and K. Y. Moravan, "NRL-APL grain size algorithm upgrade," Naval Research Laboratory, Stennis Space Center, MS. NRL/MR/7430–02-8274.

[Briggs and Tang 2002] K. B. Briggs and D. Tang, "Assessing the sediment volume contribution to scattering: bulk density fluctuations," *Proceedings OCEANS 02*, pp. 2093-2097.

[Briggs et al. 2005a] K. B Briggs, A. P. Lyons, E. Pouliquen, L. A. Mayer, and M. D. Richardson, "Seafloor roughness, sediment grain size, and temporal stability," in *Underwater Acoustic Measurements: Technologies and Results*, J. P. Papadakis and L. Bjorno (Eds.), Proceedings of a conference held in Heraklion, Crete, pp. 337-343.

[Briggs et al. 2005b] K. B. Briggs, M. Zimmer and M. D. Richardson, "Spatial and temporal variations in sediment compressional wave speed and attenuation measured at 400 kHz for SAX04," in *Boundary Influences in High-Frequency, Shallow-Water Acoustics*, N. G. Pace and P. Blondel (Eds.), University of Bath, UK, pp. 29-37.

[Broschat and Thorsos 1997] S. L. Broschat and E. I. Thorsos, "An investigation of the small-slope approximation for scattering from rough surfaces: Part II, Numerical studies," *J. Acoust. Soc. Am.* **101**, pp. 2615-2625.

[Brown 1978] G. S. Brown, "Backscattering from a Gaussian-distributed, perfectly conducting rough surface," *IEEE Trans. Antennas Propag.* **26**, pp. 472-482.

[Brown 1980] G. S. Brown, "Backscattering from a Gaussian-distributed, perfectly conducting rough surface, correction," *IEEE Trans. Antennas Propag.* **28**, pp. 943-946.

[Brown et al. 2002] C. J. Brown, K. M. Cooper, W. J. Meadows, D. S. Limpenney, and H. L. Rees, "Small-scale mapping of seabed assemblages in the eastern English Channel using sidescan sonar and remote sampling techniques, " *Estuarine Coastal Shelf Sci.* **54**, pp. 263-278.

[Brown and Barlett 2005] W. E. Brown and M. L. Barlett, "Midfrequency 'through-the-sensor' scattering measurements: a new approach," *IEEE J. Ocean. Eng.* **30**, pp. 733-747.

[Brunson and Johnson 1980] B. A. Brunson and R. K. Johnson, "Laboratory measurements of shear wave attenuation in saturated sand," *J. Acoust. Soc. Am.* **68**, pp. 1371-1375.

[Bryan and Cox 1972] K. Bryan and M.D. Cox, "An approximate equation of state for numerical models of ocean circulation," *J. Phys. Oceanogr.* **2**, pp. 510-514.

[Bryan and Stoll 1988] G. M. Bryan and R. D. Stoll, "The dynamic shear modulus of marine sediments," *J. Acoust. Soc. Am.* **83**, pp. 2159-2164.

[Bryant and Bennett 1998] W. R. Bryant and R. H. Bennett, "Origin, physical, mineralogical nature of red clays — The Pacific Ocean Basin as a model," *Geo-Mar. Lett.* **8**, pp. 189-249.

[Brydon et al. 1999] D. Brydon, S. Sun, and R. Bleck, "A new approximation of the equation of state for seawater, suitable for numerical ocean models," *J. Geophys. Res.* **104**, pp. 1537-1540.

[Buchan et al. 1967] S. Buchan, F. C. D. Dewes, D. M. McCann and D. Taylor-Smith, "Measurements of the acoustic and geotechnical properties of marine sediment cores," in *Marine Geotechnique*, A. F. Richards (Ed.), University of Illinois Press, Champaign.

[Buchan et al. 1972] S. Buchan, D. M. McCann, and D. Taylor-Smith, "Relations between the acoustic and geotechnical properties of marine sediments," *Q. J. Eng. Geol.* **5**, pp. 265-284.

[Bucker et al. 1964] H. P. Bucker, J. A. Whitney, and D. L. Keir, "Use of Stoneley waves to determine the shear velocity in ocean sediments," *J. Acoust. Soc. Am.* **36**, pp. 1595-1596.

[Buckingham 1997] M. J. Buckingham, "Theory of acoustic attenuation, dispersion, and pulse propagation in unconsolidated granular materials including marine sediments," *J. Acoust. Soc. Am.* **102**, pp. 2579-2596.

[Buckingham 1998] M. J. Buckingham, "Theory of compressional and shear waves in fluidlike marine sediments," *J. Acoust. Soc. Am.* **103**, pp. 288-299.

[Buckingham 2000] M. J. Buckingham, "Wave propagation, stress relaxation, and grain-to-grain shearing in saturated, unconsolidated marine sediments," *J. Acoust. Soc. Am.* **108**, pp. 2796-2815.

[Buckingham and Richardson 2002] M. J. Buckingham and M. D. Richardson, "On tone-burst measurements of sound speed and attenuation in sandy marine sediments," *IEEE J. Ocean. Eng.* **27**, pp. 429-453.

[Buckingham 2005] M. J. Buckingham, "Compressional and shear wave properties of marine sediments: Comparisons between theory and data," *J. Acoust. Soc. Am.* **117**, pp. 137-152.

[Bunchuk and Zhitkovskii 1980] A. V. Bunchuk and Yu. Yu. Zhitkovskii, "Sound scattering by the ocean bottom in shallow-water regions (review) ," *Sov. Phys. Acoust.* **26**, pp. 363-370.

[Bunchuk and Ivakin 1985] A. V. Bunchuk and A. N. Ivakin, "Azimuthal anisotropy of sound backscattering from the ocean bottom in shallow water," in *V Seminar on Acoustic and Statistical Models of the Ocean*, pp.47-51, (in Russian).

[Bussmann and Suess 1998] I. Bussmann and E. Suess, "Ground water seepage in Eckernförde Bay (Western Baltic Sea): effect on methane and salinity distribution of the water column, *Cont. Shelf Res.* **18**, pp. 1795-1806.

[Calvert and Veevers 1962] S. E. Calvert and J.J. Veevers, "Minor structures of unconsolidated marine sediments revealed by X-radiography," *Sedimentology* **1**, pp. 287-295.

[Camin and Isakson 2006] H. J. Camin and M. J. Isakson, "A comparison of sediment reflection coefficient measurements to elastic and poro-elastic models," *J. Acoust. Soc. Am.* (In press).

[Carbo and Molero 2002] R. Carbo and A. C. Molero, "Temperature dependence of high frequency sound attenuation in porous marine sediments," *Acust. Acta Acust.* **88**, pp. 190-194.

[Carey 1995] W. M. Carey, "Standard definitions for sound levels in the ocean," *IEEE J. Ocean. Eng.* **20**, pp. 109-113.

[Carman 1956] P. Carman, *Flow of Gases through Porous Media*, Academic Press, New York.

[Carney 1981] R. S. Carney, "Bioturbation and biodeposition," in *Principles of Benthic Marine Paleoecology*, A. J. Boucot (Ed.), Academic Press, New York, pp. 357-445.

[Caruthers and Novarini 1993] J. W. Caruthers and J. C. Novarini, "Modeling bistatic bottom scattering strength including a forward scattering lobe," *IEEE J. Ocean. Eng.* **18**, pp. 100-107.

[Castagna et al. 1995] J. P. Castagna, D.-H. Han, and M. L. Batzle, "Issues in rock physics and implications for DHI interpretation," *Leading Edge* **14**, pp. 883-886.

[Chen and Millero 1976] C. T. Chen and F. J. Millero, "Re-evaluation of Wilson's sound-speed measurements in pure water," *J. Acoust. Soc. Am.* **60** pp. 1270-1273.

[Chen and Millero 1977] C. T. Chen and F. J. Millero, "Speed of sound in seawater at high pressures," *J. Acoust. Soc. Am.* **62**, pp. 1129-1135.

[Chen and Fung 1988] M. F. Chen and A. K. Fung, "A numerical study of the regions of validity of the Kirchhoff and small-perturbation rough surface scattering models," *Radio Sci.* **23**, pp. 163-170.

[Chin-Bing et al. 1982] S. A. Chin-Bing, J. A. Davis, and R. B. Evans, "Nature of the lateral wave effect on bottom loss measurements," *J. Acoust. Soc. Am.* **71**, pp. 1433-1437.

[Choi et al. 2001] J. W. Choi, J. Na, and W. Seong, "240-kHz bistatic bottom scattering measurements in shallow water," *IEEE J. Ocean. Eng.* **26**, pp. 54-62.

[Choi et al. 2002] J. Choi, F. François-Caraillet, and B. P. Boudreau, "Lattice-automaton bioturbation simulator (LABS) implementation for small deposit feeders," *Comput. Geosci.* **28**, pp. 213-222.

[Chopra and Marfurt 2005] S. Chopra and K. J. Marfurt, "Seismic attributes — A historical perspective," *Geophysics* **70**, pp. 3SO-28SO.

[Chotiros et al. 1985] N. P. Chotiros, H. Boehme, T. G. Goldsberry, S. P. Pitt, R. A. Lamb, A. L. Garcia, and R. A. Altenberg, "Acoustic backscattering at low grazing angles from the ocean bottom. II. Statistical characterisitcs of bottom backscatter at a shallow water site," *J. Acoust. Soc. Am.* **77**, pp. 975-982.

[Chotiros and Boyle 1992] N. P. Chotiros and F. A. Boyle, *Bottom Penetration and Buried Targets*, Applied Research Laboratories, The University of Texas at Austin, ARL-TR-92-13.

[Chotiros 1994] N. P. Chotiros, "Reflection and scattering in normal incidence echo-sounding," *J. Acoust. Soc. Am.* **96**, pp. 2921-2929.

[Chotiros 1995a] N. P. Chotiros, "Biot model of sound propagation in water-saturated sand," *J. Acoust. Soc. Am.* **97**, pp. 199-214.

[Chotiros 1995b] N. P. Chotiros, "Inversion of sandy ocean sediments," in *Full Field Inversion Methods in Ocean and Seismo-Acoustics*, Diachok, Caiti, Gerstoft, and Schmidt (Eds.), Kluwer, Norwell, MA, pp. 353-358.

[Chotiros et al. 1997] N. P. Chotiros, A. M. Mautner, A. Lovik, A. Kristensen, and O. Bergem, "Acoustic penetration of a silty sand sediment in the 1-10-kHz band," *IEEE J. Ocean. Eng.* **22**, pp. 604-615.

[Chotiros 1998] N. P. Chotiros, "Response to: Comments on 'Biot model of sound propagation in water-saturated sand'," *J. Acoust. Soc. Am.* **103**, pp. 2726-2729.

[Chotiros 2002a] N. P. Chotiros, "Variability of bottom backscattering strength in the 10-500 kHz band at shallow grazing angles," in *Impact of Littoral Environmental Variability on Acoustic Predictions and Sonar Performance*, N.G. Pace and F.B. Jensen (Eds.), Kluwer Academic Publishers, Netherlands, pp. 203-210.

[Chotiros 2002b] N. P. Chotiros, "An inversion for Biot parameters in water-saturated sand," *J. Acoust. Soc. Am.* **112**, pp. 1853-1868.

[Chotiros et al. 2002a] N. P. Chotiros, D. E. Smith, and J. N. Piper, "Refraction and scattering into a sandy ocean sediment in the 30 to 40 kHz band," *IEEE J. Ocean. Eng.* **27**, pp. 362-375.

[Chotiros et al. 2002b] N. P. Chotiros, A. P. Lyons, J. Osler, and N. G. Pace,"Normal incidence reflection loss from a sandy sediment," *J. Acoust. Soc. Am.* **112**, pp. 1831-1841.

[Chotiros and Isakson 2004] N. P. Chotiros and M. J. Isakson, "A broadband model of sandy ocean sediments: Biot-Stoll with contact squirt flow and shear drag," *J. Acoust. Soc. Am.* **116**, pp. 2011-2022.

[Christensen 1996] N. I. Christensen, "Poisson's ratio and crustal seismology," *J. Geophys. Res.* **101**, pp. 3139-3156.

[Chu et al. 1997] D. Z. Chu, K. L. Williams, D. Tang, and D. R. Jackson, "High-frequency bistatic scattering by sub-bottom gas bubbles," *J. Acoust. Soc. Am.* **102**, pp. 806-814.

[Chu et al. 2001] D. Chu, D. Tang, T. C. Austin, A. A. Hinton, and R. I. Arthur, "Fine-scale acoustic tomographic imaging of shallow water sediments," *IEEE J. Ocean. Eng.* **26**, pp. 70-83.

[Clennell 1997] M. B. Clennell, "Tortuosity: a guide through the maze," in *Developments in Petrophysics*, A. M. Lovell and P. K. Harvey (Eds.), Geological Society Special Publication No. 122, pp. 299-344.

[Clifton 1976] H. E. Clifton, "Wave-formed sedimentary structures — A conceptual model," in *Beach and Nearshore Sedimentation*, R. A. Davis and R. L. Ethington (Eds.), SEPM Special Publication 24, Society for Sedimentary Geology, Tulsa, OK, pp. 126-148.

[Clifton and Dingler 1984] H. E. Clifton and J. R. Dingler, "Wave-formed structures and paleoenvironmental reconstruction," *Mar. Geol.* **60**, pp. 165-198.

[Coakley and Syvitski 1991] J. P. Coakley and J. P. M. Syvitski, "SediGraph techniques," in *Principles, Methods, and Applications of Particle Size Analysis*, J. P. M Syvitski (Ed.), Cambridge University Press, New York, pp. 127-142.

[Cox et al. 1967] R. A. Cox, F. Culkin, and J. P. Riley, "The electrical conductivity/chlorinity relationship in natural seawater," *Deep-Sea Res.* **14**, pp. 203-220.

[Cox et al. 1970] R. A. Cox, M. J. McCartney, and F. Culkin, "The specific gravity/salinity/temperature relationship in natural seawater," *Deep-Sea Res.* **17**, pp. 679-689.

[Crawford and Hay 1993] A. M. Crawford and A. E. Hay, "Determining suspended sand size and concentration from multifrequency acoustic backscatter," *J. Acoust. Soc. Am.* **94**, pp. 3312-3324.

[Crawford and Hay 2001] A. M. Crawford and A. E. Hay, "Linear transition ripple migration and wave orbital velocity skewness: observations," *J. Geophys. Res.* **106**, pp. 14113-14128.

[Crowther 1980] P. A. Crowther, "Fluctuation statistics of sea-bed acoustic backscatter," in *Bottom Interacting Ocean Acoustics,* W. A. Kuperman and F. Jensen (Eds.), Plenum, New York, pp. 609-622.

[Crowther 1983] P. A. Crowther, "Some statistics of the sea-bed and acoustic scattering therefrom," in *Acoustics and the Sea-Bed,* N. G. Pace (Ed.), Bath University, Bath, England, pp. 147-155.

[Crusius et al. 2004] J. Crusius, M. H. Bothner, and C. K. Summerfield, "Bioturbation depths, rates and processes in Massachusetts Bay sediments inferred from modeling of ^{210}Pb and $^{239+240}$Pu profiles," *Estuarine Coastal Shelf Sci.* **61**, pp. 643-655.

[Curry et al. 2004] C. W. Curry, R. H. Bennett, M. H. Hulbert, K. J. Curry, and R. W. Faas, "Comparative study of sand porosity and a technique for determining porosity of undisturbed marine sediment," *Mar. Georesourc. Geotechnol.* **22**, pp. 231-252.

[Cutter and Diaz 2000] G. R. Cutter and R. J. Diaz, "Biological alteration of physically structured flood deposits on the Eel margin, northern California," *Cont. Shelf Res.* **20**, pp. 235-253.

[Dacol and Berman 1988] D. K. Dacol and D. H. Berman, "Sound scattering from a randomly rough fluid-solid interface," *J. Acoust. Soc. Am.* **84**, pp. 292-302.

[Dacol 1990] D. K. Dacol, "The Kirchhoff approximation for acoustic scattering from a rough fluid-elastic solid interface," *J. Acoust. Soc. Am.* **88**, pp. 978-983.

[Davis and Bennell 1986] A. M. Davis and J. D. Bennell, "Dynamic properties of marine sediments," in *Ocean Seismo-Acoustics*, T. Akal and J. M. Berkson (Eds.), Plenum press, New York, pp. 501-510.

[Davis et al. 1989] A. M. Davis, J. D. Bennell, D. G. Huws, and D. Thomas, "Development of a seafloor geophysical sledge," *Mar. Geotechnol.* **8**, pp. 99-109.

[Davis et al. 1996] A. M. Davis, D. G. Huws, and R. Haynes, "Geophysical groundtruthing experiments in Eckernförde Bay," *Geo-Mar. Lett.* **16**, pp. 160-166.

[Day and Yamamoto 1999] C. M. Day and T. Yamamoto, "Low grazing angle bistatic sea floor scattering on the Florida Atlantic coastal shelf," *J. Acoust. Soc. Am.* **106**, pp. 1744-1754.

[Del Grosso 1952] V. A. Del Grosso, "The velocity of sound in seawater at zero depth," Naval Research Laboratory Report 4002.

[Del Grosso and Mader 1972a] V. A. Del Grosso and C. W. Mader, "Sound speed in sea-water samples," *J. Acoust. Soc. Am.* **52**, pp. 961-974.

[Del Grosso and Mader 1972b] V. A. Del Grosso and C. W. Mader, "Speed of sound in pure water," *J. Acoust. Soc. Am.* **52**, pp. 1442-1446.

[Del Grosso 1974] V. A. Del Grosso, "New equation for the speed of sound in natural waters (with comparisons to other equations)," *J. Acoust. Soc. Am.* **56**, pp. 1084-1091.

[de Moustier 1986] C. de Moustier, "Beyond bathymetry: Mapping acoustic scattering from the deep seafloor with Sea Beam," *J. Acoust. Soc. Am.* **79**, pp. 316-331.

[de Moustier and Alexandrou 1991] C. de Moustier and D. Alexandrou, "Angular dependence of 12 kHz seafloor acoustic backscatter," *J. Acoust. Soc. Am.* **90**, pp. 522-531.

[Deresiewicz and Skalak 1963] H. Deresiewicz and R. Skalak, "On uniqueness in dynamic poroelasticity," *Bull. Seismol. Soc. Am.* **53**, pp. 783-788.

[DeSanto 1979] J. A. DeSanto, "Theoretical methods in ocean acoustics," in *Ocean Acoustics*, J. A. DeSanto (Ed.), Springer-Verlag, New York, p. 66.

[Diaz and Cutter 2001] R. J. Diaz and G. R. Cutter, "In situ measurement of organism-sediment interaction: rates of burrow formation, abandonment and sediment oxidation, reduction," in *Organism-Sediment Interactions*, J. Y. Aller, S. A. Woodin, and R. C. Aller (Eds.), Belle W. Baruch Library in Marine Science, No. 21, University of South Carolina Press, pp. 19-33.

[Diaz et al. 2004] R. J. Diaz, M. Solan, and R. M. Valente, "A review of approaches for classifying benthic habitats and evaluating habitat quality," *J. Environ. Manage.* **73**, pp. 165-181.

[Dorfman and Dyer 1999] Y. Y. Dorfman and I. Dyer, "Monostatic and bistatic reverberation statistics west of the Mid-Atlantic Ridge," *J. Acoust. Soc. Am.* **106**, pp. 1755-1764.

[Drake 1990] T. G. Drake, "Structural features in granular flows," *J. Geophys. Res. Solid Earth Planets* **95** (B6), pp. 8681-8696.

[Drevet et al. 1999] C. Drevet, M. Brussieux, and J. P. Sessarego, "High frequency acoustic wave reflection on the surf zone seafloor," *Acust. Acta Acust.* **85**, pp. 701-706.

[Drolon et al. 2000] H. Drolon, F. Druaux, and A. Faure, "Particles shape analysis and classification using wavelet transform," *Pattern Recognition Lett.* **21**, pp. 473-482.

[Drumheller and Gragg 2001] D. M. Drumheller and R. F. Gragg, "Evaluation of a fundamental integral in rough-surface scattering theory," *J. Acoust. Soc. Am.* **110**, pp. 2270-2275.

[Dunlop 1992] J. I. Dunlop, "Measurement of acoustic attenuation in marine sediments by impedance tube," *J. Acoust. Soc. Am.* **91**, pp. 460-469.

[Dushaw et al. 1993] B. D. Dushaw, P. F. Worcester, B. D. Cornuelle, and B. M. Howe, "On equations for the speed of sound in seawater," *J. Acoust. Soc. Am.* **93**, pp. 255-275.

[Dworski and Jackson 1994] J. G. Dworski and D. R. Jackson, "Spatial and temporal variation of acoustic backscatter in the STRESS experiment," *Cont. Shelf Res.* **14**, pp. 1221-1237.

[Dyer 1970] I. Dyer, "Statistics of sound propagation in the ocean," *J. Acoust. Soc. Am.* **48**, pp. 337-345.

[Dyer and Huntley 1999] K. R. Dyer and D. A. Huntley, "The origin, classification and modeling of sand backs and ridges," *Cont. Shelf Res.* **19**, pp. 1285-1330.

[Dziak et al. 1993] R. P. Dziak, H. Matsumoto, and C. G. Fox, "Estimation of seafloor roughness spectral parameters from multibeam sonar acoustic backscatter data — Axial Seamount, Juan de Fuca Ridge," *Geophys. Res. Lett.* **20**, pp. 1863-1866.

[Eckart 1953] C. Eckart, "The scattering of sound from the sea surface," *J. Acoust. Soc. Am.* **25**, pp. 566-570.

[Eckman 1908] V. W. Eckman, "Die Zusammendrückbarkeit des Meerswassers nebst einigen Werten für Wasser und Quecksilver," (The compressibility of seawater including several estimates for water and mercury) *Publication de Circonstance, Conseil Permanent International pour l'Exploration de la Mer* **43**, pp. 1-47.

[Ehrlich and Weinberg 1970] R. Ehrlich and B. Weinberg, "An exact method for the characterization of grain shape," *J. Sedimen. Petrol.* **40**, pp. 205-212.

[Ehrlich and Chin 1980] R. Ehrlich and M. Chin, "Fourier-grain shape analysis — A new tool for sourcing and tracking abyssal silts," *Mar. Geol.* **38**, pp. 219-231.

[Elfouhaily and Guérin 2004] T. M. Elfoulhaily and C. A. Guérin, "A critical survey of approximate scattering wave theories from random rough surfaces," *Waves Random Media* **14**, pp. R1-R40.

[Erchul and Nacci 1972] R. A. Erchul and V. A. Nacci, "Electrical resistivity measuring system for porosity determination of marine sediments," *Mar. Technol. Soc. J.* **6**, pp. 47-53.

[Erickson and Jarrard 1998a] S. N. Erickson and R. D. Jarrard, "Porosity/formation-factor relationships for high-porosity siliciclastic sediments from Amazon Fan," *Geophys. Res. Lett.* **25**, pp. 2309-2312.

[Erickson and Jarrard 1998b] S. N. Erickson and R. D. Jarrard, "Velocity-porosity relationships for water-saturated siliciclastic sediments," *J. Geophys. Res. Solid Earth* **103**, pp. 30385-30406.

[Essen 1994] H. H. Essen, "Scattering from a rough sedimental seafloor containing shear and layering," *J. Acoust. Soc. Am.* **95**, pp. 1299-1310.

[Evans 2001] R. L. Evans, "Measuring the shallow porosity structure of sediments on the continental shelf. A comparison of an electromagnetic approach with cores and acoustic backscatter," *J. Geophys. Res. Oceans* **106**, pp. 27047-27060.

[Fang et al. 1993] W. W. Fang, M. G. Langseth, and P. J. Schultheiss, "Analysis and application of in situ pore pressure measurements in marine sediments," *J. Geophys. Res. Solid Earth* **98**, pp. 7921-7938.

[Feistel and Hagen 1995] R. Feistel and E. Hagen, "On the Gibbs thermodynamic potential in seawater," *Prog. Oceanogr.* **36**, pp. 249-327.

[Feistel 2003] R. Feistel, "A new extended Gibbs thermodynamic potential of seawater," *Prog. Oceanogr.* **58**, pp. 43-114.

[Fenstermacher et al. 2001] L. E. Fenstermacher, G. B. Crawford, J. C. Borgeld, T. Britt, D. A. George, M. A. Klein, N. W. Driscoll, and L. A. Mayer, "Enhanced acoustic scattering due to high abundance of sand dollars, *Dendraster excentricus*," *Mar. Georesourc. Geotechnol.* **19**, pp. 135-145.

[Fischer and Simmons 1977] F. H. Fischer and V. P. Simmons, "Sound absorption in seawater," *J. Acoust. Soc. Am.* **62**, pp. 558-564.

[Flatté et al. 1979] S. M. Flatté, R. Dashen, W. H. Munk, K. M. Watson, and F. Zachariasen, *Sound Transmission Through a Fluctuating Ocean*, Cambridge University Press, New York.

[Fleischer et al. 2001] P. Fleischer, T. H. Orsi, M. D. Richardson, and A. L. Anderson, "Distribution of free gas in marine sediments: a global overview," *Geo-Mar. Lett.* **21**, pp.103-122.

[Flemming 2000] B. W. Flemming, "A revised textural classification of gravel-free muddy sediments on the basis of ternary diagrams," *Cont. Shelf Res.* **20**, pp. 1125-1137.

[Floodgate and Judd 1992] G. D. Floodgate and A. G. Judd, "The origins of shallow gas," *Cont. Shelf Res.* **12**, pp. 1145-1156.

[Flowers and Hurdle 1972] K. D. Flowers and B. G. Hurdle, "Monostatic scattering from the ocean botttom," *J. Acoust. Soc. Am.* **51**, pp. 1109-1111.

[Fofonoff and Millard 1983] N. P. Fofonoff and R.C. Millard, "Algorithms for computation of fundamental properties of seawater," *UNESCO Tech. Pap. Mar. Sci.* **44**, pp. 1-53.

[Fofonoff 1985] N. P. Fofonoff, "Physical properties of seawater: a new salinity scale and equation of state for seawater," *J. Geophys. Res. Oceans* **90**(C2), pp. 3332-3342.

[Folk and Ward 1957] R. L. Folk and W. C. Ward, "Brazos River Bar: A study in the significance of grain size parameters," *J. Sediment. Petrol.* **27**, pp. 3-26.

[Folk 1980] R. L. Folk, *Petrology of Sedimentary Rocks*, Hemphill Publishing Company, Austin, TX.

[Fonseca et al. 2002] L. Fonseca, L. Mayer, D. Orange, and N. Driscoll, "The high-frequency backscattering angular response of gassy sediments: Model/data comparison from the Eel River Margin, California," *J. Acoust. Soc. Am.* **111**, pp. 2621-2631.

[Forch et al. 1902] C. Forch, M. Knudsen, and S. P. L. Sörensen, "Berichte ber die konstantenbestimmungen zur aufstellung der hydrographischen tabellen," *D. Kgl. Dansk. Vidensk. Selsk. Skrifter. 6, Raekke, Naturvidensk. Math. Afd* **XII.1**.

[Fox and Hayes 1985] C. G. Fox and D. E. Hayes, "Quantitative methods for analyzing the roughness of the seafloor," *Rev. Geophys.* **23**, pp. 1-48.

[Francois and Garrison 1982a] R. E. Francois and G. R. Garrision, "Sound absorption based on ocean measurements: Part I: Pure water and magnesium sulfate contributions," *J. Acoust. Soc. Am.* **72**, pp. 896-907.

[Francois and Garrison 1982b] R. E. Francois and G. R. Garrision, "Sound absorption based on ocean measurements: Part II: Boric acid contributions and equation for total absorption," *J. Acoust. Soc. Am.* **72**, pp. 1879-1890.

[François et al. 1997] F. François, J.-C. Poggiale. J.-P. Durbec, and G. Stora, "A new approach for the modeling of sediment reworking induced by a macrobenthic community," *Acta Biotheor.* **45**, pp. 295-319.

[Fraser 1935] H. J. Fraser, "Experimental study of the porosity and permeability of clastic sediments," *J. Geol.* **43**, pp. 910-1010.

[Fredsøe and Deigaard 1992] J. Fredsøe and R. Deigaard, *Mechanics of Coastal Sediment Transport*, World Scientific, Hackensack, NJ.

[Freitas et al. 2003] R. Freitas, S. Silva, V. Quintino, A. M. Rodrigues, K. Rhynas, and W. T. Collins, "Acoustic seabed classification of marine habitats: studies in the western coastal-shelf area of Portugal," *ICES J. Mar. Sci.* **60**, pp. 599-608.

[Friedrich and Levitus 1972] H. Friedrich and S. Levitus, "An approximate to the equation of state for seawater, suitable for numerical ocean models," *J. Phys. Oceanogr.* **2**, pp. 514-517.

[Frisk 1998] G. V. Frisk, *Ocean and Seabed Acoustics*, Prentice Hall, Englewood Cliffs, NJ.

[Frye and Pugh 1971] H. W. Frye and J. D. Pugh, "A new equation for the speed of sound in seawater," *J. Acoust. Soc. Am.* **50**, pp. 384-386.

[Fu et al. 1996] S. S. Fu, R. H. Wilkens, and L. N. Frazer, "Acoustic lance: New in situ seafloor velocity profiles," *J. Acoust. Soc. Am.* **99**, pp. 234-242.

[Fu et al. 2004] S. S. Fu, C. Tao, M. Prasad, R. H. Wilkens, and L. N. Frazer, "Acoustic properties of coral sands, Waikiki, Hawaii," *J. Acoust. Soc. Am.* **115**, pp. 2013-2020.

[Futterman 1962] W. I. Futterman, "Dispersive body waves," *J. Geophys. Res.* **67**, pp. 5279-5291.

[Galybin 1976] N. N. Galybin, "Back scattering of sound by a disturbed sea surface," *Sov. Phys. Acoust.* **22**, pp. 193-197.

[Gardner and Sills 2001] T. N. Gardner and G. C. Sills, "An examination of the parameters that govern the acoustic behavior of sea bed sediments containing gas bubbles," *J. Acoust. Soc. Am.* **110**, pp. 1878-1889.

[Gassmann 1951] F. Gassmann, "Über die elastizitä poröser medien," *Vierteljahresschr. Naturforsch. Ges. Zurich* **96**, pp. 1-23.

[Gensane 1989] M. Gensane, "A statistical study of acoustic signals backscattered from the sea bottom," *IEEE J. Ocean. Eng.* **14**, pp. 84-93.

[Gensane 1993] M. Gensane, "Sea bottom revergeration: The role of volume inhomogeneities of the sediment," in *Proceedings of the Ocean Reverberation Symposium*, D. D. Ellis, J. J. Preston, and H. G. Urban (Eds.), Kluwer Academic, Dordrecht, pp. 59-64.

[Gerland et al.1993] S. Gerland, M. Richter, H. Villinger, and G. Kuhn, "Nondestructive porosity determinations of Antarctic marine sediments derived from resistivity measurements with an inductive method," *Mar. Geophys. Res.* **15**, pp. 201-218.

[Goff and Jordan 1989a] J. A. Goff and T. H. Jordan, "Stochastic modeling of seafloor morphology — resolution of topographic parameters by Sea Beam data," *IEEE J. Ocean. Eng.* **14**, pp. 326-337.

[Goff and Jordan 1989b] J. A. Goff and T. H. Jordan, "Stochastic modeling of seafloor morphology — a parameterized Gaussian model," *J. Geophys. Res. Lett.* **16**, pp. 45-48.

[Goff and Tucholke 1997] J. A. Goff and B. E. Tucholke, "Multiscale spectral analysis of bathymetry on the flank of the Mid-Atlantic Ridge: Modification of the seafloor by mass wasting and sedimentation," *J. Geophys. Res. Solid Earth* **102**, pp. 15447-15462.

[Goff et al. 2004] J. A. Goff, B. J. Kraft, L. A. Mayer, S. G. Schock, C. K. Sommerfield, H. C. Olson, S. P. S. Gulick, and S. Nordfjord, "Seabed characterization on the New Jersey middle and outer shelf, correlatability and spatial variability of seafloor sediment properties," *Mar. Geol.* **209**, pp. 147-172.

[Gorgas et al. 2002] T. J. Gorgas, R. H. Wilkens, S. S. Fu, L. N. Frazer, M. D. Richardson, K. B. Briggs, and H. Lee, "In situ acoustic and laboratory ultrasonic sound speed and attenuation measured in heterogeneous soft seabed sediments: Eel River shelf, California," *Mar. Geol.* **182**, pp. 103-119.

[Gragg et al. 2001] R. F. Gragg, D. Wurmser, and R. C. Gauss, "Small-slope scattering from rough elastic ocean floors: General theory and computational algorithm," *J. Acoust. Soc. Am.* **110**, pp. 2878-2901.

[Greenlaw et al. 2004] C. F. Greenlaw, D. V. Holliday, and D. E. McGehee, "High-frequency scattering from saturated sand sediments," *J. Acoust. Soc. Am.* **115**, pp. 2818-2823.

[Greenspan and Tschiegg 1959] M. Greenspan and C. E. Tschiegg, "Tables of the speed of sound in water," *J. Acoust. Soc. Am.* **31**, pp. 75-76.

[Griffin et al. 1996] S. R. Griffin, F. B. Grosz and M. D. Richardson, "In situ sediment geoacoustic measurement system," *Sea Technol.* **37**, pp. 19-22.

[Griffiths 1967] J. C. Griffiths, *Scientific Method in Analysis of Sediments*, McGraw-Hill, New York.

[Guérin 2002] C. A. Guérin, "Scattering on rough surfaces with alpha-stable non-Gaussian height distributions," *Waves Random Media* **12**, pp. 293-306.

[Guinasso and Schink 1975] N. L. Guinasso and D. R. Schink, "Quantitative estimates of biological mixing rates in abyssal sediments," *J. Geophys. Res. Oceans Atmos.* **80**, pp. 3032-3043.

[Gunn and Best 1998] D. E. Gunn and A. I. Best, "A new automated nondestructive system for high resolution multi sensor core logging of open sediment cores," *Geo-Mar. Lett.* **18**, pp. 70-77.

[Gurevich et al. 1998] B. Gurevich, A. P. Sadovnichaja, S. L. Lopatnikov, and S. A. Shapiro, "Scattering of a compressional wave in a poroelastic medium by an ellipsoidal inclusion," *Geophys. J. Int.* **133**, pp. 91-103.

[Gurevich and Schoenberg 1999] B. Gurevich and M. Schoenberg, "Interface conditions for Biot's equations of poroelasticity," *J. Acoust. Soc. Am.* **105**, pp. 2585-2589.

[Hagen and Vogt 1999] R. A. Hagen and P. R. Vogt, "Seasonal variability of shallow biogenic gas in Chesapeake Bay," *Mar. Geol.* **158**, pp. 75-88.

[Hall 1995] M. V. Hall, "Dimensions and units of underwater acoustic parameters," *J. Acoust. Soc. Am.* **97**, pp. 3887-3889.

[Hamilton 1956] E. L. Hamilton, "Low sound velocities in high-porosity sediments," *J. Acoust. Soc. Am.* **28**, pp. 16-19.

[Hamilton et al. 1956] E. L. Hamilton, G. Shumway, H. W. Menard, and C. J. Shipek, "Acoustic and other physical properties of shallow-water sediments off San Diego," *J. Acoust. Soc. Am.* **28**, pp. 1-15.

[Hamilton 1963] E. L. Hamilton, "Sediment sound velocity measurements made in situ from bathyscaph Trieste," *J. Geophys. Res.* **68**, pp. 5991-5998.

[Hamilton et al. 1970] E. L. Hamilton, H. P. Bucker, D. L. Keir, and J. A. Whitney, "Velocities of compressional and shear waves in marine sediments determined in situ from a research submersible," *J. Geophys. Res.* **75**, pp. 4039-4049.

[Hamilton 1971a] E. L. Hamilton, "Elastic properties of marine sediments," *J. Geophys. Res.* **76**, pp. 579-604.

[Hamilton 1971b] E. L. Hamilton, "Prediction of in-situ acoustic and elastic properties of marine sediments," *Geophysics* **36**, pp. 266-284.

[Hamilton 1972] E. L. Hamilton, "Compressional wave attenuation in marine sediments," *Geophysics* **37**, pp. 620-646.

[Hamilton 1976a] E. L. Hamilton, "Shear wave velocity versus depth in marine sediments: A review," *Geophysics* **41**, pp. 985-996.

[Hamilton 1976b] E. L. Hamilton, "Sound attenuation as a function of depth in the sea floor," *J. Acoust. Soc. Am.* **59**, pp. 528-535.

[Hamilton 1976c] E. L. Hamilton, "Attenuation of shear waves in marine sediments," *J. Acoust. Soc. Am.* **60**, pp. 334-338.

[Hamilton 1978] E. L. Hamilton, "Sound velocity-density relations in sea-floor sediments and rocks," *J. Acoust. Soc. Am.* **63**, pp. 366-377.

[Hamilton 1979a] E. L. Hamilton, "Sound velocity gradients in marine sediments," *J. Acoust. Soc. Am.* **65**, pp. 909-922.

[Hamilton 1979b] E. L. Hamilton, "V_p/V_s and Poisson's ratios in marine sediments and rocks," *J. Acoust. Soc. Am.* **66**, pp. 1093-1101.

[Hamilton 1980] E. L. Hamilton, "Geoacoustic modelling of the seafloor," *J. Acoust. Soc. Am.* **68**, pp. 1313-1340.

[Hamilton and Bachman 1982] E. L. Hamilton and R. T. Bachman, "Sound velocity and related properties of marine sediments," *J. Acoust. Soc. Am.* **72**, pp. 1891-1904.

[Hamilton et al. 1982] E. L. Hamilton, R. T. Bachman, W. H. Berger, T. C. Johnson and L. A. Mayer, "Acoustic and related properties of calcareous deep-sea sediments," *J. Sediment. Petrol.* **52**, pp. 733-753.

[Hamilton 1987] E. L. Hamilton, "Acoustic properties of sediments," in *Acoustics and Ocean Bottom. II FASE Specialized Conference*, A. Lara-Saenz, C. Ranz-Guerra, and C. Carbo-Fite (Eds.), CISC, Madrid.

[Hampton 1967] L. D. Hampton, "Acoustic properties of sediments," *J. Acoust. Soc. Am.* **42**, pp. 882-890.

[Hashin and Shtrikman 1962] Z. Hashin and S. Shtrikman, "A variational approach to the theory of the elastic behavior of multiphase materials," *J. Mech. Phys. Solids* **11**, pp. 127-140.

[Hay and Mudge 2005] A. E Hay and T. Mudge, "Principal bed states during SandyDuck97: Occurrence, spectral anisotropy, and bed state storm cycle," *J. Geophys. Res. Oceans* **110**, Art. No. C03013.

[Hayes 1967] M. O. Hayes, "Relationship between coastal climate and bottom sediment type on the inner continental shelf," *Mar. Geol.* **5**, pp. 111-132.

[Haynes et al. 1997] R. Haynes, D. G. Huws, A. M. Davis, and J. D.Bennell, "Geophysical seafloor sensing in a carbonate sediment regime," *Geo-Mar. Lett.* **17**, pp. 253-259.

[Heald 2002] G. J. Heald, "High frequency seabed scattering and sediment discrimination," in *Proceedings of the Institute of Acoustics Conference* **23**, T. G. Leighton, G. J. Heald, H. D. Griffiths, and G. Griffiths (Eds.), Southampton, United Kingdom, pp. 258-267.

[Henyey and Wurmser 1990] F. S. Henyey and D. Wurmser, "Acoustic scattering from a non-Gaussian ocean surface," *J. Acoust. Soc. Am.* **87**, pp. 2477-2480.

[Henyey 1991] F. S. Henyey, "Obtaining a 2-D spectrum from a 1-D measurement," Arete Associates Tech. Rep. ARS-223-003-TR, La Jolla, CA.

[Henyey and Thorsos 1995] F. S. Henyey, E. I. Thorsos, and K. M. Nathwani, "Scattering strength cannot depend upon the length of a pulse," (Abstract) *J. Acoust. Soc. Am.* **98**, pp. 2986-2987.

[Herbert and Mayer 1991] T. D. Herbert and L. A. Mayer, "Long climatic time series from sediment physical property measurements," J. Sediment. Petrol. **61**, pp. 1089-1108.

[Herzfeld et al. 1995] U. C. Herzfeld, I.I. Kim, and J. A. Orcutt, "Is the ocean-floor a fractal?," *Math. Geol.* **27**, pp. 421-462.

[Hesselberg and Sverdrup 1914] Th. Hesselberg and H. U. Sverdrup, "Die Stabilitätsverhältnisse des Seewassers bie vertikalen Verschiebungen," *Bergens Museums Aabok* **14**, 1914.

[Hickey and Sabatier 1997] C. J. Hickey and J. M. Sabatier, "Choosing Biot parameters for modeling water-saturated sand," *J. Acoust. Soc. Am.* **102**, pp. 1480-1484.

[Hill 1963] R. Hill, "Elastic properties of reinforced solids: some theoretical principles," *J. Mech. Phys. Solids* **11**, pp. 357-372.

[Hines 1990] P. C. Hines, "Theoretical model of acoustic backscatter from a smooth seabed," *J. Acoust. Soc. Am.* **88**, pp. 324-334.

[Hines 1996] P. C. Hines, "Theoretical model of in-plane scatter from a smooth sediment seabed," *J. Acoust. Soc. Am.* **99**, pp. 836-844.

[Holland and Neumann 1998] C. W. Holland and P. Neumann, "Sub-bottom scattering: A modeling approach," *J. Acoust. Soc. Am.* **104**, pp. 1363-1373.

[Holliday et al. 2004] D. V. Holliday, C. F. Greenlaw, J. E. B. Rines, and D. Thistle, " Diel variations in acoustical scattering from a sandy seabed," Proceedings of the ICES Annual Science Conference.

[Hollister and Nowell 1991] C. D. Hollister and A. R. M. Nowell, "HEBBLE epilogue," *Mar. Geol.* **99**, pp. 445-460.

[Holyer et al. 1996] R. J. Holyer, D. K. Young, J. C. Sandidge, and K. B. Briggs, "Sediment density structure derived from textural analysis of cross-sectional X-radiographs," *Geo-Mar. Lett.* **16**, pp. 204-211.

[Horton and Melton 1970] C. W. Horton, Sr. and D. R. Melton, "Importance of the Fresnel correction in scattering from a rough surface. II. Scattering coefficient," *J. Acoust. Soc. Am.* **47**, pp. 299-303.

[Horton 1974] C. W. Horton, Sr., "Dispersion relationships in sediments and sea water," *J. Acoust. Soc. Am.* **55**, pp. 547-549.

[Horton 1981] C. W. Horton, Sr., "Comment on 'Kramers-Kronig relationship between ultrasonic attenuation and phase velocity'," *J. Acoust. Soc. Am.* **70**, p. 1182.

[Hovem and Ingram 1979] J. M. Hovem and G. D. Ingram, "Viscous attenuation of sound in saturated sand," *J. Acoust. Soc. Am.* **66**, pp. 1807-1812.

[Hovem 1980] J. M. Hovem and G. D. Ingram, "Viscous attenuation of sound in suspensions and high-porosity sediments," *J. Acoust. Soc. Am.* **67**, pp. 1559-1563.

[Howard 1968] J. D. Howard, "X-ray radiography for examination of burrowing in sediments by marine invertebrate organisms," *Sedimentology* **11**, pp. 249-258.

[Howard and Frey 1973] J. D. Howard and R. W. Frey, "Characteristic physical and biogenic sedimentary structures in Georgia estuaries," *Am. Assoc. Pet. Geol. Bull.* **57**, pp. 1169-1184.

[Hudson and Ehrlich 1980] C. B. Hudson and R. Ehrlich, "Determination of relative provenance contributions in samples of quartz sand using Q-mode factor analysis of Fourier grain shape data," *J. Sediment Petrol.* **50**, pp. 1101-1110.

[Hulbert et al. 2002] M. H. Hulbert, R. H. Bennett, R. J. Baerwald, R. L. Long, K. J. Curry, A. L. Curry, and M. T. Abril, "Observations of the sediment-water interface: Marine and fresh water environments," *Mar. Georesour. Geotechnol.* **20**, pp. 255-274.

[Huntley and Hazen 1988] D. A. Huntley and D. Z. Hazen, "Seabed stresses in combined wave and steady flow conditions on the Nova Scotia shelf — field measurements and predictions," *J. Phys. Oceanogr.* **18**, pp. 347-362.

[Huws et al. 2000] D. G. Huws, A. M. Davis, and J. R. Pyrah, "A nondestructive technique for predicting the in situ void ratio for marine sediments," *Mar. Georesour. Geotechnol.* **18**, pp. 333-346.

[Inman et al. 1993] D. L. Inman, M. H. S. Elwany, and S. A. Jenkins, "Shorerise and bar-berm profiles on ocean beaches," *J. Geophys. Res. Oceans* **98**, pp. 18181-18199.

[Irish et al. 1999] J. D. Irish, J. F. Lynch, P. A. Traykovski, A. E. Newhall, K. Prada, and A. E. Hay, "A self-contained sector-scanning sonar for bottom roughness observations as part of sediment transport studies," *J. Atmos. Oceani. Technol.* **16**, pp. 1830-1841.

[Isakson and Nielsen 2006] M. J. Isakson and T. B. Nielsen, "The viability of reflection loss measurement inversion to predict broadband acoustic behavior," *J. Acoust. Soc. Am.* **120**, pp. 135-144.

[Ishimaru 1997] A. Ishimaru, *Wave Propagation and Scattering in Random Media*, IEEE Press, New York.

[Ivakin 1981] A. N. Ivakin, "On sound scattering by multi-scale inhomogeneities of bottom sediment," *Oceanologiya* **21**, pp. 42-44.

[Ivakin and Lysanov 1981a] A. N. Ivakin and Yu. P. Lysanov, "Theory of underwater sound scattering by random inhomogeneities of the bottom," *Sov. Phys. Acoust.* **27**, pp. 61-64.

[Ivakin and Lysanov 1981b] A. N. Ivakin and Yu. P. Lysanov, "Underwater sound scattering by volume inhomogeneities of a bottom medium bounded by a rough surface," *Sov. Phys. Acoust.* **27**, pp. 212-215.

[Ivakin 1982] A. N. Ivakin, "Multiple scatttering by random inhomogeneities in seabed sediments", in *Aerophysics and Geo-Space Research*, MPTI, Moscow, pp.96-98 (in Russian).

[Ivakin 1983] A. N. Ivakin, "Sound scattering by random volume inhomogeneities and small surface roughness of an underwater ground," *Voprosy sudostroenija: Akustika*, no.17, pp. 20-25 (in Russian).

[Ivakin 1986] A. N. Ivakin, "Sound scattering by random inhomogeneities of stratified ocean sediments," *Sov. Phys. Acoust.* **32**, pp. 492-496.

[Ivakin 1989] A. N. Ivakin, "Backscattering of sound by the ocean bottom. Theory and experiment," in *Acoustics of ocean medium*, L. M. Brekhovskikh and I. B. Andreeva (Eds.), Nauka, Moscow, pp.160-169 (in Russian).

[Ivakin 1990] A. N. Ivakin, "Sound scattering by inhomogeneities of an elastic half-space," *Sov. Phys. Acoust.* **36**(4), pp. 377-380.

[Ivakin 1994a] A. N. Ivakin, "Sound scattering by rough interface and volume inhomogeneities of the sea bottom," *Acoust. Phys.* **40**, pp. 427-428.

[Ivakin 1994b] A. N. Ivakin, "Sound scattering by rough interfaces of layered media", in *Third International Congress on Air-and-Structure-Borne Sound and Vibration*, M. J. Crocker (Ed.), International Publications, Auburn University, pp.1563-1570.

[Ivakin 1998a] A. N. Ivakin, "A unified approach to volume and roughness scattering," *J. Acoust. Soc. Am.* **103**, pp. 827-837.

[Ivakin 1998b] A. N. Ivakin, "Models of seafloor roughness and volume scattering," *Proceedings OCEANS 1998* **1**, pp. 518-521.

[Ivakin and Jackson 1998] A. N. Ivakin and D. R. Jackson, "Effects of shear elasticity on sea bed scattering: Numerical examples," *J. Acoust. Soc. Am.* **103**, pp. 346-354.

[Ivakin 2001] A. N. Ivakin, "Models of scattering for remote acoustic sensing of the seafloor," *Proc. Inst. Acoust.* **23**, pp. 268-275.

[Ivakin 2004] A. Ivakin, "Scattering from discrete shell inclusions in marine sediments," *Proceedings of the 7th European Conference on Underwater Acoustics, EUCA 2004*, Delft, Netherlands, pp. 625-630.

[Ivakin 2005] A. Ivakin, "High frequency scattering from sandy sediments: roughness vs discrete inclusions," in *Boundary Influences in High-Frequency, Shallow-Water Acoustics*, N. G. Pace and P. Blondel (Eds.), University of Bath, UK, pp. 185-192.

[Jackett and McDougall 1995] D. R. Jackett and T. J. McDougall, "Minimal adjustment of hydrographic profiles to achieve static stability," *J. Atmos. Ocean. Tech.* **12**, pp. 381-389.

[D. Jackson et al. 1986a] D. R. Jackson, D. P. Winebrenner, and A. Ishimaru, "Application of the composite roughness model to high-frequency bottom backscattering," *J. Acoust. Soc. Am.* **79**, pp. 1410-1422.

[D. Jackson et al. 1986b] D. R. Jackson, A. M. Baird, J. J. Crisp, and P. A. G. Thomson, "High-frequency bottom backscatter measurements in shallow water," *J. Acoust. Soc. Am.* **80**, pp. 1188-1199.

[D. Jackson et al. 1988] D. R. Jackson, D. P. Winebrenner, and A. Ishimaru, "Comparison of perturbation theories for rough-surface scattering," *J. Acoust. Soc. Am.* **83**, pp. 961-969.

[D. Jackson and Nesbitt 1988] D. R. Jackson and E. Nesbitt, "Bottom classification using backscattering at vertical incidence," (Abstract) *J. Acoust. Soc. Am.* **83**, p. S80.

[D. Jackson and Briggs 1992] D. R. Jackson and K. B. Briggs, "High-frequency bottom backscattering: Roughness vs. sediment volume scattering," *J. Acoust. Soc. Am.* **92**, pp. 962-977.

[D. Jackson 1994] D. R. Jackson, "Models for scattering from the sea bed," *Proc. Inst. Acoust.* **16**, pp. 161-169.

[D. Jackson et al. 1996a] D. R. Jackson, K. B. Briggs, K. L. Williams, and M. D. Richardson, "Tests of models for high-frequency seafloor backscatter," *IEEE J. Ocean. Eng.* **21**, pp. 458-470.

[D. Jackson et al. 1996b] D. R. Jackson, K. L. Williams, and K. B. Briggs, "High-frequency acoustic observations of benthic spatial and temporal variability," *Geo-Mar. Lett.* **16**, pp. 212-218.

[D. Jackson et al. 1997] D. R. Jackson, E. I. Thorsos, and J. E. Moe, "Far-field considerations in boundary scattering," in *Shallow-Water Acoustics*, R. Zhang and J. Zhou (Eds.), China Ocean Press, Beijing, pp. 297-302.

[D. Jackson and Ivakin 1998] D. R. Jackson and A. N. Ivakin, "Scattering from elastic sea beds: First-order theory," *J. Acoust. Soc. Am.* **103**, pp. 336-345.

[D. Jackson et al. 1998] D. R. Jackson, K. L. Williams, T. F. Wever, C. T. Friedrichs, and L. D. Wright, "Sonar evidence for methane ebullition in Eckernförde Bay, *Cont. Shelf Res.* **18**, pp. 1893-1915.

[D. Jackson and Richardson 2002] D. R. Jackson and M. D. Richardson, "Seasonal temperature gradients within a sandy seafloor: implications for acoustic propagation and scattering," in *Proceedings of the Institute of Acoustics Conference* **23**, T. G. Leighton, G. J. Heald, H. D. Griffiths, and G. Griffiths (Eds.), Southampton, UK, 2001, pp. 361-368.

[D. Jackson et al. 2002] D. R. Jackson, K. L. Williams, E. I. Thorsos, and S. G. Kargl, "High-frequency subcritical acoustic penetration into a sandy sediment," *IEEE J. Ocean. Eng.* **27**, pp. 346-361.

[P. Jackson 1975] P. D. Jackson, "An electrical resistivity method for evaluating in-situ porosity of clean marine sands," *Mar. Geotechnol.* **1**, pp. 91-115.

[P. Jackson et al. 1978] P. D. Jackson, D. Taylor-Smith, and P. N. Stanford, "Resistivity-porosity-particle shape relationships for marine sands," *Geophysics* **43**, pp. 1250-1268.

[P. Jackson et al. 1981] P. D. Jackson, R. Baria, and D. M. McCann, "Geotechnical assessment of superficial marine sediments using in-situ geophysical probes," *Ocean Manage.* **7**, pp. 189-209.

[P. Jackson et al. 1993] P. D. Jackson, R. D. Jarrard, C. J. Pilgram, and J. M. Pearse, "Resistivity/porosity/ velocity relationships from downhole logs: an aid for evaluating pore morphology," *Proceedings of the Ocean Drilling Program, Scientific Results* **133**, pp. 661-686.

[P. Jackson et al. 1996] P. D. Jackson, K. B. Briggs, and R. C. Flint, "Evaluation of sediment heterogeneity using micro-resistivity imaging and X-radiography," *Geo-Mar. Lett.* **16**, pp. 219-225.

[P. Jackson et al. 2002] P. D. Jackson, K. B. Briggs, R. C. Flint, R. J. Holyer, and J. C. Sandidge,"Two- and three-dimensional heterogeneity in carbonate sediments using resistivity imaging," *Mar. Geol.* **182**, pp. 55-76.

[Jakeman 1982] E. Jakeman, "Fresnel scattering by a corrugated random surface with fractal slope," *J. Opt. Soc. Am.* **72**, pp. 1034-1041.

[Jenkinson 1993] I. R. Jenkinson, "Bulk-phase viscoelastic properties of seawater," *Oceanol. Acta* **16**, pp. 317-334.

[Jenkinson and Biddanda 1995] I. R. Jenkinson and B. A. Biddanda, "Bulk-phase viscoelastic properties of seawater: relationship with plankton components," *J. Plankton Res.* **17**, pp. 2251-2274.

[Jensen and Schmidt 1987] F. B. Jensen and H. Schmidt, "Subcritical penetration of narrow Gaussian beams into sediments," *J. Acoust. Soc. Am.* **82**, pp. 574-579.

[Jensen et al. 1994] F. B. Jensen, W. A. Kuperman, M. B. Porter, and H. Schmidt, *Computational Ocean Acoustics*, AIP Press, Woodbury, NY.

[Johnson and Plona 1982] D. L. Johnson and T. J. Plona, "Acoustic slow waves and the consolidation transition," *J. Acoust. Soc. Am.* **72**, pp. 556-565.

[Johnson and Olhoeft 1984] G. R. Johnson and G. R. Olhoeft, "Density of rocks and minerals," in *Handbook of Physical Properties of Rocks, Vol. III*, R. S. Carmichael (Ed.) , CRC Press, Boca Raton FL, pp. 1-38.

[Johnson et al. 1987] D. L. Johnson, J. Koplik, and R. Dashen, "Theory of dynamic permeability and tortuosity in fluid-saturated porous media," *J. Fluid Mech.* **176**, pp. 379-402.

[Johnson et al. 2002] B. D. Johnson, B. P. Boudreau, and B. S. Gardiner, "Mechanical response of sediments to bubble growth," *Mar. Geol.* **187**, pp. 347-363.

[Jones and Jackson 1997] C. D. Jones and D. R. Jackson, "Temporal fluctuation of backscattered field due to bioturbation in marine sediments," in *High Frequency Acoustics in Shallow Water*, N.G. Pace, E. Pouliquen, O. Bergem, and A.P. Lyons (Eds.), NATO SACLANT Undersea Res. Ctr., La Spezia, Italy, pp. 275-282.

[Jones 1999] C. D. Jones, "High-Frequency Acoustic Volume Scattering from Biologically Active Marine Sediments," Ph.D. dissertation, University of Washington, Seattle.

[Jones and Jackson 2000] C. D. Jones and D. R. Jackson, "Analysis of the small perturbation method for high-frequency scattering from marine sediments," in *Proceedings of the Fifth European Conference on Underwater Acoustics, Lyon,*

France, M. E. Zakharia, P. Chevret, and P. Dubail (Eds.), Lyon, France, pp. 1259-1264.

[Jones and Jackson 2001] C. D. Jones and D. R. Jackson, "Small perturbation method of high-frequency bistatic volume scattering from marine sediments," *IEEE J. Ocean. Eng.* **26**, pp. 84-93.

[Jones and Jago 1991] S. E. Jones and C. F. Jago, "Small-scale in situ measurements of S-H velocity in surficial sedimentary deposits: localized textural and biological controls," in *Shear Waves in Marine Sediments*, J. M. Hovem, M. D. Richardson, and R. D. Stoll (Eds.), Kluwer Academic Publishers, Dordrecht, pp. 313-320.

[Jones and Jago 1993] S. E. Jones and C. F. Jago, "In situ assessment of modification of sediment properties by burrowing invertebrates," *Mar. Biol.* **115**, pp. 133-142.

[Judd and Hovland 1992] A. G. Judd and M. Hovland, "The evidence of shallow gas in marine sediments," *Cont. Shelf Res.* **12**, pp. 1081-1095.

[Judd 2003] A. G. Judd, "The global importance and context of methane escape from the seabed," *Geo-Mar. Lett.* **23**, pp. 147-154.

[Judd 2004] A. G. Judd, "Natural seabed gas seeps as sources of atmospheric methane," *Environ. Geol.* **46**, pp. 988-996.

[Julien 1995] P. Y. Julien, *Erosion and Sedimentation*, Cambridge University Press, Cambridge, UK.

[Jumars et al. 1996] P. A. Jumars, D.R. Jackson, T.F. Gross, and C. Sherwood, "Acoustic remote sensing of benthic activity: a statistical approach," *Limnol. Oceanogr.* **41**, pp. 1200-1241.

[Kaczkowski and Thorsos 1994] P. J. Kaczkowski and E. I. Thorsos, "Application of the operator expansion method to scattering from one-dimensional moderately rough Dirichlet random surfaces," *J. Acoust. Soc. Am.* **96**, pp. 957-972.

[Kaplan 1974] I. R. Kaplan, *Natural Gases in Marine Sediments*, Plenum Press, New York.

[Kaye 1978] B. H. Kaye, "Specification of the ruggedness and/or texture of a fine particle profile by its fractal dimension," *Powder Technol.* **21**, pp. 1-16.

[Kearey et al. 2002] P. Kearey, M. Brooks and I. Hill, *An Introduction to Geophysical Exploration*, Blackwell Science, Oxford.

[Keller 1965] G. H. Keller, "Deep-sea nuclear sediment density probe," *Deep-Sea Res.* **12**, pp. 373-376.

[Kennett 1982] J. P. Kennett, *Marine Geology*, Prentice-Hall, Englewood Cliffs, NJ.

[Kermabon et al. 1969] A. Kermabon, C. Gehin, and P. Blavier, "A deep-sea electrical resistivity probe for measuring porosity and density of unconsolidated sediments," *Geophysics* **34**, pp. 544-571.

[Ketcham and Carlson 2001] R. A. Ketcham and W. D. Carlson, "Acquisition, optimization and interpretation of X-ray computed tomographic imagery: applications to the geosciences," *Comp. Geosci.* **27**, pp. 381-400.

[Kibblewhite 1989] A. C. Kibblewhite, "Attenuation of sound in marine sediments: A review with emphasis on new low frequency data," *J. Acoust. Soc. Am.* **86**, pp. 716-738.

[Kim et al. 2003] J. Kim, Y. Furukawa, T. Daulton, D. Lavoie, and S. Newell, "Characterization of microbially Fe(III)-reduced nontronite: Environmental cell-transmission electron microscopy study," *Clays Clay Miner.* **51**, pp. 382-389.

[Kim et al. 2004] J. Kim, H. L. Dong, J. Seabaugh, S. W. Newell, and D. D. Eberl, "Role of microbes in the smectite-to-illite reaction," *Science* **303**, pp. 830-832.

[Kinsler et al. 1999] L. E. Kinsler, A. R. Frey, A. B. Coppens, and J. V. Sanders, *Fundamentals of Acoustics*, Fourth Edition, Wiley, New York.

[Klimentos and McCann 1988] T. Klimentos and C. McCann. "Why is the Biot slow compressional wave not observed in real rocks?," *Geophysics* **53**, pp. 1605-1609.

[Knotts et al. 1993] M. E. Knotts, T. R. Michael, and K. A. O'Donnell, "Comparisons of theory and experiment in light scattering from a randomly rough surface," *J. Opt. Soc. Am. A* **10**, pp. 928-941.

[Knudsen 1901] M. Knudsen, *Hydrographic Tables*, G. E. C. Gad, Copenhagen.

[Knudsen et al. 1902] M. Knudsen, C. Forsch, and S. P. L. Sorensen, "Berichte über die konstantenbestimmungen zür aufstellung der hydrographischen tabellen," *D. Kgl. Dansk. Vidensk. Selsk. Skrifter. Naturvidensk. Math, Afd. XII* **1** pp. 1-151.

[Kraft et al. 2002] B. J. Kraft, L. A. Mayer, P. Simpkin, P. Lavoie, E. Jabs, and E. Lynskey, "Calculation of in situ acoustic wave properties in marine sediments," in *Impact of Littoral Environmental Variability on Acoustic Predictions and Sonar Performance*, N. G. Pace and F. B. Jensen (Eds.), Kluwer Academic Publishers, Netherlands, pp. 123-130.

[Kringel et al. 2003] K. Kringel, P. A. Jumars, and D. V. Holliday, "A shallow scattering layer: High-resolution acoustic analysis of nocturnal vertical migration from the seabed," *Limnol. Oceanogr.* **48**, pp. 1223-1234.

[Krinsley and Takahashi 1962] D. Krinsley and T. Takahashi, "Surface textures of sand grains — an application of electron microscopy," *Science* **135**, pp. 923-925.

[Kroner 1967] E. Kroner, "Elastic moduli of perfectly disordered composite materials, *J. Mech. Phys. Solids* **15**, pp. 319-329.

[Krumbein 1934] W. C. Krumbein, "Size frequency distributions of sediments," *J. Sediment. Petrol.* **4**, pp. 65-77.

[Krumbein 1938] W. C. Krumbein, *Manual of Sedimentary Petrography*, Appleton-Century-Crofts, New York.

[Kuo 1964] E. Y. T. Kuo, "Wave scattering and transmission at irregular surfaces," *J. Acoust. Soc. Am.* **36**, pp. 2135-2142.

[Kuo 1992] E. Y. T. Kuo, "Acoustic wave scattering from two solid boundaries at the ocean bottom: Reflection loss," *IEEE J. Ocean. Eng.* **17**, pp. 159-170.

[Kuo 1995] E. Y. T. Kuo, "Joint perturbation scattering characterization of a littoral ocean bottom reverberation: Theory, scattering strength predictions, and data comparisons," *IEEE J. Ocean. Eng.* **20**, pp. 198-210.

[Kuperman and Schmidt 1986] W. A. Kuperman and H. Schmidt, "Rough surface elastic wave scattering in a horizontally stratified ocean," *J. Acoust. Soc. Am.* **79**, pp. 1767-1777.

[Kur'yanov 1963] B. F. Kur'yanov, "The scattering of sound at a rough surface with two types of irregularities," *Sov. Phys. Acoust.* **8**, pp. 252-257.

[Kuwahara 1939] S. Kuwahara, "The velocity of sound in sea water and calculation of velocity for the use in sonic sounding," *Hydrogr. Rev., Monaco* **16**, pp. 123-140.

[Lamb and Whitman 1979] T. W. Lamb and R. V. Whitman, *Soil Mechanics, SI Version*, John Wiley and Sons, New York.

[Landau and Lifshitz 1970] L. D. Landau and E. M. Lifshitz, *Theory of Elasticity*, Pergamon, New York.

[Lapin 1964] A. D. Lapin, "Sound scattering at a rough solid surface," *Sov. Phys. Acoust.* **10**, pp. 58-64.

[Lapin 1966] A. D. Lapin, "Scattering of sound by a solid layer with rough boundaries," *Sov. Phys. Acoust.* **12**, pp. 46-51.

[Laughton 1957] A. S. Laughton, "Sound propagation in compacted ocean sediments," *Geophysics* **22**, pp. 233-260.

[Lee and Chough 1987] H. J. Lee and S. K. Chough, "Bulk density, void ratio, and porosity determined from average grain density and waterconten: an evaluation of errors," *Mar. Geotechnol.* **7**, pp. 53-62.

[Lee and Von Herzen 1994] T.-C. Lee, and R. P. Von Herzen, "In situ determination of thermal properties of sediments using a friction-heated probe source," *J. Geophys. Res. Solid Earth* **99**, pp. 12121-12132.

[Le Roux 2002] J. P. Le Roux, "Shape entropy and settling velocity of natural grains," *J. Sediment. Res.* **72**, pp. 363-366.

[Leroy 1969] C. C. Leroy, "Development of simple equations for accurate and more realistic calculation of the sound speed in sea water," *J. Acoust. Soc. Am.* **46**, pp. 216-226.

[Leurer 1997] K. C. Leurer, "Attenuation in fine-grained marine sediments: Expansion of the Biot-Stoll model by the 'effective grain model' (EGM)," *Geophysics* **62**, pp. 1465-1479.

[Levitus and Isayev 1992] S. Levitus and G. Isayev, "Polynomial approximation to the international equation of state for seawater," *J. Atmos. Ocean. Tech.* **9**, pp. 705-708.

[Liebermann 1948] L. N. Liebermann, "Reflection of sound from coastal sea bottoms," *J. Acoust. Soc. Am.* **20**, pp. 305-309.

[Lim et al. 2001] R. Lim, I. C. Paustian, and J. L. Lopes, "Acoustic transmission across a roughened fluid-fluid interface," *J. Acoust. Soc. Am.* **109**, pp. 1367-1383.

[Lim and Sammelmann 2004] R. Lim and G. S. Sammelmann, "Subcritical detection of targets buried under a rippled interface: Extended perturbation theory," *Proceedings OCEANS 2004*, pp. 1953-1961.

[Lindwall 2006] D. Lindwall, "Imaging marine geophysical environments with vector acoustics," *J. Acoust. Soc. Am. Express Lett.* **120**, pp. EL43-EL48.

[Lockwood and Willette 1973] J. C. Lockwood and J. G. Willette, "High-speed method for computing the exact solution for the pressure variations in the nearfield of a baffled piston," *J. Acoust. Soc. Am.* **53**, pp. 735-741.

[Lofi and Weber 2001] J. Lofi and O. Weber, "SCOPIX — digital processing of X-ray images for the enhancement of sedimentary structures in undisturbed core slabs," *Geo-Mar. Lett.* **20**, pp. 182-186.

[Lopes 1996] J. L. Lopes, "Observations of anomalous acoustic penetration into sediments at shallow grazing angles," *J. Acoust. Soc. Am.* **99**, p. 2473(A).

[Lopes et al. 2002] J. L. Lopes, C. L. Nesbitt, R. Lim, D. Tang, K. L. Williams, and E. I. Thorsos, "Shallow grazing angle detection of targets buried under a rippled sand interface," *Proceedings OCEANS 2002*, pp. 461-467.

[Lopes et al. 2003] J. L. Lopes, C. L. Nesbitt, R. Lim, K. L. Williams, E. I. Thorsos, and D. Tang, "Subcritical detection of targets buried under a rippled interface: calibrated levels and effects of large roughness," *Proceedings OCEANS 2003*, pp. 485-493.

[Lopes et al. 2005] J. L. Lopes, C. L. Nesbitt, R. Lim, K. L. Williams, E. I. Thorsos, and D. Tang, "Subcritical detection of spheres and elongated targets buried under a rippled interface: calibrated levels and effects of large roughness," in *Boundary Influences in High-Frequency, Shallow-Water Acoustics*, N. G. Pace and P. Blondel (Eds.), University of Bath, UK, pp. 145-158.

[Louis and McConchie 1994] D. W. Louis and D. McConchie, *Practical Sedimentology*, Chapman and Hall, New York.

[Lovell 1985a] M. A. Lovell, "Thermal conductivity and permeability assessment by electrical resistivity measurements in marine sediments," *Mar. Geotechnol.* **6**, pp. 205-240.

[Lovell 1985b] M. A. Lovell, "Thermal conductivities of marine sediments," *Q. J. Eng. Geol., London* **18**, pp. 437-441.

[Lurton 2002] X. Lurton, *An Introduction to Underwater Acoustics, Primciples and Applications*, Springer Praxis, Chichester.

[Lyons et al. 1994] A. P. Lyons, A. L. Anderson, and F. S. Dwan, "Acoustic scattering from the seafloor: Modeling and data comparison," *J. Acoust. Soc. Am.* **95**, pp. 2441-2451.

[Lyons et al. 1996] A. P. Lyons, M. E. Duncan, A. L. Anderson, and J. A. Hawkins, "Predictions of the acoustic scattering response of free methane bubbles in muddy sediments," *J. Acoust. Soc. Am.* **99**, pp. 163-172.

[Lyons et al. 1997] A. P. Lyons, D.A. Abraham, T. Akal and P. Guerrini, "Statistical evaluation of 80 kHz shallow-water seafloor reverberation," in *High Frequency Acoustics in Shallow Water*, N.G. Pace, E. Pouliquen, O. Bergem, and A.P. Lyons (Eds.), NATO SACLANT Undersea Research Centre, La Spezia, Italy, pp. 323-329.

[Lyons and Orsi 1998] A. P. Lyons and T. H. Orsi, "The effect of a layer of varying density on high-frequency reflection, forward loss, and backscatter," *IEEE J. Ocean. Eng.* **23**, pp. 411-422.

[Lyons and Abraham 1999] A. P. Lyons and D. A. Abraham, "Statistical characterization of high-frequency shallow-water seafloor backscatter," *J. Acoust. Soc. Am.* **106**, pp. 1307-1315.

[Lyons et al. 2002a] A. P. Lyons, D. A. Abraham and E. Pouliquen, "Predicting scattered envelope statistics of patchy seafloors," in *Impact of Littoral Environmental Variability on Acoustic Predictions and Sonar Performance*, N.G. Pace and F.B. Jensen (Eds.), Kluwer Academic Publishers, Netherlands, pp. 211-218.

[Lyons et al. 2002b] A. P. Lyons, W. L. J. Fox, T. Hasiotis, and E. Pouliquen, "Characterization of the two-dimensional roughness of wave-rippled sea floors using digital photogrammetry," *IEEE J. Ocen. Eng.* **27**, pp. 515-524.

[Lyons and Pouliquen 2004] A. P. Lyons and E. Pouliquen, "Advances in high-resolution seafloor characterization in support of high-frequency underwater acoustics studies: techniques and examples," *Meas. Sci. Technol.* **15**, pp. R59-R72.

[Lyons 2005] A. P. Lyons, "The potential impact of shell fragment distributions on high-frequency seafloor backscatter," *IEEE J. Ocean. Eng.* **30**, pp. 843-851.

[Lysanov 1973] Yu. P. Lysanov, "A property of the scattering coefficient in the Fraunhofer zone," *Sov. Phys. Acoust.* **18**, pp. 505-506.

[Mackenzie 1960] K. V. Mackenzie, "Reflection of sound from coastal bottoms," *J. Acoust. Soc. Am.* **32**, pp. 221-231.

[Maguer et al. 2000a] A. Maguer, W. L. J. Fox, H. Schmidt, E. Pouliquen, and E. Bovio, "Mechanisms for subcritical penetration into a sandy bottom: Experimental and modeling results," *J. Acoust. Soc. Am.* **107**, pp. 1215-1225.

[Maguer et al. 2000b] A. Maguer, E. Bovio, W. L. J. Fox, and H. Schmidt, "*In situ* esimation of sediment sound speed and critical angle," *J. Acoust. Soc. Am.* **108**, pp. 987-996.

[Maradudin et al. 1994] A. A. Maradudin, M. Nieto-Vesperinas, and E. I. Thorsos, "Enhanced backscattering of light from randomly rough surfaces and related phenomena. I. One-dimensional surfaces and angular correlation functions of scattered fields," *Comments Mod. Phys. Part B* **17**, pp. 13-17.

[Martens and Klump 1980] C. A. Martens and J. V. Klump, "Biochemical cycling in an organic-rich coastal marine basin I. Methane sediment-water exchange processes," *Geochim. Cosmochim. Acta* **44**, pp. 471-490.

[Martens et al. 1998] C. S. Martens, D. B. Albert, and M. J. Alperin, "Biogeochemical processes controlling methane in gassy coastal sediments — Part 1. A model coupling organic matter flux to gas production, oxidation and transport," *Cont. Shelf Res.* **18**, pp. 1741-1770.

[Matsumoto et al. 1993] H. Matsumoto, R. P. Dziak, and C. G. Fox, "Estimation of seafloor microtopographic roughness through modeling of acoustic backscatter data recorded by multibeam sonar systems," *J. Acoust. Soc. Am.* **94**, pp. 2776-2787.

[Matthäus 1972] W. Matthäus, "Die Viskosität des Meerswasser," (The viscosity of seawater) *Beitr. Meersekd.* **29**, pp. 93-107.

[Matthews 1927] D. J. Matthews, British Admirality Report HD 282, Hydrographic Department, First Edition.

[Matthews 1932] D. J. Matthews, *Tables of the determination of density of seawater under normal pressure, sigma-t*, Conseil Perm. Int. Expl. Mer., Copenhagen.

[Matthews 1939] D. J. Matthews, "Tables of the Velocity of Sound in Pure Water and Sea Water," British Admirality Report HD 282, Hydrographic Department.

[Matthews 1991a] M. D. Matthews, "The effects of pretreatment on size analysis," in *Principles, Methods, and Applications of Particle Size Analysis*, J. P. M. Syvitski (Ed.), Cambridge University Press, New York, pp. 34-42.

[Matthews 1991b] M. D. Matthews, "The effect of grain shape and density on size measurement," in *Principles, Methods, and Applications of Particle Size Analysis*, J. P. M. Syvitski (Ed.), Cambridge University Press, New York, pp. 22-33.

[Mavko et al. 1998] G. Mavko, T. Mukerji, and J. Dvorkin, *The Rock Physics Handbook: Tools for Seismic Analysis in Porous Media*, Cambridge University Press, New York.

[McCammon 1993] D. F. McCammon, "Low grazing angle bottom scattering strength: Survey of unclassified measurements and models and recommendations for LFA use," *U. S. Navy J. Underwater Acoust.* **43**, pp. 33-47.

[McCann and McCann 1969] C. McCann and D. M. McCann, "Attenuation of compressional waves in marine sediments," *Geophysics* **34**, pp. 882-892.

[McCann 1972] D. M. McCann, "Measurement of acoustic properties of marine sediments," *Acoustica* **26**, pp. 55-66.

[McCave and Syvitski 1991] I. N. McCave and J. P. M. Syvitski, "Principles and methods of geological particle size analysis," in *Principles, Methods, and Applications of Particle Size Analysis*, J. P. M. Syvitski (Ed.), Cambridge University Press, New York, pp. 3-21.

[McConnell and DeProspo 1994] S. O. McConnell and D. F. DeProspo, "Broadband bottom forward loss and backscattering off San Diego," *IEEE J. Ocean. Eng.* **19**, pp. 360-367.

[McDaniel and Gorman 1982] S. T. McDaniel and A. D. Gorman, "Acoustic and radar sea surface backscatter," *J. Geophys. Res. Oceans Atmos.* **87**, pp. 4127-4136.

[McDaniel and Gorman 1983] S. T. McDaniel and A. D. Gorman, "An examination of the composite-roughness scattering model," *J. Acoust. Soc. Am.* **73**, pp. 1476-1486.

[McDaniel 1986] S. T. McDaniel, "Diffractive corrections to the high-frequency Kirchhoff approximation," *J. Acoust. Soc. Am.* **79**, pp. 952-957.

[McDaniel 1990] S. T. McDaniel, "Seafloor reverberation fluctuations," *J. Acoust. Soc. Am.* **88**, pp. 1530-1535.

[McDaniel 1992] S. T. McDaniel, "Effect of surficial sediment layering on high-frequency seafloor reverberation," *J. Acoust. Soc. Am.* **91**, pp. 1353-1356.

[McDaniel 1996] S. T. McDaniel, "Fluctuations of the acoustic field Bragg-backscattered from rough surfaces," *Waves Random Media* **6**, pp. 229-249.

[McDermott 1991] I. R. McDermott, "A laboratory method to investigate shear waves in a soft soil consolidating under self weight," in *Shear Waves in Marine Sediments*, J. M. Hovem, M. D. Richardson, and R. D. Stoll (Eds.), Kluwer, Dordrecht, pp. 103-110.

[McDonald and Spindel 1971] J. F. McDonald and R. C. Spindel, "Implications of Fresnel corrections in a non-Gaussian surface scatter channel," *J. Acoust. Soc. Am.* **50**, pp. 746-757.

[McDougall et al. 2003] T. J. McDougall, D. R. Jackett, D. G. Wright and R. Feistel, "Accurate and computationally efficient algorithms for potential temperature and density of seawater," *J. Atmos. Ocean. Tech.* **20**, pp. 730-741.

[McKinney and Anderson 1964] C. M. McKinney and C. D. Anderson, "Measurements of backscattering of sound from the ocean bottom," *J. Acoust. Soc. Am.* **36**, pp. 158-163.

[McLeroy 1972] E. G. McLeroy, "Measurement and correlation of acoustic reflection and sediment properties off Panama City, Florida," Naval Coastal Systems Laboratory, AD-759 297.

[McManus 1982] D. A. McManus, "Phi and sediment size analysis: discussion," *J. Sediment. Petrol.* **52**, pp. 1011-1026.

[Medwin 1975] H. Medwin, "Speed of sound in seawater: A simple equation for realistic parameters," *J. Acoust. Soc. Am.* **58**, pp. 1318-1319.

[Medwin 1977] H. Medwin, "Acoustical determinations of bubble size spectra," *J. Acoust. Soc. Am.* **62**, pp. 1041-1044.

[Medwin and Clay 1998] H. Medwin and C. S. Clay, *Fundamentals of Acoustical Oceanography*, Academic Press, Boston.

[Mees et al. 2003] F. Mees, R. Swennen, M. Van Geet, and P. Jacobs, "Applications of X-ray Computed Tomography in the Geosciences," *Geological Society Special Publication No. 215*, The Geological Society, London.

[Meinen and Watts 1997] C. S. Meinen and D.R. Watts, "Further evidence that the sound-speed algorithm of Del Grosso is more accurate than that of Chen and Millero," *J. Acoust. Soc. Am.* **102**, pp. 2058-2062.

[Mellema 1999] G. R. Mellema, "Subcritical acoustic scattering across a rough fluid-solid interface," Doctoral thesis, University of Washington, Seattle.

[Mellor 1991] G. L. Mellor, "An equation of state for numerical models of oceans and estuaries," *J. Atmos. Ocean. Tech.* **8**, pp. 609-611.

[Meysman et al. 2003] F. J. R. Meysman, B. P. Boudreau, and J. J. Middelburg, "Relations between local, nonlocal, discrete and continuous models of bioturbation," *J. Mar. Res.* **61**, pp. 391-410.

[Michalopoulou et al. 1994] Z. H. Michalopoulou, D. Alexandrou, and C. de Moustier, "Application of a maximum-likelihood processor to acoustic backscatter for the estimation of seafloor roughness parameters," *J. Acoust. Soc. Am.* **95**, pp. 2467-2477.

[Michalopoulou and Alexandrou 1996] Z. H. Michalopoulou and D. Alexandrou, "Baysean modeling of acoustic signals for seafloor identification," *J. Acoust. Soc. Am.* **99**, pp. 223-233.

[Middleton 1984] G. Middleton, *Mechanics of Sediment Movement*, SEPM, Tulsa OK.

[Migeon et al. 1999] S. Migeon, O. Weber, J.-C. Faugeres, and J. Saint-Paul, "SCOPIX: A new X-ray imaging system for core analysis," *Geo-Mar. Lett.* **18**, pp. 251-255.

[Miller 1996] R. W. Miller, *Flow Measurement Engineering Handbook*, McGraw-Hill, New York.

[Millero and Kubinski 1975] F. J. Millero and T. Kubinski, "Speed of sound in seawater as a function of temperature and salinity at 1 atmosphere," *J. Acoust. Soc. Am.* **57**, pp. 312-319.

[Millero et al. 1977] F. J. Millero, P. Chetirkin, and F. Culkin," Relative conductivity and density of standard seawaters," *Deep-Sea Res.* **24**, pp. 315-321.

[Millero et al. 1980] F. J. Millero, C.-T. Chen, K. Schleicher, and A. Bradshaw, "A new high-pressure equation of state for seawater," *Deep-Sea Res.* **27**, pp. 255-264.

[Millero and Poisson 1981] F. J. Millero and A. Poisson, "International one-atmosphere equation of state of seawater," *Deep-Sea Res.* **28**, pp. 625-629.

[Millero and Li 1994] F. J. Millero and X. Li, "Comments on 'On equations for sound speed in seawater'," *J. Acoust. Soc. Am.* **95**, pp. 2757-2759.

[Millero 1996] F. J. Millero, *Chemical Oceanography*, Second Edition, CRC Press, Boca Raton, FL.

[Millero 2000] F. J. Millero, "Effect of changes in the composition of seawater on the density-salinity relationship," *Deep-Sea Res.* **47**, pp. 1583-1590.

[Milton et al. 1997] G. W. Milton, D. J. Eyre, and J. V. Mantese, "Finite frequency range Kramers Kronig relations: Bounds on the dispersion," *Phys. Rev. Lett.* **79**(5), pp. 3062-3065.

[Mitchell 1993] J. K. Mitchell, *Fundamentals of Soil Behavior*, Second Edition, John Wiley and Sons, New York.

[Moe and Jackson 1994] J. E. Moe and D. R. Jackson, "First order perturbation solution for rough surface scattering cross section including the effects of gradients," *J. Acoust. Soc. Am.* **96**, pp. 1748-1754.

[Moe 1996] J. E. Moe, "Near and far field acoustic scattering through and from two dimensional fluid-fluid rough interfaces," Ph.D. thesis, University of Washington, available as Applied Physics Laboratory report APL-UW TR 9606.

[Moe and Jackson 1998] J. E. Moe and D. R. Jackson, "Near field scattering through and from a two-dimensional fluid-fluid rough interface," *J. Acoust. Soc. Am.* **103**, pp. 275-287.

[Molis and Chotiros 1992] J. C. Molis and N. P. Chotiros, "A measurement of the grain bulk modulus of sands," *J. Acoust. Soc. Am.* **91** (A), p. 2463.

[Moore and Jaffe 2002] K. D. Moore and J. S. Jaffe, "Time-evolution of high-resolution topographic measurements of the sea floor using a 3-D laser line scan mapping system," *IEEE J. Ocean. Eng.* **27**, pp. 525-545.

[Morse and Ingard 1968] P. M. Morse and K. U. Ingard, *Theoretical Acoustics*, McGraw-Hill, New York.

[Mourad and Jackson 1989] P. D. Mourad and D. R. Jackson, "High frequency sonar equation models for bottom backscatter and forward loss," *Proceedings OCEANS 1989* **4**, pp. 1168-1175.

[Mourad and Jackson 1993] P. D. Mourad and D. R. Jackson, "A model/data comparison for low-frequency bottom backscatter," *J. Acoust. Soc. Am.* **94**, pp. 344-358.

[Muir et al. 1979] T. G. Muir, C. W. Horton, and L. A. Thompson, "Penetration of highly directional acoustic beams into Sediments," *J. Sound. Vib.* **64**, pp. 539-551.

[Muir et al. 1991] T. G. Muir, T. Akal, M. D. Richardson, R. D. Stoll, A. Caiti, and J. M. Hovem, "Comparison of techniques for shear wave velocity and attenuation measurements," in *Shear Waves in Marine Sediments*, J. M. Hovem, M. D. Richardson and R. D. Stoll (Eds.), Kluwer Academic Publishers, Dordrecht, pp. 283-294.

[Müller 1999] T. J. Müller, "Determination of salinity," in *Methods of Seawater Analysis*, Third Edition, K. Grasshoff, K. Kremling, and M. Ehrhardt (Eds.), Wiley-VCH, New York, pp. 41-61.

[Murad and Cushman 1997] M. A. Murad and J. H. Cushman, "A multiscale theory of swelling porous media. 2. Dual porosity models for consolidation of clays incorporating physicochemical effects," *Transp. Porous Media* **28**, pp. 69-108.

[Murray et al. 2002] J. M. H. Murray, A. Meadows, and P. S. Meadows, "Biogeomorphological implications of microscale interactions between sediment geotechnics and marine benthos: a review," *Geomorphology* **47**, pp. 15-30.

[Muzi et al. 2004] L. Muzi, A. P. Lyons, and E. Pouliquen, "Use of X-ray computed tomography for the estimation of parameters relevant to the modeling of acoustic scattering from the seafloor," *Nucl. Instrum. Methods Phys. Res. Sect. B* **213**, pp. 491-497.

[Myers et al. 1969] J. J. Myers, C. H. Holm, and R. F. McAllister, *Handbook of Ocean and Underwater Engineering*, McGraw-Hill, New York.

[Nafe and Drake 1957] J. E. Nafe and C.L. Drake, "Variation with depth in shallow and deep water marine sediments of porosity, density and velocities of compressional and shear waves," *Geophysics* **22**, pp. 523-552.

[Nafe and Drake 1963] J. E. Nafe and C. L. Drake, "Physical properties of marine sediments," in *The Sea, Vol. 3*, M. N. Hill (Ed.), Interscience, New York, pp. 794-815.

[Nesbitt and Lopes 2004] C. L. Nesbitt and J. L. Lopes, "Subcritical detection of an elongated target buried under a rippled interface," *Proceedings OCEANS 2004*, pp. 1945-1952.

[Nielsen 1992] P. Nielsen, *Coastal Bottom Boundary Layers and Sediment Transport*, World Scientific, Hackensack, NJ.

[Nilsson et al. 1991] P. Nilsson, B. Jonsson, I. L. Swanberg and K. Sundback, "Response of a marine shallow-water sediment system to an increased load of inorganic nutrients," *Mar. Ecol. Prog. Ser.* **71**, pp. 275-290.

[Nittrouer and Kravitz 1996] C. A. Nittrouer and J. A. Kravitz, "STRATAFORM: A program to study the formations and interpretation of sedimentary strata on continental margins," *Oceanography* **9**, pp. 146-152.

[Nittrouer 1999] C. A. Nittrouer, "STRATAFORM: overview of its design and synthesis of its results," *Mar. Geol.* **154**, pp. 3-12.

[Nolle et al. 1963] A. W. Nolle, J. F. Mifsud, W. A. Hoyer, W. R. Runyan, and M. B. Ward, "Acoustical properties of water-filled sands," *J. Acoust. Soc. Am.* **35**, pp. 1394-1408.

[Novarini and Caruthers 1994] J. C. Novarini and J. W. Caruthers, "The partition wavenumber in acoustic backscattering from a two-scale rough surface described by a power-law spectrum," *IEEE J. Ocean. Eng.* **19**, pp. 200-207.

[O'Donnell et al. 1981] M. O'Donnell, E. T. Jaynes, and J. G. Miller, "Kramers-Kronig relationship between ultrasonic attenuation and phase velocity," *J. Acoust. Soc. Am.* **69**, pp. 696-701.

[Ogilvy 1991] J. A. Ogilvy, *Theory of Wave Scattering from Random Rough Surfaces*, Institute of Physics Publishing, Bristol, UK.

[Ogushwitz 1985] P. R. Ogushwitz, "Applicability of the Biot theory. II. Suspensions," *J. Acoust. Soc. Am.* **77** pp. 441-452.

[Oh and Lindquist 1999] W. Oh and W. B. Lindquist, "Image thresholding by indicator kriging," *IEEE Trans. Pattern Anal. Mach. Intell.* **21**, pp. 590-602.

[Orford and Whalley 1983] J. D. Orford and W. B. Whalley, "The use of the fractal dimension to quantify the morphology of irregular-shaped particles," *Sedimentology* **30**, pp. 655-668.

[Orsi et al. 1994] T. H. Orsi, C. M. Edwards, and A. L. Anderson, "X-ray computed tomography: a non-destructive method for quantitative analysis of sediment cores," *J. Sediment. Res. Sediment. Petrol. Processes* **A64**, pp. 690-693.

[Orsi et al. 1996] T. H. Orsi, A. L. Anderson, and A. P. Lyons, "X-ray tomographic analysis of sediment macrostructure in Eckernförde Bay, western Baltic Sea," *Geo-Mar. Lett.* **16**, pp. 232-239.

[Orsi et al. 1997] T. H. Orsi, M. E. Duncan, A. P. Lyons, K. B. Briggs, M. D. Richardson, and A. L. Anderson, "High-resolution characterization of seafloor sediments for modeling acoustic backscatter," in *High Frequency Acoustics in Shallow Water*, N. G. Pace, E. Pouliquen, O. Bergem, and A. P. Lyons (Eds.), NATO SACLANT Undersea Research Centre, La Spezia, Italy, pp. 409-415.

[Orsi and Anderson 1999] T. H. Orsi and A. L. Anderson, "Bulk density calibration for X-ray tomographic analysis of marine sediments," *Geo-Mar. Lett.* **19**, pp. 270-274.

[Osler and Lyons 2004] J. C. Osler and A. P. Lyons, "Using buried directional receivers in high-frequency seafloor studies," in *High Frequency Ocean Acoustics*, M. B. Porter, M. Siderius, and W. A. Kuperman (Eds.), AIP Conference Proceedings, Melville, NY, pp. 32-39.

[Pace et al. 1985] N. G. Pace, Z. K. S. Al-Hamdani, and P. D. Thorne, "The range dependence of normal incidence acoustic backscatter from a rough surface," *J. Acoust. Soc. Am.* **77**, pp. 101-112.

[Pace and Langhorne 1993] *Acoustic Classification and Mapping of the Seabed*, N. G. Pace and D. N. Langhorne (Eds.), Institute of Acoustics, Bath.

[Pace 1994] N. G. Pace, "Low frequency acoustic backscatter from the sea bed," *Proc. Inst. Acoust.* **16**, pp. 181-188.

[Palmer 1996] D. R. Palmer, "Rayleigh scattering from non spherical particles," *J. Acoust. Soc. Am.* **99**, pp. 1901-1912.

[Panda et al. 1994] S. Panda, L. R. LeBlanc, and S. G. Schock, "Sediment classification based on impedance and attenuation estimation," *J. Acoust. Soc. Am.* **96**, pp. 3022-3035.

[Papoulis 1962] A. Papoulis, *The Fourier Integral and its Applications*, McGraw-Hill, New York.

[Parkinson 2001] R. Parkinson, *High Resolution Site Surveys*, Spon Press, New York.

[Parrott et al. 1980] D. R. Parrott, D. J. Dodds, L. H. King, and P. G. Simpkin, "Measurement and evaluation of the acoustic reflectivity of the seafloor," *Can. J. Earth Sci.* **17**, pp. 722-737.

[Piccolo et al. 1993] M. C. Piccolo, G. M. E. Perillo, and G. R. Daborn, "Soil temperature variations on a tidal flat in Minas Basin, Bay of Fundy, Canada," *Estuarine Coastal Shelf Sci.* **36**, pp. 345-357.

[Pierce 1989] A. D. Pierce, *Acoustics: An Introduction to Its Physical Principles and Applications*, Acoustical Society of America, New York.

[Pierce 1991] J. W. Pierce, "Microstructure of suspension: from stream to shelf," in *Microstructure of Fine-Grained Sediments: From Mud to Shale*, R. H. Bennett, W. R. Bryant, and M. H. Hulbert (Eds.), Frontiers in Sedimentary Geology, Springer-Verlag, New York, pp. 139-147.

[Pierce et al. 2005] A. D. Pierce, W. M. Carey, and M. Zampoli, "Low-frequency attenuation of sound in marine sediments," *Proceedings OCEANS 2005*, pp. 1270-1275.

[Piper et al. 2002] J. E. Piper, K. W. Commander, E. I. Thorsos, and K. L. Williams, "Detection of buried targets using a synthetic aperture sonar," *IEEE J. Ocean. Eng.* **27**, pp. 495-504.

[Pirard 1994] E. Pirard, "Shape processing and analysis using the calypter," *J. Microsc. Oxford* **175**, pp. 214-221.

[Plemons et al. 1972] T. D. Plemons, J. A. Shooter, and D. Middleton, "Underwater acoustic scattering from lake surfaces. II. Covariance functions and related statistics," *J. Acoust. Soc. Am.* **52**, pp. 1503-1515.

[Plona 1980] T. J. Plona, "Observation of a second bulk compressional wave in porous media at ultrasonic frequencies," *Appl. Phys. Lett.* **36**, pp. 259-261.

[Pouliquen 1992] E. Pouliquen, "Identification des fonds marins superficiels a l'aide de signaux d'echo-sounders," Doctoral thesis, University of Paris VII.

[Pouliquen et al. 1999] E. Pouliquen, O. Bergem, and N. G. Pace, "Time-evolution modeling of seafloor scatter. I. Concept," *J. Acoust. Soc. Am.* **105**, pp. 3136-3141.

[Pouliquen et al. 2000a] E. Pouliquen, A. P. Lyons, and N. G. Pace, "Penetration of acoustic waves into rippled sandy seafloors," *J. Acoust. Soc. Am.* **108**, pp. 2071-2081.

[Pouliquen et al. 2000b] E. Pouliquen, A. P. Lyons, N. G. Pace, T. H. Orsi, E. Michelozzi, L. Muzi and P. A. G. Thomson, "Backscattering from unconsolidated sediments above 100 kHz," in *Proceedings of the Fifth European Conference on Underwater Acoustics, Lyon, France*, M. E. Zakharia, P. Chevret, and P. Dubail (Eds.), Lyon, France, pp. 1283-1288.

[Pouliquen and Lyons 2002] E. Pouliquen and A. P. Lyons, "Backcattering from bioturbated sediments at very high frequency," *IEEE J. Ocean. Eng.* **27**, pp. 388-402.

[Pouliquen et al. 2003] E. Pouliquen, G. Canepa, L. Pautet, and A. P. Lyons, "Spatial and Temporal Variability of Seafloor Roughness and its Impact on Acoustic Scattering," SACLANTCEN Memorandum SM-422.

[Pouliquen et al. 2004] E. Pouliquen, G. Canepa, L. Pautet, and A. P. Lyons, "Temporal variability of seafloor roughness and its impact on acoustic scattering," *Proceedings of the 7th European Conference on Underwater Acoustics, EUCA 2004*, Delft, Netherlands, pp. 583-588.

[Preiss 1968a] K. Preiss, "Non-destructive laboratory measurements of marine sediment density in a core barrel using gamma radiation," *Deep-Sea Res.* **15**, pp. 401-407.

[Preiss 1968b] K. Preiss, "In situ measurement of marine sediment density by gamma radiation," *Deep-Sea Res.* **15**, pp. 637-641.

[Premus and Alexandrou 1994] V. Premus and D. Alexandrou, "Baysean estimation of Goff-Jordan seafloor microroughness statistics via simulated annealing," *J. Acoust. Soc. Am.* **96**, pp. 2887-2896.

[Rajan and Frisk 1992] S. D. Rajan and G. V. Frisk, "Seasonal variations of the sediment compressional wave-speed profile in the Gulf of Mexico," *J. Acoust. Soc. Am.* **91**, pp. 127-135.

[Ransom et al. 1998a] B. Ransom, K. F. Shea, P. J. Burkett, R. H. Bennett, and R. Bearwald, "Comparison of pelagic and nepheloid layer marine snow: implications for carbon cycling," *Mar. Geol.* **150**, pp. 39-50.

[Ransom et al. 1998b] B. Ransom, D. Kim, M. Kastner, and S. Wainwright, "Organic matter preservation on continental slope: Importance of mineralogy and surface area," *Geochim. Cosmochim. Acta* **62**, pp. 1329-1345.

[Rapids et al. 1998] B. Rapids, T. Nye, and T. Yamamoto, "Pilot experiment for the acquisition of marine sediment properties via small scale tomography system," *J. Acoust. Soc. Am.* **103**, pp. 212-224.

[Ratcliffe 1960] E. H. Ratcliffe, "The thermal conductivities of ocean sediments," *J. Geophys. Res.* **65**, pp. 1535-1541.

[Rauch 1980] D. Rauch, "Experimental and theoretical studies of seismic interface waves in coastal waters," in *Bottom Interacting Ocean Acoustics,* W. A. Kuperman and F. Jensen (Eds.), Plenum, New York, pp. 307-327.

[Rauch 1986] D. Rauch, "On the role of bottom interface waves in ocean seismoacoustics: A review," in *Ocean Seismo-Acoustics*, T. Akal and J. M. Berkson (Eds.), Plenum Press, New York, pp. 623-641.

[Reed et al. 2002] A. H. Reed, K. B. Briggs, and D. L. Lavoie, "Porometric properties of siliclastic marine sand: a comparison of traditional laboratory measurements with image analysis and effective medium modeling," *IEEE J. Ocean. Engr.* **27**, pp. 581-592.

[Reed et al. 2005] A. H. Reed, K. E. Thompson, C. S. Willson, K. B. Briggs, M. D. Richardson, and B. T. Hefner, "Quantification of sediment properties from pore structure and grain contacts: Microcomputed tomography analysis of SAX04 sands," in *Boundary Influences in High-Frequency, Shallow-Water Acoustics*, N. G. Pace and P. Blondel (Eds.), University of Bath, UK, pp. 83-90.

[Reineck and Singh 1973] H. E. Reineck and I. B. Singh, *Depositional Sedimentary Environments*, Springer-Verlag, New York.

[Rezak and Lavoie 1990] R. Rezak and D. L. Lavoie, "Consolidation-related fabric changes of periplatform sediments," *Geo-Mar. Lett.* **10**, pp. 101-109.

[Rezak and Lavoie 1993] *Carbonate Microfabric, Frontiers in Sedimentary Geology*, R. Rezak and D. L. Lavoie (Eds.), Springer-Verlag, New York.

[Rhoads and Cande 1971] D. C. Rhoads and S. Cande, "Sediment profile camera for in-situ study of organism-sediment relations," *Limnol. Oceanogr.* **16**, pp. 110-114.

[Rhoads and Young 1971] D. C. Rhoads and D. K. Young, "Animal-sediment relations in Cape Cod Bay, Massachustts. 2. Reworking by Molpadia-Oolitica (Holothuroidea)," *Mar. Biol.* **11**, pp. 255-261.

[Rhoads 1974] D. C. Rhoads, "Organism-sediment relationships on the muddy seafloor," *Oceanography and Mar. Biol. Annu. Rev.* **12**, pp. 263-300.

[Rhoads and Boyer 1982] D. C. Rhoads and L. F. Boyer, "The effects of marine benthos on physical properties of sediments: a successional prospective," in *Animal-Sediment Relations: The Biogenic Alteration of Sediments, Vol. 2*, P. L. McCall and M. J. S. Tevesz (Eds.), Plenum Press, New York, pp. 3-52.

[Rhoads and Germano 1982] D. C. Rhoads and J. D. Germano, "Characterization of organism-sediment relations using sediment profiles imaging — an efficient method of remote ecological monitoring of the seafloor (REMOTS system)," *Mar. Ecol. Progr. Ser.* **8**, pp. 115-128.

[Richards et al. 1974] A. F. Richards, T. J. Hirst, and J. M. Parks, "Bulk density-water content relationships in marine silts and clays," *J. Sediment. Petrol.* **44**, pp. 1004-1009.

[Richards et al. 1996] S. D. Richards, A. D. Heathershaw, and P. D. Thorne, "The effect of suspended particulate matter on sound attenuation in seawater," *J. Acoust. Soc. Am.* **100**, pp. 1447-1450.

[Richards 1998] S. D. Richards, "The effect of temperature, pressure, and salinity on sound attenuation in turbid seawater," *J. Acoust. Soc. Am.* **103**, pp. 205-211.

[Richards et al. 2003] S. D. Richards, T. G. Leighton, and N. R. Brown, "Visco-inertial absorption in dilute suspensions of irregular particles," *Proc. R. Soc. London Ser. A* **459**, pp. 2153-2167.

[Richardson and Young 1980] M. D. Richardson and D. K. Young, "Geoacoustic models and bioturbation," *Mar. Geol.* **38**, pp. 205-218.

[Richardson 1983] M. D. Richardson, "The effects of bioturbation on sediment elastic properties," *Bull. Soc. Geol. Fr.* **25**, pp.505-513.

[Richardson et al. 1983] M. D. Richardson, D. K. Young and K. B. Briggs, "Effects of hydrodynamic and biological processes on sediment geoacoustic properties in Long Island Sound, USA," *Mar. Geol.* **52**, pp. 201-226.

[Richardson 1986] M. D. Richardson, "Spatial variability of surficial shallow water sediment geoacoustic properties," in *Ocean Seismo-Acoustics*, T. Akal and J. M. Berkson (Eds.), Plenum Press, London, pp. 527-536.

[Richardson et al. 1987] M. D. Richardson, P. V. Curzi, E. Muzi, B. Miaschi, and A. Barbagelata, "Measurement of shear wave velocity in marine sediments," in *Acoustics and Ocean Bottom. II FASE Specialized Conference*, A. Lara-Saenz, C. Ranz-Guerra, and C. Carbo-Fite (Eds.), CISC, Madrid, pp. 75-84.

[Richardson et al. 1991a] M. D. Richardson, E. Muzi, L. Troiano, and B. Miaschi, "Sediment shear waves: A comparison of in situ and laboratory measurements," in *Microstructure of Fine Grained Sediments*, R. H. Bennett, W. R. Bryant, and M. H. Hurlbert (Eds.), Springer-Verlag, New York, pp. 403-415.

[Richardson et al. 1991b] M. D. Richardson, E. Muzi, B. Miaschi, and F. Turgutcan, "Shear wave velocity gradients in near-surface marine sediment," in *Shear Waves in Marine Sediments*, J. M. Hovem, M. D. Richardson, and R. D. Stoll (Eds.), Kluwer, Dordrecht, pp. 295-304.

[Richardson et al. 1992] M. D. Richardson, P. V. Curzi, E. Muzi, F. Turgutcan, and T. Akal, "A Generic Geoacoustic Model for the Adriatic Sea," SACLANTCEN Report SR-199.

[Richardson and Briggs 1993] M. D. Richardson and K. B. Briggs, "On the use of acoustic impedance values to determine sediment properties," in *Acoustic Classification and Mapping of the Seabed*, N. G. Pace and D. N. Langhorne (Eds.), Institute of Acoustics, Bath, pp.15-25.

[Richardson and Briggs 1996] M. D. Richardson and K. B. Briggs, "In-situ and laboratory geoacoustic measurements in soft mud and hard-packed sand sediments: Implications for high-frequency acoustic propagation and scattering," *Geo-Mar. Lett.* **16**, pp. 196-203.

[Richardson and Bryant 1996] M. D. Richardson and W. R. Bryant, "Benthic boundary layer processes in coastal environments: An introduction," *Geo-Mar. Lett.* **16**, pp. 133-139.

[Richardson 1997a] M. D. Richardson, "In-situ, shallow-water sediments geoacoustic properties," in *Shallow-Water Acoustics*, R. Zang and J. Zhou (Eds.), China Ocean Press, Beijing, pp. 163-170.

[Richardson 1997b] M. D. Richardson, "Attenuation of shear waves in near-surface sediments," in *High Frequency Acoustics in Shallow Water*, N. G. Pace, E. Pouliquen, O. Bergem, and A. P. Lyons (Eds.), NATO SACLANT Undersea Res. Ctr., La Spezia, Italy, pp. 451-457.

[Richardson et al. 1997] M. D. Richardson, D. M. Lavoie, and K. B. Briggs, "Geoacoustic and physical properties of carbonate sediments of the Lower Florida Keys," *Geo-Mar. Lett.* **17**, pp. 316-324.

[Richardson and Davis 1998] M. D. Richardson and A. M. Davis, "Modeling methane-rich sediments from Eckernförde Bay," *Cont. Shelf Res.* **18**, pp. 1671-1688.

[Richardson et al. 2001a] M. D. Richardson, K. B. Briggs, L. D. Bibee, P. A. Jumars, W. B. Sawyer, D. B. Albert, R. H. Bennett, T. K. Berger, M. J. Buckingham, N. P. Chotiros, P. H. Dahl, N. T. Dewitt, P. Fleischer, R. Flood, C. F. Greenlaw, D. V. Holliday, M. H. Hulbert, M. P. Hutnak, P. D. Jackson, J. S. Jaffe, H. P. Johnson, D. L. Lavoie, A. P. Lyons, C. S. Martens, D. E. McGehee, K. D. Moore, T. H. Orsi, J. N. Piper, R. I. Ray, A. H. Reed, R. F. L. Self, J. L. Schmidt, S. G. Schock, F. Simonet, R. D. Stoll, D. Tang, D. E. Thistle, E. I. Thorsos, D. J. Walter, and R. A. Wheatcroft, "Overview of SAX99: Environmental considerations," *IEEE J. Ocean. Eng.* **26**, pp. 26-53.

[Richardson et al. 2001b] M. D. Richardson, K. B. Briggs, K. L. Williams, A. P. Lyons, and D. R. Jackson, "Effects of changing roughness on acoustic scattering: (2) Anthropogenic changes," *Proc. Inst. Acoust.* **23**, pp. 383-390.

[Richardson 2002] M. D. Richardson, "Variability of shear wave speed and attenuation in surficial marine sediments," in *Impact of Littoral Environmental Variability on Acoustic Predictions and Sonar Performance*, N.G. Pace and F.B. Jensen (Eds.), Kluwer Academic Publishers, Netherlands, pp. 107-114.

[Richardson et al. 2002a] M. D. Richardson, K. L. Williams, K. B. Briggs, and E. I. Thorsos, "Dynamic measurement of sediment grain compressibility at atmospheric pressure: acoustic applications," *IEEE J. Ocean. Eng.* **27**, pp. 593-601.

[Richardson et al. 2002b] M. D. Richardson, K. B. Briggs, S. J. Bentley, D. J. Walter, and T. H. Orsi, "The effects of biological and hydrodynamic processes on physical and acoustic properties of sediments off the Eel River, California," *Mar. Geol.*. **182**, pp. 121-139.

[Richardson and Briggs 2004a] M. D. Richardson and K. B. Briggs, "Relationships among sediment physical and acoustic properties in siliciclastic and calcareous sediments," *Proceedings of the 7th European Conference on Underwater Acoustics, EUCA 2004*, Delft, Netherlands, pp. 659-664.

[Richardson and Briggs 2004b] M. D. Richardson and K. B. Briggs, "Empirical predictions of seafloor properties based on remotely measured sediment impedance," in *High Frequency Ocean Acoustic Conference*, M. B. Porter and M. Siderius (Eds.), AIP, Melville, NY, pp. 12-21.

[Richardson et al. 2005] M. D. Richardson, K. B. Briggs, A. H. Reed, W. C. Vaughan, M. A. Zimmer, L. D. Bibee, and R. I. Ray, "Characterization of the environment during SAX04: preliminary results," in *Underwater Acoustic Measurements: Technologies and Results*, J. S. Papadakis and L. Bjorno (Eds.), A conference held in Heraklion, Crete, pp. 285-292.

[Richter 1952] R. Richter, "Sediment-Gaseinien und Sedifluktion uberhaupt (Fluidal-texture in sediment)," *Notizbl. Hess. Landesamptes Bodenforsch.* **3**, pp. 67-81.

[Robertson et al. 1992] P. K. Robertson, J. P. Sully, D. J. Woeller, T. Lunne, J. J. M. Powell, and D. G. Gillespie, "Estimating coefficient of consolidation from piezocone tests," *Can. Geotech. J.* **29**, pp. 539-550.

[Robertson et al. 1995] P. K. Robertson, S. Sasitharan, J. C. Cunning, and D. C. Sego, "Shear-wave velocity to evaluate in-situ state of Ottawa sand," *J. Geotech. Eng. ASCE* **121**, pp. 262-273.

[Robertson and Wride 1998] P. K. Robertson and C. E. Wride, "Evaluating cyclic liquefaction potential using the cone penetration test," *Can. Geotech. J.* **35**, pp. 442-459.

[Rogers and Yamamoto 1999] A. K. Rogers and T. Yamamoto, "Analysis of high-frequency acoustic scattering data measured in the shallow waters of the Florida Strait," *J. Acoust. Soc. Am.* **106**, pp. 2469-2480.

[Rowden et al. 1998] A. A. Rowden, C. F. Jago, and S. E. Jones, "Influence of benthic macrofauna on geotechnical and geophysical properties of surficial sediment, North Sea," *Cont. Shelf Res.* **18**, pp. 1347-1363.

[Schaafsma and Hay 1997] A. S. Schaafsma and A. E. Hay, "Attenuation in suspensions of irregularly shaped sediment particles: A two-parameter equivalent spherical scatterer model," *J. Acoust. Soc. Am.* **102**, pp. 1485-1502.

[Schmidt 1971] P. B. Schmidt, "Monostatic and bistatic backscattering measurements from the deep ocean bottom," *J. Acoust. Soc. Am.* **50**, pp. 326-331.

[Schmidt 1999] H. Schmidt, "OASES, Version 2.2, User Guide and Reference Manual," Massachusetts Institute of Technology.

[Schock 2004] S. G. Schock, "A method for estimating the physical and acoustic properties of the sea bed using chirp sonar data," *IEEE J. Ocean. Eng.* **29**, pp. 1200-1217.

[Schüler 1952] F. Schüler, "Untersuchungen über die Mächtigkeiten von Schlickschichten mit Hilfe des Echographen," *Dtsch. Hydrogr. Z.* **5**, pp. 220-231.

[Schultheiss and Weaver 1992] P. J. Schultheiss and P. P. E. Weaver, "Multi-sensor core logging for science and industry," *Proceedings Oceans 1992*, Newport, RI.

[Sclater et al. 1969] J. G. Sclater, C. E. Corry, and V. Vacquier, "In situ measurement of the thermal conductivity of ocean-floor sediments," *J. Geophys. Res.* **74**, pp. 1070-1081.

[Self et al. 2001] R. F. L. Self, P. A'Hearn, P. A. Jumars, D. R. Jackson, M. D. Richardson, and K. B. Briggs, "Effects of macrofauna on acoustic backscatter from the seabed: field manipulations in West Sound, Orcas Island, WA, USA," *J. Mar. Res.* **59**, pp. 991-1020.

[Shepard 1954] F. P. Shepard, "Nomenclature based on sand-silt-clay ratios," *J. Sediment. Petrol.* **24**, pp. 151-158.

[Sheremet et al. 2005] A. Sheremet, A. J. Mehta, B. Liu, and G. W. Stone, "Wave-sediment interaction on a muddy inner shelf during Hurricane Claudette," *Estuarine Coastal Shelf Sci.* **63**, pp. 225-233.

[Sheriff and Geldart 1995] R. E. Sheriff and L. P. Geldart, *Exploration Seismology*, Second Edition, Cambridge University Press, Cambridge, UK.

[Shirley 1978] D. J. Shirley, "An improved shear wave transducer," *J. Acoust. Soc. Am.* **63**, pp. 1643-1645.

[Shirley and Hampton 1978] D. J. Shirley and L. D. Hampton, "Shear wave measurements in laboratory sediments," *J. Acoust. Soc. Am.* **63**, pp. 607-613.

[Shull 2001] D. H. Shull, "Transition-matrix model of bioturbation and radionuclide diagenesis," *Limnol. Oceanogr.* **46**, pp. 905-916.

[Shumway 1956] G. Shumway, "A resonant chamber method for sound speed velocity and attenuation measurements in sediments," *Geophysics* **21**, pp. 305-319.

[Shumway 1958] G. Shumway, "Sound velocity versus temperature in water saturated sediments," *Geophysics* **23**, pp. 494-505.

[Shumway 1960] G. Shumway, "Sound speed and absorption studies in marine sediments by a resonance method, I," *Geophysics* **25**, pp. 451-476.

[Simmons and Brace 1965] G. Simmons and W. F. Brace, "Comparison of static and dynamic measurements of compressibility of rocks," *J. Geophys. Res.* **70**, pp. 5649-5656.

[Simmons and Wang 1971] G. Simmons and H. Wang, *Single Crystal Elastic Constants and Calculated Aggregate Properties: A Handbook*, MIT Press, Cambridge, MA.

[Simpson and Houston 2000] H. J. Simpson and B. H. Houston, "Synthetic array measurements of acoustical waves propagating into a water-saturated sandy bottom for a smoothed and a roughened interface," *J. Acoust. Soc. Am.* **107**, pp. 2329-2337.

[Simpson et al. 2003] H. J. Simpson, B. H. Houston, S. W.Liskey, P. A. Frank, A. R. Berdoz, L. A. Kraus, C. K. Frederickson, and S. Stanic, "At-sea measurements of sound penetration into sediments using a buried vertical synthetic array," *J. Acoust. Soc. Am.* **114**, pp. 1281-1290.

[Skolnik 2001] M. I. Skolnik, *Introduction to Radar Systems*, Third Edition, McGraw-Hill, New York.

[Smeulders 2005] D. M. J. Smeulders, "Experimental evidence for slow compressional waves," *J. Eng. Mech. ASCE* **131**, pp. 908-917.

[Soukup and Gragg 2003] R. J. Soukup and R. F. Gragg, "Backscatter from a limestone seafloor at 2-3.5 kHz: Measurements and modeling," *J. Acoust. Soc. Am.* **113**, pp. 2501-2514.

[Soukup et al. 2005] R. J. Soukup, R. F. Gragg, E. L. Kunz, J. E. Summers, G. Canepa, and E. Pouliquen, "Verification of rough surface scattering predictions using an elastic scale model," in *Boundary Influences in High-Frequency,*

Shallow-Water Acoustics, N. G. Pace and P. Blondel (Eds.), University of Bath, UK, pp. 201-207.

[Spiesberger and Metzger 1991a] J. L. Spiesberger and K. Metzger, "New estimates of sound speed in water," *J. Acoust. Soc. Am.* **89**, pp. 1697-1700.

[Spiesberger and Metzger 1991b] J. L. Spiesberger and K. Metzger, "A new algorithm for sound speed in seawater," *J. Acoust. Soc. Am.* **89**, pp. 2677-2688.

[Spiesberger 1993] J. L. Spiesberger, "Is Del Grosso's sound-speed algorithm correct?," *J. Acoust. Soc. Am.* **93**, pp. 2235-2237.

[Stanic et al. 1988] S. Stanic, K. B. Briggs, P. Fleischer, R. I. Ray, and W. B. Sawyer, "Shallow-water high-frequency bottom scattering off Panama City, Florida," *J. Acoust. Soc. Am.* **83**, pp. 2134-2144.

[Stanic et al. 1989] S. Stanic, K. B. Briggs, P. Fleischer, W. B. Sawyer, and R. I. Ray, "High-frequency acoustic backscattering from a coarse shell ocean bottom," *J. Acoust. Soc. Am.* **85**, pp. 125-136.

[Stanic et al. 1991] S. Stanic, E. Kennedy, and R. I. Ray, "Variability of shallow-water bistatic bottom backscattering," *J. Acoust. Soc. Am.* **90**, pp. 547-553.

[Stanic and Kennedy 1992] S. Stanic and E. Kennedy, "Fluctuations of high-frequency shallow-water seafloor reverberation," *J. Acoust. Soc. Am.* **91**, pp. 1967-1973.

[Stanic et al. 1993] S. Stanic, E. Kennedy, and R. I. Ray, "High-frequency bistatic reverberation from a smooth ocean bottom," *J. Acoust. Soc. Am.* **93**, pp. 2633-2638.

[Stanic and Kennedy 1993] S. Stanic and E. Kennedy, "Reverberation fluctuations from a smooth seafloor," *IEEE J. Ocean. Eng.* **18**, pp. 95-99.

[Stanic et al. 1998] S. Stanic, R. R. Goodman, K. B. Briggs, N. P. Chotiros and E. T. Kennedy, "Shallow-water bottom reverberation measurements," *IEEE J. Ocean. Eng.* **23**, pp. 203-210.

[Stanton 2000] T. K. Stanton, "On acoustic scattering by a shell-covered seafloor," *J. Acoust. Soc. Am.* **108**, pp. 551-555.

[Stanton et al. 2000] T. K. Stanton, D. Chu, P. H. Weibe, R. L. Eastwood, and J. D. Warren, "Acoustic scattering by benthic and planktonic shelled animals," *J. Acoust. Soc. Am.* **108**, pp. 535-550.

[Stephens et al. 1997] K. P. Stephens, D. L. Lavoie, K. B. Briggs, Y. Furukawa, and M. D. Richardson, "Geotechnical and Geoacoustic Properties of Sediments off South Florida: Boca Raton, Indian Rocks Beach, Lower Tampa Bay, and the Lower Florida Keys," Naval Research Laboratory, NRL/MR/7431–97-8042.

[Sternlicht and de Moustier 1997] D. D. Sternlicht and C. P. de Moustier, "Temporal modeling of high frequency (30-100 kHz) acoustic seafloor backscatter: shallow water results," in *High Frequency Acoustics in Shallow Water*, N. G. Pace, E. Pouliquen, O. Bergem, and A. P. Lyons (Eds.), NATO SACLANT Undersea Res. Ctr., La Spezia, Italy, pp. 509-516.

[Sternlicht and de Moustier 2003a] D. D. Sternlicht and C. P. de Moustier, "Time-dependent seafloor acoustic backscatter (10-100 kHz)," *J. Acoust. Soc. Am.* **114**, pp. 2709-2725.

[Sternlicht and de Moustier 2003b] D. D. Sternlicht and C. de Moustier, "Remote sensing of sediment characteristics by optimized echo envelope matching," *J. Acoust. Soc. Am.* **114**, pp. 2727-2743.

[Stewart and Chotiros 1992] R. A. Stewart and N. P. Chotiros, "Estimation of sediment volume scattering cross section and absorption loss coefficient," *J. Acoust. Soc. Am.* **91**, pp. 3242-3247.

[Stewart et al. 1994] W. K. Stewart, D. Chu, S. Malik, S. Lerner, and H. Singh, "Quantitative seafloor characterization using a bathymetric sidescan sonar," *IEEE J. Ocean. Eng.* **19**, 599-610.

[Stockhausen 1963] J. H. Stockhausen, "Scattering from the Volume of an Inhomogeneous Half-Space," NRE Report, 63/9.

[Stoll 1974] R. D. Stoll, "Acoustic waves in saturated sediments," in *Physics of Sound in Marine Sediments*, L. Hampton (Ed.), Plenum, New York, pp.19-39.

[Stoll 1977] R. D. Stoll, "Acoustic waves in ocean sediments," *Geophysics* **42**, pp. 715-725.

[Stoll 1980] R. D. Stoll, "Theoretical aspects of sound-transmission in sediments," *J. Acoust. Soc. Am.* **68**, pp. 1341-1350.

[Stoll and Kan 1981] R. D. Stoll and T. K. Kan, "Reflection of acoustic waves at a water-sediment interface," *J. Acoust. Soc. Am.* **70**, pp. 149-156.

[Stoll 1985] R. D. Stoll, "Marine sediment acoustics," *J. Acoust. Soc. Am.* **77**, pp. 1789-1799.

[Stoll 1989] R. D. Stoll, *Sediment Acoustics. Lecture Notes in Earth Sciences*, Springer-Verlag, New York.

[Stoll et al. 1994] R. D. Stoll, G. M. Bryan, and E. O. Bautista, "Measuring lateral variability of sediment geoacoustic properties," *J. Acoust. Soc. Am.* **96**, pp. 427-438.

[Stoll 1998] R. D. Stoll, "Comments on 'Biot model of sound propagation in water-saturated sand'," *J. Acoust. Soc. Am.* **103**, pp. 2723-2725.

[Stoll and Bautista 1998] R. D. Stoll and E. O. Baustia, "Using the Biot theory to establish a baseline geoacoustic model for seafloor sediments," *Cont. Shelf Res.* **18**, pp. 1839-1857.

[Stoll 2002] R. D. Stoll, "Velocity dispersion in water-saturated granular sediment," *J. Acoust. Soc. Am.* **111**, pp. 785-793.

[Sumino and Anderson 1984] Y. Sumino and O. L. Anderson, "Elastic constants of minerals," in *Handbook of Physical Properties of Rocks, Vol. III*, R. S. Carmichael (Ed.) , CRC Press, Boca Raton, FL, pp. 39-138.

[Sutton et al. 1957] G. H. Sutton, H. Berckhemer, and J. E. Nafe, "Physical analysis of deep-sea sediments," *Geophysics* **22**, pp. 779-812.

[Swift et al. 1985] A. S. Swift, D. C. Hollister, and S. R. Chandler, "Close-up stereo photographs of abyssal bedforms on the Nova Scotian continental rise," *Mar. Geol.* **66**, pp. 303-322.

[Syvitski 1991] J. P. M. Syvitski, "Introduction," in *Principles, Methods, and Applications of Particle Size Analysis*, J. P. M. Syvitski (Ed.), Cambridge University Press, New York.

[Syvitski et al. 1991a] J. P. M. Syvitski, K. W. Asprey and D. A. Clattenburg, "Principles, design, and calibration of settling tubes," in *Principles, Methods, and Applications of Particle Size Analysis*, J. P. M. Syvitski (Ed.), Cambridge University Press, New York, pp. 45-63.

[Syvitski et al. 1991b] J. P. M. Syvitski, K. W. G. LeBlanc, and K. W. Asprey, "Interlaboratory, interinstrument calibration experiments," in *Principles, Methods, and Applications of Particle Size Analysis*, J. P. M. Syvitski (Ed.), Cambridge University Press, New York, pp. 174-194.

[Syvitski 1991c] J. P. M. Syvitski, "The changing microfabric of suspended particulate matter — The fluvial to marine transition: flocculation, agglomeration, and pelletization," in *Microstructure of Fine-Grained Sediments: From Mud to Shale*, R. H. Bennett, W. R. Bryant and M. H. Hulbert, (Eds.), Frontiers in Sedimentary Geology, Springer-Verlag, New York, pp. 131-138.

[Talukdar et al. 1995] K. K. Talukdar, R. C. Tyce, and C. S. Clay, "Interpretation of Sea Beam backscatter data collected at the Lauentian Fan off Nova Scotia using acoustic backscatter theory," *J. Acoust. Soc. Am.* **97**, pp. 1545-1558.

[Tang 1991] D. Tang, "Acoustic wave scattering from a random ocean bottom," Ph.D. thesis, Massachusetts Institute of Technology and Woods Hole Oceanographic Institution.

[Tang et al. 1994] D. Tang, G. Lin, D. R. Jackson, and K. L. Williams, "Analysis of high-frequency bottom and subbottom backscattering for two distinct shallow-water environments," *J. Acoust. Soc. Am.* **96**, pp. 2930-2936.

[Tang 1996a] D. Tang, "A note on scattering by a stack of rough interfaces," *J. Acoust. Soc. Am.* **99**, pp. 1414-1418.

[Tang 1996b] D. Tang, Modeling high-frequency acoustic backscattering from gas voids buried in sediments," *Geo-Mar. Lett.* **16**. pp. 261-265.

[Tang 1997] D. Tang, "Small scale volumetric inhomogeneities of shallow water sediments: Measurements and discussion," in *High Frequency Acoustics in Shallow Water*, N.G. Pace, E. Pouliquen, O. Bergem, and A.P. Lyons (Eds.), NATO SACLANT Undersea Res. Ctr., La Spezia, Italy, pp. 539-546.

[Tang and Orsi 2000a] D. Tang and T. H. Orsi, "Three-dimensional density spectra of shallow water sediments," in *Proceedings of the Fifth European Conference on Underwater Acoustics, Lyon, France*, M. E. Zakharia, P. Chevret, and P. Dubail (Eds.), Lyon, France, pp. 1277-1282.

[Tang and Orsi 2000b] D. Tang and T. H. Orsi, "Three-dimensional density spectra of sandy sediments," *J. Acoust. Soc. Am.* **107**, pp. 1953-1963.

[Tang et al. 2002] D. Tang, K. B. Briggs, K. L. Williams, D. R. Jackson, E. I. Thorsos, and D. B. Percival, "Fine-scale volume heterogeneity measurements in sand," *IEEE J. Ocean. Eng.* **27**, pp. 546-560.

[Tang 2004] D. Tang, "Fine-scale measurements of sediment roughness and subbottom variability," *IEEE J. Ocean. Eng.* **29**, pp. 929-939.

[Tappert 1977] F. D. Tappert, "The parabolic approximation method," in *Wave Propagation and Underwater Acoustics*, J. B. Keller and J. S. Papadakis (Eds.), Springer-Verlag, Berlin.

[Taylor-Smith 1971] D. Taylor-Smith, "Acoustic and electrical techniques of marine sediment identification," in *Proceedings of the International Symposium on Engineering Properties of Seafloor Soils and Their Identification*, Seattle, WA, pp. 235-267.

[Telford et al. 1990] W. M. Telford, L. P. Geldart, and R. E. Sheriff, *Applied Geophysics*, Second Edition, Cambridge University Press, Cambridge, UK.

[Thorne and Pace 1984] P. D. Thorne and N. G. Pace, "Acoustic studies of broadband scattering from a model rough surface," *J. Acoust. Soc. Am.* **75**, pp. 133-144.

[Thorne et al. 1988] P. D. Thorne, N. G. Pace, and K. S. Al-Hamdani, "Laboratory measurements of backscattering from marine sediments," *J. Acoust. Soc. Am.* **84**, pp. 303-309.

[Thorne et al. 1995] P. D. Thorne, K. R. Waters, and T.J. Brudner, "Acoustic measurements of scattering by objects of irregular shape," *J. Acoust. Soc. Am.* **97**, pp. 242-251.

[Thorne and Hanes 2002] P. D. Thorne and D. M. Hanes, "A review of acoustic measurement of small-scale sediment processes," *Cont. Shelf Res.* **22**, pp. 603-632.

[Thorsos 1988] E. I. Thorsos, "Exact numerical methods vs. the Kirchhoff approximation for rough surface scattering," in *Computational Acoustics* **II**, *Algorithms and Applications*, D. Lee, R. L. Sternberg, and M. H. Schultz (Eds.), North-Holland, Amsterdam.

[Thorsos and Jackson 1989] E. I. Thorsos and D. R. Jackson, "The validity of the perturbation approximation for rough surface scattering using a Gaussian roughness spectrum," *J. Acoust. Soc. Am.* **86**, pp. 261-277.

[Thorsos 1990] E. I. Thorsos, "Acoustic scattering from a 'Pierson-Moskowitz' sea surface," *J. Acoust. Soc. Am.* **88**, pp. 335-349.

[Thorsos and Jackson 1991] E. I. Thorsos and D. R. Jackson, "Studies of scattering theory using numerical methods," *Waves Random Media* **1**, pp. S165-S190.

[Thorsos and Broschat 1995] E. I. Thorsos and S. L. Broschat, "An investigation of the small-slope approximation for scattering from rough surfaces: Part I, Theory," *J. Acoust. Soc. Am.* **97**, pp. 2082-2093.

[Thorsos et al. 2000a] E. I. Thorsos, D. R. Jackson, and K. L. Williams, "Modeling of subcritical penetration into sediments due to interface roughness," *J. Acoust. Soc. Am.* **107**, pp. 263-277.

[Thorsos et al. 2000b] E. I. Thorsos, K. L. Williams, D. R. Jackson, and D. Tang, "High-frequency sound interaction in ocean sediments," in *Proceedings of the Fifth European Conference on Underwater Acoustics, Lyon, France*, M. E. Zakharia, P. Chevret, and P. Dubail (Eds.), Lyon, France, pp. 275-280.

[Thorsos et al. 2001a] E. I. Thorsos, K. L. Williams, N. P. Chotiros, J. T. Christoff, K. W. Commander, C. F. Greenlaw, D. V. Holliday, D. R. Jackson, J. L. Lopes, D. E. McGehee, J. E. Piper, M. D. Richardson, and D. Tang, "An overview of SAX99: Acoustic measurements," *IEEE J. Ocean. Eng.* **26**, pp. 4-25.

[Thorsos et al. 2001b] E. I. Thorsos, K. L. Williams, D. R. Jackson, M. D. Richardson, K. B. Briggs, and D. Tang, "An experiment in high-frequency sediment acoustics: SAX99," in *Proceedings of the Institute of Acoustics Conference*, T. G. Leighton, G. J. Heald, H. D. Griffiths, and G. Griffiths (Eds.), Southampton, United Kingdom.

[Tolman 1994] H. L. Tolman, "Wind waves and moveable-bed bottom friction," *J. Phys. Oceanogr.* **24**, pp. 994-1009.

[Toorman 1999] E. A. Toorman, "Sedimentation and self-weight consolidation: constitutive equations and numerical modeling." *Geotechnique* **49**, pp. 709-726.

[Traykovski et al. 1999] P. Traykovski, A. E. Hay, J. D. Irish, and J. F. Lynch, "Geometry, migration, and evolution of wave orbital ripples at LEO-15," *J. Geophys. Res. Oceans* **104(C1)**, pp. 1505-1524.

[Trevorrow and Yamamoto 1991] M. V. Trevorrow and T. Yamamoto, "Summary of marine sedimentary shear modulus and acoustic speed profile results using a gravity wave inversion technique," *J. Acoust. Soc. Am.* **90**, pp. 441-456.

[Trevorrow 2004] M. V. Trevorrow, "Statistics of fluctuations in high-frequency low-grazing-angle backscatter from a rocky seabed," *IEEE J. Ocean. Eng.* **29**, pp. 236-245.

[Turgut and Yamamoto 1990] A. Turgut and T. Yamamoto, "Measurements of acoustic wave velocities and attenuation in marine sediments," *J. Acoust. Soc. Am.* **87**, pp. 2376-2383.

[Turgut 2000] A. Turgut, "Approximate expressions for viscous attenuation in marine sediments: Relating Biot's "critical" and "peak" frequencies," *J. Acoust. Soc. Am.* **108**, pp. 513-518.

[Udden 1914] J. A. Udden, "Mechanical composition of clastic sediments," *Bull. Geol. Soc. Am.* **25**, pp. 655-744.

[Uscinski 1978] B. J. Uscinski, *The Elements of Wave Propagation in Random Media*, McGraw-Hill, New York.

[Urick and Ament 1949] R. J. Urick and W. S. Ament, "The propagation of sound in composite media," *J. Acoust. Soc. Am.* **21**, pp. 115-119.

[Urick 1954] R. J. Urick, "The backscattering of sound from a harbor bottom," *J. Acoust. Soc. Am.* **26**, pp. 231-235.

[Urick 1956] R. J. Urick, "The processes of sound scattering at the ocean surface and bottom," *J. Mar. Res.* **15**, pp. 134-148.

[Urick 1960] R. J. Urick, "Side scattering of sound in shallow water," *J. Acoust. Soc. Am.* **32**, pp. 351-355.

[Urick 1983] R.J. Urick, *Principles of Underwater Sound*, McGraw-Hill, New York.

[Van Geet et al. 2001] M. Van Geet, R. Swennen, and M. Wevers, "Towards 3-D petrography: application of microfocus computer tomography in geological science," *Comput. Geosci.* **27**, pp. 1091-1099.

[Vanorio et al. 2003] T. Vanorio, M. Prasad, and A. Nur, "Elastic properties of dry clay mineral aggregates, suspensions and sandstones," *Geophys. J. Int.* **155**, pp. 319-326.

[Vaughan et al. 2002] W. C. Vaughan, D. H. Easley, and D. L. Lavoie, "Averaging pore statistics of two-dimensional images for predicting permeability," *J. Hydraulic Eng. ASCE* **128**, pp. 1002-1007.

[Von Herzen and Maxwell 1959] R. Von Herzen and A. E. Maxwell, "The measurement of thermal conductivity of deep-sea sediments by a needle-probe method," *J. Geophys. Res.* **64**, pp. 1557-1563.

[Voronovich 1985] A. G. Voronovich, " Small-slope approximation in wave scattering by rough surfaces," *Soc. Phys. JETP* **62**, pp. 65-70.

[Voronovich 1994] A. G. Voronovich, *Wave Scattering from Rough Surfaces*, Springer-Verlag, Berlin.

[Wagner 1967] R. J. Wagner, "Shadowing of randomly rough surfaces," *J. Acoust. Soc. Am.* **41**, pp. 138-147.

[Wang et al. 2001] Z. Wang, H. Wang, and M. E. Gates, "Effective elastic properties of solid clays," *Geophysics* **66**, pp. 428-440.

[Watanabe et al. 2004] T. Watanabe, K. T. Nihei, S. Nakagawa, and L. R. Myer,"Viscoacoustic wave form inversion of transmission data for velocity and attenuation," *J. Acoust. Soc. Am.* **115**, pp. 3059-3067.

[Weaver and Schultheiss 1990] P. P. E. Weaver and P. J. Schultheiss, "Current methods for obtaining, logging and splitting marine sediment cores," *Mar. Geophys. Res.* **12**, pp. 85-100.

[Weber et al. 1997] M. E. Weber, F. Niessen, G. Kuhn, and M. Wiedicke, "Calibration and application of marine sedimentary physical properties using a multi-sensor core logger," *Mar. Geol.* **136**, pp. 151-172.

[Weinberg 1995] S. Weinberg, *The Quantum Theory of Fields, Vol. I*, Cambridge University Press, Cambridge.

[Wellington and Vinegar 1987] S. L. Wellington and H. J. Vinegar, "X-ray computerized tomography," *J. Pet. Eng.* **39**, pp. 885-898.

[Wentworth 1922] C. K. Wentworth, "A scale of grade and class terms for clastic sediments," *J. Geol.* **30**, pp. 377-392.

[Westervelt 1963] P. J. Westervelt, "Parametric acoustic array," *J. Acoust. Soc. Am.* **35**, pp. 535-537.

[Wever and Fiedler 1995] T. F. Wever and H. M. Fiedler, "Variability of acoustic turbidity in Eckernförde Bay (southwest Baltic Sea) related to the annual temperature cycle," *Mar. Geol.* **125**, pp. 21-27.

[Wever et al. 1998] T. F. Wever, F. Abegg, H. M. Fiedler, G. Fechner, and I. H. Stender, "Shallow gas in the muddy sediments of Eckernförde Bay, Germany," *Cont. Shelf Res.* **18**, pp. 1715-1739.

[Wheatcroft 1990] R. A. Wheatcroft, "Preservation potential of sedimentary event layers," *Geology* **18**, pp. 843-845.

[Wheatcroft 1991] R. A. Wheatcroft, "Conservative tracer study of horizontal sediment mixing rates in a bathyal basin, California borderland," *J. Mar. Res.* **49**, pp. 565-588.

[Wheatcroft 1994] R. A. Wheatcroft, "Temporal variation in bed configuration and one-dimensional bottom roughness at the mid-shelf STRESS site," *Cont. Shelf Res.* **14**, pp. 1167-1190.

[Wheatcroft 2002] R. A. Wheatcroft, "In situ measurements of near-surface porosity in shallow-water marine sands, *IEEE J. Ocean. Eng.* **27**, pp. 561-570.

[Wheatcroft and Drake 2003] R. A. Wheatcroft and D. E. Drake, "Post-depositional alteration and preservation of sedimentary event layers on continental margins, I. The role of episodic sedimentation," *Mar. Geol.* **199**, pp. 123-137.

[White 1998] T. E. White, "Status of measurement techniques for coastal sediment transport," *Coastal Eng.* **35**, pp. 17-45.

[Whiticar 2002] M. J. Whiticar, "Diagenetic relationships of methanogenesis, nutrients, acoustic turbidity, pockmarks and freshwater seepages in Eckernförde Bay," *Mar. Geol.* **182**, pp. 29-53.

[Wiberg and Harris 1994] P. L. Wiberg and C. K. Harris, "Ripple geometry in wave-dominated environments," *J. Geophys. Res. Oceans* **99**, pp. 775-789.

[Wilkens and Richardson 1998] R. H. Wilkens and M. D. Richardson, "The influence of gas bubbles on sediment acoustic properties: In situ, laboratory and theoretical results from Eckernförde Bay, Baltic Sea," *Cont. Shelf Res.* **18**, pp. 1859-1892.

[Williams et al. 1988] K. L. Williams, R. H. Hackman, and D. H. Trivett, "High-frequency scattering from liquid/porous sediment interfaces," *J. Acoust. Soc. Am.* **84**, pp. 760-770.

[Williams et al. 1989] K. L. Williams, L. J. Satkowiak, and D. R. Bugler, "Linear and parametric array transmission across a water-sand interface — Theory, experiment, and observation of beam displacement, *J. Acoust. Soc. Am.* **86**, pp. 311-325.

[Williams and Jackson 1997] K. L. Williams and D. R. Jackson, "Bottom bistatic scattering: Experimental results and model comparison for a carbonate sediment," in *High Frequency Acoustics in Shallow Water*, N.G. Pace, E. Pouliquen, O. Bergem, and A.P. Lyons (Eds.), NATO SACLANT Undersea Res. Ctr., La Spezia, Italy, pp. 601-605.

[Williams and Jackson 1998] K. L. Williams and D. R. Jackson, "Bistatic bottom scattering: Model, experiments, and model/data comparison," *J. Acoust. Soc. Am.* **103**, pp. 169-181.

[Williams 2001a] K. L. Williams, "An effective density fluid model for acoustic propagation in sediments derived from Biot theory," *J. Acoust. Soc. Am.* **110**, pp. 2956-2963.

[Williams 2001b] K. L. Williams, "Temporal fluctuations in the acoustic scattering from bottom-deployed objects and localized biological treatments," *IEEE J. Ocean. Eng.* **26**, pp. 63-69.

[Williams et al. 2001a] K. L. Williams, J. M. Grochocinski, and D. R. Jackson, "Interface scattering by poroelastic seafloors: First-order theory," *J. Acoust. Soc. Am.* **110**, pp. 2956-2963.

[Williams et al. 2001b] K. L. Williams, M. D. Richardson, K. B. Briggs, and D. R. Jackson, "Scattering of high-frequency acoustic energy from discrete scatterers on the seafloor: Glass spheres and shells," *Proc. Inst. Acoust.* **23**, pp. 369-374.

[Williams et al. 2002a] K. L. Williams, D. R. Jackson, E. I. Thorsos, D. Tang, and S. G. Schock, "Comparison of sound speed and attenuation measured in a sandy sediment to predictions based on the Biot theory of porous media," *IEEE J. Ocean. Eng.* **27**, pp. 413-428.

[Williams et al. 2002b] K. L. Williams, D. R. Jackson, E. I. Thorsos, D. Tang, and K. B. Briggs, "Acoustic backscattering experiments in a well characterized sand sediment: Data/model comparisons using sediment fluid and Biot models," *IEEE J. Ocean. Eng.* **27**, pp. 376-387.

[Williams et al. 2005] K. L. Williams, S. G. Kargl, E. I. Thorsos, and D. Tang, "Synthetic aperture sonar (SAS) imaging and acoustic scattering strength measurements during SAX04 (Sediment Acoustics Experiment-2004): Experimental results and associated modeling," in *Boundary Influences in High-Frequency, Shallow-Water Acoustics*, N. G. Pace and P. Blondel (Eds.), University of Bath, UK, pp. 127-135.

[Wilson 1959] W. D. Wilson, "Speed of sound in distilled water as a function of temperature and pressure," *J. Acoust. Soc. Am.* **31**, pp. 1067-1072.

[Wilson 1960a] W. D. Wilson, "Sound speed in sea water as a function of temperature, pressure, and salinity," *J. Acoust. Soc. Am.* **32**, pp. 641-644.

[Wilson 1960b] W. D. Wilson, "Equation for the speed of sound in sea water," *J. Acoust. Soc. Am.* **32**, p. 1357.

[Wilson and Powell 1983] G. R. Wilson and D. R. Powell, "Probability estimates of surface and bottom reverberation," *J. Acoust. Soc. Am.* **73**, pp. 195-200.

[Winebrenner and Ishimaru 1985] D. P. Winebrenner and A. Ishimaru, "Application of the phase-perturbation technique to randomly rough surfaces," *J. Opt. Soc. Am. A* **2**, pp. 2285-2294.

[Winebrenner and Ishimaru 1986] D. P. Winebrenner and A. Ishimaru, "On the far-field approximation for scattering from randomly rough surfaces," *IEEE Trans. Antennas Propag.* **AP-34**, pp. 847-849.

[Wingham 1985] D. J. Wingham, "The dispersion of sound in sediment, *J. Acoust. Soc. Am.* **78**, pp. 1757-1760.

[Wingham et al. 1986] D. J. Wingham, N. G. Pace, and R. V. Ceen, "An experimental study of the penetration of a water sediment interface by a parametric beam, *J. Acoust. Soc. Am.* **79**, pp. 363-374.

[Winokur and Chanesman 1966] R. S. Winokur and S. Chanesman, "A pulse method for sound speed measurements in cored ocean bottom sediments," Informal manuscript IM No. 66-5, U.S. Naval Oceanographic Office, Washington, DC.

[Winsauer et al. 1952] W. O. Winsauer, A. M. Shearin, P. H. Masson, and M. Williams, "Resistivity of brine saturated sands in relation to pore geometry," *Bull. Am. Assoc. Pet. Geol.* **36**, pp. 253-277.

[Winsborrow et al. 2003] G. Winsborrow, D. G. Huws, and E. Muyzert, "Acquisition and inversion of Love wave data to measure the lateral variability of geo-acoustic properties of marine sediments," *J. Appl. Geophys.* **54**, pp. 71-84.

[Winterweap and van Kestern 2004] J. C. Winterweap and W. G. M. van Kestern, "Introduction to the physics of cohesive sediments in the marine environment," *Developments in Sedimentology* **56**, Elsevier, Amsterdam.

[Wong and Chesterman 1968] H.-K. Wong and W. D. Chesterman, "Bottom backscattering near grazing incidence in shallow water," *J. Acoust. Soc. Am.* **44**, pp. 1713-1718.

[Wood 1964] A. B. Wood, *A Textbook of Sound*, Third Edition, G. Bell, London.

[Wood and Weston 1964] A. B. Wood and D. E. Weston, "The propagation of sound in mud," *Acoustica* **14**, pp. 156-162.

[Worley 2004] A. Worley, "The impact of experimental error and sediment models on reflection coefficient inversion," Master's thesis, University of Texas, Austin.

[Wright 1997] D. G. Wright, "An equation of state for use in ocean models: Eckart's formula revisited," *J. Atmos. Ocean. Tech.* **14**, pp. 735-740.

[Wurmser 1996] D. Wurmser, "A manifestly reciprocal theory of scattering in the presence of elastic media," *J. Math. Phys.* **37**, pp. 4434-4479.

[Wurmser 2005] D. Wurmser, "Stochastic cross-sections based on the small-slope approximation: Theory," Naval Research Laboratory NRL/MR/7140–05-8816.

[Yaglom 1987] A. M. Yaglom, *Correlation Theory of Stationary and Related Random Functions*, Springer-Verlag, New York.

[Yamamoto and Turgut 1988] T. Yamamoto and A. Turgut, "Acoustic wave propagation through porous media with arbitrary pore size distributions," *J. Acoust. Soc. Am.* **83**, pp. 1744-1751.

[Yamamoto et al. 1989] T. Yamamoto, M. V. Trevorrow, M. Badiey, and A. Turgut, "Determination of the seabed porosity and shear modulus profiles using a gravity-wave inversion," *Geophys. J. Int.* **98**, pp. 173-182.

[Yamamoto 1995] T. Yamamoto, "Velocity variabilities and other physical properties of marine sediments measured by crosswell acoustic tomography," *J. Acoust. Soc. Am.* **98**, pp. 2235-2248.

[Yamamoto 1996] T. Yamamoto, "Acoustic scattering in the ocean from velocity and density fluctuations in the sediments," *J. Acoust. Soc. Am.* **99**, pp. 866-879.

[Yamamoto 2001] T. Yamamoto, "Imaging the permeability structure within the near-surface sediments by acoustic crosswell tomography," *J. Appl. Geophys.* **47**, pp. 1-11.

[Yang and Broschat 1994] T. Yang and S. L. Broschat, "Acoustic scattering from a fluid-elastic-solid interface using the small slope approximation," *J. Acoust. Soc. Am.* **96**, pp. 1796-1804.

[Yang et al. 2002] T. Yang, S. L. Broschat, and C. Galea, "A comparison of perturbation theory and the small-slope approximation for acoustic scattering from a rough interface for a Biot medium," *IEEE J. Ocean. Eng.* **27**, pp. 403-412.

[Yao and Anandarajah 2003] M. Yao and A. Anandarajah, "Three-dimensional discrete element method of analysis of clays," *J. Eng. Mech. ASCE* **129**, pp. 585-596.

[Yargus 2003] M. W. Yargus, "Experimental Study of Sound Waves in Sandy Sediment," Ph.D. dissertation, University of Washington, Seattle, available as APL-UW TR 0301.

[Ye and Alvarez 1998] Z. Ye and A. Alvarez, "Acoustic localization in bubbly liquid media," *Phys. Rev. Lett.* **80**, pp. 3503-3506.

[Yefimov et al. 1988] A. V. Yefimov, A. N. Ivakin, and Yu. P. Lysanov, "A geoacoustic model of scattering by the ocean bottom based on deep sea drilling data," *Oceanology* **28**, pp. 290-293.

[Yelton and Chotiros 1995] D. J. Yelton and N. P. Chotiros, "New multiple scatter model of the ocean sediment," *J. Acoust. Soc. Am.* **97**, Pt. 2, p. 3387.

[Yilmaz 2001] Öz. Yilmaz, *Seismic Data Analysis: Processing, Inversion, and Interpretation of Seismic Data, 2 volumes*, Investigations in Geophysics Series No. 10, Society of Exploration Geophysics, Tulsa, OK.

[Youd et al. 2001] T. L. Youd, I. M. Idriss, R. D. Andrus, I. Arango, G. Castro, J. T. Christian, R. Dobry, W. D. L. Finn, L. F. Harder, M. E. Hynes, K. Ishihara, J. P. Koester, S. S. C. Liao, W. F. Marcuson, G. R. Martin, J. K. Mitchell, Y. Moriwaki, M. S. Power, P. K. Robertson, R. B. Seed, and K. H. Stokoe, "Liquefaction resistance of soils: Summary report from the 1996 NCEER and 1998 NCEER/NSF Workshops on Evaluation of Liquefaction Resistance of Soils," *J. Geotech. Geoenviron. Eng.* **127**, pp. 817-833.

[Zhitkovski and Lysanov 1967] Yu. Yu. Zhitkovski and Yu. P. Lysanov, "Reflection and scattering of sound from the ocean bottom (Review)," *Sov. Phys. Acoust.* **13,** pp. 1-13.

[Ziomek 1995] L. J. Ziomek, *Fundamentals of Acoustic Field Theory and Space-Time Signal Processing*, CRC Press, Boca Raton, FL.

[Zipfel and DeSanto 1972] G. G. Zipfel, Jr., and J. A. DeSanto, "Scattering of a scalar wave from a random rough surface: a diagrammatic approach," *J. Math. Phys.* **13,** pp. 1903-1911.

Index